∫ STUDENT SOLUTIONS MA

for

Stewart's

ESSENTIAL CALCULUS
EARLY TRANSCENDENTALS

THOMSON

BROOKS/COLE

AUSTRALIA · BRAZIL · CANADA · MEXICO · SINGAPORE · SPAIN · UNITED KINGDOM · UNITED STATES

Printed in the United State of America

4 5 6 7 8 9 11 10 09 08 07

Printer: Thomson/West

ISBN-13: 978-0-495-01429-4
ISBN-10: 0-495-01429-X

Cover image: Terry Why/IndexStock Imagery

Thomson Higher Education
10 Davis Drive
Belmont, CA 94002-3098
USA

For more information about our products, contact us at:
Thomson Learning Academic Resource Center
1-800-423-0563

For permission to use material from this text or product, submit a request online at
http://www.thomsonrights.com.

Any additional questions about permissions can be submitted by e-mail to **thomsonrights@thomson.com.**

Trademarks
Derive is a registered trademark of Soft Warehouse, Inc.
Maple is a registered trademark of Waterloo Maple, Inc.
Mathematica is a registered trademark of Wolfram Research, Inc.

☐ ABBREVIATIONS AND SYMBOLS

CD	concave downward
CU	concave upward
D	the domain of f
FDT	First Derivative Test
HA	horizontal asymptote(s)
I	interval of convergence
IP	inflection point(s)
R	radius of convergence
VA	vertical asymptote(s)
$\overset{CAS}{=}$	indicates the use of a computer algebra system.
$\overset{H}{=}$	indicates the use of l'Hospital's Rule.
$\overset{j}{=}$	indicates the use of Formula j in the Table of Integrals in the back endpapers.
$\overset{s}{=}$	indicates the use of the substitution $\{u = \sin x, du = \cos x\, dx\}$.
$\overset{c}{=}$	indicates the use of the substitution $\{u = \cos x, du = -\sin x\, dx\}$.

∫ CONTENTS

1 □ FUNCTIONS AND LIMITS

1.1 Functions and Their Representations

In exercises requiring estimations or approximations, your answers may vary slightly from the answers given here.

1. (a) The point $(-1, -2)$ is on the graph of f, so $f(-1) = -2$.

(b) When $x = 2$, y is about 2.8, so $f(2) \approx 2.8$.

(c) $f(x) = 2$ is equivalent to $y = 2$. When $y = 2$, we have $x = -3$ and $x = 1$.

(d) Reasonable estimates for x when $y = 0$ are $x = -2.5$ and $x = 0.3$.

(e) The domain of f consists of all x-values on the graph of f. For this function, the domain is $-3 \le x \le 3$, or $[-3, 3]$. The range of f consists of all y-values on the graph of f. For this function, the range is $-2 \le y \le 3$, or $[-2, 3]$.

(f) As x increases from -1 to 3, y increases from -2 to 3. Thus, f is increasing on the interval $[-1, 3]$.

3. No, the curve is not the graph of a function because a vertical line intersects the curve more than once. Hence, the curve fails the Vertical Line Test.

5. Yes, the curve is the graph of a function because it passes the Vertical Line Test. The domain is $[-3, 2]$ and the range is $[-3, -2) \cup [-1, 3]$.

7. The person's weight increased to about 160 pounds at age 20 and stayed fairly steady for 10 years. The person's weight dropped to about 120 pounds for the next 5 years, then increased rapidly to about 170 pounds. The next 30 years saw a gradual increase to 190 pounds. Possible reasons for the drop in weight at 30 years of age: diet, exercise, health problems.

9. The water will cool down almost to freezing as the ice melts. Then, when the ice has melted, the water will slowly warm up to room temperature.

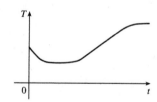

11. Of course, this graph depends strongly on the geographical location!

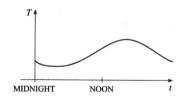

13. As the price increases, the amount sold decreases.

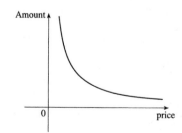

15.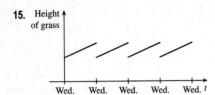

17. $f(x) = 3x^2 - x + 2$.

$f(2) = 3(2)^2 - 2 + 2 = 12 - 2 + 2 = 12$.

$f(-2) = 3(-2)^2 - (-2) + 2 = 12 + 2 + 2 = 16$.

$f(a) = 3a^2 - a + 2$.

$f(-a) = 3(-a)^2 - (-a) + 2 = 3a^2 + a + 2$.

$f(a+1) = 3(a+1)^2 - (a+1) + 2 = 3(a^2 + 2a + 1) - a - 1 + 2 = 3a^2 + 6a + 3 - a + 1 = 3a^2 + 5a + 4$.

$2f(a) = 2 \cdot f(a) = 2(3a^2 - a + 2) = 6a^2 - 2a + 4$.

$f(2a) = 3(2a)^2 - (2a) + 2 = 3(4a^2) - 2a + 2 = 12a^2 - 2a + 2$.

$f(a^2) = 3(a^2)^2 - (a^2) + 2 = 3(a^4) - a^2 + 2 = 3a^4 - a^2 + 2$.

$[f(a)]^2 = [3a^2 - a + 2]^2 = (3a^2 - a + 2)(3a^2 - a + 2)$
$= 9a^4 - 3a^3 + 6a^2 - 3a^3 + a^2 - 2a + 6a^2 - 2a + 4 = 9a^4 - 6a^3 + 13a^2 - 4a + 4$.

$f(a+h) = 3(a+h)^2 - (a+h) + 2 = 3(a^2 + 2ah + h^2) - a - h + 2 = 3a^2 + 6ah + 3h^2 - a - h + 2$.

19. $f(x) = 4 + 3x - x^2$, so $f(3+h) = 4 + 3(3+h) - (3+h)^2 = 4 + 9 + 3h - (9 + 6h + h^2) = 4 - 3h - h^2$,

and $\dfrac{f(3+h) - f(3)}{h} = \dfrac{(4 - 3h - h^2) - 4}{h} = \dfrac{h(-3 - h)}{h} = -3 - h$.

21. $\dfrac{f(x) - f(a)}{x - a} = \dfrac{\dfrac{1}{x} - \dfrac{1}{a}}{x - a} = \dfrac{\dfrac{a - x}{xa}}{x - a} = \dfrac{a - x}{xa(x - a)} = \dfrac{-1(x - a)}{xa(x - a)} = -\dfrac{1}{ax}$

23. $f(x) = x/(3x - 1)$ is defined for all x except when $0 = 3x - 1 \Leftrightarrow x = \frac{1}{3}$, so the domain

is $\left\{ x \in \mathbb{R} \mid x \neq \frac{1}{3} \right\} = \left(-\infty, \frac{1}{3} \right) \cup \left(\frac{1}{3}, \infty \right)$.

25. $f(t) = \sqrt{t} + \sqrt[3]{t}$ is defined when $t \geq 0$. These values of t give real number results for $\sqrt{t}$, whereas any value of t gives a real

number result for $\sqrt[3]{t}$. The domain is $[0, \infty)$.

27. $h(x) = 1 / \sqrt[4]{x^2 - 5x}$ is defined when

$x^2 - 5x > 0 \Leftrightarrow x(x - 5) > 0$. Note that

$x^2 - 5x \neq 0$ since that would result in division by

zero. The expression $x(x - 5)$ is positive if $x < 0$ or

$x > 5$. (See *Review of Algebra* at

www.stewartcalculus.com for methods for solving

inequalities.) Thus, the domain is $(-\infty, 0) \cup (5, \infty)$.

29. $f(x) = 5$ is defined for all real numbers, so the domain

is $\mathbb{R}$, or $(-\infty, \infty)$. The graph of f is a horizontal line

with y-intercept 5.

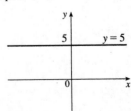

31. $f(t) = t^2 - 6t$ is defined for all real numbers, so the domain is $\mathbb{R}$, or $(-\infty, \infty)$. The graph of f is a parabola opening upward since the coefficient of t^2 is positive. To find the t-intercepts, let $y = 0$ and solve for t.

$0 = t^2 - 6t = t(t - 6) \quad \Rightarrow \quad t = 0$ and $t = 6$. The t-coordinate of the vertex is halfway between the t-intercepts, that is, at $t = 3$. Since $f(3) = 3^2 - 6 \cdot 3 = -9$, the vertex is $(3, -9)$.

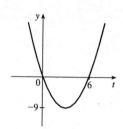

33. $g(x) = \sqrt{x - 5}$ is defined when $x - 5 \geq 0$ or $x \geq 5$, so the domain is $[5, \infty)$. Since $y = \sqrt{x - 5} \quad \Rightarrow \quad y^2 = x - 5 \quad \Rightarrow \quad x = y^2 + 5$, we see that g is the top half of a parabola.

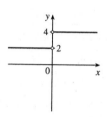

35. $G(x) = \dfrac{3x + |x|}{x}$. Since $|x| = \begin{cases} x & \text{if } x \geq 0 \\ -x & \text{if } x < 0 \end{cases}$, we have

$$G(x) = \begin{cases} \dfrac{3x + x}{x} & \text{if } x > 0 \\ \dfrac{3x - x}{x} & \text{if } x < 0 \end{cases} = \begin{cases} \dfrac{4x}{x} & \text{if } x > 0 \\ \dfrac{2x}{x} & \text{if } x < 0 \end{cases} = \begin{cases} 4 & \text{if } x > 0 \\ 2 & \text{if } x < 0 \end{cases}$$

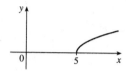

Note that G is not defined for $x = 0$. The domain is $(-\infty, 0) \cup (0, \infty)$.

37. $f(x) = \begin{cases} x + 2 & \text{if } x < 0 \\ 1 - x & \text{if } x \geq 0 \end{cases}$

The domain is $\mathbb{R}$.

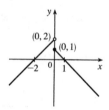

39. $f(x) = \begin{cases} x + 2 & \text{if } x \leq -1 \\ x^2 & \text{if } x > -1 \end{cases}$

Note that for $x = -1$, both $x + 2$ and x^2 are equal to 1. The domain is $\mathbb{R}$.

41. Recall that the slope m of a line between the two points (x_1, y_1) and (x_2, y_2) is $m = \dfrac{y_2 - y_1}{x_2 - x_1}$ and an equation of the line connecting those two points is $y - y_1 = m(x - x_1)$. The slope of this line segment is $\dfrac{-6 - 1}{4 - (-2)} = -\dfrac{7}{6}$, so an equation is $y - 1 = -\frac{7}{6}(x + 2)$. The function is $f(x) = -\frac{7}{6}x - \frac{4}{3}$, $-2 \leq x \leq 4$.

43. We need to solve the given equation for y. $\quad x + (y - 1)^2 = 0 \quad \Leftrightarrow \quad (y - 1)^2 = -x \quad \Leftrightarrow \quad y - 1 = \pm\sqrt{-x} \quad \Leftrightarrow$ $y = 1 \pm \sqrt{-x}$. The expression with the positive radical represents the top half of the parabola, and the one with the negative radical represents the bottom half. Hence, we want $f(x) = 1 - \sqrt{-x}$. Note that the domain is $x \leq 0$.

45. Let the length and width of the rectangle be L and W. Then the perimeter is $2L + 2W = 20$ and the area is $A = LW$. Solving the first equation for W in terms of L gives $W = \dfrac{20 - 2L}{2} = 10 - L$. Thus, $A(L) = L(10 - L) = 10L - L^2$. Since lengths are positive, the domain of A is $0 < L < 10$. If we further restrict L to be larger than W, then $5 < L < 10$ would be the domain.

47. Let the length of a side of the equilateral triangle be x. Then by the Pythagorean Theorem, the height y of the triangle satisfies

$y^2 + \left(\frac{1}{2}x\right)^2 = x^2$, so that $y^2 = x^2 - \frac{1}{4}x^2 = \frac{3}{4}x^2$ and $y = \frac{\sqrt{3}}{2}x$. Using the formula for the area A of a triangle,

$A = \frac{1}{2}(\text{base})(\text{height})$, we obtain $A(x) = \frac{1}{2}(x)\left(\frac{\sqrt{3}}{2}x\right) = \frac{\sqrt{3}}{4}x^2$, with domain $x > 0$.

49. Let each side of the base of the box have length x, and let the height of the box be h. Since the volume is 2, we know that

$2 = hx^2$, so that $h = 2/x^2$, and the surface area is $S = x^2 + 4xh$. Thus, $S(x) = x^2 + 4x(2/x^2) = x^2 + (8/x)$, with

domain $x > 0$.

51. (a)

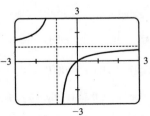

(b) On \$14,000, tax is assessed on \$4000, and $10\%(\$4000) = \$400$.

On \$26,000, tax is assessed on \$16,000, and

$10\%(\$10,000) + 15\%(\$6000) = \$1000 + \$900 = \$1900$.

(c) As in part (b), there is \$1000 tax assessed on \$20,000 of income, so the graph of T is a line segment from $(10,000, 0)$ to $(20,000, 1000)$. The tax on \$30,000 is \$2500, so the graph of T for $x > 20,000$ is the ray with initial point $(20,000, 1000)$ that passes through $(30,000, 2500)$.

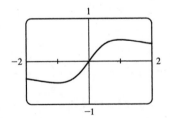

53. f is an odd function because its graph is symmetric about the origin. g is an even function because its graph is symmetric with respect to the y-axis.

55. (a) Because an even function is symmetric with respect to the y-axis, and the point $(5, 3)$ is on the graph of this even function, the point $(-5, 3)$ must also be on its graph.

(b) Because an odd function is symmetric with respect to the origin, and the point $(5, 3)$ is on the graph of this odd function, the point $(-5, -3)$ must also be on its graph.

57. $f(x) = \dfrac{x}{x^2 + 1}$.

$f(-x) = \dfrac{-x}{(-x)^2 + 1} = \dfrac{-x}{x^2 + 1} = -\dfrac{x}{x^2 + 1} = -f(x)$.

So f is an odd function.

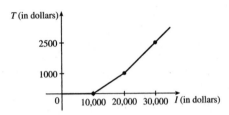

59. $f(x) = \dfrac{x}{x + 1}$, so $f(-x) = \dfrac{-x}{-x + 1} = \dfrac{x}{x - 1}$.

Since this is neither $f(x)$ nor $-f(x)$, the function f is

neither even nor odd.

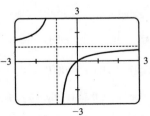

61. $f(x) = 1 + 3x^2 - x^4$.

$f(-x) = 1 + 3(-x)^2 - (-x)^4 = 1 + 3x^2 - x^4 = f(x)$.

So f is an even function.

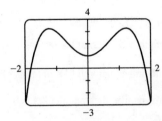

1.2   A Catalog of Essential Functions

1. (a) An equation for the family of linear functions with slope 2

is $y = f(x) = 2x + b$, where b is the y-intercept.

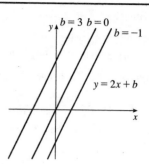

(b) $f(2) = 1$ means that the point $(2, 1)$ is on the graph of f. We can use the
point-slope form of a line to obtain an equation for the family of linear
functions through the point $(2, 1)$. $y - 1 = m(x - 2)$, which is equivalent
to $y = mx + (1 - 2m)$ in slope-intercept form.

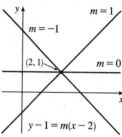

(c) To belong to both families, an equation must have slope $m = 2$, so the equation in part (b), $y = mx + (1 - 2m)$,
becomes $y = 2x - 3$. It is the *only* function that belongs to both families.

3. All members of the family of linear functions $f(x) = c - x$ have graphs
that are lines with slope -1. The y-intercept is c.

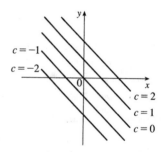

5. Since $f(-1) = f(0) = f(2) = 0$, f has zeros of -1, 0, and 2, so an equation for f is $f(x) = a[x - (-1)](x - 0)(x - 2)$,
or $f(x) = ax(x + 1)(x - 2)$. Because $f(1) = 6$, we'll substitute 1 for x and 6 for $f(x)$.

$6 = a(1)(2)(-1)$ $\Rightarrow$ $-2a = 6$ $\Rightarrow$ $a = -3$, so an equation for f is $f(x) = -3x(x + 1)(x - 2)$.

7. (a) $D = 200$, so $c = 0.0417D(a + 1) = 0.0417(200)(a + 1) = 8.34a + 8.34$. The slope is 8.34, which represents the
change in mg of the dosage for a child for each change of 1 year in age.

(b) For a newborn, $a = 0$, so $c = 8.34$ mg.

9. (a)

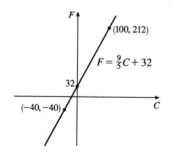

(b) The slope of $\frac{9}{5}$ means that F increases $\frac{9}{5}$ degrees for each increase
of $1°C$. (Equivalently, F increases by 9 when C increases by 5
and F decreases by 9 when C decreases by 5.) The F-intercept of
32 is the Fahrenheit temperature corresponding to a Celsius
temperature of 0.

11. (a) Using N in place of x and T in place of y, we find the slope to be $\dfrac{T_2 - T_1}{N_2 - N_1} = \dfrac{80 - 70}{173 - 113} = \dfrac{10}{60} = \dfrac{1}{6}$. So a linear

equation is $T - 80 = \frac{1}{6}(N - 173)$ $\Leftrightarrow$ $T - 80 = \frac{1}{6}N - \frac{173}{6}$ $\Leftrightarrow$ $T = \frac{1}{6}N + \frac{307}{6}$ $\left[\frac{307}{6} = 51.1\overline{6}\right]$.

(b) The slope of $\frac{1}{6}$ means that the temperature in Fahrenheit degrees increases one-sixth as rapidly as the number of cricket chirps per minute. Said differently, each increase of 6 cricket chirps per minute corresponds to an increase of $1°\text{F}$.

(c) When $N = 150$, the temperature is given approximately by $T = \frac{1}{6}(150) + \frac{307}{6} = 76.1\overline{6}°\text{F} \approx 76°\text{F}$.

13. (a) We are given $\dfrac{\text{change in pressure}}{10 \text{ feet change in depth}} = \dfrac{4.34}{10} = 0.434$. Using P for pressure and d for depth with the point

$(d, P) = (0, 15)$, we have the slope-intercept form of the line, $P = 0.434d + 15$.

(b) When $P = 100$, then $100 = 0.434d + 15$ $\Leftrightarrow$ $0.434d = 85$ $\Leftrightarrow$ $d = \frac{85}{0.434} \approx 195.85$ feet. Thus, the pressure is 100 lb/in^2 at a depth of approximately 196 feet.

15. (a) If the graph of f is shifted 3 units upward, its equation becomes $y = f(x) + 3$.

(b) If the graph of f is shifted 3 units downward, its equation becomes $y = f(x) - 3$.

(c) If the graph of f is shifted 3 units to the right, its equation becomes $y = f(x - 3)$.

(d) If the graph of f is shifted 3 units to the left, its equation becomes $y = f(x + 3)$.

(e) If the graph of f is reflected about the x-axis, its equation becomes $y = -f(x)$.

(f) If the graph of f is reflected about the y-axis, its equation becomes $y = f(-x)$.

(g) If the graph of f is stretched vertically by a factor of 3, its equation becomes $y = 3f(x)$.

(h) If the graph of f is shrunk vertically by a factor of 3, its equation becomes $y = \frac{1}{3}f(x)$.

17. (a) (graph 3) The graph of f is shifted 4 units to the right and has equation $y = f(x - 4)$.

(b) (graph 1) The graph of f is shifted 3 units upward and has equation $y = f(x) + 3$.

(c) (graph 4) The graph of f is shrunk vertically by a factor of 3 and has equation $y = \frac{1}{3}f(x)$.

(d) (graph 5) The graph of f is shifted 4 units to the left and reflected about the x-axis. Its equation is $y = -f(x + 4)$.

(e) (graph 2) The graph of f is shifted 6 units to the left and stretched vertically by a factor of 2. Its equation is
$y = 2f(x + 6)$.

19. (a) To graph $y = f(2x)$ we shrink the graph of f horizontally by a factor of 2.

(b) To graph $y = f\left(\frac{1}{2}x\right)$ we stretch the graph of f horizontally by a factor of 2.

The point $(4, -1)$ on the graph of f corresponds to the point $\left(\frac{1}{2} \cdot 4, -1\right) = (2, -1)$.

The point $(4, -1)$ on the graph of f corresponds to the point $(2 \cdot 4, -1) = (8, -1)$.

(c) To graph $y = f(-x)$ we reflect the graph of f about the y-axis.

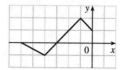

The point $(4, -1)$ on the graph of f corresponds to the point $(-1 \cdot 4, -1) = (-4, -1)$.

(d) To graph $y = -f(-x)$ we reflect the graph of f about the y-axis, then about the x-axis.

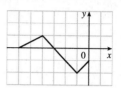

The point $(4, -1)$ on the graph of f corresponds to the point $(-1 \cdot 4, -1 \cdot -1) = (-4, 1)$.

21. $y = -x^3$: Start with the graph of $y = x^3$ and reflect about the x-axis. Note: Reflecting about the y-axis gives the same result since substituting $-x$ for x gives us $y = (-x)^3 = -x^3$.

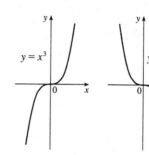

23. $y = (x + 1)^2$: Start with the graph of $y = x^2$ and shift 1 unit to the left.

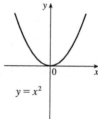

25. $y = 1 + 2\cos x$: Start with the graph of $y = \cos x$, stretch vertically by a factor of 2, and then shift 1 unit upward.

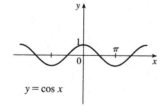 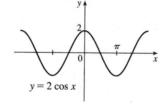 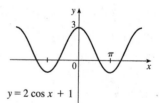

27. $y = \sin(x/2)$: Start with the graph of $y = \sin x$ and stretch horizontally by a factor of 2.

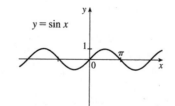 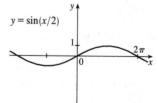

29. $y = \sqrt{x+3}$: Start with the graph of $y = \sqrt{x}$ and shift 3 units to the left.

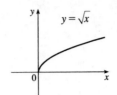

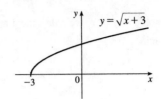

31. $y = \frac{1}{2}(x^2 + 8x) = \frac{1}{2}(x^2 + 8x + 16) - 8 = \frac{1}{2}(x+4)^2 - 8$: Start with the graph of $y = x^2$, compress vertically by a factor of 2, shift 4 units to the left, and then shift 8 units downward.

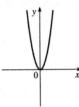
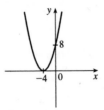
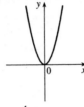
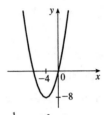

$y = x^2$ $y = \frac{1}{2}x^2$ $y = \frac{1}{2}(x+4)^2$ $y = \frac{1}{2}(x+4)^2 - 8$

33. $y = 2/(x+1)$: Start with the graph of $y = 1/x$, shift 1 unit to the left, and then stretch vertically by a factor of 2.

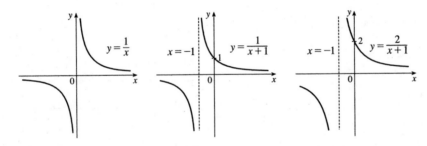

35. $f(x) = x^3 + 2x^2$; $g(x) = 3x^2 - 1$.　$D = \mathbb{R}$ for both f and g.

$(f + g)(x) = (x^3 + 2x^2) + (3x^2 - 1) = x^3 + 5x^2 - 1$, 　$D = \mathbb{R}$.

$(f - g)(x) = (x^3 + 2x^2) - (3x^2 - 1) = x^3 - x^2 + 1$, 　$D = \mathbb{R}$.

$(fg)(x) = (x^3 + 2x^2)(3x^2 - 1) = 3x^5 + 6x^4 - x^3 - 2x^2$, 　$D = \mathbb{R}$.

$\left(\dfrac{f}{g}\right)(x) = \dfrac{x^3 + 2x^2}{3x^2 - 1}$, 　$D = \left\{ x \mid x \neq \pm\dfrac{1}{\sqrt{3}} \right\}$ since $3x^2 - 1 \neq 0$.

37. $f(x) = x^2 - 1$, $D = \mathbb{R}$; $g(x) = 2x + 1$, $D = \mathbb{R}$.

(a) $(f \circ g)(x) = f(g(x)) = f(2x + 1) = (2x + 1)^2 - 1 = (4x^2 + 4x + 1) - 1 = 4x^2 + 4x$, $D = \mathbb{R}$.

(b) $(g \circ f)(x) = g(f(x)) = g(x^2 - 1) = 2(x^2 - 1) + 1 = (2x^2 - 2) + 1 = 2x^2 - 1$, $D = \mathbb{R}$.

(c) $(f \circ f)(x) = f(f(x)) = f(x^2 - 1) = (x^2 - 1)^2 - 1 = (x^4 - 2x^2 + 1) - 1 = x^4 - 2x^2$, $D = \mathbb{R}$.

(d) $(g \circ g)(x) = g(g(x)) = g(2x + 1) = 2(2x + 1) + 1 = (4x + 2) + 1 = 4x + 3$, $D = \mathbb{R}$.

39. $f(x) = \sin x$, 　$D = \mathbb{R}$; 　$g(x) = 1 - \sqrt{x}$, 　$D = [0, \infty)$.

(a) $(f \circ g)(x) = f(g(x)) = f(1 - \sqrt{x}) = \sin(1 - \sqrt{x})$, $D = [0, \infty)$.

(b) $(g \circ f)(x) = g(f(x)) = g(\sin x) = 1 - \sqrt{\sin x}$.

For $\sqrt{\sin x}$ to be defined, we must have $\sin x \geq 0 \iff$

$x \in [0, \pi] \cup [2\pi, 3\pi] \cup [-2\pi, -\pi] \cup [4\pi, 5\pi] \cup [-4\pi, -3\pi] \cup \ldots$, so $D = \{x \mid x \in [2n\pi, \pi + 2n\pi]$, where n is an integer$\}$.

(c) $(f \circ f)(x) = f(f(x)) = f(\sin x) = \sin(\sin x)$, $D = \mathbb{R}$.

(d) $(g \circ g)(x) = g(g(x)) = g(1 - \sqrt{x}) = 1 - \sqrt{1 - \sqrt{x}}$,

$D = \{x \geq 0 \mid 1 - \sqrt{x} \geq 0\} = \{x \geq 0 \mid 1 \geq \sqrt{x}\} = \{x \geq 0 \mid \sqrt{x} \leq 1\} = [0, 1]$.

41. $f(x) = x + \dfrac{1}{x}$, $\quad D = \{x \mid x \neq 0\}$; $\quad g(x) = \dfrac{x+1}{x+2}$, $\quad D = \{x \mid x \neq -2\}$.

(a) $(f \circ g)(x) = f(g(x)) = f\left(\dfrac{x+1}{x+2}\right) = \dfrac{x+1}{x+2} + \dfrac{1}{\dfrac{x+1}{x+2}} = \dfrac{x+1}{x+2} + \dfrac{x+2}{x+1}$

$\qquad = \dfrac{(x+1)(x+1) + (x+2)(x+2)}{(x+2)(x+1)} = \dfrac{(x^2 + 2x + 1) + (x^2 + 4x + 4)}{(x+2)(x+1)} = \dfrac{2x^2 + 6x + 5}{(x+2)(x+1)}$

Since $g(x)$ is not defined for $x = -2$ and $f(g(x))$ is not defined for $x = -2$ and $x = -1$, the domain of $(f \circ g)(x)$ is $D = \{x \mid x \neq -2, -1\}$.

(b) $(g \circ f)(x) = g(f(x)) = g\left(x + \dfrac{1}{x}\right) = \dfrac{\left(x + \dfrac{1}{x}\right) + 1}{\left(x + \dfrac{1}{x}\right) + 2} = \dfrac{\dfrac{x^2 + 1 + x}{x}}{\dfrac{x^2 + 1 + 2x}{x}} = \dfrac{x^2 + x + 1}{x^2 + 2x + 1} = \dfrac{x^2 + x + 1}{(x+1)^2}$

Since $f(x)$ is not defined for $x = 0$ and $g(f(x))$ is not defined for $x = -1$, the domain of $(g \circ f)(x)$ is $D = \{x \mid x \neq -1, 0\}$.

(c) $(f \circ f)(x) = f(f(x)) = f\left(x + \dfrac{1}{x}\right) = \left(x + \dfrac{1}{x}\right) + \dfrac{1}{x + \dfrac{1}{x}} = x + \dfrac{1}{x} + \dfrac{1}{\dfrac{x^2+1}{x}} = x + \dfrac{1}{x} + \dfrac{x}{x^2 + 1}$

$\qquad = \dfrac{x(x)(x^2 + 1) + 1(x^2 + 1) + x(x)}{x(x^2 + 1)} = \dfrac{x^4 + x^2 + x^2 + 1 + x^2}{x(x^2 + 1)}$

$\qquad = \dfrac{x^4 + 3x^2 + 1}{x(x^2 + 1)}$, $\quad D = \{x \mid x \neq 0\}$.

(d) $(g \circ g)(x) = g(g(x)) = g\left(\dfrac{x+1}{x+2}\right) = \dfrac{\dfrac{x+1}{x+2} + 1}{\dfrac{x+1}{x+2} + 2} = \dfrac{\dfrac{x+1+1(x+2)}{x+2}}{\dfrac{x+1+2(x+2)}{x+2}} = \dfrac{x+1+x+2}{x+1+2x+4} = \dfrac{2x+3}{3x+5}$

Since $g(x)$ is not defined for $x = -2$ and $g(g(x))$ is not defined for $x = -\frac{5}{3}$, the domain of $(g \circ g)(x)$ is $D = \{x \mid x \neq -2, -\frac{5}{3}\}$.

43. $(f \circ g \circ h)(x) = f(g(h(x))) = f(g(x+3)) = f((x+3)^2 + 2) = f(x^2 + 6x + 11) = \sqrt{(x^2 + 6x + 11) - 1} = \sqrt{x^2 + 6x + 10}$

45. Let $g(x) = x^2 + 1$ and $f(x) = x^{10}$. Then $(f \circ g)(x) = f(g(x)) = f(x^2 + 1) = (x^2 + 1)^{10} = F(x)$.

47. Let $g(t) = \cos t$ and $f(t) = \sqrt{t}$. Then $(f \circ g)(t) = f(g(t)) = f(\cos t) = \sqrt{\cos t} = u(t)$.

49. Let $h(x) = x^2$, $g(x) = 3^x$, and $f(x) = 1 - x$. Then

$(f \circ g \circ h)(x) = f(g(h(x))) = f(g(x^2)) = f\left(3^{x^2}\right) = 1 - 3^{x^2} = H(x)$.

51. Let $h(x) = \sqrt{x}$, $g(x) = \sec x$, and $f(x) = x^4$. Then

$$(f \circ g \circ h)(x) = f(g(h(x))) = f(g(\sqrt{x})) = f(\sec\sqrt{x}) = (\sec\sqrt{x})^4 = \sec^4(\sqrt{x}) = H(x).$$

53. (a) $g(2) = 5$, because the point $(2, 5)$ is on the graph of g. Thus, $f(g(2)) = f(5) = 4$, because the point $(5, 4)$ is on the graph of f.

(b) $g(f(0)) = g(0) = 3$

(c) $(f \circ g)(0) = f(g(0)) = f(3) = 0$

(d) $(g \circ f)(6) = g(f(6)) = g(6)$. This value is not defined, because there is no point on the graph of g that has x-coordinate 6.

(e) $(g \circ g)(-2) = g(g(-2)) = g(1) = 4$

(f) $(f \circ f)(4) = f(f(4)) = f(2) = -2$

55. (a) Using the relationship *distance* = *rate* · *time* with the radius r as the distance, we have $r(t) = 60t$.

(b) $A = \pi r^2 \Rightarrow (A \circ r)(t) = A(r(t)) = \pi(60t)^2 = 3600\pi t^2$. This formula gives us the extent of the rippled area (in cm^2) at any time t.

57. (a)

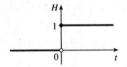

$$H(t) = \begin{cases} 0 & \text{if } t < 0 \\ 1 & \text{if } t \geq 0 \end{cases}$$

(b)

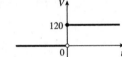

$$V(t) = \begin{cases} 0 & \text{if } t < 0 \\ 120 & \text{if } t \geq 0 \end{cases} \quad \text{so } V(t) = 120H(t).$$

(c)

Starting with the formula in part (b), we replace 120 with 240 to reflect the different voltage. Also, because we are starting 5 units to the right of $t = 0$, we replace t with $t - 5$. Thus, the formula is $V(t) = 240H(t - 5)$.

59. If $f(x) = m_1 x + b_1$ and $g(x) = m_2 x + b_2$, then

$$(f \circ g)(x) = f(g(x)) = f(m_2 x + b_2) = m_1(m_2 x + b_2) + b_1 = m_1 m_2 x + m_1 b_2 + b_1.$$

So $f \circ g$ is a linear function with slope $m_1 m_2$.

61. (a) By examining the variable terms in g and h, we deduce that we must square g to get the terms $4x^2$ and $4x$ in h. If we let $f(x) = x^2 + c$, then $(f \circ g)(x) = f(g(x)) = f(2x + 1) = (2x + 1)^2 + c = 4x^2 + 4x + (1 + c)$. Since $h(x) = 4x^2 + 4x + 7$, we must have $1 + c = 7$. So $c = 6$ and $f(x) = x^2 + 6$.

(b) We need a function g so that $f(g(x)) = 3(g(x)) + 5 = h(x)$. But $h(x) = 3x^2 + 3x + 2 = 3(x^2 + x) + 2 = 3(x^2 + x - 1) + 5$, so we see that $g(x) = x^2 + x - 1$.

63. (a) If f and g are even functions, then $f(-x) = f(x)$ and $g(-x) = g(x)$.

(i) $(f + g)(-x) = f(-x) + g(-x) = f(x) + g(x) = (f + g)(x)$, so $f + g$ is an *even* function.

(ii) $(fg)(-x) = f(-x) \cdot g(-x) = f(x) \cdot g(x) = (fg)(x)$, so fg is an *even* function.

(b) If f and g are odd functions, then $f(-x) = -f(x)$ and $g(-x) = -g(x)$.

(i) $(f + g)(-x) = f(-x) + g(-x) = -f(x) + [-g(x)] = -[f(x) + g(x)] = -(f + g)(x)$, so $f + g$ is an *odd* function.

(ii) $(fg)(-x) = f(-x) \cdot g(-x) = -f(x) \cdot [-g(x)] = f(x) \cdot g(x) = (fg)(x)$, so fg is an *even* function.

65. We need to examine $h(-x)$: $h(-x) = (f \circ g)(-x) = f(g(-x)) = f(g(x))$ [because g is even] $= h(x)$

Because $h(-x) = h(x)$, h is an even function.

1.3 The Limit of a Function

1. (a) $y = y(t) = 40t - 16t^2$. At $t = 2$, $y = 40(2) - 16(2)^2 = 16$. The average velocity between times 2 and $2 + h$ is

$$v_{\text{ave}} = \frac{y(2 + h) - y(2)}{(2 + h) - 2} = \frac{\left[40(2 + h) - 16(2 + h)^2\right] - 16}{h} = \frac{-24h - 16h^2}{h} = -24 - 16h, \text{ if } h \neq 0.$$

(i) $[2, 2.5]$: $h = 0.5$, $v_{\text{ave}} = -32$ ft/s   (ii) $[2, 2.1]$: $h = 0.1$, $v_{\text{ave}} = -25.6$ ft/s

(iii) $[2, 2.05]$: $h = 0.05$, $v_{\text{ave}} = -24.8$ ft/s   (iv) $[2, 2.01]$: $h = 0.01$, $v_{\text{ave}} = -24.16$ ft/s

(b) The instantaneous velocity when $t = 2$ (h approaches 0) is -24 ft/s.

3. (a) $f(x)$ approaches 2 as x approaches 1 from the left, so $\lim\limits_{x \to 1^-} f(x) = 2$.

(b) $f(x)$ approaches 3 as x approaches 1 from the right, so $\lim\limits_{x \to 1^+} f(x) = 3$.

(c) $\lim\limits_{x \to 1} f(x)$ does not exist because the limits in part (a) and part (b) are not equal.

(d) $f(x)$ approaches 4 as x approaches 5 from the left and from the right, so $\lim\limits_{x \to 5} f(x) = 4$.

(e) $f(5)$ is not defined, so it doesn't exist.

5. (a) $\lim\limits_{t \to 0^-} g(t) = -1$ (b) $\lim\limits_{t \to 0^+} g(t) = -2$

(c) $\lim\limits_{t \to 0} g(t)$ does not exist because the limits in part (a) and part (b) are not equal.

(d) $\lim\limits_{t \to 2^-} g(t) = 2$ (e) $\lim\limits_{t \to 2^+} g(t) = 0$

(f) $\lim\limits_{t \to 2} g(t)$ does not exist because the limits in part (d) and part (e) are not equal.

(g) $g(2) = 1$ (h) $\lim\limits_{t \to 4} g(t) = 3$

7. $\lim\limits_{x \to 1^-} f(x) = 2$, $\lim\limits_{x \to 1^+} f(x) = -2$, $f(1) = 2$

9. $\lim\limits_{x \to 3^+} f(x) = 4$, $\lim\limits_{x \to 3^-} f(x) = 2$, $\lim\limits_{x \to -2} f(x) = 2$,

$f(3) = 3$, $f(-2) = 1$

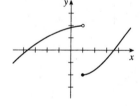

11. For $f(x) = \dfrac{x^2 - 2x}{x^2 - x - 2}$:

x	$f(x)$	x	$f(x)$
2.5	0.714286	1.9	0.655172
2.1	0.677419	1.95	0.661017
2.05	0.672131	1.99	0.665552
2.01	0.667774	1.995	0.666110
2.005	0.667221	1.999	0.666556
2.001	0.666778		

It appears that $\lim\limits_{x \to 2} \dfrac{x^2 - 2x}{x^2 - x - 2} = 0.\bar{6} = \frac{2}{3}$.

13. For $f(x) = \dfrac{\sin x}{x + \tan x}$:

x	$f(x)$
± 1	0.329033
± 0.5	0.458209
± 0.2	0.493331
± 0.1	0.498333
± 0.05	0.499583
± 0.01	0.499983

It appears that $\lim\limits_{x \to 0} \dfrac{\sin x}{x + \tan x} = 0.5 = \dfrac{1}{2}$.

15. For $f(x) = \dfrac{\sqrt{x+4}-2}{x}$:

x	$f(x)$
1	0.236068
0.5	0.242641
0.1	0.248457
0.05	0.249224
0.01	0.249844

x	$f(x)$
-1	0.267949
-0.5	0.258343
-0.1	0.251582
-0.05	0.250786
-0.01	0.250156

It appears that $\lim\limits_{x\to 0} \dfrac{\sqrt{x+4}-2}{x} = 0.25 = \frac14$.

17. For $f(x) = \dfrac{x^6 - 1}{x^{10} - 1}$:

x	$f(x)$
0.5	0.985337
0.9	0.719397
0.95	0.660186
0.99	0.612018
0.999	0.601200

x	$f(x)$
1.5	0.183369
1.1	0.484119
1.05	0.540783
1.01	0.588022
1.001	0.598800

It appears that $\lim\limits_{x\to 1} \dfrac{x^6 - 1}{x^{10} - 1} = 0.6 = \frac35$.

19. (a) From the graphs, it seems that $\lim\limits_{x\to 0} \dfrac{\tan 4x}{x} = 4$.

(b)

x	$f(x)$
± 0.1	4.227932
± 0.01	4.002135
± 0.001	4.000021
± 0.0001	4.000000

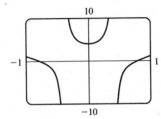
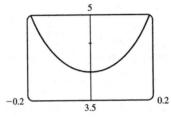

21. For $f(x) = x^2 - (2^x/1000)$:

(a)

x	$f(x)$
1	0.998000
0.8	0.638259
0.6	0.358484
0.4	0.158680
0.2	0.038851
0.1	0.008928
0.05	0.001465

It appears that $\lim\limits_{x\to 0} f(x) = 0$.

(b)

x	$f(x)$
0.04	0.000572
0.02	-0.000614
0.01	-0.000907
0.005	-0.000978
0.003	-0.000993
0.001	-0.001000

It appears that $\lim\limits_{x\to 0} f(x) = -0.001$.

23. On the left side of $x = 2$, we need $|x - 2| < \left|\frac{10}{7} - 2\right| = \frac47$. On the right side, we need $|x - 2| < \left|\frac{10}{3} - 2\right| = \frac43$. For both of these conditions to be satisfied at once, we need the more restrictive of the two to hold, that is, $|x - 2| < \frac47$. So we can choose $\delta = \frac47$, or any smaller positive number.

25. $\left|\sqrt{4x+1} - 3\right| < 0.5 \quad \Leftrightarrow \quad 2.5 < \sqrt{4x+1} < 3.5$. We plot the three parts of this inequality on the same screen and identify the x-coordinates of the points of intersection using the cursor. It appears that the inequality holds for

$1.3125 \le x \le 2.8125$. Since $|2 - 1.3125| = 0.6875$ and

$|2 - 2.8125| = 0.8125$, we choose $0 < \delta < \min\{0.6875, 0.8125\} = 0.6875$.

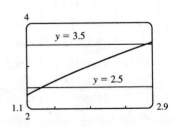

27. (a) $A = \pi r^2$ and $A = 1000$ cm^2 $\Rightarrow$ $\pi r^2 = 1000$ $\Rightarrow$ $r^2 = \frac{1000}{\pi}$ $\Rightarrow$ $r = \sqrt{\frac{1000}{\pi}}$ $[r > 0]$ ≈ 17.8412 cm.

(b) $|A - 1000| \le 5$ $\Rightarrow$ $-5 \le \pi r^2 - 1000 \le 5$ $\Rightarrow$ $1000 - 5 \le \pi r^2 \le 1000 + 5$ $\Rightarrow$

$\sqrt{\frac{995}{\pi}} \le r \le \sqrt{\frac{1005}{\pi}}$ $\Rightarrow$ $17.7966 \le r \le 17.8858$. $\sqrt{\frac{1000}{\pi}} - \sqrt{\frac{995}{\pi}} \approx 0.04466$ and $\sqrt{\frac{1005}{\pi}} - \sqrt{\frac{1000}{\pi}} \approx 0.04455$.

So if the machinist gets the radius within 0.0445 cm of 17.8412, the area will be within 5 cm^2 of 1000.

(c) x is the radius, $f(x)$ is the area, a is the target radius given in part (a), L is the target area (1000), ε is the tolerance in the area (5), and δ is the tolerance in the radius given in part (b).

29. Given $\varepsilon > 0$, we need $\delta > 0$ such that if $0 < |x - 1| < \delta$, then

$|(2x + 3) - 5| < \varepsilon$. But $|(2x + 3) - 5| < \varepsilon$ $\Leftrightarrow$ $|2x - 2| < \varepsilon$

$\Leftrightarrow$ $2|x - 1| < \varepsilon$ $\Leftrightarrow$ $|x - 1| < \varepsilon/2$. So if we choose $\delta = \varepsilon/2$,

then $0 < |x - 1| < \delta$ $\Rightarrow$ $|(2x + 3) - 5| < \varepsilon$. Thus,

$\lim_{x \to 1} (2x + 3) = 5$ by the definition of a limit.

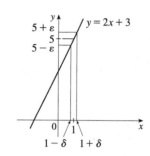

31. Given $\varepsilon > 0$, we need $\delta > 0$ such that if $0 < |x - (-3)| < \delta$, then

$|(1 - 4x) - 13| < \varepsilon$. But $|(1 - 4x) - 13| < \varepsilon$ $\Leftrightarrow$

$|-4x - 12| < \varepsilon$ $\Leftrightarrow$ $|-4| |x + 3| < \varepsilon$ $\Leftrightarrow$ $|x - (-3)| < \varepsilon/4$.

So if we choose $\delta = \varepsilon/4$, then $0 < |x - (-3)| < \delta$ $\Rightarrow$

$|(1 - 4x) - 13| < \varepsilon$. Thus, $\lim_{x \to -3} (1 - 4x) = 13$ by the definition of

a limit.

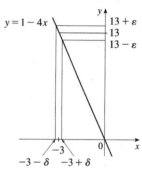

33. Given $\varepsilon > 0$, we need $\delta > 0$ such that if $0 < |x - 3| < \delta$, then $\left| \frac{x}{5} - \frac{3}{5} \right| < \varepsilon$ $\Leftrightarrow$ $\frac{1}{5} |x - 3| < \varepsilon$ $\Leftrightarrow$ $|x - 3| < 5\varepsilon$. So

choose $\delta = 5\varepsilon$. Then $0 < |x - 3| < \delta$ $\Rightarrow$ $|x - 3| < 5\varepsilon$ $\Rightarrow$ $\frac{|x - 3|}{5} < \varepsilon$ $\Rightarrow$ $\left| \frac{x}{5} - \frac{3}{5} \right| < \varepsilon$. By the definition of a

limit, $\lim_{x \to 3} \frac{x}{5} = \frac{3}{5}$.

35. Given $\varepsilon > 0$, we need $\delta > 0$ such that if $0 < |x - (-5)| < \delta$, then $\left| (4 - \frac{3}{5}x) - 7 \right| < \varepsilon$ $\Leftrightarrow$

$\left| -\frac{3}{5}x - 3 \right| < \varepsilon$ $\Leftrightarrow$ $\frac{3}{5} |x + 5| < \varepsilon$ $\Leftrightarrow$ $|x - (-5)| < \frac{5}{3}\varepsilon$. So choose $\delta = \frac{5}{3}\varepsilon$. Then $|x - (-5)| < \delta$ $\Rightarrow$

$\left| (4 - \frac{3}{5}x) - 7 \right| < \varepsilon$. Thus, $\lim_{x \to -5} \left(4 - \frac{3}{5}x \right) = 7$ by the definition of a limit.

37. Given $\varepsilon > 0$, we need $\delta > 0$ such that if $0 < |x - a| < \delta$, then $|x - a| < \varepsilon$. So $\delta = \varepsilon$ will work.

39. Given $\varepsilon > 0$, we need $\delta > 0$ such that if $0 < |x - 0| < \delta$, then $\left| x^2 - 0 \right| < \varepsilon$ $\Leftrightarrow$ $x^2 < \varepsilon$ $\Leftrightarrow$ $|x| < \sqrt{\varepsilon}$. Take $\delta = \sqrt{\varepsilon}$.

Then $0 < |x - 0| < \delta$ $\Rightarrow$ $\left| x^2 - 0 \right| < \varepsilon$. Thus, $\lim_{x \to 0} x^2 = 0$ by the definition of a limit.

41. Given $\varepsilon > 0$, we need $\delta > 0$ such that if $0 < |x - 0| < \delta$, then $||x| - 0| < \varepsilon$. But $||x|| = |x|$. So this is true if we pick $\delta = \varepsilon$.

Thus, $\lim_{x \to 0} |x| = 0$ by the definition of a limit.

43. Given $\varepsilon > 0$, we need $\delta > 0$ such that if $0 < |x - 3| < \delta$, then $|x^2 - 9| < \varepsilon$ $\Leftrightarrow$ $|(x - 3)(x + 3)| < \varepsilon$. Notice that if

$|x - 3| < 1$, then $-1 < x - 3 < 1$ $\Rightarrow$ $5 < x + 3 < 7$ $\Rightarrow$ $|x + 3| < 7$. So take $\delta = \min\{1, \varepsilon/7\}$. Then

$0 < |x - 3| < \delta$ $\Leftrightarrow$ $|(x - 3)(x + 3)| < |7(x - 3)| = 7 \cdot |x - 3| < 7\delta \le \varepsilon$. Thus, $\lim\limits_{x \to 3} x^2 = 9$ by the definition

of a limit.

45. (a) The points of intersection in the graph are $(x_1, 2.6)$ and $(x_2, 3.4)$

with $x_1 \approx 0.891$ and $x_2 \approx 1.093$. Thus, we can take δ to be the

smaller of $1 - x_1$ and $x_2 - 1$. So $\delta = x_2 - 1 \approx 0.093$.

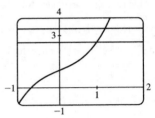

(b) Solving $x^3 + x + 1 = 3 + \varepsilon$ gives us two nonreal complex roots and one real root, which is

$$x(\varepsilon) = \frac{\left(216 + 108\varepsilon + 12\sqrt{336 + 324\varepsilon + 81\varepsilon^2}\,\right)^{2/3} - 12}{6\left(216 + 108\varepsilon + 12\sqrt{336 + 324\varepsilon + 81\varepsilon^2}\,\right)^{1/3}}. \text{ Thus, } \delta = x(\varepsilon) - 1.$$

(c) If $\varepsilon = 0.4$, then $x(\varepsilon) \approx 1.093\,272\,342$ and $\delta = x(\varepsilon) - 1 \approx 0.093$, which agrees with our answer in part (a).

1.4 Calculating Limits

1. (a) $\lim\limits_{x \to a} [f(x) + h(x)] = \lim\limits_{x \to a} f(x) + \lim\limits_{x \to a} h(x) = -3 + 8 = 5$ (b) $\lim\limits_{x \to a} [f(x)]^2 = \left[\lim\limits_{x \to a} f(x)\right]^2 = (-3)^2 = 9$

(c) $\lim\limits_{x \to a} \sqrt[3]{h(x)} = \sqrt[3]{\lim\limits_{x \to a} h(x)} = \sqrt[3]{8} = 2$ (d) $\lim\limits_{x \to a} \dfrac{1}{f(x)} = \dfrac{1}{\lim\limits_{x \to a} f(x)} = \dfrac{1}{-3} = -\dfrac{1}{3}$

(e) $\lim\limits_{x \to a} \dfrac{f(x)}{h(x)} = \dfrac{\lim\limits_{x \to a} f(x)}{\lim\limits_{x \to a} h(x)} = \dfrac{-3}{8} = -\dfrac{3}{8}$ (f) $\lim\limits_{x \to a} \dfrac{g(x)}{f(x)} = \dfrac{\lim\limits_{x \to a} g(x)}{\lim\limits_{x \to a} f(x)} = \dfrac{0}{-3} = 0$

(g) The limit does not exist, since $\lim\limits_{x \to a} g(x) = 0$ but $\lim\limits_{x \to a} f(x) \ne 0$.

(h) $\lim\limits_{x \to a} \dfrac{2f(x)}{h(x) - f(x)} = \dfrac{2\lim\limits_{x \to a} f(x)}{\lim\limits_{x \to a} h(x) - \lim\limits_{x \to a} f(x)} = \dfrac{2(-3)}{8 - (-3)} = -\dfrac{6}{11}$

3. $\lim\limits_{x \to -2} (3x^4 + 2x^2 - x + 1) = \lim\limits_{x \to -2} 3x^4 + \lim\limits_{x \to -2} 2x^2 - \lim\limits_{x \to -2} x + \lim\limits_{x \to -2} 1$ [Limit Laws 1 and 2]

$= 3 \lim\limits_{x \to -2} x^4 + 2 \lim\limits_{x \to -2} x^2 - \lim\limits_{x \to -2} x + \lim\limits_{x \to -2} 1$ [3]

$= 3(-2)^4 + 2(-2)^2 - (-2) + (1)$ [9, 8, and 7]

$= 48 + 8 + 2 + 1 = 59$

5. $\lim\limits_{x \to 8} (1 + \sqrt[3]{x})(2 - 6x^2 + x^3) = \lim\limits_{x \to 8} (1 + \sqrt[3]{x}) \cdot \lim\limits_{x \to 8} (2 - 6x^2 + x^3)$ [Limit Law 4]

$= \left(\lim\limits_{x \to 8} 1 + \lim\limits_{x \to 8} \sqrt[3]{x}\right) \cdot \left(\lim\limits_{x \to 8} 2 - 6 \lim\limits_{x \to 8} x^2 + \lim\limits_{x \to 8} x^3\right)$ [1, 2, and 3]

$= (1 + \sqrt[3]{8}) \cdot (2 - 6 \cdot 8^2 + 8^3)$ [7, 10, 9]

$= (3)(130) = 390$

7. $\lim\limits_{x \to 1} \left(\dfrac{1 + 3x}{1 + 4x^2 + 3x^4} \right)^3 = \left(\lim\limits_{x \to 1} \dfrac{1 + 3x}{1 + 4x^2 + 3x^4} \right)^3$ [6]

$$= \left[\dfrac{\lim\limits_{x \to 1} (1 + 3x)}{\lim\limits_{x \to 1} (1 + 4x^2 + 3x^4)} \right]^3 \qquad \text{[5]}$$

$$= \left[\dfrac{\lim\limits_{x \to 1} 1 + 3 \lim\limits_{x \to 1} x}{\lim\limits_{x \to 1} 1 + 4 \lim\limits_{x \to 1} x^2 + 3 \lim\limits_{x \to 1} x^4} \right]^3 \qquad \text{[2, 1, and 3]}$$

$$= \left[\dfrac{1 + 3(1)}{1 + 4(1)^2 + 3(1)^4} \right]^3 = \left[\dfrac{4}{8} \right]^3 = \left(\dfrac{1}{2} \right)^3 = \dfrac{1}{8} \qquad \text{[7, 8, and 9]}$$

9. $\lim\limits_{\theta \to \pi/2} \theta \sin\theta = \left(\lim\limits_{\theta \to \pi/2} \theta \right) \left(\lim\limits_{\theta \to \pi/2} \sin\theta \right)$ [4]

$$= \dfrac{\pi}{2} \cdot \sin\dfrac{\pi}{2} \qquad \text{[8 and Direct Substitution Property]}$$

$$= \dfrac{\pi}{2}$$

11. $\lim\limits_{x \to 2} \dfrac{x^2 + x - 6}{x - 2} = \lim\limits_{x \to 2} \dfrac{(x + 3)(x - 2)}{x - 2} = \lim\limits_{x \to 2} (x + 3) = 2 + 3 = 5$

13. $\lim\limits_{x \to 2} \dfrac{x^2 - x + 6}{x - 2}$ does not exist since $x - 2 \to 0$ but $x^2 - x + 6 \to 8$ as $x \to 2$.

15. $\lim\limits_{t \to -3} \dfrac{t^2 - 9}{2t^2 + 7t + 3} = \lim\limits_{t \to -3} \dfrac{(t + 3)(t - 3)}{(2t + 1)(t + 3)} = \lim\limits_{t \to -3} \dfrac{t - 3}{2t + 1} = \dfrac{-3 - 3}{2(-3) + 1} = \dfrac{-6}{-5} = \dfrac{6}{5}$

17. $\lim\limits_{h \to 0} \dfrac{(4 + h)^2 - 16}{h} = \lim\limits_{h \to 0} \dfrac{(16 + 8h + h^2) - 16}{h} = \lim\limits_{h \to 0} \dfrac{8h + h^2}{h} = \lim\limits_{h \to 0} \dfrac{h(8 + h)}{h} = \lim\limits_{h \to 0} (8 + h) = 8 + 0 = 8$

19. By the formula for the sum of cubes, we have

$$\lim\limits_{x \to -2} \dfrac{x + 2}{x^3 + 8} = \lim\limits_{x \to -2} \dfrac{x + 2}{(x + 2)(x^2 - 2x + 4)} = \lim\limits_{x \to -2} \dfrac{1}{x^2 - 2x + 4} = \dfrac{1}{4 + 4 + 4} = \dfrac{1}{12}.$$

21. $\lim\limits_{x \to 7} \dfrac{\sqrt{x + 2} - 3}{x - 7} = \lim\limits_{x \to 7} \dfrac{\sqrt{x + 2} - 3}{x - 7} \cdot \dfrac{\sqrt{x + 2} + 3}{\sqrt{x + 2} + 3} = \lim\limits_{x \to 7} \dfrac{(x + 2) - 9}{(x - 7)\left(\sqrt{x + 2} + 3\right)}$

$$= \lim\limits_{x \to 7} \dfrac{x - 7}{(x - 7)\left(\sqrt{x + 2} + 3\right)} = \lim\limits_{x \to 7} \dfrac{1}{\sqrt{x + 2} + 3} = \dfrac{1}{\sqrt{9} + 3} = \dfrac{1}{6}$$

23. $\lim\limits_{x \to -4} \dfrac{\frac{1}{4} + \frac{1}{x}}{4 + x} = \lim\limits_{x \to -4} \dfrac{\frac{x + 4}{4x}}{4 + x} = \lim\limits_{x \to -4} \dfrac{x + 4}{4x(4 + x)} = \lim\limits_{x \to -4} \dfrac{1}{4x} = \dfrac{1}{4(-4)} = -\dfrac{1}{16}$

25. (a)

$$\lim\limits_{x \to 0} \dfrac{x}{\sqrt{1 + 3x} - 1} \approx \dfrac{2}{3}$$

(b)

x	$f(x)$
-0.001	0.6661663
-0.0001	0.6666167
-0.00001	0.6666617
-0.000001	0.6666662
0.000001	0.6666672
0.00001	0.6666717
0.0001	0.6667167
0.001	0.6671663

The limit appears to be $\dfrac{2}{3}$.

(c) $\lim\limits_{x \to 0} \left(\dfrac{x}{\sqrt{1+3x}-1} \cdot \dfrac{\sqrt{1+3x}+1}{\sqrt{1+3x}+1} \right) = \lim\limits_{x \to 0} \dfrac{x(\sqrt{1+3x}+1)}{(1+3x)-1} = \lim\limits_{x \to 0} \dfrac{x(\sqrt{1+3x}+1)}{3x}$

$= \dfrac{1}{3} \lim\limits_{x \to 0} \left(\sqrt{1+3x}+1 \right)$ [Limit Law 3]

$= \dfrac{1}{3} \left[\sqrt{\lim\limits_{x \to 0}(1+3x)} + \lim\limits_{x \to 0} 1 \right]$ [1 and 11]

$= \dfrac{1}{3} \left(\sqrt{\lim\limits_{x \to 0} 1 + 3 \lim\limits_{x \to 0} x} + 1 \right)$ [1, 3, and 7]

$= \dfrac{1}{3} \left(\sqrt{1+3\cdot 0}+1 \right)$ [7 and 8]

$= \dfrac{1}{3}(1+1) = \dfrac{2}{3}$

27. Let $f(x) = -x^2$, $g(x) = x^2 \cos 20\pi x$, and $h(x) = x^2$.

Then $-1 \le \cos 20\pi x \le 1 \;\Rightarrow\; -x^2 \le x^2 \cos 20\pi x \le x^2 \;\Rightarrow$

$f(x) \le g(x) \le h(x)$. So since $\lim\limits_{x \to 0} f(x) = \lim\limits_{x \to 0} h(x) = 0$, by the

Squeeze Theorem we have $\lim\limits_{x \to 0} g(x) = 0$.

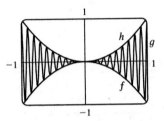

29. We have $\lim\limits_{x \to 4} (4x - 9) = 4(4) - 9 = 7$ and $\lim\limits_{x \to 4} \left(x^2 - 4x + 7 \right) = 4^2 - 4(4) + 7 = 7$. Since $4x - 9 \le f(x) \le x^2 - 4x + 7$

for $x \ge 0$, $\lim\limits_{x \to 4} f(x) = 7$ by the Squeeze Theorem.

31. $-1 \le \cos(2/x) \le 1 \;\Rightarrow\; -x^4 \le x^4 \cos(2/x) \le x^4$. Since $\lim\limits_{x \to 0} \left(-x^4 \right) = 0$ and $\lim\limits_{x \to 0} x^4 = 0$, we have

$\lim\limits_{x \to 0} \left[x^4 \cos(2/x) \right] = 0$ by the Squeeze Theorem.

33. $|x - 3| = \begin{cases} x - 3 & \text{if } x - 3 \ge 0 \\ -(x - 3) & \text{if } x - 3 < 0 \end{cases} = \begin{cases} x - 3 & \text{if } x \ge 3 \\ 3 - x & \text{if } x < 3 \end{cases}$

Thus, $\lim\limits_{x \to 3^+} (2x + |x - 3|) = \lim\limits_{x \to 3^+} (2x + x - 3) = \lim\limits_{x \to 3^+} (3x - 3) = 3(3) - 3 = 6$ and

$\lim\limits_{x \to 3^-} (2x + |x - 3|) = \lim\limits_{x \to 3^-} (2x + 3 - x) = \lim\limits_{x \to 3^-} (x + 3) = 3 + 3 = 6$. Since the left and right limits are equal,

$\lim\limits_{x \to 3} (2x + |x - 3|) = 6$.

35. Since $|x| = -x$ for $x < 0$, we have $\lim\limits_{x \to 0^-} \left(\dfrac{1}{x} - \dfrac{1}{|x|} \right) = \lim\limits_{x \to 0^-} \left(\dfrac{1}{x} - \dfrac{1}{-x} \right) = \lim\limits_{x \to 0^-} \dfrac{2}{x}$, which does not exist since the

denominator approaches 0 and the numerator does not.

37. (a) (i) If $x \to 1^+$, then $x > 1$ and $g(x) = x - 1$. Thus, $\lim\limits_{x \to 1^+} g(x) = \lim\limits_{x \to 1^+} (x - 1) = 1 - 1 = 0$.

(ii) If $x \to 1^-$, then $x < 1$ and $g(x) = 1 - x^2$. Thus, $\lim\limits_{x \to 1^-} g(x) = \lim\limits_{x \to 1^-} \left(1 - x^2 \right) = 1 - 1^2 = 0$.

Since the left- and right-hand limits of g at 1 are equal, $\lim\limits_{x \to 1} g(x) = 0$.

(iii) If $x \to 0$, then $-1 < x < 1$ and $g(x) = 1 - x^2$. Thus, $\lim\limits_{x \to 0} g(x) = \lim\limits_{x \to 0} (1 - x^2) = 1 - 0^2 = 1$.

(iv) If $x \to -1^-$, then $x < -1$ and $g(x) = -x$. Thus, $\lim\limits_{x \to -1^-} g(x) = \lim\limits_{x \to -1^-} (-x) = -(-1) = 1$.

(v) If $x \to -1^+$, then $-1 < x < 1$ and $g(x) = 1 - x^2$. Thus,

$$\lim_{x \to -1^+} g(x) = \lim_{x \to -1^+} (1 - x^2) = 1 - (-1)^2 = 1 - 1 = 0$$

(vi) $\lim\limits_{x \to -1} g(x)$ does not exist because the limits in part (iv) and part (v) are not equal.

(b)

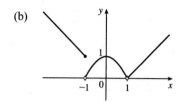

39. (a) (i) $[\![x]\!] = -2$ for $-2 \le x < -1$, so $\lim\limits_{x \to -2^+} [\![x]\!] = \lim\limits_{x \to -2^+} (-2) = -2$

(ii) $[\![x]\!] = -3$ for $-3 \le x < -2$, so $\lim\limits_{x \to -2^-} [\![x]\!] = \lim\limits_{x \to -2^-} (-3) = -3$.

The right and left limits are different, so $\lim\limits_{x \to -2} [\![x]\!]$ does not exist.

(iii) $[\![x]\!] = -3$ for $-3 \le x < -2$, so $\lim\limits_{x \to -2.4} [\![x]\!] = \lim\limits_{x \to -2.4} (-3) = -3$.

(b) (i) $[\![x]\!] = n - 1$ for $n - 1 \le x < n$, so $\lim\limits_{x \to n^-} [\![x]\!] = \lim\limits_{x \to n^-} (n - 1) = n - 1$.

(ii) $[\![x]\!] = n$ for $n \le x < n + 1$, so $\lim\limits_{x \to n^+} [\![x]\!] = \lim\limits_{x \to n^+} n = n$.

(c) $\lim\limits_{x \to a} [\![x]\!]$ exists $\Leftrightarrow$ a is not an integer.

41. The graph of $f(x) = [\![x]\!] + [\![-x]\!]$ is the same as the graph of $g(x) = -1$ with holes at each integer, since $f(a) = 0$ for any integer a. Thus, $\lim\limits_{x \to 2^-} f(x) = -1$ and $\lim\limits_{x \to 2^+} f(x) = -1$, so $\lim\limits_{x \to 2} f(x) = -1$. However,

$f(2) = [\![2]\!] + [\![-2]\!] = 2 + (-2) = 0$, so $\lim\limits_{x \to 2} f(x) \ne f(2)$.

43. $\lim\limits_{x \to 0} \dfrac{\sin 3x}{x} = \lim\limits_{x \to 0} \dfrac{3 \sin 3x}{3x}$ [multiply numerator and denominator by 3]

$\qquad = 3 \lim\limits_{3x \to 0} \dfrac{\sin 3x}{3x}$ [as $x \to 0,\ 3x \to 0$]

$\qquad = 3 \lim\limits_{\theta \to 0} \dfrac{\sin \theta}{\theta}$ [let $\theta = 3x$]

$\qquad = 3(1)$ [Equation 2]

$\qquad = 3$

45. $\lim\limits_{t \to 0} \dfrac{\tan 6t}{\sin 2t} = \lim\limits_{t \to 0} \left(\dfrac{\sin 6t}{t} \cdot \dfrac{1}{\cos 6t} \cdot \dfrac{t}{\sin 2t} \right) = \lim\limits_{t \to 0} \dfrac{6 \sin 6t}{6t} \cdot \lim\limits_{t \to 0} \dfrac{1}{\cos 6t} \cdot \lim\limits_{t \to 0} \dfrac{2t}{2 \sin 2t}$

$\qquad = 6 \lim\limits_{t \to 0} \dfrac{\sin 6t}{6t} \cdot \lim\limits_{t \to 0} \dfrac{1}{\cos 6t} \cdot \dfrac{1}{2} \lim\limits_{t \to 0} \dfrac{2t}{\sin 2t} = 6(1) \cdot \dfrac{1}{1} \cdot \dfrac{1}{2}(1) = 3$

47. Divide numerator and denominator by θ. ($\sin \theta$ also works.)

$$\lim_{\theta \to 0} \frac{\sin \theta}{\theta + \tan \theta} = \lim_{\theta \to 0} \frac{\dfrac{\sin \theta}{\theta}}{1 + \dfrac{\sin \theta}{\theta} \cdot \dfrac{1}{\cos \theta}} = \frac{\lim\limits_{\theta \to 0} \dfrac{\sin \theta}{\theta}}{1 + \lim\limits_{\theta \to 0} \dfrac{\sin \theta}{\theta} \lim\limits_{\theta \to 0} \dfrac{1}{\cos \theta}} = \frac{1}{1 + 1 \cdot 1} = \frac{1}{2}$$

49. Since $p(x)$ is a polynomial, $p(x) = a_0 + a_1 x + a_2 x^2 + \cdots + a_n x^n$. Thus, by the Limit Laws,

$$\lim_{x \to a} p(x) = \lim_{x \to a} \left(a_0 + a_1 x + a_2 x^2 + \cdots + a_n x^n\right) = a_0 + a_1 \lim_{x \to a} x + a_2 \lim_{x \to a} x^2 + \cdots + a_n \lim_{x \to a} x^n$$

$$= a_0 + a_1 a + a_2 a^2 + \cdots + a_n a^n = p(a)$$

Thus, for any polynomial p, $\lim_{x \to a} p(x) = p(a)$.

51. $\lim_{h \to 0} \sin(a + h) = \lim_{h \to 0} (\sin a \cos h + \cos a \sin h) = \lim_{h \to 0} (\sin a \cos h) + \lim_{h \to 0} (\cos a \sin h)$

$$= \left(\lim_{h \to 0} \sin a\right)\left(\lim_{h \to 0} \cos h\right) + \left(\lim_{h \to 0} \cos a\right)\left(\lim_{h \to 0} \sin h\right) = (\sin a)(1) + (\cos a)(0) = \sin a$$

53. Let $f(x) = [\![x]\!]$ and $g(x) = -[\![x]\!]$. Then $\lim_{x \to 3} f(x)$ and $\lim_{x \to 3} g(x)$ do not exist   [Example 8]

but $\lim_{x \to 3} [f(x) + g(x)] = \lim_{x \to 3} ([\![x]\!] - [\![x]\!]) = \lim_{x \to 3} 0 = 0$.

55. Since the denominator approaches 0 as $x \to -2$, the limit will exist only if the numerator also approaches

0 as $x \to -2$. In order for this to happen, we need $\lim_{x \to -2} \left(3x^2 + ax + a + 3\right) = 0$ $\Leftrightarrow$

$3(-2)^2 + a(-2) + a + 3 = 0$ $\Leftrightarrow$ $12 - 2a + a + 3 = 0$ $\Leftrightarrow$ $a = 15$. With $a = 15$, the limit becomes

$$\lim_{x \to -2} \frac{3x^2 + 15x + 18}{x^2 + x - 2} = \lim_{x \to -2} \frac{3(x + 2)(x + 3)}{(x - 1)(x + 2)} = \lim_{x \to -2} \frac{3(x + 3)}{x - 1} = \frac{3(-2 + 3)}{-2 - 1} = \frac{3}{-3} = -1.$$

1.5 Continuity

1. From Definition 1, $\lim_{x \to 4} f(x) = f(4)$.

3. (a) The following are the numbers at which f is discontinuous and the type of discontinuity at that number: -4 (removable), -2 (jump), 2 (jump), 4 (infinite).

(b) f is continuous from the left at -2 since $\lim_{x \to -2^-} f(x) = f(-2)$. f is continuous from the right at 2 and 4 since

$\lim_{x \to 2^+} f(x) = f(2)$ and $\lim_{x \to 4^+} f(x) = f(4)$. It is continuous from neither side at -4 since $f(-4)$ is undefined.

5. The graph of $y = f(x)$ must have a discontinuity at $x = 3$ and must show that $\lim_{x \to 3^-} f(x) = f(3)$.

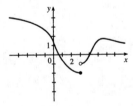

7. (a)

(b) There are discontinuities at times $t = 1, 2, 3,$ and 4. A person parking in the lot would want to keep in mind that the charge will jump at the beginning of each hour.

9. Since f and g are continuous functions,

$$\lim_{x \to 3} [2f(x) - g(x)] = 2 \lim_{x \to 3} f(x) - \lim_{x \to 3} g(x) \qquad \text{[by Limit Laws 2 and 3]}$$

$$= 2f(3) - g(3) \qquad \text{[by continuity of } f \text{ and } g \text{ at } x = 3]$$

$$= 2 \cdot 5 - g(3) = 10 - g(3)$$

Since it is given that $\lim_{x \to 3} [2f(x) - g(x)] = 4$, we have $10 - g(3) = 4$, so $g(3) = 6$.

11. $\lim_{x \to -1} f(x) = \lim_{x \to -1} (x + 2x^3)^4 = \left(\lim_{x \to -1} x + 2 \lim_{x \to -1} x^3 \right)^4 = [-1 + 2(-1)^3]^4 = (-3)^4 = 81 = f(-1)$.

By the definition of continuity, f is continuous at $a = -1$.

13. $f(x) = -\dfrac{1}{(x - 1)^2}$ is discontinuous at 1 since $f(1)$ is not defined.

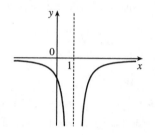

15. $f(x) = \begin{cases} 1 - x^2 & \text{if } x < 1 \\ 1/x & \text{if } x \geq 1 \end{cases}$

The left-hand limit of f at $a = 1$ is

$\lim_{x \to 1^-} f(x) = \lim_{x \to 1^-} (1 - x^2) = 0$. The right-hand limit of f at $a = 1$ is

$\lim_{x \to 1^+} f(x) = \lim_{x \to 1^+} (1/x) = 1$. Since these limits are not equal, $\lim_{x \to 1} f(x)$

does not exist and f is discontinuous at 1.

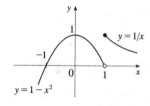

17. $F(x) = \dfrac{x}{x^2 + 5x + 6}$ is a rational function. So by Theorem 5 (or Theorem 6), F is continuous at every number in its domain,

$\{x \mid x^2 + 5x + 6 \neq 0\} = \{x \mid (x + 3)(x + 2) \neq 0\} = \{x \mid x \neq -3, \ -2\}$ or $(-\infty, -3) \cup (-3, -2) \cup (-2, \infty)$.

19. By Theorem 5, the polynomials x^2 and $2x - 1$ are continuous on $(-\infty, \infty)$. By Theorem 6, the root function $\sqrt{x}$ is

continuous on $[0, \infty)$. By Theorem 8, the composite function $\sqrt{2x - 1}$ is continuous on its domain, $[\frac{1}{2}, \infty)$. By part 1 of

Theorem 4, the sum $R(x) = x^2 + \sqrt{2x - 1}$ is continuous on $[\frac{1}{2}, \infty)$.

21. By Theorem 6, the root function $\sqrt{x}$ and the trigonometric function $\sin x$ are continuous on their domains, $[0, \infty)$ and

$(-\infty, \infty)$, respectively. Thus, the product $F(x) = \sqrt{x} \sin x$ is continuous on the intersection of those domains, $[0, \infty)$, by

part 4 of Theorem 4.

23.

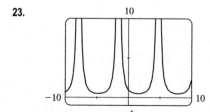

$y = \dfrac{1}{1 + \sin x}$ is undefined and hence discontinuous when

$1 + \sin x = 0 \iff \sin x = -1 \iff x = -\frac{\pi}{2} + 2\pi n$, n an

integer. The figure shows discontinuities for $n = -1$, 0, and 1; that

is, $-\dfrac{5\pi}{2} \approx -7.85$, $-\dfrac{\pi}{2} \approx -1.57$, and $\dfrac{3\pi}{2} \approx 4.71$.

25. Because we are dealing with root functions, $5 + \sqrt{x}$ is continuous on $[0, \infty)$, $\sqrt{x+5}$ is continuous on $[-5, \infty)$, so the

quotient $f(x) = \dfrac{5 + \sqrt{x}}{\sqrt{5 + x}}$ is continuous on $[0, \infty)$. Since f is continuous at $x = 4$, $\lim\limits_{x \to 4} f(x) = f(4) = \frac{7}{3}$.

27. $f(x) = \begin{cases} x^2 & \text{if } x < 1 \\ \sqrt{x} & \text{if } x \geq 1 \end{cases}$

By Theorem 5, since $f(x)$ equals the polynomial x^2 on $(-\infty, 1)$, f is continuous on $(-\infty, 1)$. By Theorem 6, since $f(x)$

equals the root function $\sqrt{x}$ on $(1, \infty)$, f is continuous on $(1, \infty)$. At $x = 1$, $\lim\limits_{x \to 1^-} f(x) = \lim\limits_{x \to 1^-} x^2 = 1$ and

$\lim\limits_{x \to 1^+} f(x) = \lim\limits_{x \to 1^+} \sqrt{x} = 1$. Thus, $\lim\limits_{x \to 1} f(x)$ exists and equals 1. Also, $f(1) = \sqrt{1} = 1$. Thus, f is continuous at $x = 1$.

We conclude that f is continuous on $(-\infty, \infty)$.

29. $f(x) = \begin{cases} x + 2 & \text{if } x < 0 \\ 2x^2 & \text{if } 0 \leq x \leq 1 \\ 2 - x & \text{if } x > 1 \end{cases}$

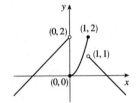

f is continuous on $(-\infty, 0)$, $(0, 1)$, and $(1, \infty)$ since on each of

these intervals it is a polynomial. Now $\lim\limits_{x \to 0^-} f(x) = \lim\limits_{x \to 0^-} (x + 2) = 2$ and

$\lim\limits_{x \to 0^+} f(x) = \lim\limits_{x \to 0^+} 2x^2 = 0$, so f is discontinuous at 0. Since $f(0) = 0$, f is continuous from the right at 0. Also

$\lim\limits_{x \to 1^-} f(x) = \lim\limits_{x \to 1^-} 2x^2 = 2$ and $\lim\limits_{x \to 1^+} f(x) = \lim\limits_{x \to 1^+} (2 - x) = 1$, so f is discontinuous at 1. Since $f(1) = 2$,

f is continuous from the left at 1.

31. $f(x) = \begin{cases} cx^2 + 2x & \text{if } x < 2 \\ x^3 - cx & \text{if } x \geq 2 \end{cases}$

f is continuous on $(-\infty, 2)$ and $(2, \infty)$. Now $\lim\limits_{x \to 2^-} f(x) = \lim\limits_{x \to 2^-} (cx^2 + 2x) = 4c + 4$ and

$\lim\limits_{x \to 2^+} f(x) = \lim\limits_{x \to 2^+} (x^3 - cx) = 8 - 2c$. So f is continuous $\iff$ $4c + 4 = 8 - 2c$ $\iff$ $6c = 4$ $\iff$ $c = \frac{2}{3}$. Thus, for f

to be continuous on $(-\infty, \infty)$, $c = \frac{2}{3}$.

33. (a) $f(x) = \dfrac{x^2 - 2x - 8}{x + 2} = \dfrac{(x - 4)(x + 2)}{x + 2}$ has a removable discontinuity at -2 because $g(x) = x - 4$ is continuous on $\mathbb{R}$

and $f(x) = g(x)$ for $x \neq -2$. [The discontinuity is removed by defining $f(-2) = -6$.]

(b) $f(x) = \dfrac{x - 7}{|x - 7|}$ $\Rightarrow$ $\lim\limits_{x \to 7^-} f(x) = -1$ and $\lim\limits_{x \to 7^+} f(x) = 1$. Thus, $\lim\limits_{x \to 7} f(x)$ does not exist, so the discontinuity is not

removable. (It is a jump discontinuity.)

(c) $f(x) = \dfrac{x^3 + 64}{x + 4} = \dfrac{(x + 4)(x^2 - 4x + 16)}{x + 4}$ has a removable discontinuity at -4 because $g(x) = x^2 - 4x + 16$ is

continuous on $\mathbb{R}$ and $f(x) = g(x)$ for $x \neq -4$. [The discontinuity is removed by defining $f(-4) = 48$.]

(d) $f(x) = \dfrac{3 - \sqrt{x}}{9 - x} = \dfrac{3 - \sqrt{x}}{(3 - \sqrt{x})(3 + \sqrt{x})}$ has a removable discontinuity at 9 because $g(x) = \dfrac{1}{3 + \sqrt{x}}$ is continuous on

$[0, \infty)$ and $f(x) = g(x)$ for $x \neq 9$. [The discontinuity is removed by defining $f(9) = \frac{1}{6}$.]

35. $f(x) = x^2 + 10\sin x$ is continuous on the interval $[31, 32]$, $f(31) \approx 957$, and $f(32) \approx 1030$. Since $957 < 1000 < 1030$, there is a number c in $(31, 32)$ such that $f(c) = 1000$ by the Intermediate Value Theorem.

37. $f(x) = x^4 + x - 3$ is continuous on the interval $[1, 2]$, $f(1) = -1$, and $f(2) = 15$. Since $-1 < 0 < 15$, there is a number c in $(1, 2)$ such that $f(c) = 0$ by the Intermediate Value Theorem. Thus, there is a root of the equation $x^4 + x - 3 = 0$ in the interval $(1, 2)$.

39. $f(x) = \cos x - x$ is continuous on the interval $[0, 1]$, $f(0) = 1$, and $f(1) = \cos 1 - 1 \approx -0.46$. Since $-0.46 < 0 < 1$, there is a number c in $(0, 1)$ such that $f(c) = 0$ by the Intermediate Value Theorem. Thus, there is a root of the equation $\cos x - x = 0$, or $\cos x = x$, in the interval $(0, 1)$.

41. (a) $f(x) = \cos x - x^3$ is continuous on the interval $[0, 1]$, $f(0) = 1 > 0$, and $f(1) = \cos 1 - 1 \approx -0.46 < 0$. Since $1 > 0 > -0.46$, there is a number c in $(0, 1)$ such that $f(c) = 0$ by the Intermediate Value Theorem. Thus, there is a root of the equation $\cos x - x^3 = 0$, or $\cos x = x^3$, in the interval $(0, 1)$.

(b) $f(0.86) \approx 0.016 > 0$ and $f(0.87) \approx -0.014 < 0$, so there is a root between 0.86 and 0.87, that is, in the interval $(0.86, 0.87)$.

43. (a) Let $f(x) = x^5 - x^2 - 4$. Then $f(1) = 1^5 - 1^2 - 4 = -4 < 0$ and $f(2) = 2^5 - 2^2 - 4 = 24 > 0$. So by the Intermediate Value Theorem, there is a number c in $(1, 2)$ such that $f(c) = c^5 - c^2 - 4 = 0$.

(b) We can see from the graphs that, correct to three decimal places, the root is $x \approx 1.434$.

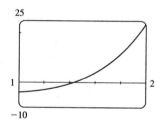

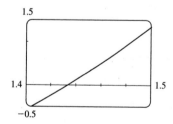

45. If there is such a number, it satisfies the equation $x^3 + 1 = x \;\Leftrightarrow\; x^3 - x + 1 = 0$. Let the left-hand side of this equation be called $f(x)$. Now $f(-2) = -5 < 0$, and $f(-1) = 1 > 0$. Note also that $f(x)$ is a polynomial, and thus continuous. So by the Intermediate Value Theorem, there is a number c between -2 and -1 such that $f(c) = 0$, so that $c = c^3 + 1$.

47. Define $u(t)$ to be the monk's distance from the monastery, as a function of time, on the first day, and define $d(t)$ to be his distance from the monastery, as a function of time, on the second day. Let D be the distance from the monastery to the top of the mountain. From the given information we know that $u(0) = 0$, $u(12) = D$, $d(0) = D$ and $d(12) = 0$. Now consider the function $u - d$, which is clearly continuous. We calculate that $(u - d)(0) = -D$ and $(u - d)(12) = D$. So by the Intermediate Value Theorem, there must be some time t_0 between 0 and 12 such that $(u - d)(t_0) = 0 \;\Leftrightarrow\; u(t_0) = d(t_0)$. So at time t_0 after 7:00 AM, the monk will be at the same place on both days.

1.6 Limits Involving Infinity

1. (a) $\lim\limits_{x\to 2} f(x) = \infty$ (b) $\lim\limits_{x\to -1^-} f(x) = \infty$ (c) $\lim\limits_{x\to -1^+} f(x) = -\infty$

   (d) $\lim\limits_{x\to \infty} f(x) = 1$ (e) $\lim\limits_{x\to -\infty} f(x) = 2$ (f) Vertical: $x = -1$, $x = 2$; Horizontal: $y = 1$, $y = 2$

3. $f(0) = 0$, $f(1) = 1$, $\lim\limits_{x\to \infty} f(x) = 0$,

    f is odd

5. $\lim\limits_{x\to 2} f(x) = -\infty$, $\lim\limits_{x\to \infty} f(x) = \infty$, $\lim\limits_{x\to -\infty} f(x) = 0$,

     $\lim\limits_{x\to 0^+} f(x) = \infty$, $\lim\limits_{x\to 0^-} f(x) = -\infty$

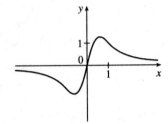

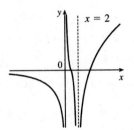

7. $f(0) = 3$, $\lim\limits_{x\to 0^-} f(x) = 4$, $\lim\limits_{x\to 0^+} f(x) = 2$, $\lim\limits_{x\to -\infty} f(x) = -\infty$, $\lim\limits_{x\to 4^-} f(x) = -\infty$, $\lim\limits_{x\to 4^+} f(x) = \infty$,

   $\lim\limits_{x\to \infty} f(x) = 3$

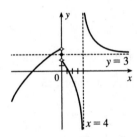

9. If $f(x) = x^2/2^x$, then a calculator gives $f(0) = 0$, $f(1) = 0.5$, $f(2) = 1$, $f(3) = 1.125$, $f(4) = 1$, $f(5) = 0.78125$,

   $f(6) = 0.5625$, $f(7) = 0.3828125$, $f(8) = 0.25$, $f(9) = 0.158203125$, $f(10) = 0.09765625$, $f(20) \approx 0.00038147$,

   $f(50) \approx 2.2204 \times 10^{-12}$, $f(100) \approx 7.8886 \times 10^{-27}$. It appears that $\lim\limits_{x\to \infty} \left(x^2/2^x \right) = 0$.

11. Vertical: $x \approx -1.62$, $x \approx 0.62$, $x = 1$;

     Horizontal: $y = 1$

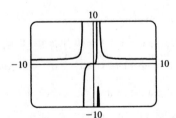

13. $\lim\limits_{x\to -3^+} \dfrac{x+2}{x+3} = -\infty$ since the numerator is negative and the denominator approaches 0 from the positive side as $x \to -3^+$.

15. $\lim\limits_{x\to 1}\dfrac{2-x}{(x-1)^2}=\infty$ since the numerator is positive and the denominator approaches 0 through positive values as $x\to 1$.

17. $\lim\limits_{x\to(-\pi/2)^-}\sec x=\lim\limits_{x\to(-\pi/2)^-}(1/\cos x)=-\infty$ since $\cos x\to 0$ as $x\to(-\pi/2)^-$ and $\cos x<0$ for $-\pi<x<-\pi/2$.

19. Divide both the numerator and denominator by x^3 (the highest power of x that occurs in the denominator).

$$\lim_{x\to\infty}\frac{x^3+5x}{2x^3-x^2+4}=\lim_{x\to\infty}\frac{\dfrac{x^3+5x}{x^3}}{\dfrac{2x^3-x^2+4}{x^3}}=\lim_{x\to\infty}\frac{1+\dfrac{5}{x^2}}{2-\dfrac{1}{x}+\dfrac{4}{x^3}}=\frac{\lim\limits_{x\to\infty}\left(1+\dfrac{5}{x^2}\right)}{\lim\limits_{x\to\infty}\left(2-\dfrac{1}{x}+\dfrac{4}{x^3}\right)}$$

$$=\frac{\lim\limits_{x\to\infty}1+5\lim\limits_{x\to\infty}\dfrac{1}{x^2}}{\lim\limits_{x\to\infty}2-\lim\limits_{x\to\infty}\dfrac{1}{x}+4\lim\limits_{x\to\infty}\dfrac{1}{x^3}}=\frac{1+5(0)}{2-0+4(0)}=\frac{1}{2}$$

21. First, multiply the factors in the denominator. Then divide both the numerator and denominator by u^4.

$$\lim_{u\to\infty}\frac{4u^4+5}{(u^2-2)(2u^2-1)}=\lim_{u\to\infty}\frac{4u^4+5}{2u^4-5u^2+2}=\lim_{u\to\infty}\frac{\dfrac{4u^4+5}{u^4}}{\dfrac{2u^4-5u^2+2}{u^4}}=\lim_{u\to\infty}\frac{4+\dfrac{5}{u^4}}{2-\dfrac{5}{u^2}+\dfrac{2}{u^4}}$$

$$=\frac{\lim\limits_{u\to\infty}\left(4+\dfrac{5}{u^4}\right)}{\lim\limits_{u\to\infty}\left(2-\dfrac{5}{u^2}+\dfrac{2}{u^4}\right)}=\frac{\lim\limits_{u\to\infty}4+5\lim\limits_{u\to\infty}\dfrac{1}{u^4}}{\lim\limits_{u\to\infty}2-5\lim\limits_{u\to\infty}\dfrac{1}{u^2}+2\lim\limits_{u\to\infty}\dfrac{1}{u^4}}=\frac{4+5(0)}{2-5(0)+2(0)}=\frac{4}{2}=2$$

23. $\lim\limits_{x\to\infty}\left(\sqrt{9x^2+x}-3x\right)=\lim\limits_{x\to\infty}\dfrac{\left(\sqrt{9x^2+x}-3x\right)\left(\sqrt{9x^2+x}+3x\right)}{\sqrt{9x^2+x}+3x}=\lim\limits_{x\to\infty}\dfrac{\left(\sqrt{9x^2+x}\right)^2-(3x)^2}{\sqrt{9x^2+x}+3x}$

$$=\lim_{x\to\infty}\frac{(9x^2+x)-9x^2}{\sqrt{9x^2+x}+3x}=\lim_{x\to\infty}\frac{x}{\sqrt{9x^2+x}+3x}\cdot\frac{1/x}{1/x}$$

$$=\lim_{x\to\infty}\frac{x/x}{\sqrt{9x^2/x^2+x/x^2}+3x/x}=\lim_{x\to\infty}\frac{1}{\sqrt{9+1/x}+3}=\frac{1}{\sqrt{9}+3}=\frac{1}{3+3}=\frac{1}{6}$$

25. $\lim\limits_{x\to\infty}\cos x$ does not exist because as x increases $\cos x$ does not approach any one value, but oscillates between 1 and -1.

27. $\lim\limits_{x\to\infty}\left(x-\sqrt{x}\right)=\lim\limits_{x\to\infty}\sqrt{x}\left(\sqrt{x}-1\right)=\infty$ since $\sqrt{x}\to\infty$ and $\sqrt{x}-1\to\infty$ as $x\to\infty$.

29. $\lim\limits_{x\to-\infty}\left(x^4+x^5\right)=\lim\limits_{x\to-\infty}x^5\left(\frac{1}{x}+1\right)$ [factor out the largest power of x] $=-\infty$ because $x^5\to-\infty$ and $1/x+1\to1$

as $x\to-\infty$.

31. $\lim\limits_{x\to\infty}\dfrac{x+x^3+x^5}{1-x^2+x^4}=\lim\limits_{x\to\infty}\dfrac{(x+x^3+x^5)/x^4}{(1-x^2+x^4)/x^4}$ [divide by the highest power of x in the denominator]

$$=\lim_{x\to\infty}\frac{1/x^3+1/x+x}{1/x^4-1/x^2+1}=\infty$$

because $(1/x^3+1/x+x)\to\infty$ and $(1/x^4-1/x^2+1)\to1$ as $x\to\infty$.

33. $\displaystyle\lim_{x\to\infty}\frac{2x^2+x-1}{x^2+x-2}=\lim_{x\to\infty}\frac{\dfrac{2x^2+x-1}{x^2}}{\dfrac{x^2+x-2}{x^2}}=\lim_{x\to\infty}\frac{2+\dfrac{1}{x}-\dfrac{1}{x^2}}{1+\dfrac{1}{x}-\dfrac{2}{x^2}}=\frac{\displaystyle\lim_{x\to\infty}\left(2+\dfrac{1}{x}-\dfrac{1}{x^2}\right)}{\displaystyle\lim_{x\to\infty}\left(1+\dfrac{1}{x}-\dfrac{2}{x^2}\right)}$

$\displaystyle=\frac{\displaystyle\lim_{x\to\infty}2+\lim_{x\to\infty}\dfrac{1}{x}-\lim_{x\to\infty}\dfrac{1}{x^2}}{\displaystyle\lim_{x\to\infty}1+\lim_{x\to\infty}\dfrac{1}{x}-2\lim_{x\to\infty}\dfrac{1}{x^2}}=\frac{2+0-0}{1+0-2(0)}=2$, so $y=2$ is a horizontal asymptote.

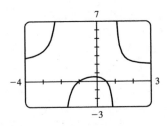

$y=f(x)=\dfrac{2x^2+x-1}{x^2+x-2}=\dfrac{(2x-1)(x+1)}{(x+2)(x-1)}$, so

$\displaystyle\lim_{x\to-2^-}f(x)=\infty,\quad\lim_{x\to-2^+}f(x)=-\infty,\quad\lim_{x\to1^-}f(x)=-\infty,$ and

$\displaystyle\lim_{x\to1^+}f(x)=\infty.$ Thus, $x=-2$ and $x=1$ are vertical asymptotes.

The graph confirms our work.

35. (a)

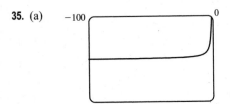

From the graph of $f(x)=\sqrt{x^2+x+1}+x$, we estimate

the value of $\displaystyle\lim_{x\to-\infty}f(x)$ to be -0.5.

(b)

x	$f(x)$
$-10{,}000$	-0.4999625
$-100{,}000$	-0.4999962
$-1{,}000{,}000$	-0.4999996

From the table, we estimate the limit

to be -0.5.

(c) $\displaystyle\lim_{x\to-\infty}\left(\sqrt{x^2+x+1}+x\right)=\lim_{x\to-\infty}\left(\sqrt{x^2+x+1}+x\right)\left[\frac{\sqrt{x^2+x+1}-x}{\sqrt{x^2+x+1}-x}\right]=\lim_{x\to-\infty}\frac{\left(x^2+x+1\right)-x^2}{\sqrt{x^2+x+1}-x}$

$\displaystyle=\lim_{x\to-\infty}\frac{(x+1)(1/x)}{\left(\sqrt{x^2+x+1}-x\right)(1/x)}=\lim_{x\to-\infty}\frac{1+(1/x)}{-\sqrt{1+(1/x)+(1/x^2)}-1}$

$\displaystyle=\frac{1+0}{-\sqrt{1+0+0}-1}=-\frac{1}{2}$

Note that for $x<0$, we have $\sqrt{x^2}=|x|=-x$, so when we divide the radical by x, with $x<0$, we get

$\dfrac{1}{x}\sqrt{x^2+x+1}=-\dfrac{1}{\sqrt{x^2}}\sqrt{x^2+x+1}=-\sqrt{1+(1/x)+(1/x^2)}.$

37. From the graph, it appears $y=1$ is a horizontal asymptote.

$\displaystyle\lim_{x\to\infty}\frac{3x^3+500x^2}{x^3+500x^2+100x+2000}=\lim_{x\to\infty}\frac{\dfrac{3x^3+500x^2}{x^3}}{\dfrac{x^3+500x^2+100x+2000}{x^3}}$

$\displaystyle=\lim_{x\to\infty}\frac{3+(500/x)}{1+(500/x)+(100/x^2)+(2000/x^3)}=\frac{3+0}{1+0+0+0}=3,$

so $y=3$ is a horizontal asymptote. The discrepancy can be explained by the

choice in the viewing window. Try $[-100{,}000,\,100{,}000]$ by $[-1,4]$ to get a

graph that lends credibility to our calculation that $y=3$ is a horizontal

asymptote.

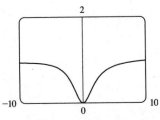

39. Let's look for a rational function.

(1) $\lim\limits_{x\to\pm\infty} f(x) = 0 \;\Rightarrow\;$ degree of numerator < degree of denominator

(2) $\lim\limits_{x\to 0} f(x) = -\infty \;\Rightarrow\;$ there is a factor of x^2 in the denominator (not just x, since that would produce a sign change at $x = 0$), and the function is negative near $x = 0$.

(3) $\lim\limits_{x\to 3^-} f(x) = \infty$ and $\lim\limits_{x\to 3^+} f(x) = -\infty \;\Rightarrow\;$ vertical asymptote at $x = 3$; there is a factor of $(x - 3)$ in the denominator.

(4) $f(2) = 0 \;\Rightarrow\;$ 2 is an x-intercept; there is at least one factor of $(x - 2)$ in the numerator.

Combining all of this information and putting in a negative sign to give us the desired left- and right-hand limits gives us

$$f(x) = \frac{2 - x}{x^2(x - 3)} \text{ as one possibility.}$$

41. Divide the numerator and the denominator by the highest power of x in $Q(x)$.

(a) If $\deg P < \deg Q$, then the numerator $\to 0$ but the denominator doesn't. So $\lim\limits_{x\to\infty} [P(x)/Q(x)] = 0$.

(b) If $\deg P > \deg Q$, then the numerator $\to \pm\infty$ but the denominator doesn't, so $\lim\limits_{x\to\infty} [P(x)/Q(x)] = \pm\infty$ (depending on the ratio of the leading coefficients of P and Q).

43. $\lim\limits_{x\to\infty} \dfrac{4x - 1}{x} = \lim\limits_{x\to\infty} \left(4 - \dfrac{1}{x}\right) = 4$ and $\lim\limits_{x\to\infty} \dfrac{4x^2 + 3x}{x^2} = \lim\limits_{x\to\infty} \left(4 + \dfrac{3}{x}\right) = 4$. Therefore, by the Squeeze Theorem,

$\lim\limits_{x\to\infty} f(x) = 4$.

45. (a) After t minutes, $25t$ liters of brine with 30 g of salt per liter has been pumped into the tank, so it contains $(5000 + 25t)$ liters of water and $25t \cdot 30 = 750t$ grams of salt. Therefore, the salt concentration at time t will be

$$C(t) = \frac{750t}{5000 + 25t} = \frac{30t}{200 + t} \, \frac{\text{g}}{\text{L}}.$$

(b) $\lim\limits_{t\to\infty} C(t) = \lim\limits_{t\to\infty} \dfrac{30t}{200 + t} = \lim\limits_{t\to\infty} \dfrac{30t/t}{200/t + t/t} = \dfrac{30}{0 + 1} = 30$. So the salt concentration approaches that of the brine being pumped into the tank.

47. $\dfrac{1}{(x + 3)^4} > 10{,}000 \;\Leftrightarrow\; (x + 3)^4 < \dfrac{1}{10{,}000} \;\Leftrightarrow\; |x + 3| < \dfrac{1}{\sqrt[4]{10{,}000}} \;\Leftrightarrow\; |x - (-3)| < \dfrac{1}{10}$

49. Let $N < 0$ be given. Then, for $x < -1$, we have $\dfrac{5}{(x + 1)^3} < N \;\Leftrightarrow\; \dfrac{5}{N} < (x + 1)^3 \;\Leftrightarrow\; \sqrt[3]{\dfrac{5}{N}} < x + 1$. Let

$\delta = -\sqrt[3]{\dfrac{5}{N}}$. Then $-1 - \delta < x < -1 \;\Rightarrow\; \sqrt[3]{\dfrac{5}{N}} < x + 1 < 0 \;\Rightarrow\; \dfrac{5}{(x + 1)^3} < N$, so $\lim\limits_{x\to -1^-} \dfrac{5}{(x + 1)^3} = -\infty$.

51. $\left| \dfrac{6x^2 + 5x - 3}{2x^2 - 1} - 3 \right| < 0.2 \;\Leftrightarrow\; 2.8 < \dfrac{6x^2 + 5x - 3}{2x^2 - 1} < 3.2$. So

we graph the three parts of this inequality on the same screen, and

find that the curve $y = \dfrac{6x^2 + 5x - 3}{2x^2 - 1}$ seems to lie between the lines

$y = 2.8$ and $y = 3.2$ whenever $x > 12.8$. So we can choose $N = 13$

(or any larger number) so that the inequality holds whenever $x \geq N$.

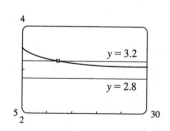

53. (a) $1/x^2 < 0.0001 \iff x^2 > 1/0.0001 = 10,000 \iff x > 100 \quad (x > 0)$

(b) If $\varepsilon > 0$ is given, then $1/x^2 < \varepsilon \iff x^2 > 1/\varepsilon \iff x > 1/\sqrt{\varepsilon}$. Let $N = 1/\sqrt{\varepsilon}$.

Then $x > N \implies x > \dfrac{1}{\sqrt{\varepsilon}} \implies \left|\dfrac{1}{x^2} - 0\right| = \dfrac{1}{x^2} < \varepsilon$, so $\lim\limits_{x \to \infty} \dfrac{1}{x^2} = 0$.

55. Suppose that $\lim\limits_{x \to \infty} f(x) = L$. Then for every $\varepsilon > 0$ there is a corresponding positive number N such that $|f(x) - L| < \varepsilon$

whenever $x > N$. If $t = 1/x$, then $x > N \iff 0 < 1/x < 1/N \iff 0 < t < 1/N$. Thus, for every $\varepsilon > 0$ there is a

corresponding $\delta > 0$ (namely $1/N$) such that $|f(1/t) - L| < \varepsilon$ whenever $0 < t < \delta$. This proves that

$\lim\limits_{t \to 0^+} f(1/t) = L = \lim\limits_{x \to \infty} f(x)$.

Now suppose that $\lim\limits_{x \to -\infty} f(x) = L$. Then for every $\varepsilon > 0$ there is a corresponding negative number N such that

$|f(x) - L| < \varepsilon$ whenever $x < N$. If $t = 1/x$, then $x < N \iff 1/N < 1/x < 0 \iff 1/N < t < 0$. Thus, for every

$\varepsilon > 0$ there is a corresponding $\delta > 0$ (namely $-1/N$) such that $|f(1/t) - L| < \varepsilon$ whenever $-\delta < t < 0$. This proves that

$\lim\limits_{t \to 0^-} f(1/t) = L = \lim\limits_{x \to -\infty} f(x)$.

1 Review

CONCEPT CHECK

1. (a) A **function** f is a rule that assigns to each element x in a set A exactly one element, called $f(x)$, in a set B. The set A is

called the **domain** of the function. The **range** of f is the set of all possible values of $f(x)$ as x varies throughout the

domain.

(b) If f is a function with domain A, then its **graph** is the set of ordered pairs $\{(x, f(x)) \mid x \in A\}$.

(c) Use the Vertical Line Test on page 4.

2. The four ways to represent a function are: verbally, numerically, visually, and algebraically. An example of each is given

below.

Verbally: An assignment of students to chairs in a classroom (a description in words)

Numerically: A tax table that assigns an amount of tax to an income (a table of values)

Visually: A graphical history of the Dow Jones average (a graph)

Algebraically: A relationship between distance, rate, and time: $d = rt$ (an explicit formula)

3. (a) An **even function** f satisfies $f(-x) = f(x)$ for every number x in its domain. It is symmetric with respect to the y-axis.

(b) An **odd function** g satisfies $g(-x) = -g(x)$ for every number x in its domain. It is symmetric with respect to the origin.

4. A **mathematical model** is a mathematical description (often by means of a function or an equation) of a real-world

phenomenon.

5. (a) Linear function: $f(x) = 2x + 1$, $f(x) = ax + b$ (b) Power function: $f(x) = x^2$, $f(x) = x^a$

(c) Exponential function: $f(x) = 2^x$, $f(x) = a^x$ (d) Quadratic function: $f(x) = x^2 + x + 1$, $f(x) = ax^2 + bx + c$

(e) Polynomial of degree 5: $f(x) = x^5 + 2$

(f) Rational function: $f(x) = \dfrac{x}{x + 2}$, $f(x) = \dfrac{P(x)}{Q(x)}$ where $P(x)$ and $Q(x)$ are polynomials

6.

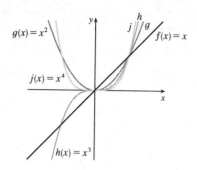

7. (a)

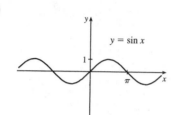

(b)

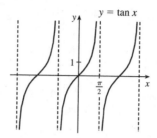

(c)

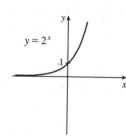

(d)

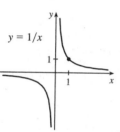

(e)

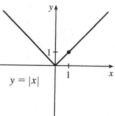

(f)

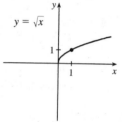

8. (a) The domain of $f + g$ is the intersection of the domain of f and the domain of g; that is, $A \cap B$.

(b) The domain of fg is also $A \cap B$.

(c) The domain of f/g must exclude values of x that make g equal to 0; that is, $\{x \in A \cap B \mid g(x) \neq 0\}$.

9. Given two functions f and g, the **composite function** $f \circ g$ is defined by $(f \circ g)(x) = f(g(x))$. The domain of $f \circ g$ is the set of all x in the domain of g such that $g(x)$ is in the domain of f.

10. (a) If the graph of f is shifted 2 units upward, its equation becomes $y = f(x) + 2$.

(b) If the graph of f is shifted 2 units downward, its equation becomes $y = f(x) - 2$.

(c) If the graph of f is shifted 2 units to the right, its equation becomes $y = f(x - 2)$.

(d) If the graph of f is shifted 2 units to the left, its equation becomes $y = f(x + 2)$.

(e) If the graph of f is reflected about the x-axis, its equation becomes $y = -f(x)$.

(f) If the graph of f is reflected about the y-axis, its equation becomes $y = f(-x)$.

(g) If the graph of f is stretched vertically by a factor of 2, its equation becomes $y = 2f(x)$.

(h) If the graph of f is shrunk vertically by a factor of 2, its equation becomes $y = \frac{1}{2}f(x)$.

(i) If the graph of f is stretched horizontally by a factor of 2, its equation becomes $y = f(\frac{1}{2})x$.

(j) If the graph of f is shrunk horizontally by a factor of 2, its equation becomes $y = f(2x)$.

11. (a) $\lim\limits_{x \to a} f(x) = L$: See Definition 1.3.1 and Figures 1 and 2 in Section 1.3.

(b) $\lim\limits_{x \to a^+} f(x) = L$: See the paragraph after Definition 1.3.2 and Figure 9(b) in Section 1.3.

(c) $\lim\limits_{x \to a^-} f(x) = L$: See Definition 1.3.2 and Figure 9(a) in Section 1.3.

(d) $\lim_{x \to a} f(x) = \infty$: See Definition 1.6.1 and Figure 2 in Section 1.6.

(e) $\lim_{x \to \infty} f(x) = L$: See Definition 1.6.3 and Figure 8 in Section 1.6.

12. In general, the limit of a function fails to exist when the function does not approach a fixed number. For each of the following functions, the limit fails to exist at $x = 2$.

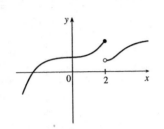

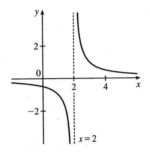

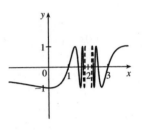

The left- and right-hand limits are not equal.

There is an infinite discontinuity.

There are an infinite number of oscillations.

13. (a)–(g) See the statements of Limit Laws 1–6 and 11 in Section 1.4.

14. See Theorem 1.4.4.

15. (a) A function f is continuous at a number a if $f(x)$ approaches $f(a)$ as x approaches a; that is, $\lim_{x \to a} f(x) = f(a)$.

(b) A function f is continuous on the interval $(-\infty, \infty)$ if f is continuous at every real number a. The graph of such a function has no break and every vertical line crosses it.

16. See Theorem 1.5.9.

17. (a) See Definition 1.6.2 and Figures 2–4 in Section 1.6.

(b) See Definition 1.6.4 and Figures 8 and 9 in Section 1.6.

TRUE-FALSE QUIZ

1. **False.** Let $f(x) = x^2$, $s = -1$, and $t = 1$. Then $f(s+t) = (-1+1)^2 = 0^2 = 0$, but

$$f(s) + f(t) = (-1)^2 + 1^2 = 2 \neq 0 = f(s+t).$$

3. **True.** See the Vertical Line Test.

5. **False.** Limit Law 2 applies only if the individual limits exist (these don't).

7. **True.** Limit Law 5 applies.

9. **False.** Consider $\lim_{x \to 5} \dfrac{x(x-5)}{x-5}$ or $\lim_{x \to 5} \dfrac{\sin(x-5)}{x-5}$. The first limit exists and is equal to 5. By Equation 1.4.5, we know that the latter limit exists (and it is equal to 1).

11. **True.** A polynomial is continuous everywhere, so $\lim_{x \to b} p(x)$ exists and is equal to $p(b)$.

13. **True.** See Figure 10 in Section 1.6.

15. **False.** Consider $f(x) = \begin{cases} 1/(x-1) & \text{if } x \neq 1 \\ 2 & \text{if } x = 1 \end{cases}$

17. **True.** Use Theorem 1.5.7 with $a = 2$, $b = 5$, and $g(x) = 4x^2 - 11$. Note that $f(4) = 3$ is not needed.

19. **True,** by the definition of a limit with $\varepsilon = 1$.

EXERCISES

1. (a) When $x = 2$, $y \approx 2.7$. Thus, $f(2) \approx 2.7$.

  (b) $f(x) = 3 \quad \Rightarrow \quad x \approx 2.3, 5.6$

  (c) The domain of f is $-6 \le x \le 6$, or $[-6, 6]$.

  (d) The range of f is $-4 \le y \le 4$, or $[-4, 4]$.

  (e) f is increasing on $[-4, 4]$, that is, on $-4 \le x \le 4$.

  (f) f is odd since its graph is symmetric about the origin.

3. $f(x) = \sqrt{4 - 3x^2}$.   Domain: $4 - 3x^2 \ge 0 \quad \Rightarrow \quad 3x^2 \le 4 \quad \Rightarrow \quad x^2 \le \frac{4}{3} \quad \Rightarrow \quad |x| \le \frac{2}{\sqrt{3}}$.

  Range: $y \ge 0$ and $y \le \sqrt{4} \quad \Rightarrow \quad 0 \le y \le 2$.

5. $y = 1 + \sin x$. Domain: $\mathbb{R}$. Range: $-1 \le \sin x \le 1 \quad \Rightarrow \quad 0 \le 1 + \sin x \le 2 \quad \Rightarrow \quad 0 \le y \le 2$.

7. (a) To obtain the graph of $y = f(x) + 8$, we shift the graph of $y = f(x)$ up 8 units.

  (b) To obtain the graph of $y = f(x + 8)$, we shift the graph of $y = f(x)$ left 8 units.

  (c) To obtain the graph of $y = 1 + 2f(x)$, we stretch the graph of $y = f(x)$ vertically by a factor of 2, and then shift the resulting graph 1 unit upward.

  (d) To obtain the graph of $y = f(x - 2) - 2$, we shift the graph of $y = f(x)$ right 2 units (for the "-2" inside the parentheses), and then shift the resulting graph 2 units downward.

  (e) To obtain the graph of $y = -f(x)$, we reflect the graph of $y = f(x)$ about the x-axis.

  (f) To obtain the graph of $y = 3 - f(x)$, we reflect the graph of $y = f(x)$ about the x-axis, and then shift the resulting graph 3 units upward.

9. $y = -\sin 2x$: Start with the graph of $y = \sin x$, compress horizontally by a factor of 2, and reflect about the x-axis.

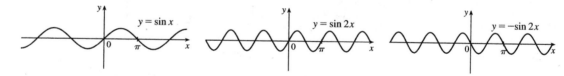

11. $y = 1 + \frac{1}{2}x^3$: Start with the graph of $y = x^3$, compress vertically by a factor of 2, and shift 1 unit upward.

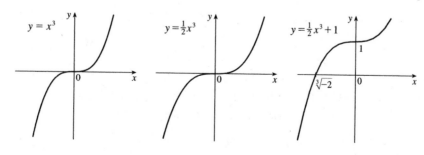

13. $f(x) = \dfrac{1}{x+2}$:

Start with the graph of $f(x) = 1/x$ and shift 2 units to the left.

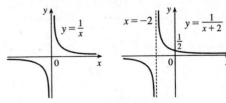

15. (a) The terms of f are a mixture of odd and even powers of x, so f is neither even nor odd.

(b) The terms of f are all odd powers of x, so f is odd.

(c) $f(-x) = \cos\big((-x)^2\big) = \cos(x^2) = f(x)$, so f is even.

(d) $f(-x) = 1 + \sin(-x) = 1 - \sin x$. Now $f(-x) \neq f(x)$ and $f(-x) \neq -f(x)$, so f is neither even nor odd.

17. $f(x) = \sqrt{x}$, $D = [0, \infty)$; $g(x) = \sin x$, $D = \mathbb{R}$.

(a) $(f \circ g)(x) = f(g(x)) = f(\sin x) = \sqrt{\sin x}$. For $\sqrt{\sin x}$ to be defined, we must have $\sin x \geq 0 \iff x \in [0, \pi]$,

$[2\pi, 3\pi]$, $[-2\pi, -\pi]$, $[4\pi, 5\pi]$, $[-4\pi, -3\pi]$, ..., so $D = \{x \mid x \in [2n\pi, \pi + 2n\pi]$, where n is an integer$\}$.

(b) $(g \circ f)(x) = g(f(x)) = g(\sqrt{x}) = \sin \sqrt{x}$. x must be greater than or equal to 0 for $\sqrt{x}$ to be defined, so $D = [0, \infty)$.

(c) $(f \circ f)(x) = f(f(x)) = f(\sqrt{x}) = \sqrt{\sqrt{x}} = \sqrt[4]{x}$. $D = [0, \infty)$.

(d) $(g \circ g)(x) = g(g(x)) = g(\sin x) = \sin(\sin x)$. $D = \mathbb{R}$.

19. The graphs of $f(x) = \sin^n x$, where n is a positive integer, all have domain $\mathbb{R}$. For odd n, the range is $[-1, 1]$ and for even n, the range is $[0, 1]$. For odd n, the functions are odd and symmetric with respect to the origin. For even n, the functions are even and symmetric with respect to the y-axis. As n becomes large, the graphs become less rounded and more "spiky."

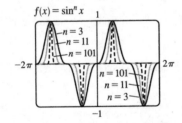 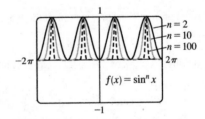

21. (a) (i) $\displaystyle\lim_{x \to 2^+} f(x) = 3$ (ii) $\displaystyle\lim_{x \to -3^+} f(x) = 0$

(iii) $\displaystyle\lim_{x \to -3} f(x)$ does not exist since the left and right limits are not equal. (The left limit is -2.)

(iv) $\displaystyle\lim_{x \to 4} f(x) = 2$

(v) $\displaystyle\lim_{x \to 0} f(x) = \infty$ (vi) $\displaystyle\lim_{x \to 2^-} f(x) = -\infty$

(vii) $\displaystyle\lim_{x \to \infty} f(x) = 4$ (viii) $\displaystyle\lim_{x \to -\infty} f(x) = -1$

(b) The equations of the horizontal asymptotes are $y = -1$ and $y = 4$.

(c) The equations of the vertical asymptotes are $x = 0$ and $x = 2$.

(d) f is discontinuous at $x = -3, 0, 2,$ and 4. The discontinuities are jump, infinite, infinite, and removable, respectively.

23. $\displaystyle\lim_{x \to 0} \cos(x + \sin x) = \cos\left[\lim_{x \to 0}(x + \sin x)\right]$ [by Theorem 1.5.7] $= \cos 0 = 1$

25. $\displaystyle\lim_{x \to -3} \frac{x^2 - 9}{x^2 + 2x - 3} = \lim_{x \to -3} \frac{(x+3)(x-3)}{(x+3)(x-1)} = \lim_{x \to -3} \frac{x-3}{x-1} = \frac{-3-3}{-3-1} = \frac{-6}{-4} = \frac{3}{2}$

27. $\lim\limits_{h\to 0} \dfrac{(h-1)^3+1}{h} = \lim\limits_{h\to 0} \dfrac{(h^3-3h^2+3h-1)+1}{h} = \lim\limits_{h\to 0} \dfrac{h^3-3h^2+3h}{h} = \lim\limits_{h\to 0}\left(h^2-3h+3\right) = 3$

Another solution: Factor the numerator as a sum of two cubes and then simplify.

$\lim\limits_{h\to 0} \dfrac{(h-1)^3+1}{h} = \lim\limits_{h\to 0} \dfrac{(h-1)^3+1^3}{h} = \lim\limits_{h\to 0} \dfrac{[(h-1)+1]\left[(h-1)^2-1(h-1)+1^2\right]}{h}$

$\qquad\qquad = \lim\limits_{h\to 0}\left[(h-1)^2-h+2\right] = 1-0+2 = 3$

29. $\lim\limits_{r\to 9} \dfrac{\sqrt{r}}{(r-9)^4} = \infty$ since $(r-9)^4 \to 0$ as $r\to 9$ and $\dfrac{\sqrt{r}}{(r-9)^4} > 0$ for $r\neq 9$.

31. $\lim\limits_{s\to 16} \dfrac{4-\sqrt{s}}{s-16} = \lim\limits_{s\to 16} \dfrac{4-\sqrt{s}}{(\sqrt{s}+4)(\sqrt{s}-4)} = \lim\limits_{s\to 16} \dfrac{-1}{\sqrt{s}+4} = \dfrac{-1}{\sqrt{16}+4} = -\dfrac{1}{8}$

33. $\lim\limits_{x\to\infty} \dfrac{1+2x-x^2}{1-x+2x^2} = \lim\limits_{x\to\infty} \dfrac{\left(1+2x-x^2\right)/x^2}{\left(1-x+2x^2\right)/x^2} = \lim\limits_{x\to\infty} \dfrac{1/x^2+2/x-1}{1/x^2-1/x+2} = \dfrac{0+0-1}{0-0+2} = -\dfrac{1}{2}$

35. $\lim\limits_{x\to\infty}\left(\sqrt{x^2+4x+1}-x\right) = \lim\limits_{x\to\infty}\left[\dfrac{\sqrt{x^2+4x+1}-x}{1}\cdot\dfrac{\sqrt{x^2+4x+1}+x}{\sqrt{x^2+4x+1}+x}\right] = \lim\limits_{x\to\infty} \dfrac{(x^2+4x+1)-x^2}{\sqrt{x^2+4x+1}+x}$

$\qquad\qquad = \lim\limits_{x\to\infty} \dfrac{(4x+1)/x}{\left(\sqrt{x^2+4x+1}+x\right)/x} \qquad \left[\text{divide by } x = \sqrt{x^2} \text{ for } x > 0\right]$

$\qquad\qquad = \lim\limits_{x\to\infty} \dfrac{4+1/x}{\sqrt{1+4/x+1/x^2}+1} = \dfrac{4+0}{\sqrt{1+0+0}+1} = \dfrac{4}{2} = 2$

37. $\lim\limits_{x\to 0} \dfrac{\cot 2x}{\csc x} = \lim\limits_{x\to 0} \dfrac{\cos 2x\,\sin x}{\sin 2x} = \lim\limits_{x\to 0}\cos 2x\left[\dfrac{(\sin x)/x}{(\sin 2x)/x}\right] = \lim\limits_{x\to 0}\cos 2x\left[\dfrac{\lim\limits_{x\to 0}[(\sin x)/x]}{2\lim\limits_{x\to 0}[(\sin 2x)/2x]}\right] = 1\cdot\dfrac{1}{2\cdot 1} = \dfrac{1}{2}$

39. From the graph of $y = (\cos^2 x)/x^2$, it appears that $y = 0$ is the horizontal

asymptote and $x = 0$ is the vertical asymptote. Now $0 \le (\cos x)^2 \le 1 \;\Rightarrow$

$\dfrac{0}{x^2} \le \dfrac{\cos^2 x}{x^2} \le \dfrac{1}{x^2} \;\Rightarrow\; 0 \le \dfrac{\cos^2 x}{x^2} \le \dfrac{1}{x^2}$. But $\lim\limits_{x\to\pm\infty} 0 = 0$ and

$\lim\limits_{x\to\pm\infty} \dfrac{1}{x^2} = 0$, so by the Squeeze Theorem, $\lim\limits_{x\to\pm\infty} \dfrac{\cos^2 x}{x^2} = 0$.

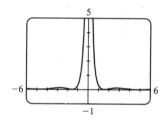

Thus, $y = 0$ is the horizontal asymptote. $\lim\limits_{x\to 0} \dfrac{\cos^2 x}{x^2} = \infty$ because $\cos^2 x \to 1$ and $x^2 \to 0$ as $x\to 0$, so $x = 0$ is the

vertical asymptote.

41. Since $2x-1 \le f(x) \le x^2$ for $0 < x < 3$ and $\lim\limits_{x\to 1}(2x-1) = 1 = \lim\limits_{x\to 1}x^2$, we have $\lim\limits_{x\to 1}f(x) = 1$ by the Squeeze Theorem.

43. Given $\varepsilon > 0$, we need $\delta > 0$ so that if $0 < |x-5| < \delta$, then $|(7x-27)-8| < \varepsilon \;\Leftrightarrow\; |7x-35| < \varepsilon \;\Leftrightarrow$

$|x-5| < \varepsilon/7$. So take $\delta = \varepsilon/7$. Then $0 < |x-5| < \delta \;\Rightarrow\; |(7x-27)-8| < \varepsilon$. Thus, $\lim\limits_{x\to 5}(7x-27) = 8$ by the

definition of a limit.

45. If $\varepsilon > 0$ is given, then $1/x^4 < \varepsilon \;\Leftrightarrow\; x^4 > 1/\varepsilon \;\Leftrightarrow\; x > 1/\sqrt[4]{\varepsilon}$. Let $N = 1/\sqrt[4]{\varepsilon}$.

Then $x > N \;\Rightarrow\; x > \dfrac{1}{\sqrt[4]{\varepsilon}} \;\Rightarrow\; \left|\dfrac{1}{x^4}-0\right| = \dfrac{1}{x^4} < \varepsilon$, so $\lim\limits_{x\to\infty} \dfrac{1}{x^4} = 0$.

47. (a) $f(x) = \sqrt{-x}$ if $x < 0$, $f(x) = 3 - x$ if $0 \le x < 3$, $f(x) = (x-3)^2$ if $x > 3$.

   (i) $\lim\limits_{x \to 0^+} f(x) = \lim\limits_{x \to 0^+} (3 - x) = 3$

   (iii) Because of (i) and (ii), $\lim\limits_{x \to 0} f(x)$ does not exist.

   (v) $\lim\limits_{x \to 3^+} f(x) = \lim\limits_{x \to 3^+} (x - 3)^2 = 0$

   (ii) $\lim\limits_{x \to 0^-} f(x) = \lim\limits_{x \to 0^-} \sqrt{-x} = 0$

   (iv) $\lim\limits_{x \to 3^-} f(x) = \lim\limits_{x \to 3^-} (3 - x) = 0$

   (vi) Because of (iv) and (v), $\lim\limits_{x \to 3} f(x) = 0$.

(b) f is discontinuous at 0 since $\lim\limits_{x \to 0} f(x)$ does not exist.

f is discontinuous at 3 since $f(3)$ does not exist.

(c)

49. $f(x) = 2x^3 + x^2 + 2$ is a polynomial, so it is continuous on $[-2, -1]$ and $f(-2) = -10 < 0 < 1 = f(-1)$. So by the Intermediate Value Theorem there is a number c in $(-2, -1)$ such that $f(c) = 0$, that is, the equation $2x^3 + x^2 + 2 = 0$ has a root in $(-2, -1)$.

2 ☐ DERIVATIVES

2.1 Derivatives and Rates of Change

1. (a) (i) Using Definition 1,

$$m = \lim_{x \to a} \frac{f(x) - f(a)}{x - a} \quad \lim_{x \to -3} \frac{f(x) - f(-3)}{x - (-3)} = \lim_{x \to -3} \frac{(x^2 + 2x) - (3)}{x - (-3)} = \lim_{x \to -3} \frac{(x + 3)(x - 1)}{x + 3}$$

$$= \lim_{x \to -3} (x - 1) = -4$$

(ii) Using Equation 2,

$$m = \lim_{h \to 0} \frac{f(a + h) - f(a)}{h} = \lim_{h \to 0} \frac{f(-3 + h) - f(-3)}{h} = \lim_{h \to 0} \frac{\left[(-3 + h)^2 + 2(-3 + h)\right] - (3)}{h}$$

$$= \lim_{h \to 0} \frac{9 - 6h + h^2 - 6 + 2h - 3}{h} = \lim_{h \to 0} \frac{h(h - 4)}{h} = \lim_{h \to 0} (h - 4) = -4$$

(b) Using the point-slope form of the equation of a line, an equation of the tangent line is $y - 3 = -4(x + 3)$. Solving for y gives us $y = -4x - 9$, which is the slope-intercept form of the equation of the tangent line.

(c)

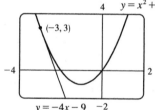
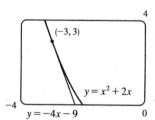

3. Using (1) with $f(x) = \dfrac{x - 1}{x - 2}$ and $P(3, 2)$,

$$m = \lim_{x \to a} \frac{f(x) - f(a)}{x - a} = \lim_{x \to 3} \frac{\dfrac{x - 1}{x - 2} - 2}{x - 3} = \lim_{x \to 3} \frac{\dfrac{x - 1 - 2(x - 2)}{x - 2}}{x - 3} = \lim_{x \to 3} \frac{3 - x}{(x - 2)(x - 3)} = \lim_{x \to 3} \frac{-1}{x - 2} = \frac{-1}{1} = -1.$$

Tangent line: $y - 2 = -1(x - 3) \quad \Leftrightarrow \quad y - 2 = -x + 3 \quad \Leftrightarrow \quad y = -x + 5$

5. Using (1), $m = \lim\limits_{x \to 1} \dfrac{\sqrt{x} - \sqrt{1}}{x - 1} = \lim\limits_{x \to 1} \dfrac{(\sqrt{x} - 1)(\sqrt{x} + 1)}{(x - 1)(\sqrt{x} + 1)} = \lim\limits_{x \to 1} \dfrac{x - 1}{(x - 1)(\sqrt{x} + 1)} = \lim\limits_{x \to 1} \dfrac{1}{\sqrt{x} + 1} = \dfrac{1}{2}.$

Tangent line: $y - 1 = \frac{1}{2}(x - 1) \quad \Leftrightarrow \quad y = \frac{1}{2}x + \frac{1}{2}$

7. (a) Using (2) with $y = f(x) = 3 + 4x^2 - 2x^3$,

$$m = \lim_{h \to 0} \frac{f(a+h) - f(a)}{h} = \lim_{h \to 0} \frac{3 + 4(a+h)^2 - 2(a+h)^3 - (3 + 4a^2 - 2a^3)}{h}$$

$$= \lim_{h \to 0} \frac{3 + 4(a^2 + 2ah + h^2) - 2(a^3 + 3a^2h + 3ah^2 + h^3) - 3 - 4a^2 + 2a^3}{h}$$

$$= \lim_{h \to 0} \frac{3 + 4a^2 + 8ah + 4h^2 - 2a^3 - 6a^2h - 6ah^2 - 2h^3 - 3 - 4a^2 + 2a^3}{h}$$

$$= \lim_{h \to 0} \frac{8ah + 4h^2 - 6a^2h - 6ah^2 - 2h^3}{h} = \lim_{h \to 0} \frac{h(8a + 4h - 6a^2 - 6ah - 2h^2)}{h}$$

$$= \lim_{h \to 0} (8a + 4h - 6a^2 - 6ah - 2h^2) = 8a - 6a^2$$

(b) At $(1, 5)$: $m = 8(1) - 6(1)^2 = 2$, so an equation of the tangent line

   is $y - 5 = 2(x - 1) \iff y = 2x + 3$.

   At $(2, 3)$: $m = 8(2) - 6(2)^2 = -8$, so an equation of the tangent

   line is $y - 3 = -8(x - 2) \iff y = -8x + 19$.

(c)

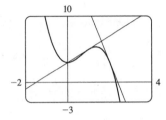

9. (a) Since the slope of the tangent at $t = 0$ is 0, the car's initial velocity was 0.

(b) The slope of the tangent is greater at C than at B, so the car was going faster at C.

(c) Near A, the tangent lines are becoming steeper as x increases, so the velocity was increasing, so the car was speeding up.

   Near B, the tangent lines are becoming less steep, so the car was slowing down. The steepest tangent near C is the one at

   C, so at C the car had just finished speeding up, and was about to start slowing down.

(d) Between D and E, the slope of the tangent is 0, so the car did not move during that time.

11. Let $s(t) = 40t - 16t^2$.

$$v(2) = \lim_{t \to 2} \frac{s(t) - s(2)}{t - 2} = \lim_{t \to 2} \frac{(40t - 16t^2) - 16}{t - 2} = \lim_{t \to 2} \frac{-16t^2 + 40t - 16}{t - 2} = \lim_{t \to 2} \frac{-8(2t^2 - 5t + 2)}{t - 2}$$

$$= \lim_{t \to 2} \frac{-8(t - 2)(2t - 1)}{t - 2} = -8 \lim_{t \to 2} (2t - 1) = -8(3) = -24$$

Thus, the instantaneous velocity when $t = 2$ is -24 ft/s.

13. $v(a) = \lim_{h \to 0} \dfrac{s(a+h) - s(a)}{h} = \lim_{h \to 0} \dfrac{\dfrac{1}{(a+h)^2} - \dfrac{1}{a^2}}{h} = \lim_{h \to 0} \dfrac{\dfrac{a^2 - (a+h)^2}{a^2(a+h)^2}}{h} = \lim_{h \to 0} \dfrac{a^2 - (a^2 + 2ah + h^2)}{ha^2(a+h)^2}$

$$= \lim_{h \to 0} \frac{-(2ah + h^2)}{ha^2(a+h)^2} = \lim_{h \to 0} \frac{-h(2a + h)}{ha^2(a+h)^2} = \lim_{h \to 0} \frac{-(2a+h)}{a^2(a+h)^2} = \frac{-2a}{a^2 \cdot a^2} = \frac{-2}{a^3} \text{ m/s}$$

So $v(1) = \dfrac{-2}{1^3} = -2$ m/s, $v(2) = \dfrac{-2}{2^3} = -\dfrac{1}{4}$ m/s, and $v(3) = \dfrac{-2}{3^3} = -\dfrac{2}{27}$ m/s.

15. $g'(0)$ is the only negative value. The slope at $x = 4$ is smaller than the slope at $x = 2$ and both are smaller than the slope at

   $x = -2$. Thus, $g'(0) < 0 < g'(4) < g'(2) < g'(-2)$.

17. We begin by drawing a curve through the origin with a slope of 3 to satisfy $f(0) = 0$ and $f'(0) = 3$. Since $f'(1) = 0$, we will round off our figure so that there is a horizontal tangent directly over $x = 1$. Last, we make sure that the curve has a slope of -1 as we pass over $x = 2$. Two of the many possibilities are shown.

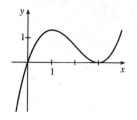 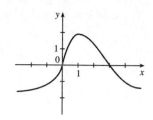

19. Using Definition 4 with $f(x) = 3x^2 - 5x$ and the point $(2, 2)$, we have

$$f'(2) = \lim_{h \to 0} \frac{f(2+h) - f(2)}{h} = \lim_{h \to 0} \frac{[3(2+h)^2 - 5(2+h)] - 2}{h}$$

$$= \lim_{h \to 0} \frac{(12 + 12h + 3h^2 - 10 - 5h) - 2}{h} = \lim_{h \to 0} \frac{3h^2 + 7h}{h} = \lim_{h \to 0} (3h + 7) = 7$$

So an equation of the tangent line at $(2, 2)$ is $y - 2 = 7(x - 2)$ or $y = 7x - 12$.

21. (a) Using Definition 4 with $F(x) = 5x/(1 + x^2)$ and the point $(2, 2)$, we have

(b)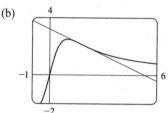

$$F'(2) = \lim_{h \to 0} \frac{F(2+h) - F(2)}{h} = \lim_{h \to 0} \frac{\dfrac{5(2+h)}{1 + (2+h)^2} - 2}{h}$$

$$= \lim_{h \to 0} \frac{\dfrac{5h + 10}{h^2 + 4h + 5} - 2}{h} = \lim_{h \to 0} \frac{\dfrac{5h + 10 - 2(h^2 + 4h + 5)}{h^2 + 4h + 5}}{h}$$

$$= \lim_{h \to 0} \frac{-2h^2 - 3h}{h(h^2 + 4h + 5)} = \lim_{h \to 0} \frac{h(-2h - 3)}{h(h^2 + 4h + 5)} = \lim_{h \to 0} \frac{-2h - 3}{h^2 + 4h + 5} = \frac{-3}{5}$$

So an equation of the tangent line at $(2, 2)$ is $y - 2 = -\frac{3}{5}(x - 2)$ or $y = -\frac{3}{5}x + \frac{16}{5}$.

23. Use Definition 4 with $f(x) = 3 - 2x + 4x^2$.

$$f'(a) = \lim_{h \to 0} \frac{f(a+h) - f(a)}{h} = \lim_{h \to 0} \frac{[3 - 2(a+h) + 4(a+h)^2] - (3 - 2a + 4a^2)}{h}$$

$$= \lim_{h \to 0} \frac{(3 - 2a - 2h + 4a^2 + 8ah + 4h^2) - (3 - 2a + 4a^2)}{h}$$

$$= \lim_{h \to 0} \frac{-2h + 8ah + 4h^2}{h} = \lim_{h \to 0} \frac{h(-2 + 8a + 4h)}{h} = \lim_{h \to 0} (-2 + 8a + 4h) = -2 + 8a$$

25. Use Definition 4 with $f(t) = (2t + 1)/(t + 3)$.

$$f'(a) = \lim_{h \to 0} \frac{f(a+h) - f(a)}{h} = \lim_{h \to 0} \frac{\dfrac{2(a+h) + 1}{(a+h) + 3} - \dfrac{2a + 1}{a + 3}}{h} = \lim_{h \to 0} \frac{(2a + 2h + 1)(a + 3) - (2a + 1)(a + h + 3)}{h(a + h + 3)(a + 3)}$$

$$= \lim_{h \to 0} \frac{(2a^2 + 6a + 2ah + 6h + a + 3) - (2a^2 + 2ah + 6a + a + h + 3)}{h(a + h + 3)(a + 3)}$$

$$= \lim_{h \to 0} \frac{5h}{h(a + h + 3)(a + 3)} = \lim_{h \to 0} \frac{5}{(a + h + 3)(a + 3)} = \frac{5}{(a + 3)^2}$$

27. Use Definition 4 with $f(x) = 1/\sqrt{x+2}$.

$$f'(a) = \lim_{h \to 0} \frac{f(a+h) - f(a)}{h} = \lim_{h \to 0} \frac{\dfrac{1}{\sqrt{(a+h)+2}} - \dfrac{1}{\sqrt{a+2}}}{h} = \lim_{h \to 0} \frac{\dfrac{\sqrt{a+2} - \sqrt{a+h+2}}{\sqrt{a+h+2}\,\sqrt{a+2}}}{h}$$

$$= \lim_{h \to 0} \left[\frac{\sqrt{a+2} - \sqrt{a+h+2}}{h\,\sqrt{a+h+2}\,\sqrt{a+2}} \cdot \frac{\sqrt{a+2} + \sqrt{a+h+2}}{\sqrt{a+2} + \sqrt{a+h+2}} \right] = \lim_{h \to 0} \frac{(a+2) - (a+h+2)}{h\sqrt{a+h+2}\,\sqrt{a+2}\,\left(\sqrt{a+2} + \sqrt{a+h+2}\right)}$$

$$= \lim_{h \to 0} \frac{-h}{h\sqrt{a+h+2}\,\sqrt{a+2}\,\left(\sqrt{a+2} + \sqrt{a+h+2}\right)} = \lim_{h \to 0} \frac{-1}{\sqrt{a+h+2}\,\sqrt{a+2}\,\left(\sqrt{a+2} + \sqrt{a+h+2}\right)}$$

$$= \frac{-1}{\left(\sqrt{a+2}\right)^2 \left(2\sqrt{a+2}\right)} = -\frac{1}{2(a+2)^{3/2}}$$

Note that the answers to Exercises 29–34 are not unique.

29. By Definition 4, $\displaystyle\lim_{h \to 0} \frac{(1+h)^{10} - 1}{h} = f'(1)$, where $f(x) = x^{10}$ and $a = 1$.

Or: By Definition 4, $\displaystyle\lim_{h \to 0} \frac{(1+h)^{10} - 1}{h} = f'(0)$, where $f(x) = (1+x)^{10}$ and $a = 0$.

31. By Equation 5, $\displaystyle\lim_{x \to 5} \frac{2^x - 32}{x - 5} = f'(5)$, where $f(x) = 2^x$ and $a = 5$.

33. By Definition 4, $\displaystyle\lim_{h \to 0} \frac{\cos(\pi + h) + 1}{h} = f'(\pi)$, where $f(x) = \cos x$ and $a = \pi$.

Or: By Definition 4, $\displaystyle\lim_{h \to 0} \frac{\cos(\pi + h) + 1}{h} = f'(0)$, where $f(x) = \cos(\pi + x)$ and $a = 0$.

35. The sketch shows the graph for a room temperature of $72°$ and a refrigerator temperature of $38°$. The initial rate of change is greater in magnitude than the rate of change after an hour.

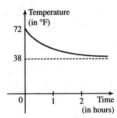

37. (a) (i) $[2000, 2002]$: $\dfrac{P(2002) - P(2000)}{2002 - 2000} = \dfrac{77 - 55}{2} = \dfrac{22}{2} = 11$ percent/year

    (ii) $[2000, 2001]$: $\dfrac{P(2001) - P(2000)}{2001 - 2000} = \dfrac{68 - 55}{1} = 13$ percent/year

    (iii) $[1999, 2000]$: $\dfrac{P(2000) - P(1999)}{2000 - 1999} = \dfrac{55 - 39}{1} = 16$ percent/year

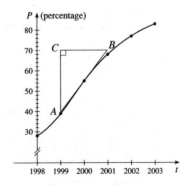

(b) Using the values from (ii) and (iii), we have $\dfrac{13 + 16}{2} = 14.5$ percent/year.

(c) Estimating A as $(1999, 40)$ and B as $(2001, 70)$, the slope at 2000 is

$$\frac{70 - 40}{2001 - 1999} = \frac{30}{2} = 15 \text{ percent/year.}$$

39. (a) (i) $\dfrac{\Delta C}{\Delta x} = \dfrac{C(105) - C(100)}{105 - 100} = \dfrac{6601.25 - 6500}{5} = \$20.25/\text{unit.}$

    (ii) $\dfrac{\Delta C}{\Delta x} = \dfrac{C(101) - C(100)}{101 - 100} = \dfrac{6520.05 - 6500}{1} = \$20.05/\text{unit.}$

(b) $\dfrac{C(100+h)-C(100)}{h} = \dfrac{\left[5000+10(100+h)+0.05(100+h)^2\right]-6500}{h} = \dfrac{20h+0.05h^2}{h}$

$= 20+0.05h,\ h \neq 0$

So the instantaneous rate of change is $\displaystyle\lim_{h\to 0}\dfrac{C(100+h)-C(100)}{h} = \lim_{h\to 0}(20+0.05h) = \$20/\text{unit}.$

41. (a) $f'(x)$ is the rate of change of the production cost with respect to the number of ounces of gold produced. Its units are dollars per ounce.

(b) After 800 ounces of gold have been produced, the rate at which the production cost is increasing is $\$17/\text{ounce}$. So the cost of producing the 800th (or 801st) ounce is about \$17.

(c) In the short term, the values of $f'(x)$ will decrease because more efficient use is made of start-up costs as x increases. But eventually $f'(x)$ might increase due to large-scale operations.

43. $T'(10)$ is the rate at which the temperature is changing at 10:00 AM. To estimate the value of $T'(10)$, we will average the

difference quotients obtained using the times $t = 8$ and $t = 12$. Let $A = \dfrac{T(8)-T(10)}{8-10} = \dfrac{72-81}{-2} = 4.5$ and

$B = \dfrac{T(12)-T(10)}{12-10} = \dfrac{88-81}{2} = 3.5.$ Then $T'(10) = \displaystyle\lim_{t\to 10}\dfrac{T(t)-T(10)}{t-10} \approx \dfrac{A+B}{2} = \dfrac{4.5+3.5}{2} = 4^\circ\text{F/h}.$

45. (a) $S'(T)$ is the rate at which the oxygen solubility changes with respect to the water temperature. Its units are $(\text{mg/L})/^\circ\text{C}$.

(b) For $T = 16^\circ\text{C}$, it appears that the tangent line to the curve goes through the points $(0, 14)$ and $(32, 6)$. So

$S'(16) \approx \dfrac{6-14}{32-0} = -\dfrac{8}{32} = -0.25\ (\text{mg/L})/^\circ\text{C}.$ This means that as the temperature increases past 16°C, the oxygen

solubility is decreasing at a rate of $0.25\ (\text{mg/L})/^\circ\text{C}.$

47. Since $f(x) = x\sin(1/x)$ when $x \neq 0$ and $f(0) = 0$, we have

$f'(0) = \displaystyle\lim_{h\to 0}\dfrac{f(0+h)-f(0)}{h} = \lim_{h\to 0}\dfrac{h\sin(1/h)-0}{h} = \lim_{h\to 0}\sin(1/h).$ This limit does not exist since $\sin(1/h)$ takes the

values -1 and 1 on any interval containing 0. (Compare with Example 5 in Section 1.3.)

2.2 The Derivative as a Function

1. It appears that f is an odd function, so f' will be an even function—that is, $f'(-a) = f'(a)$.

(a) $f'(-3) \approx 1.5$ (b) $f'(-2) \approx 1$

(c) $f'(-1) \approx 0$ (d) $f'(0) \approx -4$

(e) $f'(1) \approx 0$ (f) $f'(2) \approx 1$

(g) $f'(3) \approx 1.5$

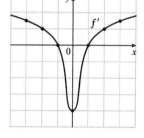

3. (a)$' = $ II, since from left to right, the slopes of the tangents to graph (a) start out negative, become 0, then positive, then 0, then negative again. The actual function values in graph II follow the same pattern.

(b)$' = $ IV, since from left to right, the slopes of the tangents to graph (b) start out at a fixed positive quantity, then suddenly become negative, then positive again. The discontinuities in graph IV indicate sudden changes in the slopes of the tangents.

(c)$' = $ I, since the slopes of the tangents to graph (c) are negative for $x < 0$ and positive for $x > 0$, as are the function values of graph I.

(d)$' = $ III, since from left to right, the slopes of the tangents to graph (d) are positive, then 0, then negative, then 0, then positive, then 0, then negative again, and the function values in graph III follow the same pattern.

Hints for Exercises 4–11: First plot x-intercepts on the graph of f' for any horizontal tangents on the graph of f. Look for any corners on the graph of f—there will be a discontinuity on the graph of f'. On any interval where f has a tangent with positive (or negative) slope, the graph of f' will be positive (or negative). If the graph of the function is linear, the graph of f' will be a horizontal line.

5.

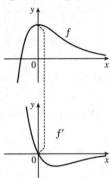

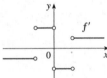

7.

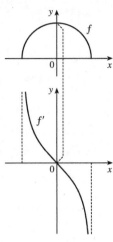

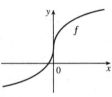

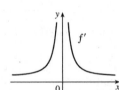

9.

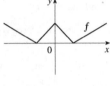

11.

13. It appears that there are horizontal tangents on the graph of M for $t = 1963$ and $t = 1971$. Thus, there are zeros for those values of t on the graph of M'. The derivative is negative for the years 1963 to 1971.

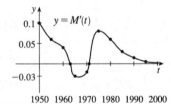

15. (a) By zooming in, we estimate that $f'(0) = 0$, $f'\left(\frac{1}{2}\right) = 1$, $f'(1) = 2$, and $f'(2) = 4$.

   (b) By symmetry, $f'(-x) = -f'(x)$. So $f'\left(-\frac{1}{2}\right) = -1$, $f'(-1) = -2$, and $f'(-2) = -4$.

   (c) It appears that $f'(x)$ is twice the value of x, so we guess that $f'(x) = 2x$.

   (d) $f'(x) = \lim\limits_{h \to 0} \dfrac{f(x+h) - f(x)}{h} = \lim\limits_{h \to 0} \dfrac{(x+h)^2 - x^2}{h}$

$\qquad = \lim\limits_{h \to 0} \dfrac{(x^2 + 2hx + h^2) - x^2}{h} = \lim\limits_{h \to 0} \dfrac{2hx + h^2}{h} = \lim\limits_{h \to 0} \dfrac{h(2x+h)}{h} = \lim\limits_{h \to 0} (2x + h) = 2x$

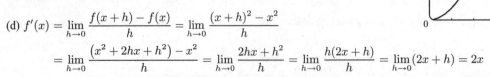

17. $f'(x) = \lim\limits_{h \to 0} \dfrac{f(x+h) - f(x)}{h} = \lim\limits_{h \to 0} \dfrac{\left[\frac{1}{2}(x+h) - \frac{1}{3}\right] - \left(\frac{1}{2}x - \frac{1}{3}\right)}{h} = \lim\limits_{h \to 0} \dfrac{\frac{1}{2}x + \frac{1}{2}h - \frac{1}{3} - \frac{1}{2}x + \frac{1}{3}}{h}$

$\qquad = \lim\limits_{h \to 0} \dfrac{\frac{1}{2}h}{h} = \lim\limits_{h \to 0} \frac{1}{2} = \frac{1}{2}$

Domain of f = domain of f' = $\mathbb{R}$.

19. $f'(x) = \lim\limits_{h \to 0} \dfrac{f(x+h) - f(x)}{h} = \lim\limits_{h \to 0} \dfrac{\left[(x+h)^3 - 3(x+h) + 5\right] - (x^3 - 3x + 5)}{h}$

$\qquad = \lim\limits_{h \to 0} \dfrac{(x^3 + 3x^2h + 3xh^2 + h^3 - 3x - 3h + 5) - (x^3 - 3x + 5)}{h} = \lim\limits_{h \to 0} \dfrac{3x^2h + 3xh^2 + h^3 - 3h}{h}$

$\qquad = \lim\limits_{h \to 0} \dfrac{h(3x^2 + 3xh + h^2 - 3)}{h} = \lim\limits_{h \to 0} (3x^2 + 3xh + h^2 - 3) = 3x^2 - 3$

Domain of f = domain of f' = $\mathbb{R}$.

21. $g'(x) = \lim\limits_{h \to 0} \dfrac{g(x+h) - g(x)}{h} = \lim\limits_{h \to 0} \dfrac{\sqrt{1 + 2(x+h)} - \sqrt{1 + 2x}}{h} \left[\dfrac{\sqrt{1 + 2(x+h)} + \sqrt{1 + 2x}}{\sqrt{1 + 2(x+h)} + \sqrt{1 + 2x}}\right]$

$\qquad = \lim\limits_{h \to 0} \dfrac{(1 + 2x + 2h) - (1 + 2x)}{h\left[\sqrt{1 + 2(x+h)} + \sqrt{1 + 2x}\right]} = \lim\limits_{h \to 0} \dfrac{2}{\sqrt{1 + 2x + 2h} + \sqrt{1 + 2x}} = \dfrac{2}{2\sqrt{1 + 2x}} = \dfrac{1}{\sqrt{1 + 2x}}$

Domain of g = $\left[-\frac{1}{2}, \infty\right)$, domain of g' = $\left(-\frac{1}{2}, \infty\right)$.

23. $G'(t) = \lim\limits_{h \to 0} \dfrac{G(t+h) - G(t)}{h} = \lim\limits_{h \to 0} \dfrac{\dfrac{4(t+h)}{(t+h)+1} - \dfrac{4t}{t+1}}{h} = \lim\limits_{h \to 0} \dfrac{\dfrac{4(t+h)(t+1) - 4t(t+h+1)}{(t+h+1)(t+1)}}{h}$

$\qquad = \lim\limits_{h \to 0} \dfrac{(4t^2 + 4ht + 4t + 4h) - (4t^2 + 4ht + 4t)}{h(t+h+1)(t+1)} = \lim\limits_{h \to 0} \dfrac{4h}{h(t+h+1)(t+1)}$

$\qquad = \lim\limits_{h \to 0} \dfrac{4}{(t+h+1)(t+1)} = \dfrac{4}{(t+1)^2}$

Domain of G = domain of G' = $(-\infty, -1) \cup (-1, \infty)$.

25. (a) $f'(x) = \lim\limits_{h \to 0} \dfrac{f(x+h) - f(x)}{h} = \lim\limits_{h \to 0} \dfrac{\left[(x+h)^4 + 2(x+h)\right] - (x^4 + 2x)}{h}$

$\qquad = \lim\limits_{h \to 0} \dfrac{x^4 + 4x^3h + 6x^2h^2 + 4xh^3 + h^4 + 2x + 2h - x^4 - 2x}{h}$

$\qquad = \lim\limits_{h \to 0} \dfrac{4x^3h + 6x^2h^2 + 4xh^3 + h^4 + 2h}{h} = \lim\limits_{h \to 0} \dfrac{h(4x^3 + 6x^2h + 4xh^2 + h^3 + 2)}{h}$

$\qquad = \lim\limits_{h \to 0} (4x^3 + 6x^2h + 4xh^2 + h^3 + 2) = 4x^3 + 2$

(b) Notice that $f'(x) = 0$ when f has a horizontal tangent, $f'(x)$ is

positive when the tangents have positive slope, and $f'(x)$ is

negative when the tangents have negative slope

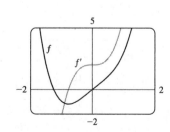

27. f is not differentiable at $x = -4$, because the graph has a corner there, and at $x = 0$, because there is a discontinuity there.

29. f is not differentiable at $x = -1$, because the graph has a vertical tangent there, and at $x = 4$, because the graph has a corner there.

31. As we zoom in toward $(-1, 0)$, the curve appears more and more like a straight line, so $f(x) = x + \sqrt{|x|}$ is differentiable at $x = -1$. But no matter how much we zoom in toward the origin, the curve doesn't straighten out—we can't eliminate the sharp point (a cusp). So f is not differentiable at $x = 0$.

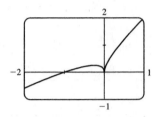

33. $a = f$, $b = f'$, $c = f''$. We can see this because where a has a horizontal tangent, $b = 0$, and where b has a horizontal tangent, $c = 0$. We can immediately see that c can be neither f nor f', since at the points where c has a horizontal tangent, neither a nor b is equal to 0.

35. We can immediately see that a is the graph of the acceleration function, since at the points where a has a horizontal tangent, neither c nor b is equal to 0. Next, we note that $a = 0$ at the point where b has a horizontal tangent, so b must be the graph of the velocity function, and hence, $b' = a$. We conclude that c is the graph of the position function.

37. $f'(x) = \lim\limits_{h \to 0} \dfrac{f(x+h) - f(x)}{h} = \lim\limits_{h \to 0} \dfrac{[1 + 4(x+h) - (x+h)^2] - (1 + 4x - x^2)}{h}$

$= \lim\limits_{h \to 0} \dfrac{(1 + 4x + 4h - x^2 - 2xh - h^2) - (1 + 4x - x^2)}{h} = \lim\limits_{h \to 0} \dfrac{4h - 2xh - h^2}{h} = \lim\limits_{h \to 0} (4 - 2x - h) = 4 - 2x$

$f''(x) = \lim\limits_{h \to 0} \dfrac{f'(x+h) - f'(x)}{h} = \lim\limits_{h \to 0} \dfrac{[4 - 2(x+h)] - (4 - 2x)}{h} = \lim\limits_{h \to 0} \dfrac{-2h}{h} = \lim\limits_{h \to 0} (-2) = -2$

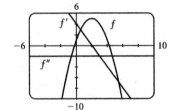

We see from the graph that our answers are reasonable because the graph of f' is that of a linear function and the graph of f'' is that of a constant function.

39. (a) Note that we have factored $x - a$ as the difference of two cubes in the third step.

$f'(a) = \lim\limits_{x \to a} \dfrac{f(x) - f(a)}{x - a} = \lim\limits_{x \to a} \dfrac{x^{1/3} - a^{1/3}}{x - a} = \lim\limits_{x \to a} \dfrac{x^{1/3} - a^{1/3}}{(x^{1/3} - a^{1/3})(x^{2/3} + x^{1/3}a^{1/3} + a^{2/3})}$

$= \lim\limits_{x \to a} \dfrac{1}{x^{2/3} + x^{1/3}a^{1/3} + a^{2/3}} = \dfrac{1}{3a^{2/3}}$ or $\tfrac{1}{3}a^{-2/3}$

(b) $f'(0) = \lim\limits_{h \to 0} \dfrac{f(0+h) - f(0)}{h} = \lim\limits_{h \to 0} \dfrac{\sqrt[3]{h} - 0}{h} = \lim\limits_{h \to 0} \dfrac{1}{h^{2/3}}$. This function increases without bound, so the limit does not exist, and therefore $f'(0)$ does not exist.

(c) $\lim\limits_{x \to 0} |f'(x)| = \lim\limits_{x \to 0} \dfrac{1}{3x^{2/3}} = \infty$ and f is continuous at $x = 0$ (root function), so f has a vertical tangent at $x = 0$.

41. $f(x) = |x - 6| = \begin{cases} x - 6 & \text{if } x - 6 \geq 6 \\ -(x - 6) & \text{if } x - 6 < 0 \end{cases} = \begin{cases} x - 6 & \text{if } x \geq 6 \\ 6 - x & \text{if } x < 6 \end{cases}$

So the right-hand limit is $\displaystyle\lim_{x \to 6^+} \frac{f(x) - f(6)}{x - 6} = \lim_{x \to 6^+} \frac{|x - 6| - 0}{x - 6} = \lim_{x \to 6^+} \frac{x - 6}{x - 6} = \lim_{x \to 6^+} 1 = 1$, and the left-hand limit

is $\displaystyle\lim_{x \to 6^-} \frac{f(x) - f(6)}{x - 6} = \lim_{x \to 6^-} \frac{|x - 6| - 0}{x - 6} = \lim_{x \to 6^-} \frac{6 - x}{x - 6} = \lim_{x \to 6^-} (-1) = -1$. Since these limits are not equal,

$f'(6) = \displaystyle\lim_{x \to 6} \frac{f(x) - f(6)}{x - 6}$ does not exist and f is not differentiable at 6.

However, a formula for f' is $f'(x) = \begin{cases} 1 & \text{if } x > 6 \\ -1 & \text{if } x < 6 \end{cases}$

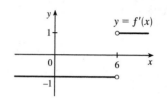

Another way of writing the formula is $f'(x) = \dfrac{x - 6}{|x - 6|}$.

43. (a) If f is even, then

$$f'(-x) = \lim_{h \to 0} \frac{f(-x + h) - f(-x)}{h} = \lim_{h \to 0} \frac{f[-(x - h)] - f(-x)}{h}$$

$$= \lim_{h \to 0} \frac{f(x - h) - f(x)}{h} = -\lim_{h \to 0} \frac{f(x - h) - f(x)}{-h} \quad [\text{let } \Delta x = -h]$$

$$= -\lim_{\Delta x \to 0} \frac{f(x + \Delta x) - f(x)}{\Delta x} = -f'(x)$$

Therefore, f' is odd.

(b) If f is odd, then

$$f'(-x) = \lim_{h \to 0} \frac{f(-x + h) - f(-x)}{h} = \lim_{h \to 0} \frac{f[-(x - h)] - f(-x)}{h}$$

$$= \lim_{h \to 0} \frac{-f(x - h) + f(x)}{h} = \lim_{h \to 0} \frac{f(x - h) - f(x)}{-h} \quad [\text{let } \Delta x = -h]$$

$$= \lim_{\Delta x \to 0} \frac{f(x + \Delta x) - f(x)}{\Delta x} = f'(x)$$

Therefore, f' is even.

45.

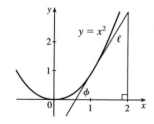

In the right triangle in the diagram, let Δy be the side opposite angle ϕ and Δx the side adjacent angle ϕ. Then the slope of the tangent line ℓ is $m = \Delta y / \Delta x = \tan \phi$. Note that $0 < \phi < \frac{\pi}{2}$. We know (see Exercise 15) that the derivative of $f(x) = x^2$ is $f'(x) = 2x$. So the slope of the tangent to the curve at the point $(1, 1)$ is 2. Thus, ϕ is the angle between 0 and $\frac{\pi}{2}$ whose tangent is 2; that is, $\phi = \tan^{-1} 2 \approx 63°$.

2.3 Basic Differentiation Formulas

1. $f(x) = 186.5$ is a constant function, so its derivative is 0, that is, $f'(x) = 0$.

3. $f(x) = 5x - 1 \quad \Rightarrow \quad f'(x) = 5 - 0 = 5$

5. $f(x) = x^3 - 4x + 6 \;\Rightarrow\; f'(x) = 3x^2 - 4(1) + 0 = 3x^2 - 4$

7. $f(x) = x - 3\sin x \;\Rightarrow\; f'(x) = 1 - 3\cos x$

9. $f(t) = \frac{1}{4}(t^4 + 8) \;\Rightarrow\; f'(t) = \frac{1}{4}(t^4 + 8)' = \frac{1}{4}(4t^{4-1} + 0) = t^3$

11. $y = x^{-2/5} \;\Rightarrow\; y' = -\frac{2}{5}x^{(-2/5)-1} = -\frac{2}{5}x^{-7/5} = -\dfrac{2}{5x^{7/5}}$

13. $V(r) = \frac{4}{3}\pi r^3 \;\Rightarrow\; V'(r) = \frac{4}{3}\pi(3r^2) = 4\pi r^2$

15. $F(x) = \left(\frac{1}{2}x\right)^5 = \left(\frac{1}{2}\right)^5 x^5 = \frac{1}{32}x^5 \;\Rightarrow\; F'(x) = \frac{1}{32}(5x^4) = \frac{5}{32}x^4$

17. $y = 4\pi^2 \;\Rightarrow\; y' = 0$ since $4\pi^2$ is a constant.

19. $y = \dfrac{x^2 + 4x + 3}{\sqrt{x}} = x^{3/2} + 4x^{1/2} + 3x^{-1/2} \;\Rightarrow$

$y' = \frac{3}{2}x^{1/2} + 4\left(\frac{1}{2}\right)x^{-1/2} + 3\left(-\frac{1}{2}\right)x^{-3/2} = \frac{3}{2}\sqrt{x} + \dfrac{2}{\sqrt{x}} - \dfrac{3}{2x\sqrt{x}} \;\Rightarrow$

$\left[\text{note that } x^{3/2} = x^{2/2} \cdot x^{1/2} = x\sqrt{x}\right]$

21. $v = t^2 - \dfrac{1}{\sqrt[4]{t^3}} = t^2 - t^{-3/4} \;\Rightarrow\; v' = 2t - \left(-\frac{3}{4}\right)t^{-7/4} = 2t + \dfrac{3}{4t^{7/4}} = 2t + \dfrac{3}{4t\sqrt[4]{t^3}}$

23. $z = \dfrac{A}{y^{10}} + B\cos y = Ay^{-10} + B\cos y \;\Rightarrow\; \dfrac{dz}{dy} = A(-10)y^{-11} + B(-\sin y) = -\dfrac{10A}{y^{11}} - B\sin y$

25. $y = 6\cos x \;\Rightarrow\; y' = -6\sin x$. At $(\pi/3, 3)$, $y' = -6\sin(\pi/3) = -6\left(\sqrt{3}/2\right) = -3\sqrt{3}$ and an equation of the tangent

line is $y - 3 = -3\sqrt{3}\,(x - \pi/3)$ or $y = -3\sqrt{3}x + 3 + \pi\sqrt{3}$. The slope of the normal line is $1/(3\sqrt{3})$ (the negative

reciprocal of $-3\sqrt{3}$) and an equation of the normal line is $y - 3 = \dfrac{1}{3\sqrt{3}}\left(x - \dfrac{\pi}{3}\right)$ or $y = \dfrac{1}{3\sqrt{3}}x + 3 - \dfrac{\pi}{9\sqrt{3}}$.

27. $y = f(x) = x + \sqrt{x} \;\Rightarrow\; f'(x) = 1 + \frac{1}{2}x^{-1/2}$.

So the slope of the tangent line at $(1, 2)$ is $f'(1) = 1 + \frac{1}{2}(1) = \frac{3}{2}$

and its equation is $y - 2 = \frac{3}{2}(x - 1)$ or $y = \frac{3}{2}x + \frac{1}{2}$.

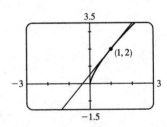

29. $f(x) = x^4 - 3x^3 + 16x \;\Rightarrow\; f'(x) = 4x^3 - 9x^2 + 16 \;\Rightarrow\; f''(x) = 12x^2 - 18x$

31. $g(t) = 2\cos t - 3\sin t \;\Rightarrow\; g'(t) = -2\sin t - 3\cos t \;\Rightarrow\; g''(t) = -2\cos t + 3\sin t$

33. $\dfrac{d}{dx}(\sin x) = \cos x \;\Rightarrow\; \dfrac{d^2}{dx^2}(\sin x) = -\sin x \;\Rightarrow\; \dfrac{d^3}{dx^3}(\sin x) = -\cos x \;\Rightarrow\; \dfrac{d^4}{dx^4}(\sin x) = \sin x$.

The derivatives of $\sin x$ occur in a cycle of four. Since $99 = 4(24) + 3$, we have $\dfrac{d^{99}}{dx^{99}}(\sin x) = \dfrac{d^3}{dx^3}(\sin x) = -\cos x$.

35. $f(x) = x + 2\sin x$ has a horizontal tangent when $f'(x) = 0 \;\Leftrightarrow\; 1 + 2\cos x = 0 \;\Leftrightarrow\; \cos x = -\frac{1}{2} \;\Leftrightarrow$

$x = \frac{2\pi}{3} + 2\pi n$ or $\frac{4\pi}{3} + 2\pi n$, where n is an integer. Note that $\frac{4\pi}{3}$ and $\frac{2\pi}{3}$ are $\pm\frac{\pi}{3}$ units from π. This allows us to write the

solutions in the more compact equivalent form $(2n + 1)\pi \pm \frac{\pi}{3}$, n an integer.

37. $y = 6x^3 + 5x - 3$ $\Rightarrow$ $m = y' = 18x^2 + 5$, but $x^2 \geq 0$ for all x, so $m \geq 5$ for all x.

39. The slope of $y = x^2 - 5x + 4$ is given by $m = y' = 2x - 5$. The slope of $x - 3y = 5$ $\Leftrightarrow$ $y = \frac{1}{3}x - \frac{5}{3}$ is $\frac{1}{3}$,

so the desired normal line must have slope $\frac{1}{3}$, and hence, the tangent line to the parabola must have slope -3. This occurs if

$2x - 5 = -3$ $\Rightarrow$ $2x = 2$ $\Rightarrow$ $x = 1$. When $x = 1$, $y = 1^2 - 5(1) + 4 = 0$, and an equation of the normal line is

$y - 0 = \frac{1}{3}(x - 1)$ or $y = \frac{1}{3}x - \frac{1}{3}$.

41. (a) $s = t^3 - 3t$ $\Rightarrow$ $v(t) = s'(t) = 3t^2 - 3$ $\Rightarrow$ $a(t) = v'(t) = 6t$

(b) $a(2) = 6(2) = 12 \text{ m/s}^2$

(c) $v(t) = 3t^2 - 3 = 0$ when $t^2 = 1$, that is, $t = 1$ and $a(1) = 6 \text{ m/s}^2$.

43. (a) $s = f(t) = t^3 - 12t^2 + 36t$ $\Rightarrow$ $v(t) = f'(t) = 3t^2 - 24t + 36$

(b) $v(3) = 27 - 72 + 36 = -9 \text{ ft/s}$

(c) The particle is at rest when $v(t) = 0$. $3t^2 - 24t + 36 = 0$ $\Rightarrow$ $3(t - 2)(t - 6) = 0$ $\Rightarrow$ $t = 2$ s or 6 s.

(d) The particle is moving in the positive direction when $v(t) > 0$. $3(t - 2)(t - 6) > 0$ $\Leftrightarrow$ $0 \leq t < 2$ or $t > 6$.

(e) Since the particle is moving in the positive direction and in the
negative direction, we need to calculate the distance traveled in the
intervals $[0, 2]$, $[2, 6]$, and $[6, 8]$ separately.

$|f(2) - f(0)| = |32 - 0| = 32$.

$|f(6) - f(2)| = |0 - 32| = 32$.

$|f(8) - f(6)| = |32 - 0| = 32$.

The total distance is $32 + 32 + 32 = 96$ ft.

(f)

45. (a) $s(t) = t^3 - 4.5t^2 - 7t$ $\Rightarrow$ $v(t) = s'(t) = 3t^2 - 9t - 7 = 5$ $\Leftrightarrow$ $3t^2 - 9t - 12 = 0$ $\Leftrightarrow$
$3(t - 4)(t + 1) = 0$ $\Leftrightarrow$ $t = 4$ or -1. Since $t \geq 0$, the particle reaches a velocity of 5 m/s at $t = 4$ s.

(b) $a(t) = v'(t) = 6t - 9 = 0$ $\Leftrightarrow$ $t = 1.5$. The acceleration changes from negative to positive, so the velocity changes
from decreasing to increasing. Thus, at $t = 1.5$ s, the velocity has its minimum value.

47. (a) $h = 10t - 0.83t^2$ $\Rightarrow$ $v(t) = \dfrac{dh}{dt} = 10 - 1.66t$, so $v(3) = 10 - 1.66(3) = 5.02 \text{ m/s}$.

(b) $h = 25$ $\Rightarrow$ $10t - 0.83t^2 = 25$ $\Rightarrow$ $0.83t^2 - 10t + 25 = 0$ $\Rightarrow$ $t = \dfrac{10 \pm \sqrt{17}}{1.66} \approx 3.54$ or 8.51.

The value $t_1 = (10 - \sqrt{17})/1.66$ corresponds to the time it takes for the stone to rise 25 m and

$t_2 = (10 + \sqrt{17})/1.66$ corresponds to the time when the stone is 25 m high on the way down. Thus,

$v(t_1) = 10 - 1.66[(10 - \sqrt{17})/1.66] = \sqrt{17} \approx 4.12 \text{ m/s}$.

49. (a) $C(x) = 2000 + 3x + 0.01x^2 + 0.0002x^3$ $\Rightarrow$ $C'(x) = 3 + 0.02x + 0.0006x^2$

(b) $C'(100) = 3 + 0.02(100) + 0.0006(10,000) = 3 + 2 + 6 = \$11/\text{pair}$. $C'(100)$ is the rate at which the cost is increasing
as the 100th pair of jeans is produced. It predicts the cost of the 101st pair.

(c) The cost of manufacturing the 101st pair of jeans is

$C(101) - C(100) = (2000 + 303 + 102.01 + 206.0602) - (2000 + 300 + 100 + 200) = 11.0702 \approx \$11.07$.

51. $S(r) = 4\pi r^2 \;\Rightarrow\; S'(r) = 8\pi r \;\Rightarrow$

(a) $S'(1) = 8\pi$ ft^2/ft

(b) $S'(2) = 16\pi$ ft^2/ft

(c) $S'(3) = 24\pi$ ft^2/ft

As the radius increases, the surface area grows at an increasing rate. In fact, the rate of change is linear with respect to the radius.

53. (a) To find the rate of change of volume with respect to pressure, we first solve for V in terms of P.

$$PV = C \;\Rightarrow\; V = \frac{C}{P} \;\Rightarrow\; \frac{dV}{dP} = -\frac{C}{P^2}.$$

(b) From the formula for dV/dP in part (a), we see that as P increases, the absolute value of dV/dP decreases. Thus, the volume is decreasing more rapidly at the beginning.

55. $f'(x) = \lim\limits_{h \to 0} \dfrac{f(x+h) - f(x)}{h} = \lim\limits_{h \to 0} \dfrac{\frac{1}{x+h} - \frac{1}{x}}{h} = \lim\limits_{h \to 0} \dfrac{x - (x+h)}{hx(x+h)} = \lim\limits_{h \to 0} \dfrac{-h}{hx(x+h)} = \lim\limits_{h \to 0} \dfrac{-1}{x(x+h)} = -\dfrac{1}{x^2}$

57. $y = A \sin x + B \cos x \;\Rightarrow\; y' = A \cos x - B \sin x \;\Rightarrow\; y'' = -A \sin x - B \cos x$. Substituting these

expressions for y, y', and y'' into the given differential equation $y'' + y' - 2y = \sin x$ gives us

$(-A \sin x - B \cos x) + (A \cos x - B \sin x) - 2(A \sin x + B \cos x) = \sin x \;\Leftrightarrow$

$-3A \sin x - B \sin x + A \cos x - 3B \cos x = \sin x \;\Leftrightarrow\; (-3A - B) \sin x + (A - 3B) \cos x = 1 \sin x$, so we must have

$-3A - B = 1$ and $A - 3B = 0$ (since 0 is the coefficient of $\cos x$ on the right side). Solving for A and B, we add the first

equation to three times the second to get $B = -\frac{1}{10}$ and $A = -\frac{3}{10}$.

59.

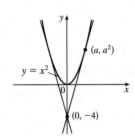

Let (a, a^2) be a point on the parabola at which the tangent line passes through the point $(0, -4)$. The tangent line has slope $2a$ and equation $y - (-4) = 2a(x - 0)$ $\;\Leftrightarrow\; y = 2ax - 4$. Since (a, a^2) also lies on the line, $a^2 = 2a(a) - 4$, or $a^2 = 4$. So $a = \pm 2$ and the points are $(2, 4)$ and $(-2, 4)$.

61. $y = f(x) = ax^2 \;\Rightarrow\; f'(x) = 2ax$. So the slope of the tangent to the parabola at $x = 2$ is $m = 2a(2) = 4a$. The slope of

the given line, $2x + y = b \;\Leftrightarrow\; y = -2x + b$, is seen to be -2, so we must have $4a = -2 \;\Leftrightarrow\; a = -\frac{1}{2}$. So when $x = 2$,

the point in question has y-coordinate $-\frac{1}{2} \cdot 2^2 = -2$. Now we simply require that the given line, whose equation is

$2x + y = b$, pass through the point $(2, -2)$: $2(2) + (-2) = b \;\Leftrightarrow\; b = 2$. So we must have $a = -\frac{1}{2}$ and $b = 2$.

63. $y = f(x) = ax^3 + bx^2 + cx + d \;\Rightarrow\; f'(x) = 3ax^2 + 2bx + c$. The point $(-2, 6)$ is on f, so $f(-2) = 6 \;\Rightarrow$

$-8a + 4b - 2c + d = 6$ **(1)**. The point $(2, 0)$ is on f, so $f(2) = 0 \;\Rightarrow\; 8a + 4b + 2c + d = 0$ **(2)**. Since there are

horizontal tangents at $(-2, 6)$ and $(2, 0)$, $f'(\pm 2) = 0$. $f'(-2) = 0 \;\Rightarrow\; 12a - 4b + c = 0$ **(3)** and $f'(2) = 0 \;\Rightarrow$

$12a + 4b + c = 0$ **(4)**. Subtracting equation **(3)** from **(4)** gives $8b = 0 \;\Rightarrow\; b = 0$. Adding **(1)** and **(2)** gives $8b + 2d = 6$,

so $d = 3$ since $b = 0$. From **(3)** we have $c = -12a$, so **(2)** becomes $8a + 4(0) + 2(-12a) + 3 = 0 \;\Rightarrow\; 3 = 16a \;\Rightarrow$

$a = \frac{3}{16}$. Now $c = -12a = -12\left(\frac{3}{16}\right) = -\frac{9}{4}$ and the desired cubic function is $y = \frac{3}{16}x^3 - \frac{9}{4}x + 3$.

65. *Solution 1:*   Let $f(x) = x^{1000}$. Then, by the definition of a derivative, $f'(1) = \lim\limits_{x \to 1} \dfrac{f(x) - f(1)}{x - 1} = \lim\limits_{x \to 1} \dfrac{x^{1000} - 1}{x - 1}$.

But this is just the limit we want to find, and we know (from the Power Rule) that $f'(x) = 1000x^{999}$, so

$f'(1) = 1000(1)^{999} = 1000$. So $\lim\limits_{x \to 1} \dfrac{x^{1000} - 1}{x - 1} = 1000$.

Solution 2:   Note that $(x^{1000} - 1) = (x - 1)(x^{999} + x^{998} + x^{997} + \cdots + x^2 + x + 1)$. So

$$\lim_{x \to 1} \frac{x^{1000} - 1}{x - 1} = \lim_{x \to 1} \frac{(x - 1)(x^{999} + x^{998} + x^{997} + \cdots + x^2 + x + 1)}{x - 1} = \lim_{x \to 1} (x^{999} + x^{998} + x^{997} + \cdots + x^2 + x + 1)$$

$$= \underbrace{1 + 1 + 1 + \cdots + 1 + 1 + 1}_{1000 \text{ ones}} = 1000, \text{ as above.}$$

67. $y = x^2 \;\Rightarrow\; y' = 2x$, so the slope of a tangent line at the point (a, a^2) is $y' = 2a$ and the slope of a normal line is $-1/(2a)$,

for $a \neq 0$. The slope of the normal line through the points (a, a^2) and $(0, c)$ is $\dfrac{a^2 - c}{a - 0}$, so $\dfrac{a^2 - c}{a} = -\dfrac{1}{2a} \;\Rightarrow$

$a^2 - c = -\frac{1}{2} \;\Rightarrow\; a^2 = c - \frac{1}{2}$. The last equation has two solutions if $c > \frac{1}{2}$, one solution if $c = \frac{1}{2}$, and no solution if

$c < \frac{1}{2}$. Since the y-axis is normal to $y = x^2$ regardless of the value of c (this is the case for $a = 0$), we have three normal lines

if $c > \frac{1}{2}$ and one normal line if $c \leq \frac{1}{2}$.

2.4   The Product and Quotient Rules

1. Product Rule: $y = (x^2 + 1)(x^3 + 1) \;\Rightarrow$

$$y' = (x^2 + 1)(3x^2) + (x^3 + 1)(2x) = 3x^4 + 3x^2 + 2x^4 + 2x = 5x^4 + 3x^2 + 2x.$$

Multiplying first: $y = (x^2 + 1)(x^3 + 1) = x^5 + x^3 + x^2 + 1 \;\Rightarrow\; y' = 5x^4 + 3x^2 + 2x$ (equivalent).

3. $g(t) = t^3 \cos t \;\Rightarrow\; g'(t) = t^3(-\sin t) + (\cos t) \cdot 3t^2 = 3t^2 \cos t - t^3 \sin t$ or $t^2(3 \cos t - t \sin t)$

5. $F(y) = \left(\dfrac{1}{y^2} - \dfrac{3}{y^4} \right)(y + 5y^3) = (y^{-2} - 3y^{-4})(y + 5y^3) \;\overset{\text{PR}}{\Rightarrow}$

$F'(y) = (y^{-2} - 3y^{-4})(1 + 15y^2) + (y + 5y^3)(-2y^{-3} + 12y^{-5})$

$\quad = (y^{-2} + 15 - 3y^{-4} - 45y^{-2}) + (-2y^{-2} + 12y^{-4} - 10 + 60y^{-2})$

$\quad = 5 + 14y^{-2} + 9y^{-4}$ or $5 + 14/y^2 + 9/y^4$

7. $f(x) = \sin x + \frac{1}{2} \cot x \;\Rightarrow\; f'(x) = \cos x - \frac{1}{2} \csc^2 x$

9. $h(\theta) = \theta \csc \theta - \cot \theta \;\Rightarrow\; h'(\theta) = \theta(-\csc \theta \cot \theta) + (\csc \theta) \cdot 1 - (-\csc^2 \theta) = \csc \theta - \theta \csc \theta \cot \theta + \csc^2 \theta$

11. $g(x) = \dfrac{3x - 1}{2x + 1} \;\overset{\text{QR}}{\Rightarrow}\; g'(x) = \dfrac{(2x + 1)(3) - (3x - 1)(2)}{(2x + 1)^2} = \dfrac{6x + 3 - 6x + 2}{(2x + 1)^2} = \dfrac{5}{(2x + 1)^2}$

13. $y = \dfrac{t^2}{3t^2 - 2t + 1} \;\overset{\text{QR}}{\Rightarrow}$

$$y' = \frac{(3t^2 - 2t + 1)(2t) - t^2(6t - 2)}{(3t^2 - 2t + 1)^2} = \frac{2t[3t^2 - 2t + 1 - t(3t - 1)]}{(3t^2 - 2t + 1)^2} = \frac{2t(3t^2 - 2t + 1 - 3t^2 + t)}{(3t^2 - 2t + 1)^2} = \frac{2t(1 - t)}{(3t^2 - 2t + 1)^2}$$

15. $y = \dfrac{v^3 - 2v\sqrt{v}}{v} = v^2 - 2\sqrt{v} = v^2 - 2v^{1/2} \quad\Rightarrow\quad y' = 2v - 2\left(\frac{1}{2}\right)v^{-1/2} = 2v - v^{-1/2}.$

We can change the form of the answer as follows: $2v - v^{-1/2} = 2v - \dfrac{1}{\sqrt{v}} = \dfrac{2v\sqrt{v}-1}{\sqrt{v}} = \dfrac{2v^{3/2}-1}{\sqrt{v}}$

17. $y = \dfrac{r^2}{1+\sqrt{r}} \quad\Rightarrow$

$y' = \dfrac{(1+\sqrt{r})(2r) - r^2\left(\frac{1}{2}r^{-1/2}\right)}{(1+\sqrt{r})^2} = \dfrac{2r + 2r^{3/2} - \frac{1}{2}r^{3/2}}{(1+\sqrt{r})^2} = \dfrac{2r + \frac{3}{2}r^{3/2}}{(1+\sqrt{r})^2} = \dfrac{\frac{1}{2}r(4+3r^{1/2})}{(1+\sqrt{r})^2} = \dfrac{r(4+3\sqrt{r})}{2(1+\sqrt{r})^2}$

19. $y = \dfrac{x}{\cos x} \quad\Rightarrow\quad y' = \dfrac{(\cos x)(1) - (x)(-\sin x)}{(\cos x)^2} = \dfrac{\cos x + x\sin x}{\cos^2 x}$

21. $f(\theta) = \dfrac{\sec\theta}{1+\sec\theta} \quad\Rightarrow$

$f'(\theta) = \dfrac{(1+\sec\theta)(\sec\theta\tan\theta) - (\sec\theta)(\sec\theta\tan\theta)}{(1+\sec\theta)^2} = \dfrac{(\sec\theta\tan\theta)[(1+\sec\theta)-\sec\theta]}{(1+\sec\theta)^2} = \dfrac{\sec\theta\tan\theta}{(1+\sec\theta)^2}$

23. $y = \dfrac{\sin x}{x^2} \quad\Rightarrow\quad y' = \dfrac{x^2\cos x - (\sin x)(2x)}{(x^2)^2} = \dfrac{x(x\cos x - 2\sin x)}{x^4} = \dfrac{x\cos x - 2\sin x}{x^3}$

25. $f(x) = \dfrac{x}{x+c/x} \quad\Rightarrow\quad f'(x) = \dfrac{(x+c/x)(1) - x(1 - c/x^2)}{\left(x + \frac{c}{x}\right)^2} = \dfrac{x + c/x - x + c/x}{\left(\dfrac{x^2+c}{x}\right)^2} = \dfrac{2c/x}{\dfrac{(x^2+c)^2}{x^2}} \cdot \dfrac{x^2}{x^2} = \dfrac{2cx}{(x^2+c)^2}$

27. $y = \dfrac{2x}{x+1} \quad\Rightarrow\quad y' = \dfrac{(x+1)(2) - (2x)(1)}{(x+1)^2} = \dfrac{2}{(x+1)^2}.$ At $(1,1)$, $y' = \frac{1}{2}$, and an equation of the tangent line is

$y - 1 = \frac{1}{2}(x-1)$, or $y = \frac{1}{2}x + \frac{1}{2}$.

29. $y = \tan x \quad\Rightarrow\quad y' = \sec^2 x \quad\Rightarrow\quad$ the slope of the tangent line at $\left(\frac{\pi}{4}, 1\right)$ is $\sec^2\frac{\pi}{4} = \left(\sqrt{2}\right)^2 = 2$ and an equation of the

tangent line is $y - 1 = 2\left(x - \frac{\pi}{4}\right)$ or $y = 2x + 1 - \frac{\pi}{2}$.

31. (a) $y = f(x) = \dfrac{1}{1+x^2} \quad\Rightarrow$

$f'(x) = \dfrac{(1+x^2)(0) - 1(2x)}{(1+x^2)^2} = \dfrac{-2x}{(1+x^2)^2}.$ So the slope of the

tangent line at the point $\left(-1, \frac{1}{2}\right)$ is $f'(-1) = \dfrac{2}{2^2} = \frac{1}{2}$ and its

equation is $y - \frac{1}{2} = \frac{1}{2}(x+1)$ or $y = \frac{1}{2}x + 1$.

(b)

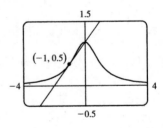

33. $f(x) = \dfrac{x^2}{1+x} \quad\Rightarrow\quad f'(x) = \dfrac{(1+x)(2x) - x^2(1)}{(1+x)^2} = \dfrac{2x + 2x^2 - x^2}{(1+x)^2} = \dfrac{x^2 + 2x}{x^2 + 2x + 1} \quad\Rightarrow$

$f''(x) = \dfrac{(x^2+2x+1)(2x+2) - (x^2+2x)(2x+2)}{(x^2+2x+1)^2} = \dfrac{(2x+2)(x^2+2x+1 - x^2 - 2x)}{[(x+1)^2]^2}$

$= \dfrac{2(x+1)(1)}{(x+1)^4} = \dfrac{2}{(x+1)^3},$

so $f''(1) = \dfrac{2}{(1+1)^3} = \dfrac{2}{8} = \dfrac{1}{4}$.

35. $H(\theta) = \theta \sin \theta \;\Rightarrow\; H'(\theta) = \theta \, (\cos \theta) + (\sin \theta) \cdot 1 = \theta \cos \theta + \sin \theta \;\Rightarrow$

$H''(\theta) = \theta \, (-\sin \theta) + (\cos \theta) \cdot 1 + \cos \theta = -\theta \sin \theta + 2 \cos \theta$

37. $\dfrac{d}{dx}(\csc x) = \dfrac{d}{dx}\left(\dfrac{1}{\sin x}\right) = \dfrac{(\sin x)(0) - 1(\cos x)}{\sin^2 x} = \dfrac{-\cos x}{\sin^2 x} = -\dfrac{1}{\sin x} \cdot \dfrac{\cos x}{\sin x} = -\csc x \cot x$

39. $\dfrac{d}{dx}(\cot x) = \dfrac{d}{dx}\left(\dfrac{\cos x}{\sin x}\right) = \dfrac{(\sin x)(-\sin x) - (\cos x)(\cos x)}{\sin^2 x} = -\dfrac{\sin^2 x + \cos^2 x}{\sin^2 x} = -\dfrac{1}{\sin^2 x} = -\csc^2 x$

41. We are given that $f(5) = 1$, $f'(5) = 6$, $g(5) = -3$, and $g'(5) = 2$.

(a) $(fg)'(5) = f(5)g'(5) + g(5)f'(5) = (1)(2) + (-3)(6) = 2 - 18 = -16$

(b) $\left(\dfrac{f}{g}\right)'(5) = \dfrac{g(5)f'(5) - f(5)g'(5)}{[g(5)]^2} = \dfrac{(-3)(6) - (1)(2)}{(-3)^2} = -\dfrac{20}{9}$

(c) $\left(\dfrac{g}{f}\right)'(5) = \dfrac{f(5)g'(5) - g(5)f'(5)}{[f(5)]^2} = \dfrac{(1)(2) - (-3)(6)}{(1)^2} = 20$

43. (a) From the graphs of f and g, we obtain the following values: $f(1) = 2$ since the point $(1, 2)$ is on the graph of f;

$g(1) = 1$ since the point $(1, 1)$ is on the graph of g; $f'(1) = 2$ since the slope of the line segment between $(0, 0)$ and $(2, 4)$

is $\dfrac{4 - 0}{2 - 0} = 2$; $g'(1) = -1$ since the slope of the line segment between $(-2, 4)$ and $(2, 0)$ is $\dfrac{0 - 4}{2 - (-2)} = -1$.

Now $u(x) = f(x)g(x)$, so $u'(1) = f(1)g'(1) + g(1)\,f'(1) = 2 \cdot (-1) + 1 \cdot 2 = 0$.

(b) $v(x) = f(x)/g(x)$, so $v'(5) = \dfrac{g(5)f'(5) - f(5)g'(5)}{[g(5)]^2} = \dfrac{2\left(-\frac{1}{3}\right) - 3 \cdot \frac{2}{3}}{2^2} = \dfrac{-\frac{8}{3}}{4} = -\dfrac{2}{3}$

45. (a) $y = xg(x) \;\Rightarrow\; y' = xg'(x) + g(x) \cdot 1 = xg'(x) + g(x)$

(b) $y = \dfrac{x}{g(x)} \;\Rightarrow\; y' = \dfrac{g(x) \cdot 1 - xg'(x)}{[g(x)]^2} = \dfrac{g(x) - xg'(x)}{[g(x)]^2}$

(c) $y = \dfrac{g(x)}{x} \;\Rightarrow\; y' = \dfrac{xg'(x) - g(x) \cdot 1}{(x)^2} = \dfrac{xg'(x) - g(x)}{x^2}$

47. (a) $x(t) = 8 \sin t \;\Rightarrow\; v(t) = x'(t) = 8 \cos t \;\Rightarrow\; a(t) = x''(t) = -8 \sin t$

(b) The mass at time $t = \frac{2\pi}{3}$ has position $x\left(\frac{2\pi}{3}\right) = 8 \sin \frac{2\pi}{3} = 8\left(\frac{\sqrt{3}}{2}\right) = 4\sqrt{3}$, velocity $v\left(\frac{2\pi}{3}\right) = 8 \cos \frac{2\pi}{3} = 8\left(-\frac{1}{2}\right) = -4$,

and acceleration $a\left(\frac{2\pi}{3}\right) = -8 \sin \frac{2\pi}{3} = -8\left(\frac{\sqrt{3}}{2}\right) = -4\sqrt{3}$. Since $v\left(\frac{2\pi}{3}\right) < 0$, the particle is moving to the left. Because

v and a have the same sign, the particle is speeding up.

49. $PV = nRT \;\Rightarrow\; T = \dfrac{PV}{nR} = \dfrac{PV}{(10)(0.0821)} = \dfrac{1}{0.821}(PV)$. Using the Product Rule, we have

$\dfrac{dT}{dt} = \dfrac{1}{0.821}[P(t)V'(t) + V(t)P'(t)] = \dfrac{1}{0.821}[(8)(-0.15) + (10)(0.10)] \approx -0.2436$ K/min.

51. If $y = f(x) = \dfrac{x}{x+1}$, then $f'(x) = \dfrac{(x+1)(1) - x(1)}{(x+1)^2} = \dfrac{1}{(x+1)^2}$. When $x = a$, the equation of the tangent line is

$y - \dfrac{a}{a+1} = \dfrac{1}{(a+1)^2}(x - a)$. This line passes through $(1, 2)$ when $2 - \dfrac{a}{a+1} = \dfrac{1}{(a+1)^2}(1 - a)$ $\Leftrightarrow$

$2(a+1)^2 - a(a+1) = 1 - a$ $\Leftrightarrow$ $2a^2 + 4a + 2 - a^2 - a - 1 + a = 0$ $\Leftrightarrow$ $a^2 + 4a + 1 = 0$.

The quadratic formula gives the roots of this equation as $a = \dfrac{-4 \pm \sqrt{4^2 - 4(1)(1)}}{2(1)} = \dfrac{-4 \pm \sqrt{12}}{2} = -2 \pm \sqrt{3}$,

so there are two such tangent lines. Since

$$f\left(-2 \pm \sqrt{3}\right) = \dfrac{-2 \pm \sqrt{3}}{-2 \pm \sqrt{3} + 1} = \dfrac{-2 \pm \sqrt{3}}{-1 \pm \sqrt{3}} \cdot \dfrac{-1 \mp \sqrt{3}}{-1 \mp \sqrt{3}}$$

$$= \dfrac{2 \pm 2\sqrt{3} \mp \sqrt{3} - 3}{1 - 3} = \dfrac{-1 \pm \sqrt{3}}{-2} = \dfrac{1 \mp \sqrt{3}}{2},$$

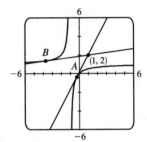

the lines touch the curve at $A\left(-2 + \sqrt{3}, \frac{1-\sqrt{3}}{2}\right) \approx (-0.27, -0.37)$

and $B\left(-2 - \sqrt{3}, \frac{1+\sqrt{3}}{2}\right) \approx (-3.73, 1.37)$.

53. (a) $(fgh)' = [(fg)h]' = (fg)'h + (fg)h' = (f'g + fg')h + (fg)h' = f'gh + fg'h + fgh'$

(b) $y = x \sin x \cos x$ $\Rightarrow$ $\dfrac{dy}{dx} = \sin x \cos x + x \cos x \cos x + x \sin x (-\sin x) = \sin x \cos x + x \cos^2 x - x \sin^2 x$

55. (a) $\dfrac{d}{dx}\left(\dfrac{1}{g(x)}\right) = \dfrac{g(x) \cdot \frac{d}{dx}(1) - 1 \cdot \frac{d}{dx}[g(x)]}{[g(x)]^2}$ [Quotient Rule] $= \dfrac{g(x) \cdot 0 - 1 \cdot g'(x)}{[g(x)]^2} = \dfrac{0 - g'(x)}{[g(x)]^2} = -\dfrac{g'(x)}{[g(x)]^2}$

(b) $y = \dfrac{1}{x^4 + x^2 + 1}$ $\Rightarrow$ $y' = -\dfrac{2x(2x^2 + 1)}{(x^4 + x^2 + 1)^2}$

(c) $\dfrac{d}{dx}(x^{-n}) = \dfrac{d}{dx}\left(\dfrac{1}{x^n}\right) = -\dfrac{(x^n)'}{(x^n)^2}$ [by the Reciprocal Rule] $= -\dfrac{nx^{n-1}}{x^{2n}} = -nx^{n-1-2n} = -nx^{-n-1}$

2.5 The Chain Rule

1. Let $u = g(x) = 4x$ and $y = f(u) = \sin u$. Then $\dfrac{dy}{dx} = \dfrac{dy}{du}\dfrac{du}{dx} = (\cos u)(4) = 4 \cos 4x$.

3. Let $u = g(x) = 1 - x^2$ and $y = f(u) = u^{10}$. Then $\dfrac{dy}{dx} = \dfrac{dy}{du}\dfrac{du}{dx} = (10u^9)(-2x) = -20x(1 - x^2)^9$.

5. Let $u = g(x) = \sin x$ and $y = f(u) = \sqrt{u}$. Then $\dfrac{dy}{dx} = \dfrac{dy}{du}\dfrac{du}{dx} = \frac{1}{2}u^{-1/2} \cos x = \dfrac{\cos x}{2\sqrt{u}} = \dfrac{\cos x}{2\sqrt{\sin x}}$.

7. $F(x) = \sqrt[4]{1 + 2x + x^3} = (1 + 2x + x^3)^{1/4}$ $\Rightarrow$

$$F'(x) = \frac{1}{4}(1 + 2x + x^3)^{-3/4} \cdot \dfrac{d}{dx}(1 + 2x + x^3) = \dfrac{1}{4(1 + 2x + x^3)^{3/4}} \cdot (2 + 3x^2)$$

$$= \dfrac{2 + 3x^2}{4(1 + 2x + x^3)^{3/4}} = \dfrac{2 + 3x^2}{4\sqrt[4]{(1 + 2x + x^3)^3}}$$

9. $g(t) = \dfrac{1}{(t^4 + 1)^3} = (t^4 + 1)^{-3}$ $\Rightarrow$ $g'(t) = -3(t^4 + 1)^{-4}(4t^3) = -12t^3(t^4 + 1)^{-4} = \dfrac{-12t^3}{(t^4 + 1)^4}$

11. $y = \cos(a^3 + x^3)$ $\Rightarrow$ $y' = -\sin(a^3 + x^3) \cdot 3x^2$ [a^3 is just a constant] $= -3x^2 \sin(a^3 + x^3)$

13. $y = \cot(x/2) \;\Rightarrow\; y' = -\csc^2(x/2) \cdot \frac{1}{2} = -\frac{1}{2}\csc^2(x/2)$

15. $g(x) = (1 + 4x)^5(3 + x - x^2)^8 \;\Rightarrow$

$$g'(x) = (1 + 4x)^5 \cdot 8(3 + x - x^2)^7(1 - 2x) + (3 + x - x^2)^8 \cdot 5(1 + 4x)^4 \cdot 4$$
$$= 4(1 + 4x)^4(3 + x - x^2)^7[2(1 + 4x)(1 - 2x) + 5(3 + x - x^2)]$$
$$= 4(1 + 4x)^4(3 + x - x^2)^7[(2 + 4x - 16x^2) + (15 + 5x - 5x^2)] = 4(1 + 4x)^4(3 + x - x^2)^7(17 + 9x - 21x^2)$$

17. $y = (2x - 5)^4(8x^2 - 5)^{-3} \;\Rightarrow$

$$y' = 4(2x - 5)^3(2)(8x^2 - 5)^{-3} + (2x - 5)^4(-3)(8x^2 - 5)^{-4}(16x)$$
$$= 8(2x - 5)^3(8x^2 - 5)^{-3} - 48x(2x - 5)^4(8x^2 - 5)^{-4}$$

[This simplifies to $8(2x - 5)^3(8x^2 - 5)^{-4}(-4x^2 + 30x - 5)$.]

19. $y = x^3\cos nx \;\Rightarrow\; y' = x^3(-\sin nx)(n) + \cos nx\,(3x^2) = x^2(3\cos nx - nx\sin nx)$

21. $y = \sin(x\cos x) \;\Rightarrow\; y' = \cos(x\cos x) \cdot [x(-\sin x) + \cos x \cdot 1] = (\cos x - x\sin x)\cos(x\cos x)$

23. $F(z) = \sqrt{\dfrac{z-1}{z+1}} = \left(\dfrac{z-1}{z+1}\right)^{1/2} \;\Rightarrow$

$$F'(z) = \frac{1}{2}\left(\frac{z-1}{z+1}\right)^{-1/2} \cdot \frac{d}{dz}\left(\frac{z-1}{z+1}\right) = \frac{1}{2}\left(\frac{z+1}{z-1}\right)^{1/2} \cdot \frac{(z+1)(1) - (z-1)(1)}{(z+1)^2}$$
$$= \frac{1}{2}\frac{(z+1)^{1/2}}{(z-1)^{1/2}} \cdot \frac{z+1-z+1}{(z+1)^2} = \frac{1}{2}\frac{(z+1)^{1/2}}{(z-1)^{1/2}} \cdot \frac{2}{(z+1)^2} = \frac{1}{(z-1)^{1/2}(z+1)^{3/2}}$$

25. $y = \dfrac{r}{\sqrt{r^2 + 1}} \;\Rightarrow$

$$y' = \frac{\sqrt{r^2+1}\,(1) - r \cdot \frac{1}{2}(r^2+1)^{-1/2}(2r)}{\left(\sqrt{r^2+1}\right)^2} = \frac{\sqrt{r^2+1} - \dfrac{r^2}{\sqrt{r^2+1}}}{\left(\sqrt{r^2+1}\right)^2} = \frac{\dfrac{\sqrt{r^2+1}\sqrt{r^2+1} - r^2}{\sqrt{r^2+1}}}{\left(\sqrt{r^2+1}\right)^2}$$
$$= \frac{(r^2+1) - r^2}{\left(\sqrt{r^2+1}\right)^3} = \frac{1}{(r^2+1)^{3/2}} \;\text{ or }\; (r^2+1)^{-3/2}$$

Another solution: Write y as a product and make use of the Product Rule. $y = r(r^2+1)^{-1/2} \;\Rightarrow$

$y' = r \cdot -\frac{1}{2}(r^2+1)^{-3/2}(2r) + (r^2+1)^{-1/2} \cdot 1 = (r^2+1)^{-3/2}[-r^2 + (r^2+1)^1] = (r^2+1)^{-3/2}(1) = (r^2+1)^{-3/2}$.

The step that students usually have trouble with is factoring out $(r^2+1)^{-3/2}$. But this is no different than factoring out x^2 from $x^2 + x^5$; that is, we are just factoring out a factor with the *smallest* exponent that appears on it. In this case, $-\frac{3}{2}$ is smaller than $-\frac{1}{2}$.

27. $y = \tan(\cos x) \;\Rightarrow\; y' = \sec^2(\cos x) \cdot (-\sin x) = -\sin x\sec^2(\cos x)$

29. $y = \sin\sqrt{1+x^2} \;\Rightarrow\; y' = \cos\sqrt{1+x^2} \cdot \frac{1}{2}(1+x^2)^{-1/2} \cdot 2x = \left(x\cos\sqrt{1+x^2}\,\right)/\sqrt{1+x^2}$

31. $y = (1 + \cos^2 x)^6 \;\Rightarrow\; y' = 6(1 + \cos^2 x)^5 \cdot 2\cos x\,(-\sin x) = -12\cos x\sin x\,(1 + \cos^2 x)^5$

33. $y = \sec^2 x + \tan^2 x = (\sec x)^2 + (\tan x)^2 \;\Rightarrow$

$$y' = 2(\sec x)(\sec x\tan x) + 2(\tan x)(\sec^2 x) = 2\sec^2 x\tan x + 2\sec^2 x\tan x = 4\sec^2 x\tan x$$

35. $y = \cot^2(\sin\theta) = [\cot(\sin\theta)]^2 \quad\Rightarrow$

$y' = 2[\cot(\sin\theta)] \cdot \dfrac{d}{d\theta}[\cot(\sin\theta)] = 2\cot(\sin\theta) \cdot [-\csc^2(\sin\theta) \cdot \cos\theta] = -2\cos\theta\,\cot(\sin\theta)\,\csc^2(\sin\theta)$

37. $y = \sin\left(\tan\sqrt{\sin x}\right) \quad\Rightarrow$

$y' = \cos\left(\tan\sqrt{\sin x}\right) \cdot \dfrac{d}{dx}\left(\tan\sqrt{\sin x}\right) = \cos\left(\tan\sqrt{\sin x}\right)\sec^2\sqrt{\sin x} \cdot \dfrac{d}{dx}(\sin x)^{1/2}$

$= \cos\left(\tan\sqrt{\sin x}\right)\sec^2\sqrt{\sin x} \cdot \tfrac{1}{2}(\sin x)^{-1/2} \cdot \cos x = \cos\left(\tan\sqrt{\sin x}\right)\left(\sec^2\sqrt{\sin x}\right)\left(\dfrac{1}{2\sqrt{\sin x}}\right)(\cos x)$

39. $y = (1+2x)^{10} \quad\Rightarrow\quad y' = 10(1+2x)^9 \cdot 2 = 20(1+2x)^9$. At $(0,1)$, $y' = 20(1+0)^9 = 20$, and an equation of the tangent

line is $y - 1 = 20(x-0)$, or $y = 20x + 1$.

41. (a) $y = f(x) = \tan(\tfrac{\pi}{4}x^2) \quad\Rightarrow\quad f'(x) = \sec^2(\tfrac{\pi}{4}x^2)(2 \cdot \tfrac{\pi}{4}x)$.   (b)

    The slope of the tangent at $(1,1)$ is thus

    $f'(1) = \sec^2\tfrac{\pi}{4}\left(\tfrac{\pi}{2}\right) = 2 \cdot \tfrac{\pi}{2} = \pi$, and its equation

    is $y - 1 = \pi(x-1)$ or $y = \pi x - \pi + 1$.

43. $F(t) = (1-7t)^6 \quad\Rightarrow\quad F'(t) = 6(1-7t)^5(-7) = -42(1-7t)^5 \quad\Rightarrow\quad F''(t) = -42 \cdot 5(1-7t)^4(-7) = 1470(1-7t)^4$

45. $y = (x^3+1)^{2/3} \quad\Rightarrow\quad y' = \tfrac{2}{3}(x^3+1)^{-1/3}(3x^2) = 2x^2(x^3+1)^{-1/3} \quad\Rightarrow$

$y'' = 2x^2\left(-\tfrac{1}{3}\right)(x^3+1)^{-4/3}(3x^2) + (x^3+1)^{-1/3}(4x) = 4x(x^3+1)^{-1/3} - 2x^4(x^3+1)^{-4/3}$

47. $F(x) = f(g(x)) \quad\Rightarrow\quad F'(x) = f'(g(x)) \cdot g'(x)$, so $F'(5) = f'(g(5)) \cdot g'(5) = f'(-2) \cdot 6 = 4 \cdot 6 = 24$

49. (a) $h(x) = f(g(x)) \quad\Rightarrow\quad h'(x) = f'(g(x)) \cdot g'(x)$, so $h'(1) = f'(g(1)) \cdot g'(1) = f'(2) \cdot 6 = 5 \cdot 6 = 30$.

    (b) $H(x) = g(f(x)) \quad\Rightarrow\quad H'(x) = g'(f(x)) \cdot f'(x)$, so $H'(1) = g'(f(1)) \cdot f'(1) = g'(3) \cdot 4 = 9 \cdot 4 = 36$.

51. (a) $u(x) = f(g(x)) \quad\Rightarrow\quad u'(x) = f'(g(x))g'(x)$. So $u'(1) = f'(g(1))g'(1) = f'(3)g'(1)$. To find $f'(3)$, note that f is

    linear from $(2,4)$ to $(6,3)$, so its slope is $\dfrac{3-4}{6-2} = -\dfrac{1}{4}$. To find $g'(1)$, note that g is linear from $(0,6)$ to $(2,0)$, so its slope

    is $\dfrac{0-6}{2-0} = -3$. Thus, $f'(3)g'(1) = \left(-\tfrac{1}{4}\right)(-3) = \tfrac{3}{4}$.

    (b) $v(x) = g(f(x)) \quad\Rightarrow\quad v'(x) = g'(f(x))f'(x)$. So $v'(1) = g'(f(1))f'(1) = g'(2)f'(1)$, which does not exist since

    $g'(2)$ does not exist.

    (c) $w(x) = g(g(x)) \quad\Rightarrow\quad w'(x) = g'(g(x))g'(x)$. So $w'(1) = g'(g(1))g'(1) = g'(3)g'(1)$. To find $g'(3)$, note that g is

    linear from $(2,0)$ to $(5,2)$, so its slope is $\dfrac{2-0}{5-2} = \dfrac{2}{3}$. Thus, $g'(3)g'(1) = \left(\tfrac{2}{3}\right)(-3) = -2$.

53. (a) $F(x) = f(\cos x) \quad\Rightarrow\quad F'(x) = f'(\cos x)\dfrac{d}{dx}(\cos x) = -\sin x\, f'(\cos x)$

    (b) $G(x) = \cos(f(x)) \quad\Rightarrow\quad G'(x) = -\sin(f(x))\,f'(x)$

55. $r(x) = f(g(h(x))) \quad\Rightarrow\quad r'(x) = f'(g(h(x))) \cdot g'(h(x)) \cdot h'(x)$, so

    $r'(1) = f'(g(h(1))) \cdot g'(h(1)) \cdot h'(1) = f'(g(2)) \cdot g'(2) \cdot 4 = f'(3) \cdot 5 \cdot 4 = 6 \cdot 5 \cdot 4 = 120$

57. For the tangent line to be horizontal, $f'(x) = 0$. $f(x) = 2\sin x + \sin^2 x \Rightarrow f'(x) = 2\cos x + 2\sin x \cos x = 0 \Leftrightarrow$

$2\cos x\,(1 + \sin x) = 0 \Leftrightarrow \cos x = 0$ or $\sin x = -1$, so $x = \frac{\pi}{2} + 2n\pi$ or $\frac{3\pi}{2} + 2n\pi$, where n is any integer. Now

$f\!\left(\frac{\pi}{2}\right) = 3$ and $f\!\left(\frac{3\pi}{2}\right) = -1$, so the points on the curve with a horizontal tangent are $\left(\frac{\pi}{2} + 2n\pi, 3\right)$ and $\left(\frac{3\pi}{2} + 2n\pi, -1\right)$,

where n is any integer.

59. $s(t) = 10 + \frac{1}{4}\sin(10\pi t) \Rightarrow$ the velocity after t seconds is $v(t) = s'(t) = \frac{1}{4}\cos(10\pi t)(10\pi) = \frac{5\pi}{2}\cos(10\pi t)$ cm/s.

61. (a) $B(t) = 4.0 + 0.35\sin\dfrac{2\pi t}{5.4} \Rightarrow \dfrac{dB}{dt} = \left(0.35\cos\dfrac{2\pi t}{5.4}\right)\!\left(\dfrac{2\pi}{5.4}\right) = \dfrac{0.7\pi}{5.4}\cos\dfrac{2\pi t}{5.4} = \dfrac{7\pi}{54}\cos\dfrac{2\pi t}{5.4}$

(b) At $t = 1$, $\dfrac{dB}{dt} = \dfrac{7\pi}{54}\cos\dfrac{2\pi}{5.4} \approx 0.16$.

63. By the Chain Rule, $a(t) = \dfrac{dv}{dt} = \dfrac{dv}{ds}\dfrac{ds}{dt} = \dfrac{dv}{ds}v(t) = v(t)\dfrac{dv}{ds}$. The derivative dv/dt is the rate of change of the velocity with

respect to time (in other words, the acceleration) whereas the derivative dv/ds is the rate of change of the velocity with respect

to the displacement.

65. (a) $\dfrac{d}{dx}\left(\sin^n x \cos nx\right) = n\sin^{n-1} x \cos x \cos nx + \sin^n x\,(-n\sin nx)$ [Product Rule]

$= n\sin^{n-1} x\,(\cos nx \cos x - \sin nx \sin x)$ [factor out $n\sin^{n-1} x$]

$= n\sin^{n-1} x \cos(nx + x)$ [Addition Formula for cosine]

$= n\sin^{n-1} x \cos[(n+1)x]$ [factor out x]

(b) $\dfrac{d}{dx}\left(\cos^n x \cos nx\right) = n\cos^{n-1} x\,(-\sin x)\cos nx + \cos^n x\,(-n\sin nx)$ [Product Rule]

$= -n\cos^{n-1} x\,(\cos nx \sin x + \sin nx \cos x)$ [factor out $-n\cos^{n-1} x$]

$= -n\cos^{n-1} x \sin(nx + x)$ [Addition Formula for sine]

$= -n\cos^{n-1} x \sin[(n+1)x]$ [factor out x]

67. Since $\theta^\circ = \left(\frac{\pi}{180}\right)\theta$ rad, we have $\dfrac{d}{d\theta}\left(\sin\theta^\circ\right) = \dfrac{d}{d\theta}\left(\sin\frac{\pi}{180}\theta\right) = \frac{\pi}{180}\cos\frac{\pi}{180}\theta = \frac{\pi}{180}\cos\theta^\circ$.

69. $\dfrac{d^2 y}{dx^2} = \dfrac{d}{dx}\left(\dfrac{dy}{dx}\right)$ [Leibniz notation for the second derivative]

$= \dfrac{d}{dx}\left(\dfrac{dy}{du}\dfrac{du}{dx}\right)$ [Chain Rule]

$= \dfrac{dy}{du}\cdot\dfrac{d}{dx}\left(\dfrac{du}{dx}\right) + \dfrac{du}{dx}\cdot\dfrac{d}{dx}\left(\dfrac{dy}{du}\right)$ [Product Rule]

$= \dfrac{dy}{du}\cdot\dfrac{d^2 u}{dx^2} + \dfrac{du}{dx}\cdot\dfrac{d}{du}\left(\dfrac{dy}{du}\right)\cdot\dfrac{du}{dx}$ [dy/du is a function of u]

$= \dfrac{dy}{du}\dfrac{d^2 u}{dx^2} + \dfrac{d^2 y}{du^2}\left(\dfrac{du}{dx}\right)^2$

Or: Using function notation for $y = f(u)$ and $u = g(x)$, we have $y = f(g(x))$, so

$y' = f'(g(x))\cdot g'(x)$ [by the Chain Rule] $\Rightarrow$

$(y')' = [f'(g(x))\cdot g'(x)]' = f'(g(x))\cdot g''(x) + g'(x)\cdot f''(g(x))\cdot g'(x) = f'(g(x))\cdot g''(x) + f''(g(x))\cdot [g'(x)]^2$.

2.6 Implicit Differentiation

1. (a) $\frac{d}{dx}(xy + 2x + 3x^2) = \frac{d}{dx}(4) \;\Rightarrow\; (x \cdot y' + y \cdot 1) + 2 + 6x = 0 \;\Rightarrow\; xy' = -y - 2 - 6x \;\Rightarrow$

$y' = \frac{-y - 2 - 6x}{x}$ or $y' = -6 - \frac{y+2}{x}$.

(b) $xy + 2x + 3x^2 = 4 \;\Rightarrow\; xy = 4 - 2x - 3x^2 \;\Rightarrow\; y = \frac{4 - 2x - 3x^2}{x} = \frac{4}{x} - 2 - 3x$, so $y' = -\frac{4}{x^2} - 3$.

(c) From part (a), $y' = \frac{-y - 2 - 6x}{x} = \frac{-(4/x - 2 - 3x) - 2 - 6x}{x} = \frac{-4/x - 3x}{x} = -\frac{4}{x^2} - 3$.

3. $\frac{d}{dx}(x^3 + x^2y + 4y^2) = \frac{d}{dx}(6) \;\Rightarrow\; 3x^2 + (x^2y' + y \cdot 2x) + 8yy' = 0 \;\Rightarrow\; x^2y' + 8yy' = -3x^2 - 2xy \;\Rightarrow$

$(x^2 + 8y)y' = -3x^2 - 2xy \;\Rightarrow\; y' = -\frac{3x^2 + 2xy}{x^2 + 8y} = -\frac{x(3x + 2y)}{x^2 + 8y}$

5. $\frac{d}{dx}(x^2y + xy^2) = \frac{d}{dx}(3x) \;\Rightarrow\; (x^2y' + y \cdot 2x) + (x \cdot 2yy' + y^2 \cdot 1) = 3 \;\Rightarrow\; x^2y' + 2xyy' = 3 - 2xy - y^2 \;\Rightarrow$

$y'(x^2 + 2xy) = 3 - 2xy - y^2 \;\Rightarrow\; y' = \frac{3 - 2xy - y^2}{x^2 + 2xy}$

7. $\frac{d}{dx}(x^2y^2 + x\sin y) = \frac{d}{dx}(4) \;\Rightarrow\; x^2 \cdot 2yy' + y^2 \cdot 2x + x\cos y \cdot y' + \sin y \cdot 1 = 0 \;\Rightarrow$

$2x^2yy' + x\cos y \cdot y' = -2xy^2 - \sin y \;\Rightarrow\; (2x^2y + x\cos y)y' = -2xy^2 - \sin y \;\Rightarrow\; y' = \frac{-2xy^2 - \sin y}{2x^2y + x\cos y}$

9. $\frac{d}{dx}(4\cos x \sin y) = \frac{d}{dx}(1) \;\Rightarrow\; 4[\cos x \cdot \cos y \cdot y' + \sin y \cdot (-\sin x)] = 0 \;\Rightarrow$

$y'(4\cos x \cos y) = 4\sin x \sin y \;\Rightarrow\; y' = \frac{4\sin x \sin y}{4\cos x \cos y} = \tan x \tan y$

11. $\frac{d}{dx}[\tan(x/y)] = \frac{d}{dx}(x + y) \;\Rightarrow\; \sec^2(x/y) \cdot \frac{y \cdot 1 - x \cdot y'}{y^2} = 1 + y' \;\Rightarrow$

$y\sec^2(x/y) - x\sec^2(x/y) \cdot y' = y^2 + y^2y' \;\Rightarrow\; y\sec^2(x/y) - y^2 = y^2y' + x\sec^2(x/y) \;\Rightarrow$

$y\sec^2(x/y) - y^2 = [y^2 + x\sec^2(x/y)] \cdot y' \;\Rightarrow\; y' = \frac{y\sec^2(x/y) - y^2}{y^2 + x\sec^2(x/y)}$

13. $\sqrt{xy} = 1 + x^2y \;\Rightarrow\; \frac{1}{2}(xy)^{-1/2}(xy' + y \cdot 1) = 0 + x^2y' + y \cdot 2x \;\Rightarrow\; \frac{x}{2\sqrt{xy}}y' + \frac{y}{2\sqrt{xy}} = x^2y' + 2xy \;\Rightarrow$

$y'\left(\frac{x}{2\sqrt{xy}} - x^2\right) = 2xy - \frac{y}{2\sqrt{xy}} \;\Rightarrow\; y'\left(\frac{x - 2x^2\sqrt{xy}}{2\sqrt{xy}}\right) = \frac{4xy\sqrt{xy} - y}{2\sqrt{xy}} \;\Rightarrow\; y' = \frac{4xy\sqrt{xy} - y}{x - 2x^2\sqrt{xy}}$

15. $\frac{d}{dx}\{f(x) + x^2[f(x)]^3\} = \frac{d}{dx}(10) \;\Rightarrow\; f'(x) + x^2 \cdot 3[f(x)]^2 \cdot f'(x) + [f(x)]^3 \cdot 2x = 0$. If $x = 1$, we have

$f'(1) + 1^2 \cdot 3[f(1)]^2 \cdot f'(1) + [f(1)]^3 \cdot 2(1) = 0 \;\Rightarrow\; f'(1) + 1 \cdot 3 \cdot 2^2 \cdot f'(1) + 2^3 \cdot 2 = 0 \;\Rightarrow$

$f'(1) + 12f'(1) = -16 \;\Rightarrow\; 13f'(1) = -16 \;\Rightarrow\; f'(1) = -\frac{16}{13}$.

17. $x^2 + xy + y^2 = 3 \ \Rightarrow \ 2x + xy' + y \cdot 1 + 2yy' = 0 \ \Rightarrow \ xy' + 2yy' = -2x - y \ \Rightarrow \ y'(x + 2y) = -2x - y \ \Rightarrow$

$y' = \dfrac{-2x - y}{x + 2y}$. When $x = 1$ and $y = 1$, we have $y' = \dfrac{-2 - 1}{1 + 2 \cdot 1} = \dfrac{-3}{3} = -1$, so an equation of the tangent line is

$y - 1 = -1(x - 1)$ or $y = -x + 2$.

19. $x^2 + y^2 = (2x^2 + 2y^2 - x)^2 \ \Rightarrow \ 2x + 2yy' = 2(2x^2 + 2y^2 - x)(4x + 4yy' - 1)$. When $x = 0$ and $y = \frac{1}{2}$, we have

$0 + y' = 2(\frac{1}{2})(2y' - 1) \ \Rightarrow \ y' = 2y' - 1 \ \Rightarrow \ y' = 1$, so an equation of the tangent line is $y - \frac{1}{2} = 1(x - 0)$

or $y = x + \frac{1}{2}$.

21. $2(x^2 + y^2)^2 = 25(x^2 - y^2) \ \Rightarrow \ 4(x^2 + y^2)(2x + 2yy') = 25(2x - 2yy') \ \Rightarrow$

$4(x + yy')(x^2 + y^2) = 25(x - yy') \ \Rightarrow \ 4yy'(x^2 + y^2) + 25yy' = 25x - 4x(x^2 + y^2) \ \Rightarrow$

$y' = \dfrac{25x - 4x(x^2 + y^2)}{25y + 4y(x^2 + y^2)}$. When $x = 3$ and $y = 1$, we have $y' = \frac{75 - 120}{25 + 40} = -\frac{45}{65} = -\frac{9}{13}$, so an equation of the tangent line

is $y - 1 = -\frac{9}{13}(x - 3)$ or $y = -\frac{9}{13}x + \frac{40}{13}$.

23. $9x^2 + y^2 = 9 \ \Rightarrow \ 18x + 2yy' = 0 \ \Rightarrow \ 2yy' = -18x \ \Rightarrow \ y' = -9x/y \ \Rightarrow$

$y'' = -9\left(\dfrac{y \cdot 1 - x \cdot y'}{y^2}\right) = -9\left(\dfrac{y - x(-9x/y)}{y^2}\right) = -9 \cdot \dfrac{y^2 + 9x^2}{y^3} = -9 \cdot \dfrac{9}{y^3}$ [since x and y must satisfy the original

equation, $9x^2 + y^2 = 9$]. Thus, $y'' = -81/y^3$.

25. $x^3 + y^3 = 1 \ \Rightarrow \ 3x^2 + 3y^2y' = 0 \ \Rightarrow \ y' = -\dfrac{x^2}{y^2} \ \Rightarrow$

$y'' = -\dfrac{y^2(2x) - x^2 \cdot 2yy'}{(y^2)^2} = -\dfrac{2xy^2 - 2x^2y(-x^2/y^2)}{y^4} = -\dfrac{2xy^4 + 2x^4y}{y^6} = -\dfrac{2xy(y^3 + x^3)}{y^6} = -\dfrac{2x}{y^5}$,

since x and y must satisfy the original equation, $x^3 + y^3 = 1$.

27. (a) $y^2 = 5x^4 - x^2 \ \Rightarrow \ 2yy' = 5(4x^3) - 2x \ \Rightarrow \ y' = \dfrac{10x^3 - x}{y}$.

(b)

So at the point $(1, 2)$ we have $y' = \dfrac{10(1)^3 - 1}{2} = \dfrac{9}{2}$, and an equation

of the tangent line is $y - 2 = \frac{9}{2}(x - 1)$ or $y = \frac{9}{2}x - \frac{5}{2}$.

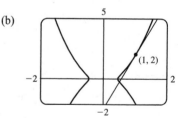

29. (a) There are eight points with horizontal tangents: four at $x \approx 1.57735$ and

four at $x \approx 0.42265$.

(b) $y' = \dfrac{3x^2 - 6x + 2}{2(2y^3 - 3y^2 - y + 1)} \ \Rightarrow \ y' = -1$ at $(0, 1)$ and $y' = \frac{1}{3}$ at $(0, 2)$.

Equations of the tangent lines are $y = -x + 1$ and $y = \frac{1}{3}x + 2$.

(c) $y' = 0 \ \Rightarrow \ 3x^2 - 6x + 2 = 0 \ \Rightarrow \ x = 1 \pm \frac{1}{3}\sqrt{3}$

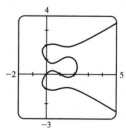

(d) By multiplying the right side of the equation by $x - 3$, we obtain the first graph. By modifying the equation in other ways, we can generate the other graphs.

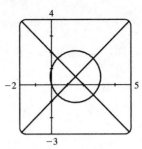

$$y(y^2 - 1)(y - 2)$$
$$= x(x - 1)(x - 2)(x - 3)$$

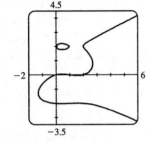

$$y(y^2 - 4)(y - 2)$$
$$= x(x - 1)(x - 2)$$

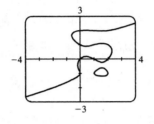

$$y(y + 1)(y^2 - 1)(y - 2)$$
$$= x(x - 1)(x - 2)$$

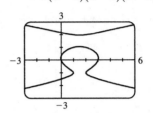

$$(y + 1)(y^2 - 1)(y - 2)$$
$$= (x - 1)(x - 2)$$

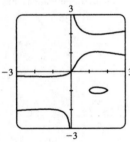

$$x(y + 1)(y^2 - 1)(y - 2)$$
$$= y(x - 1)(x - 2)$$

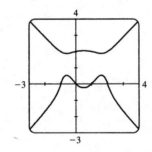

$$y(y^2 + 1)(y - 2)$$
$$= x(x^2 - 1)(x - 2)$$

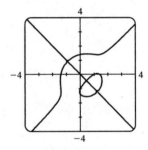

$$y(y + 1)(y^2 - 2)$$
$$= x(x - 1)(x^2 - 2)$$

31. From Exercise 21, a tangent to the lemniscate will be horizontal if $y' = 0 \Rightarrow 25x - 4x(x^2 + y^2) = 0 \Rightarrow x[25 - 4(x^2 + y^2)] = 0 \Rightarrow x^2 + y^2 = \frac{25}{4}$ **(1)**. (Note that when x is 0, y is also 0, and there is no horizontal tangent at the origin.) Substituting $\frac{25}{4}$ for $x^2 + y^2$ in the equation of the lemniscate, $2(x^2 + y^2)^2 = 25(x^2 - y^2)$, we get $x^2 - y^2 = \frac{25}{8}$ **(2)**. Solving **(1)** and **(2)**, we have $x^2 = \frac{75}{16}$ and $y^2 = \frac{25}{16}$, so the four points are $\left(\pm\frac{5\sqrt{3}}{4}, \pm\frac{5}{4} \right)$.

33. $x^2 + y^2 = r^2$ is a circle with center O and $ax + by = 0$ is a line through O.

$x^2 + y^2 = r^2 \Rightarrow 2x + 2yy' = 0 \Rightarrow y' = -x/y$, so the slope of the tangent line at P_0 (x_0, y_0) is $-x_0/y_0$. The slope of the line OP_0 is y_0/x_0, which is the negative reciprocal of $-x_0/y_0$. Hence, the curves are orthogonal, and the families of curves are orthogonal trajectories of each other.

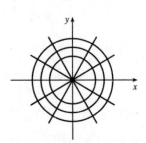

35. $y = cx^2 \Rightarrow y' = 2cx$ and $x^2 + 2y^2 = k \Rightarrow 2x + 4yy' = 0 \Rightarrow$

$2yy' = -x \Rightarrow y' = -\dfrac{x}{2(y)} = -\dfrac{x}{2(cx^2)} = -\dfrac{1}{2cx}$, so the curves

are orthogonal.

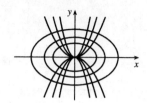

37. If the circle has radius r, its equation is $x^2 + y^2 = r^2 \Rightarrow 2x + 2yy' = 0 \Rightarrow y' = -\dfrac{x}{y}$, so the slope of the tangent line

at $P(x_0, y_0)$ is $-\dfrac{x_0}{y_0}$. The negative reciprocal of that slope is $\dfrac{-1}{-x_0/y_0} = \dfrac{y_0}{x_0}$, which is the slope of OP, so the tangent line at

P is perpendicular to the radius OP.

39. To find the points at which the ellipse $x^2 - xy + y^2 = 3$ crosses the x-axis, let $y = 0$ and solve for x.

$y = 0 \Rightarrow x^2 - x(0) + 0^2 = 3 \Leftrightarrow x = \pm\sqrt{3}$. So the graph of the ellipse crosses the x-axis at the points $(\pm\sqrt{3}, 0)$.

Using implicit differentiation to find y', we get $2x - xy' - y + 2yy' = 0 \Rightarrow y'(2y - x) = y - 2x \Leftrightarrow y' = \dfrac{y - 2x}{2y - x}$.

So y' at $(\sqrt{3}, 0)$ is $\dfrac{0 - 2\sqrt{3}}{2(0) - \sqrt{3}} = 2$ and y' at $(-\sqrt{3}, 0)$ is $\dfrac{0 + 2\sqrt{3}}{2(0) + \sqrt{3}} = 2$. Thus, the tangent lines at these points are parallel.

41. $x^2y^2 + xy = 2 \Rightarrow x^2 \cdot 2yy' + y^2 \cdot 2x + x \cdot y' + y \cdot 1 = 0 \Leftrightarrow y'(2x^2y + x) = -2xy^2 - y \Leftrightarrow$

$y' = -\dfrac{2xy^2 + y}{2x^2y + x}$. So $-\dfrac{2xy^2 + y}{2x^2y + x} = -1 \Leftrightarrow 2xy^2 + y = 2x^2y + x \Leftrightarrow y(2xy + 1) = x(2xy + 1) \Leftrightarrow$

$y(2xy + 1) - x(2xy + 1) = 0 \Leftrightarrow (2xy + 1)(y - x) = 0 \Leftrightarrow xy = -\frac{1}{2}$ or $y = x$. But $xy = -\frac{1}{2} \Rightarrow$

$x^2y^2 + xy = \frac{1}{4} - \frac{1}{2} \neq 2$, so we must have $x = y$. Then $x^2y^2 + xy = 2 \Rightarrow x^4 + x^2 = 2 \Leftrightarrow x^4 + x^2 - 2 = 0 \Leftrightarrow$

$(x^2 + 2)(x^2 - 1) = 0$. So $x^2 = -2$, which is impossible, or $x^2 = 1 \Leftrightarrow x = \pm 1$. Since $x = y$, the points on the curve

where the tangent line has a slope of -1 are $(-1, -1)$ and $(1, 1)$.

43. (a) $y = J(x)$ and $xy'' + y' + xy = 0 \Rightarrow xJ''(x) + J'(x) + xJ(x) = 0$. If $x = 0$, we have $0 + J'(0) + 0 = 0$,

so $J'(0) = 0$.

(b) Differentiating $xy'' + y' + xy = 0$ implicitly, we get $xy''' + y'' \cdot 1 + y'' + xy' + y \cdot 1 = 0 \Rightarrow$

$xy''' + 2y'' + xy' + y = 0$, so $xJ'''(x) + 2J''(x) + xJ'(x) + J(x) = 0$. If $x = 0$, we have

$0 + 2J''(0) + 0 + 1 \quad [J(0) = 1 \text{ is given}] = 0 \Rightarrow 2J''(0) = -1 \Rightarrow J''(0) = -\frac{1}{2}$.

2.7 Related Rates

1. $V = x^3 \Rightarrow \dfrac{dV}{dt} = \dfrac{dV}{dx}\dfrac{dx}{dt} = 3x^2\dfrac{dx}{dt}$

3. Let s denote the side of a square. The square's area A is given by $A = s^2$. Differentiating with respect to t gives us

$\dfrac{dA}{dt} = 2s\dfrac{ds}{dt}$. When $A = 16$, $s = 4$. Substitution 4 for s and 6 for $\dfrac{ds}{dt}$ gives us $\dfrac{dA}{dt} = 2(4)(6) = 48 \text{ cm}^2/\text{s}$.

5. $y = x^3 + 2x \Rightarrow \dfrac{dy}{dt} = \dfrac{dy}{dx}\dfrac{dx}{dt} = (3x^2 + 2)(5) = 5(3x^2 + 2)$. When $x = 2$, $\dfrac{dy}{dt} = 5(14) = 70$.

7. $z^2 = x^2 + y^2 \;\Rightarrow\; 2z\dfrac{dz}{dt} = 2x\dfrac{dx}{dt} + 2y\dfrac{dy}{dt} \;\Rightarrow\; \dfrac{dz}{dt} = \dfrac{1}{z}\left(x\dfrac{dx}{dt} + y\dfrac{dy}{dt}\right)$. When $x = 5$ and $y = 12$,

$z^2 = 5^2 + 12^2 \;\Rightarrow\; z^2 = 169 \;\Rightarrow\; z = \pm 13$. For $\dfrac{dx}{dt} = 2$ and $\dfrac{dy}{dt} = 3$, $\dfrac{dz}{dt} = \dfrac{1}{\pm 13}(5\cdot 2 + 12\cdot 3) = \pm\dfrac{46}{13}$.

9. (a) Given: the rate of decrease of the surface area is $1 \text{ cm}^2/\text{min}$. If we let t be

time (in minutes) and S be the surface area (in cm^2), then we are given that

$dS/dt = -1 \text{ cm}^2/\text{s}$.

(b) Unknown: the rate of decrease of the diameter when the diameter is 10 cm.

If we let x be the diameter, then we want to find dx/dt when $x = 10$ cm.

(c)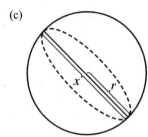

(d) If the radius is r and the diameter $x = 2r$, then $r = \frac{1}{2}x$ and

$$S = 4\pi r^2 = 4\pi\left(\tfrac{1}{2}x\right)^2 = \pi x^2 \;\Rightarrow\; \dfrac{dS}{dt} = \dfrac{dS}{dx}\dfrac{dx}{dt} = 2\pi x\dfrac{dx}{dt}.$$

(e) $-1 = \dfrac{dS}{dt} = 2\pi x\dfrac{dx}{dt} \;\Rightarrow\; \dfrac{dx}{dt} = -\dfrac{1}{2\pi x}$. When $x = 10$, $\dfrac{dx}{dt} = -\dfrac{1}{20\pi}$. So the rate of decrease is $\frac{1}{20\pi}$ cm/min.

11. (a) Given: a plane flying horizontally at an altitude of 1 mi and a speed of 500 mi/h passes directly over a radar station. If we

let t be time (in hours) and x be the horizontal distance traveled by the plane (in mi), then we are given that

$dx/dt = 500 \text{ mi/h}$.

(b) Unknown: the rate at which the distance from the plane to the station is increasing

when it is 2 mi from the station. If we let y be the distance from the plane to the station,

then we want to find dy/dt when $y = 2$ mi.

(c)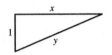

(d) By the Pythagorean Theorem, $y^2 = x^2 + 1 \;\Rightarrow\; 2y\,(dy/dt) = 2x\,(dx/dt)$.

(e) $\dfrac{dy}{dt} = \dfrac{x}{y}\dfrac{dx}{dt} = \dfrac{x}{y}(500)$. Since $y^2 = x^2 + 1$, when $y = 2$, $x = \sqrt{3}$, so $\dfrac{dy}{dt} = \dfrac{\sqrt{3}}{2}(500) = 250\sqrt{3} \approx 433 \text{ mi/h}$.

13.

We are given that $\dfrac{dx}{dt} = 60 \text{ mi/h}$ and $\dfrac{dy}{dt} = 25 \text{ mi/h}$. $z^2 = x^2 + y^2 \;\Rightarrow\;$

$2z\dfrac{dz}{dt} = 2x\dfrac{dx}{dt} + 2y\dfrac{dy}{dt} \;\Rightarrow\; z\dfrac{dz}{dt} = x\dfrac{dx}{dt} + y\dfrac{dy}{dt} \;\Rightarrow\; \dfrac{dz}{dt} = \dfrac{1}{z}\left(x\dfrac{dx}{dt} + y\dfrac{dy}{dt}\right)$.

After 2 hours, $x = 2\,(60) = 120$ and $y = 2\,(25) = 50 \;\Rightarrow\; z = \sqrt{120^2 + 50^2} = 130$,

so $\dfrac{dz}{dt} = \dfrac{1}{z}\left(x\dfrac{dx}{dt} + y\dfrac{dy}{dt}\right) = \dfrac{120(60) + 50(25)}{130} = 65 \text{ mi/h}$.

15.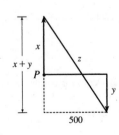

We are given that $\dfrac{dx}{dt} = 4 \text{ ft/s}$ and $\dfrac{dy}{dt} = 5 \text{ ft/s}$. $z^2 = (x + y)^2 + 500^2 \;\Rightarrow\;$

$2z\dfrac{dz}{dt} = 2(x + y)\left(\dfrac{dx}{dt} + \dfrac{dy}{dt}\right)$. 15 minutes after the woman starts, we have

$x = (4 \text{ ft/s})(20 \text{ min})(60 \text{ s/min}) = 4800 \text{ ft}$ and $y = 5\cdot 15\cdot 60 = 4500 \;\Rightarrow\;$

$z = \sqrt{(4800 + 4500)^2 + 500^2} = \sqrt{86,740,000}$, so

$\dfrac{dz}{dt} = \dfrac{x + y}{z}\left(\dfrac{dx}{dt} + \dfrac{dy}{dt}\right) = \dfrac{4800 + 4500}{\sqrt{86,740,000}}(4 + 5) = \dfrac{837}{\sqrt{8674}} \approx 8.99 \text{ ft/s}$.

17. $A = \frac{1}{2}bh$, where b is the base and h is the altitude. We are given that $\frac{dh}{dt} = 1$ cm/min and $\frac{dA}{dt} = 2$ cm²/min. Using the

Product Rule, we have $\frac{dA}{dt} = \frac{1}{2}\left(b\,\frac{dh}{dt} + h\,\frac{db}{dt}\right)$. When $h = 10$ and $A = 100$, we have $100 = \frac{1}{2}b(10) \;\Rightarrow\; \frac{1}{2}b = 10 \;\Rightarrow$

$b = 20$, so $2 = \frac{1}{2}\left(20 \cdot 1 + 10\,\frac{db}{dt}\right) \;\Rightarrow\; 4 = 20 + 10\,\frac{db}{dt} \;\Rightarrow\; \frac{db}{dt} = \frac{4 - 20}{10} = -1.6$ cm/min.

19.

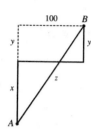

We are given that $\frac{dx}{dt} = 35$ km/h and $\frac{dy}{dt} = 25$ km/h. $z^2 = (x + y)^2 + 100^2 \;\Rightarrow$

$2z\,\frac{dz}{dt} = 2(x + y)\left(\frac{dx}{dt} + \frac{dy}{dt}\right)$. At 4:00 PM, $x = 4(35) = 140$ and

$y = 4(25) = 100 \;\Rightarrow\; z = \sqrt{(140 + 100)^2 + 100^2} = \sqrt{67{,}600} = 260$, so

$\frac{dz}{dt} = \frac{x + y}{z}\left(\frac{dx}{dt} + \frac{dy}{dt}\right) = \frac{140 + 100}{260}(35 + 25) = \frac{720}{13} \approx 55.4$ km/h.

21. Using Q for the origin, we are given $\frac{dx}{dt} = -2$ ft/s and need to find $\frac{dy}{dt}$ when $x = -5$.

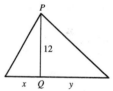

Using the Pythagorean Theorem twice, we have $\sqrt{x^2 + 12^2} + \sqrt{y^2 + 12^2} = 39$,
the total length of the rope. Differentiating with respect to t, we get

$\frac{x}{\sqrt{x^2 + 12^2}}\,\frac{dx}{dt} + \frac{y}{\sqrt{y^2 + 12^2}}\,\frac{dy}{dt} = 0$, so $\frac{dy}{dt} = -\frac{x\,\sqrt{y^2 + 12^2}}{y\,\sqrt{x^2 + 12^2}}\,\frac{dx}{dt}$.

Now when $x = -5$, $39 = \sqrt{(-5)^2 + 12^2} + \sqrt{y^2 + 12^2} = 13 + \sqrt{y^2 + 12^2} \;\Leftrightarrow\; \sqrt{y^2 + 12^2} = 26$, and

$y = \sqrt{26^2 - 12^2} = \sqrt{532}$. So when $x = -5$, $\frac{dy}{dt} = -\frac{(-5)(26)}{\sqrt{532}\,(13)}(-2) = -\frac{10}{\sqrt{133}} \approx -0.87$ ft/s. So cart B is moving

towards Q at about 0.87 ft/s.

23. By similar triangles, $\frac{3}{1} = \frac{b}{h}$, so $b = 3h$. The trough has volume

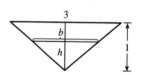

$V = \frac{1}{2}bh(10) = 5(3h)h = 15h^2 \;\Rightarrow\; 12 = \frac{dV}{dt} = 30h\,\frac{dh}{dt} \;\Rightarrow\; \frac{dh}{dt} = \frac{2}{5h}$.

When $h = \frac{1}{2}$, $\frac{dh}{dt} = \frac{2}{5 \cdot \frac{1}{2}} = \frac{4}{5}$ ft/min.

25. We are given that $\frac{dV}{dt} = 30$ ft³/min. $V = \frac{1}{3}\pi r^2 h = \frac{1}{3}\pi\left(\frac{h}{2}\right)^2 h = \frac{\pi h^3}{12} \;\Rightarrow$

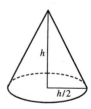

$\frac{dV}{dt} = \frac{dV}{dh}\,\frac{dh}{dt} \;\Rightarrow\; 30 = \frac{\pi h^2}{4}\,\frac{dh}{dt} \;\Rightarrow\; \frac{dh}{dt} = \frac{120}{\pi h^2}$. When $h = 10$ ft,

$\frac{dh}{dt} = \frac{120}{10^2\pi} = \frac{6}{5\pi} \approx 0.38$ ft/min.

27. $A = \frac{1}{2}bh$, but $b = 5$ m and $\sin\theta = \frac{h}{4} \;\Rightarrow\; h = 4\sin\theta$, so $A = \frac{1}{2}(5)(4\sin\theta) = 10\sin\theta$.

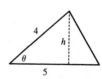

We are given $\frac{d\theta}{dt} = 0.06$ rad/s, so $\frac{dA}{dt} = \frac{dA}{d\theta}\,\frac{d\theta}{dt} = (10\cos\theta)(0.06) = 0.6\cos\theta$.

When $\theta = \frac{\pi}{3}$, $\frac{dA}{dt} = 0.6\left(\cos\frac{\pi}{3}\right) = (0.6)\left(\frac{1}{2}\right) = 0.3$ m²/s.

29. Differentiating both sides of $PV = C$ with respect to t and using the Product Rule gives us $P \dfrac{dV}{dt} + V \dfrac{dP}{dt} = 0 \implies$

$\dfrac{dV}{dt} = -\dfrac{V}{P} \dfrac{dP}{dt}$. When $V = 600$, $P = 150$ and $\dfrac{dP}{dt} = 20$, so we have $\dfrac{dV}{dt} = -\dfrac{600}{150}(20) = -80$. Thus, the volume is

decreasing at a rate of 80 cm³/min.

31. With $R_1 = 80$ and $R_2 = 100$, $\dfrac{1}{R} = \dfrac{1}{R_1} + \dfrac{1}{R_2} = \dfrac{1}{80} + \dfrac{1}{100} = \dfrac{180}{8000} = \dfrac{9}{400}$, so $R = \dfrac{400}{9}$. Differentiating $\dfrac{1}{R} = \dfrac{1}{R_1} + \dfrac{1}{R_2}$

with respect to t, we have $-\dfrac{1}{R^2} \dfrac{dR}{dt} = -\dfrac{1}{R_1^2} \dfrac{dR_1}{dt} - \dfrac{1}{R_2^2} \dfrac{dR_2}{dt} \implies \dfrac{dR}{dt} = R^2 \left(\dfrac{1}{R_1^2} \dfrac{dR_1}{dt} + \dfrac{1}{R_2^2} \dfrac{dR_2}{dt} \right)$. When $R_1 = 80$ and

$R_2 = 100$, $\dfrac{dR}{dt} = \dfrac{400^2}{9^2} \left[\dfrac{1}{80^2}(0.3) + \dfrac{1}{100^2}(0.2) \right] = \dfrac{107}{810} \approx 0.132 \ \Omega/\text{s}$.

33. (a) By the Pythagorean Theorem, $4000^2 + y^2 = \ell^2$. Differentiating with respect to t,

we obtain $2y \dfrac{dy}{dt} = 2\ell \dfrac{d\ell}{dt}$. We know that $\dfrac{dy}{dt} = 600$ ft/s, so when $y = 3000$ ft,

$\ell = \sqrt{4000^2 + 3000^2} = \sqrt{25{,}000{,}000} = 5000$ ft

and $\dfrac{d\ell}{dt} = \dfrac{y}{\ell} \dfrac{dy}{dt} = \dfrac{3000}{5000}(600) = \dfrac{1800}{5} = 360$ ft/s.

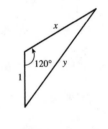

(b) Here $\tan\theta = \dfrac{y}{4000} \implies \dfrac{d}{dt}(\tan\theta) = \dfrac{d}{dt}\left(\dfrac{y}{4000}\right) \implies \sec^2\theta \dfrac{d\theta}{dt} = \dfrac{1}{4000} \dfrac{dy}{dt} \implies \dfrac{d\theta}{dt} = \dfrac{\cos^2\theta}{4000} \dfrac{dy}{dt}$. When

$y = 3000$ ft, $\dfrac{dy}{dt} = 600$ ft/s, $\ell = 5000$ and $\cos\theta = \dfrac{4000}{\ell} = \dfrac{4000}{5000} = \dfrac{4}{5}$, so $\dfrac{d\theta}{dt} = \dfrac{(4/5)^2}{4000}(600) = 0.096$ rad/s.

35. We are given that $\dfrac{dx}{dt} = 300$ km/h. By the Law of Cosines,

$y^2 = x^2 + 1^2 - 2(1)(x)\cos 120° = x^2 + 1 - 2x\left(-\tfrac{1}{2}\right) = x^2 + x + 1$, so

$2y \dfrac{dy}{dt} = 2x \dfrac{dx}{dt} + \dfrac{dx}{dt} \implies \dfrac{dy}{dt} = \dfrac{2x + 1}{2y} \dfrac{dx}{dt}$. After 1 minute, $x = \dfrac{300}{60} = 5$ km $\implies$

$y = \sqrt{5^2 + 5 + 1} = \sqrt{31}$ km $\implies \dfrac{dy}{dt} = \dfrac{2(5) + 1}{2\sqrt{31}}(300) = \dfrac{1650}{\sqrt{31}} \approx 296$ km/h.

37. Let the distance between the runner and the friend be ℓ. Then by the Law of Cosines,

$\ell^2 = 200^2 + 100^2 - 2 \cdot 200 \cdot 100 \cdot \cos\theta = 50{,}000 - 40{,}000\cos\theta$ ($\star$). Differentiating

implicitly with respect to t, we obtain $2\ell \dfrac{d\ell}{dt} = -40{,}000(-\sin\theta) \dfrac{d\theta}{dt}$. Now if D is the

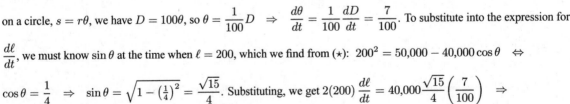

distance run when the angle is θ radians, then by the formula for the length of an arc

on a circle, $s = r\theta$, we have $D = 100\theta$, so $\theta = \dfrac{1}{100}D \implies \dfrac{d\theta}{dt} = \dfrac{1}{100} \dfrac{dD}{dt} = \dfrac{7}{100}$. To substitute into the expression for

$\dfrac{d\ell}{dt}$, we must know $\sin\theta$ at the time when $\ell = 200$, which we find from ($\star$): $200^2 = 50{,}000 - 40{,}000\cos\theta \iff$

$\cos\theta = \dfrac{1}{4} \implies \sin\theta = \sqrt{1 - \left(\tfrac{1}{4}\right)^2} = \dfrac{\sqrt{15}}{4}$. Substituting, we get $2(200)\dfrac{d\ell}{dt} = 40{,}000 \dfrac{\sqrt{15}}{4}\left(\dfrac{7}{100}\right) \implies$

$d\ell/dt = \dfrac{7\sqrt{15}}{4} \approx 6.78$ m/s. Whether the distance between them is increasing or decreasing depends on the direction in which

the runner is running.

2.8 Linear Approximations and Differentials

1. $f(x) = x^4 + 3x^2 \;\Rightarrow\; f'(x) = 4x^3 + 6x$, so $f(-1) = 4$ and $f'(-1) = -10$.

Thus, $L(x) = f(-1) + f'(-1)(x - (-1)) = 4 + (-10)(x + 1) = -10x - 6$.

3. $f(x) = \cos x \;\Rightarrow\; f'(x) = -\sin x$, so $f\left(\frac{\pi}{2}\right) = 0$ and $f'\left(\frac{\pi}{2}\right) = -1$.

Thus, $L(x) = f\left(\frac{\pi}{2}\right) + f'\left(\frac{\pi}{2}\right)\left(x - \frac{\pi}{2}\right) = 0 - 1\left(x - \frac{\pi}{2}\right) = -x + \frac{\pi}{2}$.

5. $f(x) = \sqrt{1-x} \;\Rightarrow\; f'(x) = \dfrac{-1}{2\sqrt{1-x}}$, so $f(0) = 1$ and $f'(0) = -\frac{1}{2}$.

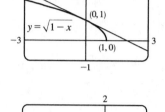

Therefore,

$\sqrt{1-x} = f(x) \approx f(0) + f'(0)(x - 0) = 1 + \left(-\frac{1}{2}\right)(x - 0) = 1 - \frac{1}{2}x$.

So $\sqrt{0.9} = \sqrt{1 - 0.1} \approx 1 - \frac{1}{2}(0.1) = 0.95$

and $\sqrt{0.99} = \sqrt{1 - 0.01} \approx 1 - \frac{1}{2}(0.01) = 0.995$.

7. $f(x) = \sqrt[3]{1-x} = (1-x)^{1/3} \;\Rightarrow\; f'(x) = -\frac{1}{3}(1-x)^{-2/3}$, so $f(0) = 1$

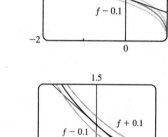

and $f'(0) = -\frac{1}{3}$. Thus, $f(x) \approx f(0) + f'(0)(x - 0) = 1 - \frac{1}{3}x$. We need

$\sqrt[3]{1-x} - 0.1 < 1 - \frac{1}{3}x < \sqrt[3]{1-x} + 0.1$, which is true when

$-1.204 < x < 0.706$.

9. $f(x) = \dfrac{1}{(1+2x)^4} = (1+2x)^{-4} \;\Rightarrow$

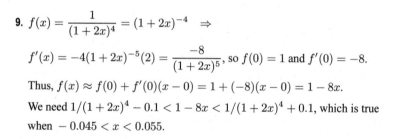

$f'(x) = -4(1+2x)^{-5}(2) = \dfrac{-8}{(1+2x)^5}$, so $f(0) = 1$ and $f'(0) = -8$.

Thus, $f(x) \approx f(0) + f'(0)(x - 0) = 1 + (-8)(x - 0) = 1 - 8x$.

We need $1/(1+2x)^4 - 0.1 < 1 - 8x < 1/(1+2x)^4 + 0.1$, which is true

when $-0.045 < x < 0.055$.

11. To estimate $(2.001)^5$, we'll find the linearization of $f(x) = x^5$ at $a = 2$. Since $f'(x) = 5x^4$, $f(2) = 32$, and $f'(2) = 80$, we

have $L(x) = 32 + 80(x - 2) = 80x - 128$. Thus, $x^5 \approx 80x - 128$ when x is near 2 , so

$(2.001)^5 \approx 80(2.001) - 128 = 160.08 - 128 = 32.08$.

13. To estimate $(8.06)^{2/3}$, we'll find the linearization of $f(x) = x^{2/3}$ at $a = 8$. Since $f'(x) = \frac{2}{3}x^{-1/3} = 2/(3\sqrt[3]{x})$, $f(8) = 4$,

and $f'(8) = \frac{1}{3}$, we have $L(x) = 4 + \frac{1}{3}(x - 8) = \frac{1}{3}x + \frac{4}{3}$. Thus, $x^{2/3} \approx \frac{1}{3}x + \frac{4}{3}$ when x is near 8, so

$(8.06)^{2/3} \approx \frac{1}{3}(8.06) + \frac{4}{3} = \frac{12.06}{3} = 4.02$.

15. $y = f(x) = \sec x \;\Rightarrow\; f'(x) = \sec x \tan x$, so $f(0) = 1$ and $f'(0) = 1 \cdot 0 = 0$. The linear approximation of f at 0 is

$f(0) + f'(0)(x - 0) = 1 + 0(x) = 1$. Since 0.08 is close to 0, approximating $\sec 0.08$ with 1 is reasonable.

17. (a) The differential dy is defined in terms of dx by the equation $dy = f'(x)\,dx$. For $y = f(x) = x^2 \sin 2x$,

$f'(x) = x^2 \cos 2x \cdot 2 + \sin 2x \cdot 2x = 2x(x \cos 2x + \sin 2x)$, so $dy = 2x(x \cos 2x + \sin 2x)\,dx$.

(b) $y = \sqrt{4 + 5x}$ $\Rightarrow$ $dy = \frac{1}{2}(4 + 5x)^{-1/2} \cdot 5\,dx = \dfrac{5}{2\sqrt{4+5x}}\,dx$

19. (a) $y = \tan x$ $\Rightarrow$ $dy = \sec^2 x\,dx$

(b) When $x = \pi/4$ and $dx = -0.1$, $dy = [\sec(\pi/4)]^2\,(-0.1) = \left(\sqrt{2}\right)^2(-0.1) = -0.2$.

$$\Delta y = f(x + \Delta x) - f(x) = \tan\left(\frac{\pi}{4} - 0.1\right) - \tan\left(\frac{\pi}{4}\right) \approx -0.18237.$$

21. (a) If x is the edge length, then $V = x^3$ $\Rightarrow$ $dV = 3x^2\,dx$. When $x = 30$ and $dx = 0.1$, $dV = 3(30)^2(0.1) = 270$, so the

maximum possible error in computing the volume of the cube is about 270 cm^3. The relative error is calculated by dividing

the change in V, ΔV, by V. We approximate ΔV with dV.

Relative error $= \dfrac{\Delta V}{V} \approx \dfrac{dV}{V} = \dfrac{3x^2\,dx}{x^3} = 3\dfrac{dx}{x} = 3\left(\dfrac{0.1}{30}\right) = 0.01$.

Percentage error $=$ relative error $\times 100\% = 0.01 \times 100\% = 1\%$.

(b) $S = 6x^2$ $\Rightarrow$ $dS = 12x\,dx$. When $x = 30$ and $dx = 0.1$, $dS = 12(30)(0.1) = 36$, so the maximum possible error in

computing the surface area of the cube is about 36 cm^2.

Relative error $= \dfrac{\Delta S}{S} \approx \dfrac{dS}{S} = \dfrac{12x\,dx}{6x^2} = 2\dfrac{dx}{x} = 2\left(\dfrac{0.1}{30}\right) = 0.00\overline{6}$.

Percentage error $=$ relative error $\times 100\% = 0.00\overline{6} \times 100\% = 0.\overline{6}\%$.

23. (a) For a sphere of radius r, the circumference is $C = 2\pi r$ and the surface area is $S = 4\pi r^2$, so

$$r = \frac{C}{2\pi} \Rightarrow S = 4\pi\left(\frac{C}{2\pi}\right)^2 = \frac{C^2}{\pi} \Rightarrow dS = \frac{2}{\pi}C\,dC. \text{ When } C = 84 \text{ and } dC = 0.5, dS = \frac{2}{\pi}(84)(0.5) = \frac{84}{\pi},$$

so the maximum error is about $\dfrac{84}{\pi} \approx 27\text{ cm}^2$. Relative error $\approx \dfrac{dS}{S} = \dfrac{84/\pi}{84^2/\pi} = \dfrac{1}{84} \approx 0.012$

(b) $V = \frac{4}{3}\pi r^3 = \frac{4}{3}\pi\left(\dfrac{C}{2\pi}\right)^3 = \dfrac{C^3}{6\pi^2}$ $\Rightarrow$ $dV = \dfrac{1}{2\pi^2}C^2\,dC$. When $C = 84$ and $dC = 0.5$,

$dV = \dfrac{1}{2\pi^2}(84)^2(0.5) = \dfrac{1764}{\pi^2}$, so the maximum error is about $\dfrac{1764}{\pi^2} \approx 179\text{ cm}^3$.

The relative error is approximately $\dfrac{dV}{V} = \dfrac{1764/\pi^2}{(84)^3/(6\pi^2)} = \dfrac{1}{56} \approx 0.018$.

25. $F = kR^4$ $\Rightarrow$ $dF = 4kR^3\,dR$ $\Rightarrow$ $\dfrac{dF}{F} = \dfrac{4kR^3\,dR}{kR^4} = 4\left(\dfrac{dR}{R}\right)$. Thus, the relative change in F is about 4 times the

relative change in R. So a 5% increase in the radius corresponds to a 20% increase in blood flow.

27. (a) The graph shows that $f'(1) = 2$, so $L(x) = f(1) + f'(1)(x - 1) = 5 + 2(x - 1) = 2x + 3$.

$f(0.9) \approx L(0.9) = 4.8$ and $f(1.1) \approx L(1.1) = 5.2$.

(b) From the graph, we see that $f'(x)$ is positive and decreasing. This means that the slopes of the tangent lines are positive,

but the tangents are becoming less steep. So the tangent lines lie *above* the curve. Thus, the estimates in part (a) are too

large.

2   Review

CONCEPT CHECK

1. See Definition 2.1.1.

2. See the paragraph containing Formula 3 in Section 2.1.

3. See Definition 2.1.4. The pages following the definition discuss interpretations of $f'(a)$ as the slope of the tangent line to the graph of f at $x = a$ and as the instantaneous rate of change of $f(x)$ with respect to x when $x = a$.

4. (a) The average rate of change of y with respect to x over the interval $[x_1, x_2]$ is $\dfrac{f(x_2) - f(x_1)}{x_2 - x_1}$.

   (b) The instantaneous rate of change of y with respect to x at $x = x_1$ is $\displaystyle\lim_{x_2 \to x_1} \dfrac{f(x_2) - f(x_1)}{x_2 - x_1}$.

5. See the paragraphs before and after Example 7 in Section 2.2.

6. (a) A function f is differentiable at a number a if its derivative f' exists at $x = a$; that is, if $f'(a)$ exists.

   (b) See Theorem 2.2.4. This theorem also tells us that if f is *not* continuous at a, then f is *not* differentiable at a.

   (c)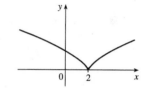

7. See the discussion and Figure 7 on page 89.

8. (a) The Power Rule: If n is any real number, then $\dfrac{d}{dx}(x^n) = nx^{n-1}$. The derivative of a variable base raised to a constant power is the power times the base raised to the power minus one.

   (b) The Constant Multiple Rule: If c is a constant and f is a differentiable function, then $\dfrac{d}{dx}[cf(x)] = c\dfrac{d}{dx}f(x)$. The derivative of a constant times a function is the constant times the derivative of the function.

   (c) The Sum Rule: If f and g are both differentiable, then $\dfrac{d}{dx}[f(x) + g(x)] = \dfrac{d}{dx}f(x) + \dfrac{d}{dx}g(x)$. The derivative of a sum of functions is the sum of the derivatives.

   (d) The Difference Rule: If f and g are both differentiable, then $\dfrac{d}{dx}[f(x) - g(x)] = \dfrac{d}{dx}f(x) - \dfrac{d}{dx}g(x)$. The derivative of a difference of functions is the difference of the derivatives.

   (e) The Product Rule: If f and g are both differentiable, then $\dfrac{d}{dx}[f(x)g(x)] = f(x)\dfrac{d}{dx}g(x) + g(x)\dfrac{d}{dx}f(x)$. The derivative of a product of two functions is the first function times the derivative of the second function plus the second function times the derivative of the first function.

   (f) The Quotient Rule: If f and g are both differentiable, then $\dfrac{d}{dx}\left[\dfrac{f(x)}{g(x)}\right] = \dfrac{g(x)\dfrac{d}{dx}f(x) - f(x)\dfrac{d}{dx}g(x)}{[g(x)]^2}$.
   The derivative of a quotient of functions is the denominator times the derivative of the numerator minus the numerator times the derivative of the denominator, all divided by the square of the denominator.

   (g) The Chain Rule: If f and g are both differentiable and $F = f \circ g$ is the composite function defined by $F(x) = f(g(x))$, then F is differentiable and F' is given by the product $F'(x) = f'(g(x))g'(x)$. The derivative of a composite function is the derivative of the outer function evaluated at the inner function times the derivative of the inner function.

9. (a) $y = x^n \Rightarrow y' = nx^{n-1}$

    (c) $y = \cos x \Rightarrow y' = -\sin x$

    (e) $y = \csc x \Rightarrow y' = -\csc x \cot x$

    (g) $y = \cot x \Rightarrow y' = -\csc^2 x$

    (b) $y = \sin x \Rightarrow y' = \cos x$

    (d) $y = \tan x \Rightarrow y' = \sec^2 x$

    (f) $y = \sec x \Rightarrow y' = \sec x \tan x$

10. Implicit differentiation consists of differentiating both sides of an equation involving x and y with respect to x, and then solving the resulting equation for y'.

11. (a) The linearization L of f at $x = a$ is $L(x) = f(a) + f'(a)(x - a)$.

    (b) If $y = f(x)$, then the differential dy is given by $dy = f'(x)\,dx$.

    (c) See Figure 5 in Section 2.8.

TRUE-FALSE QUIZ

1. False.   See the note on page 88.

3. False.   See the warning before the Product Rule on page 106.

5. True by the Chain Rule.

7. False. $f(x) = |x^2 + x| = x^2 + x$ for $x \geq 0$ or $x \leq -1$ and $|x^2 + x| = -(x^2 + x)$ for $-1 < x < 0$.

    So $f'(x) = 2x + 1$ for $x > 0$ or $x < -1$ and $f'(x) = -(2x + 1)$ for $-1 < x < 0$. But $|2x + 1| = 2x + 1$

    for $x \geq -\frac{1}{2}$ and $|2x + 1| = -2x - 1$ for $x < -\frac{1}{2}$.

9. True. $g(x) = x^5 \Rightarrow g'(x) = 5x^4 \Rightarrow g'(2) = 5(2)^4 = 80$, and by the definition of the derivative,

$$\lim_{x \to 2} \frac{g(x) - g(2)}{x - 2} = g'(2) = 80.$$

11. False.   A tangent line to the parabola $y = x^2$ has slope $dy/dx = 2x$, so at $(-2, 4)$ the slope of the tangent is $2(-2) = -4$

    and an equation of the tangent line is $y - 4 = -4(x + 2)$. [The given equation, $y - 4 = 2x(x + 2)$, is not even

    linear!]

EXERCISES

1. Estimating the slopes of the tangent lines at $x = 2, 3$, and 5, we obtain approximate values 0.4, 2, and 0.1. Since the

    graph is concave downward at $x = 5$, $f''(5)$ is negative. Arranging the numbers in increasing order, we have:

    $f''(5) < 0 < f'(5) < f'(2) < 1 < f'(3)$.

3. (a) $f'(r)$ is the rate at which the total cost changes with respect to the interest rate. Its units are dollars/(percent per year).

    (b) The total cost of paying off the loan is increasing by $1200/(percent per year) as the interest rate reaches 10%. So if the

    interest rate goes up from 10% to 11%, the cost goes up approximately $1200.

    (c) As r increases, C increases. So $f'(r)$ will always be positive.

5.

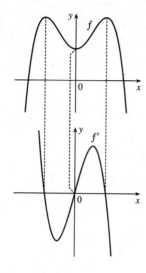

7. The graph of a has tangent lines with positive slope for $x < 0$ and negative slope for $x > 0$, and the values of c fit this pattern, so c must be the graph of the derivative of the function for a. The graph of c has horizontal tangent lines to the left and right of the x-axis and b has zeros at these points. Hence, b is the graph of the derivative of the function for c. Therefore, a is the graph of f, c is the graph of f', and b is the graph of f''.

9. $C'(1990)$ is the rate at which the total value of US currency in circulation is changing in billions of dollars per year. To estimate the value of $C'(1990)$, we will average the difference quotients obtained using the times $t = 1985$ and $t = 1995$.

Let $A = \dfrac{C(1985) - C(1990)}{1985 - 1990} = \dfrac{187.3 - 271.9}{-5} = \dfrac{-84.6}{-5} = 16.92$ and

$B = \dfrac{C(1995) - C(1990)}{1995 - 1990} = \dfrac{409.3 - 271.9}{5} = \dfrac{137.4}{5} = 27.48$. Then

$C'(1990) = \lim\limits_{t \to 1990} \dfrac{C(t) - C(1990)}{t - 1990} \approx \dfrac{A + B}{2} = \dfrac{16.92 + 27.48}{2} = \dfrac{44.4}{2} = 22.2$ billion dollars/year.

11. $f(x) = x^3 + 5x + 4 \;\Rightarrow$

$f'(x) = \lim\limits_{h \to 0} \dfrac{f(x + h) - f(x)}{h} = \lim\limits_{h \to 0} \dfrac{(x + h)^3 + 5(x + h) + 4 - (x^3 + 5x + 4)}{h}$

$= \lim\limits_{h \to 0} \dfrac{3x^2 h + 3xh^2 + h^3 + 5h}{h} = \lim\limits_{h \to 0} (3x^2 + 3xh + h^2 + 5) = 3x^2 + 5$

13. $y = (x^4 - 3x^2 + 5)^3 \;\Rightarrow$

$y' = 3(x^4 - 3x^2 + 5)^2 \dfrac{d}{dx}(x^4 - 3x^2 + 5) = 3(x^4 - 3x^2 + 5)^2 (4x^3 - 6x) = 6x(x^4 - 3x^2 + 5)^2 (2x^2 - 3)$

15. $y = \sqrt{x} + \dfrac{1}{\sqrt[3]{x^4}} = x^{1/2} + x^{-4/3} \;\Rightarrow\; y' = \tfrac{1}{2}x^{-1/2} - \tfrac{4}{3}x^{-7/3} = \dfrac{1}{2\sqrt{x}} - \dfrac{4}{3\sqrt[3]{x^7}}$

17. $y = 2x\sqrt{x^2 + 1} \;\Rightarrow$

$y' = 2x \cdot \tfrac{1}{2}(x^2 + 1)^{-1/2}(2x) + \sqrt{x^2 + 1}\,(2) = \dfrac{2x^2}{\sqrt{x^2 + 1}} + 2\sqrt{x^2 + 1} = \dfrac{2x^2 + 2(x^2 + 1)}{\sqrt{x^2 + 1}} = \dfrac{2(2x^2 + 1)}{\sqrt{x^2 + 1}}$

19. $y = \dfrac{t}{1 - t^2} \;\Rightarrow\; y' = \dfrac{(1 - t^2)(1) - t(-2t)}{(1 - t^2)^2} = \dfrac{1 - t^2 + 2t^2}{(1 - t^2)^2} = \dfrac{t^2 + 1}{(1 - t^2)^2}$

21. $y = \tan\sqrt{1 - x} \;\Rightarrow\; y' = \left(\sec^2\sqrt{1 - x}\right)\left(\dfrac{1}{2\sqrt{1 - x}}\right)(-1) = -\dfrac{\sec^2\sqrt{1 - x}}{2\sqrt{1 - x}}$

23. $\dfrac{d}{dx}\left(xy^4 + x^2 y\right) = \dfrac{d}{dx}\left(x + 3y\right) \;\;\Rightarrow\;\; x \cdot 4y^3 y' + y^4 \cdot 1 + x^2 \cdot y' + y \cdot 2x = 1 + 3y' \;\;\Rightarrow$

$y'\left(4xy^3 + x^2 - 3\right) = 1 - y^4 - 2xy \;\;\Rightarrow\;\; y' = \dfrac{1 - y^4 - 2xy}{4xy^3 + x^2 - 3}$

25. $y = \dfrac{\sec 2\theta}{1 + \tan 2\theta} \;\;\Rightarrow$

$y' = \dfrac{(1 + \tan 2\theta)(\sec 2\theta \tan 2\theta \cdot 2) - (\sec 2\theta)(\sec^2 2\theta \cdot 2)}{(1 + \tan 2\theta)^2} = \dfrac{2 \sec 2\theta \left[(1 + \tan 2\theta)\tan 2\theta - \sec^2 2\theta\right]}{(1 + \tan 2\theta)^2}$

$= \dfrac{2 \sec 2\theta \left(\tan 2\theta + \tan^2 2\theta - \sec^2 2\theta\right)}{(1 + \tan 2\theta)^2} = \dfrac{2 \sec 2\theta \left(\tan 2\theta - 1\right)}{(1 + \tan 2\theta)^2} \qquad \left[1 + \tan^2 x = \sec^2 x\right]$

27. $y = \left(1 - x^{-1}\right)^{-1} \;\;\Rightarrow$

$y' = -1(1 - x^{-1})^{-2}\left[-(-1x^{-2})\right] = -(1 - 1/x)^{-2}x^{-2} = -((x-1)/x)^{-2}x^{-2} = -(x-1)^{-2}$

29. $\sin(xy) = x^2 - y \;\;\Rightarrow\;\; \cos(xy)(xy' + y \cdot 1) = 2x - y' \;\;\Rightarrow\;\; x\cos(xy)y' + y' = 2x - y\cos(xy) \;\;\Rightarrow$

$y'[x\cos(xy) + 1] = 2x - y\cos(xy) \;\;\Rightarrow\;\; y' = \dfrac{2x - y\cos(xy)}{x\cos(xy) + 1}$

31. $y = \cot(3x^2 + 5) \;\;\Rightarrow\;\; y' = -\csc^2(3x^2 + 5)(6x) = -6x\csc^2(3x^2 + 5)$

33. $y = \sin\left(\tan\sqrt{1 + x^3}\right) \;\;\Rightarrow\;\; y' = \cos\left(\tan\sqrt{1 + x^3}\right)\left(\sec^2\sqrt{1 + x^3}\right)\left[3x^2/\left(2\sqrt{1 + x^3}\right)\right]$

35. $y = \tan^2(\sin\theta) = [\tan(\sin\theta)]^2 \;\;\Rightarrow\;\; y' = [\tan(\sin\theta)] \cdot \sec^2(\sin\theta) \cdot \cos\theta = 2\cos\theta\,\tan(\sin\theta)\,\sec^2(\sin\theta)$

37. $y = (x\tan x)^{1/5} \;\;\Rightarrow\;\; y' = \tfrac{1}{5}(x\tan x)^{-4/5}(\tan x + x\sec^2 x)$

39. $f(t) = \sqrt{4t + 1} \;\;\Rightarrow\;\; f'(t) = \tfrac{1}{2}(4t + 1)^{-1/2} \cdot 4 = 2(4t + 1)^{-1/2} \;\;\Rightarrow$

$f''(t) = 2\left(-\tfrac{1}{2}\right)(4t + 1)^{-3/2} \cdot 4 = -4/(4t + 1)^{3/2}$, so $f''(2) = -4/9^{3/2} = -\tfrac{4}{27}$.

41. $x^6 + y^6 = 1 \;\;\Rightarrow\;\; 6x^5 + 6y^5 y' = 0 \;\;\Rightarrow\;\; y' = -x^5/y^5 \;\;\Rightarrow$

$y'' = -\dfrac{y^5(5x^4) - x^5(5y^4 y')}{(y^5)^2} = -\dfrac{5x^4 y^4\left[y - x(-x^5/y^5)\right]}{y^{10}} = -\dfrac{5x^4\left[(y^6 + x^6)/y^5\right]}{y^6} = -\dfrac{5x^4}{y^{11}}$

43. $y = 4\sin^2 x \;\;\Rightarrow\;\; y' = 4 \cdot 2\sin x\cos x$. At $\left(\tfrac{\pi}{6}, 1\right)$, $y' = 8 \cdot \tfrac{1}{2} \cdot \tfrac{\sqrt{3}}{2} = 2\sqrt{3}$, so an equation of the tangent line

is $y - 1 = 2\sqrt{3}\left(x - \tfrac{\pi}{6}\right)$, or $y = 2\sqrt{3}\,x + 1 - \pi\sqrt{3}/3$.

45. $y = \sqrt{1 + 4\sin x} \;\;\Rightarrow\;\; y' = \tfrac{1}{2}(1 + 4\sin x)^{-1/2} \cdot 4\cos x = \dfrac{2\cos x}{\sqrt{1 + 4\sin x}}$. At $(0, 1)$, $y' = \dfrac{2}{\sqrt{1}} = 2$, so an equation of the

tangent line is $y - 1 = 2(x - 0)$, or $y = 2x + 1$.

47. $y = \sin x + \cos x \;\;\Rightarrow\;\; y' = \cos x - \sin x = 0 \;\;\Leftrightarrow\;\; \cos x = \sin x$ and $0 \leq x \leq 2\pi \;\;\Leftrightarrow\;\; x = \tfrac{\pi}{4}$ or $\tfrac{5\pi}{4}$, so the points

are $\left(\tfrac{\pi}{4}, \sqrt{2}\right)$ and $\left(\tfrac{5\pi}{4}, -\sqrt{2}\right)$.

49. (a) $h(x) = f(x)g(x) \implies h'(x) = f(x)g'(x) + g(x)f'(x) \implies$

$h'(2) = f(2)g'(2) + g(2)f'(2) = (3)(4) + (5)(-2) = 12 - 10 = 2$

(b) $F(x) = f(g(x)) \implies F'(x) = f'(g(x))g'(x) \implies F'(2) = f'(g(2))g'(2) = f'(5)(4) = 11 \cdot 4 = 44$

51. $f(x) = x^2 g(x) \implies f'(x) = x^2 g'(x) + g(x)(2x) = x[xg'(x) + 2g(x)]$

53. $f(x) = [g(x)]^2 \implies f'(x) = 2[g(x)]^1 \cdot g'(x) = 2g(x)g'(x)$

55. $f(x) = g(g(x)) \implies f'(x) = g'(g(x))g'(x)$

57. $f(x) = g(\sin x) \implies f'(x) = g'(\sin x) \cdot \cos x$

59. $h(x) = \dfrac{f(x)g(x)}{f(x) + g(x)} \implies$

$h'(x) = \dfrac{[f(x) + g(x)] [f(x)g'(x) + g(x)f'(x)] - f(x)g(x) [f'(x) + g'(x)]}{[f(x) + g(x)]^2}$

$= \dfrac{[f(x)]^2 g'(x) + f(x)g(x)f'(x) + f(x)g(x)g'(x) + [g(x)]^2 f'(x) - f(x)g(x)f'(x) - f(x)g(x)g'(x)}{[f(x) + g(x)]^2}$

$= \dfrac{f'(x)[g(x)]^2 + g'(x)[f(x)]^2}{[f(x) + g(x)]^2}$

61. f is not differentiable: at $x = -4$ because f is not continuous, at $x = -1$ because f has a corner, at $x = 2$ because f is not continuous, and at $x = 5$ because f has a vertical tangent.

63. (a) $y = t^3 - 12t + 3 \implies v(t) = y' = 3t^2 - 12 \implies a(t) = v'(t) = 6t$

(b) $v(t) = 3(t^2 - 4) > 0$ when $t > 2$, so it moves upward when $t > 2$ and downward when $0 \le t < 2$.

(c) Distance upward $= y(3) - y(2) = -6 - (-13) = 7$, distance downward $= y(0) - y(2) = 3 - (-13) = 16$.
Total distance $= 7 + 16 = 23$.

65. If $x = $ edge length, then $V = x^3 \implies dV/dt = 3x^2\, dx/dt = 10 \implies dx/dt = 10/(3x^2)$ and $S = 6x^2 \implies$

$dS/dt = (12x)\, dx/dt = 12x[10/(3x^2)] = 40/x$. When $x = 30$, $dS/dt = \frac{40}{30} = \frac{4}{3}$ cm^2/min.

67. Given $dh/dt = 5$ and $dx/dt = 15$, find dz/dt. $z^2 = x^2 + h^2 \implies$

$2z\dfrac{dz}{dt} = 2x\dfrac{dx}{dt} + 2h\dfrac{dh}{dt} \implies \dfrac{dz}{dt} = \dfrac{1}{z}(15x + 5h)$. When $t = 3$,

$h = 45 + 3(5) = 60$ and $x = 15(3) = 45 \implies z = \sqrt{45^2 + 60^2} = 75$,

so $\dfrac{dz}{dt} = \frac{1}{75}[15(45) + 5(60)] = 13$ ft/s.

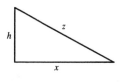

69. We are given $d\theta/dt = -0.25$ rad/h. $\tan \theta = 400/x \implies$

$x = 400 \cot \theta \implies \dfrac{dx}{dt} = -400 \csc^2 \theta \dfrac{d\theta}{dt}$. When $\theta = \frac{\pi}{6}$,

$\dfrac{dx}{dt} = -400(2)^2(-0.25) = 400$ ft/h.

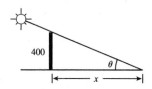

71. (a) $f(x) = \sqrt[3]{1 + 3x} = (1 + 3x)^{1/3}$ ⇒ $f'(x) = (1 + 3x)^{-2/3}$, so the linearization of f at $a = 0$ is

$L(x) = f(0) + f'(0)(x - 0) = 1^{1/3} + 1^{-2/3}x = 1 + x$. Thus, $\sqrt[3]{1 + 3x} \approx 1 + x$ ⇒

$\sqrt[3]{1.03} = \sqrt[3]{1 + 3(0.01)} \approx 1 + (0.01) = 1.01$.

(b) The linear approximation is $\sqrt[3]{1 + 3x} \approx 1 + x$, so for the required accuracy

we want $\sqrt[3]{1 + 3x} - 0.1 < 1 + x < \sqrt[3]{1 + 3x} + 0.1$. From the graph,

it appears that this is true when $-0.23 < x < 0.40$.

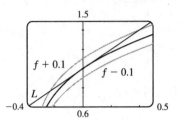

73. $A = x^2 + \frac{1}{2}\pi\left(\frac{1}{2}x\right)^2 = \left(1 + \frac{\pi}{8}\right)x^2$ ⇒ $dA = \left(2 + \frac{\pi}{4}\right)x\,dx$. When

$x = 60$ and $dx = 0.1$, $dA = \left(2 + \frac{\pi}{4}\right)60(0.1) = 12 + \frac{3\pi}{2}$, so the maximum

error is approximately $12 + \frac{3\pi}{2} \approx 16.7$ cm^2.

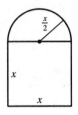

75. $\displaystyle\lim_{h \to 0} \frac{\sqrt[4]{16 + h} - 2}{h} = \left[\frac{d}{dx}\sqrt[4]{x}\right]_{x=16} = \left.\frac{1}{4}x^{-3/4}\right|_{x=16} = \frac{1}{4\left(\sqrt[4]{16}\right)^3} = \frac{1}{32}$

77. $\displaystyle\lim_{x \to 0} \frac{\sqrt{1 + \tan x} - \sqrt{1 + \sin x}}{x^3} = \lim_{x \to 0} \frac{\left(\sqrt{1 + \tan x} - \sqrt{1 + \sin x}\right)\left(\sqrt{1 + \tan x} + \sqrt{1 + \sin x}\right)}{x^3\left(\sqrt{1 + \tan x} + \sqrt{1 + \sin x}\right)}$

$\displaystyle = \lim_{x \to 0} \frac{(1 + \tan x) - (1 + \sin x)}{x^3\left(\sqrt{1 + \tan x} + \sqrt{1 + \sin x}\right)} = \lim_{x \to 0} \frac{\sin x\,(1/\cos x - 1)}{x^3\left(\sqrt{1 + \tan x} + \sqrt{1 + \sin x}\right)} \cdot \frac{\cos x}{\cos x}$

$\displaystyle = \lim_{x \to 0} \frac{\sin x\,(1 - \cos x)}{x^3\left(\sqrt{1 + \tan x} + \sqrt{1 + \sin x}\right)\cos x} \cdot \frac{1 + \cos x}{1 + \cos x}$

$\displaystyle = \lim_{x \to 0} \frac{\sin x \cdot \sin^2 x}{x^3\left(\sqrt{1 + \tan x} + \sqrt{1 + \sin x}\right)\cos x\,(1 + \cos x)}$

$\displaystyle = \left(\lim_{x \to 0} \frac{\sin x}{x}\right)^3 \lim_{x \to 0} \frac{1}{\left(\sqrt{1 + \tan x} + \sqrt{1 + \sin x}\right)\cos x\,(1 + \cos x)}$

$\displaystyle = 1^3 \cdot \frac{1}{\left(\sqrt{1} + \sqrt{1}\right) \cdot 1 \cdot (1 + 1)} = \frac{1}{4}$

3 ☐ INVERSE FUNCTIONS:
Exponential, Logarithmic, and Inverse Trigonometric Functions

3.1 Exponential Functions

1. (a) $f(x) = a^x$, $a > 0$ (b) $\mathbb{R}$ (c) $(0, \infty)$ (d) See Figures 6(c), 6(b), and 6(a), respectively.

3. All of these graphs approach 0 as $x \to -\infty$, all of them pass through the point $(0, 1)$, and all of them are increasing and approach ∞ as $x \to \infty$. The larger the base, the faster the function increases for $x > 0$, and the faster it approaches 0 as $x \to -\infty$.

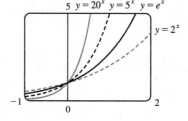

5. The functions with bases greater than 1 (3^x and 10^x) are increasing, while those with bases less than 1 $\left[\left(\frac{1}{3}\right)^x \text{ and } \left(\frac{1}{10}\right)^x \right]$ are decreasing. The graph of $\left(\frac{1}{3}\right)^x$ is the reflection of that of 3^x about the y-axis, and the graph of $\left(\frac{1}{10}\right)^x$ is the reflection of that of 10^x about the y-axis. The graph of 10^x increases more quickly than that of 3^x for $x > 0$, and approaches 0 faster as $x \to -\infty$.

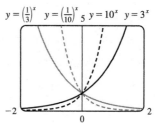

7. We start with the graph of $y = 4^x$ (Figure 3) and then shift 3 units downward. This shift doesn't affect the domain, but the range of $y = 4^x - 3$ is $(-3, \infty)$. There is a horizontal asymptote of $y = -3$.

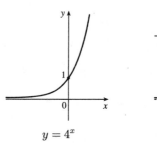

$y = 4^x$ $y = 4^x - 3$

9. We start with the graph of $y = 2^x$ (Figure 3), reflect it about the y-axis, and then about the x-axis (or just rotate $180°$ to handle both reflections) to obtain the graph of $y = -2^{-x}$. In each graph, $y = 0$ is the horizontal asymptote.

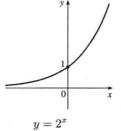
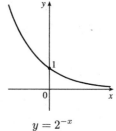
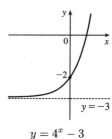

$y = 2^x$ $y = 2^{-x}$ $y = -2^{-x}$

11. We start with the graph of $y = e^x$ (Figure 10) and reflect about the y-axis to get the graph of $y = e^{-x}$. Then we compress the graph vertically by a factor of 2 to obtain the graph of $y = \frac{1}{2}e^{-x}$ and then reflect about the x-axis to get the graph of $y = -\frac{1}{2}e^{-x}$. Finally, we shift the graph upward one unit to get the graph of $y = 1 - \frac{1}{2}e^{-x}$.

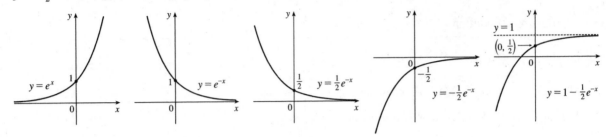

13. (a) To find the equation of the graph that results from shifting the graph of $y = e^x$ 2 units downward, we subtract 2 from the original function to get $y = e^x - 2$.

(b) To find the equation of the graph that results from shifting the graph of $y = e^x$ 2 units to the right, we replace x with $x - 2$ in the original function to get $y = e^{x-2}$.

(c) To find the equation of the graph that results from reflecting the graph of $y = e^x$ about the x-axis, we multiply the original function by -1 to get $y = -e^x$.

(d) To find the equation of the graph that results from reflecting the graph of $y = e^x$ about the y-axis, we replace x with $-x$ in the original function to get $y = e^{-x}$.

(e) To find the equation of the graph that results from reflecting the graph of $y = e^x$ about the x-axis and then about the y-axis, we first multiply the original function by -1 (to get $y = -e^x$) and then replace x with $-x$ in this equation to get $y = -e^{-x}$.

15. (a) The denominator $1 + e^x$ is never equal to zero because $e^x > 0$, so the domain of $f(x) = 1/(1 + e^x)$ is $\mathbb{R}$.

(b) $1 - e^x = 0 \Leftrightarrow e^x = 1 \Leftrightarrow x = 0$, so the domain of $f(x) = 1/(1 - e^x)$ is $(-\infty, 0) \cup (0, \infty)$.

17. Use $y = Ca^x$ with the points $(1, 6)$ and $(3, 24)$. $6 = Ca^1 \quad \left[C = \frac{6}{a}\right]$ and $24 = Ca^3 \Rightarrow 24 = \left(\frac{6}{a}\right)a^3 \Rightarrow$ $4 = a^2 \Rightarrow a = 2$ [since $a > 0$] and $C = \frac{6}{2} = 3$. The function is $f(x) = 3 \cdot 2^x$.

19. 2 ft = 24 in, $f(24) = 24^2$ in = 576 in = 48 ft. $g(24) = 2^{24}$ in = $2^{24}/(12 \cdot 5280)$ mi ≈ 265 mi

21. The graph of g finally surpasses that of f at $x \approx 35.8$.

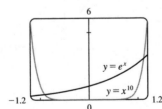 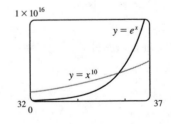

23. $\lim\limits_{x\to\infty} (1.001)^x = \infty$ by (3), since $1.001 > 1$.

25. Divide numerator and denominator by e^{3x}: $\lim\limits_{x\to\infty} \dfrac{e^{3x} - e^{-3x}}{e^{3x} + e^{-3x}} = \lim\limits_{x\to\infty} \dfrac{1 - e^{-6x}}{1 + e^{-6x}} = \dfrac{1 - 0}{1 + 0} = 1$

27. Let $t = 3/(2 - x)$. As $x \to 2^+$, $t \to -\infty$. So $\lim\limits_{x\to 2^+} e^{3/(2-x)} = \lim\limits_{t\to -\infty} e^t = 0$ by (5).

29. Since $-1 \le \cos x \le 1$ and $e^{-2x} > 0$, we have $-e^{-2x} \le e^{-2x} \cos x \le e^{-2x}$. We know that $\lim\limits_{x\to\infty} (-e^{-2x}) = 0$ and

$\lim\limits_{x\to\infty} (e^{-2x}) = 0$, so by the Squeeze Theorem, $\lim\limits_{x\to\infty} (e^{-2x} \cos x) = 0$.

31.

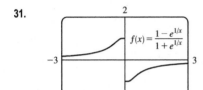

From the graph, it appears that f is an odd function (f is undefined for $x = 0$).
To prove this, we must show that $f(-x) = -f(x)$.

$$f(-x) = \frac{1 - e^{1/(-x)}}{1 + e^{1/(-x)}} = \frac{1 - e^{(-1/x)}}{1 + e^{(-1/x)}} = \frac{1 - \dfrac{1}{e^{1/x}}}{1 + \dfrac{1}{e^{1/x}}} \cdot \frac{e^{1/x}}{e^{1/x}} = \frac{e^{1/x} - 1}{e^{1/x} + 1}$$

$$= -\frac{1 - e^{1/x}}{1 + e^{1/x}} = -f(x)$$

So f is an odd function.

3.2 Inverse Functions and Logarithms

1. (a) See Definition 1.

(b) It must pass the Horizontal Line Test.

3. f is not one-to-one because $2 \ne 6$, but $f(2) = 2.0 = f(6)$.

5. No horizontal line intersects the graph of f more than once. Thus, by the Horizontal Line Test, f is one-to-one.

7. The horizontal line $y = 0$ (the x-axis) intersects the graph of f in more than one point. Thus, by the Horizontal Line Test,

f is not one-to-one.

9. The graph of $f(x) = x^2 - 2x$ is a parabola with axis of symmetry $x = -\dfrac{b}{2a} = -\dfrac{-2}{2(1)} = 1$. Pick any x-values equidistant

from 1 to find two equal function values. For example, $f(0) = 0$ and $f(2) = 0$, so f is not one-to-one.

11. $g(x) = 1/x$. $x_1 \ne x_2 \;\Rightarrow\; 1/x_1 \ne 1/x_2 \;\Rightarrow\; g(x_1) \ne g(x_2)$, so g is one-to-one.

Geometric solution: The graph of g is the hyperbola shown in Figure 9 in Section 1.2. It passes the Horizontal Line Test,

so g is one-to-one.

13. A football will attain every height h up to its maximum height twice: once on the way up, and again on the way down. Thus,

even if t_1 does not equal t_2, $f(t_1)$ may equal $f(t_2)$, so f is not 1-1.

15. Since $f(2) = 9$ and f is 1-1, we know that $f^{-1}(9) = 2$. Remember, if the point $(2, 9)$ is on the graph of f, then the point

$(9, 2)$ is on the graph of f^{-1}.

17. First, we must determine x such that $g(x) = 4$. By inspection, we see that if $x = 0$, then $g(x) = 4$. Since g is 1-1 (g is an

increasing function), it has an inverse, and $g^{-1}(4) = 0$.

19. We solve $C = \frac{5}{9}(F - 32)$ for F: $\frac{9}{5}C = F - 32 \;\Rightarrow\; F = \frac{9}{5}C + 32$. This gives us a formula for the inverse function, that is, the Fahrenheit temperature F as a function of the Celsius temperature C.

$F \geq -459.67 \;\Rightarrow\; \frac{9}{5}C + 32 \geq -459.67 \;\Rightarrow\; \frac{9}{5}C \geq -491.67 \;\Rightarrow\; C \geq -273.15$, the domain of the inverse function.

21. $f(x) = \sqrt{10 - 3x} \;\Rightarrow\; y = \sqrt{10 - 3x} \;(y \geq 0) \;\Rightarrow\; y^2 = 10 - 3x \;\Rightarrow\; 3x = 10 - y^2 \;\Rightarrow\; x = -\frac{1}{3}y^2 + \frac{10}{3}$.

Interchange x and y: $y = -\frac{1}{3}x^2 + \frac{10}{3}$. So $f^{-1}(x) = -\frac{1}{3}x^2 + \frac{10}{3}$. Note that the domain of f^{-1} is $x \geq 0$.

23. $y = f(x) = e^{x^3} \;\Rightarrow\; \ln y = x^3 \;\Rightarrow\; x = \sqrt[3]{\ln y}$. Interchange x and y: $y = \sqrt[3]{\ln x}$. So $f^{-1}(x) = \sqrt[3]{\ln x}$.

25. $y = f(x) = \ln(x + 3) \;\Rightarrow\; x + 3 = e^y \;\Rightarrow\; x = e^y - 3$. Interchange x and y: $y = e^x - 3$. So $f^{-1}(x) = e^x - 3$.

27. $y = f(x) = x^4 + 1 \;\Rightarrow\; y - 1 = x^4 \;\Rightarrow\; x = \sqrt[4]{y - 1}$ (not $\pm$ since

$x \geq 0$). Interchange x and y: $y = \sqrt[4]{x - 1}$. So $f^{-1}(x) = \sqrt[4]{x - 1}$. The

graph of $y = \sqrt[4]{x - 1}$ is just the graph of $y = \sqrt[4]{x}$ shifted right one unit.

From the graph, we see that f and f^{-1} are reflections about the line $y = x$.

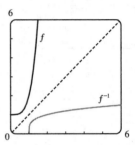

29. Reflect the graph of f about the line $y = x$. The points $(-1, -2)$, $(1, -1)$,

$(2, 2)$, and $(3, 3)$ on f are reflected to $(-2, -1)$, $(-1, 1)$, $(2, 2)$, and $(3, 3)$

on f^{-1}.

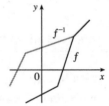

31. (a) $x_1 \neq x_2 \;\Rightarrow\; x_1^3 \neq x_2^3 \;\Rightarrow\; f(x_1) \neq f(x_2)$, so f is one-to-one.

(b) $f'(x) = 3x^2$ and $f(2) = 8 \;\Rightarrow\; f^{-1}(8) = 2$, so $(f^{-1})'(8) = 1/f'(f^{-1}(8)) = 1/f'(2) = \frac{1}{12}$.

(c) $y = x^3 \;\Rightarrow\; x = y^{1/3}$. Interchanging x and y gives $y = x^{1/3}$, $\qquad$ (e)

so $f^{-1}(x) = x^{1/3}$. Domain(f^{-1}) = range$(f) = \mathbb{R}$.

Range(f^{-1}) = domain$(f) = \mathbb{R}$.

(d) $f^{-1}(x) = x^{1/3} \;\Rightarrow\; (f^{-1})'(x) = \frac{1}{3}x^{-2/3} \;\Rightarrow\;$

$(f^{-1})'(8) = \frac{1}{3}\left(\frac{1}{4}\right) = \frac{1}{12}$ as in part (b).

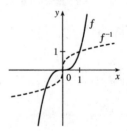

33. (a) Since $x \geq 0$, $x_1 \neq x_2 \;\Rightarrow\; x_1^2 \neq x_2^2 \;\Rightarrow\; 9 - x_1^2 \neq 9 - x_2^2 \;\Rightarrow\; f(x_1) \neq f(x_2)$, so f is 1-1.

(b) $f'(x) = -2x$ and $f(1) = 8 \;\Rightarrow\; f^{-1}(8) = 1$, so $(f^{-1})'(8) = \dfrac{1}{f'(f^{-1}(8))} = \dfrac{1}{f'(1)} = \dfrac{1}{(-2)} = -\dfrac{1}{2}$.

(c) $y = 9 - x^2 \;\Rightarrow\; x^2 = 9 - y \;\Rightarrow\; x = \sqrt{9 - y}$. $\qquad$ (e)

Interchange x and y: $y = \sqrt{9 - x}$, so $f^{-1}(x) = \sqrt{9 - x}$.

Domain(f^{-1}) = range $(f) = [0, 9]$.

Range(f^{-1}) = domain $(f) = [0, 3]$.

(d) $(f^{-1})'(x) = -1/(2\sqrt{9 - x}) \;\Rightarrow\; (f^{-1})'(8) = -\frac{1}{2}$ as in part (b).

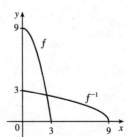

35. $f(0) = 1 \;\Rightarrow\; f^{-1}(1) = 0$, and $f(x) = x^3 + x + 1 \;\Rightarrow\; f'(x) = 3x^2 + 1$ and $f'(0) = 1$. Thus,

$$\left(f^{-1}\right)'(1) = \frac{1}{f'(f^{-1}(1))} = \frac{1}{f'(0)} = \frac{1}{1} = 1.$$

37. $f(0) = 3 \;\Rightarrow\; f^{-1}(3) = 0$, and $f(x) = 3 + x^2 + \tan(\pi x/2) \;\Rightarrow\; f'(x) = 2x + \frac{\pi}{2}\sec^2(\pi x/2)$ and

$f'(0) = \frac{\pi}{2} \cdot 1 = \frac{\pi}{2}$. Thus, $\left(f^{-1}\right)'(3) = 1/f'\left(f^{-1}(3)\right) = 1/f'(0) = 2/\pi.$

39. $f(4) = 5 \;\Rightarrow\; f^{-1}(5) = 4$. Thus, $g'(5) = \dfrac{1}{f'(f^{-1}(5))} = \dfrac{1}{f'(4)} = \dfrac{1}{2/3} = \dfrac{3}{2}.$

41. (a) It is defined as the inverse of the exponential function with base a, that is, $\log_a x = y \;\Leftrightarrow\; a^y = x$.

(b) $(0, \infty)$ (c) $\mathbb{R}$ (d) See Figure 13.

43. (a) $\log_2 64 = 6$ since $2^6 = 64$.   (b) $\log_6 \frac{1}{36} = -2$ since $6^{-2} = \frac{1}{36}$.

45. (a) $\log_{10} 1.25 + \log_{10} 80 = \log_{10}(1.25 \cdot 80) = \log_{10} 100 = \log_{10} 10^2 = 2$

(b) $\log_5 10 + \log_5 20 - 3\log_5 2 = \log_5 (10 \cdot 20) - \log_5 2^3 = \log_5 \frac{200}{8} = \log_5 25 = \log_5 5^2 = 2$

47. $\log_2 \left(\dfrac{x^3 y}{z^2}\right) = \log_2(x^3 y) - \log_2 z^2 = \log_2 x^3 + \log_2 y - \log_2 z^2 = 3\log_2 x + \log_2 y - 2\log_2 z$

(assuming that the variables are positive)

49. $\ln(uv)^{10} = 10\ln(uv) = 10(\ln u + \ln v) = 10\ln u + 10\ln v$

51. $2\ln 4 - \ln 2 = \ln 4^2 - \ln 2 = \ln 16 - \ln 2 = \ln \frac{16}{2} = \ln 8$

53. $\ln(1 + x^2) + \frac{1}{2}\ln x - \ln \sin x = \ln(1 + x^2) + \ln x^{1/2} - \ln \sin x = \ln[(1 + x^2)\sqrt{x}] - \ln \sin x = \ln \dfrac{(1 + x^2)\sqrt{x}}{\sin x}$

55. To graph these functions, we use $\log_{1.5} x = \dfrac{\ln x}{\ln 1.5}$ and $\log_{50} x = \dfrac{\ln x}{\ln 50}$.

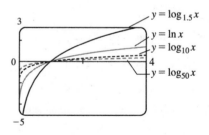

These graphs all approach $-\infty$ as $x \to 0^+$, and they all pass through the
point $(1, 0)$. Also, they are all increasing, and all approach ∞ as $x \to \infty$.
The functions with larger bases increase extremely slowly, and the ones with
smaller bases do so somewhat more quickly. The functions with large bases
approach the y-axis more closely as $x \to 0^+$.

57. 3 ft = 36 in, so we need x such that $\log_2 x = 36 \;\Leftrightarrow\; x = 2^{36} = 68,719,476,736$. In miles, this is

$68,719,476,736 \text{ in} \cdot \dfrac{1 \text{ ft}}{12 \text{ in}} \cdot \dfrac{1 \text{ mi}}{5280 \text{ ft}} \approx 1,084,587.7 \text{ mi}.$

59. (a) Shift the graph of $y = \log_{10} x$ five units to the left to   (b) Reflect the graph of $y = \ln x$ about the x-axis to obtain
obtain the graph of $y = \log_{10}(x + 5)$. Note the vertical   the graph of $y = -\ln x$.
asymptote of $x = -5$.

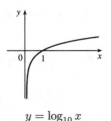

$y = \log_{10} x$

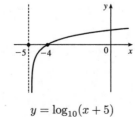

$y = \log_{10}(x + 5)$

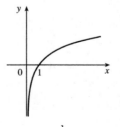

$y = \ln x$ $y = -\ln x$

61. (a) $2\ln x = 1 \Rightarrow \ln x = \frac{1}{2} \Rightarrow x = e^{1/2} = \sqrt{e}$

(b) $e^{-x} = 5 \Rightarrow -x = \ln 5 \Rightarrow x = -\ln 5$

63. (a) $2^{x-5} = 3 \Leftrightarrow \log_2 3 = x - 5 \Leftrightarrow x = 5 + \log_2 3$.

Or: $2^{x-5} = 3 \Leftrightarrow \ln(2^{x-5}) = \ln 3 \Leftrightarrow (x-5)\ln 2 = \ln 3 \Leftrightarrow x - 5 = \dfrac{\ln 3}{\ln 2} \Leftrightarrow x = 5 + \dfrac{\ln 3}{\ln 2}$

(b) $\ln x + \ln(x-1) = \ln(x(x-1)) = 1 \Leftrightarrow x(x-1) = e^1 \Leftrightarrow x^2 - x - e = 0$. The quadratic formula (with $a = 1$, $b = -1$, and $c = -e$) gives $x = \frac{1}{2}(1 \pm \sqrt{1+4e})$, but we reject the negative root since the natural logarithm is not defined for $x < 0$. So $x = \frac{1}{2}(1 + \sqrt{1+4e})$.

65. (a) $e^x < 10 \Rightarrow \ln e^x < \ln 10 \Rightarrow x < \ln 10 \Rightarrow x \in (-\infty, \ln 10)$

(b) $\ln x > -1 \Rightarrow e^{\ln x} > e^{-1} \Rightarrow x > e^{-1} \Rightarrow x \in (1/e, \infty)$

67. (a) For $f(x) = \sqrt{3 - e^{2x}}$, we must have $3 - e^{2x} \geq 0 \Rightarrow e^{2x} \leq 3 \Rightarrow 2x \leq \ln 3 \Rightarrow x \leq \frac{1}{2}\ln 3$. Thus, the domain of f is $(-\infty, \frac{1}{2}\ln 3]$.

(b) $y = f(x) = \sqrt{3 - e^{2x}}$ [note that $y \geq 0$] $\Rightarrow y^2 = 3 - e^{2x} \Rightarrow e^{2x} = 3 - y^2 \Rightarrow 2x = \ln(3 - y^2) \Rightarrow x = \frac{1}{2}\ln(3 - y^2)$. Interchange x and y: $y = \frac{1}{2}\ln(3 - x^2)$. So $f^{-1}(x) = \frac{1}{2}\ln(3 - x^2)$. For the domain of f^{-1}, we must have $3 - x^2 > 0 \Rightarrow x^2 < 3 \Rightarrow |x| < \sqrt{3} \Rightarrow -\sqrt{3} < x < \sqrt{3} \Rightarrow 0 \leq x < \sqrt{3}$ since $x \geq 0$. Note that the domain of f^{-1}, $[0, \sqrt{3})$, equals the range of f.

69. Let $t = 2 - x$. As $x \to 2^-, t \to 0^+$. $\displaystyle\lim_{x \to 2^-} \ln(2 - x) = \lim_{t \to 0^+} \ln t = -\infty$ by (15).

71. $\displaystyle\lim_{x \to 0} \ln(\cos x) = \ln 1 = 0$. [$\ln(\cos x)$ is continuous at $x = 0$ since it is the composite of two continuous functions.]

73. $\displaystyle\lim_{x \to \infty} [\ln(1 + x^2) - \ln(1 + x)] = \lim_{x \to \infty} \ln \frac{1 + x^2}{1 + x} = \ln\left(\lim_{x \to \infty} \frac{1 + x^2}{1 + x}\right) = \ln\left(\lim_{x \to \infty} \frac{\frac{1}{x} + x}{\frac{1}{x} + 1}\right) = \infty$, since the limit in parentheses is ∞.

75. We see that the graph of $y = f(x) = \sqrt{x^3 + x^2 + x + 1}$ is increasing, so f is 1-1.

Enter $x = \sqrt{y^3 + y^2 + y + 1}$ and use your CAS to solve the equation for y.

Using Derive, we get two (irrelevant) solutions involving imaginary expressions, as well as one which can be simplified to the following:

$$y = f^{-1}(x) = -\frac{\sqrt[3]{4}}{6}\left(\sqrt[3]{D - 27x^2 + 20} - \sqrt[3]{D + 27x^2 - 20} + \sqrt[3]{2}\right)$$

where $D = 3\sqrt{3}\sqrt{27x^4 - 40x^2 + 16}$.

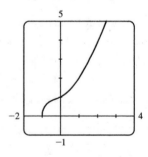

Maple and Mathematica each give two complex expressions and one real expression, and the real expression is equivalent to that given by Derive. For example, Maple's expression simplifies to $\dfrac{1}{6}\dfrac{M^{2/3} - 8 - 2M^{1/3}}{2M^{1/3}}$, where

$M = 108x^2 + 12\sqrt{48 - 120x^2 + 81x^4} - 80$.

77. (a) Let $\varepsilon > 0$ be given. We need N such that $|a^x - 0| < \varepsilon$ when $x < N$. But $a^x < \varepsilon \Leftrightarrow x < \log_a \varepsilon$. Let $N = \log_a \varepsilon$. Then $x < N \Rightarrow x < \log_a \varepsilon \Rightarrow |a^x - 0| = a^x < \varepsilon$, so $\displaystyle\lim_{x \to -\infty} a^x = 0$.

(b) Let $M > 0$ be given. We need N such that $a^x > M$ when $x > N$. But $a^x > M \Leftrightarrow x > \log_a M$. Let $N = \log_a M$. Then $x > N \Rightarrow x > \log_a M \Rightarrow a^x > M$, so $\displaystyle\lim_{x \to \infty} a^x = \infty$.

3.3   Derivatives of Logarithmic and Exponential Functions

1. $f(x) = \log_2(1 - 3x) \;\Rightarrow\; f'(x) = \dfrac{1}{(1-3x)\ln 2}\dfrac{d}{dx}(1-3x) = \dfrac{-3}{(1-3x)\ln 2}$ or $\dfrac{3}{(3x-1)\ln 2}$

3. $f(\theta) = \ln(\cos\theta) \;\Rightarrow\; f'(\theta) = \dfrac{1}{\cos\theta}\dfrac{d}{d\theta}(\cos\theta) = \dfrac{-\sin\theta}{\cos\theta} = -\tan\theta$

5. $f(x) = \sqrt[5]{\ln x} = (\ln x)^{1/5} \;\Rightarrow\; f'(x) = \tfrac{1}{5}(\ln x)^{-4/5}\dfrac{d}{dx}(\ln x) = \dfrac{1}{5(\ln x)^{4/5}}\cdot\dfrac{1}{x} = \dfrac{1}{5x\sqrt[5]{(\ln x)^4}}$

7. $f(x) = \sin x \ln(5x) \;\Rightarrow$

$f'(x) = \sin x \cdot \dfrac{1}{5x}\cdot\dfrac{d}{dx}(5x) + \ln(5x)\cdot\cos x = \dfrac{\sin x \cdot 5}{5x} + \cos x \ln(5x) = \dfrac{\sin x}{x} + \cos x \ln(5x)$

9. $g(x) = \ln\dfrac{a-x}{a+x} = \ln(a-x) - \ln(a+x) \;\Rightarrow$

$g'(x) = \dfrac{1}{a-x}(-1) - \dfrac{1}{a+x} = \dfrac{-(a+x) - (a-x)}{(a-x)(a+x)} = \dfrac{-2a}{a^2 - x^2}$

11. $F(t) = \ln\dfrac{(2t+1)^3}{(3t-1)^4} = \ln(2t+1)^3 - \ln(3t-1)^4 = 3\ln(2t+1) - 4\ln(3t-1) \;\Rightarrow$

$F'(t) = 3\cdot\dfrac{1}{2t+1}\cdot 2 - 4\cdot\dfrac{1}{3t-1}\cdot 3 = \dfrac{6}{2t+1} - \dfrac{12}{3t-1}$, or combined, $\dfrac{-6(t+3)}{(2t+1)(3t-1)}$.

13. $f(u) = \dfrac{\ln u}{1 + \ln(2u)} \;\Rightarrow$

$f'(u) = \dfrac{[1 + \ln(2u)]\cdot\frac{1}{u} - \ln u\cdot\frac{1}{2u}\cdot 2}{[1 + \ln(2u)]^2} = \dfrac{\frac{1}{u}[1 + \ln(2u) - \ln u]}{[1 + \ln(2u)]^2} = \dfrac{1 + (\ln 2 + \ln u) - \ln u}{u[1 + \ln(2u)]^2} = \dfrac{1 + \ln 2}{u[1 + \ln(2u)]^2}$

15. $y = \ln|2 - x - 5x^2| \;\Rightarrow\; y' = \dfrac{1}{2 - x - 5x^2}\cdot(-1 - 10x) = \dfrac{-10x - 1}{2 - x - 5x^2}$ or $\dfrac{10x + 1}{5x^2 + x - 2}$

17. By the Product Rule, $f(x) = x^2 e^x \;\Rightarrow\; f'(x) = x^2\dfrac{d}{dx}(e^x) + e^x\dfrac{d}{dx}(x^2) = x^2 e^x + e^x(2x) = xe^x(x+2)$.

19. By the Quotient Rule, $y = \dfrac{e^x}{x^2} \;\Rightarrow$

$y' = \dfrac{x^2\dfrac{d}{dx}(e^x) - e^x\dfrac{d}{dx}(x^2)}{(x^2)^2} = \dfrac{x^2(e^x) - e^x(2x)}{x^4} = \dfrac{xe^x(x-2)}{x^4} = \dfrac{e^x(x-2)}{x^3}$.

21. $y = xe^{-x^2} \;\Rightarrow\; y' = xe^{-x^2}(-2x) + e^{-x^2}\cdot 1 = e^{-x^2}(-2x^2 + 1) = e^{-x^2}(1 - 2x^2)$

23. $y = e^{x\cos x} \;\Rightarrow\; y' = e^{x\cos x}\cdot\dfrac{d}{dx}(x\cos x) = e^{x\cos x}[x(-\sin x) + (\cos x)\cdot 1] = e^{x\cos x}(\cos x - x\sin x)$

25. $h(t) = t^3 - 3^t \;\Rightarrow\; h'(t) = 3t^2 - 3^t\ln 3$

27. By the Quotient Rule, $y = \dfrac{ae^x + b}{ce^x + d} \;\Rightarrow$

$y' = \dfrac{(ce^x + d)(ae^x) - (ae^x + b)(ce^x)}{(ce^x + d)^2} = \dfrac{(ace^x + ad - ace^x - bc)e^x}{(ce^x + d)^2} = \dfrac{(ad - bc)e^x}{(ce^x + d)^2}$.

The notations $\overset{PR}{\Rightarrow}$ and $\overset{QR}{\Rightarrow}$ indicate the use of the Product and Quotient Rules, respectively.

29. Using Theorem 6 and the Chain Rule, $y = 2^{\sin \pi x}$ $\Rightarrow$

$$y' = 2^{\sin \pi x}(\ln 2) \cdot \frac{d}{dx}(\sin \pi x) = 2^{\sin \pi x}(\ln 2) \cdot \cos \pi x \cdot \pi = 2^{\sin \pi x}(\pi \ln 2) \cos \pi x$$

31. $f(u) = e^{1/u}$ $\Rightarrow$ $f'(u) = e^{1/u} \cdot \dfrac{d}{du}\left(\dfrac{1}{u}\right) = e^{1/u}\left(\dfrac{-1}{u^2}\right) = \left(\dfrac{-1}{u^2}\right)e^{1/u}$

33. $y = \ln\left(e^{-x} + xe^{-x}\right) = \ln\left(e^{-x}(1+x)\right) = \ln\left(e^{-x}\right) + \ln(1+x) = -x + \ln(1+x)$ $\Rightarrow$

$$y' = -1 + \frac{1}{1+x} = \frac{-1-x+1}{1+x} = -\frac{x}{1+x}$$

35. $F(t) = e^{t \sin 2t}$ $\Rightarrow$ $F'(t) = e^{t \sin 2t}(t \sin 2t)' = e^{t \sin 2t}(t \cdot 2 \cos 2t + \sin 2t \cdot 1) = e^{t \sin 2t}(2t \cos 2t + \sin 2t)$

37. $y = e^{\alpha x} \sin \beta x$ $\Rightarrow$ $y' = e^{\alpha x} \cdot \beta \cos \beta x + \sin \beta x \cdot \alpha e^{\alpha x} = e^{\alpha x}(\beta \cos \beta x + \alpha \sin \beta x)$ $\Rightarrow$

$$y'' = e^{\alpha x}(-\beta^2 \sin \beta x + \alpha\beta \cos \beta x) + (\beta \cos \beta x + \alpha \sin \beta x) \cdot \alpha e^{\alpha x}$$
$$= e^{\alpha x}(-\beta^2 \sin \beta x + \alpha\beta \cos \beta x + \alpha\beta \cos \beta x + \alpha^2 \sin \beta x) = e^{\alpha x}(\alpha^2 \sin \beta x - \beta^2 \sin \beta x + 2\alpha\beta \cos \beta x)$$
$$= e^{\alpha x}\left[(\alpha^2 - \beta^2) \sin \beta x + 2\alpha\beta \cos \beta x\right]$$

39. $y = x \ln x$ $\Rightarrow$ $y' = x(1/x) + (\ln x) \cdot 1 = 1 + \ln x$ $\Rightarrow$ $y'' = 1/x$

41. $y = f(x) = \ln \ln x$ $\Rightarrow$ $f'(x) = \dfrac{1}{\ln x}\left(\dfrac{1}{x}\right)$ $\Rightarrow$ $f'(e) = \dfrac{1}{e}$, so an equation of the tangent line at $(e, 0)$ is

$$y - 0 = \frac{1}{e}(x - e), \text{ or } y = \frac{1}{e}x - 1, \text{ or } x - ey = e.$$

43. $f(x) = \dfrac{x}{1 - \ln(x-1)}$ $\Rightarrow$

$$f'(x) = \frac{[1 - \ln(x-1)] \cdot 1 - x \cdot \dfrac{-1}{x-1}}{[1 - \ln(x-1)]^2} = \frac{\dfrac{(x-1)[1 - \ln(x-1)] + x}{x-1}}{[1 - \ln(x-1)]^2} = \frac{x - 1 - (x-1)\ln(x-1) + x}{(x-1)[1 - \ln(x-1)]^2}$$

$$= \frac{2x - 1 - (x-1)\ln(x-1)}{(x-1)[1 - \ln(x-1)]^2}$$

$$\text{Dom}(f) = \{x \mid x - 1 > 0 \quad \text{and} \quad 1 - \ln(x-1) \neq 0\} = \{x \mid x > 1 \quad \text{and} \quad \ln(x-1) \neq 1\}$$
$$= \{x \mid x > 1 \quad \text{and} \quad x - 1 \neq e^1\} = \{x \mid x > 1 \quad \text{and} \quad x \neq 1 + e\} = (1, 1 + e) \cup (1 + e, \infty)$$

45. $y = (2x + 1)^5(x^4 - 3)^6$ $\Rightarrow$ $\ln y = \ln\left((2x+1)^5(x^4-3)^6\right)$ $\Rightarrow$ $\ln y = 5\ln(2x+1) + 6\ln(x^4 - 3)$ $\Rightarrow$

$$\frac{1}{y}y' = 5 \cdot \frac{1}{2x+1} \cdot 2 + 6 \cdot \frac{1}{x^4 - 3} \cdot 4x^3 \Rightarrow$$

$$y' = y\left(\frac{10}{2x+1} + \frac{24x^3}{x^4 - 3}\right) = (2x+1)^5(x^4-3)^6\left(\frac{10}{2x+1} + \frac{24x^3}{x^4 - 3}\right).$$

[The answer could be simplified to $y' = 2(2x+1)^4(x^4 - 3)^5(29x^4 + 12x^3 - 15)$, but this is unnecessary.]

47. $y = \dfrac{\sin^2 x \,\tan^4 x}{(x^2+1)^2}$ $\Rightarrow$ $\ln y = \ln(\sin^2 x \,\tan^4 x) - \ln(x^2+1)^2$ $\Rightarrow$

$\ln y = \ln(\sin x)^2 + \ln(\tan x)^4 - \ln(x^2+1)^2$ $\Rightarrow$ $\ln y = 2\ln|\sin x| + 4\ln|\tan x| - 2\ln(x^2+1)$ $\Rightarrow$

$\dfrac{1}{y}\,y' = 2\cdot\dfrac{1}{\sin x}\cdot\cos x + 4\cdot\dfrac{1}{\tan x}\cdot\sec^2 x - 2\cdot\dfrac{1}{x^2+1}\cdot 2x$ $\Rightarrow$ $y' = \dfrac{\sin^2 x \,\tan^4 x}{(x^2+1)^2}\left(2\cot x + \dfrac{4\sec^2 x}{\tan x} - \dfrac{4x}{x^2+1}\right)$

49. $y = x^x$ $\Rightarrow$ $\ln y = \ln x^x$ $\Rightarrow$ $\ln y = x\ln x$ $\Rightarrow$ $y'/y = x(1/x) + (\ln x)\cdot 1$ $\Rightarrow$ $y' = y(1+\ln x)$ $\Rightarrow$
$y' = x^x(1+\ln x)$

51. $y = (\cos x)^x$ $\Rightarrow$ $\ln y = \ln(\cos x)^x$ $\Rightarrow$ $\ln y = x\ln\cos x$ $\Rightarrow$ $\dfrac{1}{y}\,y' = x\cdot\dfrac{1}{\cos x}\cdot(-\sin x) + \ln\cos x\cdot 1$ $\Rightarrow$

$y' = y\left(\ln\cos x - \dfrac{x\sin x}{\cos x}\right)$ $\Rightarrow$ $y' = (\cos x)^x(\ln\cos x - x\tan x)$

53. $y = (\tan x)^{1/x}$ $\Rightarrow$ $\ln y = \ln(\tan x)^{1/x}$ $\Rightarrow$ $\ln y = \dfrac{1}{x}\ln\tan x$ $\Rightarrow$

$\dfrac{1}{y}\,y' = \dfrac{1}{x}\cdot\dfrac{1}{\tan x}\cdot\sec^2 x + \ln\tan x\cdot\left(-\dfrac{1}{x^2}\right)$ $\Rightarrow$ $y' = y\left(\dfrac{\sec^2 x}{x\tan x} - \dfrac{\ln\tan x}{x^2}\right)$ $\Rightarrow$

$y' = (\tan x)^{1/x}\left(\dfrac{\sec^2 x}{x\tan x} - \dfrac{\ln\tan x}{x^2}\right)$ or $y' = (\tan x)^{1/x}\cdot\dfrac{1}{x}\left(\csc x\sec x - \dfrac{\ln\tan x}{x}\right)$

55. $\dfrac{d}{dx}\left(e^{x^2 y}\right) = \dfrac{d}{dx}(x+y)$ $\Rightarrow$ $e^{x^2 y}(x^2 y' + y\cdot 2x) = 1 + y'$ $\Rightarrow$ $x^2 e^{x^2 y} y' + 2xy e^{x^2 y} = 1 + y'$ $\Rightarrow$

$x^2 e^{x^2 y} y' - y' = 1 - 2xy e^{x^2 y}$ $\Rightarrow$ $y'(x^2 e^{x^2 y} - 1) = 1 - 2xy e^{x^2 y}$ $\Rightarrow$ $y' = \dfrac{1 - 2xy e^{x^2 y}}{x^2 e^{x^2 y} - 1}$

57. $y = \ln(x^2 + y^2)$ $\Rightarrow$ $y' = \dfrac{1}{x^2+y^2}\dfrac{d}{dx}(x^2+y^2)$ $\Rightarrow$ $y' = \dfrac{2x+2yy'}{x^2+y^2}$ $\Rightarrow$ $x^2 y' + y^2 y' = 2x + 2yy'$ $\Rightarrow$

$x^2 y' + y^2 y' - 2yy' = 2x$ $\Rightarrow$ $(x^2 + y^2 - 2y)y' = 2x$ $\Rightarrow$ $y' = \dfrac{2x}{x^2 + y^2 - 2y}$

59. $s(t) = 2e^{-1.5t}\sin 2\pi t$ $\Rightarrow$

$v(t) = s'(t) = 2[e^{-1.5t}(\cos 2\pi t)(2\pi) + (\sin 2\pi t)e^{-1.5t}(-1.5)] = 2e^{-1.5t}(2\pi\cos 2\pi t - 1.5\sin 2\pi t)$

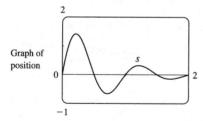

Graph of position

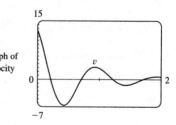

Graph of velocity

61. $y = Ae^{-x} + Bxe^{-x}$ $\Rightarrow$ $y' = A(-e^{-x}) + B[x(-e^{-x}) + e^{-x}\cdot 1]$
$= -Ae^{-x} + Be^{-x} - Bxe^{-x} = (B - A)e^{-x} - Bxe^{-x}$

$\Rightarrow$ $y'' = (B - A)(-e^{-x}) - B[x(-e^{-x}) + e^{-x}\cdot 1]$
$= (A - B)e^{-x} - Be^{-x} + Bxe^{-x} = (A - 2B)e^{-x} + Bxe^{-x}$,

so $y'' + 2y' + y = (A - 2B)e^{-x} + Bxe^{-x} + 2[(B - A)e^{-x} - Bxe^{-x}] + Ae^{-x} + Bxe^{-x}$
$= [(A - 2B) + 2(B - A) + A]e^{-x} + [B - 2B + B]xe^{-x} = 0.$

63. $f(x) = e^{2x}$ $\Rightarrow$ $f'(x) = 2e^{2x}$ $\Rightarrow$ $f''(x) = 2 \cdot 2e^{2x} = 2^2 e^{2x}$ $\Rightarrow$

$f'''(x) = 2^2 \cdot 2e^{2x} = 2^3 e^{2x}$ $\Rightarrow$ $\cdots$ $\Rightarrow$ $f^{(n)}(x) = 2^n e^{2x}$

65. $f(x) = \ln(x-1)$ $\Rightarrow$ $f'(x) = 1/(x-1) = (x-1)^{-1}$ $\Rightarrow$ $f''(x) = -(x-1)^{-2}$ $\Rightarrow$ $f'''(x) = 2(x-1)^{-3}$ $\Rightarrow$

$f^{(4)}(x) = -2 \cdot 3(x-1)^{-4}$ $\Rightarrow$ $\cdots$ $\Rightarrow$ $f^{(n)}(x) = (-1)^{n-1} \cdot 2 \cdot 3 \cdot 4 \cdot \cdots \cdot (n-1)(x-1)^{-n} = (-1)^{n-1} \dfrac{(n-1)!}{(x-1)^n}$

67. We use Theorem 3.2.7. Note that $f(0) = 3 + 0 + e^0 = 4$, so $f^{-1}(4) = 0$. Also $f'(x) = 1 + e^x$. Therefore,

$$\left(f^{-1}\right)'(4) = \frac{1}{f'(f^{-1}(4))} = \frac{1}{f'(0)} = \frac{1}{1+e^0} = \frac{1}{2}.$$

3.4 Exponential Growth and Decay

1. The relative growth rate is $\dfrac{1}{P}\dfrac{dP}{dt} = 0.7944$, so $\dfrac{dP}{dt} = 0.7944P$ and, by Theorem 2, $P(t) = P(0)e^{0.7944t} = 2e^{0.7944t}$.

Thus, $P(6) = 2e^{0.7944(6)} \approx 234.99$ or about 235 members.

3. (a) By Theorem 2, $P(t) = P(0)e^{kt} = 100e^{kt}$. Now $P(1) = 100e^{k(1)} = 420$ $\Rightarrow$ $e^k = \frac{420}{100}$ $\Rightarrow$ $k = \ln 4.2$.

So $P(t) = 100e^{(\ln 4.2)t} = 100(4.2)^t$.

(b) $P(3) = 100(4.2)^3 = 7408.8 \approx 7409$ bacteria

(c) $dP/dt = kP$ $\Rightarrow$ $P'(3) = k \cdot P(3) = (\ln 4.2)\left(100(4.2)^3\right)$ [from part (a)] $\approx 10{,}632$ bacteria/hour

(d) $P(t) = 100(4.2)^t = 10{,}000$ $\Rightarrow$ $(4.2)^t = 100$ $\Rightarrow$ $t = (\ln 100)/(\ln 4.2) \approx 3.2$ hours

5. (a) Let the population (in millions) in the year t be $P(t)$. Since the initial time is the year 1750, we substitute $t - 1750$ for t in

Theorem 2, so the exponential model gives $P(t) = P(1750)e^{k(t-1750)}$. Then $P(1800) = 980 = 790e^{k(1800-1750)}$ $\Rightarrow$

$\frac{980}{790} = e^{k(50)}$ $\Rightarrow$ $\ln \frac{980}{790} = 50k$ $\Rightarrow$ $k = \frac{1}{50} \ln \frac{980}{790} \approx 0.0043104$. So with this model, we have

$P(1900) = 790e^{k(1900-1750)} \approx 1508$ million and $P(1950) = 790e^{k(1950-1750)} \approx 1871$ million. Both of these estimates

are much too low.

(b) In this case, the exponential model gives $P(t) = P(1850)e^{k(t-1850)}$ $\Rightarrow$ $P(1900) = 1650 = 1260e^{k(1900-1850)}$ $\Rightarrow$

$\ln \frac{1650}{1260} = k(50)$ $\Rightarrow$ $k = \frac{1}{50} \ln \frac{1650}{1260} \approx 0.005393$. So with this model, we estimate

$P(1950) = 1260e^{k(1950-1850)} \approx 2161$ million. This is still too low, but closer than the estimate of $P(1950)$ in part (a).

(c) The exponential model gives $P(t) = P(1900)e^{k(t-1900)}$ $\Rightarrow$ $P(1950) = 2560 = 1650e^{k(1950-1900)}$ $\Rightarrow$

$\ln \frac{2560}{1650} = k(50)$ $\Rightarrow$ $k = \frac{1}{50} \ln \frac{2560}{1650} \approx 0.008785$. With this model, we estimate

$P(2000) = 1650e^{k(2000-1900)} \approx 3972$ million. This is much too low. The discrepancy is explained by the wars in the

first part of the 20th century and the lower mortality rate in the latter part of the century due to advances in medical science.

The exponential model assumes, among other things, that the birth and mortality rates will remain constant.

7. (a) If $y = [\text{N}_2\text{O}_5]$ then by Theorem 2, $\dfrac{dy}{dt} = -0.0005y$ $\Rightarrow$ $y(t) = y(0)e^{-0.0005t} = Ce^{-0.0005t}$.

(b) $y(t) = Ce^{-0.0005t} = 0.9C$ $\Rightarrow$ $e^{-0.0005t} = 0.9$ $\Rightarrow$ $-0.0005t = \ln 0.9$ $\Rightarrow$ $t = -2000 \ln 0.9 \approx 211$ s

9. (a) If $y(t)$ is the mass (in mg) remaining after t years, then $y(t) = y(0)e^{kt} = 100e^{kt}$. $y(30) = 100e^{30k} = \frac{1}{2}(100)$ $\Rightarrow$

$e^{30k} = \frac{1}{2}$ $\Rightarrow$ $k = -(\ln 2)/30$ $\Rightarrow$ $y(t) = 100e^{-(\ln 2)t/30} = 100 \cdot 2^{-t/30}$

(b) $y(100) = 100 \cdot 2^{-100/30} \approx 9.92$ mg

(c) $100e^{-(\ln 2)t/30} = 1$ $\Rightarrow$ $-(\ln 2)t/30 = \ln\frac{1}{100}$ $\Rightarrow$ $t = -30\,\frac{\ln 0.01}{\ln 2} \approx 199.3$ years

11. Let $y(t)$ be the level of radioactivity. Thus, $y(t) = y(0)e^{-kt}$ and k is determined by using the half-life:

$y(5730) = \frac{1}{2}y(0)$ $\Rightarrow$ $y(0)e^{-k(5730)} = \frac{1}{2}y(0)$ $\Rightarrow$ $e^{-5730k} = \frac{1}{2}$ $\Rightarrow$ $-5730k = \ln\frac{1}{2}$ $\Rightarrow$

$k = -\dfrac{\ln\frac{1}{2}}{5730} = \dfrac{\ln 2}{5730}$. If 74% of the ^{14}C remains, then we know that $y(t) = 0.74y(0)$ $\Rightarrow$ $0.74 = e^{-t(\ln 2)/5730}$ $\Rightarrow$

$\ln 0.74 = -\dfrac{t\ln 2}{5730}$ $\Rightarrow$ $t = -\dfrac{5730(\ln 0.74)}{\ln 2} \approx 2489 \approx 2500$ years.

13. (a) Using Newton's Law of Cooling, $\dfrac{dT}{dt} = k(T - T_s)$, we have $\dfrac{dT}{dt} = k(T - 75)$. Now let $y = T - 75$, so

$y(0) = T(0) - 75 = 185 - 75 = 110$, so y is a solution of the initial-value problem $dy/dt = ky$ with $y(0) = 110$ and by

Theorem 2 we have $y(t) = y(0)e^{kt} = 110e^{kt}$.

$y(30) = 110e^{30k} = 150 - 75$ $\Rightarrow$ $e^{30k} = \frac{75}{110} = \frac{15}{22}$ $\Rightarrow$ $k = \frac{1}{30}\ln\frac{15}{22}$, so $y(t) = 110e^{\frac{1}{30}t\ln\left(\frac{15}{22}\right)}$ and

$y(45) = 110e^{\frac{45}{30}\ln\left(\frac{15}{22}\right)} \approx 62\,°F$. Thus, $T(45) \approx 62 + 75 = 137\,°F$.

(b) $T(t) = 100$ $\Rightarrow$ $y(t) = 25$. $y(t) = 110e^{\frac{1}{30}t\ln\left(\frac{15}{22}\right)} = 25$ $\Rightarrow$ $e^{\frac{1}{30}t\ln\left(\frac{15}{22}\right)} = \frac{25}{110}$ $\Rightarrow$ $\frac{1}{30}t\ln\frac{15}{22} = \ln\frac{25}{110}$ $\Rightarrow$

$t = \dfrac{30\ln\frac{25}{110}}{\ln\frac{15}{22}} \approx 116$ min.

15. $\dfrac{dT}{dt} = k(T - 20)$. Letting $y = T - 20$, we get $\dfrac{dy}{dt} = ky$, so $y(t) = y(0)e^{kt}$. $y(0) = T(0) - 20 = 5 - 20 = -15$, so

$y(25) = y(0)e^{25k} = -15e^{25k}$, and $y(25) = T(25) - 20 = 10 - 20 = -10$, so $-15e^{25k} = -10$ $\Rightarrow$ $e^{25k} = \frac{2}{3}$. Thus,

$25k = \ln\left(\frac{2}{3}\right)$ and $k = \frac{1}{25}\ln\left(\frac{2}{3}\right)$, so $y(t) = y(0)e^{kt} = -15e^{(1/25)\ln(2/3)t}$. More simply, $e^{25k} = \frac{2}{3}$ $\Rightarrow$ $e^k = \left(\frac{2}{3}\right)^{1/25}$ $\Rightarrow$

$e^{kt} = \left(\frac{2}{3}\right)^{t/25}$ $\Rightarrow$ $y(t) = -15 \cdot \left(\frac{2}{3}\right)^{t/25}$.

(a) $T(50) = 20 + y(50) = 20 - 15 \cdot \left(\frac{2}{3}\right)^{50/25} = 20 - 15 \cdot \left(\frac{2}{3}\right)^2 = 20 - \frac{20}{3} = 13.\overline{3}\,°C$

(b) $15 = T(t) = 20 + y(t) = 20 - 15 \cdot \left(\frac{2}{3}\right)^{t/25}$ $\Rightarrow$ $15 \cdot \left(\frac{2}{3}\right)^{t/25} = 5$ $\Rightarrow$ $\left(\frac{2}{3}\right)^{t/25} = \frac{1}{3}$ $\Rightarrow$

$(t/25)\ln\left(\frac{2}{3}\right) = \ln\left(\frac{1}{3}\right)$ $\Rightarrow$ $t = 25\ln\left(\frac{1}{3}\right)/\ln\left(\frac{2}{3}\right) \approx 67.74$ min.

17. (a) Let $P(h)$ be the pressure at altitude h. Then $dP/dh = kP$ $\Rightarrow$ $P(h) = P(0)e^{kh} = 101.3e^{kh}$.

$P(1000) = 101.3e^{1000k} = 87.14$ $\Rightarrow$ $1000k = \ln\left(\frac{87.14}{101.3}\right)$ $\Rightarrow$ $k = \frac{1}{1000}\ln\left(\frac{87.14}{101.3}\right)$ $\Rightarrow$

$P(h) = 101.3\,e^{\frac{1}{1000}h\ln\left(\frac{87.14}{101.3}\right)}$, so $P(3000) = 101.3e^{3\ln\left(\frac{87.14}{101.3}\right)} \approx 64.5$ kPa.

(b) $P(6187) = 101.3\,e^{\frac{6187}{1000}\ln\left(\frac{87.14}{101.3}\right)} \approx 39.9$ kPa

19. Using $A = A_0\left(1 + \dfrac{r}{n}\right)^{nt}$ with $A_0 = 3000$, $r = 0.05$, and $t = 5$, we have:

(a) Annually: $n = 1$; $A = 3000\left(1 + \dfrac{0.05}{1}\right)^{1\cdot5} = \$3828.84$

(b) Semiannually: $n = 2$; $A = 3000\left(1 + \dfrac{0.05}{2}\right)^{2\cdot5} = \$3840.25$

(c) Monthly: $n = 12$; $A = 3000\left(1 + \dfrac{0.05}{12}\right)^{12\cdot5} = \$3850.08$

(d) Weekly: $n = 52$; $A = 3000\left(1 + \dfrac{0.05}{52}\right)^{52\cdot5} = \$3851.61$

(e) Daily: $n = 365$; $A = 3000\left(1 + \dfrac{0.05}{365}\right)^{365\cdot5} = \$3852.01$

(f) Continuously: $A = 3000e^{(0.05)5} = \$3852.08$

3.5   Inverse Trigonometric Functions

1. (a) $\sin^{-1}\left(\dfrac{\sqrt{3}}{2}\right) = \dfrac{\pi}{3}$ since $\sin\dfrac{\pi}{3} = \dfrac{\sqrt{3}}{2}$ and $\dfrac{\pi}{3}$ is in $\left[-\dfrac{\pi}{2}, \dfrac{\pi}{2}\right]$.

(b) $\cos^{-1}(-1) = \pi$ since $\cos\pi = -1$ and π is in $[0, \pi]$.

3. (a) $\tan^{-1}\sqrt{3} = \dfrac{\pi}{3}$ since $\tan\dfrac{\pi}{3} = \sqrt{3}$ and $\dfrac{\pi}{3}$ is in $\left(-\dfrac{\pi}{2}, \dfrac{\pi}{2}\right)$.

(b) $\arcsin\left(-\dfrac{1}{\sqrt{2}}\right) = -\dfrac{\pi}{4}$ since $\sin\left(-\dfrac{\pi}{4}\right) = -\dfrac{1}{\sqrt{2}}$ and $-\dfrac{\pi}{4}$ is in $\left[-\dfrac{\pi}{2}, \dfrac{\pi}{2}\right]$.

5. (a) $\sin(\sin^{-1}(0.7)) = 0.7$ since 0.7 is in $[-1, 1]$.

(b) $\tan^{-1}\left(\tan\dfrac{4\pi}{3}\right) = \tan^{-1}\sqrt{3} = \dfrac{\pi}{3}$ since $\dfrac{\pi}{3}$ is in $\left[-\dfrac{\pi}{2}, \dfrac{\pi}{2}\right]$.

7. Let $y = \sin^{-1}x$. Then $-\dfrac{\pi}{2} \le y \le \dfrac{\pi}{2}$ $\Rightarrow$ $\cos y \ge 0$, so $\cos(\sin^{-1}x) = \cos y = \sqrt{1 - \sin^2 y} = \sqrt{1 - x^2}$.

9. Let $y = \tan^{-1}x$. Then $\tan y = x$, so from the triangle we see that $\sin(\tan^{-1}x) = \sin y = \dfrac{x}{\sqrt{1 + x^2}}$.

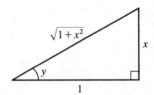

11. Let $y = \cos^{-1}x$. Then $\cos y = x$ and $0 \le y \le \pi$ $\Rightarrow$ $-\sin y\,\dfrac{dy}{dx} = 1$ $\Rightarrow$

$\dfrac{dy}{dx} = -\dfrac{1}{\sin y} = -\dfrac{1}{\sqrt{1 - \cos^2 y}} = -\dfrac{1}{\sqrt{1 - x^2}}$. [Note that $\sin y \ge 0$ for $0 \le y \le \pi$.]

13. Let $y = \cot^{-1}x$. Then $\cot y = x$ $\Rightarrow$ $-\csc^2 y\,\dfrac{dy}{dx} = 1$ $\Rightarrow$ $\dfrac{dy}{dx} = -\dfrac{1}{\csc^2 y} = -\dfrac{1}{1 + \cot^2 y} = -\dfrac{1}{1 + x^2}$.

15. Let $y = \csc^{-1}x$. Then $\csc y = x$ $\Rightarrow$ $-\csc y \cot y\,\dfrac{dy}{dx} = 1$ $\Rightarrow$

$\dfrac{dy}{dx} = -\dfrac{1}{\csc y \cot y} = -\dfrac{1}{\csc y \sqrt{\csc^2 y - 1}} = -\dfrac{1}{x\sqrt{x^2 - 1}}$. Note that $\cot y \ge 0$ on the domain of $\csc^{-1}x$.

17. $y = \tan^{-1}\sqrt{x} \;\Rightarrow\; y' = \dfrac{1}{1+(\sqrt{x})^2}\cdot\dfrac{d}{dx}(\sqrt{x}) = \dfrac{1}{1+x}\left(\tfrac{1}{2}x^{-1/2}\right) = \dfrac{1}{2\sqrt{x}\,(1+x)}$

19. $y = \sin^{-1}(2x+1) \;\Rightarrow$

$$y' = \frac{1}{\sqrt{1-(2x+1)^2}}\cdot\frac{d}{dx}(2x+1) = \frac{1}{\sqrt{1-(4x^2+4x+1)}}\cdot 2 = \frac{2}{\sqrt{-4x^2-4x}} = \frac{1}{\sqrt{-x^2-x}}$$

21. $H(x) = (1+x^2)\arctan x \;\Rightarrow\; H'(x) = (1+x^2)\dfrac{1}{1+x^2} + (\arctan x)(2x) = 1 + 2x\arctan x$

23. $y = \cos^{-1}(e^{2x}) \;\Rightarrow\; y' = -\dfrac{1}{\sqrt{1-(e^{2x})^2}}\cdot\dfrac{d}{dx}(e^{2x}) = -\dfrac{2e^{2x}}{\sqrt{1-e^{4x}}}$

25. $y = \arctan(\cos\theta) \;\Rightarrow\; y' = \dfrac{1}{1+(\cos\theta)^2}(-\sin\theta) = -\dfrac{\sin\theta}{1+\cos^2\theta}$

27. $h(t) = \cot^{-1}(t) + \cot^{-1}(1/t) \;\Rightarrow$

$$h'(t) = -\frac{1}{1+t^2} - \frac{1}{1+(1/t)^2}\cdot\frac{d}{dt}\frac{1}{t} = -\frac{1}{1+t^2} - \frac{t^2}{t^2+1}\cdot\left(-\frac{1}{t^2}\right) = -\frac{1}{1+t^2} + \frac{1}{t^2+1} = 0.$$

Note that this makes sense because $h(t) = \frac{\pi}{2}$ for $t > 0$ and $h(t) = -\frac{\pi}{2}$ for $t < 0$.

29. $y = \arccos\left(\dfrac{b+a\cos x}{a+b\cos x}\right) \;\Rightarrow$

$$y' = -\frac{1}{\sqrt{1-\left(\dfrac{b+a\cos x}{a+b\cos x}\right)^2}}\,\frac{(a+b\cos x)(-a\sin x) - (b+a\cos x)(-b\sin x)}{(a+b\cos x)^2}$$

$$= \frac{1}{\sqrt{a^2+b^2\cos^2 x - b^2 - a^2\cos^2 x}}\,\frac{(a^2-b^2)\sin x}{|a+b\cos x|}$$

$$= \frac{1}{\sqrt{a^2-b^2}\sqrt{1-\cos^2 x}}\,\frac{(a^2-b^2)\sin x}{|a+b\cos x|} = \frac{\sqrt{a^2-b^2}}{|a+b\cos x|}\,\frac{\sin x}{|\sin x|}$$

But $0 \le x \le \pi$, so $|\sin x| = \sin x$. Also $a > b > 0 \;\Rightarrow\; b\cos x \ge -b > -a$, so $a+b\cos x > 0$. Thus $y' = \dfrac{\sqrt{a^2-b^2}}{a+b\cos x}$.

31. $g(x) = \cos^{-1}(3-2x) \;\Rightarrow\; g'(x) = -\dfrac{1}{\sqrt{1-(3-2x)^2}}(-2) = \dfrac{2}{\sqrt{1-(3-2x)^2}}$.

Domain$(g) = \{x \mid -1 \le 3-2x \le 1\} = \{x \mid -4 \le -2x \le -2\} = \{x \mid 2 \ge x \ge 1\} = [1,2]$.

Domain$(g') = \{x \mid 1-(3-2x)^2 > 0\} = \{x \mid (3-2x)^2 < 1\} = \{x \mid |3-2x| < 1\}$
$\qquad = \{x \mid -1 < 3-2x < 1\} = \{x \mid -4 < -2x < -2\} = \{x \mid 2 > x > 1\} = (1,2)$

33. $g(x) = x\sin^{-1}\left(\dfrac{x}{4}\right) + \sqrt{16-x^2} \;\Rightarrow\; g'(x) = \sin^{-1}\left(\dfrac{x}{4}\right) + \dfrac{x}{4\sqrt{1-(x/4)^2}} - \dfrac{x}{\sqrt{16-x^2}} = \sin^{-1}\left(\dfrac{x}{4}\right) \;\Rightarrow$

$g'(2) = \sin^{-1}\left(\tfrac{1}{2}\right) = \tfrac{\pi}{6}$

35. $\displaystyle\lim_{x\to-1^+}\sin^{-1}x = \sin^{-1}(-1) = -\tfrac{\pi}{2}$

37. Let $t = e^x$. As $x \to \infty$, $t \to \infty$. $\displaystyle\lim_{x \to \infty} \arctan(e^x) = \lim_{t \to \infty} \arctan t = \frac{\pi}{2}$ by (8).

39.

$$\frac{dx}{dt} = 2 \text{ ft/s}, \ \sin\theta = \frac{x}{10} \ \Rightarrow \ \theta = \sin^{-1}\left(\frac{x}{10}\right), \ \frac{d\theta}{dx} = \frac{1/10}{\sqrt{1 - (x/10)^2}},$$

$$\frac{d\theta}{dt} = \frac{d\theta}{dx}\frac{dx}{dt} = \frac{1/10}{\sqrt{1 - (x/10)^2}}(2) \text{ rad/s}, \ \frac{d\theta}{dt}\bigg]_{x=6} = \frac{2/10}{\sqrt{1 - (6/10)^2}} \text{ rad/s} = \frac{1}{4} \text{ rad/s}$$

41. $y = \sec^{-1}x \ \Rightarrow \ \sec y = x \ \Rightarrow \ \sec y \tan y \dfrac{dy}{dx} = 1 \ \Rightarrow \ \dfrac{dy}{dx} = \dfrac{1}{\sec y \tan y}$. Now $\tan^2 y = \sec^2 y - 1 = x^2 - 1$, so

$\tan y = \pm\sqrt{x^2 - 1}$. For $y \in \left[0, \frac{\pi}{2}\right)$, $x \geq 1$, so $\sec y = x = |x|$ and $\tan y \geq 0 \ \Rightarrow \ \dfrac{dy}{dx} = \dfrac{1}{x\sqrt{x^2 - 1}} = \dfrac{1}{|x|\sqrt{x^2 - 1}}$.

For $y \in \left(\frac{\pi}{2}, \pi\right]$, $x \leq -1$, so $|x| = -x$ and $\tan y = -\sqrt{x^2 - 1} \ \Rightarrow$

$$\frac{dy}{dx} = \frac{1}{\sec y \tan y} = \frac{1}{x\left(-\sqrt{x^2 - 1}\right)} = \frac{1}{(-x)\sqrt{x^2 - 1}} = \frac{1}{|x|\sqrt{x^2 - 1}}$$

3.6　Hyperbolic Functions

1. (a) $\sinh 0 = \frac{1}{2}\left(e^0 - e^0\right) = 0$　　　　　　(b) $\cosh 0 = \frac{1}{2}\left(e^0 + e^0\right) = \frac{1}{2}(1 + 1) = 1$

3. (a) $\sinh(\ln 2) = \dfrac{e^{\ln 2} - e^{-\ln 2}}{2} = \dfrac{e^{\ln 2} - \left(e^{\ln 2}\right)^{-1}}{2} = \dfrac{2 - 2^{-1}}{2} = \dfrac{2 - \frac{1}{2}}{2} = \dfrac{3}{4}$

(b) $\sinh 2 = \frac{1}{2}\left(e^2 - e^{-2}\right) \approx 3.62686$

5. (a) $\operatorname{sech} 0 = \dfrac{1}{\cosh 0} = \dfrac{1}{1} = 1$　　　　　　(b) $\cosh^{-1} 1 = 0$ because $\cosh 0 = 1$.

7. $\sinh(-x) = \frac{1}{2}\left[e^{-x} - e^{-(-x)}\right] = \frac{1}{2}\left(e^{-x} - e^x\right) = -\frac{1}{2}\left(e^x - e^{-x}\right) = -\sinh x$

9. $\cosh x + \sinh x = \frac{1}{2}\left(e^x + e^{-x}\right) + \frac{1}{2}\left(e^x - e^{-x}\right) = \frac{1}{2}(2e^x) = e^x$

11. $\sinh x \cosh y + \cosh x \sinh y = \left[\frac{1}{2}(e^x - e^{-x})\right]\left[\frac{1}{2}(e^y + e^{-y})\right] + \left[\frac{1}{2}(e^x + e^{-x})\right]\left[\frac{1}{2}(e^y - e^{-y})\right]$

$$= \frac{1}{4}[(e^{x+y} + e^{x-y} - e^{-x+y} - e^{-x-y}) + (e^{x+y} - e^{x-y} + e^{-x+y} - e^{-x-y})]$$

$$= \frac{1}{4}(2e^{x+y} - 2e^{-x-y}) = \frac{1}{2}[e^{x+y} - e^{-(x+y)}] = \sinh(x + y)$$

13. Putting $y = x$ in the result from Exercise 11, we have

$\sinh 2x = \sinh(x + x) = \sinh x \cosh x + \cosh x \sinh x = 2 \sinh x \cosh x$.

15. By Exercise 9, $(\cosh x + \sinh x)^n = (e^x)^n = e^{nx} = \cosh nx + \sinh nx$.

17. $\tanh x = \frac{4}{5} > 0$, so $x > 0$. $\coth x = 1/\tanh x = \frac{5}{4}$, $\operatorname{sech}^2 x = 1 - \tanh^2 x = 1 - \left(\frac{4}{5}\right)^2 = \frac{9}{25} \ \Rightarrow \ \operatorname{sech} x = \frac{3}{5}$ (since

$\operatorname{sech} x > 0$), $\cosh x = 1/\operatorname{sech} x = \frac{5}{3}$, $\sinh x = \tanh x \cosh x = \frac{4}{5} \cdot \frac{5}{3} = \frac{4}{3}$, and $\operatorname{csch} x = 1/\sinh x = \frac{3}{4}$.

19. (a) $\displaystyle\lim_{x \to \infty} \tanh x = \lim_{x \to \infty} \frac{e^x - e^{-x}}{e^x + e^{-x}} \cdot \frac{e^{-x}}{e^{-x}} = \lim_{x \to \infty} \frac{1 - e^{-2x}}{1 + e^{-2x}} = \frac{1 - 0}{1 + 0} = 1$

(b) $\displaystyle\lim_{x\to-\infty} \tanh x = \lim_{x\to-\infty}\frac{e^x - e^{-x}}{e^x + e^{-x}}\cdot\frac{e^x}{e^x} = \lim_{x\to-\infty}\frac{e^{2x} - 1}{e^{2x} + 1} = \frac{0-1}{0+1} = -1$

(c) $\displaystyle\lim_{x\to\infty}\sinh x = \lim_{x\to\infty}\frac{e^x - e^{-x}}{2} = \infty$

(d) $\displaystyle\lim_{x\to-\infty}\sinh x = \lim_{x\to-\infty}\frac{e^x - e^{-x}}{2} = -\infty$

(e) $\displaystyle\lim_{x\to\infty}\operatorname{sech} x = \lim_{x\to\infty}\frac{2}{e^x + e^{-x}} = 0$

(f) $\displaystyle\lim_{x\to\infty}\coth x = \lim_{x\to\infty}\frac{e^x + e^{-x}}{e^x - e^{-x}}\cdot\frac{e^{-x}}{e^{-x}} = \lim_{x\to\infty}\frac{1 + e^{-2x}}{1 - e^{-2x}} = \frac{1+0}{1-0} = 1$ [*Or:* Use part (a)]

(g) $\displaystyle\lim_{x\to 0^+}\coth x = \lim_{x\to 0^+}\frac{\cosh x}{\sinh x} = \infty$, since $\sinh x \to 0$ through positive values and $\cosh x \to 1$.

(h) $\displaystyle\lim_{x\to 0^-}\coth x = \lim_{x\to 0^-}\frac{\cosh x}{\sinh x} = -\infty$, since $\sinh x \to 0$ through negative values and $\cosh x \to 1$.

(i) $\displaystyle\lim_{x\to-\infty}\operatorname{csch} x = \lim_{x\to-\infty}\frac{2}{e^x - e^{-x}} = 0$

21. Let $y = \sinh^{-1} x$. Then $\sinh y = x$ and, by Example 1(a), $\cosh^2 y - \sinh^2 y = 1$ $\Rightarrow$ [with $\cosh y > 0$]

$\cosh y = \sqrt{1 + \sinh^2 y} = \sqrt{1 + x^2}$. So by Exercise 9, $e^y = \sinh y + \cosh y = x + \sqrt{1 + x^2}$ $\Rightarrow$ $y = \ln\!\left(x + \sqrt{1 + x^2}\,\right)$.

23. (a) Let $y = \tanh^{-1} x$. Then $x = \tanh y = \dfrac{\sinh y}{\cosh y} = \dfrac{(e^y - e^{-y})/2}{(e^y + e^{-y})/2}\cdot\dfrac{e^y}{e^y} = \dfrac{e^{2y} - 1}{e^{2y} + 1}$ $\Rightarrow$

$xe^{2y} + x = e^{2y} - 1$ $\Rightarrow$ $1 + x = e^{2y} - xe^{2y}$ $\Rightarrow$ $1 + x = e^{2y}(1 - x)$ $\Rightarrow$

$e^{2y} = \dfrac{1+x}{1-x}$ $\Rightarrow$ $2y = \ln\!\left(\dfrac{1+x}{1-x}\right)$ $\Rightarrow$ $y = \frac{1}{2}\ln\!\left(\dfrac{1+x}{1-x}\right)$.

(b) Let $y = \tanh^{-1} x$. Then $x = \tanh y$, so from Exercise 14 we have

$e^{2y} = \dfrac{1 + \tanh y}{1 - \tanh y} = \dfrac{1+x}{1-x}$ $\Rightarrow$ $2y = \ln\!\left(\dfrac{1+x}{1-x}\right)$ $\Rightarrow$ $y = \frac{1}{2}\ln\!\left(\dfrac{1+x}{1-x}\right)$.

25. (a) Let $y = \cosh^{-1} x$. Then $\cosh y = x$ and $y \geq 0$ $\Rightarrow$ $\sinh y\,\dfrac{dy}{dx} = 1$ $\Rightarrow$

$\dfrac{dy}{dx} = \dfrac{1}{\sinh y} = \dfrac{1}{\sqrt{\cosh^2 y - 1}} = \dfrac{1}{\sqrt{x^2 - 1}}$ (since $\sinh y \geq 0$ for $y \geq 0$). *Or:* Use Formula 4.

(b) Let $y = \tanh^{-1} x$. Then $\tanh y = x$ $\Rightarrow$ $\operatorname{sech}^2 y\,\dfrac{dy}{dx} = 1$ $\Rightarrow$ $\dfrac{dy}{dx} = \dfrac{1}{\operatorname{sech}^2 y} = \dfrac{1}{1 - \tanh^2 y} = \dfrac{1}{1 - x^2}$.

Or: Use Formula 5.

(c) Let $y = \operatorname{sech}^{-1} x$. Then $\operatorname{sech} y = x$ $\Rightarrow$ $-\operatorname{sech} y \tanh y\,\dfrac{dy}{dx} = 1$ $\Rightarrow$

$\dfrac{dy}{dx} = -\dfrac{1}{\operatorname{sech} y \tanh y} = -\dfrac{1}{\operatorname{sech} y \sqrt{1 - \operatorname{sech}^2 y}} = -\dfrac{1}{x\sqrt{1 - x^2}}$. (Note that $y > 0$ and so $\tanh y > 0$.)

27. $f(x) = x\cosh x$ $\Rightarrow$ $f'(x) = x\,(\cosh x)' + (\cosh x)(x)' = x\sinh x + \cosh x$

29. $h(x) = \sinh(x^2)$ $\Rightarrow$ $h'(x) = \cosh(x^2)\cdot 2x = 2x\cosh(x^2)$

31. $h(t) = \coth\sqrt{1 + t^2}$ $\Rightarrow$ $h'(t) = -\operatorname{csch}^2\sqrt{1 + t^2}\cdot\frac{1}{2}(1 + t^2)^{-1/2}(2t) = -\dfrac{t\operatorname{csch}^2\sqrt{1 + t^2}}{\sqrt{1 + t^2}}$

33. $H(t) = \tanh(e^t)$ $\Rightarrow$ $H'(t) = \operatorname{sech}^2(e^t)\cdot e^t = e^t\operatorname{sech}^2(e^t)$

35. $y = e^{\cosh 3x}$ $\Rightarrow$ $y' = e^{\cosh 3x}\cdot\sinh 3x\cdot 3 = 3e^{\cosh 3x}\sinh 3x$

37. $y = \tanh^{-1}\sqrt{x} \;\Rightarrow\; y' = \dfrac{1}{1 - (\sqrt{x})^2} \cdot \tfrac{1}{2}x^{-1/2} = \dfrac{1}{2\sqrt{x}(1-x)}$

39. $y = x\sinh^{-1}(x/3) - \sqrt{9 + x^2} \;\Rightarrow\;$

$y' = \sinh^{-1}\left(\dfrac{x}{3}\right) + x\dfrac{1/3}{\sqrt{1 + (x/3)^2}} - \dfrac{2x}{2\sqrt{9 + x^2}} = \sinh^{-1}\left(\dfrac{x}{3}\right) + \dfrac{x}{\sqrt{9 + x^2}} - \dfrac{x}{\sqrt{9 + x^2}} = \sinh^{-1}\left(\dfrac{x}{3}\right)$

41. $y = \coth^{-1}\sqrt{x^2 + 1} \;\Rightarrow\; y' = \dfrac{1}{1 - (x^2 + 1)}\dfrac{2x}{2\sqrt{x^2 + 1}} = -\dfrac{1}{x\sqrt{x^2 + 1}}$

43. (a) $y = 20\cosh(x/20) - 15 \;\Rightarrow\; y' = 20\sinh(x/20) \cdot \tfrac{1}{20} = \sinh(x/20)$. Since the right pole is positioned at $x = 7$,

we have $y'(7) = \sinh\tfrac{7}{20} \approx 0.3572$.

(b) If α is the angle between the tangent line and the x-axis, then $\tan\alpha =$ slope of the line $= \sinh\tfrac{7}{20}$, so

$\alpha = \tan^{-1}\left(\sinh\tfrac{7}{20}\right) \approx 0.343$ rad $\approx 19.66°$. Thus, the angle between the line and the pole is $\theta = 90° - \alpha \approx 70.34°$.

45. (a) $y = A\sinh mx + B\cosh mx \;\Rightarrow\; y' = mA\cosh mx + mB\sinh mx \;\Rightarrow\;$

$y'' = m^2 A\sinh mx + m^2 B\cosh mx = m^2(A\sinh mx + B\cosh mx) = m^2 y$

(b) From part (a), a solution of $y'' = 9y$ is $y(x) = A\sinh 3x + B\cosh 3x$. So $-4 = y(0) = A\sinh 0 + B\cosh 0 = B$, so

$B = -4$. Now $y'(x) = 3A\cosh 3x - 12\sinh 3x \;\Rightarrow\; 6 = y'(0) = 3A \;\Rightarrow\; A = 2$, so $y = 2\sinh 3x - 4\cosh 3x$.

47. The tangent to $y = \cosh x$ has slope 1 when $y' = \sinh x = 1 \;\Rightarrow\; x = \sinh^{-1} 1 = \ln(1 + \sqrt{2})$, by Equation 3.

Since $\sinh x = 1$ and $y = \cosh x = \sqrt{1 + \sinh^2 x}$, we have $\cosh x = \sqrt{2}$. The point is $\left(\ln(1 + \sqrt{2}), \sqrt{2}\right)$.

49. If $ae^x + be^{-x} = \alpha\cosh(x + \beta)$ [or $\alpha\sinh(x + \beta)$], then

$ae^x + be^{-x} = \tfrac{\alpha}{2}\left(e^{x+\beta} \pm e^{-x-\beta}\right) = \tfrac{\alpha}{2}\left(e^x e^\beta \pm e^{-x}e^{-\beta}\right) = \left(\tfrac{\alpha}{2}e^\beta\right)e^x \pm \left(\tfrac{\alpha}{2}e^{-\beta}\right)e^{-x}$. Comparing coefficients

of e^x and e^{-x}, we have $a = \tfrac{\alpha}{2}e^\beta$ **(1)** and $b = \pm\tfrac{\alpha}{2}e^{-\beta}$ **(2)**. We need to find α and β. Dividing equation **(1)** by

equation **(2)** gives us $\dfrac{a}{b} = \pm e^{2\beta} \;\Rightarrow\;$ **(∗)** $2\beta = \ln(\pm\tfrac{a}{b}) \;\Rightarrow\; \beta = \tfrac{1}{2}\ln(\pm\tfrac{a}{b})$. Solving equations **(1)** and **(2)** for e^β gives

us $e^\beta = \dfrac{2a}{\alpha}$ and $e^\beta = \pm\dfrac{\alpha}{2b}$, so $\dfrac{2a}{\alpha} = \pm\dfrac{\alpha}{2b} \;\Rightarrow\; \alpha^2 = \pm 4ab \;\Rightarrow\; \alpha = 2\sqrt{\pm ab}$.

(∗) If $\dfrac{a}{b} > 0$, we use the $+$ sign and obtain a cosh function, whereas if $\dfrac{a}{b} < 0$, we use the $-$ sign and obtain a sinh

function.

In summary, if a and b have the same sign, we have $ae^x + be^{-x} = 2\sqrt{ab}\cosh\left(x + \tfrac{1}{2}\ln\tfrac{a}{b}\right)$, whereas, if a and b have the

opposite sign, then $ae^x + be^{-x} = 2\sqrt{-ab}\sinh\left(x + \tfrac{1}{2}\ln\left(-\tfrac{a}{b}\right)\right)$.

3.7 Indeterminate Forms and l'Hospital's Rule

Note: The use of l'Hospital's Rule is indicated by an H above the equal sign: $\overset{\text{H}}{=}$

1. This limit has the form $\tfrac{0}{0}$. We can simply factor the numerator to evaluate this limit.

$\lim\limits_{x \to -1} \dfrac{x^2 - 1}{x + 1} = \lim\limits_{x \to -1} \dfrac{(x + 1)(x - 1)}{x + 1} = \lim\limits_{x \to -1} (x - 1) = -2$

3. This limit has the form $\tfrac{0}{0}$. $\lim\limits_{x \to (\pi/2)^+} \dfrac{\cos x}{1 - \sin x} \overset{\text{H}}{=} \lim\limits_{x \to (\pi/2)^+} \dfrac{-\sin x}{-\cos x} = \lim\limits_{x \to (\pi/2)^+} \tan x = -\infty$.

5. This limit has the form $\tfrac{0}{0}$. $\lim\limits_{t \to 0} \dfrac{e^t - 1}{t^3} \overset{\text{H}}{=} \lim\limits_{t \to 0} \dfrac{e^t}{3t^2} = \infty$ since $e^t \to 1$ and $3t^2 \to 0^+$ as $t \to 0$.

7. This limit has the form $\frac{0}{0}$. $\lim\limits_{x\to 0} \dfrac{\tan px}{\tan qx} \overset{\text{H}}{=} \lim\limits_{x\to 0} \dfrac{p\sec^2 px}{q\sec^2 qx} = \dfrac{p(1)^2}{q(1)^2} = \dfrac{p}{q}$

9. $\lim\limits_{x\to 0^+} [(\ln x)/x] = -\infty$ since $\ln x \to -\infty$ as $x \to 0^+$ and dividing by small values of x just increases the magnitude of the quotient $(\ln x)/x$.   L'Hospital's Rule does not apply.

11. This limit has the form $\frac{0}{0}$. $\lim\limits_{t\to 0} \dfrac{5^t - 3^t}{t} \overset{\text{H}}{=} \lim\limits_{t\to 0} \dfrac{5^t \ln 5 - 3^t \ln 3}{1} = \ln 5 - \ln 3 = \ln \frac{5}{3}$

13. This limit has the form $\frac{0}{0}$. $\lim\limits_{x\to 0} \dfrac{e^x - 1 - x}{x^2} \overset{\text{H}}{=} \lim\limits_{x\to 0} \dfrac{e^x - 1}{2x} \overset{\text{H}}{=} \lim\limits_{x\to 0} \dfrac{e^x}{2} = \dfrac{1}{2}$

15. This limit has the form $\frac{\infty}{\infty}$. $\lim\limits_{x\to\infty} \dfrac{x}{\ln(1 + 2e^x)} \overset{\text{H}}{=} \lim\limits_{x\to\infty} \dfrac{1}{\dfrac{1}{1 + 2e^x}\cdot 2e^x} = \lim\limits_{x\to\infty} \dfrac{1 + 2e^x}{2e^x} \overset{\text{H}}{=} \lim\limits_{x\to\infty} \dfrac{2e^x}{2e^x} = 1$

17. This limit has the form $\frac{0}{0}$. $\lim\limits_{x\to 1} \dfrac{1 - x + \ln x}{1 + \cos \pi x} \overset{\text{H}}{=} \lim\limits_{x\to 1} \dfrac{-1 + 1/x}{-\pi \sin \pi x} \overset{\text{H}}{=} \lim\limits_{x\to 1} \dfrac{-1/x^2}{-\pi^2 \cos \pi x} = \dfrac{-1}{-\pi^2(-1)} = -\dfrac{1}{\pi^2}$

19. This limit has the form $\frac{0}{0}$. $\lim\limits_{x\to 1} \dfrac{x^a - ax + a - 1}{(x-1)^2} \overset{\text{H}}{=} \lim\limits_{x\to 1} \dfrac{ax^{a-1} - a}{2(x-1)} \overset{\text{H}}{=} \lim\limits_{x\to 1} \dfrac{a(a-1)x^{a-2}}{2} = \dfrac{a(a-1)}{2}$

21. This limit has the form $0 \cdot (-\infty)$. We need to write this product as a quotient, but keep in mind that we will have to differentiate both the numerator and the denominator. If we differentiate $\dfrac{1}{\ln x}$, we get a complicated expression that results in a more difficult limit. Instead we write the quotient as $\dfrac{\ln x}{x^{-1/2}}$.

$$\lim\limits_{x\to 0^+} \sqrt{x}\,\ln x = \lim\limits_{x\to 0^+} \dfrac{\ln x}{x^{-1/2}} \overset{\text{H}}{=} \lim\limits_{x\to 0^+} \dfrac{1/x}{-\frac{1}{2}x^{-3/2}} \cdot \dfrac{-2x^{3/2}}{-2x^{3/2}} = \lim\limits_{x\to 0^+} (-2\sqrt{x}) = 0$$

23. This limit has the form $\infty \cdot 0$. We'll change it to the form $\frac{0}{0}$.

$$\lim\limits_{x\to 0} \cot 2x \sin 6x = \lim\limits_{x\to 0} \dfrac{\sin 6x}{\tan 2x} \overset{\text{H}}{=} \lim\limits_{x\to 0} \dfrac{6\cos 6x}{2\sec^2 2x} = \dfrac{6(1)}{2(1)^2} = 3$$

25. This limit has the form $\infty \cdot 0$. $\lim\limits_{x\to\infty} x^3 e^{-x^2} = \lim\limits_{x\to\infty} \dfrac{x^3}{e^{x^2}} \overset{\text{H}}{=} \lim\limits_{x\to\infty} \dfrac{3x^2}{2xe^{x^2}} = \lim\limits_{x\to\infty} \dfrac{3x}{2e^{x^2}} \overset{\text{H}}{=} \lim\limits_{x\to\infty} \dfrac{3}{4xe^{x^2}} = 0$

27. As $x \to \infty$, $1/x \to 0$, and $e^{1/x} \to 1$. So the limit has the form $\infty - \infty$ and we will change the form to a product by factoring out x.

$$\lim\limits_{x\to\infty} (xe^{1/x} - x) = \lim\limits_{x\to\infty} x(e^{1/x} - 1) = \lim\limits_{x\to\infty} \dfrac{e^{1/x} - 1}{1/x} \overset{\text{H}}{=} \lim\limits_{x\to\infty} \dfrac{e^{1/x}(-1/x^2)}{-1/x^2} = \lim\limits_{x\to\infty} e^{1/x} = e^0 = 1$$

29. The limit has the form $\infty - \infty$ and we will change the form to a product by factoring out x.

$$\lim\limits_{x\to\infty} (x - \ln x) = \lim\limits_{x\to\infty} x\left(1 - \dfrac{\ln x}{x}\right) = \infty \text{ since } \lim\limits_{x\to\infty} \dfrac{\ln x}{x} \overset{\text{H}}{=} \lim\limits_{x\to\infty} \dfrac{1/x}{1} = 0$$

31. $y = x^{x^2} \;\Rightarrow\; \ln y = x^2 \ln x$, so $\lim\limits_{x\to 0^+} \ln y = \lim\limits_{x\to 0^+} x^2 \ln x = \lim\limits_{x\to 0^+} \dfrac{\ln x}{1/x^2} \overset{\text{H}}{=} \lim\limits_{x\to 0^+} \dfrac{1/x}{-2/x^3} = \lim\limits_{x\to 0^+} \left(-\frac{1}{2}x^2\right) = 0 \;\Rightarrow$

$\lim\limits_{x\to 0^+} x^{x^2} = \lim\limits_{x\to 0^+} e^{\ln y} = e^0 = 1$.

33. $y = (1 - 2x)^{1/x}$ $\Rightarrow$ $\ln y = \dfrac{1}{x}\ln(1 - 2x)$, so $\lim\limits_{x \to 0} \ln y = \lim\limits_{x \to 0} \dfrac{\ln(1 - 2x)}{x} \overset{\text{H}}{=} \lim\limits_{x \to 0} \dfrac{-2/(1 - 2x)}{1} = -2$ $\Rightarrow$

$\lim\limits_{x \to 0}(1 - 2x)^{1/x} = \lim\limits_{x \to 0} e^{\ln y} = e^{-2}$.

35. $y = (\cos x)^{1/x^2}$ $\Rightarrow$ $\ln y = \dfrac{1}{x^2}\ln\cos x$ $\Rightarrow$

$\lim\limits_{x \to 0^+} \ln y = \lim\limits_{x \to 0^+} \dfrac{\ln\cos x}{x^2} \overset{\text{H}}{=} \lim\limits_{x \to 0^+} \dfrac{-\tan x}{2x} \overset{\text{H}}{=} \lim\limits_{x \to 0^+} \dfrac{-\sec^2 x}{2} = -\dfrac{1}{2}$ $\Rightarrow$

$\lim\limits_{x \to 0^+} (\cos x)^{1/x^2} = \lim\limits_{x \to 0^+} e^{\ln y} = e^{-1/2} = 1/\sqrt{e}$

37. From the graph, it appears that $\lim\limits_{x \to \infty} x\left[\ln(x + 5) - \ln x\right] = 5$. To prove this, we first note that

$\ln(x + 5) - \ln x = \ln\dfrac{x + 5}{x} = \ln\left(1 + \dfrac{5}{x}\right) \to \ln 1 = 0$ as $x \to \infty$. Thus,

$\lim\limits_{x \to \infty} x\left[\ln(x + 5) - \ln x\right] = \lim\limits_{x \to \infty} \dfrac{\ln(x + 5) - \ln x}{1/x} \overset{\text{H}}{=} \lim\limits_{x \to \infty} \dfrac{\dfrac{1}{x + 5} - \dfrac{1}{x}}{-1/x^2}$

$= \lim\limits_{x \to \infty} \left[\dfrac{x - (x + 5)}{x(x + 5)} \cdot \dfrac{-x^2}{1}\right] = \lim\limits_{x \to \infty} \dfrac{5x^2}{x^2 + 5x} = 5$

39. $\lim\limits_{x \to \infty} \dfrac{e^x}{x^n} \overset{\text{H}}{=} \lim\limits_{x \to \infty} \dfrac{e^x}{nx^{n-1}} \overset{\text{H}}{=} \lim\limits_{x \to \infty} \dfrac{e^x}{n(n - 1)x^{n-2}} \overset{\text{H}}{=} \cdots \overset{\text{H}}{=} \lim\limits_{x \to \infty} \dfrac{e^x}{n!} = \infty$

41. First we will find $\lim\limits_{n \to \infty}\left(1 + \dfrac{r}{n}\right)^{nt}$, which is of the form 1^∞. $y = \left(1 + \dfrac{r}{n}\right)^{nt}$ $\Rightarrow$ $\ln y = nt\ln\left(1 + \dfrac{r}{n}\right)$, so

$\lim\limits_{n \to \infty} \ln y = \lim\limits_{n \to \infty} nt\ln\left(1 + \dfrac{r}{n}\right) = t\lim\limits_{n \to \infty} \dfrac{\ln(1 + r/n)}{1/n} \overset{\text{H}}{=} t\lim\limits_{n \to \infty} \dfrac{(-r/n^2)}{(1 + r/n)(-1/n^2)} = t\lim\limits_{n \to \infty} \dfrac{r}{1 + i/n} = tr$ $\Rightarrow$

$\lim\limits_{n \to \infty} y = e^{rt}$. Thus, as $n \to \infty$, $A = A_0\left(1 + \dfrac{r}{n}\right)^{nt} \to A_0 e^{rt}$.

43. $\lim\limits_{E \to 0^+} P(E) = \lim\limits_{E \to 0^+}\left(\dfrac{e^E + e^{-E}}{e^E - e^{-E}} - \dfrac{1}{E}\right)$

$= \lim\limits_{E \to 0^+} \dfrac{E\left(e^E + e^{-E}\right) - 1\left(e^E - e^{-E}\right)}{\left(e^E - e^{-E}\right)E} = \lim\limits_{E \to 0^+} \dfrac{Ee^E + Ee^{-E} - e^E + e^{-E}}{Ee^E - Ee^{-E}}$ [form is $\frac{0}{0}$]

$\overset{\text{H}}{=} \lim\limits_{E \to 0^+} \dfrac{Ee^E + e^E \cdot 1 + E\left(-e^{-E}\right) + e^{-E} \cdot 1 - e^E + \left(-e^{-E}\right)}{Ee^E + e^E \cdot 1 - \left[E(-e^{-E}) + e^{-E} \cdot 1\right]}$

$= \lim\limits_{E \to 0^+} \dfrac{Ee^E - Ee^{-E}}{Ee^E + e^E + Ee^{-E} - e^{-E}} = \lim\limits_{E \to 0^+} \dfrac{e^E - e^{-E}}{e^E + \dfrac{e^E}{E} + e^{-E} - \dfrac{e^{-E}}{E}}$ [divide by E]

$= \dfrac{0}{2 + L}$, where $L = \lim\limits_{E \to 0^+} \dfrac{e^E - e^{-E}}{E}$ [form is $\frac{0}{0}$]

$\overset{\text{H}}{=} \lim\limits_{E \to 0^+} \dfrac{e^E + e^{-E}}{1} = \dfrac{1 + 1}{1} = 2$

Thus, $\lim\limits_{E \to 0^+} P(E) = \dfrac{0}{2 + 2} = 0$.

45. We see that both numerator and denominator approach 0, so we can use l'Hospital's Rule:

$$\lim_{x \to a} \frac{\sqrt{2a^3x - x^4} - a\sqrt[3]{aax}}{a - \sqrt[4]{ax^3}} \overset{\text{H}}{=} \lim_{x \to a} \frac{\frac{1}{2}(2a^3x - x^4)^{-1/2}(2a^3 - 4x^3) - a\left(\frac{1}{3}\right)(aax)^{-2/3}a^2}{-\frac{1}{4}(ax^3)^{-3/4}(3ax^2)}$$

$$= \frac{\frac{1}{2}(2a^3a - a^4)^{-1/2}(2a^3 - 4a^3) - \frac{1}{3}a^3(a^2a)^{-2/3}}{-\frac{1}{4}(aa^3)^{-3/4}(3aa^2)}$$

$$= \frac{(a^4)^{-1/2}(-a^3) - \frac{1}{3}a^3(a^3)^{-2/3}}{-\frac{3}{4}a^3(a^4)^{-3/4}} = \frac{-a - \frac{1}{3}a}{-\frac{3}{4}} = \frac{4}{3}\left(\frac{4}{3}a\right) = \frac{16}{9}a$$

47. Since $f(2) = 0$, the given limit has the form $\frac{0}{0}$.

$$\lim_{x \to 0} \frac{f(2 + 3x) + f(2 + 5x)}{x} \overset{\text{H}}{=} \lim_{x \to 0} \frac{f'(2 + 3x) \cdot 3 + f'(2 + 5x) \cdot 5}{1} = f'(2) \cdot 3 + f'(2) \cdot 5 = 8f'(2) = 8 \cdot 7 = 56$$

49. Since $\lim_{h \to 0}[f(x + h) - f(x - h)] = f(x) - f(x) = 0$ (f is differentiable and hence continuous) and $\lim_{h \to 0} 2h = 0$, we use

l'Hospital's Rule:

$$\lim_{h \to 0} \frac{f(x + h) - f(x - h)}{2h} \overset{\text{H}}{=} \lim_{h \to 0} \frac{f'(x + h)(1) - f'(x - h)(-1)}{2}$$

$$= \frac{f'(x) + f'(x)}{2} = \frac{2f'(x)}{2} = f'(x)$$

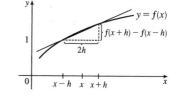

$\dfrac{f(x + h) - f(x - h)}{2h}$ is the slope of the secant line between

$(x - h, f(x - h))$ and $(x + h, f(x + h))$. As $h \to 0$, this line gets

closer to the tangent line and its slope approaches $f'(x)$.

51. (a) We show that $\lim_{x \to 0} \dfrac{f(x)}{x^n} = 0$ for every integer $n \geq 0$. Let $y = \dfrac{1}{x^2}$. Then

$$\lim_{x \to 0} \frac{f(x)}{x^{2n}} = \lim_{x \to 0} \frac{e^{-1/x^2}}{(x^2)^n} = \lim_{y \to \infty} \frac{y^n}{e^y} \overset{\text{H}}{=} \lim_{y \to \infty} \frac{ny^{n-1}}{e^y} \overset{\text{H}}{=} \cdots \overset{\text{H}}{=} \lim_{y \to \infty} \frac{n!}{e^y} = 0 \;\Rightarrow$$

$$\lim_{x \to 0} \frac{f(x)}{x^n} = \lim_{x \to 0} x^n \frac{f(x)}{x^{2n}} = \lim_{x \to 0} x^n \lim_{x \to 0} \frac{f(x)}{x^{2n}} = 0. \text{ Thus, } f'(0) = \lim_{x \to 0} \frac{f(x) - f(0)}{x - 0} = \lim_{x \to 0} \frac{f(x)}{x} = 0.$$

(b) Using the Chain Rule and the Quotient Rule we see that $f^{(n)}(x)$ exists for $x \neq 0$. In fact, we prove by induction that for

each $n \geq 0$, there is a polynomial p_n and a non-negative integer k_n with $f^{(n)}(x) = p_n(x)f(x)/x^{k_n}$ for $x \neq 0$. This is

true for $n = 0$; suppose it is true for the nth derivative. Then $f'(x) = f(x)(2/x^3)$, so

$$f^{(n+1)}(x) = \left[x^{k_n}\left[p'_n(x)f(x) + p_n(x)f'(x)\right] - k_n x^{k_n - 1}p_n(x)f(x)\right]x^{-2k_n}$$

$$= \left[x^{k_n}p'_n(x) + p_n(x)(2/x^3) - k_n x^{k_n - 1}p_n(x)\right]f(x)x^{-2k_n}$$

$$= \left[x^{k_n + 3}p'_n(x) + 2p_n(x) - k_n x^{k_n + 2}p_n(x)\right]f(x)x^{-(2k_n + 3)}$$

which has the desired form.

Now we show by induction that $f^{(n)}(0) = 0$ for all n. By part (a), $f'(0) = 0$. Suppose that $f^{(n)}(0) = 0$. Then

$$f^{(n+1)}(0) = \lim_{x \to 0} \frac{f^{(n)}(x) - f^{(n)}(0)}{x - 0} = \lim_{x \to 0} \frac{f^{(n)}(x)}{x} = \lim_{x \to 0} \frac{p_n(x)f(x)/x^{k_n}}{x} = \lim_{x \to 0} \frac{p_n(x)f(x)}{x^{k_n + 1}}$$

$$= \lim_{x \to 0} p_n(x) \lim_{x \to 0} \frac{f(x)}{x^{k_n + 1}} = p_n(0) \cdot 0 = 0$$

3 Review

CONCEPT CHECK

---

1. (a) A function f is called a *one-to-one function* if it never takes on the same value twice; that is, if $f(x_1) \neq f(x_2)$ whenever $x_1 \neq x_2$. (Or, f is 1-1 if each output corresponds to only one input.)

   Use the Horizontal Line Test: A function is one-to-one if and only if no horizontal line intersects its graph more than once.

   (b) If f is a one-to-one function with domain A and range B, then its *inverse function* f^{-1} has domain B and range A and is defined by $f^{-1}(y) = x \iff f(x) = y$ for any y in B. The graph of f^{-1} is obtained by reflecting the graph of f about the line $y = x$.

   (c) $(f^{-1})'(a) = \dfrac{1}{f'(f^{-1}(a))}$

2. (a) $e = \lim\limits_{x \to 0} (1+x)^{1/x}$

   (b) $e \approx 2.71828$

   (c) The differentiation formula for $y = a^x$ $[y' = a^x \ln a]$ is simplest when $a = e$ because $\ln e = 1$.

   (d) The differentiation formula for $y = \log_a x$ $[y' = 1/(x \ln a)]$ is simplest when $a = e$ because $\ln e = 1$.

3. (a) The function $f(x) = e^x$ has domain $\mathbb{R}$ and range $(0, \infty)$.

   (b) The function $f(x) = \ln x$ has domain $(0, \infty)$ and range $\mathbb{R}$.

   (c) The graphs are reflections of one another about the line $y = x$. See Figure 3.2.15.

   (d) $\log_a x = \dfrac{\ln x}{\ln a}$

4. (a) The inverse sine function $f(x) = \sin^{-1} x$ is defined as follows:

$$\sin^{-1} x = y \iff \sin y = x \quad \text{and} \quad -\frac{\pi}{2} \le y \le \frac{\pi}{2}$$

   Its domain is $-1 \le x \le 1$ and its range is $-\frac{\pi}{2} \le y \le \frac{\pi}{2}$.

   (b) The inverse cosine function $f(x) = \cos^{-1} x$ is defined as follows:

$$\cos^{-1} x = y \iff \cos y = x \quad \text{and} \quad 0 \le y \le \pi$$

   Its domain is $-1 \le x \le 1$ and its range is $0 \le y \le \pi$.

   (c) See Definition 3.5.7. Domain $= \mathbb{R}$, Range $= \left(-\frac{\pi}{2}, \frac{\pi}{2}\right)$. See Figure 10 in Section 3.5.

5. $\sinh x = \dfrac{e^x - e^{-x}}{2}$, $\cosh x = \dfrac{e^x + e^{-x}}{2}$, $\tanh x = \dfrac{\sinh x}{\cosh x} = \dfrac{e^x - e^{-x}}{e^x + e^{-x}}$

6. (a) $y = e^x \implies y' = e^x$ 　　　　　　　　　　　　　　(b) $y = a^x \implies y' = a^x \ln a$

   (c) $y = \ln x \implies y' = 1/x$ 　　　　　　　　　　　　(d) $y = \log_a x \implies y' = 1/(x \ln a)$

   (e) $y = \sin^{-1} x \implies y' = 1/\sqrt{1 - x^2}$ 　　　　(f) $y = \cos^{-1} x \implies y' = -1/\sqrt{1 - x^2}$

   (g) $y = \tan^{-1} x \implies y' = 1/(1 + x^2)$ 　　　　　(h) $y = \sinh x \implies y' = \cosh x$

   (i) $y = \cosh x \implies y' = \sinh x$ 　　　　　　　　　(j) $y = \tanh x \implies y' = \operatorname{sech}^2 x$

(k) $y = \sinh^{-1} x \;\Rightarrow\; y' = 1/\sqrt{1 + x^2}$ (l) $y = \cosh^{-1} x \;\Rightarrow\; y' = 1/\sqrt{x^2 - 1}$

(m) $y = \tanh^{-1} x \;\Rightarrow\; y' = 1/(1 - x^2)$

7. (a) $\dfrac{dy}{dt} = ky$; the relative growth rate, $\dfrac{1}{y}\dfrac{dy}{dt}$, is constant.

   (b) The equation in part (a) is an appropriate model for population growth if we assume that there is enough room and nutrition to support the growth.

   (c) If $y(0) = y_0$, then the solution is $y(t) = y_0 e^{kt}$.

8. (a) See l'Hospital's Rule and the three notes that follow it in Section 3.7.

   (b) Write fg as $\dfrac{f}{1/g}$ or $\dfrac{g}{1/f}$.

   (c) Convert the difference into a quotient using a common denominator, rationalizing, factoring, or some other method.

   (d) Convert the power to a product by taking the natural logarithm of both sides of $y = f^g$ or by writing f^g as $e^{g \ln f}$.

TRUE-FALSE QUIZ

1. True.   If f is one-to-one, with domain $\mathbb{R}$, then $f^{-1}(f(6)) = 6$ by the first cancellation equation in (3.2.4).

3. False.   For example, $\cos \frac{\pi}{2} = \cos\left(-\frac{\pi}{2}\right)$, so $\cos x$ is not 1-1.

5. True, since $\ln x$ is an increasing function on $(0, \infty)$.

7. True.   We can divide by e^x since $e^x \neq 0$ for every x.

9. False.   Let $x = e$. Then $(\ln x)^6 = (\ln e)^6 = 1^6 = 1$, but $6 \ln x = 6 \ln e = 6 \cdot 1 = 6 \neq 1 = (\ln x)^6$.

11. False. $\ln 10$ is a constant, so its derivative is 0.

13. False.   The "-1" is not an exponent; it is an indication of an inverse function.

15. True.   See Figure 3.6.2.

EXERCISES

1. No. f is not 1-1 because the graph of f fails the Horizontal Line Test.

3. (a) $f^{-1}(3) = 7$ since $f(7) = 3$.

   (b) $(f^{-1})'(3) = \dfrac{1}{f'(f^{-1}(3))} = \dfrac{1}{f'(7)} = \dfrac{1}{8}$

5.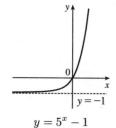

$y = 5^x - 1$

7. Reflect the graph of $y = \ln x$ about the x-axis to obtain the graph of $y = -\ln x$.

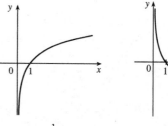

$y = \ln x$ $y = -\ln x$

9.

$$y = 2\arctan x$$

11. (a) $e^{2\ln 3} = \left(e^{\ln 3}\right)^2 = 3^2 = 9$

(b) $\log_{10} 25 + \log_{10} 4 = \log_{10}(25 \cdot 4) = \log_{10} 100 = \log_{10} 10^2 = 2$

13. $\ln x = \frac{1}{3} \iff \log_e x = \frac{1}{3} \implies x = e^{1/3}$

15. $e^{e^x} = 17 \implies \ln e^{e^x} = \ln 17 \implies e^x = \ln 17 \implies \ln e^x = \ln(\ln 17) \implies x = \ln\ln 17$

17. $\ln(x+1) + \ln(x-1) = 1 \implies \ln[(x+1)(x-1)] = 1 \implies \ln(x^2-1) = \ln e \implies x^2 - 1 = e \implies$
$x^2 = e + 1 \implies x = \sqrt{e+1}$ since $\ln(x-1)$ is defined only when $x > 1$.

19. $\tan^{-1} x = 1 \implies \tan\tan^{-1} x = \tan 1 \implies x = \tan 1 \ (\approx 1.5574)$

21. $f(t) = t^2 \ln t \implies f'(t) = t^2 \cdot \frac{1}{t} + (\ln t)(2t) = t + 2t \ln t \text{ or } t(1 + 2\ln t)$

23. $h(\theta) = e^{\tan 2\theta} \implies h'(\theta) = e^{\tan 2\theta} \cdot \sec^2 2\theta \cdot 2 = 2\sec^2(2\theta) e^{\tan 2\theta}$

25. $y = \ln|\sec 5x + \tan 5x| \implies$

$$y' = \frac{1}{\sec 5x + \tan 5x}(\sec 5x \tan 5x \cdot 5 + \sec^2 5x \cdot 5) = \frac{5\sec 5x \, (\tan 5x + \sec 5x)}{\sec 5x + \tan 5x} = 5\sec 5x$$

27. $y = e^{cx}(c\sin x - \cos x) \implies y' = ce^{cx}(c\sin x - \cos x) + e^{cx}(c\cos x + \sin x) = (c^2+1)e^{cx}\sin x$

29. $y = \ln(\sec^2 x) = 2\ln|\sec x| \implies y' = (2/\sec x)(\sec x \tan x) = 2\tan x$

31. $y = xe^{-1/x} \implies y' = e^{-1/x} + xe^{-1/x}(1/x^2) = e^{-1/x}(1 + 1/x)$

33. $y = 2^{-t^2} \implies y' = 2^{-t^2}(\ln 2)(-2t) = (-2\ln 2)t\, 2^{-t^2}$

35. $H(v) = v\tan^{-1} v \implies H'(v) = v \cdot \frac{1}{1+v^2} + \tan^{-1} v \cdot 1 = \frac{v}{1+v^2} + \tan^{-1} v$

37. $y = x\sinh(x^2) \implies y' = x\cosh(x^2) \cdot 2x + \sinh(x^2) \cdot 1 = 2x^2 \cosh(x^2) + \sinh(x^2)$

39. $y = \ln\sin x - \frac{1}{2}\sin^2 x \implies y' = \frac{1}{\sin x} \cdot \cos x - \frac{1}{2} \cdot 2\sin x \cdot \cos x = \cot x - \sin x \cos x$

41. $y = \ln\left(\frac{1}{x}\right) + \frac{1}{\ln x} = \ln x^{-1} + (\ln x)^{-1} = -\ln x + (\ln x)^{-1} \implies y' = -1 \cdot \frac{1}{x} + (-1)(\ln x)^{-2} \cdot \frac{1}{x} = -\frac{1}{x} - \frac{1}{x(\ln x)^2}$

43. $y = \ln(\cosh 3x) \implies y' = (1/\cosh 3x)(\sinh 3x)(3) = 3\tanh 3x$

45. $y = \cosh^{-1}(\sinh x) \implies y' = (\cosh x)/\sqrt{\sinh^2 x - 1}$

47. $f(x) = e^{\sin^3(\ln(x^2+1))} \implies$

$$f'(x) = e^{\sin^3(\ln(x^2+1))} \cdot 3\sin^2(\ln(x^2+1)) \cdot \cos(\ln(x^2+1)) \cdot \frac{1}{x^2+1} \cdot 2x$$

$$= \frac{6x}{x^2+1}\sin^2(\ln(x^2+1)) \cdot \cos(\ln(x^2+1)) \cdot e^{\sin^3(\ln(x^2+1))}$$

49. $f(x) = e^{g(x)} \implies f'(x) = e^{g(x)}g'(x)$

51. $f(x) = \ln|g(x)|$ $\Rightarrow$ $f'(x) = \dfrac{1}{g(x)}g'(x) = \dfrac{g'(x)}{g(x)}$

53. $f(x) = 2^x$ $\Rightarrow$ $f'(x) = 2^x \ln 2$ $\Rightarrow$ $f''(x) = 2^x(\ln 2)^2$ $\Rightarrow$ $\cdots$ $\Rightarrow$ $f^{(n)}(x) = 2^x(\ln 2)^n$

55. We first show it is true for $n = 1$: $f'(x) = e^x + xe^x = (x+1)e^x$. We now assume it is true for $n = k$:

$f^{(k)}(x) = (x+k)e^x$. With this assumption, we must show it is true for $n = k+1$:

$f^{(k+1)}(x) = \dfrac{d}{dx}\left[f^{(k)}(x)\right] = \dfrac{d}{dx}\left[(x+k)e^x\right] = e^x + (x+k)e^x = [x+(k+1)]e^x.$

Therefore, $f^{(n)}(x) = (x+n)e^x$ by mathematical induction.

57. $y = (2+x)e^{-x}$ $\Rightarrow$ $y' = (2+x)(-e^{-x}) + e^{-x} \cdot 1 = e^{-x}[-(2+x)+1] = e^{-x}(-x-1)$. At $(0,2)$, $y' = 1(-1) = -1$,

so an equation of the tangent line is $y - 2 = -1(x-0)$, or $y = -x+2$.

59. $y = [\ln(x+4)]^2$ $\Rightarrow$ $y' = 2[\ln(x+4)]^1 \cdot \dfrac{1}{x+4} \cdot 1 = 2\dfrac{\ln(x+4)}{x+4}$ and $y' = 0$ $\Leftrightarrow$ $\ln(x+4) = 0$ $\Leftrightarrow$

$x+4 = e^0$ $\Rightarrow$ $x+4 = 1$ $\Leftrightarrow$ $x = -3$, so the tangent is horizontal at the point $(-3,0)$.

61. (a) The line $x - 4y = 1$ has slope $\frac{1}{4}$. A tangent to $y = e^x$ has slope $\frac{1}{4}$ when $y' = e^x = \frac{1}{4}$ $\Rightarrow$ $x = \ln\frac{1}{4} = -\ln 4$.

Since $y = e^x$, the y-coordinate is $\frac{1}{4}$ and the point of tangency is $\left(-\ln 4, \frac{1}{4}\right)$. Thus, an equation of the tangent line

is $y - \frac{1}{4} = \frac{1}{4}(x + \ln 4)$ or $y = \frac{1}{4}x + \frac{1}{4}(\ln 4 + 1)$.

(b) The slope of the tangent at the point (a, e^a) is $\dfrac{d}{dx}e^x\bigg|_{x=a} = e^a$. Thus, an equation of the tangent line is

$y - e^a = e^a(x-a)$. We substitute $x = 0$, $y = 0$ into this equation, since we want the line to pass through the origin:

$0 - e^a = e^a(0-a)$ $\Leftrightarrow$ $-e^a = e^a(-a)$ $\Leftrightarrow$ $a = 1$. So an equation of the tangent line at the point $(a, e^a) = (1, e)$

is $y - e = e(x-1)$ or $y = ex$.

63. (a) $y(t) = y(0)e^{kt} = 200e^{kt}$ $\Rightarrow$ $y(0.5) = 200e^{0.5k} = 360$ $\Rightarrow$ $e^{0.5k} = 1.8$ $\Rightarrow$ $0.5k = \ln 1.8$ $\Rightarrow$

$k = 2\ln 1.8 = \ln(1.8)^2 = \ln 3.24$ $\Rightarrow$ $y(t) = 200e^{(\ln 3.24)t} = 200(3.24)^t$

(b) $y(4) = 200(3.24)^4 \approx 22{,}040$ bacteria

(c) $y'(t) = 200(3.24)^t \cdot \ln 3.24$, so $y'(4) = 200(3.24)^4 \cdot \ln 3.24 \approx 25{,}910$ bacteria per hour

(d) $200(3.24)^t = 10{,}000$ $\Rightarrow$ $(3.24)^t = 50$ $\Rightarrow$ $t\ln 3.24 = \ln 50$ $\Rightarrow$ $t = \ln 50/\ln 3.24 \approx 3.33$ hours

65. (a) $C'(t) = -kC(t)$ $\Rightarrow$ $C(t) = C(0)e^{-kt}$ by Theorem 3.4.2. But $C(0) = C_0$, so $C(t) = C_0 e^{-kt}$.

(b) $C(30) = \frac{1}{2}C_0$ since the concentration is reduced by half. Thus, $\frac{1}{2}C_0 = C_0 e^{-30k}$ $\Rightarrow$ $\ln\frac{1}{2} = -30k$ $\Rightarrow$

$k = -\frac{1}{30}\ln\frac{1}{2} = \frac{1}{30}\ln 2$. Since 10% of the original concentration remains if 90% is eliminated, we want the value of t

such that $C(t) = \frac{1}{10}C_0$. Therefore, $\frac{1}{10}C_0 = C_0 e^{-t(\ln 2)/30}$ $\Rightarrow$ $\ln 0.1 = -t(\ln 2)/30$ $\Rightarrow$ $t = -\frac{30}{\ln 2}\ln 0.1 \approx 100$ h.

67. $\displaystyle\lim_{x\to\infty} e^{-3x} = 0$ since $-3x \to -\infty$ as $x \to \infty$ and $\displaystyle\lim_{t\to-\infty} e^t = 0$.

69. Let $t = 2/(x-3)$. As $x \to 3^-$, $t \to -\infty$. $\displaystyle\lim_{x\to 3^-} e^{2/(x-3)} = \lim_{t\to-\infty} e^t = 0$

71. Let $t = \sinh x$. As $x \to 0^+$, $t \to 0^+$. $\displaystyle\lim_{x\to 0^+} \ln(\sinh x) = \lim_{t\to 0^+} \ln t = -\infty$

73. $\lim\limits_{x \to \infty} \dfrac{(1 + 2^x)/2^x}{(1 - 2^x)/2^x} = \lim\limits_{x \to \infty} \dfrac{1/2^x + 1}{1/2^x - 1} = \dfrac{0 + 1}{0 - 1} = -1$

75. $\lim\limits_{x \to 0} \dfrac{\tan \pi x}{\ln(1 + x)} \overset{\text{H}}{=} \lim\limits_{x \to 0} \dfrac{\pi \sec^2 \pi x}{1/(1 + x)} = \dfrac{\pi \cdot 1^2}{1/1} = \pi$

77. $\lim\limits_{x \to 0} \dfrac{e^{4x} - 1 - 4x}{x^2} \overset{\text{H}}{=} \lim\limits_{x \to 0} \dfrac{4e^{4x} - 4}{2x} \overset{\text{H}}{=} \lim\limits_{x \to 0} \dfrac{16e^{4x}}{2} = \lim\limits_{x \to 0} 8e^{4x} = 8 \cdot 1 = 8$

79. $\lim\limits_{x \to \infty} x^3 e^{-x} = \lim\limits_{x \to \infty} \dfrac{x^3}{e^x} \overset{\text{H}}{=} \lim\limits_{x \to \infty} \dfrac{3x^2}{e^x} \overset{\text{H}}{=} \lim\limits_{x \to \infty} \dfrac{6x}{e^x} \overset{\text{H}}{=} \lim\limits_{x \to \infty} \dfrac{6}{e^x} = 0$

81. $\lim\limits_{x \to 1^+} \left(\dfrac{x}{x - 1} - \dfrac{1}{\ln x} \right) = \lim\limits_{x \to 1^+} \left(\dfrac{x \ln x - x + 1}{(x - 1) \ln x} \right) \overset{\text{H}}{=} \lim\limits_{x \to 1^+} \dfrac{x \cdot (1/x) + \ln x - 1}{(x - 1) \cdot (1/x) + \ln x}$

$\qquad\qquad = \lim\limits_{x \to 1^+} \dfrac{\ln x}{1 - 1/x + \ln x} \overset{\text{H}}{=} \lim\limits_{x \to 1^+} \dfrac{1/x}{1/x^2 + 1/x} = \dfrac{1}{1 + 1} = \dfrac{1}{2}$

83. $f(x) = \ln x + \tan^{-1} x \;\Rightarrow\; f(1) = \ln 1 + \tan^{-1} 1 = \dfrac{\pi}{4} \;\Rightarrow\; f^{-1}\left(\dfrac{\pi}{4}\right) = 1.$

$\qquad f'(x) = \dfrac{1}{x} + \dfrac{1}{1 + x^2}$, so $(f^{-1})'\left(\dfrac{\pi}{4}\right) = \dfrac{1}{f'(1)} = \dfrac{1}{3/2} = \dfrac{2}{3}$.

4 ☐ APPLICATIONS OF DIFFERENTIATION

4.1 Maximum and Minimum Values

1. A function f has an **absolute minimum** at $x = c$ if $f(c)$ is the smallest function value on the entire domain of f, whereas f has a **local minimum** at c if $f(c)$ is the smallest function value when x is near c.

3. Absolute maximum at b; absolute minimum at d; local maxima at b and e; local minima at d and s; neither a maximum nor a minimum at a, c, r, and t.

5. Absolute maximum value is $f(4) = 4$; absolute minimum value is $f(7) = 0$; local maximum values are $f(4) = 4$ and $f(6) = 3$; local minimum values are $f(2) = 1$ and $f(5) = 2$.

7. Absolute minimum at 2, absolute maximum at 3, local minimum at 4

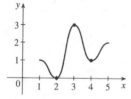

9. Absolute maximum at 5, absolute minimum at 2, local maximum at 3, local minima at 2 and 4

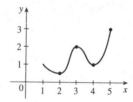

11. (a)

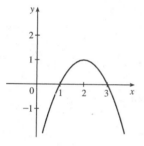

(b)

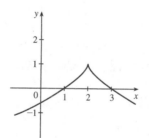

(c)

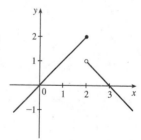

13. (a) *Note:* By the Extreme Value Theorem, f must *not* be continuous; because if it were, it would attain an absolute minimum.

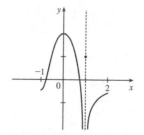

(b)

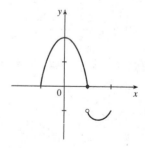

15. $f(x) = 8 - 3x$, $x \geq 1$. Absolute maximum $f(1) = 5$; no local maximum. No absolute or local minimum.

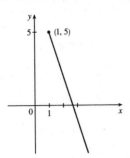

17. $f(x) = x^2$, $0 < x < 2$. No absolute or local maximum or minimum value.

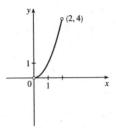

19. $f(\theta) = \sin\theta$, $-2\pi \leq \theta \leq 2\pi$. Absolute and local maxima $f\left(-\frac{3\pi}{2}\right) = f\left(\frac{\pi}{2}\right) = 1$. Absolute and local minima $f\left(-\frac{\pi}{2}\right) = f\left(\frac{3\pi}{2}\right) = -1$.

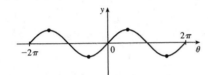

21. $f(x) = 1 - \sqrt{x}$. Absolute maximum $f(0) = 1$; no local maximum. No absolute or local minimum.

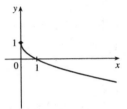

23. $f(x) = 5x^2 + 4x$ $\Rightarrow$ $f'(x) = 10x + 4$. $f'(x) = 0$ $\Rightarrow$ $x = -\frac{2}{5}$, so $-\frac{2}{5}$ is the only critical number.

25. $f(x) = x^3 + 3x^2 - 24x$ $\Rightarrow$ $f'(x) = 3x^2 + 6x - 24 = 3(x^2 + 2x - 8)$.

$f'(x) = 0$ $\Rightarrow$ $3(x + 4)(x - 2) = 0$ $\Rightarrow$ $x = -4, 2$. These are the only critical numbers.

27. $s(t) = 3t^4 + 4t^3 - 6t^2$ $\Rightarrow$ $s'(t) = 12t^3 + 12t^2 - 12t$. $s'(t) = 0$ $\Rightarrow$ $12t(t^2 + t - 1)$ $\Rightarrow$

$t = 0$ or $t^2 + t - 1 = 0$. Using the quadratic formula to solve the latter equation gives us

$t = \dfrac{-1 \pm \sqrt{1^2 - 4(1)(-1)}}{2(1)} = \dfrac{-1 \pm \sqrt{5}}{2} \approx 0.618, -1.618$. The three critical numbers are $0, \dfrac{-1 \pm \sqrt{5}}{2}$.

29. $g(y) = \dfrac{y-1}{y^2 - y + 1}$ $\Rightarrow$

$g'(y) = \dfrac{(y^2 - y + 1)(1) - (y - 1)(2y - 1)}{(y^2 - y + 1)^2} = \dfrac{y^2 - y + 1 - (2y^2 - 3y + 1)}{(y^2 - y + 1)^2} = \dfrac{-y^2 + 2y}{(y^2 - y + 1)^2} = \dfrac{y(2 - y)}{(y^2 - y + 1)^2}$.

$g'(y) = 0$ $\Rightarrow$ $y = 0, 2$. The expression $y^2 - y + 1$ is never equal to 0, so $g'(y)$ exists for all real numbers. The critical numbers are 0 and 2.

31. $F(x) = x^{4/5}(x - 4)^2$ $\Rightarrow$

$F'(x) = x^{4/5} \cdot 2(x - 4) + (x - 4)^2 \cdot \frac{4}{5}x^{-1/5} = \frac{1}{5}x^{-1/5}(x - 4)\left[5 \cdot x \cdot 2 + (x - 4) \cdot 4\right]$

$= \dfrac{(x - 4)(14x - 16)}{5x^{1/5}} = \dfrac{2(x - 4)(7x - 8)}{5x^{1/5}}$

$F'(x) = 0$ $\Rightarrow$ $x = 4, \frac{8}{7}$. $F'(0)$ does not exist. Thus, the three critical numbers are $0, \frac{8}{7}$, and 4.

33. $f(\theta) = 2\cos\theta + \sin^2\theta \;\Rightarrow\; f'(\theta) = -2\sin\theta + 2\sin\theta\cos\theta.$ $f'(\theta) = 0 \;\Rightarrow\; 2\sin\theta(\cos\theta - 1) = 0 \;\Rightarrow\; \sin\theta = 0$ or $\cos\theta = 1 \;\Rightarrow\; \theta = n\pi$ (n an integer) or $\theta = 2n\pi$. The solutions $\theta = n\pi$ include the solutions $\theta = 2n\pi$, so the critical numbers are $\theta = n\pi$.

35. $f(x) = x\ln x \;\Rightarrow\; f'(x) = x(1/x) + (\ln x)\cdot 1 = \ln x + 1.$ $f'(x) = 0 \;\Leftrightarrow\; \ln x = -1 \;\Leftrightarrow\; x = e^{-1} = 1/e.$ Therefore, the only critical number is $x = 1/e$.

37. $f(x) = 3x^2 - 12x + 5,\; [0, 3].$ $f'(x) = 6x - 12 = 0 \;\Leftrightarrow\; x = 2.$ Applying the Closed Interval Method, we find that $f(0) = 5,\; f(2) = -7,$ and $f(3) = -4.$ So $f(0) = 5$ is the absolute maximum value and $f(2) = -7$ is the absolute minimum value.

39. $f(x) = 2x^3 - 3x^2 - 12x + 1,\; [-2, 3].$ $f'(x) = 6x^2 - 6x - 12 = 6(x^2 - x - 2) = 6(x - 2)(x + 1) = 0 \;\Leftrightarrow\; x = 2, -1.$ $f(-2) = -3,\; f(-1) = 8,\; f(2) = -19,$ and $f(3) = -8.$ So $f(-1) = 8$ is the absolute maximum value and $f(2) = -19$ is the absolute minimum value.

41. $f(x) = x^4 - 2x^2 + 3,\; [-2, 3].$ $f'(x) = 4x^3 - 4x = 4x(x^2 - 1) = 4x(x + 1)(x - 1) = 0 \;\Leftrightarrow\; x = -1, 0, 1.$ $f(-2) = 11,\; f(-1) = 2,\; f(0) = 3,\; f(1) = 2,\; f(3) = 66.$ So $f(3) = 66$ is the absolute maximum value and $f(\pm 1) = 2$ is the absolute minimum value.

43. $f(t) = t\sqrt{4 - t^2},\; [-1, 2].$

$$f'(t) = t\cdot\tfrac{1}{2}(4 - t^2)^{-1/2}(-2t) + (4 - t^2)^{1/2}\cdot 1 = \frac{-t^2}{\sqrt{4 - t^2}} + \sqrt{4 - t^2} = \frac{-t^2 + (4 - t^2)}{\sqrt{4 - t^2}} = \frac{4 - 2t^2}{\sqrt{4 - t^2}}.$$

$f'(t) = 0 \;\Rightarrow\; 4 - 2t^2 = 0 \;\Rightarrow\; t^2 = 2 \;\Rightarrow\; t = \pm\sqrt{2},$ but $t = -\sqrt{2}$ is not in the given interval, $[-1, 2].$ $f'(t)$ does not exist if $4 - t^2 = 0 \;\Rightarrow\; t = \pm 2,$ but -2 is not in the given interval. $f(-1) = -\sqrt{3},\; f(\sqrt{2}) = 2,$ and $f(2) = 0.$ So $f(\sqrt{2}) = 2$ is the absolute maximum value and $f(-1) = -\sqrt{3}$ is the absolute minimum value.

45. $f(x) = \sin x + \cos x,\; \left[0, \tfrac{\pi}{3}\right].$ $f'(x) = \cos x - \sin x = 0 \;\Leftrightarrow\; \sin x = \cos x \;\Rightarrow\; \dfrac{\sin x}{\cos x} = 1 \;\Rightarrow\; \tan x = 1 \;\Rightarrow\; x = \tfrac{\pi}{4}.$ $f(0) = 1,\; f(\tfrac{\pi}{4}) = \sqrt{2} \approx 1.41,\; f(\tfrac{\pi}{3}) = \dfrac{\sqrt{3}+1}{2} \approx 1.37.$ So $f(\tfrac{\pi}{4}) = \sqrt{2}$ is the absolute maximum value and $f(0) = 1$ is the absolute minimum value.

47. $f(x) = xe^{-x^2/8},\; [-1, 4].$ $f'(x) = x\cdot e^{-x^2/8}\cdot(-\tfrac{x}{4}) + e^{-x^2/8}\cdot 1 = e^{-x^2/8}(-\tfrac{x^2}{4} + 1).$ Since $e^{-x^2/8}$ is never 0, $f'(x) = 0 \;\Rightarrow\; -x^2/4 + 1 = 0 \;\Rightarrow\; 1 = x^2/4 \;\Rightarrow\; x^2 = 4 \;\Rightarrow\; x = \pm 2,$ but -2 is not in the given interval, $[-1, 4].$ $f(-1) = -e^{-1/8} \approx -0.88,\; f(2) = 2e^{-1/2} \approx 1.21,$ and $f(4) = 4e^{-2} \approx 0.54.$ So $f(2) = 2e^{-1/2}$ is the absolute maximum value and $f(-1) = -e^{-1/8}$ is the absolute minimum value.

49. $f(x) = x^a(1 - x)^b,\; 0 \le x \le 1, a > 0, b > 0.$

$$f'(x) = x^a\cdot b(1 - x)^{b-1}(-1) + (1 - x)^b\cdot ax^{a-1} = x^{a-1}(1 - x)^{b-1}[x\cdot b(-1) + (1 - x)\cdot a]$$
$$= x^{a-1}(1 - x)^{b-1}(a - ax - bx)$$

At the endpoints, we have $f(0) = f(1) = 0$ [the minimum value of f]. In the interval $(0, 1),$ $f'(x) = 0 \;\Leftrightarrow\; x = \dfrac{a}{a + b}.$

$$f\left(\frac{a}{a+b}\right) = \left(\frac{a}{a+b}\right)^a\left(1 - \frac{a}{a+b}\right)^b = \frac{a^a}{(a+b)^a}\left(\frac{a+b-a}{a+b}\right)^b = \frac{a^a}{(a+b)^a}\cdot\frac{b^b}{(a+b)^b} = \frac{a^a b^b}{(a+b)^{a+b}}.$$

So $f\left(\dfrac{a}{a+b}\right) = \dfrac{a^a b^b}{(a+b)^{a+b}}$ is the absolute maximum value.

51. (a)

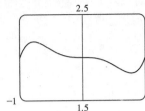

From the graph, it appears that the absolute maximum value is about

$f(-0.77) = 2.19$, and the absolute minimum value is about

$f(0.77) = 1.81$.

(b) $f(x) = x^5 - x^3 + 2 \Rightarrow f'(x) = 5x^4 - 3x^2 = x^2(5x^2 - 3)$. So $f'(x) = 0 \Rightarrow x = 0, \pm\sqrt{\frac{3}{5}}$.

$f\left(-\sqrt{\frac{3}{5}}\right) = \left(-\sqrt{\frac{3}{5}}\right)^5 - \left(-\sqrt{\frac{3}{5}}\right)^3 + 2 = -\left(\frac{3}{5}\right)^2\sqrt{\frac{3}{5}} + \frac{3}{5}\sqrt{\frac{3}{5}} + 2 = \left(\frac{3}{5} - \frac{9}{25}\right)\sqrt{\frac{3}{5}} + 2 = \frac{6}{25}\sqrt{\frac{3}{5}} + 2$ (maximum)

and similarly, $f\left(\sqrt{\frac{3}{5}}\right) = -\frac{6}{25}\sqrt{\frac{3}{5}} + 2$ (minimum).

53. (a)

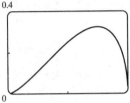

From the graph, it appears that the absolute maximum value is about

$f(0.75) = 0.32$, and the absolute minimum value is $f(0) = f(1) = 0$;

that is, at both endpoints.

(b) $f(x) = x\sqrt{x - x^2} \Rightarrow f'(x) = x \cdot \dfrac{1 - 2x}{2\sqrt{x - x^2}} + \sqrt{x - x^2} = \dfrac{(x - 2x^2) + (2x - 2x^2)}{2\sqrt{x - x^2}} = \dfrac{3x - 4x^2}{2\sqrt{x - x^2}}$.

So $f'(x) = 0 \Rightarrow 3x - 4x^2 = 0 \Rightarrow x(3 - 4x) = 0 \Rightarrow x = 0$ or $\frac{3}{4}$.

$f(0) = f(1) = 0$ (minimum), and $f\left(\frac{3}{4}\right) = \frac{3}{4}\sqrt{\frac{3}{4} - \left(\frac{3}{4}\right)^2} = \frac{3}{4}\sqrt{\frac{3}{16}} = \frac{3\sqrt{3}}{16}$ (maximum).

55. The density is defined as $\rho = \dfrac{\text{mass}}{\text{volume}} = \dfrac{1000}{V(T)}$ (in g/cm³). But a critical point of ρ will also be a critical point of V

[since $\dfrac{d\rho}{dT} = -1000V^{-2}\dfrac{dV}{dT}$ and V is never 0], and V is easier to differentiate than ρ.

$V(T) = 999.87 - 0.06426T + 0.0085043T^2 - 0.0000679T^3 \Rightarrow V'(T) = -0.06426 + 0.0170086T - 0.0002037T^2$.

Setting this equal to 0 and using the quadratic formula to find T, we get

$T = \dfrac{-0.0170086 \pm \sqrt{0.0170086^2 - 4 \cdot 0.0002037 \cdot 0.06426}}{2(-0.0002037)} \approx 3.9665°\text{C}$ or $79.5318°\text{C}$. Since we are only interested in

the region $0°\text{C} \le T \le 30°\text{C}$, we check the density ρ at the endpoints and at $3.9665°\text{C}$: $\rho(0) \approx \dfrac{1000}{999.87} \approx 1.00013$;

$\rho(30) \approx \dfrac{1000}{1003.7628} \approx 0.99625$; $\rho(3.9665) \approx \dfrac{1000}{999.7447} \approx 1.000255$. So water has its maximum density at

about $3.9665°\text{C}$.

57. We apply the Closed Interval Method to the continuous function

$I(t) = 0.00009045t^5 + 0.001438t^4 - 0.06561t^3 + 0.4598t^2 - 0.6270t + 99.33$ on $[0, 10]$. Its derivative is

$I'(t) = 0.00045225t^4 + 0.005752t^3 - 0.19683t^2 + 0.9196t - 0.6270$. Since I' exists for all t, the only critical numbers of I

occur when $I'(t) = 0$. We use a rootfinder on a computer algebra system (or a graphing device) to find that $I'(t) = 0$ when

$t \approx -29.7186, 0.8231, 5.1309$, or 11.0459, but only the second and third roots lie in the interval $[0, 10]$. The values of I at

these critical numbers are $I(0.8231) \approx 99.09$ and $I(5.1309) \approx 100.67$. The values of I at the endpoints of the interval are

$I(0) = 99.33$ and $I(10) \approx 96.86$. Comparing these four numbers, we see that food was most expensive at $t \approx 5.1309$

(corresponding roughly to August, 1989) and cheapest at $t = 10$ (midyear 1994).

59. (a) $v(r) = k(r_0 - r)r^2 = kr_0 r^2 - kr^3 \Rightarrow v'(r) = 2kr_0 r - 3kr^2.$ $v'(r) = 0 \Rightarrow kr(2r_0 - 3r) = 0 \Rightarrow$
$r = 0$ or $\frac{2}{3}r_0$ (but 0 is not in the interval). Evaluating v at $\frac{1}{2}r_0$, $\frac{2}{3}r_0$, and r_0, we get $v\left(\frac{1}{2}r_0\right) = \frac{1}{8}kr_0^3$, $v\left(\frac{2}{3}r_0\right) = \frac{4}{27}kr_0^3$,
and $v(r_0) = 0$. Since $\frac{4}{27} > \frac{1}{8}$, v attains its maximum value at $r = \frac{2}{3}r_0$. This supports the statement in the text.

(b) From part (a), the maximum value of v is $\frac{4}{27}kr_0^3$.

(c)

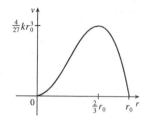

61. $f(x) = x^{101} + x^{51} + x + 1 \Rightarrow f'(x) = 101x^{100} + 51x^{50} + 1 \geq 1$ for all x, so $f'(x) = 0$ has no solution. Thus, $f(x)$
has no critical number, so $f(x)$ can have no local maximum or minimum.

63. If f has a local minimum at c, then $g(x) = -f(x)$ has a local maximum at c, so $g'(c) = 0$ by the case of Fermat's Theorem
proved in the text. Thus, $f'(c) = -g'(c) = 0$.

4.2 The Mean Value Theorem

1. $f(x) = x^2 - 4x + 1$, $[0, 4]$. Since f is a polynomial, it is continuous and differentiable on $\mathbb{R}$, so it is continuous on $[0, 4]$
and differentiable on $(0, 4)$. Also, $f(0) = 1 = f(4)$. $f'(c) = 0 \Leftrightarrow 2c - 4 = 0 \Leftrightarrow c = 2$, which is in the open interval
$(0, 4)$, so $c = 2$ satisfies the conclusion of Rolle's Theorem.

3. $f(x) = \sin 2\pi x$, $[-1, 1]$. f, being the composite of the sine function and the polynomial $2\pi x$, is continuous and
differentiable on $\mathbb{R}$, so it is continuous on $[-1, 1]$ and differentiable on $(-1, 1)$. Also, $f(-1) = 0 = f(1)$.
$f'(c) = 0 \Leftrightarrow 2\pi \cos 2\pi c = 0 \Leftrightarrow \cos 2\pi c = 0 \Leftrightarrow 2\pi c = \pm\frac{\pi}{2} + 2\pi n \Leftrightarrow c = \pm\frac{1}{4} + n$. If $n = 0$ or ± 1, then
$c = \pm\frac{1}{4}, \pm\frac{3}{4}$ is in $(-1, 1)$.

5. $f(x) = 1 - x^{2/3}$.

$f(-1) = 1 - (-1)^{2/3} = 1 - 1 = 0 = f(1)$.

$f'(x) = -\frac{2}{3}x^{-1/3}$, so $f'(c) = 0$ has no solution.

This does not contradict Rolle's Theorem, since
$f'(0)$ does not exist, and so f is not differentiable
on $(-1, 1)$.

7. $\dfrac{f(8) - f(0)}{8 - 0} = \dfrac{6 - 4}{8} = \dfrac{1}{4}$. The values of c which satisfy
$f'(c) = \frac{1}{4}$ seem to be about $c = 0.8$, 3.2, 4.4, and 6.1.

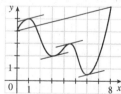

9. (a), (b) The equation of the secant line is $y - 5 = \dfrac{8.5 - 5}{8 - 1}(x - 1) \Leftrightarrow y = \frac{1}{2}x + \frac{9}{2}$.

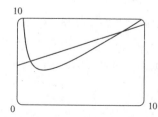

(c) $f(x) = x + 4/x \Rightarrow f'(x) = 1 - 4/x^2$.

So $f'(c) = \frac{1}{2} \Rightarrow c^2 = 8 \Rightarrow c = 2\sqrt{2}$, and $f(c) = 2\sqrt{2} + \frac{4}{2\sqrt{2}} = 3\sqrt{2}$. Thus, an equation of the tangent line

is $y - 3\sqrt{2} = \frac{1}{2}(x - 2\sqrt{2}) \Leftrightarrow y = \frac{1}{2}x + 2\sqrt{2}$.

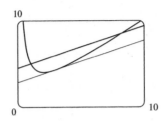

11. $f(x) = 3x^2 + 2x + 5$, $[-1, 1]$. f is continuous on $[-1, 1]$ and differentiable on $(-1, 1)$ since polynomials are continuous

and differentiable on $\mathbb{R}$. $f'(c) = \dfrac{f(b) - f(a)}{b - a} \Leftrightarrow 6c + 2 = \dfrac{f(1) - f(-1)}{1 - (-1)} = \dfrac{10 - 6}{2} = 2 \Leftrightarrow 6c = 0 \Leftrightarrow c = 0$,

which is in $(-1, 1)$.

13. $f(x) = e^{-2x}$, $[0, 3]$. f is continuous and differentiable on $\mathbb{R}$, so it is continuous on $[0, 3]$ and differentiable on $(0, 3)$.

$f'(c) = \dfrac{f(b) - f(a)}{b - a} \Leftrightarrow -2e^{-2c} = \dfrac{e^{-6} - e^0}{3 - 0} \Leftrightarrow e^{-2c} = \dfrac{1 - e^{-6}}{6} \Leftrightarrow -2c = \ln\left(\dfrac{1 - e^{-6}}{6}\right) \Leftrightarrow$

$c = -\dfrac{1}{2}\ln\left(\dfrac{1 - e^{-6}}{6}\right) \approx 0.897$, which is in $(0, 3)$.

15. $f(x) = |x - 1|$. $f(3) - f(0) = |3 - 1| - |0 - 1| = 1$. Since $f'(c) = -1$ if $c < 1$ and $f'(c) = 1$ if $c > 1$,

$f'(c)(3 - 0) = \pm 3$ and so is never equal to 1. This does not contradict the Mean Value Theorem since $f'(1)$ does not exist.

17. Let $f(x) = 1 + 2x + x^3 + 4x^5$. Then $f(-1) = -6 < 0$ and $f(0) = 1 > 0$. Since f is a polynomial, it is continuous, so the

Intermediate Value Theorem says that there is a number c between -1 and 0 such that $f(c) = 0$. Thus, the given equation has

a real root. Suppose the equation has distinct real roots a and b with $a < b$. Then $f(a) = f(b) = 0$. Since f is a polynomial, it

is differentiable on (a, b) and continuous on $[a, b]$. By Rolle's Theorem, there is a number r in (a, b) such that $f'(r) = 0$. But

$f'(x) = 2 + 3x^2 + 20x^4 \geq 2$ for all x, so $f'(x)$ can never be 0. This contradiction shows that the equation can't have two

distinct real roots. Hence, it has exactly one real root.

19. Let $f(x) = x^3 - 15x + c$ for x in $[-2, 2]$. If f has two real roots a and b in $[-2, 2]$, with $a < b$, then $f(a) = f(b) = 0$. Since

the polynomial f is continuous on $[a, b]$ and differentiable on (a, b), Rolle's Theorem implies that there is a number r in (a, b)

such that $f'(r) = 0$. Now $f'(r) = 3r^2 - 15$. Since r is in (a, b), which is contained in $[-2, 2]$, we have $|r| < 2$, so $r^2 < 4$. It

follows that $3r^2 - 15 < 3 \cdot 4 - 15 = -3 < 0$. This contradicts $f'(r) = 0$, so the given equation can't have two real roots in

$[-2, 2]$. Hence, it has at most one real root in $[-2, 2]$.

21. (a) Suppose that a cubic polynomial $P(x)$ has roots $a_1 < a_2 < a_3 < a_4$, so $P(a_1) = P(a_2) = P(a_3) = P(a_4)$.

By Rolle's Theorem there are numbers c_1, c_2, c_3 with $a_1 < c_1 < a_2$, $a_2 < c_2 < a_3$ and $a_3 < c_3 < a_4$ and

$P'(c_1) = P'(c_2) = P'(c_3) = 0$. Thus, the second-degree polynomial $P'(x)$ has three distinct real roots, which is

impossible.

(b) We prove by induction that a polynomial of degree n has at most n real roots. This is certainly true for $n = 1$. Suppose that the result is true for all polynomials of degree n and let $P(x)$ be a polynomial of degree $n + 1$. Suppose that $P(x)$ has more than $n + 1$ real roots, say $a_1 < a_2 < a_3 < \cdots < a_{n+1} < a_{n+2}$. Then $P(a_1) = P(a_2) = \cdots = P(a_{n+2}) = 0$. By Rolle's Theorem there are real numbers $c_1, \ldots, c_{n+1}$ with $a_1 < c_1 < a_2, \ldots, a_{n+1} < c_{n+1} < a_{n+2}$ and $P'(c_1) = \cdots = P'(c_{n+1}) = 0$. Thus, the nth degree polynomial $P'(x)$ has at least $n + 1$ roots. This contradiction shows that $P(x)$ has at most $n + 1$ real roots.

23. By the Mean Value Theorem, $f(4) - f(1) = f'(c)(4 - 1)$ for some $c \in (1, 4)$. But for every $c \in (1, 4)$ we have $f'(c) \geq 2$. Putting $f'(c) \geq 2$ into the above equation and substituting $f(1) = 10$, we get
$f(4) = f(1) + f'(c)(4 - 1) = 10 + 3f'(c) \geq 10 + 3 \cdot 2 = 16$. So the smallest possible value of $f(4)$ is 16.

25. Suppose that such a function f exists. By the Mean Value Theorem there is a number $0 < c < 2$ with
$$f'(c) = \frac{f(2) - f(0)}{2 - 0} = \frac{5}{2}.$$ But this is impossible since $f'(x) \leq 2 < \frac{5}{2}$ for all x, so no such function can exist.

27. We use Exercise 26 with $f(x) = \sqrt{1 + x}$, $g(x) = 1 + \frac{1}{2}x$, and $a = 0$. Notice that $f(0) = 1 = g(0)$ and
$$f'(x) = \frac{1}{2\sqrt{1 + x}} < \frac{1}{2} = g'(x) \text{ for } x > 0.$$ So by Exercise 26, $f(b) < g(b) \;\Rightarrow\; \sqrt{1 + b} < 1 + \frac{1}{2}b$ for $b > 0$.

Another method: Apply the Mean Value Theorem directly to either $f(x) = 1 + \frac{1}{2}x - \sqrt{1 + x}$ or $g(x) = \sqrt{1 + x}$ on $[0, b]$.

29. Let $f(x) = \sin x$ and let $b < a$. Then $f(x)$ is continuous on $[b, a]$ and differentiable on (b, a). By the Mean Value Theorem, there is a number $c \in (b, a)$ with $\sin a - \sin b = f(a) - f(b) = f'(c)(a - b) = (\cos c)(a - b)$. Thus, $|\sin a - \sin b| \leq |\cos c|\,|b - a| \leq |a - b|$. If $a < b$, then $|\sin a - \sin b| = |\sin b - \sin a| \leq |b - a| = |a - b|$. If $a = b$, both sides of the inequality are 0.

31. For $x > 0$, $f(x) = g(x)$, so $f'(x) = g'(x)$. For $x < 0$, $f'(x) = (1/x)' = -1/x^2$ and $g'(x) = (1 + 1/x)' = -1/x^2$, so again $f'(x) = g'(x)$. However, the domain of $g(x)$ is not an interval [it is $(-\infty, 0) \cup (0, \infty)$] so we cannot conclude that $f - g$ is constant (in fact it is not).

33. Let $f(x) = \arcsin\left(\dfrac{x - 1}{x + 1}\right) - 2\arctan\sqrt{x} + \dfrac{\pi}{2}$. Note that the domain of f is $[0, \infty)$. Thus,
$$f'(x) = \frac{1}{\sqrt{1 - \left(\dfrac{x - 1}{x + 1}\right)^2}} \cdot \frac{(x + 1) - (x - 1)}{(x + 1)^2} - \frac{2}{1 + x} \cdot \frac{1}{2\sqrt{x}} = \frac{1}{\sqrt{x}\,(x + 1)} - \frac{1}{\sqrt{x}\,(x + 1)} = 0. \text{ Then}$$
$f(x) = C$ on $(0, \infty)$ by Theorem 5. By continuity of f, $f(x) = C$ on $[0, \infty)$. To find C, we let $x = 0 \;\Rightarrow$
$\arcsin(-1) - 2\arctan(0) + \frac{\pi}{2} = C \;\Rightarrow\; -\frac{\pi}{2} - 0 + \frac{\pi}{2} = 0 = C$. Thus, $f(x) = 0 \;\Rightarrow$
$$\arcsin\left(\frac{x - 1}{x + 1}\right) = 2\arctan\sqrt{x} - \frac{\pi}{2}.$$

35. Let $g(t)$ and $h(t)$ be the position functions of the two runners and let $f(t) = g(t) - h(t)$. By hypothesis,
$f(0) = g(0) - h(0) = 0$ and $f(b) = g(b) - h(b) = 0$, where b is the finishing time. Then by the Mean Value Theorem, there
is a time c, with $0 < c < b$, such that $f'(c) = \dfrac{f(b) - f(0)}{b - 0}$. But $f(b) = f(0) = 0$, so $f'(c) = 0$. Since
$f'(c) = g'(c) - h'(c) = 0$, we have $g'(c) = h'(c)$. So at time c, both runners have the same speed $g'(c) = h'(c)$.

4.3 Derivatives and the Shapes of Graphs

Abbreviations: CU = Concave upward, CD = Concave downward, IP = Inflection point, HA = Horizontal asymptote, VA = Vertical asymptote.

1. (a) $f(x) = x^3 - 12x + 1 \Rightarrow f'(x) = 3x^2 - 12 = 3(x+2)(x-2)$.

We don't need to include "3" in the chart to determine the sign of $f'(x)$.

Interval	$x+2$	$x-2$	$f'(x)$	f
$x < -2$	$-$	$-$	$+$	increasing on $(-\infty, -2)$
$-2 < x < 2$	$+$	$-$	$-$	decreasing on $(-2, 2)$
$x > 2$	$+$	$+$	$+$	increasing on $(2, \infty)$

So f is increasing on $(-\infty, -2)$ and $(2, \infty)$ and f is decreasing on $(-2, 2)$.

(b) f changes from increasing to decreasing at $x = -2$ and from decreasing to increasing at $x = 2$. Thus, $f(-2) = 17$ is a local maximum value and $f(2) = -15$ is a local minimum value.

(c) $f''(x) = 6x$. $f''(x) > 0 \Leftrightarrow x > 0$ and $f''(x) < 0 \Leftrightarrow x < 0$. Thus, f is CU on $(0, \infty)$ and CD on $(-\infty, 0)$.

There is an IP where the concavity changes, at $(0, f(0)) = (0, 1)$.

3. (a) $f(x) = x - 2\sin x$ on $(0, 3\pi) \Rightarrow f'(x) = 1 - 2\cos x$. $f'(x) > 0 \Leftrightarrow 1 - 2\cos x > 0 \Leftrightarrow \cos x < \frac{1}{2} \Leftrightarrow$
$\frac{\pi}{3} < x < \frac{5\pi}{3}$ or $\frac{7\pi}{3} < x < 3\pi$. $f'(x) < 0 \Leftrightarrow \cos x > \frac{1}{2} \Leftrightarrow 0 < x < \frac{\pi}{3}$ or $\frac{5\pi}{3} < x < \frac{7\pi}{3}$. So f is increasing on $\left(\frac{\pi}{3}, \frac{5\pi}{3}\right)$ and $\left(\frac{7\pi}{3}, 3\pi\right)$, and f is decreasing on $\left(0, \frac{\pi}{3}\right)$ and $\left(\frac{5\pi}{3}, \frac{7\pi}{3}\right)$.

(b) f changes from increasing to decreasing at $x = \frac{5\pi}{3}$, and from decreasing to increasing at $x = \frac{\pi}{3}$ and at $x = \frac{7\pi}{3}$. Thus,
$f\left(\frac{5\pi}{3}\right) = \frac{5\pi}{3} + \sqrt{3} \approx 6.97$ is a local maximum value and $f\left(\frac{\pi}{3}\right) = \frac{\pi}{3} - \sqrt{3} \approx -0.68$ and $f\left(\frac{7\pi}{3}\right) = \frac{7\pi}{3} - \sqrt{3} \approx 5.60$ are local minimum values.

(c) $f''(x) = 2\sin x > 0 \Leftrightarrow 0 < x < \pi$ and $2\pi < x < 3\pi$, $f''(x) < 0 \Leftrightarrow \pi < x < 2\pi$. Thus, f is CU on $(0, \pi)$ and $(2\pi, 3\pi)$, and f is CD on $(\pi, 2\pi)$. There are IPs at (π, π) and $(2\pi, 2\pi)$.

5. (a) $y = f(x) = xe^x \Rightarrow f'(x) = xe^x + e^x = e^x(x+1)$. So $f'(x) > 0 \Leftrightarrow x + 1 > 0 \Leftrightarrow x > -1$. Thus, f is increasing on $(-1, \infty)$ and decreasing on $(-\infty, -1)$.

(b) f changes from decreasing to increasing at its only critical number, $x = -1$. Thus, $f(-1) = -e^{-1}$ is a local minimum value.

(c) $f'(x) = e^x(x+1) \Rightarrow f''(x) = e^x(1) + (x+1)e^x = e^x(x+2)$. So $f''(x) > 0 \Leftrightarrow x + 2 > 0 \Leftrightarrow x > -2$. Thus, f is CU on $(-2, \infty)$ and CD on $(-\infty, -2)$. Since the concavity changes direction at $x = -2$, the point $\left(-2, -2e^{-2}\right)$ is an IP.

7. (a) $y = f(x) = \dfrac{\ln x}{\sqrt{x}}$. (Note that f is only defined for $x > 0$.)

$f'(x) = \dfrac{\sqrt{x}\,(1/x) - \ln x \left(\frac{1}{2}x^{-1/2}\right)}{x} = \dfrac{\dfrac{1}{\sqrt{x}} - \dfrac{\ln x}{2\sqrt{x}}}{x} \cdot \dfrac{2\sqrt{x}}{2\sqrt{x}} = \dfrac{2 - \ln x}{2x^{3/2}} > 0 \Leftrightarrow 2 - \ln x > 0 \Leftrightarrow$

$\ln x < 2 \Leftrightarrow x < e^2$. Therefore f is increasing on $(0, e^2)$ and decreasing on (e^2, ∞).

(b) f changes from increasing to decreasing at $x = e^2$, so $f\left(e^2\right) = \dfrac{\ln e^2}{\sqrt{e^2}} = \dfrac{2}{e}$ is a local maximum value.

(c) $f''(x) = \dfrac{2x^{3/2}(-1/x) - (2 - \ln x)(3x^{1/2})}{\left(2x^{3/2}\right)^2} = \dfrac{-2x^{1/2} + 3x^{1/2}(\ln x - 2)}{4x^3} = \dfrac{x^{1/2}(-2 + 3\ln x - 6)}{4x^3} = \dfrac{3\ln x - 8}{4x^{5/2}}.$

$f''(x) = 0 \iff \ln x = \tfrac{8}{3} \iff x = e^{8/3}.$ $f''(x) > 0 \iff x > e^{8/3}$, so f is CU on $\left(e^{8/3}, \infty\right)$ and CD on $\left(0, e^{8/3}\right)$.

There is an IP at $\left(e^{8/3}, \tfrac{8}{3}e^{-4/3}\right) \approx (14.39, 0.70).$

9. $f(x) = x + \sqrt{1 - x} \implies f'(x) = 1 + \tfrac{1}{2}(1 - x)^{-1/2}(-1) = 1 - \dfrac{1}{2\sqrt{1 - x}}$. Note that f is defined for $1 - x \geq 0$; that is,

for $x \leq 1$. $f'(x) = 0 \implies 2\sqrt{1 - x} = 1 \implies \sqrt{1 - x} = \tfrac{1}{2} \implies 1 - x = \tfrac{1}{4} \implies x = \tfrac{3}{4}$. f' does not exist at $x = 1$,

but we can't have a local maximum or minimum at an endpoint.

First Derivative Test: $f'(x) > 0 \implies x < \tfrac{3}{4}$ and $f'(x) < 0 \implies \tfrac{3}{4} < x < 1$. Since f' changes from positive to

negative at $x = \tfrac{3}{4}$, $f\left(\tfrac{3}{4}\right) = \tfrac{5}{4}$ is a local maximum value.

Second Derivative Test: $f''(x) = -\tfrac{1}{2}\left(-\tfrac{1}{2}\right)(1 - x)^{-3/2}(-1) = -\dfrac{1}{4\left(\sqrt{1 - x}\right)^3}.$

$f''\left(\tfrac{3}{4}\right) = -2 < 0 \implies f\left(\tfrac{3}{4}\right) = \tfrac{5}{4}$ is a local maximum value.

Preference: The First Derivative Test may be slightly easier to apply in this case.

11. (a) By the Second Derivative Test, if $f'(2) = 0$ and $f''(2) = -5 < 0$, f has a local maximum at $x = 2$.

(b) If $f'(6) = 0$, we know that f has a horizontal tangent at $x = 6$. Knowing that $f''(6) = 0$ does not provide any additional information since the Second Derivative Test fails. For example, the first and second derivatives of $y = (x - 6)^4$, $y = -(x - 6)^4$, and $y = (x - 6)^3$ all equal zero for $x = 6$, but the first has a local minimum at $x = 6$, the second has a local maximum at $x = 6$, and the third has an inflection point at $x = 6$.

13. (a) There is an IP at $x = 3$ because the graph of f changes from CD to CU there. There is an IP at $x = 5$ because the graph of f changes from CU to CD there.

(b) There is an IP at $x = 2$ and at $x = 6$ because $f'(x)$ has a maximum value there, and so $f''(x)$ changes from positive to negative there. There is an IP at $x = 4$ because $f'(x)$ has a minimum value there and so $f''(x)$ changes from negative to positive there.

(c) There is an inflection point at $x = 1$ because $f''(x)$ changes from negative to positive there, and so the graph of f changes from CD to CU. There is an IP at $x = 7$ because $f''(x)$ changes from positive to negative there, and so the graph of f changes from CU to CD.

15. The function must be always decreasing (since the first derivative is always negative) and concave downward (since the second derivative is always negative).

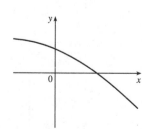

17. $f'(0) = f'(2) = f'(4) = 0 \Rightarrow$ horizontal tangents at $x = 0, 2, 4$.

$f'(x) > 0$ if $x < 0$ or $2 < x < 4 \Rightarrow f$ is increasing on $(-\infty, 0)$ and $(2, 4)$.

$f'(x) < 0$ if $0 < x < 2$ or $x > 4 \Rightarrow f$ is decreasing on $(0, 2)$ and $(4, \infty)$.

$f''(x) > 0$ if $1 < x < 3 \Rightarrow f$ is CU on $(1, 3)$. $f''(x) < 0$ if $x < 1$ or

$x > 3 \Rightarrow f$ is CD on $(-\infty, 1)$ and $(3, \infty)$. There are IPs when $x = 1$ and 3.

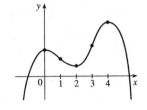

19. $f'(x) > 0$ if $|x| < 2 \Rightarrow f$ is increasing on $(-2, 2)$.

$f'(x) < 0$ if $|x| > 2 \Rightarrow f$ is decreasing on $(-\infty, -2)$

and $(2, \infty)$. $f'(-2) = 0 \Rightarrow$ horizontal tangent

at $x = -2$. $\lim\limits_{x \to 2} |f'(x)| = \infty \Rightarrow$ there is a vertical

asymptote or vertical tangent (cusp) at $x = 2$.

$f''(x) > 0$ if $x \neq 2 \Rightarrow f$ is CU on $(-\infty, 2)$ and $(2, \infty)$.

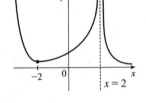

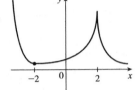

21. (a) f is increasing where f' is positive, that is, on $(0, 2)$, $(4, 6)$, and $(8, \infty)$; and decreasing where f' is negative, that is, on $(2, 4)$ and $(6, 8)$.

(b) f has local maxima where f' changes from positive to negative, at $x = 2$ and at $x = 6$, and local minima where f' changes from negative to positive, at $x = 4$ and at $x = 8$.

(c) f is CU where f' is increasing, that is, on $(3, 6)$ and $(6, \infty)$, and CD where f' is decreasing, that is, on $(0, 3)$.

(e)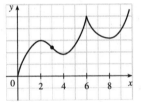

(d) There is an IP where f changes from being CD to being CU, that is, at $x = 3$.

23. (a) $f(x) = 2x^3 - 3x^2 - 12x \Rightarrow f'(x) = 6x^2 - 6x - 12 = 6(x^2 - x - 2) = 6(x - 2)(x + 1)$.

$f'(x) > 0 \Leftrightarrow x < -1$ or $x > 2$ and $f'(x) < 0 \Leftrightarrow -1 < x < 2$.

So f is increasing on $(-\infty, -1)$ and $(2, \infty)$, and f is decreasing on $(-1, 2)$.

(b) Since f changes from increasing to decreasing at $x = -1$, $f(-1) = 7$ is a local

maximum value. Since f changes from decreasing to increasing at $x = 2$,

$f(2) = -20$ is a local minimum value.

(d)

(c) $f''(x) = 6(2x - 1) \Rightarrow f''(x) > 0$ on $\left(\frac{1}{2}, \infty\right)$ and $f''(x) < 0$ on $\left(-\infty, \frac{1}{2}\right)$.

So f is CU on $\left(\frac{1}{2}, \infty\right)$ and CD on $\left(-\infty, \frac{1}{2}\right)$. There is a change in concavity at

$x = \frac{1}{2}$, and we have an IP at $\left(\frac{1}{2}, -\frac{13}{2}\right)$.

25. (a) $f(x) = x^4 - 6x^2 \Rightarrow f'(x) = 4x^3 - 12x = 4x(x^2 - 3) = 0$ when $x = 0, \pm\sqrt{3}$.

Interval	$4x$	$x^2 - 3$	$f'(x)$	f
$x < -\sqrt{3}$	$-$	$+$	$-$	decreasing on $\left(-\infty, -\sqrt{3}\right)$
$-\sqrt{3} < x < 0$	$-$	$-$	$+$	increasing on $\left(-\sqrt{3}, 0\right)$
$0 < x < \sqrt{3}$	$+$	$-$	$-$	decreasing on $\left(0, \sqrt{3}\right)$
$x > \sqrt{3}$	$+$	$+$	$+$	increasing on $\left(\sqrt{3}, \infty\right)$

(b) Local minimum values $f(\pm\sqrt{3}) = -9$,

local maximum value $f(0) = 0$

(c) $f''(x) = 12x^2 - 12 = 12(x^2 - 1) > 0 \iff x^2 > 1 \iff |x| > 1 \iff$ $x > 1$ or $x < -1$, so f is CU on $(-\infty, -1)$, $(1, \infty)$ and CD on $(-1, 1)$.

There are IPs at $(\pm 1, -5)$.

(d)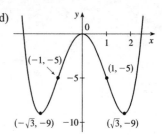

27. (a) $h(x) = 3x^5 - 5x^3 + 3 \Rightarrow h'(x) = 15x^4 - 15x^2 = 15x^2(x^2 - 1) = 0$ when $x = 0, \pm 1$. Since $15x^2$ is nonnegative,

$h'(x) > 0 \iff x^2 > 1 \iff |x| > 1 \iff x > 1$ or $x < -1$, so h is increasing on $(-\infty, -1)$ and $(1, \infty)$ and

decreasing on $(-1, 1)$, with a horizontal tangent at $x = 0$.

(b) Local maximum value $h(-1) = 5$, local minimum value $h(1) = 1$

(c) $h''(x) = 60x^3 - 30x = 30x(2x^2 - 1) = 60x\left(x + \frac{1}{\sqrt{2}}\right)\left(x - \frac{1}{\sqrt{2}}\right) \Rightarrow$

$h''(x) > 0$ when $x > \frac{1}{\sqrt{2}}$ or $-\frac{1}{\sqrt{2}} < x < 0$, so h is CU on $\left(-\frac{1}{\sqrt{2}}, 0\right)$ and

$\left(\frac{1}{\sqrt{2}}, \infty\right)$ and CD on $\left(-\infty, -\frac{1}{\sqrt{2}}\right)$ and $\left(0, \frac{1}{\sqrt{2}}\right)$. There are IPs at $(0, 3)$ and

$\left(\pm\frac{1}{\sqrt{2}}, 3 \mp \frac{7}{8}\sqrt{2}\right)$ [about $(-0.71, 4.24)$ and $(0.71, 1.76)$].

(d)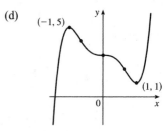

29. (a) $A(x) = x\sqrt{x+3} \Rightarrow A'(x) = x \cdot \frac{1}{2}(x+3)^{-1/2} + \sqrt{x+3} \cdot 1 = \dfrac{x}{2\sqrt{x+3}} + \sqrt{x+3} = \dfrac{x + 2(x+3)}{2\sqrt{x+3}} = \dfrac{3x+6}{2\sqrt{x+3}}$.

The domain of A is $[-3, \infty)$. $A'(x) > 0$ for $x > -2$ and $A'(x) < 0$ for $-3 < x < -2$, so A is increasing on $(-2, \infty)$

and decreasing on $(-3, -2)$.

(b) $A(-2) = -2$ is a local minimum value.

(c) $A''(x) = \dfrac{2\sqrt{x+3} \cdot 3 - (3x+6) \cdot \dfrac{1}{\sqrt{x+3}}}{\left(2\sqrt{x+3}\right)^2}$

$= \dfrac{6(x+3) - (3x+6)}{4(x+3)^{3/2}} = \dfrac{3x+12}{4(x+3)^{3/2}} = \dfrac{3(x+4)}{4(x+3)^{3/2}}$

$A''(x) > 0$ for all $x > -3$, so A is CU on $(-3, \infty)$. No IP

(d)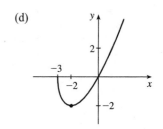

31. (a) $C(x) = x^{1/3}(x+4) = x^{4/3} + 4x^{1/3} \Rightarrow C'(x) = \frac{4}{3}x^{1/3} + \frac{4}{3}x^{-2/3} = \frac{4}{3}x^{-2/3}(x+1) = \dfrac{4(x+1)}{3\sqrt[3]{x^2}}$. $C'(x) > 0$ if

$-1 < x < 0$ or $x > 0$ and $C'(x) < 0$ for $x < -1$, so C is increasing on $(-1, \infty)$ and C is decreasing on $(-\infty, -1)$.

(b) $C(-1) = -3$ is a local minimum value.

(c) $C''(x) = \frac{4}{9}x^{-2/3} - \frac{8}{9}x^{-5/3} = \frac{4}{9}x^{-5/3}(x-2) = \dfrac{4(x-2)}{9\sqrt[3]{x^5}}$.

$C''(x) < 0$ for $0 < x < 2$ and $C''(x) > 0$ for $x < 0$ and $x > 2$,

so C is CD on $(0, 2)$ and CU on $(-\infty, 0)$ and $(2, \infty)$.

There are IPs at $(0, 0)$ and $\left(2, 6\sqrt[3]{2}\right) \approx (2, 7.56)$.

(d)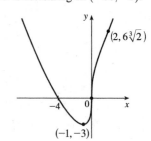

33. (a) $f(\theta) = 2\cos\theta - \cos 2\theta$, $0 \le \theta \le 2\pi$.

$f'(\theta) = -2\sin\theta + 2\sin 2\theta = -2\sin\theta + 2(2\sin\theta\cos\theta) = 2\sin\theta\,(2\cos\theta - 1)$.

Interval	$\sin\theta$	$2\cos\theta - 1$	$f'(\theta)$	f
$0 < \theta < \frac{\pi}{3}$	+	+	+	increasing on $(0, \frac{\pi}{3})$
$\frac{\pi}{3} < \theta < \pi$	+	−	−	decreasing on $(\frac{\pi}{3}, \pi)$
$\pi < \theta < \frac{5\pi}{3}$	−	−	+	increasing on $(\pi, \frac{5\pi}{3})$
$\frac{5\pi}{3} < \theta < 2\pi$	−	+	−	decreasing on $(\frac{5\pi}{3}, 2\pi)$

(b) $f\left(\frac{\pi}{3}\right) = \frac{3}{2}$ and $f\left(\frac{5\pi}{3}\right) = \frac{3}{2}$ are local maximum values and $f(\pi) = -3$ is a local minimum value.

(c) $f'(\theta) = -2\sin\theta + 2\sin 2\theta$ $\Rightarrow$

$\qquad f''(\theta) = -2\cos\theta + 4\cos 2\theta = -2\cos\theta + 4(2\cos^2\theta - 1)$

$\qquad\qquad = 2(4\cos^2\theta - \cos\theta - 2)$

$\qquad f''(\theta) = 0 \iff \cos\theta = \dfrac{1 \pm \sqrt{33}}{8} \iff \theta = \cos^{-1}\left(\dfrac{1 \pm \sqrt{33}}{8}\right) \iff$

(d)

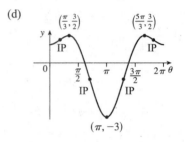

$\theta = \cos^{-1}\left(\dfrac{1 + \sqrt{33}}{8}\right) \approx 0.5678$, $2\pi - \cos^{-1}\left(\dfrac{1 + \sqrt{33}}{8}\right) \approx 5.7154$,

$\cos^{-1}\left(\dfrac{1 - \sqrt{33}}{8}\right) \approx 2.2057$, or $2\pi - \cos^{-1}\left(\dfrac{1 - \sqrt{33}}{8}\right) \approx 4.0775$. Denote these four values of θ by $\theta_1, \theta_4, \theta_2$, and θ_3,

respectively. Then f is CU on $(0, \theta_1)$, CD on (θ_1, θ_2), CU on (θ_2, θ_3), CD on (θ_3, θ_4), and CU on $(\theta_4, 2\pi)$. To find the *exact* y-coordinate for $\theta = \theta_1$, we have

$$f(\theta_1) = 2\cos\theta_1 - \cos 2\theta_1 = 2\cos\theta_1 - \left(2\cos^2\theta_1 - 1\right) = 2\left(\frac{1 + \sqrt{33}}{8}\right) - 2\left(\frac{1 + \sqrt{33}}{8}\right)^2 + 1$$

$$= \tfrac{1}{4} + \tfrac{1}{4}\sqrt{33} - \tfrac{1}{32} - \tfrac{1}{16}\sqrt{33} - \tfrac{33}{32} + 1 = \tfrac{3}{16} + \tfrac{3}{16}\sqrt{33} = \tfrac{3}{16}\left(1 + \sqrt{33}\right) = y_1 \approx 1.26.$$

Similarly, $f(\theta_2) = \frac{3}{16}\left(1 - \sqrt{33}\right) = y_2 \approx -0.89$. So f has IPs at (θ_1, y_1), (θ_2, y_2), (θ_3, y_2), and (θ_4, y_1).

35. $f(x) = \dfrac{x^2}{x^2 - 1} = \dfrac{x^2}{(x+1)(x-1)}$ has domain $(-\infty, -1) \cup (-1, 1) \cup (1, \infty)$.

(a) $\displaystyle\lim_{x \to \pm\infty} f(x) = \lim_{x \to \pm\infty} \frac{x^2/x^2}{(x^2 - 1)/x^2} = \lim_{x \to \pm\infty} \frac{1}{1 - 1/x^2} = \frac{1}{1 - 0} = 1$, so $y = 1$ is a HA.

$\displaystyle\lim_{x \to -1^-} \frac{x^2}{x^2 - 1} = \infty$ since $x^2 \to 1$ and $(x^2 - 1) \to 0^+$ as $x \to -1^-$, so $x = -1$ is a VA.

$\displaystyle\lim_{x \to 1^+} \frac{x^2}{x^2 - 1} = \infty$ since $x^2 \to 1$ and $(x^2 - 1) \to 0^+$ as $x \to 1^+$, so $x = 1$ is a VA.

(b) $f(x) = \dfrac{x^2}{x^2 - 1}$ $\Rightarrow$ $f'(x) = \dfrac{(x^2 - 1)(2x) - x^2(2x)}{(x^2 - 1)^2} = \dfrac{2x[(x^2 - 1) - x^2]}{(x^2 - 1)^2} = \dfrac{-2x}{(x^2 - 1)^2}$. Since $(x^2 - 1)^2$ is

positive for all x in the domain of f, the sign of the derivative is determined by the sign of $-2x$. Thus, $f'(x) > 0$ if $x < 0$ $(x \ne -1)$ and $f'(x) < 0$ if $x > 0$ $(x \ne 1)$. So f is increasing on $(-\infty, -1)$ and $(-1, 0)$, and f is decreasing on $(0, 1)$ and $(1, \infty)$.

(c) $f'(x) = 0 \Rightarrow x = 0$ and $f(0) = 0$ is a local maximum value.

(e)

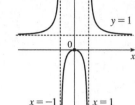

(d) $f''(x) = \dfrac{(x^2 - 1)^2(-2) - (-2x) \cdot 2(x^2 - 1)(2x)}{[(x^2 - 1)^2]^2}$

$= \dfrac{2(x^2 - 1)[-(x^2 - 1) + 4x^2]}{(x^2 - 1)^4} = \dfrac{2(3x^2 + 1)}{(x^2 - 1)^3}.$

The sign of $f''(x)$ is determined by the denominator; that is, $f''(x) > 0$ if

$|x| > 1$ and $f''(x) < 0$ if $|x| < 1$. Thus, f is CU on $(-\infty, -1)$ and $(1, \infty)$,

and f is CD on $(-1, 1)$. No IP

37. (a) $\lim\limits_{x \to -\infty} \left(\sqrt{x^2 + 1} - x\right) = \infty$ and

$\lim\limits_{x \to \infty} \left(\sqrt{x^2 + 1} - x\right) = \lim\limits_{x \to \infty} \left(\sqrt{x^2 + 1} - x\right) \dfrac{\sqrt{x^2 + 1} + x}{\sqrt{x^2 + 1} + x} = \lim\limits_{x \to \infty} \dfrac{1}{\sqrt{x^2 + 1} + x} = 0$, so $y = 0$ is a HA.

(b) $f(x) = \sqrt{x^2 + 1} - x \Rightarrow f'(x) = \dfrac{x}{\sqrt{x^2 + 1}} - 1$. Since $\dfrac{x}{\sqrt{x^2 + 1}} < 1$ for all x, $f'(x) < 0$, so f is decreasing on $\mathbb{R}$.

(c) No minimum or maximum

(d) $f''(x) = \dfrac{\left(x^2 + 1\right)^{1/2}(1) - x \cdot \frac{1}{2}\left(x^2 + 1\right)^{-1/2}(2x)}{\left(\sqrt{x^2 + 1}\right)^2}$

(e)

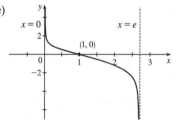

$= \dfrac{\left(x^2 + 1\right)^{1/2} - \dfrac{x^2}{\left(x^2 + 1\right)^{1/2}}}{x^2 + 1} = \dfrac{\left(x^2 + 1\right) - x^2}{\left(x^2 + 1\right)^{3/2}} = \dfrac{1}{\left(x^2 + 1\right)^{3/2}} > 0,$

so f is CU on $\mathbb{R}$. No IP

39. $f(x) = \ln(1 - \ln x)$ is defined when $x > 0$ (so that $\ln x$ is defined) and $1 - \ln x > 0$ [so that $\ln(1 - \ln x)$ is defined]. The

second condition is equivalent to $1 > \ln x \Leftrightarrow x < e$, so f has domain $(0, e)$.

(a) As $x \to 0^+$, $\ln x \to -\infty$, so $1 - \ln x \to \infty$ and $f(x) \to \infty$. As $x \to e^-$, $\ln x \to 1^-$, so $1 - \ln x \to 0^+$ and

$f(x) \to -\infty$. Thus, $x = 0$ and $x = e$ are VAs. There is no HA.

(b) $f'(x) = \dfrac{1}{1 - \ln x}\left(-\dfrac{1}{x}\right) = -\dfrac{1}{x(1 - \ln x)} < 0$ on $(0, e)$. Thus, f is decreasing on its domain, $(0, e)$.

(c) $f'(x) \neq 0$ on $(0, e)$, so f has no local maximum or minimum value.

(e)

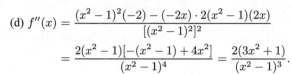

(d) $f''(x) = -\dfrac{-[x(1 - \ln x)]'}{[x(1 - \ln x)]^2} = \dfrac{x(-1/x) + (1 - \ln x)}{x^2(1 - \ln x)^2} = -\dfrac{\ln x}{x^2(1 - \ln x)^2}$

so $f''(x) > 0 \Leftrightarrow \ln x < 0 \Leftrightarrow 0 < x < 1$. Thus, f is CU on $(0, 1)$ and

CD on $(1, e)$. There is an IP at $(1, 0)$.

41. (a) $\lim\limits_{x \to \pm\infty} e^{-1/(x+1)} = 1$ since $-1/(x + 1) \to 0$, so $y = 1$ is a HA. $\lim\limits_{x \to -1^+} e^{-1/(x+1)} = 0$ since $-1/(x + 1) \to -\infty$,

$\lim\limits_{x \to -1^-} e^{-1/(x+1)} = \infty$ since $-1/(x + 1) \to \infty$, so $x = -1$ is a VA.

(b) $f(x) = e^{-1/(x+1)} \Rightarrow f'(x) = e^{-1/(x+1)}\left[-(-1)\dfrac{1}{(x + 1)^2}\right]$ [Reciprocal Rule] $= e^{-1/(x+1)}/(x + 1)^2 \Rightarrow$

$f'(x) > 0$ for all x except -1, so f is increasing on $(-\infty, -1)$ and $(-1, \infty)$.

(c) No local maximum or minimum

(d) $f''(x) = \dfrac{(x+1)^2 e^{-1/(x+1)} \left[1/(x+1)^2\right] - e^{-1/(x+1)} \left[2(x+1)\right]}{\left[(x+1)^2\right]^2}$

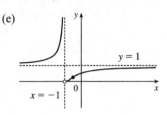

(e)

$\qquad = \dfrac{e^{-1/(x+1)} \left[1 - (2x+2)\right]}{(x+1)^4} = -\dfrac{e^{-1/(x+1)}(2x+1)}{(x+1)^4} \;\Rightarrow$

$f''(x) > 0 \;\Leftrightarrow\; 2x + 1 < 0 \;\Leftrightarrow\; x < -\frac{1}{2}$, so f is CU on $(-\infty, -1)$

and $\left(-1, -\frac{1}{2}\right)$, and CD on $\left(-\frac{1}{2}, \infty\right)$. There is an IP at $\left(-\frac{1}{2}, e^{-2}\right)$.

43. The nonnegative factors $(x+1)^2$ and $(x-6)^4$ do not affect the sign of $f'(x) = (x+1)^2(x-3)^5(x-6)^4$.

So $f'(x) > 0 \;\Rightarrow\; (x-3)^5 > 0 \;\Rightarrow\; x - 3 > 0 \;\Rightarrow\; x > 3$. Thus, f is increasing on the interval $(3, \infty)$.

45. (a)

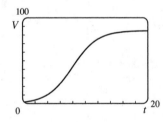

From the graph, we get an estimate of $f(1) \approx 1.41$ as a local maximum value,

and no local minimum value. $\quad f(x) = \dfrac{x+1}{\sqrt{x^2+1}} \;\Rightarrow\; f'(x) = \dfrac{1-x}{(x^2+1)^{3/2}}$.

$f'(x) = 0 \;\Leftrightarrow\; x = 1. \quad f(1) = \frac{2}{\sqrt{2}} = \sqrt{2}$ is the exact value.

(b) From the graph in part (a), f increases most rapidly somewhere between $x = -\frac{1}{2}$ and $x = -\frac{1}{4}$. To find the exact value,

we need to find the maximum value of f', which we can do by finding the critical numbers of f'.

$f''(x) = \dfrac{2x^2 - 3x - 1}{(x^2+1)^{5/2}} = 0 \;\Leftrightarrow\; x = \dfrac{3 \pm \sqrt{17}}{4}. \quad x = \dfrac{3 + \sqrt{17}}{4}$ corresponds to the *minimum* value of f'.

The maximum value of f' occurs at $\frac{3 - \sqrt{17}}{4} \approx -0.28$.

47.

From the graph, we estimate that the most rapid increase in the percentage of

households in the United States with at least one VCR occurs at about $t = 8$.

To maximize the first derivative, we need to determine the values for which the

second derivative is 0. We'll use $V(t) = \dfrac{a}{1 + be^{ct}}$, and substitute $a = 85$,

$b = 53$, and $c = -0.5$ later.

$V'(t) = -\dfrac{a(bce^{ct})}{(1 + be^{ct})^2}$ [by the Reciprocal Rule] and

$V''(t) = -abc \cdot \dfrac{(1 + be^{ct})^2 \cdot ce^{ct} - e^{ct} \cdot 2(1 + be^{ct}) \cdot bce^{ct}}{\left[(1 + be^{ct})^2\right]^2} = \dfrac{-abc \cdot ce^{ct}(1 + be^{ct})[(1 + be^{ct}) - 2be^{ct}]}{(1 + be^{ct})^4}$

$\qquad = \dfrac{-abc^2 e^{ct}(1 - be^{ct})}{(1 + be^{ct})^3}$

So $V''(t) = 0 \;\Leftrightarrow\; 1 = be^{ct} \;\Leftrightarrow\; e^{ct} = 1/b \;\Leftrightarrow\; ct = \ln(1/b) \;\Leftrightarrow\; t = (1/c)\ln(1/b) = -2\ln\frac{1}{53} \approx 7.94$ years,

which corresponds to roughly midyear 1988.

49. $f(x) = ax^3 + bx^2 + cx + d \Rightarrow f'(x) = 3ax^2 + 2bx + c$. We are given that

$f(1) = 0$ and $f(-2) = 3$, so $f(1) = a + b + c + d = 0$ and

$f(-2) = -8a + 4b - 2c + d = 3$. Also $f'(1) = 3a + 2b + c = 0$ and

$f'(-2) = 12a - 4b + c = 0$ by Fermat's Theorem. Solving these four equations, we get

$a = \frac{2}{9}, b = \frac{1}{3}, c = -\frac{4}{3}, d = \frac{7}{9}$, so the function is $f(x) = \frac{1}{9}(2x^3 + 3x^2 - 12x + 7)$.

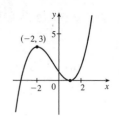

51. $f(x) = \tan x - x \Rightarrow f'(x) = \sec^2 x - 1 > 0$ for $0 < x < \frac{\pi}{2}$ since $\sec^2 x > 1$ for $0 < x < \frac{\pi}{2}$. So f is increasing on

$\left(0, \frac{\pi}{2}\right)$. Thus, $f(x) > f(0) = 0$ for $0 < x < \frac{\pi}{2} \Rightarrow \tan x - x > 0 \Rightarrow \tan x > x$ for $0 < x < \frac{\pi}{2}$.

53. Let the cubic function be $f(x) = ax^3 + bx^2 + cx + d \Rightarrow f'(x) = 3ax^2 + 2bx + c \Rightarrow f''(x) = 6ax + 2b$. So f is CU

when $6ax + 2b > 0 \Leftrightarrow x > -b/(3a)$, CD when $x < -b/(3a)$, and so the only IP occurs when $x = -b/(3a)$. If the graph

has three x-intercepts x_1, x_2 and x_3, then the expression for $f(x)$ must factor as $f(x) = a(x - x_1)(x - x_2)(x - x_3)$.

Multiplying these factors together gives us $f(x) = a\left[x^3 - (x_1 + x_2 + x_3)x^2 + (x_1x_2 + x_1x_3 + x_2x_3)x - x_1x_2x_3\right]$.

Equating the coefficients of the x^2-terms for the two forms of f gives us $b = -a(x_1 + x_2 + x_3)$. Hence, the x-coordinate of

the point of inflection is $-\dfrac{b}{3a} = -\dfrac{-a(x_1 + x_2 + x_3)}{3a} = \dfrac{x_1 + x_2 + x_3}{3}$.

55. By hypothesis $g = f'$ is differentiable on an open interval containing c. Since $(c, f(c))$ is an IP, the concavity changes at

$x = c$, so $f''(x)$ changes signs at $x = c$. Hence, by the First Derivative Test, f' has a local extremum at $x = c$. Thus, by

Fermat's Theorem $f''(c) = 0$.

57. Using the fact that $|x| = \sqrt{x^2}$, we have that $g(x) = x\sqrt{x^2} \Rightarrow g'(x) = \sqrt{x^2} + \sqrt{x^2} = 2\sqrt{x^2} = 2|x| \Rightarrow$

$g''(x) = 2x(x^2)^{-1/2} = \dfrac{2x}{|x|} < 0$ for $x < 0$ and $g''(x) > 0$ for $x > 0$, so $(0,0)$ is an IP. But $g''(0)$ does not exist.

4.4 Curve Sketching

1. $y = f(x) = x^3 + x = x(x^2 + 1)$ **A.** f is a polynomial, so $D = \mathbb{R}$.

B. x-intercept $= 0$, y-intercept $= f(0) = 0$ **C.** $f(-x) = -f(x)$, so f is

odd; the curve is symmetric about the origin.   **D.** f is a polynomial, so there is

no asymptote.   **E.** $f'(x) = 3x^2 + 1 > 0$, so f is increasing on $(-\infty, \infty)$.

F. There is no critical number and hence, no local maximum or minimum value.

G. $f''(x) = 6x > 0$ on $(0, \infty)$ and $f''(x) < 0$ on $(-\infty, 0)$, so f is CU on

$(0, \infty)$ and CD on $(-\infty, 0)$. Since the concavity changes at $x = 0$, there is an

IP at $(0, 0)$.

H.

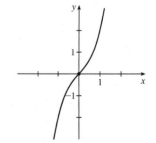

3. $y = f(x) = 2 - 15x + 9x^2 - x^3 = -(x - 2)(x^2 - 7x + 1)$ **A.** $D = \mathbb{R}$ **B.** y-intercept: $f(0) = 2$;

x-intercepts: $f(x) = 0 \Rightarrow x = 2$ or   [by the quadratic formula] $x = \frac{7 \pm \sqrt{45}}{2} \approx 0.15, 6.85$

C. No symmetry   **D.** No asymptote

E. $f'(x) = -15 + 18x - 3x^2 = -3(x^2 - 6x + 5)$

$\qquad = -3(x - 1)(x - 5) > 0 \Leftrightarrow 1 < x < 5,$

so f is increasing on $(1, 5)$ and decreasing on $(-\infty, 1)$ and $(5, \infty)$.

F. Local maximum value $f(5) = 27$, local minimum value $f(1) = -5$

G. $f''(x) = 18 - 6x = -6(x - 3) > 0 \Leftrightarrow x < 3$, so f is CU on $(-\infty, 3)$

and CD on $(3, \infty)$.   IP at $(3, 11)$

H.
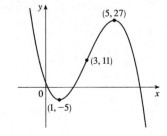

5. $y = f(x) = x^4 + 4x^3 = x^3(x + 4)$ **A.** $D = \mathbb{R}$ **B.** y-intercept: $f(0) = 0$;

x-intercepts: $f(x) = 0 \Leftrightarrow x = -4, 0$ **C.** No symmetry

D. No asymptote   **E.** $f'(x) = 4x^3 + 12x^2 = 4x^2(x + 3) > 0 \Leftrightarrow$

$x > -3$, so f is increasing on $(-3, \infty)$ and decreasing on $(-\infty, -3)$.

F. Local minimum value $f(-3) = -27$, no local maximum

G. $f''(x) = 12x^2 + 24x = 12x(x + 2) < 0 \Leftrightarrow -2 < x < 0$, so f is CD

on $(-2, 0)$ and CU on $(-\infty, -2)$ and $(0, \infty)$.   IP at $(0, 0)$ and $(-2, -16)$

H.
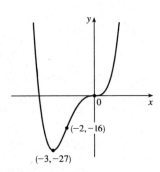

7. $y = f(x) = 2x^5 - 5x^2 + 1$ **A.** $D = \mathbb{R}$ **B.** y-intercept: $f(0) = 1$ **C.** No symmetry   **D.** No asymptote

E. $f'(x) = 10x^4 - 10x = 10x(x^3 - 1) = 10x(x - 1)(x^2 + x + 1)$, so $f'(x) < 0 \Leftrightarrow 0 < x < 1$ and $f'(x) > 0 \Leftrightarrow$

$x < 0$ or $x > 1$. Thus, f is increasing on $(-\infty, 0)$ and $(1, \infty)$ and decreasing on $(0, 1)$.   **F.** Local maximum

value $f(0) = 1$, local minimum value $f(1) = -2$

G. $f''(x) = 40x^3 - 10 = 10(4x^3 - 1)$ so $f''(x) = 0 \Leftrightarrow x = 1/\sqrt[3]{4}$.

$f''(x) > 0 \Leftrightarrow x > 1/\sqrt[3]{4}$ and $f''(x) < 0 \Leftrightarrow x < 1/\sqrt[3]{4}$, so f is CD

on $\left(-\infty, 1/\sqrt[3]{4}\right)$ and CU on $\left(1/\sqrt[3]{4}, \infty\right)$.

IP at $\left(\dfrac{1}{\sqrt[3]{4}}, 1 - \dfrac{9}{2\left(\sqrt[3]{4}\right)^2}\right) \approx (0.630, -0.786)$.

H.
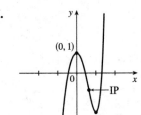

9. $y = f(x) = x/(x - 1)$ **A.** $D = \{x \mid x \neq 1\} = (-\infty, 1) \cup (1, \infty)$ **B.** x-intercept $= 0$, y-intercept $= f(0) = 0$

C. No symmetry   **D.** $\displaystyle\lim_{x \to \pm\infty} \frac{x}{x - 1} = 1$, so $y = 1$ is a HA. $\displaystyle\lim_{x \to 1^-} \frac{x}{x - 1} = -\infty$, $\displaystyle\lim_{x \to 1^+} \frac{x}{x - 1} = \infty$, so $x = 1$ is a VA.

E. $f'(x) = \dfrac{(x - 1) - x}{(x - 1)^2} = \dfrac{-1}{(x - 1)^2} < 0$ for $x \neq 1$, so f is decreasing on

$(-\infty, 1)$ and $(1, \infty)$.   **F.** No extreme values   **G.** $f''(x) = \dfrac{2}{(x - 1)^3} > 0$

$\Leftrightarrow x > 1$, so f is CU on $(1, \infty)$ and CD on $(-\infty, 1)$.   No IP

H.
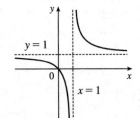

11. $y = f(x) = 1/(x^2 - 9)$ **A.** $D = \{x \mid x \neq \pm 3\} = (-\infty, -3) \cup (-3, 3) \cup (3, \infty)$ **B.** y-intercept $= f(0) = -\frac{1}{9}$,

no x-intercept   **C.** $f(-x) = f(x) \Rightarrow f$ is even; the curve is symmetric about the y-axis.   **D.** $\lim\limits_{x \to \pm\infty} \dfrac{1}{x^2 - 9} = 0$,

so $y = 0$ is a HA. $\lim\limits_{x \to 3^-} \dfrac{1}{x^2 - 9} = -\infty$, $\lim\limits_{x \to 3^+} \dfrac{1}{x^2 - 9} = \infty$, $\lim\limits_{x \to -3^-} \dfrac{1}{x^2 - 9} = \infty$, $\lim\limits_{x \to -3^+} \dfrac{1}{x^2 - 9} = -\infty$,

so $x = 3$ and $x = -3$ are VAs.   **E.** $f'(x) = -\dfrac{2x}{(x^2 - 9)^2} > 0 \Leftrightarrow x < 0 \ (x \neq -3)$ so f is increasing on

$(-\infty, -3)$ and $(-3, 0)$ and decreasing on $(0, 3)$ and $(3, \infty)$.

F. Local maximum value $f(0) = -\frac{1}{9}$.

G. $y'' = \dfrac{-2(x^2 - 9)^2 + (2x)2(x^2 - 9)(2x)}{(x^2 - 9)^4} = \dfrac{6(x^2 + 3)}{(x^2 - 9)^3} > 0 \Leftrightarrow$

$x^2 > 9 \Leftrightarrow x > 3$ or $x < -3$, so f is CU on $(-\infty, -3)$ and $(3, \infty)$ and

CD on $(-3, 3)$.   No IP

H.

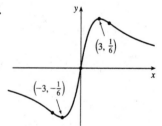

13. $y = f(x) = x/(x^2 + 9)$ **A.** $D = \mathbb{R}$ **B.** y-intercept: $f(0) = 0$; x-intercept: $f(x) = 0 \Leftrightarrow x = 0$

C. $f(-x) = -f(x)$, so f is odd and the curve is symmetric about the origin.   **D.** $\lim\limits_{x \to \pm\infty} \dfrac{x}{x^2 + 9} = 0$, so $y = 0$ is a HA;

no VA   **E.** $f'(x) = \dfrac{(x^2 + 9)(1) - x(2x)}{(x^2 + 9)^2} = \dfrac{9 - x^2}{(x^2 + 9)^2} = \dfrac{(3 + x)(3 - x)}{(x^2 + 9)^2} > 0 \Leftrightarrow -3 < x < 3$,

so f is increasing on $(-3, 3)$ and decreasing on $(-\infty, -3)$ and $(3, \infty)$.   **F.** Local minimum value $f(-3) = -\frac{1}{6}$,

local maximum value $f(3) = \frac{1}{6}$

G. $f''(x) = \dfrac{(x^2 + 9)^2(-2x) - (9 - x^2) \cdot 2(x^2 + 9)(2x)}{[(x^2 + 9)^2]^2}$

$= \dfrac{(2x)(x^2 + 9)[-(x^2 + 9) - 2(9 - x^2)]}{(x^2 + 9)^4}$

$= \dfrac{2x(x^2 - 27)}{(x^2 + 9)^3} = 0 \Leftrightarrow x = 0, \pm\sqrt{27} = \pm 3\sqrt{3}$

$f''(x) > 0 \Leftrightarrow -3\sqrt{3} < x < 0$ or $x > 3\sqrt{3}$, so f is CU on $\left(-3\sqrt{3}, 0\right)$ and

$\left(3\sqrt{3}, \infty\right)$, and CD on $\left(-\infty, -3\sqrt{3}\right)$ and $\left(0, 3\sqrt{3}\right)$. There are three IPs:

$(0, 0)$ and $\left(\pm 3\sqrt{3}, \pm\frac{1}{12}\sqrt{3}\right)$.

H.

15. $y = f(x) = (x - 1)/x^2$ **A.** $D = \{x \mid x \neq 0\} = (-\infty, 0) \cup (0, \infty)$ **B.** No y-intercept; x-intercept: $f(x) = 0 \Leftrightarrow$

$x = 1$ **C.** No symmetry   **D.** $\lim\limits_{x \to \pm\infty} \dfrac{x - 1}{x^2} = 0$, so $y = 0$ is a HA. $\lim\limits_{x \to 0} \dfrac{x - 1}{x^2} = -\infty$, so $x = 0$ is a VA.

E. $f'(x) = \dfrac{x^2 \cdot 1 - (x - 1) \cdot 2x}{(x^2)^2} = \dfrac{-x^2 + 2x}{x^4} = \dfrac{-(x - 2)}{x^3}$, so $f'(x) > 0 \Leftrightarrow 0 < x < 2$ and $f'(x) < 0 \Leftrightarrow$

$x < 0$ or $x > 2$. Thus, f is increasing on $(0, 2)$ and decreasing on $(-\infty, 0)$

and $(2, \infty)$.   **F.** No local minimum, local maximum value $f(2) = \frac{1}{4}$.

G. $f''(x) = \dfrac{x^3 \cdot (-1) - [-(x - 2)] \cdot 3x^2}{(x^3)^2} = \dfrac{2x^3 - 6x^2}{x^6} = \dfrac{2(x - 3)}{x^4}$.

$f''(x)$ is negative on $(-\infty, 0)$ and $(0, 3)$ and positive on $(3, \infty)$, so f is

CD on $(-\infty, 0)$ and $(0, 3)$ and CU on $(3, \infty)$.   IP at $\left(3, \frac{2}{9}\right)$

H.

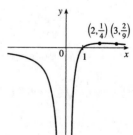

17. $y = f(x) = x\sqrt{5 - x}$ **A.** The domain is $\{x \mid 5 - x \geq 0\} = (-\infty, 5]$ **B.** y-intercept: $f(0) = 0$;
x-intercepts: $f(x) = 0 \Leftrightarrow x = 0, 5$ **C.** No symmetry **D.** No asymptote

E. $f'(x) = x \cdot \frac{1}{2}(5 - x)^{-1/2}(-1) + (5 - x)^{1/2} \cdot 1 = \frac{1}{2}(5 - x)^{-1/2}[-x + 2(5 - x)] = \dfrac{10 - 3x}{2\sqrt{5 - x}} > 0 \Leftrightarrow x < \frac{10}{3}$,

so f is increasing on $\left(-\infty, \frac{10}{3}\right)$ and decreasing on $\left(\frac{10}{3}, 5\right)$.

F. Local maximum value $f\left(\frac{10}{3}\right) = \frac{10}{9}\sqrt{15} \approx 4.3$; no local minimum

H.

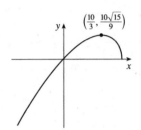

G. $f''(x) = \dfrac{2(5 - x)^{1/2}(-3) - (10 - 3x) \cdot 2\left(\frac{1}{2}\right)(5 - x)^{-1/2}(-1)}{\left(2\sqrt{5 - x}\right)^2}$

$= \dfrac{(5 - x)^{-1/2}[-6(5 - x) + (10 - 3x)]}{4(5 - x)} = \dfrac{3x - 20}{4(5 - x)^{3/2}}$

$f''(x) < 0$ for $x < 5$, so f is CD on $(-\infty, 5)$. No IP

19. $y = f(x) = x/\sqrt{x^2 + 1}$ **A.** $D = \mathbb{R}$ **B.** y-intercept: $f(0) = 0$; x-intercept: $f(x) = 0 \Rightarrow x = 0$
C. $f(-x) = -f(x)$, so f is odd; the graph is symmetric about the origin.

D. $\displaystyle\lim_{x \to \infty} f(x) = \lim_{x \to \infty} \frac{x}{\sqrt{x^2 + 1}} = \lim_{x \to \infty} \frac{x/x}{\sqrt{x^2 + 1}/x} = \lim_{x \to \infty} \frac{x/x}{\sqrt{x^2 + 1}/\sqrt{x^2}} = \lim_{x \to \infty} \frac{1}{\sqrt{1 + 1/x^2}} = \frac{1}{\sqrt{1 + 0}} = 1$ and

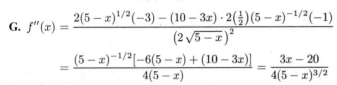

$\displaystyle\lim_{x \to -\infty} f(x) = \lim_{x \to -\infty} \frac{x}{\sqrt{x^2 + 1}} = \lim_{x \to -\infty} \frac{x/x}{\sqrt{x^2 + 1}/x} = \lim_{x \to -\infty} \frac{x/x}{\sqrt{x^2 + 1}/\left(-\sqrt{x^2}\right)}$

$= \displaystyle\lim_{x \to -\infty} \frac{1}{-\sqrt{1 + 1/x^2}} = \frac{1}{-\sqrt{1 + 0}} = -1$

so $y = \pm 1$ are HAs. No VA. **E.** $f'(x) = \dfrac{\sqrt{x^2 + 1} - x \cdot \dfrac{2x}{2\sqrt{x^2 + 1}}}{[(x^2 + 1)^{1/2}]^2} = \dfrac{x^2 + 1 - x^2}{(x^2 + 1)^{3/2}} = \dfrac{1}{(x^2 + 1)^{3/2}} > 0$ for all x,

so f is increasing on $\mathbb{R}$. **F.** No extreme values

H.

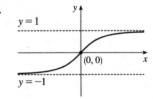

G. $f''(x) = -\frac{3}{2}(x^2 + 1)^{-5/2} \cdot 2x = \dfrac{-3x}{(x^2 + 1)^{5/2}}$, so $f''(x) > 0$ for $x < 0$

and $f''(x) < 0$ for $x > 0$. Thus, f is CU on $(-\infty, 0)$ and CD on $(0, \infty)$.
IP at $(0, 0)$

21. $y = f(x) = \sqrt{1 - x^2}/x$ **A.** $D = \{x \mid |x| \leq 1, x \neq 0\} = [-1, 0) \cup (0, 1]$ **B.** x-intercepts ± 1, no y-intercept

C. $f(-x) = -f(x)$, so the curve is symmetric about $(0, 0)$. **D.** $\displaystyle\lim_{x \to 0^+} \frac{\sqrt{1 - x^2}}{x} = \infty$, $\displaystyle\lim_{x \to 0^-} \frac{\sqrt{1 - x^2}}{x} = -\infty$,

so $x = 0$ is a VA. **E.** $f'(x) = \dfrac{(-x^2/\sqrt{1 - x^2}) - \sqrt{1 - x^2}}{x^2} = -\dfrac{1}{x^2\sqrt{1 - x^2}} < 0$, so f is decreasing on $(-1, 0)$

and $(0, 1)$. **F.** No extreme values **G.** $f''(x) = \dfrac{2 - 3x^2}{x^3(1 - x^2)^{3/2}} > 0 \Leftrightarrow$

H.

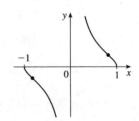

$-1 < x < -\sqrt{\frac{2}{3}}$ or $0 < x < \sqrt{\frac{2}{3}}$, so f is CU on $\left(-1, -\sqrt{\frac{2}{3}}\right)$ and $\left(0, \sqrt{\frac{2}{3}}\right)$

and CD on $\left(-\sqrt{\frac{2}{3}}, 0\right)$ and $\left(\sqrt{\frac{2}{3}}, 1\right)$. IP at $\left(\pm\sqrt{\frac{2}{3}}, \pm\frac{1}{\sqrt{2}}\right)$

23. $y = f(x) = x - 3x^{1/3}$ **A.** $D = \mathbb{R}$ **B.** y-intercept: $f(0) = 0$; x-intercepts: $f(x) = 0 \Rightarrow x = 3x^{1/3} \Rightarrow$

$x^3 = 27x \Rightarrow x^3 - 27x = 0 \Rightarrow x(x^2 - 27) = 0 \Rightarrow x = 0, \pm 3\sqrt{3}$ **C.** $f(-x) = -f(x)$, so f is odd;

the graph is symmetric about the origin.   **D.** No asymptote   **E.** $f'(x) = 1 - x^{-2/3} = 1 - \dfrac{1}{x^{2/3}} = \dfrac{x^{2/3} - 1}{x^{2/3}}$.

$f'(x) > 0$ when $|x| > 1$ and $f'(x) < 0$ when $0 < |x| < 1$, so f is increasing on $(-\infty, -1)$ and $(1, \infty)$, and

decreasing on $(-1, 0)$ and $(0, 1)$ [hence decreasing on $(-1, 1)$ since f is

continuous on $(-1, 1)$].   **F.** Local maximum value $f(-1) = 2$, local minimum

value $f(1) = -2$ **G.** $f''(x) = \frac{2}{3}x^{-5/3} < 0$ when $x < 0$ and $f''(x) > 0$

when $x > 0$, so f is CD on $(-\infty, 0)$ and CU on $(0, \infty)$.   IP at $(0, 0)$

H.
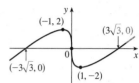

25. $y = f(x) = x + \sqrt{|x|}$ **A.** $D = \mathbb{R}$ **B.** x-intercepts $0, -1$; y-intercept 0 **C.** No symmetry   **D.** $\displaystyle\lim_{x \to \infty} \left(x + \sqrt{|x|}\right) = \infty$,

$\displaystyle\lim_{x \to -\infty} \left(x + \sqrt{|x|}\right) = -\infty$.   No asymptote   **E.** For $x > 0$, $f(x) = x + \sqrt{x} \Rightarrow f'(x) = 1 + \dfrac{1}{2\sqrt{x}} > 0$, so f increases

on $(0, \infty)$. For $x < 0$, $f(x) = x + \sqrt{-x} \Rightarrow f'(x) = 1 - \dfrac{1}{2\sqrt{-x}} > 0 \Leftrightarrow 2\sqrt{-x} > 1 \Leftrightarrow -x > \frac{1}{4} \Leftrightarrow$

$x < -\frac{1}{4}$, so f increases on $\left(-\infty, -\frac{1}{4}\right)$ and decreases on $\left(-\frac{1}{4}, 0\right)$.   **H.**

F. Local maximum value $f\left(-\frac{1}{4}\right) = \frac{1}{4}$, local minimum value $f(0) = 0$

G. For $x > 0$, $f''(x) = -\frac{1}{4}x^{-3/2} \Rightarrow f''(x) < 0$, so f is CD on $(0, \infty)$.

For $x < 0$, $f''(x) = -\frac{1}{4}(-x)^{-3/2} \Rightarrow f''(x) < 0$, so f is CD on $(-\infty, 0)$.

No IP

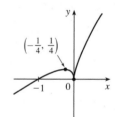

27. $y = f(x) = 3\sin x - \sin^3 x$ **A.** $D = \mathbb{R}$ **B.** y-intercept: $f(0) = 0$; x-intercepts: $f(x) = 0 \Rightarrow$

$\sin x \left(3 - \sin^2 x\right) = 0 \Rightarrow \sin x = 0$ [since $\sin^2 x \le 1 < 3$] $\Rightarrow x = n\pi$, n an integer.   **C.** $f(-x) = -f(x)$, so f is

odd; the graph (shown for $-2\pi \le x \le 2\pi$) is symmetric about the origin and periodic with period 2π.   **D.** No asymptote

E. $f'(x) = 3\cos x - 3\sin^2 x \cos x = 3\cos x \left(1 - \sin^2 x\right) = 3\cos^3 x$. $f'(x) > 0 \Leftrightarrow \cos x > 0 \Leftrightarrow$

$x \in \left(2n\pi - \frac{\pi}{2}, 2n\pi + \frac{\pi}{2}\right)$ for each integer n, and $f'(x) < 0 \Leftrightarrow \cos x < 0 \Leftrightarrow x \in \left(2n\pi + \frac{\pi}{2}, 2n\pi + \frac{3\pi}{2}\right)$ for each

integer n. Thus, f is increasing on $\left(2n\pi - \frac{\pi}{2}, 2n\pi + \frac{\pi}{2}\right)$ for each integer n, and f is decreasing on $\left(2n\pi + \frac{\pi}{2}, 2n\pi + \frac{3\pi}{2}\right)$

for each integer n.   **F.** f has local maximum values $f\left(2n\pi + \frac{\pi}{2}\right) = 2$ and local minimum values $f\left(2n\pi + \frac{3\pi}{2}\right) = -2$.

G. $f''(x) = -9\sin x \cos^2 x = -9\sin x \left(1 - \sin^2 x\right) = -9\sin x \left(1 - \sin x\right)(1 + \sin x)$.

$f''(x) < 0 \Leftrightarrow \sin x > 0$ and $\sin x \ne \pm 1 \Leftrightarrow x \in \left(2n\pi, 2n\pi + \frac{\pi}{2}\right) \cup \left(2n\pi + \frac{\pi}{2}, 2n\pi + \pi\right)$ for some integer n.

$f''(x) > 0 \Leftrightarrow \sin x < 0$ and $\sin x \ne \pm 1 \Leftrightarrow x \in \left((2n - 1)\pi, (2n - 1)\pi + \frac{\pi}{2}\right) \cup \left((2n - 1)\pi + \frac{\pi}{2}, 2n\pi\right)$

for some integer n. Thus, f is CD on the intervals $\left(2n\pi, \left(2n + \frac{1}{2}\right)\pi\right)$ and   **H.**

$\left(\left(2n + \frac{1}{2}\right)\pi, (2n + 1)\pi\right)$ [hence CD on the intervals $(2n\pi, (2n + 1)\pi)$] for

each integer n, and f is CU on the intervals $\left((2n - 1)\pi, \left(2n - \frac{1}{2}\right)\pi\right)$ and

$\left(\left(2n - \frac{1}{2}\right)\pi, 2n\pi\right)$ [hence CU on the intervals $((2n - 1)\pi, 2n\pi)$] for each

integer n. f has IPs at $(n\pi, 0)$ for each integer n.

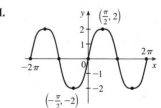

29. $y = f(x) = x \tan x$, $-\frac{\pi}{2} < x < \frac{\pi}{2}$ **A.** $D = \left(-\frac{\pi}{2}, \frac{\pi}{2}\right)$ **B.** Intercepts are 0   **C.** $f(-x) = f(x)$, so the curve is

symmetric about the y-axis.   **D.** $\lim\limits_{x \to (\pi/2)^-} x \tan x = \infty$ and $\lim\limits_{x \to -(\pi/2)^+} x \tan x = \infty$, so $x = \frac{\pi}{2}$ and $x = -\frac{\pi}{2}$ are VAs.

E. $f'(x) = \tan x + x \sec^2 x > 0$ $\Leftrightarrow$ $0 < x < \frac{\pi}{2}$, so f increases on $\left(0, \frac{\pi}{2}\right)$ **H.**

and decreases on $\left(-\frac{\pi}{2}, 0\right)$.   **F.** Absolute and local minimum value $f(0) = 0$.

G. $y'' = 2\sec^2 x + 2x \tan x \sec^2 x > 0$ for $-\frac{\pi}{2} < x < \frac{\pi}{2}$, so f is CU

on $\left(-\frac{\pi}{2}, \frac{\pi}{2}\right)$.   No IP

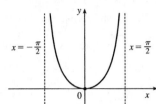

31. $y = f(x) = \frac{1}{2}x - \sin x$, $0 < x < 3\pi$ **A.** $D = (0, 3\pi)$ **B.** No y-intercept. The x-intercept, approximately 1.9, can be

found using Newton's method.   **C.** No symmetry   **D.** No asymptote   **E.** $f'(x) = \frac{1}{2} - \cos x > 0$ $\Leftrightarrow$ $\cos x < \frac{1}{2}$ $\Leftrightarrow$

$\frac{\pi}{3} < x < \frac{5\pi}{3}$ or $\frac{7\pi}{3} < x < 3\pi$, so f is increasing on $\left(\frac{\pi}{3}, \frac{5\pi}{3}\right)$ and $\left(\frac{7\pi}{3}, 3\pi\right)$ and decreasing on $\left(0, \frac{\pi}{3}\right)$ and $\left(\frac{5\pi}{3}, \frac{7\pi}{3}\right)$.

F. Local minimum value $f\left(\frac{\pi}{3}\right) = \frac{\pi}{6} - \frac{\sqrt{3}}{2}$, local maximum value   **H.**

$f\left(\frac{5\pi}{3}\right) = \frac{5\pi}{6} + \frac{\sqrt{3}}{2}$, local minimum value $f\left(\frac{7\pi}{3}\right) = \frac{7\pi}{6} - \frac{\sqrt{3}}{2}$

G. $f''(x) = \sin x > 0$ $\Leftrightarrow$ $0 < x < \pi$ or $2\pi < x < 3\pi$, so f is CU

on $(0, \pi)$ and $(2\pi, 3\pi)$ and CD on $(\pi, 2\pi)$.   IPs at $\left(\pi, \frac{\pi}{2}\right)$ and $(2\pi, \pi)$

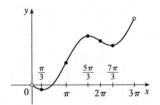

33. $y = f(x) = \dfrac{\sin x}{1 + \cos x}$ $\left[\substack{\text{when} \\ \cos x \neq 1} \; = \; \dfrac{\sin x}{1 + \cos x} \cdot \dfrac{1 - \cos x}{1 - \cos x} = \dfrac{\sin x \,(1 - \cos x)}{\sin^2 x} = \dfrac{1 - \cos x}{\sin x} = \csc x - \cot x\right]$

A. The domain of f is the set of all real numbers except odd integer multiples of π.   **B.** y-intercept: $f(0) = 0$;

x-intercepts: $x = n\pi$, n an even integer.   **C.** $f(-x) = -f(x)$, so f is an odd function; the graph is symmetric about the

origin and has period 2π.   **D.** When n is an odd integer, $\lim\limits_{x \to (n\pi)^-} f(x) = \infty$ and $\lim\limits_{x \to (n\pi)^+} f(x) = -\infty$, so $x = n\pi$ is a

VA for each odd integer n. No HA.   **E.** $f'(x) = \dfrac{(1 + \cos x) \cdot \cos x - \sin x(-\sin x)}{(1 + \cos x)^2} = \dfrac{1 + \cos x}{(1 + \cos x)^2} = \dfrac{1}{1 + \cos x}$.

$f'(x) > 0$ for all x except odd multiples of π, so f is increasing on $((2k-1)\pi, (2k+1)\pi)$ for each integer k.

F. No extreme values   **G.** $f''(x) = \dfrac{\sin x}{(1 + \cos x)^2} > 0$ $\Rightarrow$ **H.**

$\sin x > 0$ $\Rightarrow$ $x \in (2k\pi, (2k+1)\pi)$ and $f''(x) < 0$ on

$((2k-1)\pi, 2k\pi)$ for each integer k. f is CU on $(2k\pi, (2k+1)\pi)$

and CD on $((2k-1)\pi, 2k\pi)$ for each integer k. f has IPs

at $(2k\pi, 0)$ for each integer k.

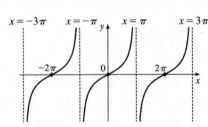

35. $y = 1/(1 + e^{-x})$ **A.** $D = \mathbb{R}$ **B.** No x-intercept; y-intercept $= f(0) = \frac{1}{2}$.   **C.** No symmetry

D. $\lim\limits_{x \to \infty} 1/(1 + e^{-x}) = \frac{1}{1+0} = 1$ and $\lim\limits_{x \to -\infty} 1/(1 + e^{-x}) = 0$ (since $\lim\limits_{x \to -\infty} e^{-x} = \infty$), so f has HAs $y = 0$ and $y = 1$.

E. $f'(x) = -(1 + e^{-x})^{-2}(-e^{-x}) = e^{-x}/(1 + e^{-x})^2$. This is positive for all x, so f is increasing on $\mathbb{R}$.

F. No extreme values

G. $f''(x) = \dfrac{(1+e^{-x})^2(-e^{-x}) - e^{-x}(2)(1+e^{-x})(-e^{-x})}{(1+e^{-x})^4} = \dfrac{e^{-x}(e^{-x}-1)}{(1+e^{-x})^3}$ **H.**

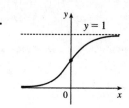

The second factor in the numerator is negative for $x > 0$ and positive for $x < 0$, and the other factors are always positive, so f is CU on $(-\infty, 0)$ and CD on $(0, \infty)$. f has an IP at $\left(0, \frac{1}{2}\right)$.

---

37. $y = f(x) = x\ln x$ **A.** $D = (0, \infty)$ **B.** x-intercept when $\ln x = 0 \;\Leftrightarrow\; x = 1$, no y-intercept   **C.** No symmetry

D. $\displaystyle\lim_{x\to\infty} x\ln x = \infty,\; \lim_{x\to 0^+} x\ln x = \lim_{x\to 0^+} \dfrac{\ln x}{1/x} \overset{\text{H}}{=} \lim_{x\to 0^+} \dfrac{1/x}{-1/x^2} = \lim_{x\to 0^+}(-x) = 0$, no asymptote.

E. $f'(x) = \ln x + 1 = 0$ when $\ln x = -1 \;\Leftrightarrow\; x = e^{-1}$.

$f'(x) > 0 \;\Leftrightarrow\; \ln x > -1 \;\Leftrightarrow\; x > e^{-1}$, so f is increasing on $(1/e, \infty)$ and decreasing on $(0, 1/e)$.   **F.** $f(1/e) = -1/e$ is an absolute and local minimum value.   **G.** $f''(x) = 1/x > 0$, so f is CU on $(0, \infty)$.   No IP

H.

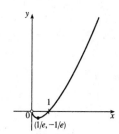

---

39. $y = f(x) = xe^{-x}$ **A.** $D = \mathbb{R}$ **B.** Intercepts are 0   **C.** No symmetry

D. $\displaystyle\lim_{x\to\infty} xe^{-x} = \lim_{x\to\infty} \dfrac{x}{e^x} \overset{\text{H}}{=} \lim_{x\to\infty} \dfrac{1}{e^x} = 0$, so $y = 0$ is a HA. $\displaystyle\lim_{x\to -\infty} xe^{-x} = -\infty$

E. $f'(x) = e^{-x} - xe^{-x} = e^{-x}(1-x) > 0 \;\Leftrightarrow\; x < 1$, so f is increasing on $(-\infty, 1)$ and decreasing on $(1, \infty)$.   **F.** Absolute and local maximum value $f(1) = 1/e$.   **G.** $f''(x) = e^{-x}(x-2) > 0 \;\Leftrightarrow\; x > 2$, so f is CU on $(2, \infty)$ and CD on $(-\infty, 2)$.   IP at $\left(2, 2/e^2\right)$

H.

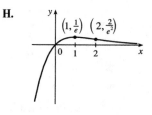

---

41. $y = f(x) = \ln(\sin x)$

A. $D = \{x \text{ in } \mathbb{R} \mid \sin x > 0\} = \displaystyle\bigcup_{n=-\infty}^{\infty} (2n\pi, (2n+1)\pi) = \cdots \cup (-4\pi, -3\pi) \cup (-2\pi, -\pi) \cup (0, \pi) \cup (2\pi, 3\pi) \cup \cdots$

B. No y-intercept; x-intercepts: $f(x) = 0 \;\Leftrightarrow\; \ln(\sin x) = 0 \;\Leftrightarrow\; \sin x = e^0 = 1 \;\Leftrightarrow\; x = 2n\pi + \frac{\pi}{2}$ for each integer n.   **C.** f is periodic with period 2π.   **D.** $\displaystyle\lim_{x\to(2n\pi)^+} f(x) = -\infty$ and $\displaystyle\lim_{x\to[(2n+1)\pi]^-} f(x) = -\infty$, so the lines $x = n\pi$ are VAs for all integers n.   **E.** $f'(x) = \frac{\cos x}{\sin x} = \cot x$, so $f'(x) > 0$ when $2n\pi < x < 2n\pi + \frac{\pi}{2}$ for each integer n, and $f'(x) < 0$ when $2n\pi + \frac{\pi}{2} < x < (2n+1)\pi$. Thus, f is increasing on $\left(2n\pi, 2n\pi + \frac{\pi}{2}\right)$ and decreasing on $\left(2n\pi + \frac{\pi}{2}, (2n+1)\pi\right)$ for each integer n.   **F.** Local maximum values $f\left(2n\pi + \frac{\pi}{2}\right) = 0$, no local minimum.

G. $f''(x) = -\csc^2 x < 0$, so f is CD on $(2n\pi, (2n+1)\pi)$ for each integer n.   No IP

H.

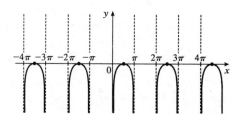

43. $y = f(x) = xe^{-x^2}$ **A.** $D = \mathbb{R}$ **B.** Intercepts are 0 **C.** $f(-x) = -f(x)$, so the curve is symmetric

about the origin. **D.** $\lim\limits_{x \to \pm\infty} xe^{-x^2} = \lim\limits_{x \to \pm\infty} \dfrac{x}{e^{x^2}} \overset{H}{=} \lim\limits_{x \to \pm\infty} \dfrac{1}{2xe^{x^2}} = 0$, so $y = 0$ is a HA.

E. $f'(x) = e^{-x^2} - 2x^2 e^{-x^2} = e^{-x^2}\left(1 - 2x^2\right) > 0 \iff x^2 < \frac{1}{2} \iff |x| < \frac{1}{\sqrt{2}}$, so f is increasing on $\left(-\frac{1}{\sqrt{2}}, \frac{1}{\sqrt{2}}\right)$

and decreasing on $\left(-\infty, -\frac{1}{\sqrt{2}}\right)$ and $\left(\frac{1}{\sqrt{2}}, \infty\right)$. **F.** Local maximum value $f\left(\frac{1}{\sqrt{2}}\right) = 1/\sqrt{2e}$, local minimum value

$f\left(-\frac{1}{\sqrt{2}}\right) = -1/\sqrt{2e}$ **G.** $f''(x) = -2xe^{-x^2}\left(1 - 2x^2\right) - 4xe^{-x^2} = 2xe^{-x^2}\left(2x^2 - 3\right) > 0 \iff$

$x > \sqrt{\frac{3}{2}}$ or $-\sqrt{\frac{3}{2}} < x < 0$, so f is CU on $\left(\sqrt{\frac{3}{2}}, \infty\right)$ **H.**

and $\left(-\sqrt{\frac{3}{2}}, 0\right)$ and CD on $\left(-\infty, -\sqrt{\frac{3}{2}}\right)$ and $\left(0, \sqrt{\frac{3}{2}}\right)$.

IPs at $(0, 0)$ and $\left(\pm\sqrt{\frac{3}{2}}, \pm\sqrt{\frac{3}{2}}\,e^{-3/2}\right)$

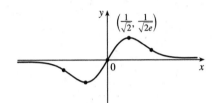

45. $y = -\dfrac{W}{24EI}x^4 + \dfrac{WL}{12EI}x^3 - \dfrac{WL^2}{24EI}x^2 = -\dfrac{W}{24EI}x^2\left(x^2 - 2Lx + L^2\right) = \dfrac{-W}{24EI}x^2(x - L)^2 = cx^2(x - L)^2$

where $c = -\dfrac{W}{24EI}$ is a negative constant and $0 \le x \le L$. We sketch $f(x) = cx^2(x - L)^2$ for $c = -1$.

$f(0) = f(L) = 0$

$f'(x) = cx^2[2(x - L)] + (x - L)^2(2cx) = 2cx(x - L)[x + (x - L)]$

$\quad = 2cx(x - L)(2x - L)$

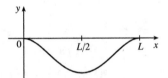

So for $0 < x < L$, $f'(x) > 0 \iff x(x - L)(2x - L) < 0$ [since $c < 0$] $\iff L/2 < x < L$ and $f'(x) < 0 \iff$

$0 < x < L/2$. Thus, f is increasing on $(L/2, L)$ and decreasing on $(0, L/2)$, and there is a local and absolute minimum at the

point $(L/2, f(L/2)) = (L/2, cL^4/16)$.

$f'(x) = 2c[x(x - L)(2x - L)] \Rightarrow$

$f''(x) = 2c[1(x - L)(2x - L) + x(1)(2x - L) + x(x - L)(2)] = 2c(6x^2 - 6Lx + L^2) = 0 \iff$

$x = \dfrac{6L \pm \sqrt{12L^2}}{12} = \frac{1}{2}L \pm \frac{\sqrt{3}}{6}L$, and these are the x-coordinates of the two inflection points.

47. $y = f(x) = \dfrac{-2x^2 + 5x - 1}{2x - 1} = -x + 2 + \dfrac{1}{2x - 1}$ **A.** $D = \left\{x \in \mathbb{R} \mid x \ne \frac{1}{2}\right\} = \left(-\infty, \frac{1}{2}\right) \cup \left(\frac{1}{2}, \infty\right)$

B. y-intercept: $f(0) = 1$; x-intercepts: $f(x) = 0 \Rightarrow -2x^2 + 5x - 1 = 0 \Rightarrow x = \dfrac{-5 \pm \sqrt{17}}{-4} \Rightarrow x \approx 0.22, 2.28$.

C. No symmetry **D.** $\lim\limits_{x \to (1/2)^-} f(x) = -\infty$ and $\lim\limits_{x \to (1/2)^+} f(x) = \infty$, so $x = \frac{1}{2}$ is a VA.

$\lim\limits_{x \to \pm\infty}[f(x) - (-x + 2)] = \lim\limits_{x \to \pm\infty} \dfrac{1}{2x - 1} = 0$, so the line $y = -x + 2$ is a SA.

E. $f'(x) = -1 - \dfrac{2}{(2x-1)^2} < 0$ for $x \neq \frac{1}{2}$, so f is decreasing on $\left(-\infty, \frac{1}{2}\right)$

and $\left(\frac{1}{2}, \infty\right)$.   **F.** No extreme values   **G.** $f'(x) = -1 - 2(2x-1)^{-2}$ $\Rightarrow$

$f''(x) = -2(-2)(2x-1)^{-3}(2) = \dfrac{8}{(2x-1)^3}$, so $f''(x) > 0$ when $x > \frac{1}{2}$ and

$f''(x) < 0$ when $x < \frac{1}{2}$. Thus, f is CU on $\left(\frac{1}{2}, \infty\right)$ and CD on $\left(-\infty, \frac{1}{2}\right)$.
No IP

H.

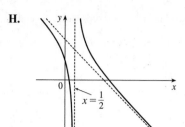

49. $y = f(x) = (x^2 + 4)/x = x + 4/x$ **A.** $D = \{x \mid x \neq 0\} = (-\infty, 0) \cup (0, \infty)$ **B.** No intercept

C. $f(-x) = -f(x)$ $\Rightarrow$ symmetry about the origin   **D.** $\lim\limits_{x \to \infty} (x + 4/x) = \infty$ but $f(x) - x = 4/x \to 0$ as $x \to \pm\infty$,

so $y = x$ is a SA. $\lim\limits_{x \to 0^+} (x + 4/x) = \infty$ and $\lim\limits_{x \to 0^-} (x + 4/x) = -\infty$, so $x = 0$ is a VA.

E. $f'(x) = 1 - 4/x^2 > 0$ $\Leftrightarrow$ $x^2 > 4$ $\Leftrightarrow$ $x > 2$ or $x < -2$, so f is

increasing on $(-\infty, -2)$ and $(2, \infty)$ and decreasing on $(-2, 0)$ and $(0, 2)$.

F. Local maximum value $f(-2) = -4$, local minimum value $f(2) = 4$

G. $f''(x) = 8/x^3 > 0$ $\Leftrightarrow$ $x > 0$ so f is CU on $(0, \infty)$ and CD

on $(-\infty, 0)$. No IP

H.

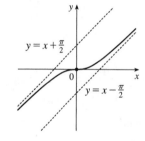

51. $y = f(x) = x - \tan^{-1} x$, $f'(x) = 1 - \dfrac{1}{1+x^2} = \dfrac{1+x^2-1}{1+x^2} = \dfrac{x^2}{1+x^2}$,

$f''(x) = \dfrac{(1+x^2)(2x) - x^2(2x)}{(1+x^2)^2} = \dfrac{2x(1+x^2-x^2)}{(1+x^2)^2} = \dfrac{2x}{(1+x^2)^2}$.

$\lim\limits_{x \to \infty} \left[f(x) - \left(x - \frac{\pi}{2}\right)\right] = \lim\limits_{x \to \infty} \left(\frac{\pi}{2} - \tan^{-1} x\right) = \frac{\pi}{2} - \frac{\pi}{2} = 0$, so

$y = x - \frac{\pi}{2}$ is a SA. Also,

$\lim\limits_{x \to -\infty} \left[f(x) - \left(x + \frac{\pi}{2}\right)\right] = \lim\limits_{x \to -\infty} \left(-\frac{\pi}{2} - \tan^{-1} x\right) = -\frac{\pi}{2} - \left(-\frac{\pi}{2}\right) = 0$

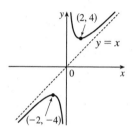

so $y = x + \frac{\pi}{2}$ is also a SA. $f'(x) \geq 0$ for all x, with equality $\Leftrightarrow$ $x = 0$, so f is increasing on $\mathbb{R}$. $f''(x)$ has the same sign

as x, so f is CD on $(-\infty, 0)$ and CU on $(0, \infty)$. $f(-x) = -f(x)$, so f is an odd function; its graph is symmetric about the

origin. f has no local extreme values. Its only IP is at $(0, 0)$.

53. $f(x) = 4x^4 - 32x^3 + 89x^2 - 95x + 29$ $\Rightarrow$ $f'(x) = 16x^3 - 96x^2 + 178x - 95$ $\Rightarrow$ $f''(x) = 48x^2 - 192x + 178$.

$f(x) = 0$ $\Leftrightarrow$ $x \approx 0.5,\ 1.60$; $f'(x) = 0$ $\Leftrightarrow$ $x \approx 0.92,\ 2.5,\ 2.58$ and $f''(x) = 0$ $\Leftrightarrow$ $x \approx 1.46,\ 2.54$.

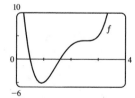

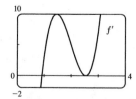

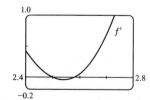

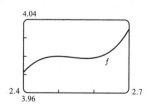

From the graphs of f', we estimate that $f' < 0$ and that f is decreasing on $(-\infty, 0.92)$ and $(2.5, 2.58)$, and that $f' > 0$ and

f is increasing on $(0.92, 2.5)$ and $(2.58, \infty)$ with local minimum values $f(0.92) \approx -5.12$ and $f(2.58) \approx 3.998$ and local maximum value $f(2.5) \approx 4$. The graphs of f' make it clear that f has a maximum and a minimum near $x = 2.5$, shown more clearly in the fourth graph.

From the graph of f'', we estimate that $f'' > 0$ and that f is CU on $(-\infty, 1.46)$ and $(2.54, \infty)$, and that $f'' < 0$ and f is CD on $(1.46, 2.54)$. There are IPs at about $(1.46, -1.40)$ and $(2.54, 3.999)$.

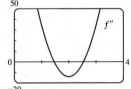

55. $f(x) = x^2 - 4x + 7 \cos x, \quad -4 \le x \le 4. \quad f'(x) = 2x - 4 - 7 \sin x \quad \Rightarrow \quad f''(x) = 2 - 7 \cos x.$

$f(x) = 0 \quad \Leftrightarrow \quad x \approx 1.10; f'(x) = 0 \quad \Leftrightarrow \quad x \approx -1.49, -1.07, \text{ or } 2.89; f''(x) = 0 \quad \Leftrightarrow \quad x = \pm \cos^{-1}\left(\frac{2}{7}\right) \approx \pm 1.28.$

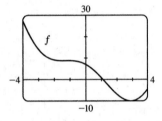
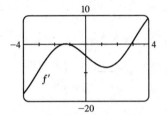
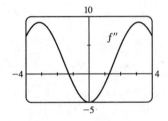

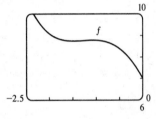
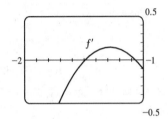

From the graphs of f', we estimate that f is decreasing ($f' < 0$) on $(-4, -1.49)$, increasing on $(-1.49, -1.07)$, decreasing on $(-1.07, 2.89)$, and increasing on $(2.89, 4)$, with local minimum values $f(-1.49) \approx 8.75$ and $f(2.89) \approx -9.99$ and local maximum value $f(-1.07) \approx 8.79$ (notice the second graph of f). From the graph of f'', we estimate that f is CU ($f'' > 0$) on $(-4, -1.28)$, CD on $(-1.28, 1.28)$, and CU on $(1.28, 4)$. There are IPs at about $(-1.28, 8.77)$ and $(1.28, -1.48)$.

57. $f(x) = 1 + \dfrac{1}{x} + \dfrac{8}{x^2} + \dfrac{1}{x^3} \quad \Rightarrow \quad f'(x) = -\dfrac{1}{x^2} - \dfrac{16}{x^3} - \dfrac{3}{x^4} = -\dfrac{1}{x^4}(x^2 + 16x + 3) \quad \Rightarrow$

$f''(x) = \dfrac{2}{x^3} + \dfrac{48}{x^4} + \dfrac{12}{x^5} = \dfrac{2}{x^5}(x^2 + 24x + 6).$

From the graphs, it appears that f increases on $(-15.8, -0.2)$ and decreases on $(-\infty, -15.8)$, $(-0.2, 0)$, and $(0, \infty)$; that f has a local minimum value of $f(-15.8) \approx 0.97$ and a local maximum value of $f(-0.2) \approx 72$; that f is CD on $(-\infty, -24)$ and $(-0.25, 0)$ and is CU on $(-24, -0.25)$ and $(0, \infty)$; and that f has IPs at $(-24, 0.97)$ and $(-0.25, 60)$.

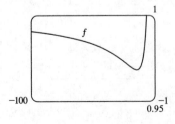
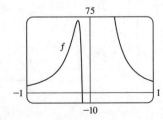

To find the exact values, note that $f' = 0 \Rightarrow x = \dfrac{-16 \pm \sqrt{256 - 12}}{2} = -8 \pm \sqrt{61}$ [≈ -0.19 and -15.81].

f' is positive (f is increasing) on $\left(-8 - \sqrt{61}, -8 + \sqrt{61}\right)$ and f' is negative (f is decreasing) on $\left(-\infty, -8 - \sqrt{61}\right)$,

$\left(-8 + \sqrt{61}, 0\right)$, and $(0, \infty)$. $f'' = 0 \Rightarrow x = \dfrac{-24 \pm \sqrt{576 - 24}}{2} = -12 \pm \sqrt{138}$ [≈ -0.25 and -23.75].

f'' is positive (f is CU) on $\left(-12 - \sqrt{138}, -12 + \sqrt{138}\right)$ and $(0, \infty)$ and f'' is negative (f is CD) on $\left(-\infty, -12 - \sqrt{138}\right)$ and $\left(-12 + \sqrt{138}, 0\right)$.

59. $f(x) = x^4 + cx^2 = x^2(x^2 + c)$. Note that f is an even function. For $c \geq 0$, the only x-intercept is the point $(0, 0)$. We

calculate $f'(x) = 4x^3 + 2cx = 4x\left(x^2 + \frac{1}{2}c\right) \Rightarrow f''(x) = 12x^2 + 2c$. If $c \geq 0$, $x = 0$ is the only critical point and there

is no inflection point. As we can see from the examples, there is no change in the basic shape of the graph for $c \geq 0$; it merely

becomes steeper as c increases. For $c = 0$, the graph is the simple curve $y = x^4$. For $c < 0$, there are x-intercepts at 0

and at $\pm\sqrt{-c}$. Also, there is a maximum at $(0, 0)$, and there are minima

at $\left(\pm\sqrt{-\frac{1}{2}c}, -\frac{1}{4}c^2\right)$. As $c \to -\infty$, the x-coordinates of these minima

get larger in absolute value, and the minimum points move downward.

There are IPs at $\left(\pm\sqrt{-\frac{1}{6}c}, -\frac{5}{36}c^2\right)$, which also move away from the

origin as $c \to -\infty$.

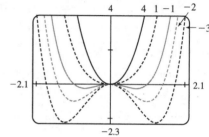

61.

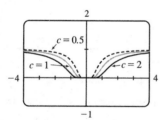 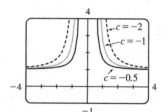

$c = 0$ is a transitional value—we get the graph of $y = 1$. For $c > 0$, we see that there is a HA at $y = 1$, and that the graph

spreads out as c increases. At first glance there appears to be a minimum at $(0, 0)$, but $f(0)$ is undefined, so there is no

minimum or maximum. For $c < 0$, we still have the HA at $y = 1$, but the range is $(1, \infty)$ rather than $(0, 1)$. We also have

a VA at $x = 0$. $f(x) = e^{-c/x^2} \Rightarrow f'(x) = e^{-c/x^2}\left(\dfrac{2c}{x^3}\right) \Rightarrow f''(x) = \dfrac{2c(2c - 3x^2)}{x^6 e^{c/x^2}}$. $f'(x) \neq 0$ and $f'(x)$

exists for all $x \neq 0$ (and 0 is not in the domain of f), so there are no maxima or minima. $f''(x) = 0 \Rightarrow x = \pm\sqrt{2c/3}$,

so if $c > 0$, the IPs spread out as c increases, and if $c < 0$, there are no IP. For $c > 0$, there are IPs at $\left(\pm\sqrt{2c/3}, e^{-3/2}\right)$.

Note that the y-coordinate of the IP is constant.

63. $f(x) = cx + \sin x \Rightarrow f'(x) = c + \cos x \Rightarrow f''(x) = -\sin x$

$f(-x) = -f(x)$, so f is an odd function and its graph is symmetric with respect to the origin.

$f(x) = 0 \Leftrightarrow \sin x = -cx$, so 0 is always an x-intercept.

$f'(x) = 0 \Leftrightarrow \cos x = -c$, so there is no critical number when $|c| > 1$. If $|c| \leq 1$, then there are infinitely many critical numbers. If x_1 is the unique solution of $\cos x = -c$ in the interval $[0, \pi]$, then the critical numbers are $2n\pi \pm x_1$, where n ranges over the integers. (Special cases: When $c = 1$, $x_1 = 0$; when $c = 0$, $x = \frac{\pi}{2}$; and when $c = -1$, $x_1 = \pi$.)

$f''(x) < 0 \iff \sin x > 0$, so f is CD on intervals of the form $(2n\pi, (2n+1)\pi)$. f is CU on intervals of the form $((2n-1)\pi, 2n\pi)$. The IPs of f are the points $(2n\pi, 2n\pi c)$, where n is an integer.

If $c \geq 1$, then $f'(x) \geq 0$ for all x, so f is increasing and has no extremum. If $c \leq -1$, then $f'(x) \leq 0$ for all x, so f is decreasing and has no extremum. If $|c| < 1$, then $f'(x) > 0 \iff \cos x > -c \iff x$ is in an interval of the form $(2n\pi - x_1, 2n\pi + x_1)$ for some integer n. These are the intervals on which f is increasing. Similarly, we find that f is decreasing on the intervals of the form $(2n\pi + x_1, 2(n+1)\pi - x_1)$. Thus, f has local maxima at the points $2n\pi + x_1$, where f has the values $c(2n\pi + x_1) + \sin x_1 = c(2n\pi + x_1) + \sqrt{1 - c^2}$, and f has local minima at the points $2n\pi - x_1$, where we have $f(2n\pi - x_1) = c(2n\pi - x_1) - \sin x_1 = c(2n\pi - x_1) - \sqrt{1 - c^2}$.

The transitional values of c are -1 and 1. The IPs move vertically, but not horizontally, when c changes. When $|c| \geq 1$, there is no extremum. For $|c| < 1$, the maxima are spaced 2π apart horizontally, as are the minima. The horizontal spacing between maxima and adjacent minima is regular (and equals π) when $c = 0$, but the horizontal space between a local maximum and the nearest local minimum shrinks as $|c|$ approaches 1.

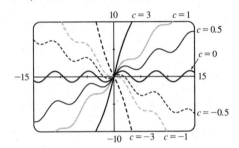

4.5 Optimization Problems

1. (a)

First Number	Second Number	Product
1	22	22
2	21	42
3	20	60
4	19	76
5	18	90
6	17	102
7	16	112
8	15	120
9	14	126
10	13	130
11	12	132

We needn't consider pairs where the first number is larger than the second, since we can just interchange the numbers in such cases. The answer appears to be 11 and 12, but we have considered only integers in the table.

(b) Call the two numbers x and y. Then $x + y = 23$, so $y = 23 - x$. Call the product P. Then

$P = xy = x(23 - x) = 23x - x^2$, so we wish to maximize the function $P(x) = 23x - x^2$. Since $P'(x) = 23 - 2x$, we see that $P'(x) = 0 \iff x = \frac{23}{2} = 11.5$. Thus, the maximum value of P is $P(11.5) = (11.5)^2 = 132.25$ and it occurs when $x = y = 11.5$.

Or: Note that $P''(x) = -2 < 0$ for all x, so P is everywhere concave downward and the local maximum at $x = 11.5$ must be an absolute maximum.

3. The two numbers are x and $\dfrac{100}{x}$, where $x > 0$. Minimize $f(x) = x + \dfrac{100}{x}$. $f'(x) = 1 - \dfrac{100}{x^2} = \dfrac{x^2 - 100}{x^2}$. The critical

number is $x = 10$. Since $f'(x) < 0$ for $0 < x < 10$ and $f'(x) > 0$ for $x > 10$, there is an absolute minimum at $x = 10$. The

numbers are 10 and 10.

5. If the rectangle has dimensions x and y, then its perimeter is $2x + 2y = 100$ m, so $y = 50 - x$. Thus, the area is

$A = xy = x(50 - x)$. We wish to maximize the function $A(x) = x(50 - x) = 50x - x^2$, where $0 < x < 50$. Since

$A'(x) = 50 - 2x = -2(x - 25)$, $A'(x) > 0$ for $0 < x < 25$ and $A'(x) < 0$ for $25 < x < 50$. Thus, A has an absolute

maximum at $x = 25$, and $A(25) = 25^2 = 625$ m^2. The dimensions of the rectangle that maximize its area are $x = y = 25$ m.

(The rectangle is a square.)

7. (a)

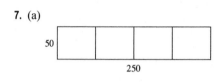
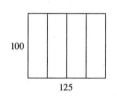

The areas of the three figures are

12,500, 12,500, and 9000 ft^2.

There appears to be a maximum

area of at least 12,500 ft^2.

(b) Let x denote the length of each of two sides and three dividers.

Let y denote the length of the other two sides.

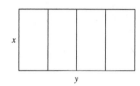

(c) Area $A = \text{length} \times \text{width} = y \cdot x$

(d) Length of fencing $= 750 \quad \Rightarrow \quad 5x + 2y = 750$

(e) $5x + 2y = 750 \quad \Rightarrow \quad y = 375 - \frac{5}{2}x \quad \Rightarrow \quad A(x) = \left(375 - \frac{5}{2}x\right)x = 375x - \frac{5}{2}x^2$

(f) $A'(x) = 375 - 5x = 0 \quad \Rightarrow \quad x = 75$. Since $A''(x) = -5 < 0$ there is an absolute maximum when $x = 75$. Then

$y = \frac{375}{2} = 187.5$. The largest area is $75\left(\frac{375}{2}\right) = 14{,}062.5$ ft^2. These values of x and y are between the values in the first

and second figures in part (a). Our original estimate was low.

9. Let b be the length of the base of the box and h the height. The surface area is $1200 = b^2 + 4hb \quad \Rightarrow \quad h = (1200 - b^2)/(4b)$.

The volume is $V = b^2h = b^2(1200 - b^2)/4b = 300b - b^3/4 \quad \Rightarrow \quad V'(b) = 300 - \frac{3}{4}b^2$.

$V'(b) = 0 \quad \Rightarrow \quad 300 = \frac{3}{4}b^2 \quad \Rightarrow \quad b^2 = 400 \quad \Rightarrow \quad b = \sqrt{400} = 20$. Since $V'(b) > 0$ for $0 < b < 20$ and $V'(b) < 0$ for

$b > 20$, there is an absolute maximum when $b = 20$ by the First Derivative Test for Absolute Extreme Values (see page 229).

If $b = 20$, then $h = (1200 - 20^2)/(4 \cdot 20) = 10$, so the largest possible volume is $b^2h = (20)^2(10) = 4000$ cm^3.

11. (a) Let the rectangle have sides x and y and area A, so $A = xy$ or $y = A/x$. The problem is to minimize the

perimeter $= 2x + 2y = 2x + 2A/x = P(x)$. Now $P'(x) = 2 - 2A/x^2 = 2(x^2 - A)/x^2$. So the critical number is

$x = \sqrt{A}$. Since $P'(x) < 0$ for $0 < x < \sqrt{A}$ and $P'(x) > 0$ for $x > \sqrt{A}$, there is an absolute minimum at $x = \sqrt{A}$.

The sides of the rectangle are $\sqrt{A}$ and $A/\sqrt{A} = \sqrt{A}$, so the rectangle is a square.

(b) Let p be the perimeter and x and y the lengths of the sides, so $p = 2x + 2y$ ⟹ $2y = p - 2x$ ⟹ $y = \frac{1}{2}p - x$.

The area is $A(x) = x\left(\frac{1}{2}p - x\right) = \frac{1}{2}px - x^2$. Now $A'(x) = 0$ ⟹ $\frac{1}{2}p - 2x = 0$ ⟹ $2x = \frac{1}{2}p$ ⟹ $x = \frac{1}{4}p$.

Since $A''(x) = -2 < 0$, there is an absolute maximum for A when $x = \frac{1}{4}p$ by the Second Derivative Test. The sides of

the rectangle are $\frac{1}{4}p$ and $\frac{1}{2}p - \frac{1}{4}p = \frac{1}{4}p$, so the rectangle is a square.

13.

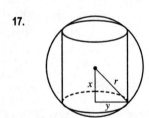

From the figure, we see that there are two points that are farthest away from $A(1, 0)$.

The distance d from A to an arbitrary point $P(x, y)$ on the ellipse is

$d = \sqrt{(x-1)^2 + (y-0)^2}$ and the square of the distance is

$S = d^2 = x^2 - 2x + 1 + y^2 = x^2 - 2x + 1 + (4 - 4x^2) = -3x^2 - 2x + 5$.

$S' = -6x - 2$ and $S' = 0$ ⟹ $x = -\frac{1}{3}$. Now $S'' = -6 < 0$, so we know that S

has a maximum at $x = -\frac{1}{3}$. Since $-1 \le x \le 1$, $S(-1) = 4$, $S\left(-\frac{1}{3}\right) = \frac{16}{3}$,

and $S(1) = 0$, we see that the maximum distance is $\sqrt{\frac{16}{3}}$. The corresponding y-values are

$y = \pm\sqrt{4 - 4\left(-\frac{1}{3}\right)^2} = \pm\sqrt{\frac{32}{9}} = \pm\frac{4}{3}\sqrt{2} \approx \pm 1.89$. The points are $\left(-\frac{1}{3}, \pm\frac{4}{3}\sqrt{2}\right)$.

15.

The height h of the equilateral triangle with sides of length L is $\frac{\sqrt{3}}{2}L$,

since $h^2 + (L/2)^2 = L^2$ ⟹ $h^2 = L^2 - \frac{1}{4}L^2 = \frac{3}{4}L^2$ ⟹ $h = \frac{\sqrt{3}}{2}L$. Using

similar triangles, $\dfrac{\frac{\sqrt{3}}{2}L - y}{x} = \dfrac{\frac{\sqrt{3}}{2}L}{L/2} = \sqrt{3}$ ⟹ $\sqrt{3}\,x = \frac{\sqrt{3}}{2}L - y$ ⟹

$y = \frac{\sqrt{3}}{2}L - \sqrt{3}\,x$ ⟹ $y = \frac{\sqrt{3}}{2}(L - 2x)$.

The area of the inscribed rectangle is $A(x) = (2x)y = \sqrt{3}\,x(L - 2x) = \sqrt{3}\,Lx - 2\sqrt{3}\,x^2$, where $0 \le x \le L/2$. Now

$0 = A'(x) = \sqrt{3}\,L - 4\sqrt{3}\,x$ ⟹ $x = \sqrt{3}\,L/(4\sqrt{3}) = L/4$. Since $A(0) = A(L/2) = 0$, the maximum occurs when

$x = L/4$, and $y = \frac{\sqrt{3}}{2}L - \frac{\sqrt{3}}{4}L = \frac{\sqrt{3}}{4}L$, so the dimensions are $L/2$ and $\frac{\sqrt{3}}{4}L$.

17.

The cylinder has volume $V = \pi y^2(2x)$. Also $x^2 + y^2 = r^2$ ⟹ $y^2 = r^2 - x^2$,

so $V(x) = \pi(r^2 - x^2)(2x) = 2\pi(r^2 x - x^3)$, where $0 \le x \le r$.

$V'(x) = 2\pi(r^2 - 3x^2) = 0$ ⟹ $x = r/\sqrt{3}$. Now $V(0) = V(r) = 0$, so there is a

maximum when $x = r/\sqrt{3}$ and

$V(r/\sqrt{3}) = \pi(r^2 - r^2/3)(2r/\sqrt{3}) = 4\pi r^3/(3\sqrt{3})$.

19.

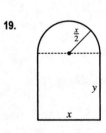

Perimeter $= 30$ ⟹ $2y + x + \pi\left(\dfrac{x}{2}\right) = 30$ ⟹

$y = \dfrac{1}{2}\left(30 - x - \dfrac{\pi x}{2}\right) = 15 - \dfrac{x}{2} - \dfrac{\pi x}{4}$. The area is the area of the rectangle

plus the area of the semicircle, or $xy + \frac{1}{2}\pi\left(\dfrac{x}{2}\right)^2$,

so $A(x) = x\left(15 - \dfrac{x}{2} - \dfrac{\pi x}{4}\right) + \frac{1}{8}\pi x^2 = 15x - \frac{1}{2}x^2 - \frac{\pi}{8}x^2$.

$A'(x) = 15 - \left(1 + \frac{\pi}{4}\right)x = 0 \Rightarrow x = \dfrac{15}{1 + \pi/4} = \dfrac{60}{4 + \pi}$. $A''(x) = -\left(1 + \frac{\pi}{4}\right) < 0$, so this gives a maximum.

The dimensions are $x = \dfrac{60}{4 + \pi}$ ft and $y = 15 - \dfrac{30}{4 + \pi} - \dfrac{15\pi}{4 + \pi} = \dfrac{60 + 15\pi - 30 - 15\pi}{4 + \pi} = \dfrac{30}{4 + \pi}$ ft, so the height of the rectangle is half the base.

21. 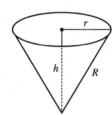 Let x be the length of the wire used for the square. The total area is

$$A(x) = \left(\frac{x}{4}\right)^2 + \frac{1}{2}\left(\frac{10-x}{3}\right)\frac{\sqrt{3}}{2}\left(\frac{10-x}{3}\right)$$

$$= \tfrac{1}{16}x^2 + \tfrac{\sqrt{3}}{36}(10 - x)^2, \quad 0 \le x \le 10$$

$A'(x) = \tfrac{1}{8}x - \tfrac{\sqrt{3}}{18}(10 - x) = 0 \Leftrightarrow \tfrac{9}{72}x + \tfrac{4\sqrt{3}}{72}x - \tfrac{40\sqrt{3}}{72} = 0 \Leftrightarrow x = \tfrac{40\sqrt{3}}{9 + 4\sqrt{3}}$.

Now $A(0) = \left(\tfrac{\sqrt{3}}{36}\right)100 \approx 4.81$, $A(10) = \tfrac{100}{16} = 6.25$ and $A\left(\tfrac{40\sqrt{3}}{9 + 4\sqrt{3}}\right) \approx 2.72$, so

(a) The maximum area occurs when $x = 10$ m, and all the wire is used for the square.

(b) The minimum area occurs when $x = \tfrac{40\sqrt{3}}{9 + 4\sqrt{3}} \approx 4.35$ m.

23. 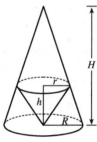 $h^2 + r^2 = R^2 \Rightarrow V = \tfrac{\pi}{3}r^2 h = \tfrac{\pi}{3}(R^2 - h^2)h = \tfrac{\pi}{3}(R^2 h - h^3)$.

$V'(h) = \tfrac{\pi}{3}(R^2 - 3h^2) = 0$ when $h = \tfrac{1}{\sqrt{3}}R$. This gives an absolute maximum, since

$V'(h) > 0$ for $0 < h < \tfrac{1}{\sqrt{3}}R$ and $V'(h) < 0$ for $h > \tfrac{1}{\sqrt{3}}R$. The maximum volume

is $V\left(\tfrac{1}{\sqrt{3}}R\right) = \tfrac{\pi}{3}\left(\tfrac{1}{\sqrt{3}}R^3 - \tfrac{1}{3\sqrt{3}}R^3\right) = \tfrac{2}{9\sqrt{3}}\pi R^3$.

25. By similar triangles, $\dfrac{H}{R} = \dfrac{H - h}{r}$ **(1)**. The volume of the inner cone is $V = \tfrac{1}{3}\pi r^2 h$, so

we'll solve **(1)** for h. $\dfrac{Hr}{R} = H - h \Rightarrow h = H - \dfrac{Hr}{R} = \dfrac{HR - Hr}{R} = \dfrac{H}{R}(R - r)$ **(2)**.

Thus, $V(r) = \dfrac{\pi}{3}r^2 \cdot \dfrac{H}{R}(R - r) = \dfrac{\pi H}{3R}(Rr^2 - r^3) \Rightarrow$

$V'(r) = \dfrac{\pi H}{3R}(2Rr - 3r^2) = \dfrac{\pi H}{3R}r(2R - 3r)$.

$V'(r) = 0 \Rightarrow r = 0$ or $2R = 3r \Rightarrow r = \tfrac{2}{3}R$ and from **(2)**, $h = \dfrac{H}{R}\left(R - \tfrac{2}{3}R\right) = \dfrac{H}{R}\left(\tfrac{1}{3}R\right) = \tfrac{1}{3}H$.

$V'(r)$ changes from positive to negative at $r = \tfrac{2}{3}R$, so the inner cone has a maximum volume of

$V = \tfrac{1}{3}\pi r^2 h = \tfrac{1}{3}\pi\left(\tfrac{2}{3}R\right)^2\left(\tfrac{1}{3}H\right) = \tfrac{4}{27}\cdot\tfrac{1}{3}\pi R^2 H$, which is approximately 15% of the volume of the larger cone.

27. $P(R) = \dfrac{E^2 R}{(R + r)^2} \Rightarrow$

$$P'(R) = \dfrac{(R + r)^2 \cdot E^2 - E^2 R \cdot 2(R + r)}{[(R + r)^2]^2} = \dfrac{(R^2 + 2Rr + r^2)E^2 - 2E^2 R^2 - 2E^2 Rr}{(R + r)^4}$$

$$= \dfrac{E^2 r^2 - E^2 R^2}{(R + r)^4} = \dfrac{E^2(r^2 - R^2)}{(R + r)^4} = \dfrac{E^2(r + R)(r - R)}{(R + r)^4} = \dfrac{E^2(r - R)}{(R + r)^3}$$

$P'(R) = 0 \Rightarrow R = r \Rightarrow P(r) = \dfrac{E^2 r}{(r + r)^2} = \dfrac{E^2 r}{4r^2} = \dfrac{E^2}{4r}$. The expression for $P'(R)$ shows that $P'(R) > 0$ for

$R < r$ and $P'(R) < 0$ for $R > r$. Thus, the maximum value of the power is $E^2/(4r)$, and this occurs when $R = r$.

29. $S = 6sh - \frac{3}{2}s^2 \cot\theta + 3s^2 \frac{\sqrt{3}}{2} \csc\theta$

(a) $\dfrac{dS}{d\theta} = \frac{3}{2}s^2 \csc^2\theta - 3s^2 \frac{\sqrt{3}}{2} \csc\theta \cot\theta$ or $\frac{3}{2}s^2 \csc\theta \left(\csc\theta - \sqrt{3}\cot\theta\right)$.

(b) $\dfrac{dS}{d\theta} = 0$ when $\csc\theta - \sqrt{3}\cot\theta = 0$ $\Rightarrow$ $\dfrac{1}{\sin\theta} - \sqrt{3}\dfrac{\cos\theta}{\sin\theta} = 0$ $\Rightarrow$ $\cos\theta = \frac{1}{\sqrt{3}}$. The First Derivative Test shows

that the minimum surface area occurs when $\theta = \cos^{-1}\left(\frac{1}{\sqrt{3}}\right) \approx 55°$.

(c) If $\cos\theta = \frac{1}{\sqrt{3}}$, then $\cot\theta = \frac{1}{\sqrt{2}}$ and $\csc\theta = \frac{\sqrt{3}}{\sqrt{2}}$, so the surface area is

$$S = 6sh - \frac{3}{2}s^2\frac{1}{\sqrt{2}} + 3s^2\frac{\sqrt{3}}{2}\frac{\sqrt{3}}{\sqrt{2}} = 6sh - \frac{3}{2\sqrt{2}}s^2 + \frac{9}{2\sqrt{2}}s^2$$

$$= 6sh + \frac{6}{2\sqrt{2}}s^2 = 6s\left(h + \frac{1}{2\sqrt{2}}s\right)$$

31.

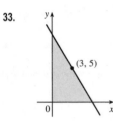

The total illumination is $I(x) = \dfrac{3k}{x^2} + \dfrac{k}{(10-x)^2}$, $0 < x < 10$. Then

$$I'(x) = \frac{-6k}{x^3} + \frac{2k}{(10-x)^3} = 0 \Rightarrow 6k(10-x)^3 = 2kx^3 \Rightarrow$$

$3(10-x)^3 = x^3 \Rightarrow \sqrt[3]{3}(10-x) = x \Rightarrow 10\sqrt[3]{3} - \sqrt[3]{3}x = x \Rightarrow 10\sqrt[3]{3} = x + \sqrt[3]{3}x \Rightarrow$

$10\sqrt[3]{3} = (1+\sqrt[3]{3})x \Rightarrow x = \dfrac{10\sqrt[3]{3}}{1+\sqrt[3]{3}} \approx 5.9$ ft. This gives a minimum since $I''(x) > 0$ for $0 < x < 10$.

33.

The line with slope m (where $m < 0$) through $(3,5)$ has equation $y - 5 = m(x-3)$ or

$y = mx + (5-3m)$. The y-intercept is $5-3m$ and the x-intercept is $-5/m+3$. So the triangle

has area $A(m) = \frac{1}{2}(5-3m)(-5/m+3) = 15 - 25/(2m) - \frac{9}{2}m$. Now

$$A'(m) = \frac{25}{2m^2} - \frac{9}{2} = 0 \Leftrightarrow m^2 = \frac{25}{9} \Rightarrow m = -\frac{5}{3} \text{ (since } m < 0\text{)}.$$

$A''(m) = -\dfrac{25}{m^3} > 0$, so there is an absolute minimum when $m = -\frac{5}{3}$. Thus, an equation of the

line is $y - 5 = -\frac{5}{3}(x-3)$ or $y = -\frac{5}{3}x + 10$.

35. (a) If $c(x) = \dfrac{C(x)}{x}$, then, by Quotient Rule, we have $c'(x) = \dfrac{xC'(x) - C(x)}{x^2}$. Now $c'(x) = 0$ when $xC'(x) - C(x) = 0$

and this gives $C'(x) = \dfrac{C(x)}{x} = c(x)$. Therefore, the marginal cost equals the average cost.

(b) (i) $C(x) = 16{,}000 + 200x + 4x^{3/2}$, $C(1000) = 16{,}000 + 200{,}000 + 40{,}000\sqrt{10} \approx 216{,}000 + 126{,}491$, so

$C(1000) \approx \$342{,}491$. $c(x) = C(x)/x = \dfrac{16{,}000}{x} + 200 + 4x^{1/2}$, $c(1000) \approx \$342.49/\text{unit}$.

$C'(x) = 200 + 6x^{1/2}$, $C'(1000) = 200 + 60\sqrt{10} \approx \$389.74/\text{unit}$.

(ii) We must have $C'(x) = c(x) \Leftrightarrow 200 + 6x^{1/2} = \dfrac{16{,}000}{x} + 200 + 4x^{1/2} \Leftrightarrow 2x^{3/2} = 16{,}000 \Leftrightarrow$

$x = (8{,}000)^{2/3} = 400$ units. To check that this is a minimum, we calculate

$c'(x) = \dfrac{-16{,}000}{x^2} + \dfrac{2}{\sqrt{x}} = \dfrac{2}{x^2}(x^{3/2} - 8000)$. This is negative for $x < (8000)^{2/3} = 400$, zero at $x = 400$, and

positive for $x > 400$, so c is decreasing on $(0, 400)$ and increasing on $(400, \infty)$. Thus, c has an absolute minimum

at $x = 400$. [*Note: $c''(x)$ is not positive for all $x > 0$.*]

(iii) The minimum average cost is $c(400) = 40 + 200 + 80 = \$320/\text{unit}$.

37. (a) We are given that the demand function p is linear and $p(27{,}000) = 10$, $p(33{,}000) = 8$, so the slope is

$\frac{10-8}{27{,}000-33{,}000} = -\frac{1}{3000}$ and an equation of the line is $y - 10 = \left(-\frac{1}{3000}\right)(x - 27{,}000)$ ⇒

$y = p(x) = -\frac{1}{3000}x + 19 = 19 - (x/3000)$.

(b) The revenue is $R(x) = xp(x) = 19x - (x^2/3000)$ ⇒ $R'(x) = 19 - (x/1500) = 0$ when $x = 28{,}500$. Since

$R''(x) = -1/1500 < 0$, the maximum revenue occurs when $x = 28{,}500$ ⇒ the price is $p(28{,}500) = \$9.50$.

39. (a) As in Example 6, we see that the demand function p is linear. We are given that $p(1000) = 450$ and deduce that

$p(1100) = 440$, since a \$10 reduction in price increases sales by 100 per week. The slope for p is $\frac{440-450}{1100-1000} = -\frac{1}{10}$,

so an equation is $p - 450 = -\frac{1}{10}(x - 1000)$ or $p(x) = -\frac{1}{10}x + 550$.

(b) $R(x) = xp(x) = -\frac{1}{10}x^2 + 550x$. $R'(x) = -\frac{1}{5}x + 550 = 0$ when $x = 5(550) = 2750$.

$p(2750) = 275$, so the rebate should be $450 - 275 = \$175$.

(c) $C(x) = 68{,}000 + 150x$ ⇒ $P(x) = R(x) - C(x) = -\frac{1}{10}x^2 + 550x - 68{,}000 - 150x = -\frac{1}{10}x^2 + 400x - 68{,}000$,

$P'(x) = -\frac{1}{5}x + 400 = 0$ when $x = 2000$. $p(2000) = 350$. Therefore, the rebate to maximize profits should be

$450 - 350 = \$100$.

41.

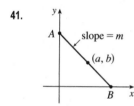

Every line segment in the first quadrant passing through (a, b) with endpoints on the x- and y-axes

satisfies an equation of the form $y - b = m(x - a)$, where $m < 0$. By setting $x = 0$ and then

$y = 0$, we find its endpoints, $A(0, b - am)$ and $B\left(a - \frac{b}{m}, 0\right)$. The distance d from A to B is

given by $d = \sqrt{[(a - \frac{b}{m}) - 0]^2 + [0 - (b - am)]^2}$. It follows that the square of the length of

the line segment, as a function of m, is given by $S(m) = \left(a - \frac{b}{m}\right)^2 + (am - b)^2 = a^2 - \frac{2ab}{m} + \frac{b^2}{m^2} + a^2m^2 - 2abm + b^2$.

Thus, $\qquad S'(m) = \frac{2ab}{m^2} - \frac{2b^2}{m^3} + 2a^2m - 2ab = \frac{2}{m^3}(abm - b^2 + a^2m^4 - abm^3)$

$\qquad\qquad\qquad = \frac{2}{m^3}[b(am - b) + am^3(am - b)] = \frac{2}{m^3}(am - b)(b + am^3)$

Thus, $S'(m) = 0$ ⇔ $m = b/a$ or $m = -\sqrt[3]{\frac{b}{a}}$. Since $b/a > 0$ and $m < 0$, m must equal $-\sqrt[3]{\frac{b}{a}}$. Since $\frac{2}{m^3} < 0$, we see

that $S'(m) < 0$ for $m < -\sqrt[3]{\frac{b}{a}}$ and $S'(m) > 0$ for $m > -\sqrt[3]{\frac{b}{a}}$. Thus, S has its absolute minimum value when $m = -\sqrt[3]{\frac{b}{a}}$.

That value is

$$S\left(-\sqrt[3]{\tfrac{b}{a}}\right) = \left(a + b\sqrt[3]{\tfrac{a}{b}}\right)^2 + \left(-a\sqrt[3]{\tfrac{b}{a}} - b\right)^2 = \left(a + \sqrt[3]{ab^2}\right)^2 + \left(\sqrt[3]{a^2b} + b\right)^2$$

$$= a^2 + 2a^{4/3}b^{2/3} + a^{2/3}b^{4/3} + a^{4/3}b^{2/3} + 2a^{2/3}b^{4/3} + b^2 = a^2 + 3a^{4/3}b^{2/3} + 3a^{2/3}b^{4/3} + b^2$$

The last expression is of the form $x^3 + 3x^2y + 3xy^2 + y^3$ $[= (x + y)^3]$ with $x = a^{2/3}$ and $y = b^{2/3}$, so we can write it

as $(a^{2/3} + b^{2/3})^3$ and the shortest such line segment has length $\sqrt{S} = (a^{2/3} + b^{2/3})^{3/2}$.

43.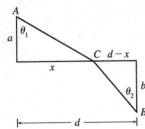

The total time is

$$(x) = \begin{pmatrix} \text{time from} \\ A \text{ to } C \end{pmatrix} + \begin{pmatrix} \text{time from} \\ C \text{ to } B \end{pmatrix} = \frac{\sqrt{a^2 + x^2}}{v_1} + \frac{\sqrt{b^2 + (d-x)^2}}{v_2}, \quad 0 < x < d.$$

$$T'(x) = \frac{x}{v_1\sqrt{a^2 + x^2}} - \frac{d-x}{v_2\sqrt{b^2 + (d-x)^2}} = \frac{\sin\theta_1}{v_1} - \frac{\sin\theta_2}{v_2}$$

The minimum occurs when $T'(x) = 0 \Rightarrow \dfrac{\sin\theta_1}{v_1} = \dfrac{\sin\theta_2}{v_2}$. [*Note:* $T''(x) > 0$]

45.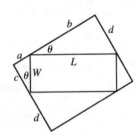

$y^2 = x^2 + z^2$, but triangles CDE and BCA are similar, so $z/8 = x/(4\sqrt{x-4}) \Rightarrow$

$z = 2x/\sqrt{x-4}$. Thus, we minimize $f(x) = y^2 = x^2 + 4x^2/(x-4) = x^3/(x-4)$,

$4 < x \le 8.$ $f'(x) = \dfrac{(x-4)(3x^2) - x^3}{(x-4)^2} = \dfrac{x^2[3(x-4) - x]}{(x-4)^2} = \dfrac{2x^2(x-6)}{(x-4)^2} = 0$

when $x = 6.$ $f'(x) < 0$ when $x < 6$, $f'(x) > 0$ when $x > 6$, so the minimum

occurs when $x = 6$ in.

47.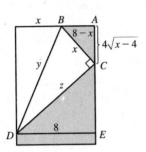

In the small triangle with sides a and c and hypotenuse W, $\sin\theta = \dfrac{a}{W}$ and

$\cos\theta = \dfrac{c}{W}$. In the triangle with sides b and d and hypotenuse L, $\sin\theta = \dfrac{d}{L}$ and

$\cos\theta = \dfrac{b}{L}$. Thus, $a = W\sin\theta$, $c = W\cos\theta$, $d = L\sin\theta$, and $b = L\cos\theta$, so the

area of the circumscribed rectangle is

$$
\begin{aligned}
A(\theta) &= (a+b)(c+d) = (W\sin\theta + L\cos\theta)(W\cos\theta + L\sin\theta) \\
&= W^2\sin\theta\cos\theta + WL\sin^2\theta + LW\cos^2\theta + L^2\sin\theta\cos\theta = LW\sin^2\theta + LW\cos^2\theta + (L^2 + W^2)\sin\theta\cos\theta \\
&= LW(\sin^2\theta + \cos^2\theta) + (L^2 + W^2) \cdot \tfrac{1}{2} \cdot 2\sin\theta\cos\theta = LW + \tfrac{1}{2}(L^2 + W^2)\sin 2\theta, \quad 0 \le \theta \le \tfrac{\pi}{2}
\end{aligned}
$$

This expression shows, without calculus, that the maximum value of $A(\theta)$ occurs when $\sin 2\theta = 1 \Leftrightarrow 2\theta = \tfrac{\pi}{2} \Rightarrow$

$\theta = \tfrac{\pi}{4}$. So the maximum area is $A\left(\tfrac{\pi}{4}\right) = LW + \tfrac{1}{2}(L^2 + W^2) = \tfrac{1}{2}(L^2 + 2LW + W^2) = \tfrac{1}{2}(L+W)^2$.

49.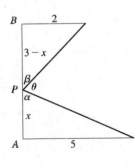

From the figure, $\tan\alpha = \dfrac{5}{x}$ and $\tan\beta = \dfrac{2}{3-x}$. Since $\alpha + \beta + \theta = 180° = \pi$,

$$\theta = \pi - \tan^{-1}\left(\frac{5}{x}\right) - \tan^{-1}\left(\frac{2}{3-x}\right) \Rightarrow$$

$$\frac{d\theta}{dx} = -\frac{1}{1 + \left(\dfrac{5}{x}\right)^2}\left(-\frac{5}{x^2}\right) - \frac{1}{1 + \left(\dfrac{2}{3-x}\right)^2}\left[\frac{2}{(3-x)^2}\right]$$

$$= \frac{x^2}{x^2 + 25} \cdot \frac{5}{x^2} - \frac{(3-x)^2}{(3-x)^2 + 4} \cdot \frac{2}{(3-x)^2}.$$

Now $\dfrac{d\theta}{dx} = 0 \Rightarrow \dfrac{5}{x^2 + 25} = \dfrac{2}{x^2 - 6x + 13} \Rightarrow 2x^2 + 50 = 5x^2 - 30x + 65 \Rightarrow 3x^2 - 30x + 15 = 0 \Rightarrow$

$x^2 - 10x + 5 = 0 \;\Rightarrow\; x = 5 \pm 2\sqrt{5}$. We reject the root with the $+$ sign, since it is larger than 3.

$d\theta/dx > 0$ for $x < 5 - 2\sqrt{5}$ and $d\theta/dx < 0$ for $x > 5 - 2\sqrt{5}$, so θ is maximized when $|AP| = x = 5 - 2\sqrt{5} \approx 0.53$.

4.6   Newton's Method

1. (a)

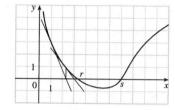

The tangent line at $x = 1$ intersects the x-axis at $x \approx 2.3$, so $x_2 \approx 2.3$. The tangent line at $x = 2.3$ intersects the x-axis at $x \approx 3$, so $x_3 \approx 3.0$.

(b) $x_1 = 5$ would *not* be a better first approximation than $x_1 = 1$ since the tangent line is nearly horizontal. In fact, the second approximation for $x_1 = 5$ appears to be to the left of $x = 1$.

3. Since $x_1 = 3$ and $y = 5x - 4$ is tangent to $y = f(x)$ at $x = 3$, we simply need to find where the tangent line intersects the x-axis. $y = 0 \;\Rightarrow\; 5x_2 - 4 = 0 \;\Rightarrow\; x_2 = \frac{4}{5}$.

5. $f(x) = x^3 + 2x - 4 \;\Rightarrow\; f'(x) = 3x^2 + 2$, so $x_{n+1} = x_n - \dfrac{x_n^3 + 2x_n - 4}{3x_n^2 + 2}$.

Now $x_1 = 1 \;\Rightarrow\; x_2 = 1 - \dfrac{1 + 2 - 4}{3 \cdot 1^2 + 2} = 1 - \dfrac{-1}{5} = 1.2 \;\Rightarrow\; x_3 = 1.2 - \dfrac{(1.2)^3 + 2(1.2) - 4}{3(1.2)^2 + 2} \approx 1.1797$.

7. $f(x) = x^3 + x + 3 \;\Rightarrow\; f'(x) = 3x^2 + 1$, so $x_{n+1} = x_n - \dfrac{x_n^3 + x_n + 3}{3x_n^2 + 1}$.

Now $x_1 = -1 \;\Rightarrow\;$

$x_2 = -1 - \dfrac{(-1)^3 + (-1) + 3}{3(-1)^2 + 1} = -1 - \dfrac{-1 - 1 + 3}{3 + 1} = -1 - \dfrac{1}{4} = -1.25$.

Newton's method follows the tangent line at $(-1, 1)$ down to its intersection with the x-axis at $(-1.25, 0)$, giving the second approximation $x_2 = -1.25$.

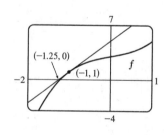

9. To approximate $x = \sqrt[3]{30}$ (so that $x^3 = 30$), we can take $f(x) = x^3 - 30$. So $f'(x) = 3x^2$, and thus,

$x_{n+1} = x_n - \dfrac{x_n^3 - 30}{3x_n^2}$. Since $\sqrt[3]{27} = 3$ and 27 is close to 30, we'll use $x_1 = 3$. We need to find approximations until they

agree to eight decimal places. $x_1 = 3 \;\Rightarrow\; x_2 \approx 3.11111111,\; x_3 \approx 3.10723734,\; x_4 \approx 3.10723251 \approx x_5$. So $\sqrt[3]{30} \approx 3.10723251$, to eight decimal places.

Here is a quick and easy method for finding the iterations for Newton's method on a programmable calculator.

(The screens shown are from the TI-83 Plus, but the method is similar on other calculators.) Assign $f(x) = x^3 - 30$

to Y_1, and $f'(x) = 3x^2$ to Y_2. Now store $x_1 = 3$ in X and then enter $X - Y_1/Y_2 \to X$ to get $x_2 = 3.\overline{1}$. By successively pressing the ENTER key, you get the approximations $x_3, x_4, \ldots$.

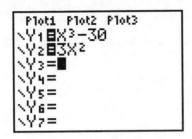
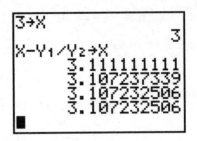

In Derive, load the utility file SOLVE. Enter NEWTON(x^3-30,x,3) and then APPROXIMATE to get

[3, 3.11111111, 3.10723733, 3.10723250, 3.10723250]. You can request a specific iteration by adding a fourth argument. For

example, NEWTON(x^3-30,x,3,2) gives [3, 3.11111111, 3.10723733].

In Maple, make the assignments $f := x \to x^3 - 30;$, $g := x \to x - f(x)/D(f)(x);$, and $x := 3.;$. Repeatedly execute

the command $x := g(x);$ to generate successive approximations.

In Mathematica, make the assignments $f[x\_] := x^3 - 30$, $g[x\_] := x - f[x]/f'[x]$, and $x = 3$. Repeatedly execute the

command $x = g[x]$ to generate successive approximations.

11. $\sin x = x^2$, so $f(x) = \sin x - x^2$ $\Rightarrow$ $f'(x) = \cos x - 2x$ $\Rightarrow$

$x_{n+1} = x_n - \dfrac{\sin x_n - x_n^2}{\cos x_n - 2x_n}$. From the figure, the positive root of $\sin x = x^2$ is

near 1. $x_1 = 1$ $\Rightarrow$ $x_2 \approx 0.891396$, $x_3 \approx 0.876985$, $x_4 \approx 0.876726 \approx x_5$. So

the positive root is 0.876726, to six decimal places.

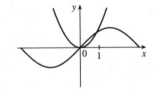

13. $f(x) = x^5 - x^4 - 5x^3 - x^2 + 4x + 3$ $\Rightarrow$ $f'(x) = 5x^4 - 4x^3 - 15x^2 - 2x + 4$ $\Rightarrow$

$x_{n+1} = x_n - \dfrac{x_n^5 - x_n^4 - 5x_n^3 - x_n^2 + 4x_n + 3}{5x_n^4 - 4x_n^3 - 15x_n^2 - 2x_n + 4}$. From the graph of f, there appear to be roots near -1.4, 1.1, and 2.7.

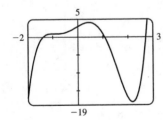

$x_1 = -1.4$	$x_1 = 1.1$	$x_1 = 2.7$
$x_2 \approx -1.39210970$	$x_2 \approx 1.07780402$	$x_2 \approx 2.72046250$
$x_3 \approx -1.39194698$	$x_3 \approx 1.07739442$	$x_3 \approx 2.71987870$
$x_4 \approx -1.39194691 \approx x_5$	$x_4 \approx 1.07739428 \approx x_5$	$x_4 \approx 2.71987822 \approx x_5$

To eight decimal places, the roots of the equation are -1.39194691, 1.07739428,

and 2.71987822.

15. Solving $e^{-x} = 2 + x$ is the same as solving $f(x) = e^{-x} - x - 2 = 0$.

$f'(x) = -e^{-x} - 1$ $\Rightarrow$ $x_{n+1} = x_n - \dfrac{e^{-x_n} - x_n - 2}{-e^{-x_n} - 1}$. From the graph of $y = e^{-x}$

and $y = 2 + x$, there appears to be a root near $x = -0.5$. Now $x_1 = -0.5$ $\Rightarrow$

$x_2 \approx -0.44385167$, $x_3 \approx -0.44285470$, $x_4 \approx -0.44285440 \approx x_5$. To eight

decimal places, the root of the equation is -0.44285440.

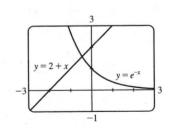

17. From the graph, $y = x^2\sqrt{2 - x - x^2}$ and $y = 1$ intersect twice, at $x \approx -2$ and at $x \approx -1$.

$$f(x) = x^2\sqrt{2 - x - x^2} - 1 \quad \Rightarrow \quad f'(x) = x^2 \cdot \tfrac{1}{2}(2 - x - x^2)^{-1/2}(-1 - 2x) + (2 - x - x^2)^{1/2} \cdot 2x$$

$$= \tfrac{1}{2}x(2 - x - x^2)^{-1/2}\left[x(-1 - 2x) + 4(2 - x - x^2)\right] = \frac{x(8 - 5x - 6x^2)}{2\sqrt{(2 + x)(1 - x)}},$$

so $x_{n+1} = x_n - \dfrac{x_n^2\sqrt{2 - x_n - x_n^2} - 1}{\dfrac{x_n(8 - 5x_n - 6x_n^2)}{2\sqrt{(2 + x_n)(1 - x_n)}}}$. Trying $x_1 = -2$ won't work because $f'(-2)$ is undefined, so we try $x_1 = -1.95$.

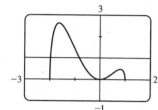

$x_1 = -1.95$	$x_1 = -0.8$
$x_2 \approx -1.98580357$	$x_2 \approx -0.82674444$
$x_3 \approx -1.97899778$	$x_3 \approx -0.82646236$
$x_4 \approx -1.97807848$	$x_4 \approx -0.82646233 \approx x_5$
$x_5 \approx -1.97806682$	
$x_6 \approx -1.97806681 \approx x_7$	

To eight decimal places, the roots of the equation are -1.97806681 and -0.82646233.

19.

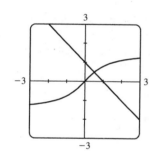

From the graph of $y = \tan^{-1} x$ and $y = 1 - x$, there appears to be a point of intersection near $x = 0.5$. Solving $\tan^{-1} x = 1 - x$ is the same as solving $f(x) = \tan^{-1} x + x - 1 = 0$. $f(x) = \tan^{-1} x + x - 1 \quad \Rightarrow$

$$f'(x) = \frac{1}{1 + x^2} + 1, \text{ so } x_{n+1} = x_n - \frac{\tan^{-1} x_n + x_n - 1}{1/(1 + x_n^2) + 1}.$$

Now $x_1 = 0.5 \quad \Rightarrow \quad x_2 \approx 0.52019577, x_3 \approx 0.52026899 \approx x_4$.

To eight decimal places, the root of the equation is 0.52026899.

21. (a) $f(x) = x^2 - a \quad \Rightarrow \quad f'(x) = 2x$, so Newton's method gives

$$x_{n+1} = x_n - \frac{x_n^2 - a}{2x_n} = x_n - \frac{1}{2}x_n + \frac{a}{2x_n} = \frac{1}{2}x_n + \frac{a}{2x_n} = \frac{1}{2}\left(x_n + \frac{a}{x_n}\right).$$

(b) Using (a) with $a = 1000$ and $x_1 = \sqrt{900} = 30$, we get $x_2 \approx 31.666667$, $x_3 \approx 31.622807$, and $x_4 \approx 31.622777 \approx x_5$.

So $\sqrt{1000} \approx 31.622777$.

23. $f(x) = x^3 - 3x + 6 \quad \Rightarrow \quad f'(x) = 3x^2 - 3$. If $x_1 = 1$, then $f'(x_1) = 0$ and the tangent line used for approximating x_2 is horizontal. Attempting to find x_2 results in trying to divide by zero.

25. For $f(x) = x^{1/3}$, $f'(x) = \tfrac{1}{3}x^{-2/3}$ and

$$x_{n+1} = x_n - \frac{f(x_n)}{f'(x_n)} = x_n - \frac{x_n^{1/3}}{\tfrac{1}{3}x_n^{-2/3}} = x_n - 3x_n = -2x_n.$$

Therefore, each successive approximation becomes twice as large as the previous one in absolute value, so the sequence of approximations fails to converge to the root, which is 0. In the figure, we have $x_1 = 0.5$, $x_2 = -2(0.5) = -1$, and $x_3 = -2(-1) = 2$.

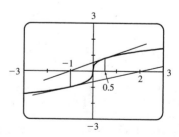

27.

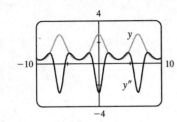

From the figure, we see that $y = f(x) = e^{\cos x}$ is periodic with period 2π. To find the x-coordinates of the IP, we only need to approximate the zeros of y'' on $[0, \pi]$. $f'(x) = -e^{\cos x} \sin x \Rightarrow f''(x) = e^{\cos x}(\sin^2 x - \cos x)$. Since $e^{\cos x} \neq 0$, we will use Newton's method with $g(x) = \sin^2 x - \cos x$, $g'(x) = 2 \sin x \cos x + \sin x$, and $x_1 = 1$. $x_2 \approx 0.904173$, $x_3 \approx 0.904557 \approx x_4$. Thus, $(0.904557, 1.855277)$ is the IP.

29. In this case, $A = 18,000$, $R = 375$, and $n = 5(12) = 60$. So the formula $A = \dfrac{R}{i}\left[1 - (1 + i)^{-n}\right]$ becomes

$$18,000 = \frac{375}{x}\left[1 - (1 + x)^{-60}\right] \quad \Leftrightarrow \quad 48x = 1 - (1 + x)^{-60} \quad \text{[multiply each term by } (1 + x)^{60}\text{]} \quad \Leftrightarrow$$

$48x(1 + x)^{60} - (1 + x)^{60} + 1 = 0$. Let the LHS be called $f(x)$, so that

$$f'(x) = 48x(60)(1 + x)^{59} + 48(1 + x)^{60} - 60(1 + x)^{59}$$

$$= 12(1 + x)^{59}\left[4x(60) + 4(1 + x) - 5\right] = 12(1 + x)^{59}(244x - 1)$$

$x_{n+1} = x_n - \dfrac{48x_n(1 + x_n)^{60} - (1 + x_n)^{60} + 1}{12(1 + x_n)^{59}(244x_n - 1)}$. An interest rate of 1% per month seems like a reasonable estimate for

$x = i$. So let $x_1 = 1\% = 0.01$, and we get $x_2 \approx 0.0082202$, $x_3 \approx 0.0076802$, $x_4 \approx 0.0076291$, $x_5 \approx 0.0076286 \approx x_6$.

Thus, the dealer is charging a monthly interest rate of 0.76286% (or 9.55% per year, compounded monthly).

4.7 Antiderivatives

1. $f(x) = 6x^2 - 8x + 3 \Rightarrow F(x) = 6\dfrac{x^{2+1}}{2+1} - 8\dfrac{x^{1+1}}{1+1} + 3x + C = 2x^3 - 4x^2 + 3x + C$

Check: $F'(x) = 2 \cdot 3x^2 - 4 \cdot 2x + 3 + 0 = 6x^2 - 8x + 3 = f(x)$

3. $f(x) = 5x^{1/4} - 7x^{3/4} \Rightarrow F(x) = 5\dfrac{x^{1/4+1}}{\frac{1}{4}+1} - 7\dfrac{x^{3/4+1}}{\frac{3}{4}+1} + C = 5\dfrac{x^{5/4}}{5/4} - 7\dfrac{x^{7/4}}{7/4} + C = 4x^{5/4} - 4x^{7/4} + C$

5. $f(x) = \sqrt[3]{x} + \dfrac{5}{x^6} = x^{1/3} + 5x^{-6}$ has domain $(-\infty, 0) \cup (0, \infty)$, so

$$F(x) = \begin{cases} \dfrac{x^{1/3+1}}{\frac{1}{3}+1} + 5\dfrac{x^{-6+1}}{-6+1} + C_1 = \frac{3}{4}x^{4/3} - x^{-5} + C_1 & \text{if } x < 0 \\[3mm] \frac{3}{4}x^{4/3} - x^{-5} + C_2 & \text{if } x > 0 \end{cases}$$

See Example 1(b) for a similar problem.

7. $f(u) = \dfrac{u^4 + 3\sqrt{u}}{u^2} = \dfrac{u^4}{u^2} + \dfrac{3u^{1/2}}{u^2} = u^2 + 3u^{-3/2} \Rightarrow$

$F(u) = \dfrac{u^3}{3} + 3\dfrac{u^{-3/2+1}}{-3/2+1} + C = \dfrac{1}{3}u^3 + 3\dfrac{u^{-1/2}}{-1/2} + C = \dfrac{1}{3}u^3 - \dfrac{6}{\sqrt{u}} + C$

9. $g(\theta) = \cos\theta - 5\sin\theta \Rightarrow G(\theta) = \sin\theta - 5(-\cos\theta) + C = \sin\theta + 5\cos\theta + C$

11. $f(x) = 2x + 5(1 - x^2)^{-1/2} = 2x + \dfrac{5}{\sqrt{1 - x^2}} \Rightarrow F(x) = x^2 + 5\sin^{-1} x + C$

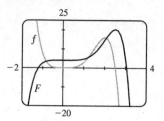

13. $f(x) = 5x^4 - 2x^5 \quad \Rightarrow \quad F(x) = 5 \cdot \dfrac{x^5}{5} - 2 \cdot \dfrac{x^6}{6} + C = x^5 - \frac{1}{3}x^6 + C.$

$F(0) = 4 \quad \Rightarrow \quad 0^5 - \frac{1}{3} \cdot 0^6 + C = 4 \quad \Rightarrow \quad C = 4,$ so $F(x) = x^5 - \frac{1}{3}x^6 + 4.$

The graph confirms our answer since $f(x) = 0$ when F has a local maximum, f is

positive when F is increasing, and f is negative when F is decreasing.

15. $f''(x) = 6x + 12x^2 \quad \Rightarrow \quad f'(x) = 6 \cdot \dfrac{x^2}{2} + 12 \cdot \dfrac{x^3}{3} + C = 3x^2 + 4x^3 + C \quad \Rightarrow$

$f(x) = 3 \cdot \dfrac{x^3}{3} + 4 \cdot \dfrac{x^4}{4} + Cx + D = x^3 + x^4 + Cx + D \quad [C \text{ and } D \text{ are just arbitrary constants}]$

17. $f''(x) = 1 + x^{4/5} \quad \Rightarrow \quad f'(x) = x + \frac{5}{9}x^{9/5} + C \quad \Rightarrow$

$f(x) = \frac{1}{2}x^2 + \frac{5}{9} \cdot \frac{5}{14}x^{14/5} + Cx + D = \frac{1}{2}x^2 + \frac{25}{126}x^{14/5} + Cx + D$

19. $f'(x) = \sqrt{x}(6 + 5x) = 6x^{1/2} + 5x^{3/2} \quad \Rightarrow \quad f(x) = 4x^{3/2} + 2x^{5/2} + C.$

$f(1) = 6 + C \text{ and } f(1) = 10 \quad \Rightarrow \quad C = 4,$ so $f(x) = 4x^{3/2} + 2x^{5/2} + 4.$

21. $f'(t) = 2\cos t + \sec^2 t \quad \Rightarrow \quad f(t) = 2\sin t + \tan t + C \text{ because } -\pi/2 < t < \pi/2.$

$f\left(\frac{\pi}{3}\right) = 2\left(\sqrt{3}/2\right) + \sqrt{3} + C = 2\sqrt{3} + C \text{ and } f\left(\frac{\pi}{3}\right) = 4 \quad \Rightarrow \quad C = 4 - 2\sqrt{3},$ so $f(t) = 2\sin t + \tan t + 4 - 2\sqrt{3}.$

23. $f''(x) = 24x^2 + 2x + 10 \quad \Rightarrow \quad f'(x) = 8x^3 + x^2 + 10x + C. \quad f'(1) = 8 + 1 + 10 + C \text{ and } f'(1) = -3 \quad \Rightarrow$

$19 + C = -3 \quad \Rightarrow \quad C = -22,$ so $f'(x) = 8x^3 + x^2 + 10x - 22$ and hence, $f(x) = 2x^4 + \frac{1}{3}x^3 + 5x^2 - 22x + D.$

$f(1) = 2 + \frac{1}{3} + 5 - 22 + D \text{ and } f(1) = 5 \quad \Rightarrow \quad D = 22 - \frac{7}{3} = \frac{59}{3},$ so $f(x) = 2x^4 + \frac{1}{3}x^3 + 5x^2 - 22x + \frac{59}{3}.$

25. $f''(\theta) = \sin\theta + \cos\theta \quad \Rightarrow \quad f'(\theta) = -\cos\theta + \sin\theta + C. \quad f'(0) = -1 + C \text{ and } f'(0) = 4 \quad \Rightarrow \quad C = 5,$ so

$f'(\theta) = -\cos\theta + \sin\theta + 5$ and hence, $f(\theta) = -\sin\theta - \cos\theta + 5\theta + D. \quad f(0) = -1 + D \text{ and } f(0) = 3 \quad \Rightarrow \quad D = 4,$ so

$f(\theta) = -\sin\theta - \cos\theta + 5\theta + 4.$

27. $f''(x) = x^{-2}, x > 0 \quad \Rightarrow \quad f'(x) = -1/x + C \quad \Rightarrow \quad f(x) = -\ln|x| + Cx + D = -\ln x + Cx + D \text{ (since } x > 0).$

$f(1) = 0 \quad \Rightarrow \quad C + D = 0 \text{ and } f(2) = 0 \quad \Rightarrow \quad -\ln 2 + 2C + D = 0 \quad \Rightarrow \quad -\ln 2 + 2C - C = 0 \text{ [since } D = -C] \quad \Rightarrow$

$-\ln 2 + C = 0 \quad \Rightarrow \quad C = \ln 2 \text{ and } D = -\ln 2. \text{ So } f(x) = -\ln x + (\ln 2)x - \ln 2.$

29. Given $f'(x) = 2x + 1,$ we have $f(x) = x^2 + x + C.$ Since f passes through $(1,6), f(1) = 6 \quad \Rightarrow \quad 1^2 + 1 + C = 6 \quad \Rightarrow$

$C = 4.$ Therefore, $f(x) = x^2 + x + 4$ and $f(2) = 2^2 + 2 + 4 = 10.$

31. b is the antiderivative of $f.$ For small x, f is negative, so the graph of its antiderivative must be decreasing. But both a and c

are increasing for small $x,$ so only b can be f's antiderivative. Also, f is positive where b is increasing, which supports our

conclusion.

33. $v(t) = s'(t) = \sin t - \cos t \quad \Rightarrow \quad s(t) = -\cos t - \sin t + C. \quad s(0) = -1 + C \text{ and } s(0) = 0 \quad \Rightarrow \quad C = 1,$ so

$s(t) = -\cos t - \sin t + 1.$

35. $a(t) = v'(t) = 10\sin t + 3\cos t \quad \Rightarrow \quad v(t) = -10\cos t + 3\sin t + C \quad \Rightarrow \quad s(t) = -10\sin t - 3\cos t + Ct + D.$

$s(0) = -3 + D = 0 \text{ and } s(2\pi) = -3 + 2\pi C + D = 12 \quad \Rightarrow \quad D = 3 \text{ and } C = \frac{6}{\pi}. \text{ Thus,}$

$s(t) = -10\sin t - 3\cos t + \frac{6}{\pi}t + 3.$

37. (a) We first observe that since the stone is dropped 450 m above the ground, $v(0) = 0$ and $s(0) = 450$.

$v'(t) = a(t) = -9.8 \Rightarrow v(t) = -9.8t + C$. Now $v(0) = 0 \Rightarrow C = 0$, so $v(t) = -9.8t \Rightarrow$

$s(t) = -4.9t^2 + D$. Last, $s(0) = 450 \Rightarrow D = 450 \Rightarrow s(t) = 450 - 4.9t^2$.

(b) The stone reaches the ground when $s(t) = 0$. $450 - 4.9t^2 = 0 \Rightarrow t^2 = 450/4.9 \Rightarrow t_1 = \sqrt{450/4.9} \approx 9.58$ s.

(c) The velocity with which the stone strikes the ground is $v(t_1) = -9.8\sqrt{450/4.9} \approx -93.9$ m/s.

(d) This is just reworking parts (a) and (b) with $v(0) = -5$. Using $v(t) = -9.8t + C$, $v(0) = -5 \Rightarrow 0 + C = -5 \Rightarrow$

$v(t) = -9.8t - 5$. So $s(t) = -4.9t^2 - 5t + D$ and $s(0) = 450 \Rightarrow D = 450 \Rightarrow s(t) = -4.9t^2 - 5t + 450$.

Solving $s(t) = 0$ by using the quadratic formula gives us $t = \left(5 \pm \sqrt{8845}\right)/(-9.8) \Rightarrow t_1 \approx 9.09$ s.

39. By Exercise 38 with $a = -9.8$, $s(t) = -4.9t^2 + v_0 t + s_0$ and $v(t) = s'(t) = -9.8t + v_0$. So

$[v(t)]^2 = (-9.8t + v_0)^2 = (9.8)^2 t^2 - 19.6 v_0 t + v_0^2 = v_0^2 + 96.04t^2 - 19.6 v_0 t = v_0^2 - 19.6\left(-4.9t^2 + v_0 t\right)$.

But $-4.9t^2 + v_0 t$ is just $s(t)$ without the s_0 term; that is, $s(t) - s_0$. Thus, $[v(t)]^2 = v_0^2 - 19.6\left[s(t) - s_0\right]$.

41. Using Exercise 38 with $a = -32$, $v_0 = 0$, and $s_0 = h$ (the height of the cliff), we know that the height at time t is

$s(t) = -16t^2 + h$. $v(t) = s'(t) = -32t$ and $v(t) = -120 \Rightarrow -32t = -120 \Rightarrow t = 3.75$, so

$0 = s(3.75) = -16(3.75)^2 + h \Rightarrow h = 16(3.75)^2 = 225$ ft.

43. Taking the upward direction to be positive we have that for $0 \le t \le 10$ (using the subscript 1 to refer to $0 \le t \le 10$),

$a_1(t) = -(9 - 0.9t) = v_1'(t) \Rightarrow v_1(t) = -9t + 0.45t^2 + v_0$, but $v_1(0) = v_0 = -10 \Rightarrow$

$v_1(t) = -9t + 0.45t^2 - 10 = s_1'(t) \Rightarrow s_1(t) = -\frac{9}{2}t^2 + 0.15t^3 - 10t + s_0$. But $s_1(0) = 500 = s_0 \Rightarrow$

$s_1(t) = -\frac{9}{2}t^2 + 0.15t^3 - 10t + 500$. $s_1(10) = -450 + 150 - 100 + 500 = 100$, so it takes

more than 10 seconds for the raindrop to fall. Now for $t > 10$, $a(t) = 0 = v'(t) \Rightarrow$

$v(t) = \text{constant} = v_1(10) = -9(10) + 0.45(10)^2 - 10 = -55 \Rightarrow v(t) = -55$.

At 55 m/s, it will take $100/55 \approx 1.8$ s to fall the last 100 m. Hence, the total time is $10 + \frac{100}{55} = \frac{130}{11} \approx 11.8$ s.

45. $a(t) = k$, the initial velocity is 30 mi/h $= 30 \cdot \frac{5280}{3600} = 44$ ft/s, and the final velocity (after 5 seconds) is

50 mi/h $= 50 \cdot \frac{5280}{3600} = \frac{220}{3}$ ft/s. So $v(t) = kt + C$ and $v(0) = 44 \Rightarrow C = 44$. Thus, $v(t) = kt + 44 \Rightarrow$

$v(5) = 5k + 44$. But $v(5) = \frac{220}{3}$, so $5k + 44 = \frac{220}{3} \Rightarrow 5k = \frac{88}{3} \Rightarrow k = \frac{88}{15} \approx 5.87$ ft/s².

47. Let the acceleration be $a(t) = k$ km/h². We have $v(0) = 100$ km/h and we can take the initial position $s(0)$ to be 0.

We want the time t_f for which $v(t) = 0$ to satisfy $s(t) < 0.08$ km. In general, $v'(t) = a(t) = k$, so $v(t) = kt + C$,

where $C = v(0) = 100$. Now $s'(t) = v(t) = kt + 100$, so $s(t) = \frac{1}{2}kt^2 + 100t + D$, where $D = s(0) = 0$.

Thus, $s(t) = \frac{1}{2}kt^2 + 100t$. Since $v(t_f) = 0$, we have $kt_f + 100 = 0$ or $t_f = -100/k$, so

$s(t_f) = \frac{1}{2}k\left(-\frac{100}{k}\right)^2 + 100\left(-\frac{100}{k}\right) = 10{,}000\left(\frac{1}{2k} - \frac{1}{k}\right) = -\frac{5{,}000}{k}$. The condition $s(t_f)$ must satisfy is

$-\frac{5{,}000}{k} < 0.08 \Rightarrow -\frac{5{,}000}{0.08} > k$ [k is negative] $\Rightarrow k < -62{,}500$ km/h², or equivalently,

$k < -\frac{3125}{648} \approx -4.82$ m/s². Thus, a constant deceleration of 4.82 m/s² is required.

49. (a) First note that 90 mi/h $= 90 \times \frac{5280}{3600}$ ft/s $= 132$ ft/s. Then $a(t) = 4$ ft/s² $\Rightarrow$ $v(t) = 4t + C$, but $v(0) = 0$ $\Rightarrow$

$C = 0$. Now $4t = 132$ when $t = \frac{132}{4} = 33$ s, so it takes 33 s to reach 132 ft/s. Therefore, taking $s(0) = 0$, we have

$s(t) = 2t^2$, $0 \leq t \leq 33$. So $s(33) = 2178$ ft. 15 minutes $= 15(60) = 900$ s, so for $33 < t \leq 933$ we have

$v(t) = 132$ ft/s $\Rightarrow$ $s(933) = 132(900) + 2178 = 120{,}978$ ft $= 22.9125$ mi.

(b) As in part (a), the train accelerates for 33 s and travels 2178 ft while doing so. Similarly, it decelerates for 33 s and travels

2178 ft at the end of its trip. During the remaining $900 - 66 = 834$ s it travels at 132 ft/s, so the distance traveled is

$132 \cdot 834 = 110{,}088$ ft. Thus, the total distance is $2178 + 110{,}088 + 2178 = 114{,}444$ ft $= 21.675$ mi.

(c) 45 mi $= 45(5280) = 237{,}600$ ft. Subtract $2(2178)$ to take care of the speeding up and slowing down, and we have

233,244 ft at 132 ft/s for a trip of $233{,}244/132 = 1767$ s at 90 mi/h. The total time is

$1767 + 2(33) = 1833$ s $= 30$ min 33 s $= 30.55$ min.

(d) $37.5(60) = 2250$ s. $2250 - 2(33) = 2184$ s at maximum speed. $2184(132) + 2(2178) = 292{,}644$ total feet or

$292{,}644/5280 = 55.425$ mi.

4   Review

CONCEPT CHECK

1. A function f has an **absolute maximum** at $x = c$ if $f(c)$ is the largest function value on the entire domain of f, whereas f has a **local maximum** at c if $f(c)$ is the largest function value when x is near c. See Figure 4 in Section 4.1.

2. (a) See Theorem 4.1.3.

(b) See the Closed Interval Method before Example 6 in Section 4.1.

3. (a) See Theorem 4.1.4.

(b) See Definition 4.1.6.

4. (a) See Rolle's Theorem at the beginning of Section 4.2.

(b) See the Mean Value Theorem in Section 4.2. Geometric interpretation—there is some point P on the graph of a function f [on the interval (a, b)] where the tangent line is parallel to the secant line that connects $(a, f(a))$ and $(b, f(b))$.

5. (a) See the I/D Test before Example 1 in Section 4.3.

(b) A function f is concave upward on an interval I if the graph of f lies above all of its tangent lines on I.

(c) See the Concavity Test before Example 4 in Section 4.3.

(d) An inflection point is a point where a curve changes its direction of concavity. They can be found by determining the points at which the second derivative changes sign.

6. (a) See the First Derivative Test after Example 1 in Section 4.3.

(b) See the Second Derivative Test before Example 5 in Section 4.3.

(c) See the note before Example 6 in Section 4.3.

7. Without calculus you could get misleading graphs that fail to show the most interesting features of a function. See the third paragraph in Section 4.4.

8. (a) See Figure 3 in Section 4.6.

(b) $x_2 = x_1 - \dfrac{f(x_1)}{f'(x_1)}$

(c) $x_{n+1} = x_n - \dfrac{f(x_n)}{f'(x_n)}$

(d) Newton's method is likely to fail or to work very slowly when $f'(x_1)$ is close to 0.

9. (a) See the definition at the beginning of Section 4.7.

(b) If F_1 and F_2 are both antiderivatives of f on an interval I, then they differ by a constant.

TRUE-FALSE QUIZ

1. False.   For example, take $f(x) = x^3$, then $f'(x) = 3x^2$ and $f'(0) = 0$, but $f(0) = 0$ is not a maximum or minimum; $(0,0)$ is an inflection point.

3. False.   For example, $f(x) = x$ is continuous on $(0,1)$ but attains neither a maximum nor a minimum value on $(0,1)$. Don't confuse this with f being continuous on the *closed* interval $[a,b]$, which would make the statement true.

5. True.   This is an example of part (b) of the I/D Test.

7. False. $f'(x) = g'(x) \;\Rightarrow\; f(x) = g(x) + C$. For example, if $f(x) = x + 2$ and $g(x) = x + 1$, then $f'(x) = g'(x) = 1$, but $f(x) \neq g(x)$.

9. True.   The graph of one such function is sketched.

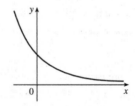

11. True.   Let $x_1 < x_2$ where $x_1, x_2 \in I$. Then $f(x_1) < f(x_2)$ and $g(x_1) < g(x_2)$ (since f and g are increasing on I), so $(f + g)(x_1) = f(x_1) + g(x_1) < f(x_2) + g(x_2) = (f + g)(x_2)$.

13. False.   Take $f(x) = x$ and $g(x) = x - 1$. Then both f and g are increasing on $(0,1)$. But $f(x)g(x) = x(x - 1)$ is not increasing on $(0,1)$.

15. True.   Let $x_1, x_2 \in I$ and $x_1 < x_2$. Then $f(x_1) < f(x_2)$ (f is increasing) $\Rightarrow$

$\dfrac{1}{f(x_1)} > \dfrac{1}{f(x_2)}$ (f is positive) $\;\Rightarrow\; g(x_1) > g(x_2) \;\Rightarrow\; g(x) = 1/f(x)$ is decreasing on I.

17. True.   By the Mean Value Theorem, there exists a number c in $(0,1)$ such that $f(1) - f(0) = f'(c)(1 - 0) = f'(c)$. Since $f'(c)$ is nonzero, $f(1) - f(0) \neq 0$, so $f(1) \neq f(0)$.

EXERCISES

1. $f(x) = 10 + 27x - x^3$, $0 \le x \le 4$. $f'(x) = 27 - 3x^2 = -3(x^2 - 9) = -3(x + 3)(x - 3) = 0$ only when $x = 3$ (since -3 is not in the domain). $f'(x) > 0$ for $x < 3$ and $f'(x) < 0$ for $x > 3$, so $f(3) = 64$ is a local maximum value. Checking the endpoints, we find $f(0) = 10$ and $f(4) = 54$. Thus, $f(0) = 10$ is the absolute minimum value and $f(3) = 64$ is the absolute maximum value.

3. $f(x) = \dfrac{x}{x^2 + x + 1}$, $-2 \le x \le 0$. $f'(x) = \dfrac{(x^2 + x + 1)(1) - x(2x + 1)}{(x^2 + x + 1)^2} = \dfrac{1 - x^2}{(x^2 + x + 1)^2} = 0$ $\Leftrightarrow$ $x = -1$ (since 1

is not in the domain). $f'(x) < 0$ for $-2 < x < -1$ and $f'(x) > 0$ for $-1 < x < 0$, so $f(-1) = -1$ is a local and absolute

minimum value. $f(-2) = -\frac{2}{3}$ and $f(0) = 0$, so $f(0) = 0$ is an absolute maximum value.

5. $f(0) = 0$, $f'(-2) = f'(1) = f'(9) = 0$, $\lim\limits_{x \to \infty} f(x) = 0$, $\lim\limits_{x \to 6} f(x) = -\infty$,

$f'(x) < 0$ on $(-\infty, -2)$, $(1, 6)$, and $(9, \infty)$, $f'(x) > 0$ on $(-2, 1)$ and $(6, 9)$,

$f''(x) > 0$ on $(-\infty, 0)$ and $(12, \infty)$, $f''(x) < 0$ on $(0, 6)$ and $(6, 12)$

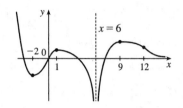

7. f is odd, $f'(x) < 0$ for $0 < x < 2$,

$f'(x) > 0$ for $x > 2$, $f''(x) > 0$ for $0 < x < 3$,

$f''(x) < 0$ for $x > 3$, $\lim_{x \to \infty} f(x) = -2$

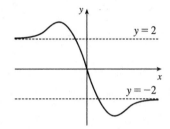

9. (a) $f(x) = 2 - 2x - x^3$ is a polynomial, so there is no asymptote.

(b) $f'(x) = -2 - 3x^2 = -1(3x^2 + 2) < 0$, so f is decreasing on $\mathbb{R}$.

(c) No local extrema

(d) $f''(x) = -6x < 0$ on $(0, \infty)$ and $f''(x) > 0$ on $(-\infty, 0)$, so f is

CD on $(0, \infty)$ and CU on $(-\infty, 0)$. IP at $(0, 2)$

(e)

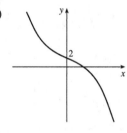

11. (a) $y = f(x) = \sin^2 x - 2 \cos x$ has no asymptote.

(b) $y' = 2 \sin x \cos x + 2 \sin x = 2 \sin x (\cos x + 1)$. $y' = 0$ $\Leftrightarrow$ $\sin x = 0$ or $\cos x = -1$ $\Leftrightarrow$ $x = n\pi$ or

$x = (2n + 1)\pi$. $y' > 0$ when $\sin x > 0$, since $\cos x + 1 \ge 0$ for all x. Therefore, $y' > 0$ (and so f is increasing) on

$(2n\pi, (2n + 1)\pi)$; $y' < 0$ (and so f is decreasing) on $((2n - 1)\pi, 2n\pi)$ or equivalently, $((2n + 1)\pi, (2n + 2)\pi)$.

(c) Local maxima are $f((2n + 1)\pi) = 2$; local minima are $f(2n\pi) = -2$.

(d) $y' = \sin 2x + 2 \sin x$ $\Rightarrow$

$\qquad y'' = 2 \cos 2x + 2 \cos x = 2(2 \cos^2 x - 1) + 2 \cos x$

$\qquad\quad = 4 \cos^2 x + 2 \cos x - 2 = 2(2 \cos^2 x + \cos x - 1)$

$\qquad\quad = 2(2 \cos x - 1)(\cos x + 1)$

$y'' = 0$ $\Leftrightarrow$ $\cos x = \frac{1}{2}$ or -1 $\Leftrightarrow$ $x = 2n\pi \pm \frac{\pi}{3}$ or $x = (2n + 1)\pi$.

$y'' > 0$ (and so f is CU) on $(2n\pi - \frac{\pi}{3}, 2n\pi + \frac{\pi}{3})$; $y'' \le 0$ (and so f is CD) on $(2n\pi + \frac{\pi}{3}, 2n\pi + \frac{5\pi}{3})$.

IPs at $(2n\pi \pm \frac{\pi}{3}, -\frac{1}{4})$

(e)

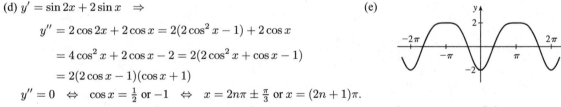

13. (a) $\lim\limits_{x \to \pm\infty} (e^x + e^{-3x}) = \infty$, no asymptote.

(b) $y = f(x) = e^x + e^{-3x}$ $\Rightarrow$ $f'(x) = e^x - 3e^{-3x} = e^{-3x}(e^{4x} - 3) > 0$

$\Leftrightarrow$ $e^{4x} > 3$ $\Leftrightarrow$ $4x > \ln 3$ $\Leftrightarrow$ $x > \frac{1}{4} \ln 3$, so f is increasing on

$\left(\frac{1}{4} \ln 3, \infty\right)$ and decreasing on $\left(-\infty, \frac{1}{4} \ln 3\right)$.

(c) $f\left(\frac{1}{4} \ln 3\right) = 3^{1/4} + 3^{-3/4} \approx 1.75$ is a local and absolute minimum.

(d) $f''(x) = e^x + 9e^{-3x} > 0$, so f is CU on $(-\infty, \infty)$. No IP

(e)

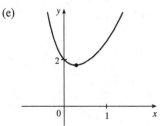

15. $y = f(x) = x^4 - 3x^3 + 3x^2 - x = x(x-1)^3$ **A.** $D = \mathbb{R}$ **B.** y-intercept: $f(0) = 0$; x-intercepts: $f(x) = 0$ ⇔

$x = 0$ or $x = 1$ **C.** No symmetry **D.** f is a polynomial function and hence, it has no asymptote.

E. $f'(x) = 4x^3 - 9x^2 + 6x - 1$. Since the sum of the coefficients is 0, 1 is a root of f', so

$f'(x) = (x-1)(4x^2 - 5x + 1) = (x-1)^2(4x-1)$. $f'(x) < 0$ ⇒ $x < \frac{1}{4}$, so f is decreasing on $\left(-\infty, \frac{1}{4}\right)$

and f is increasing on $\left(\frac{1}{4}, \infty\right)$. **F.** $f'(x)$ does not change sign at $x = 1$, so there

is not a local extremum there. $f\left(\frac{1}{4}\right) = -\frac{27}{256}$ is a local minimum value.

G. $f''(x) = 12x^2 - 18x + 6 = 6(2x-1)(x-1)$.

$f''(x) = 0$ ⇔ $x = \frac{1}{2}$ or 1. $f''(x) < 0$ ⇔ $\frac{1}{2} < x < 1$ ⇒

f is CD on $\left(\frac{1}{2}, 1\right)$ and CU on $\left(-\infty, \frac{1}{2}\right)$ and $(1, \infty)$. IPs at $\left(\frac{1}{2}, -\frac{1}{16}\right)$ and $(1, 0)$

H.

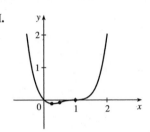

17. $y = f(x) = \dfrac{1}{x(x-3)^2}$ **A.** $D = \{x \mid x \neq 0, 3\} = (-\infty, 0) \cup (0, 3) \cup (3, \infty)$ **B.** No intercepts. **C.** No symmetry.

D. $\displaystyle\lim_{x \to \pm\infty} \frac{1}{x(x-3)^2} = 0$, so $y = 0$ is a HA. $\displaystyle\lim_{x \to 0^+} \frac{1}{x(x-3)^2} = \infty$, $\displaystyle\lim_{x \to 0^-} \frac{1}{x(x-3)^2} = -\infty$, $\displaystyle\lim_{x \to 3} \frac{1}{x(x-3)^2} = \infty$,

so $x = 0$ and $x = 3$ are VA. **E.** $f'(x) = -\dfrac{(x-3)^2 + 2x(x-3)}{x^2(x-3)^4} = \dfrac{3(1-x)}{x^2(x-3)^3}$ ⇒ $f'(x) > 0$ ⇔ $1 < x < 3$,

so f is increasing on $(1, 3)$ and decreasing on $(-\infty, 0)$, $(0, 1)$, and $(3, \infty)$. **H.**

F. Local minimum value $f(1) = \frac{1}{4}$ **G.** $f''(x) = \dfrac{6(2x^2 - 4x + 3)}{x^3(x-3)^4}$. Note that

$2x^2 - 4x + 3 > 0$ for all x since it has negative discriminant. So $f''(x) > 0$ ⇔

$x > 0$ ⇒ f is CU on $(0, 3)$ and $(3, \infty)$ and CD on $(-\infty, 0)$. No IP

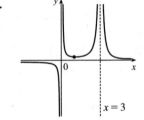

19. $y = f(x) = x\sqrt{2 + x}$ **A.** $D = [-2, \infty)$ **B.** y-intercept: $f(0) = 0$; x-intercepts: -2 and 0 **C.** No symmetry

D. No asymptote **E.** $f'(x) = \dfrac{x}{2\sqrt{2+x}} + \sqrt{2+x} = \dfrac{1}{2\sqrt{2+x}}[x + 2(2+x)] = \dfrac{3x+4}{2\sqrt{2+x}} = 0$ when $x = -\frac{4}{3}$, so f is

decreasing on $\left(-2, -\frac{4}{3}\right)$ and increasing on $\left(-\frac{4}{3}, \infty\right)$. **F.** Local minimum value **H.**

$f\left(-\frac{4}{3}\right) = -\frac{4}{3}\sqrt{\frac{2}{3}} = -\frac{4\sqrt{6}}{9} \approx -1.09$, no local maximum

G. $f''(x) = \dfrac{2\sqrt{2+x} \cdot 3 - (3x+4)\dfrac{1}{\sqrt{2+x}}}{4(2+x)} = \dfrac{6(2+x) - (3x+4)}{4(2+x)^{3/2}} = \dfrac{3x+8}{4(2+x)^{3/2}}$

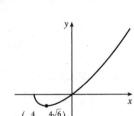

$f''(x) > 0$ for $x > -2$, so f is CU on $(-2, \infty)$. No IP

21. $y = f(x) = \sin^{-1}(1/x)$ **A.** $D = \{x \mid -1 \le 1/x \le 1\} = (-\infty, -1] \cup [1, \infty)$. **B.** No intercept **C.** $f(-x) = -f(x)$,

symmetric about the origin **D.** $\lim\limits_{x \to \pm\infty} \sin^{-1}(1/x) = \sin^{-1}(0) = 0$, so $y = 0$ is a HA.

E. $f'(x) = \dfrac{1}{\sqrt{1 - (1/x)^2}}\left(-\dfrac{1}{x^2}\right) = \dfrac{-1}{\sqrt{x^4 - x^2}} < 0$, so f is decreasing on $(-\infty, -1)$ and $(1, \infty)$.

F. No local extreme value, but $f(1) = \frac{\pi}{2}$ is the absolute maximum value and

$f(-1) = -\frac{\pi}{2}$ is the absolute minimum value.

H.

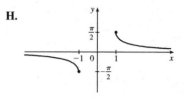

G. $f''(x) = \dfrac{4x^3 - 2x}{2(x^4 - x^2)^{3/2}} = \dfrac{x(2x^2 - 1)}{(x^4 - x^2)^{3/2}} > 0$ for $x > 1$ and $f''(x) < 0$

for $x < -1$, so f is CU on $(1, \infty)$ and CD on $(-\infty, -1)$.   No IP

23. $f(x) = \dfrac{x^2 - 1}{x^3} \ \Rightarrow \ f'(x) = \dfrac{x^3(2x) - (x^2 - 1)3x^2}{x^6} = \dfrac{3 - x^2}{x^4} \ \Rightarrow$

$f''(x) = \dfrac{x^4(-2x) - (3 - x^2)4x^3}{x^8} = \dfrac{2x^2 - 12}{x^5}$

Estimates: From the graphs of f' and f'', it appears that f is increasing on

$(-1.73, 0)$ and $(0, 1.73)$ and decreasing on $(-\infty, -1.73)$ and $(1.73, \infty)$;

f has a local maximum of about $f(1.73) = 0.38$ and a local minimum of about

$f(-1.7) = -0.38$; f is CU on $(-2.45, 0)$ and $(2.45, \infty)$, and CD on

$(-\infty, -2.45)$ and $(0, 2.45)$; and f has IPs at about $(-2.45, -0.34)$

and $(2.45, 0.34)$.

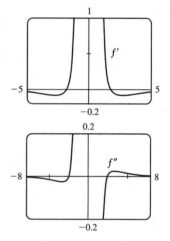

Exact: Now $f'(x) = \dfrac{3 - x^2}{x^4}$ is positive for $0 < x^2 < 3$, that is, f is increasing

on $(-\sqrt{3}, 0)$ and $(0, \sqrt{3})$; and $f'(x)$ is negative (and so f is decreasing) on

$(-\infty, -\sqrt{3})$ and $(\sqrt{3}, \infty)$. $f'(x) = 0$ when $x = \pm\sqrt{3}$.

f' goes from positive to negative at $x = \sqrt{3}$, so f has a local maximum of

$f(\sqrt{3}) = \dfrac{(\sqrt{3})^2 - 1}{(\sqrt{3})^3} = \dfrac{2\sqrt{3}}{9}$; and since f is odd, we know that maxima on the

interval $(0, \infty)$ correspond to minima on $(-\infty, 0)$, so f has a local minimum of

$f(-\sqrt{3}) = -\dfrac{2\sqrt{3}}{9}$. Also, $f''(x) = \dfrac{2x^2 - 12}{x^5}$ is positive (so f is CU) on

$(-\sqrt{6}, 0)$ and $(\sqrt{6}, \infty)$, and negative (so f is CD) on $(-\infty, -\sqrt{6})$ and

$(0, \sqrt{6})$. There are IP at $\left(\sqrt{6}, \dfrac{5\sqrt{6}}{36}\right)$ and $\left(-\sqrt{6}, -\dfrac{5\sqrt{6}}{36}\right)$.

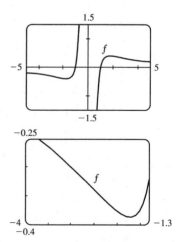

25. $f(x) = 3x^6 - 5x^5 + x^4 - 5x^3 - 2x^2 + 2 \Rightarrow f'(x) = 18x^5 - 25x^4 + 4x^3 - 15x^2 - 4x \Rightarrow$

$f''(x) = 90x^4 - 100x^3 + 12x^2 - 30x - 4$

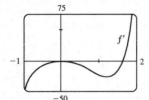 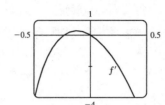 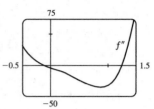

From the graphs of f' and f'', it appears that f is increasing on $(-0.23, 0)$ and $(1.62, \infty)$ and decreasing on $(-\infty, -0.23)$

and $(0, 1.62)$; f has a local maximum of about $f(0) = 2$ and local minima of about $f(-0.23) = 1.96$ and $f(1.62) = -19.2$;

f is CU on $(-\infty, -0.12)$ and $(1.24, \infty)$ and CD on $(-0.12, 1.24)$; and f has inflection points at about $(-0.12, 1.98)$ and

$(1.24, -12.1)$.

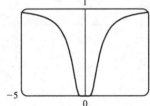 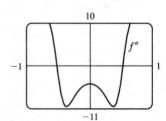

27.

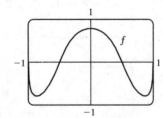

From the graph, we estimate the points of inflection to be about $(\pm 0.82, 0.22)$.

$f(x) = e^{-1/x^2} \Rightarrow f'(x) = 2x^{-3}e^{-1/x^2} \Rightarrow$

$f''(x) = 2[x^{-3}(2x^{-3})e^{-1/x^2} + e^{-1/x^2}(-3x^{-4})] = 2x^{-6}e^{-1/x^2}(2 - 3x^2)$.

This is 0 when $2 - 3x^2 = 0 \Leftrightarrow x = \pm\sqrt{\frac{2}{3}}$, so the IPs are $\left(\pm\sqrt{\frac{2}{3}}, e^{-3/2}\right)$.

29. $f(x) = \arctan(\cos(3 \arcsin x))$. We use a CAS to compute f' and f'', and to graph f, f', and f'':

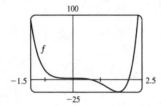 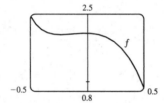 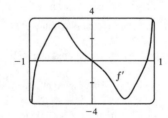

From the graph of f', it appears that the only maximum occurs at $x = 0$ and there are minima at $x = \pm 0.87$.

From the graph of f'', it appears that there are inflection points at $x = \pm 0.52$.

31. The family of functions $f(x) = \ln(\sin x + C)$ all have the same period and all have maximum values at $x = \frac{\pi}{2} + 2\pi n$. Since the domain of ln is $(0, \infty)$, f has a graph only if $\sin x + C > 0$ somewhere. Since $-1 \le \sin x \le 1$, this happens if $C > -1$, that is, f has no graph if $C \le -1$. Similarly, if $C > 1$, then $\sin x + C > 0$ and f is continuous on $(-\infty, \infty)$. As C increases, the graph of f is shifted vertically upward and flattens out. If $-1 < C \le 1$, f is

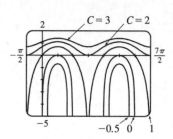

defined where $\sin x + C > 0$ $\Leftrightarrow$ $\sin x > -C$ $\Leftrightarrow$ $\sin^{-1}(-C) < x < \pi - \sin^{-1}(-C)$. Since the period is 2π, the domain of f is $\big(2n\pi + \sin^{-1}(-C), (2n+1)\pi - \sin^{-1}(-C)\big)$, n an integer.

33. $f(x) = x^{101} + x^{51} + x - 1 = 0$. Since f is continuous and $f(0) = -1$ and $f(1) = 2$, the equation has at least one root in $(0, 1)$, by the Intermediate Value Theorem. Suppose the equation has two roots, a and b, with $a < b$. Then $f(a) = 0 = f(b)$, so by the Mean Value Theorem, there is a number x in (a, b) such that

$$f'(x) = \frac{f(b) - f(a)}{b - a} = \frac{0}{b - a} = 0, \text{ so } f' \text{ has a root in } (a, b). \text{ But this is impossible since}$$

$f'(x) = 101x^{100} + 51x^{50} + 1 \ge 1$ for all x.

35. Since f is continuous on $[32, 33]$ and differentiable on $(32, 33)$, then by the Mean Value Theorem there exists a number c in

$(32, 33)$ such that $f'(c) = \frac{1}{5}c^{-4/5} = \dfrac{\sqrt[5]{33} - \sqrt[5]{32}}{33 - 32} = \sqrt[5]{33} - 2$, but $\frac{1}{5}c^{-4/5} > 0$ $\Rightarrow$ $\sqrt[5]{33} - 2 > 0$ $\Rightarrow$ $\sqrt[5]{33} > 2$. Also

f' is decreasing, so that $f'(c) < f'(32) = \frac{1}{5}(32)^{-4/5} = 0.0125$ $\Rightarrow$ $0.0125 > f'(c) = \sqrt[5]{33} - 2$ $\Rightarrow$ $\sqrt[5]{33} < 2.0125$. Therefore, $2 < \sqrt[5]{33} < 2.0125$.

37. Call the two integers x and y. Then $x + 4y = 1000$, so $x = 1000 - 4y$. Their product is $P = xy = (1000 - 4y)y$, so our problem is to maximize the function $P(y) = 1000y - 4y^2$, where $0 < y < 250$ and y is an integer. $P'(y) = 1000 - 8y$, so $P'(y) = 0$ $\Leftrightarrow$ $y = 125$. $P''(y) = -8 < 0$, so $P(125) = 62{,}500$ is an absolute maximum. Since the optimal y turned out to be an integer, we have found the desired pair of numbers, namely $x = 1000 - 4(125) = 500$ and $y = 125$.

39.

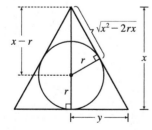

By similar triangles, $\dfrac{y}{x} = \dfrac{r}{\sqrt{x^2 - 2rx}}$, so the area of the triangle is

$$A(x) = \tfrac{1}{2}(2y)x = xy = \frac{rx^2}{\sqrt{x^2 - 2rx}} \quad \Rightarrow$$

$$A'(x) = \frac{2rx\sqrt{x^2 - 2rx} - rx^2(x - r)/\sqrt{x^2 - 2rx}}{x^2 - 2rx} = \frac{rx^2(x - 3r)}{(x^2 - 2rx)^{3/2}} = 0$$

when $x = 3r$. $A'(x) < 0$ when $2r < x < 3r$, $A'(x) > 0$ when $x > 3r$. So $x = 3r$ gives a minimum and

$A(3r) = r(9r^2)/(\sqrt{3}\,r) = 3\sqrt{3}\,r^2$.

41.

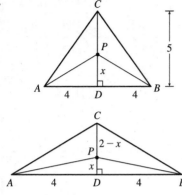

We minimize $L(x) = |PA| + |PB| + |PC| = 2\sqrt{x^2 + 16} + (5 - x)$,

$0 \leq x \leq 5$. $L'(x) = 2x/\sqrt{x^2 + 16} - 1 = 0 \iff 2x = \sqrt{x^2 + 16} \iff$

$4x^2 = x^2 + 16 \iff x = \frac{4}{\sqrt{3}}$. $L(0) = 13, L\left(\frac{4}{\sqrt{3}}\right) \approx 11.9, L(5) \approx 12.8$, so the

minimum occurs when $x = \frac{4}{\sqrt{3}} \approx 2.3$.

If $|CD| = 2$, $L(x)$ changes from $(5 - x)$ to $(2 - x)$ with $0 \leq x \leq 2$. But we still

get $L'(x) = 0 \iff x = \frac{4}{\sqrt{3}}$, which isn't in the interval $[0, 2]$. Now $L(0) = 10$

and $L(2) = 2\sqrt{20} = 4\sqrt{5} \approx 8.9$. The minimum occurs when $P = C$.

43. $v = K\sqrt{\dfrac{L}{C} + \dfrac{C}{L}} \Rightarrow \dfrac{dv}{dL} = \dfrac{K}{2\sqrt{(L/C) + (C/L)}}\left(\dfrac{1}{C} - \dfrac{C}{L^2}\right) = 0 \iff \dfrac{1}{C} = \dfrac{C}{L^2} \iff L^2 = C^2 \iff L = C.$

This gives the minimum velocity since $v' < 0$ for $0 < L < C$ and $v' > 0$ for $L > C$.

45. Let x denote the number of \$1 decreases in ticket price. Then the ticket price is \$12 − \$1(x), and the average attendance is

$11{,}000 + 1000(x)$. Now the revenue per game is

$$R(x) = (\text{price per person}) \times (\text{number of people per game})$$
$$= (12 - x)(11{,}000 + 1000x) = -1000x^2 + 1000x + 132{,}000$$

for $0 \leq x \leq 4$ [since the seating capacity is 15,000] $\Rightarrow R'(x) = -2000x + 1000 = 0 \iff x = 0.5$. This is a

maximum since $R''(x) = -2000 < 0$ for all x. Now we must check the value of $R(x) = (12 - x)(11{,}000 + 1000x)$ at

$x = 0.5$ and at the endpoints of the domain to see which value of x gives the maximum value of R.

$R(0) = (12)(11{,}000) = 132{,}000$, $R(0.5) = (11.5)(11{,}500) = 132{,}250$, and $R(4) = (8)(15{,}000) = 120{,}000$. Thus, the

maximum revenue of \$132,250 per game occurs when the average attendance is 11,500 and the ticket price is \$11.50.

47. $f(t) = \cos t + t - t^2 \Rightarrow f'(t) = -\sin t + 1 - 2t$. $f'(t)$ exists for all t, so

to find the maximum of f, we can examine the zeros of f'. From the graph of

f', we see that a good choice for t_1 is $t_1 = 0.3$. Use $g(t) = -\sin t + 1 - 2t$

and $g'(t) = -\cos t - 2$ to obtain $t_2 \approx 0.33535293$, $t_3 \approx 0.33541803 \approx t_4$.

Since $f''(t) = -\cos t - 2 < 0$ for all t, $f(0.33541803) \approx 1.16718557$ is the

absolute maximum.

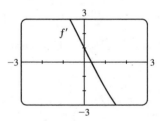

49. $f(x) = e^x - (2/\sqrt{x}) = e^x - 2x^{-1/2} \Rightarrow F(x) = e^x - 2\dfrac{x^{-1/2+1}}{-1/2 + 1} + C = e^x - 2\dfrac{x^{1/2}}{1/2} + C = e^x - 4\sqrt{x} + C$

51. $f'(x) = 2/(1 + x^2) \Rightarrow f(x) = 2\arctan x + C$. $f(0) = 2\arctan 0 + C = 0 + C = C$ and $f(0) = -1 \Rightarrow$

$C = -1$. Therefore, $f(x) = 2\arctan x - 1$.

53. $f''(x) = 1 - 6x + 48x^2$ $\Rightarrow$ $f'(x) = x - 3x^2 + 16x^3 + C$. $f'(0) = C$ and $f'(0) = 2$ $\Rightarrow$ $C = 2$,

so $f'(x) = x - 3x^2 + 16x^3 + 2$ and hence, $f(x) = \frac{1}{2}x^2 - x^3 + 4x^4 + 2x + D$.

$f(0) = D$ and $f(0) = 1$ $\Rightarrow$ $D = 1$, so $f(x) = \frac{1}{2}x^2 - x^3 + 4x^4 + 2x + 1$.

55. $a(t) = v'(t) = t - 2$ $\Rightarrow$ $v(t) = \frac{1}{2}t^2 - 2t + C$. $v(0) = C$ and $v(0) = 3$ $\Rightarrow$ $C = 3$, so $v(t) = \frac{1}{2}t^2 - 2t + 3$

and $s(t) = \frac{1}{6}t^3 - t^2 + 3t + D$. $s(0) = D$ and $s(0) = 1$ $\Rightarrow$ $D = 1$, and $s(t) = \frac{1}{6}t^3 - t^2 + 3t + 1$.

57. Choosing the positive direction to be upward, we have $a(t) = -9.8$ $\Rightarrow$ $v(t) = -9.8t + v_0$, but $v(0) = 0 = v_0$ $\Rightarrow$

$v(t) = -9.8t = s'(t)$ $\Rightarrow$ $s(t) = -4.9t^2 + s_0$, but $s(0) = s_0 = 500$ $\Rightarrow$ $s(t) = -4.9t^2 + 500$. When $s = 0$,

$-4.9t^2 + 500 = 0$ $\Rightarrow$ $t_1 = \sqrt{\frac{500}{4.9}} \approx 10.1$ $\Rightarrow$ $v(t_1) = -9.8\sqrt{\frac{500}{4.9}} \approx -98.995$ m/s. Since the canister has been

designed to withstand an impact velocity of 100 m/s, the canister will *not burst*.

59. (a)

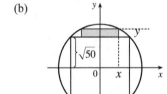

The cross-sectional area of the rectangular beam is

$A = 2x \cdot 2y = 4xy = 4x\sqrt{100 - x^2}, 0 \le x \le 10$, so

$\dfrac{dA}{dx} = 4x\left(\frac{1}{2}\right)\left(100 - x^2\right)^{-1/2}(-2x) + \left(100 - x^2\right)^{1/2} \cdot 4$

$= \dfrac{-4x^2}{(100 - x^2)^{1/2}} + 4(100 - x^2)^{1/2} = \dfrac{4[-x^2 + (100 - x^2)]}{(100 - x^2)^{1/2}}$

$\dfrac{dA}{dx} = 0$ when $-x^2 + (100 - x^2) = 0$ $\Rightarrow$ $x^2 = 50$ $\Rightarrow$ $x = \sqrt{50} \approx 7.07$ $\Rightarrow$ $y = \sqrt{100 - \left(\sqrt{50}\right)^2} = \sqrt{50}$.

Since $A(0) = A(10) = 0$, the rectangle of maximum area is a square.

(b)

The cross-sectional area of each rectangular plank (shaded in the figure) is

$A = 2x\left(y - \sqrt{50}\right) = 2x\left[\sqrt{100 - x^2} - \sqrt{50}\right], 0 \le x \le \sqrt{50}$, so

$\dfrac{dA}{dx} = 2\left(\sqrt{100 - x^2} - \sqrt{50}\right) + 2x\left(\frac{1}{2}\right)(100 - x^2)^{-1/2}(-2x)$

$= 2(100 - x^2)^{1/2} - 2\sqrt{50} - \dfrac{2x^2}{(100 - x^2)^{1/2}}$

Set $\dfrac{dA}{dx} = 0$: $(100 - x^2) - \sqrt{50}(100 - x^2)^{1/2} - x^2 = 0$ $\Rightarrow$ $100 - 2x^2 = \sqrt{50}(100 - x^2)^{1/2}$ $\Rightarrow$

$10{,}000 - 400x^2 + 4x^4 = 50(100 - x^2)$ $\Rightarrow$ $4x^4 - 350x^2 + 5000 = 0$ $\Rightarrow$ $2x^4 - 175x^2 + 2500 = 0$ $\Rightarrow$

$x^2 = \dfrac{175 \pm \sqrt{10{,}625}}{4} \approx 69.52$ or 17.98 $\Rightarrow$ $x \approx 8.34$ or 4.24. But $8.34 > \sqrt{50}$, so $x_1 \approx 4.24$ $\Rightarrow$

$y - \sqrt{50} = \sqrt{100 - x_1^2} - \sqrt{50} \approx 1.99$. Each plank should have dimensions about $8\frac{1}{2}$ inches by 2 inches.

(c) From the figure in part (a), the width is $2x$ and the depth is $2y$, so the strength is

$S = k(2x)(2y)^2 = 8kxy^2 = 8kx(100 - x^2) = 800kx - 8kx^3, 0 \le x \le 10$. $dS/dx = 800k - 24kx^2 = 0$ when

$24kx^2 = 800k$ $\Rightarrow$ $x^2 = \frac{100}{3}$ $\Rightarrow$ $x = \frac{10}{\sqrt{3}}$ $\Rightarrow$ $y = \sqrt{\frac{200}{3}} = \frac{10\sqrt{2}}{\sqrt{3}} = \sqrt{2}\,x$. Since $S(0) = S(10) = 0$, the

maximum strength occurs when $x = \frac{10}{\sqrt{3}}$. The dimensions should be $\frac{20}{\sqrt{3}} \approx 11.55$ inches by $\frac{20\sqrt{2}}{\sqrt{3}} \approx 16.33$ inches.

5 □ INTEGRALS

5.1 Areas and Distances

1. (a) Since f is *increasing*, we can obtain a *lower* estimate by using

left endpoints. We are instructed to use five rectangles, so $n = 5$.

$$L_5 = \sum_{i=1}^{5} f(x_{i-1})\,\Delta x \quad [\Delta x = \tfrac{b-a}{n} = \tfrac{10-0}{5} = 2]$$

$$= f(x_0) \cdot 2 + f(x_1) \cdot 2 + f(x_2) \cdot 2 + f(x_3) \cdot 2 + f(x_4) \cdot 2$$

$$= 2\,[f(0) + f(2) + f(4) + f(6) + f(8)]$$

$$\approx 2(1 + 3 + 4.3 + 5.4 + 6.3) = 2(20) = 40$$

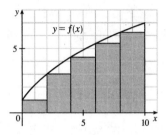

Since f is *increasing*, we can obtain an *upper* estimate by using

right endpoints.

$$R_5 = \sum_{i=1}^{5} f(x_i)\,\Delta x$$

$$= 2\,[f(x_1) + f(x_2) + f(x_3) + f(x_4) + f(x_5)]$$

$$= 2\,[f(2) + f(4) + f(6) + f(8) + f(10)]$$

$$\approx 2(3 + 4.3 + 5.4 + 6.3 + 7) = 2(26) = 52$$

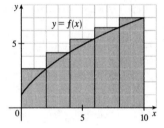

Comparing R_5 to L_5, we see that we have added the area of the rightmost upper rectangle, $f(10) \cdot 2$, to the sum and
subtracted the area of the leftmost lower rectangle, $f(0) \cdot 2$, from the sum.

(b) $L_{10} = \sum_{i=1}^{10} f(x_{i-1})\,\Delta x \quad [\Delta x = \tfrac{10-0}{10} = 1]$

$$= 1\,[f(x_0) + f(x_1) + \cdots + f(x_9)]$$

$$= f(0) + f(1) + \cdots + f(9)$$

$$\approx 1 + 2.1 + 3 + 3.7 + 4.3 + 4.9 + 5.4 + 5.8 + 6.3 + 6.7$$

$$= 43.2$$

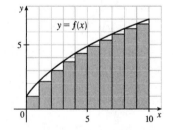

$$R_{10} = \sum_{i=1}^{10} f(x_i)\,\Delta x = f(1) + f(2) + \cdots + f(10)$$

$$= L_{10} + 1 \cdot f(10) - 1 \cdot f(0) \quad \begin{bmatrix} \text{add rightmost upper rectangle,} \\ \text{subtract leftmost lower rectangle} \end{bmatrix}$$

$$= 43.2 + 7 - 1 = 49.2$$

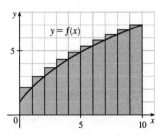

3. (a) $R_4 = \sum\limits_{i=1}^{4} f(x_i)\,\Delta x \quad [\Delta x = \frac{5-1}{4} = 1]$

$\quad = f(x_1) \cdot 1 + f(x_2) \cdot 1 + f(x_3) \cdot 1 + f(x_4) \cdot 1$

$\quad = f(2) + f(3) + f(4) + f(5)$

$\quad = \frac{1}{2} + \frac{1}{3} + \frac{1}{4} + \frac{1}{5} = \frac{77}{60} = 1.28\overline{3}$

Since f is *decreasing* on $[1, 5]$, an *underestimate* is obtained by using the *right* endpoint approximation, R_4.

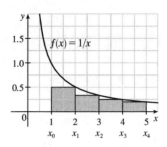

(b) $L_4 = \sum\limits_{i=1}^{4} f(x_{i-1})\,\Delta x$

$\quad = f(1) + f(2) + f(3) + f(4)$

$\quad = 1 + \frac{1}{2} + \frac{1}{3} + \frac{1}{4} = \frac{25}{12} = 2.08\overline{3}$

L_4 is an overestimate. Alternatively, we could just add the area of the leftmost upper rectangle and subtract the area of the rightmost lower rectangle; that is, $L_4 = R_4 + f(1) \cdot 1 - f(5) \cdot 1$.

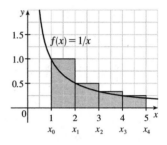

5. (a) $f(x) = 1 + x^2$ and $\Delta x = \dfrac{2 - (-1)}{3} = 1 \quad \Rightarrow$

$R_3 = 1 \cdot f(0) + 1 \cdot f(1) + 1 \cdot f(2) = 1 \cdot 1 + 1 \cdot 2 + 1 \cdot 5 = 8.$

$\Delta x = \dfrac{2 - (-1)}{6} = 0.5 \quad \Rightarrow$

$R_6 = 0.5[f(-0.5) + f(0) + f(0.5) + f(1) + f(1.5) + f(2)]$

$\quad = 0.5(1.25 + 1 + 1.25 + 2 + 3.25 + 5)$

$\quad = 0.5(13.75) = 6.875$

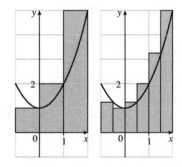

(b) $L_3 = 1 \cdot f(-1) + 1 \cdot f(0) + 1 \cdot f(1) = 1 \cdot 2 + 1 \cdot 1 + 1 \cdot 2 = 5$

$L_6 = 0.5[f(-1) + f(-0.5) + f(0) + f(0.5) + f(1) + f(1.5)]$

$\quad = 0.5(2 + 1.25 + 1 + 1.25 + 2 + 3.25)$

$\quad = 0.5(10.75) = 5.375$

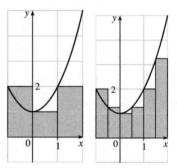

(c) $M_3 = 1 \cdot f(-0.5) + 1 \cdot f(0.5) + 1 \cdot f(1.5)$

$\quad = 1 \cdot 1.25 + 1 \cdot 1.25 + 1 \cdot 3.25 = 5.75$

$M_6 = 0.5[f(-0.75) + f(-0.25) + f(0.25) + f(0.75) + f(1.25) + f(1.75)]$

$\quad = 0.5(1.5625 + 1.0625 + 1.0625 + 1.5625 + 2.5625 + 4.0625)$

$\quad = 0.5(11.875) = 5.9375$

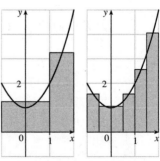

(d) M_6 appears to be the best estimate.

7. Since v is an increasing function, L_6 will give us a lower estimate and R_6 will give us an upper estimate.

$L_6 = (0 \text{ ft/s})(0.5 \text{ s}) + (6.2)(0.5) + (10.8)(0.5) + (14.9)(0.5) + (18.1)(0.5) + (19.4)(0.5) = 0.5(69.4) = 34.7 \text{ ft}$

$R_6 = 0.5(6.2 + 10.8 + 14.9 + 18.1 + 19.4 + 20.2) = 0.5(89.6) = 44.8 \text{ ft}$

9. Lower estimate for oil leakage: $R_5 = (7.6 + 6.8 + 6.2 + 5.7 + 5.3)(2) = (31.6)(2) = 63.2 \text{ L}.$

Upper estimate for oil leakage: $L_5 = (8.7 + 7.6 + 6.8 + 6.2 + 5.7)(2) = (35)(2) = 70 \text{ L}.$

11. For a decreasing function, using left endpoints gives us an overestimate and using right endpoints results in an underestimate.
We will use M_6 to get an estimate. $\Delta t = 1$, so

$$M_6 = 1[v(0.5) + v(1.5) + v(2.5) + v(3.5) + v(4.5) + v(5.5)]$$

$$\approx 55 + 40 + 28 + 18 + 10 + 4 = 155 \text{ ft}$$

For a very rough check on the above calculation, we can draw a line from $(0, 70)$ to $(6, 0)$ and calculate the area of the triangle: $\frac{1}{2}(70)(6) = 210$. This is clearly an overestimate, so our midpoint estimate of 155 is reasonable.

13. $f(x) = \sqrt[4]{x}$, $1 \le x \le 16$. $\Delta x = (16 - 1)/n = 15/n$ and $x_i = 1 + i\,\Delta x = 1 + 15i/n$.

$A = \lim\limits_{n \to \infty} R_n = \lim\limits_{n \to \infty} \sum\limits_{i=1}^{n} f(x_i)\,\Delta x = \lim\limits_{n \to \infty} \sum\limits_{i=1}^{n} \sqrt[4]{1 + \frac{15i}{n}} \cdot \frac{15}{n}.$

15. $\lim\limits_{n \to \infty} \sum\limits_{i=1}^{n} \dfrac{\pi}{4n} \tan \dfrac{i\pi}{4n}$ can be interpreted as the area of the region lying under the graph of $y = \tan x$ on the interval $\left[0, \frac{\pi}{4}\right]$,

since for $y = \tan x$ on $\left[0, \frac{\pi}{4}\right]$ with $\Delta x = \dfrac{\pi/4 - 0}{n} = \dfrac{\pi}{4n}$, $x_i = 0 + i\,\Delta x = \dfrac{i\pi}{4n}$, and $x_i^* = x_i$, the expression for the area is

$A = \lim\limits_{n \to \infty} \sum\limits_{i=1}^{n} f(x_i^*)\,\Delta x = \lim\limits_{n \to \infty} \sum\limits_{i=1}^{n} \tan\left(\dfrac{i\pi}{4n}\right) \dfrac{\pi}{4n}$. Note that this answer is not unique, since the expression for the area is

the same for the function $y = \tan(x - k\pi)$ on the interval $\left[k\pi, k\pi + \frac{\pi}{4}\right]$, where k is any integer.

17. (a) $y = f(x) = x^5$. $\Delta x = \dfrac{2 - 0}{n} = \dfrac{2}{n}$ and $x_i = 0 + i\,\Delta x = \dfrac{2i}{n}$.

$$A = \lim\limits_{n \to \infty} R_n = \lim\limits_{n \to \infty} \sum\limits_{i=1}^{n} f(x_i)\,\Delta x = \lim\limits_{n \to \infty} \sum\limits_{i=1}^{n} \left(\dfrac{2i}{n}\right)^5 \cdot \dfrac{2}{n} = \lim\limits_{n \to \infty} \sum\limits_{i=1}^{n} \dfrac{32i^5}{n^5} \cdot \dfrac{2}{n} = \lim\limits_{n \to \infty} \dfrac{64}{n^6} \sum\limits_{i=1}^{n} i^5.$$

(b) $\sum\limits_{i=1}^{n} i^5 \overset{\text{CAS}}{=} \dfrac{n^2(n+1)^2(2n^2 + 2n - 1)}{12}$

(c) $\lim\limits_{n \to \infty} \dfrac{64}{n^6} \cdot \dfrac{n^2(n+1)^2(2n^2 + 2n - 1)}{12} = \dfrac{64}{12} \lim\limits_{n \to \infty} \dfrac{(n^2 + 2n + 1)(2n^2 + 2n - 1)}{n^2 \cdot n^2}$

$$= \dfrac{16}{3} \lim\limits_{n \to \infty} \left(1 + \dfrac{2}{n} + \dfrac{1}{n^2}\right)\left(2 + \dfrac{2}{n} - \dfrac{1}{n^2}\right) = \dfrac{16}{3} \cdot 1 \cdot 2 = \dfrac{32}{3}$$

19. $y = f(x) = \cos x.$ $\Delta x = \dfrac{b-0}{n} = \dfrac{b}{n}$ and $x_i = 0 + i\,\Delta x = \dfrac{bi}{n}.$

$$A = \lim_{n \to \infty} R_n = \lim_{n \to \infty} \sum_{i=1}^{n} f(x_i)\,\Delta x = \lim_{n \to \infty} \sum_{i=1}^{n} \cos\left(\frac{bi}{n}\right) \cdot \frac{b}{n} \overset{\text{CAS}}{=} \lim_{n \to \infty} \left[\frac{b\sin\left(b\left(\frac{1}{2n}+1\right)\right)}{2n\sin\left(\frac{b}{2n}\right)} - \frac{b}{2n} \right] \overset{\text{CAS}}{=} \sin b$$

If $b = \frac{\pi}{2}$, then $A = \sin\frac{\pi}{2} = 1.$

5.2 The Definite Integral

1. $R_4 = \displaystyle\sum_{i=1}^{4} f(x_i)\,\Delta x$ $\quad [x_i^* = x_i$ is a right endpoint and $\Delta x = 0.5]$

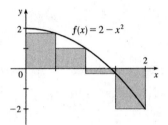

$\quad = 0.5\,[f(0.5) + f(1) + f(1.5) + f(2)] \quad [f(x) = 2 - x^2]$

$\quad = 0.5\,[1.75 + 1 + (-0.25) + (-2)]$

$\quad = 0.5(0.5) = 0.25$

The Riemann sum represents the sum of the areas of the two rectangles above the x-axis minus the sum of the areas of the two rectangles below the x-axis; that is, the *net area* of the rectangles with respect to the x-axis.

3. $M_5 = \displaystyle\sum_{i=1}^{5} f(\overline{x}_i)\,\Delta x$ $\quad [x_i^* = \overline{x}_i = \frac{1}{2}(x_{i-1} + x_i)$ is a midpoint and $\Delta x = 1]$

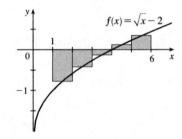

$\quad = 1\,[f(1.5) + f(2.5) + f(3.5)$

$\qquad\qquad + f(4.5) + f(5.5)] \quad [f(x) = \sqrt{x} - 2]$

$\quad \approx -0.856759$

The Riemann sum represents the sum of the areas of the two rectangles above the x-axis minus the sum of the areas of the three rectangles below the x-axis.

5. $f(x) = x^3.$ $\displaystyle\sum_{i=1}^{n} f(x_i^*)\,\Delta x_i = \frac{1}{2}\sum_{i=1}^{n} f(x_i^*) = \frac{1}{2}[(-1)^3 + (-0.4)^3 + (0.2)^3 + 1^3] = -0.028$

7. $\Delta x = (b-a)/n = (8-0)/4 = 8/4 = 2.$

(a) Using the right endpoints to approximate $\int_0^8 f(x)\,dx$, we have

$\quad \displaystyle\sum_{i=1}^{4} f(x_i)\,\Delta x = 2[f(2) + f(4) + f(6) + f(8)] \approx 2[1 + 2 + (-2) + 1] = 4.$

(b) Using the left endpoints to approximate $\int_0^8 f(x)\,dx$, we have

$\quad \displaystyle\sum_{i=1}^{4} f(x_{i-1})\,\Delta x = 2[f(0) + f(2) + f(4) + f(6)] \approx 2[2 + 1 + 2 + (-2)] = 6.$

(c) Using the midpoint of each subinterval to approximate $\int_0^8 f(x)\,dx$, we have

$\quad \displaystyle\sum_{i=1}^{4} f(\overline{x}_i)\,\Delta x = 2[f(1) + f(3) + f(5) + f(7)] \approx 2[3 + 2 + 1 + (-1)] = 10.$

9. Since f is increasing, $L_5 \leq \int_0^{25} f(x)\, dx \leq R_5$.

$$\text{Lower estimate} = L_5 = \sum_{i=1}^{5} f(x_{i-1})\, \Delta x = 5[f(0) + f(5) + f(10) + f(15) + f(20)]$$

$$= 5(-42 - 37 - 25 - 6 + 15) = 5(-95) = -475$$

$$\text{Upper estimate} = R_5 = \sum_{i=1}^{5} f(x_i)\, \Delta x = 5[f(5) + f(10) + f(15) + f(20) + f(25)]$$

$$= 5(-37 - 25 - 6 + 15 + 36) = 5(-17) = -85$$

11. $\Delta x = (10 - 2)/4 = 2$, so the endpoints are 2, 4, 6, 8, and 10, and the midpoints are 3, 5, 7, and 9. The Midpoint Rule

gives $\int_2^{10} \sqrt{x^3 + 1}\, dx \approx \sum_{i=1}^{4} f(\overline{x}_i)\, \Delta x = 2\left(\sqrt{3^3 + 1} + \sqrt{5^3 + 1} + \sqrt{7^3 + 1} + \sqrt{9^3 + 1}\right) \approx 124.1644$.

13. $\Delta x = (1 - 0)/5 = 0.2$, so the endpoints are 0, 0.2, 0.4, 0.6, 0.8, and 1, and the midpoints are 0.1, 0.3, 0.5, 0.7, and 0.9.

The Midpoint Rule gives

$$\int_0^1 \sin(x^2)\, dx \approx \sum_{i=1}^{5} f(\overline{x}_i)\, \Delta x = 0.2\left[\sin(0.1)^2 + \sin(0.3)^2 + \sin(0.5)^2 + \sin(0.7)^2 + \sin(0.9)^2\right] \approx 0.3084.$$

15. On $[0, \pi]$, $\displaystyle \lim_{n \to \infty} \sum_{i=1}^{n} x_i \sin x_i\, \Delta x = \int_0^\pi x \sin x\, dx$.

17. On $[1, 8]$, $\displaystyle \lim_{\max \Delta x_i \to 0} \sum_{i=1}^{n} \sqrt{2x_i^* + (x_i^*)^2}\, \Delta x_i = \int_1^8 \sqrt{2x + x^2}\, dx$.

19. Note that $\Delta x = \dfrac{5 - (-1)}{n} = \dfrac{6}{n}$ and $x_i = -1 + i\, \Delta x = -1 + \dfrac{6i}{n}$.

$$\int_{-1}^5 (1 + 3x)\, dx = \lim_{n \to \infty} \sum_{i=1}^{n} f(x_i)\, \Delta x = \lim_{n \to \infty} \sum_{i=1}^{n}\left[1 + 3\left(-1 + \frac{6i}{n}\right)\right]\frac{6}{n} = \lim_{n \to \infty} \frac{6}{n} \sum_{i=1}^{n}\left[-2 + \frac{18i}{n}\right]$$

$$= \lim_{n \to \infty} \frac{6}{n}\left[\sum_{i=1}^{n}(-2) + \sum_{i=1}^{n}\frac{18i}{n}\right] = \lim_{n \to \infty} \frac{6}{n}\left[-2n + \frac{18}{n}\sum_{i=1}^{n} i\right]$$

$$= \lim_{n \to \infty} \frac{6}{n}\left[-2n + \frac{18}{n} \cdot \frac{n(n+1)}{2}\right] = \lim_{n \to \infty}\left[-12 + \frac{108}{n^2} \cdot \frac{n(n+1)}{2}\right]$$

$$= \lim_{n \to \infty}\left[-12 + 54\frac{n+1}{n}\right] = \lim_{n \to \infty}\left[-12 + 54\left(1 + \frac{1}{n}\right)\right] = -12 + 54 \cdot 1 = 42$$

21. Note that $\Delta x = \dfrac{2 - 0}{n} = \dfrac{2}{n}$ and $x_i = 0 + i\, \Delta x = \dfrac{2i}{n}$.

$$\int_0^2 (2 - x^2)\, dx = \lim_{n \to \infty} \sum_{i=1}^{n} f(x_i)\, \Delta x = \lim_{n \to \infty} \sum_{i=1}^{n}\left(2 - \frac{4i^2}{n^2}\right)\left(\frac{2}{n}\right) = \lim_{n \to \infty} \frac{2}{n}\left[\sum_{i=1}^{n} 2 - \frac{4}{n^2}\sum_{i=1}^{n} i^2\right]$$

$$= \lim_{n \to \infty} \frac{2}{n}\left(2n - \frac{4}{n^2}\sum_{i=1}^{n} i^2\right) = \lim_{n \to \infty}\left[4 - \frac{8}{n^3} \cdot \frac{n(n+1)(2n+1)}{6}\right]$$

$$= \lim_{n \to \infty}\left(4 - \frac{4}{3} \cdot \frac{n+1}{n} \cdot \frac{2n+1}{n}\right) = \lim_{n \to \infty}\left[4 - \frac{4}{3}\left(1 + \frac{1}{n}\right)\left(2 + \frac{1}{n}\right)\right]$$

$$= 4 - \tfrac{4}{3} \cdot 1 \cdot 2 = \tfrac{4}{3}$$

23. Note that $\Delta x = \dfrac{2-1}{n} = \dfrac{1}{n}$ and $x_i = 1 + i\,\Delta x = 1 + i(1/n) = 1 + i/n$.

$$\int_1^2 x^3 \, dx = \lim_{n\to\infty} \sum_{i=1}^n f(x_i)\,\Delta x = \lim_{n\to\infty} \sum_{i=1}^n \left(1 + \frac{i}{n}\right)^3 \left(\frac{1}{n}\right) = \lim_{n\to\infty} \frac{1}{n} \sum_{i=1}^n \left(\frac{n+i}{n}\right)^3$$

$$= \lim_{n\to\infty} \frac{1}{n^4} \sum_{i=1}^n (n^3 + 3n^2 i + 3ni^2 + i^3) = \lim_{n\to\infty} \frac{1}{n^4} \left[\sum_{i=1}^n n^3 + \sum_{i=1}^n 3n^2 i + \sum_{i=1}^n 3ni^2 + \sum_{i=1}^n i^3\right]$$

$$= \lim_{n\to\infty} \frac{1}{n^4} \left[n \cdot n^3 + 3n^2 \sum_{i=1}^n i + 3n \sum_{i=1}^n i^2 + \sum_{i=1}^n i^3\right]$$

$$= \lim_{n\to\infty} \left[1 + \frac{3}{n^2} \cdot \frac{n(n+1)}{2} + \frac{3}{n^3} \cdot \frac{n(n+1)(2n+1)}{6} + \frac{1}{n^4} \cdot \frac{n^2(n+1)^2}{4}\right]$$

$$= \lim_{n\to\infty} \left[1 + \frac{3}{2} \cdot \frac{n+1}{n} + \frac{1}{2} \cdot \frac{n+1}{n} \cdot \frac{2n+1}{n} + \frac{1}{4} \cdot \frac{(n+1)^2}{n^2}\right]$$

$$= \lim_{n\to\infty} \left[1 + \frac{3}{2}\left(1 + \frac{1}{n}\right) + \frac{1}{2}\left(1 + \frac{1}{n}\right)\left(2 + \frac{1}{n}\right) + \frac{1}{4}\left(1 + \frac{1}{n}\right)^2\right] = 1 + \frac{3}{2} + \frac{1}{2} \cdot 2 + \frac{1}{4} = 3.75$$

25. $f(x) = \dfrac{x}{1+x^5}$, $a = 2$, $b = 6$, and $\Delta x = \dfrac{6-2}{n} = \dfrac{4}{n}$. Using Equation 3, we get $x_i^* = x_i = 2 + i\,\Delta x = 2 + \dfrac{4i}{n}$,

so $\displaystyle\int_2^6 \frac{x}{1+x^5}\, dx = \lim_{n\to\infty} R_n = \lim_{n\to\infty} \sum_{i=1}^n \frac{2 + \dfrac{4i}{n}}{1 + \left(2 + \dfrac{4i}{n}\right)^5} \cdot \frac{4}{n}$.

27. $\Delta x = (\pi - 0)/n = \pi/n$ and $x_i^* = x_i = \pi i/n$.

$$\int_0^\pi \sin 5x \, dx = \lim_{n\to\infty} \sum_{i=1}^n (\sin 5x_i)\left(\frac{\pi}{n}\right) = \lim_{n\to\infty} \sum_{i=1}^n \left(\sin \frac{5\pi i}{n}\right)\frac{\pi}{n} \overset{\text{CAS}}{=} \pi \lim_{n\to\infty} \frac{1}{n} \cot\left(\frac{5\pi}{2n}\right) \overset{\text{CAS}}{=} \pi \left(\frac{2}{5\pi}\right) = \frac{2}{5}$$

29. (a) Think of $\int_0^2 f(x)\,dx$ as the area of a trapezoid with bases 1 and 3 and height 2. The area of a trapezoid is $A = \frac{1}{2}(b+B)h$,

so $\int_0^2 f(x)\,dx = \frac{1}{2}(1+3)2 = 4$.

(b) $\int_0^5 f(x)\,dx = \int_0^2 f(x)\,dx + \int_2^3 f(x)\,dx + \int_3^5 f(x)\,dx$

             trapezoid   rectangle   triangle

       $= \frac{1}{2}(1+3)2 + \quad 3\cdot 1 \quad + \quad \frac{1}{2}\cdot 2\cdot 3 \quad = 4 + 3 + 3 = 10$

(c) $\int_5^7 f(x)\,dx$ is the negative of the area of the triangle with base 2 and height 3. $\int_5^7 f(x)\,dx = -\frac{1}{2}\cdot 2\cdot 3 = -3$.

(d) $\int_7^9 f(x)\,dx$ is the negative of the area of a trapezoid with bases 3 and 2 and height 2, so it equals

$-\frac{1}{2}(B+b)h = -\frac{1}{2}(3+2)2 = -5$. Thus,

$\int_0^9 f(x)\,dx = \int_0^5 f(x)\,dx + \int_5^7 f(x)\,dx + \int_7^9 f(x)\,dx = 10 + (-3) + (-5) = 2$.

31. $\int_0^3 \left(\frac{1}{2}x - 1\right) dx$ can be interpreted as the area of the triangle above the x-axis

minus the area of the triangle below the x-axis; that is,

$\frac{1}{2}(1)\left(\frac{1}{2}\right) - \frac{1}{2}(2)(1) = \frac{1}{4} - 1 = -\frac{3}{4}$.

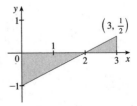

33. $\int_{-3}^{0}\left(1+\sqrt{9-x^2}\right)dx$ can be interpreted as the area under the graph of

$f(x) = 1 + \sqrt{9 - x^2}$ between $x = -3$ and $x = 0$. This is equal to one-quarter

the area of the circle with radius 3, plus the area of the rectangle, so

$\int_{-3}^{0}\left(1+\sqrt{9-x^2}\right)dx = \frac{1}{4}\pi \cdot 3^2 + 1 \cdot 3 = 3 + \frac{9}{4}\pi.$

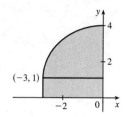

35. $\int_{-1}^{2} |x| \, dx$ can be interpreted as the sum of the areas of the two shaded

triangles; that is, $\frac{1}{2}(1)(1) + \frac{1}{2}(2)(2) = \frac{1}{2} + \frac{4}{2} = \frac{5}{2}.$

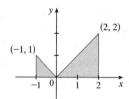

37. $\int_{9}^{4} \sqrt{t} \, dt = -\int_{4}^{9} \sqrt{t} \, dt$ [because we reversed the limits of integration]

$\qquad\qquad = -\int_{4}^{9} \sqrt{x} \, dx$ [we can use any letter without changing the value of the integral]

$\qquad\qquad = -\frac{38}{3}$

39. $\int_{-2}^{2} f(x) \, dx + \int_{2}^{5} f(x) \, dx - \int_{-2}^{-1} f(x) \, dx = \int_{-2}^{5} f(x) \, dx + \int_{-1}^{-2} f(x) \, dx$ [by Property 5 and reversing limits]

$\qquad\qquad\qquad\qquad\qquad\qquad\qquad\qquad\qquad\quad = \int_{-1}^{5} f(x) \, dx$ [Property 5]

41. $\int_{0}^{9} [2f(x) + 3g(x)] \, dx = 2\int_{0}^{9} f(x) \, dx + 3\int_{0}^{9} g(x) \, dx = 2(37) + 3(16) = 122$

43. $\int_{0}^{1}\left(5 - 6x^2\right) dx = \int_{0}^{1} 5 \, dx - 6\int_{0}^{1} x^2 \, dx = 5(1 - 0) - 6\left(\frac{1}{3}\right) = 5 - 2 = 3$

45. On $[1, 3]$, $\ln 1 \le \ln x \le \ln 3$ $\Rightarrow$ $0(3 - 1) \le \int_{1}^{3} \ln x \, dx \le (\ln 3)(3 - 1)$ $\Rightarrow$ $0 \le \int_{1}^{3} \ln x \, dx \le 2 \ln 3.$

47. If $1 \le x \le 2$, then $\frac{1}{2} \le \frac{1}{x} \le 1$, so $\frac{1}{2}(2 - 1) \le \int_{1}^{2} \frac{1}{x} \, dx \le 1(2 - 1)$ or $\frac{1}{2} \le \int_{1}^{2} \frac{1}{x} \, dx \le 1.$

49. If $\frac{\pi}{4} \le x \le \frac{\pi}{3}$, then $1 \le \tan x \le \sqrt{3}$, so $1\left(\frac{\pi}{3} - \frac{\pi}{4}\right) \le \int_{\pi/4}^{\pi/3} \tan x \, dx \le \sqrt{3}\left(\frac{\pi}{3} - \frac{\pi}{4}\right)$ or $\frac{\pi}{12} \le \int_{\pi/4}^{\pi/3} \tan x \, dx \le \frac{\pi}{12}\sqrt{3}.$

51. $\lim\limits_{n\to\infty} \sum\limits_{i=1}^{n} \frac{i^4}{n^5} = \lim\limits_{n\to\infty} \sum\limits_{i=1}^{n} \frac{i^4}{n^4} \cdot \frac{1}{n} = \lim\limits_{n\to\infty} \sum\limits_{i=1}^{n} \left(\frac{i}{n}\right)^4 \frac{1}{n}.$ At this point, we need to recognize the limit as being of the form

$\lim\limits_{n\to\infty} \sum\limits_{i=1}^{n} f(x_i) \, \Delta x$, where $\Delta x = (1 - 0)/n = 1/n$, $x_i = 0 + i\,\Delta x = i/n$, and $f(x) = x^4.$ Thus, the definite integral

is $\int_{0}^{1} x^4 \, dx.$

5.3   Evaluating Definite Integrals

1. $\displaystyle\int_{-1}^{3} x^5 \, dx = \left[\frac{x^6}{6}\right]_{-1}^{3} = \frac{3^6}{6} - \frac{(-1)^6}{6} = \frac{729 - 1}{6} = \frac{364}{3}$

3. $\int_{0}^{2}\left(6x^2 - 4x + 5\right) dx = \left[6 \cdot \frac{1}{3}x^3 - 4 \cdot \frac{1}{2}x^2 + 5x\right]_{0}^{2} = \left[2x^3 - 2x^2 + 5x\right]_{0}^{2} = (16 - 8 + 10) - 0 = 18$

5. $\int_{0}^{1} x^{4/5} \, dx = \left[\frac{5}{9}x^{9/5}\right]_{0}^{1} = \frac{5}{9} - 0 = \frac{5}{9}$

7. $\int_{-1}^{0}(2x - e^x)\, dx = \left[x^2 - e^x\right]_{-1}^{0} = (0 - 1) - (1 - e^{-1}) = -2 + 1/e$

9. $\int_{-2}^{2}(3u + 1)^2\, du = \int_{-2}^{2}\left(9u^2 + 6u + 1\right)\, du = \left[9 \cdot \frac{1}{3}u^3 + 6 \cdot \frac{1}{2}u^2 + u\right]_{-2}^{2} = \left[3u^3 + 3u^2 + u\right]_{-2}^{2}$

$\qquad = (24 + 12 + 2) - (-24 + 12 - 2) = 38 - (-14) = 52$

11. $\int_{-2}^{-1}\left(4y^3 + \frac{2}{y^3}\right)\, dy = \left[4 \cdot \frac{1}{4}y^4 + 2 \cdot \frac{1}{-2}y^{-2}\right]_{-2}^{-1} = \left[y^4 - \frac{1}{y^2}\right]_{-2}^{-1} = (1 - 1) - \left(16 - \frac{1}{4}\right) = -\frac{63}{4}$

13. $\int_{0}^{1}x(\sqrt[3]{x} + \sqrt[4]{x})\, dx = \int_{0}^{1}(x^{4/3} + x^{5/4})\, dx = \left[\frac{3}{7}x^{7/3} + \frac{4}{9}x^{9/4}\right]_{0}^{1} = \left(\frac{3}{7} + \frac{4}{9}\right) - 0 = \frac{55}{63}$

15. $\int_{0}^{\pi/4}\sec^2 t\, dt = \left[\tan t\right]_{0}^{\pi/4} = \tan\frac{\pi}{4} - \tan 0 = 1 - 0 = 1$

17. $\int_{1}^{9}\frac{1}{2x}\, dx = \frac{1}{2}\int_{1}^{9}\frac{1}{x}\, dx = \frac{1}{2}\left[\ln|x|\right]_{1}^{9} = \frac{1}{2}(\ln 9 - \ln 1) = \frac{1}{2}\ln 9 - 0 = \ln 9^{1/2} = \ln 3$

19. $\int_{1/2}^{\sqrt{3}/2}\frac{6}{\sqrt{1 - t^2}}\, dt = 6\int_{1/2}^{\sqrt{3}/2}\frac{1}{\sqrt{1 - t^2}}\, dt = 6\left[\sin^{-1} t\right]_{1/2}^{\sqrt{3}/2} = 6\left[\sin^{-1}\left(\frac{\sqrt{3}}{2}\right) - \sin^{-1}\left(\frac{1}{2}\right)\right]$

$\qquad = 6\left(\frac{\pi}{3} - \frac{\pi}{6}\right) = 6\left(\frac{\pi}{6}\right) = \pi$

21. $\int_{1}^{64}\frac{1 + \sqrt[3]{x}}{\sqrt{x}}\, dx = \int_{1}^{64}\left(\frac{1}{x^{1/2}} + \frac{x^{1/3}}{x^{1/2}}\right)\, dx = \int_{1}^{64}\left(x^{-1/2} + x^{(1/3) - (1/2)}\right)\, dx = \int_{1}^{64}\left(x^{-1/2} + x^{-1/6}\right)\, dx$

$\qquad = \left[2x^{1/2} + \frac{6}{5}x^{5/6}\right]_{1}^{64} = \left(16 + \frac{192}{5}\right) - \left(2 + \frac{6}{5}\right) = 14 + \frac{186}{5} = \frac{256}{5}$

23. $\int_{-1}^{1}e^{u+1}\, du = \left[e^{u+1}\right]_{-1}^{1} = e^2 - e^0 = e^2 - 1 \quad$ [or start with $e^{u+1} = e^u e^1$]

25. $\int_{0}^{\pi/4}\frac{1 + \cos^2\theta}{\cos^2\theta}\, d\theta = \int_{0}^{\pi/4}\left(\frac{1}{\cos^2\theta} + \frac{\cos^2\theta}{\cos^2\theta}\right)\, d\theta = \int_{0}^{\pi/4}\left(\sec^2\theta + 1\right)\, d\theta$

$\qquad = \left[\tan\theta + \theta\right]_{0}^{\pi/4} = \left(\tan\frac{\pi}{4} + \frac{\pi}{4}\right) - (0 + 0) = 1 + \frac{\pi}{4}$

27. $\int_{1}^{e}\frac{x^2 + x + 1}{x}\, dx = \int_{1}^{e}\left(x + 1 + \frac{1}{x}\right)\, dx = \left[\frac{1}{2}x^2 + x + \ln|x|\right]_{1}^{e} = \left(\frac{1}{2}e^2 + e + \ln e\right) - \left(\frac{1}{2} + 1 + \ln 1\right) = \frac{1}{2}e^2 + e - \frac{1}{2}$

29. $f(x) = 1/x^2$ is not continuous on the interval $[-1, 3]$, so the Evaluation Theorem does not apply. In fact, f has an infinite

discontinuity at $x = 0$, so $\int_{-1}^{3}(1/x^2)\, dx$ does not exist.

31. It appears that the area under the graph is about $\frac{2}{3}$ of the area of the

viewing rectangle, or about $\frac{2}{3}\pi \approx 2.1$. The actual area is

$\int_{0}^{\pi}\sin x\, dx = \left[-\cos x\right]_{0}^{\pi} = (-\cos\pi) - (-\cos 0) = -(-1) + 1 = 2.$

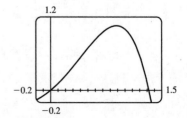

33. The graph shows that $y = x + x^2 - x^4$ has x-intercepts at $x = 0$ and at

$x = a \approx 1.32$. So the area of the region that lies under the curve and

above the x-axis is

$\int_{0}^{a}\left(x + x^2 - x^4\right)\, dx = \left[\frac{1}{2}x^2 + \frac{1}{3}x^3 - \frac{1}{5}x^5\right]_{0}^{a} = \left(\frac{1}{2}a^2 + \frac{1}{3}a^3 - \frac{1}{5}a^5\right) - 0$

$\qquad \approx 0.84$

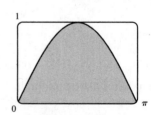

35. $\int_{-1}^{2} x^3 \, dx = \left[\frac{1}{4}x^4\right]_{-1}^{2} = 4 - \frac{1}{4} = \frac{15}{4} = 3.75$

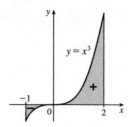

37. $\dfrac{d}{dx}\left[\sqrt{x^2+1}+C\right] = \dfrac{d}{dx}\left[(x^2+1)^{1/2}+C\right] = \frac{1}{2}(x^2+1)^{-1/2} \cdot 2x = \dfrac{x}{\sqrt{x^2+1}}$

39. $\int x\sqrt{x}\,dx = \int x^{3/2}\,dx = \frac{2}{5}x^{5/2}+C.$

The members of the family in the figure correspond to $C = 5, 3, 0, -2,$ and $-4.$

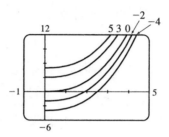

41. $\displaystyle\int (1-t)(2+t^2)\,dt = \int (2 - 2t + t^2 - t^3)\,dt = 2t - 2\frac{t^2}{2} + \frac{t^3}{3} - \frac{t^4}{4} + C = 2t - t^2 + \frac{1}{3}t^3 - \frac{1}{4}t^4 + C$

43. $\displaystyle\int \frac{\sin x}{1 - \sin^2 x}\,dx = \int \frac{\sin x}{\cos^2 x}\,dx = \int \frac{1}{\cos x} \cdot \frac{\sin x}{\cos x}\,dx = \int \sec x \tan x \, dx = \sec x + C$

45. $A = \int_{0}^{2}(2y - y^2)\,dy = \left[y^2 - \frac{1}{3}y^3\right]_{0}^{2} = \left(4 - \frac{8}{3}\right) - 0 = \frac{4}{3}$

47. If $w'(t)$ is the rate of change of weight in pounds per year, then $w(t)$ represents the weight in pounds of the child at age t. We know from the Net Change Theorem that $\int_{5}^{10} w'(t)\,dt = w(10) - w(5)$, so the integral represents the increase in the child's weight (in pounds) between the ages of 5 and 10.

49. Since $r(t)$ is the rate at which oil leaks, we can write $r(t) = -V'(t)$, where $V(t)$ is the volume of oil at time t. [Note that the minus sign is needed because V is decreasing, so $V'(t)$ is negative, but $r(t)$ is positive.] Thus, by the Net Change Theorem, $\int_{0}^{120} r(t)\,dt = -\int_{0}^{120} V'(t)\,dt = -[V(120) - V(0)] = V(0) - V(120)$, which is the number of gallons of oil that leaked from the tank in the first two hours (120 minutes).

51. By the Net Change Theorem, $\int_{1000}^{5000} R'(x)\,dx = R(5000) - R(1000)$, so it represents the increase in revenue when production is increased from 1000 units to 5000 units.

53. In general, the unit of measurement for $\int_{a}^{b} f(x)\,dx$ is the product of the unit for $f(x)$ and the unit for x. Since $f(x)$ is measured in newtons and x is measured in meters, the units for $\int_{0}^{100} f(x)\,dx$ are newton-meters.

55. (a) Displacement $= \int_0^3 (3t - 5) \, dt = \left[\frac{3}{2}t^2 - 5t\right]_0^3 = \frac{27}{2} - 15 = -\frac{3}{2}$ m

(b) Distance traveled $= \int_0^3 |3t - 5| \, dt = \int_0^{5/3} (5 - 3t) \, dt + \int_{5/3}^3 (3t - 5) \, dt$

$$= \left[5t - \frac{3}{2}t^2\right]_0^{5/3} + \left[\frac{3}{2}t^2 - 5t\right]_{5/3}^3 = \frac{25}{3} - \frac{3}{2} \cdot \frac{25}{9} + \frac{27}{2} - 15 - \left(\frac{3}{2} \cdot \frac{25}{9} - \frac{25}{3}\right) = \frac{41}{6}$$ m

57. (a) $v'(t) = a(t) = t + 4 \;\Rightarrow\; v(t) = \frac{1}{2}t^2 + 4t + C \;\Rightarrow\; v(0) = C = 5 \;\Rightarrow\; v(t) = \frac{1}{2}t^2 + 4t + 5$ m/s

(b) Distance traveled $= \int_0^{10} |v(t)| \, dt = \int_0^{10} \left|\frac{1}{2}t^2 + 4t + 5\right| \, dt = \int_0^{10} \left(\frac{1}{2}t^2 + 4t + 5\right) dt$

$$= \left[\frac{1}{6}t^3 + 2t^2 + 5t\right]_0^{10} = \frac{500}{3} + 200 + 50 = 416\frac{2}{3}$$ m

59. Let s be the position of the car. We know from Equation 2 that $s(100) - s(0) = \int_0^{100} v(t) \, dt$. We use the

Midpoint Rule for $0 \le t \le 100$ with $n = 5$. Note that the length of each of the five time intervals is

20 seconds $= \frac{20}{3600}$ hour $= \frac{1}{180}$ hour. So the distance traveled is

$$\int_0^{100} v(t) \, dt \approx \frac{1}{180}[v(10) + v(30) + v(50) + v(70) + v(90)] = \frac{1}{180}(38 + 58 + 51 + 53 + 47) = \frac{247}{180} \approx 1.4 \text{ miles.}$$

61. By the Net Change Theorem, the amount of water that flows from the tank during the first 10 minutes is

$$\int_0^{10} r(t) \, dt = \int_0^{10} (200 - 4t) \, dt = \left[200t - 2t^2\right]_0^{10} = (2000 - 200) - 0 = 1800 \text{ liters.}$$

63. The second derivative is the derivative of the first derivative, so we'll apply the Net Change Theorem with $F = h'$.

$\int_1^2 h''(u) \, du = \int_1^2 (h')'(u) \, du = h'(2) - h'(1) = 5 - 2 = 3$. The other information is unnecessary.

5.4   The Fundamental Theorem of Calculus

1. (a) $g(x) = \int_0^x f(t) \, dt$.

$g(0) = \int_0^0 f(t) \, dt = 0$ $g(1) = \int_0^1 f(t) \, dt = 1 \cdot 2 = 2$ [rectangle]

$g(2) = \int_0^2 f(t) \, dt = \int_0^1 f(t) \, dt + \int_1^2 f(t) \, dt$ $g(3) = \int_0^3 f(t) \, dt = g(2) + \int_2^3 f(t) \, dt$

$\quad = g(1) + \int_1^2 f(t) \, dt$ $\quad = 5 + \frac{1}{2} \cdot 1 \cdot 4 = 7$

$\quad = 2 + 1 \cdot 2 + \frac{1}{2} \cdot 1 \cdot 2 = 5$ [rectangle plus triangle]

$g(6) = g(3) + \int_3^6 f(t) \, dt$ [the integral is negative since f lies under the x-axis]

$\quad = 7 + \left[-\left(\frac{1}{2} \cdot 2 \cdot 2 + 1 \cdot 2\right)\right] = 7 - 4 = 3$

(b) g is increasing on $(0, 3)$ because as x increases from 0 to 3,   (d)

we keep adding more area.

(c) g has a maximum value when we start subtracting area;

that is, at $x = 3$.

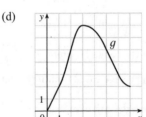

3.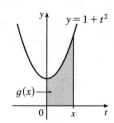

(a) By FTC1, $g(x) = \int_0^x (1 + t^2)\, dt$ $\Rightarrow$ $g'(x) = f(x) = 1 + x^2$.

(b) By FTC2, $g(x) = \int_0^x (1 + t^2)\, dt = \left[t + \frac{1}{3} t^3 \right]_0^x = \left(x + \frac{1}{3} x^3 \right) - 0$ $\Rightarrow$ $g'(x) = 1 + x^2$.

5. $f(t) = \sqrt{1 + 2t}$ and $g(x) = \int_0^x \sqrt{1 + 2t}\, dt$, so by FTC1, $g'(x) = f(x) = \sqrt{1 + 2x}$.

7. $f(t) = t^2 \sin t$ and $g(y) = \int_2^y t^2 \sin t\, dt$, so by FTC1, $g'(y) = f(y) = y^2 \sin y$.

9. Let $u = \dfrac{1}{x}$. Then $\dfrac{du}{dx} = -\dfrac{1}{x^2}$. Also, $\dfrac{dh}{dx} = \dfrac{dh}{du}\dfrac{du}{dx}$, so

$$h'(x) = \frac{d}{dx} \int_2^{1/x} \arctan t\, dt = \frac{d}{du} \int_2^u \arctan t\, dt \cdot \frac{du}{dx} = \arctan u \frac{du}{dx} = -\frac{\arctan(1/x)}{x^2}.$$

11. Let $u = \sqrt{x}$. Then $\dfrac{du}{dx} = \dfrac{1}{2\sqrt{x}}$. Also, $\dfrac{dy}{dx} = \dfrac{dy}{du}\dfrac{du}{dx}$, so

$$y' = \frac{d}{dx} \int_3^{\sqrt{x}} \frac{\cos t}{t}\, dt = \frac{d}{du} \int_3^u \frac{\cos t}{t}\, dt \cdot \frac{du}{dx} = \frac{\cos u}{u} \cdot \frac{1}{2\sqrt{x}} = \frac{\cos \sqrt{x}}{\sqrt{x}} \cdot \frac{1}{2\sqrt{x}} = \frac{\cos \sqrt{x}}{2x}.$$

13. $g(x) = \displaystyle\int_{2x}^{3x} \frac{u^2 - 1}{u^2 + 1}\, du = \int_{2x}^0 \frac{u^2 - 1}{u^2 + 1}\, du + \int_0^{3x} \frac{u^2 - 1}{u^2 + 1}\, du = -\int_0^{2x} \frac{u^2 - 1}{u^2 + 1}\, du + \int_0^{3x} \frac{u^2 - 1}{u^2 + 1}\, du$ $\Rightarrow$

$$g'(x) = -\frac{(2x)^2 - 1}{(2x)^2 + 1} \cdot \frac{d}{dx}(2x) + \frac{(3x)^2 - 1}{(3x)^2 + 1} \cdot \frac{d}{dx}(3x) = -2 \cdot \frac{4x^2 - 1}{4x^2 + 1} + 3 \cdot \frac{9x^2 - 1}{9x^2 + 1}$$

15. $f_{\text{ave}} = \dfrac{1}{b - a} \displaystyle\int_a^b f(x)\, dx = \dfrac{1}{1 - (-1)} \int_{-1}^1 x^2\, dx = \tfrac{1}{2} \cdot 2 \int_0^1 x^2\, dx = \left[\tfrac{1}{3} x^3 \right]_0^1 = \tfrac{1}{3}$

17. $g_{\text{ave}} = \dfrac{1}{\frac{\pi}{2} - 0} \displaystyle\int_0^{\pi/2} \cos x\, dx = \dfrac{2}{\pi} \left[\sin x \right]_0^{\pi/2} = \dfrac{2}{\pi} (1 - 0) = \dfrac{2}{\pi}$

19. (a) $f_{\text{ave}} = \dfrac{1}{5 - 2} \displaystyle\int_2^5 (x - 3)^2\, dx = \dfrac{1}{3} \int_2^5 (x^2 - 6x + 9)\, dx$

$= \tfrac{1}{3} \left[\tfrac{1}{3} x^3 - 3x^2 + 9x \right]_2^5$

$= \tfrac{1}{3} \left[\tfrac{1}{3}(125) - 3(25) + 9(5) - \tfrac{1}{3}(8) + 3(4) - 9(2) \right] = 1$

(b) $f(c) = f_{\text{ave}}$ $\Leftrightarrow$ $(c - 3)^2 = 1$ $\Leftrightarrow$

$c - 3 = \pm 1$ $\Leftrightarrow$ $c = 2$ or 4

(c)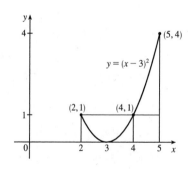

21. $f_{\text{ave}} = \dfrac{1}{50 - 20} \displaystyle\int_{20}^{50} f(x)\, dx \approx \dfrac{1}{30} M_3 = \dfrac{1}{30} \cdot \dfrac{50 - 20}{3} [f(25) + f(35) + f(45)] = \tfrac{1}{3}(38 + 29 + 48) = \tfrac{115}{3} = 38\tfrac{1}{3}$

23. $F(x) = \displaystyle\int_1^x f(t)\, dt$ $\Rightarrow$ $F'(x) = f(x) = \displaystyle\int_1^{x^2} \frac{\sqrt{1 + u^4}}{u}\, du$ $\left[\text{since } f(t) = \displaystyle\int_1^{t^2} \frac{\sqrt{1 + u^4}}{u}\, du \right]$ $\Rightarrow$

$$F''(x) = f'(x) = \frac{\sqrt{1 + (x^2)^4}}{x^2} \cdot \frac{d}{dx}(x^2) = \frac{\sqrt{1 + x^8}}{x^2} \cdot 2x = \frac{2\sqrt{1 + x^8}}{x}. \text{ So } F''(2) = \sqrt{1 + 2^8} = \sqrt{257}.$$

25. (a) By FTC1, $g'(x) = f(x)$. So $g'(x) = f(x) = 0$ at $x = 1, 3, 5, 7$, and 9. g has local maxima at $x = 1$ and 5 (since $f = g'$ changes from positive to negative there) and local minima at $x = 3$ and 7. There is no local maximum or minimum at $x = 9$, since f is not defined for $x > 9$.

(b) We can see from the graph that $\left|\int_0^1 f\,dt\right| < \left|\int_1^3 f\,dt\right| < \left|\int_3^5 f\,dt\right| < \left|\int_5^7 f\,dt\right| < \left|\int_7^9 f\,dt\right|$. So $g(1) = \left|\int_0^1 f\,dt\right|$,

$g(5) = \int_0^5 f\,dt = g(1) - \left|\int_1^3 f\,dt\right| + \left|\int_3^5 f\,dt\right|$, and $g(9) = \int_0^9 f\,dt = g(5) - \left|\int_5^7 f\,dt\right| + \left|\int_7^9 f\,dt\right|$. Thus,

$g(1) < g(5) < g(9)$, and so the absolute maximum of $g(x)$ occurs at $x = 9$.

(c) g is concave downward on those intervals where $g'' < 0$.

But $g'(x) = f(x)$, so $g''(x) = f'(x)$, which is negative on (approximately) $\left(\frac{1}{2}, 2\right)$, $(4, 6)$ and $(8, 9)$. So g is concave downward on these intervals.

(d)

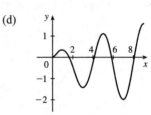

27. By FTC2, $\int_1^4 f'(x)\,dx = f(4) - f(1)$, so $17 = f(4) - 12 \implies f(4) = 17 + 12 = 29$.

29. (a) The Fresnel function $S(x) = \int_0^x \sin\left(\frac{\pi}{2}t^2\right)dt$ has local maximum values where $0 = S'(x) = \sin\left(\frac{\pi}{2}x^2\right)$ and S' changes from positive to negative. For $x > 0$, this happens when $\frac{\pi}{2}x^2 = (2n - 1)\pi$ [odd multiples of π] $\Leftrightarrow$ $x^2 = 2(2n - 1) \Leftrightarrow x = \sqrt{4n - 2}$, n any positive integer. For $x < 0$, S' changes from positive to negative where $\frac{\pi}{2}x^2 = 2n\pi$ [even multiples of π] $\Leftrightarrow$ $x^2 = 4n \Leftrightarrow x = -2\sqrt{n}$. S' does not change sign at $x = 0$.

(b) S is concave upward on those intervals where $S''(x) > 0$. Differentiating our expression for $S'(x)$, we get $S''(x) = \cos\left(\frac{\pi}{2}x^2\right)\left(2\frac{\pi}{2}x\right) = \pi x \cos\left(\frac{\pi}{2}x^2\right)$. For $x > 0$, $S''(x) > 0$ where $\cos\left(\frac{\pi}{2}x^2\right) > 0 \Leftrightarrow 0 < \frac{\pi}{2}x^2 < \frac{\pi}{2}$ or $\left(2n - \frac{1}{2}\right)\pi < \frac{\pi}{2}x^2 < \left(2n + \frac{1}{2}\right)\pi$, n any integer $\Leftrightarrow 0 < x < 1$ or $\sqrt{4n - 1} < x < \sqrt{4n + 1}$, n any positive integer. For $x < 0$, $S''(x) > 0$ where $\cos\left(\frac{\pi}{2}x^2\right) < 0 \Leftrightarrow \left(2n - \frac{3}{2}\right)\pi < \frac{\pi}{2}x^2 < \left(2n - \frac{1}{2}\right)\pi$, n any integer $\Leftrightarrow 4n - 3 < x^2 < 4n - 1 \Leftrightarrow \sqrt{4n - 3} < |x| < \sqrt{4n - 1} \implies \sqrt{4n - 3} < -x < \sqrt{4n - 1} \implies -\sqrt{4n - 3} > x > -\sqrt{4n - 1}$, so the intervals of upward concavity for $x < 0$ are $\left(-\sqrt{4n - 1}, -\sqrt{4n - 3}\right)$, n any positive integer. To summarize: S is concave upward on the intervals $(0, 1)$, $\left(-\sqrt{3}, -1\right)$, $\left(\sqrt{3}, \sqrt{5}\right)$, $\left(-\sqrt{7}, -\sqrt{5}\right)$, $\left(\sqrt{7}, 3\right)$, ...

(c) In Maple, we use `plot({int(sin(Pi*t^2/2),t=0..x),0.2},x=0..2);`. Note that Maple recognizes the Fresnel function, calling it `FresnelS(x)`. In Mathematica, we use `Plot[{Integrate[Sin[Pi*t^2/2],{t,0,x}],0.2},{x,0,2}]`. In Derive, we load the utility file FRESNEL and plot `FRESNEL_SIN(x)`. From the graphs, we see that $\int_0^x \sin\left(\frac{\pi}{2}t^2\right)dt = 0.2$ at $x \approx 0.74$.

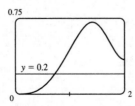

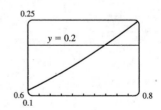

31. Using FTC1, we differentiate both sides of $6 + \int_a^x \frac{f(t)}{t^2}\, dt = 2\sqrt{x}$ to get $\frac{f(x)}{x^2} = 2\frac{1}{2\sqrt{x}}$ ⇒ $f(x) = x^{3/2}$.

To find a, we substitute $x = a$ in the original equation to obtain $6 + \int_a^a \frac{f(t)}{t^2}\, dt = 2\sqrt{a}$ ⇒ $6 + 0 = 2\sqrt{a}$ ⇒

$3 = \sqrt{a}$ ⇒ $a = 9$.

33. (a) Let $F(t) = \int_0^t f(s)\, ds$. Then, by FTC1, $F'(t) = f(t) =$ rate of depreciation, so $F(t)$ represents the loss in value over the interval $[0, t]$.

(b) $C(t) = \frac{1}{t}\left[A + \int_0^t f(s)\, ds\right] = \frac{A + F(t)}{t}$ represents the average expenditure per unit of t during the interval $[0, t]$, assuming that there has been only one overhaul during that time period. The company wants to minimize average expenditure.

(c) $C(t) = \frac{1}{t}\left[A + \int_0^t f(s)\, ds\right]$. Using FTC1, we have $C'(t) = -\frac{1}{t^2}\left[A + \int_0^t f(s)\, ds\right] + \frac{1}{t}f(t)$.

$C'(t) = 0$ ⇒ $t\, f(t) = A + \int_0^t f(s)\, ds$ ⇒ $f(t) = \frac{1}{t}\left[A + \int_0^t f(s)\, ds\right] = C(t)$.

5.5 The Substitution Rule

1. Let $u = 3x$. Then $du = 3\, dx$, so $dx = \frac{1}{3}du$. Thus,

$\int \cos 3x\, dx = \int \cos u\left(\frac{1}{3}\, du\right) = \frac{1}{3}\int \cos u\, du = \frac{1}{3}\sin u + C = \frac{1}{3}\sin 3x + C$. Don't forget that it is often very easy to check

an indefinite integration by differentiating your answer. In this case, $\frac{d}{dx}\left(\frac{1}{3}\sin 3x + C\right) = \frac{1}{3}(\cos 3x) \cdot 3 = \cos 3x$, the

desired result.

3. Let $u = x^3 + 1$. Then $du = 3x^2\, dx$ and $x^2\, dx = \frac{1}{3}\, du$, so

$$\int x^2\sqrt{x^3 + 1}\, dx = \int \sqrt{u}\left(\frac{1}{3}\, du\right) = \frac{1}{3}\frac{u^{3/2}}{3/2} + C = \frac{1}{3}\cdot\frac{2}{3}u^{3/2} + C = \frac{2}{9}(x^3 + 1)^{3/2} + C.$$

5. Let $u = 1 + 2x$. Then $du = 2\, dx$ and $dx = \frac{1}{2}\, du$, so

$$\int \frac{4}{(1 + 2x)^3}\, dx = 4\int u^{-3}\left(\frac{1}{2}\, du\right) = 2\frac{u^{-2}}{-2} + C = -\frac{1}{u^2} + C = -\frac{1}{(1 + 2x)^2} + C.$$

7. Let $u = x^2 + 3$. Then $du = 2x\, dx$, so $\int 2x(x^2 + 3)^4\, dx = \int u^4\, du = \frac{1}{5}u^5 + C = \frac{1}{5}(x^2 + 3)^5 + C$.

9. Let $u = 3x - 2$. Then $du = 3\, dx$ and $dx = \frac{1}{3}\, du$, so $\int (3x - 2)^{20}\, dx = \int u^{20}\left(\frac{1}{3}\, du\right) = \frac{1}{3}\cdot\frac{1}{21}u^{21} + C = \frac{1}{63}(3x - 2)^{21} + C$.

11. Let $u = \ln x$. Then $du = \frac{dx}{x}$, so $\int \frac{(\ln x)^2}{x}\, dx = \int u^2\, du = \frac{1}{3}u^3 + C = \frac{1}{3}(\ln x)^3 + C$.

13. Let $u = 5 - 3x$. Then $du = -3\, dx$ and $dx = -\frac{1}{3}\, du$, so

$$\int \frac{dx}{5 - 3x} = \int \frac{1}{u}\left(-\frac{1}{3}\, du\right) = -\frac{1}{3}\ln|u| + C = -\frac{1}{3}\ln|5 - 3x| + C.$$

15. Let $u = 3ax + bx^3$. Then $du = (3a + 3bx^2)\,dx = 3(a + bx^2)\,dx$, so

$$\int \frac{a + bx^2}{\sqrt{3ax + bx^3}}\,dx = \int \frac{\frac{1}{3}\,du}{u^{1/2}} = \frac{1}{3}\int u^{-1/2}\,du = \frac{1}{3}\cdot 2u^2 + C = \frac{2}{3}\sqrt{3ax + bx^3} + C.$$

17. Let $u = \pi t$. Then $du = \pi\,dt$ and $dt = \frac{1}{\pi}\,du$, so $\int \sin \pi t\,dt = \int \sin u \left(\frac{1}{\pi}\,du\right) = \frac{1}{\pi}(-\cos u) + C = -\frac{1}{\pi}\cos \pi t + C.$

19. Let $u = 1 + e^x$. Then $du = e^x\,dx$, so $\int e^x \sqrt{1 + e^x}\,dx = \int \sqrt{u}\,du = \frac{2}{3}u^{3/2} + C = \frac{2}{3}(1 + e^x)^{3/2} + C.$

Or: Let $u = \sqrt{1 + e^x}$. Then $u^2 = 1 + e^x$ and $2u\,du = e^x\,dx$, so

$\int e^x \sqrt{1 + e^x}\,dx = \int u \cdot 2u\,du = \frac{2}{3}u^3 + C = \frac{2}{3}(1 + e^x)^{3/2} + C.$

21. Let $u = \sin \theta$. Then $du = \cos \theta\,d\theta$, so $\int \cos \theta \sin^6 \theta\,d\theta = \int u^6\,du = \frac{1}{7}u^7 + C = \frac{1}{7}\sin^7 \theta + C.$

23. Let $u = \cot x$. Then $du = -\csc^2 x\,dx$ and $\csc^2 x\,dx = -du$, so

$$\int \sqrt{\cot x}\,\csc^2 x\,dx = \int \sqrt{u}\,(-du) = -\frac{u^{3/2}}{3/2} + C = -\frac{2}{3}(\cot x)^{3/2} + C.$$

25. Let $u = \sin^{-1} x$. Then $du = \frac{1}{\sqrt{1 - x^2}}\,dx$, so $\int \frac{dx}{\sqrt{1 - x^2}\,\sin^{-1} x} = \int \frac{1}{u}\,du = \ln |u| + C = \ln \left|\sin^{-1} x\right| + C.$

27. Let $u = \sec x$. Then $du = \sec x \tan x\,dx$, so

$\int \sec^3 x \tan x\,dx = \int \sec^2 x\,(\sec x \tan x)\,dx = \int u^2\,du = \frac{1}{3}u^3 + C = \frac{1}{3}\sec^3 x + C.$

29. $\displaystyle \int \frac{e^x + 1}{e^x}\,dx = \int \left(\frac{e^x}{e^x} + \frac{1}{e^x}\right)dx = \int (1 + e^{-x})\,dx = x - e^{-x} + C$ [Substitute $u = -x$.]

31. $\displaystyle \int \frac{\sin 2x}{1 + \cos^2 x}\,dx = 2\int \frac{\sin x \cos x}{1 + \cos^2 x}\,dx = 2I.$ Let $u = \cos x$. Then $du = -\sin x\,dx$, so

$2I = -2\int \frac{u\,du}{1 + u^2} = -2 \cdot \frac{1}{2}\ln(1 + u^2) + C = -\ln(1 + u^2) + C = -\ln(1 + \cos^2 x) + C.$

Or: Let $u = 1 + \cos^2 x$.

33. Let $u = 1 + x^2$. Then $du = 2x\,dx$, so

$$\int \frac{1 + x}{1 + x^2}\,dx = \int \frac{1}{1 + x^2}\,dx + \int \frac{x}{1 + x^2}\,dx = \tan^{-1} x + \int \frac{\frac{1}{2}\,du}{u} = \tan^{-1} x + \frac{1}{2}\ln|u| + C$$
$$= \tan^{-1} x + \frac{1}{2}\ln\left|1 + x^2\right| + C = \tan^{-1} x + \frac{1}{2}\ln(1 + x^2) + C \text{ [since } 1 + x^2 > 0\text{]}.$$

35. Let $u = x - 1$, so $du = dx$. When $x = 0$, $u = -1$; when $x = 2$, $u = 1$. Thus, $\int_0^2 (x - 1)^{25}\,dx = \int_{-1}^1 u^{25}\,du = 0$

by Theorem 7(b), since $f(u) = u^{25}$ is an odd function.

37. Let $u = 1 + 2x^3$, so $du = 6x^2\,dx$. When $x = 0$, $u = 1$; when $x = 1$, $u = 3$. Thus,

$\int_0^1 x^2 (1 + 2x^3)^5\,dx = \int_1^3 u^5 \left(\frac{1}{6}\,du\right) = \frac{1}{6}\left[\frac{1}{6}u^6\right]_1^3 = \frac{1}{36}(3^6 - 1^6) = \frac{1}{36}(729 - 1) = \frac{728}{36} = \frac{182}{9}.$

39. Let $u = t/4$, so $du = \frac{1}{4}\,dt$. When $t = 0$, $u = 0$; when $t = \pi$, $u = \pi/4$. Thus,

$\int_0^\pi \sec^2(t/4)\,dt = \int_0^{\pi/4} \sec^2 u\,(4\,du) = 4\left[\tan u\right]_0^{\pi/4} = 4\left(\tan \frac{\pi}{4} - \tan 0\right) = 4(1 - 0) = 4.$

41. Let $u = \sqrt{x}$, so $du = \dfrac{1}{2\sqrt{x}}\,dx$. When $x = 1$, $u = 1$; when $x = 4$, $u = 2$.

Thus, $\displaystyle\int_1^4 \dfrac{e^{\sqrt{x}}}{\sqrt{x}}\,dx = \int_1^2 e^u(2\,du) = 2\left[e^u\right]_1^2 = 2(e^2 - e)$.

43. Let $u = x - 1$, so $u + 1 = x$ and $du = dx$. When $x = 1$, $u = 0$; when $x = 2$, $u = 1$. Thus,

$\displaystyle\int_1^2 x\sqrt{x-1}\,dx = \int_0^1 (u+1)\sqrt{u}\,du = \int_0^1 (u^{3/2} + u^{1/2})\,du = \left[\frac{2}{5}u^{5/2} + \frac{2}{3}u^{3/2}\right]_0^1 = \frac{2}{5} + \frac{2}{3} = \frac{16}{15}$.

45. Let $u = e^z + z$, so $du = (e^z + 1)\,dz$. When $z = 0$, $u - 1$; when $z = 1$, $u = e + 1$. Thus,

$\displaystyle\int_0^1 \dfrac{e^z + 1}{e^z + z}\,dz = \int_1^{e+1} \dfrac{1}{u}\,du = \Big[\ln|u|\Big]_1^{e+1} = \ln|e+1| - \ln|1| = \ln(e+1)$.

47. $\displaystyle\int_{-\pi/6}^{\pi/6} \tan^3\theta\,d\theta = 0$ by Theorem 7(b), since $f(\theta) = \tan^3\theta$ is an odd function.

49. Let $u = \ln x$, so $du = \dfrac{dx}{x}$. When $x = e$, $u = 1$; when $x = e^4$; $u = 4$.

Thus, $\displaystyle\int_e^{e^4} \dfrac{dx}{x\sqrt{\ln x}} = \int_1^4 u^{-1/2}\,du = 2\left[u^{1/2}\right]_1^4 = 2(2-1) = 2$.

51. $f_{\text{ave}} = \dfrac{1}{5-0}\displaystyle\int_0^5 te^{-t^2}\,dt = \dfrac{1}{5}\int_0^{-25} e^u - \left(\tfrac{1}{2}\,du\right)$ $[u = -t^2,\, du = -2t\,dt,\, t\,dt = -\tfrac{1}{2}\,du]$

$= -\dfrac{1}{10}\left[e^u\right]_0^{-25} = -\dfrac{1}{10}(e^{-25} - 1) = \dfrac{1}{10}(1 - e^{-25})$

53. $h_{\text{ave}} = \dfrac{1}{\pi - 0}\displaystyle\int_0^\pi \cos^4 x \sin x\,dx = \dfrac{1}{\pi}\int_1^{-1} u^4(-du)$ $[u = \cos x,\, du = -\sin x\,dx]$

$= \dfrac{1}{\pi}\displaystyle\int_{-1}^1 u^4\,du = \dfrac{1}{\pi}\cdot 2\int_0^1 u^4\,du = \dfrac{2}{\pi}\left[\tfrac{1}{5}u^5\right]_0^1 = \dfrac{2}{5\pi}$

55. First write the integral as a sum of two integrals:

$I = \displaystyle\int_{-2}^2 (x+3)\sqrt{4-x^2}\,dx = I_1 + I_2 = \int_{-2}^2 x\sqrt{4-x^2}\,dx + \int_{-2}^2 3\sqrt{4-x^2}\,dx.$ $I_1 = 0$ by Theorem 7(b), since

$f(x) = x\sqrt{4-x^2}$ is an odd function and we are integrating from $x = -2$ to $x = 2$. We interpret I_2 as three times the area of

a semicircle with radius 2, so $I = 0 + 3\cdot\frac{1}{2}\left(\pi\cdot 2^2\right) = 6\pi$.

57. First figure: Let $u = \sqrt{x}$, so $x = u^2$ and $dx = 2u\,du$. When $x = 0$, $u = 0$; when $x = 1$, $u = 1$.

Thus, $A_1 = \displaystyle\int_0^1 e^{\sqrt{x}}\,dx = \int_0^1 e^u(2u\,du) = 2\int_0^1 ue^u\,du$.

Second figure: $A_2 = \displaystyle\int_0^1 2xe^x\,dx = 2\int_0^1 ue^u\,du$.

Third figure: Let $u = \sin x$, so $du = \cos x\,dx$. When $x = 0$, $u = 0$; when $x = \frac{\pi}{2}$, $u = 1$.

Thus, $A_3 = \displaystyle\int_0^{\pi/2} e^{\sin x}\sin 2x\,dx = \int_0^{\pi/2} e^{\sin x}(2\sin x\ \cos x)\,dx = \int_0^1 e^u(2u\,du) =$

$2\displaystyle\int_0^1 ue^u\,du$.

Since $A_1 = A_2 = A_3$, all three areas are equal.

59. The volume of inhaled air in the lungs at time t is

$$V(t) = \int_0^t f(u)\, du = \int_0^t \tfrac{1}{2} \sin\!\left(\tfrac{2\pi}{5} u\right) du = \int_0^{2\pi t/5} \tfrac{1}{2} \sin v \left(\tfrac{5}{2\pi}\, dv\right) \qquad \left[\text{substitute } v = \tfrac{2\pi}{5} u, \, dv = \tfrac{2\pi}{5}\, du\right]$$

$$= \tfrac{5}{4\pi} \Big[-\cos v \Big]_0^{2\pi t/5} = \tfrac{5}{4\pi} \Big[-\cos\!\left(\tfrac{2\pi}{5} t\right) + 1 \Big] = \tfrac{5}{4\pi} \Big[1 - \cos\!\left(\tfrac{2\pi}{5} t\right) \Big] \text{ liters}$$

61. Let $u = 2x$. Then $du = 2\, dx$, so $\int_0^2 f(2x)\, dx = \int_0^4 f(u)\left(\tfrac{1}{2}\, du\right) = \tfrac{1}{2} \int_0^4 f(u)\, du = \tfrac{1}{2}(10) = 5$.

63. Let $u = -x$. Then $du = -dx$, so

$$\int_a^b f(-x)\, dx = \int_{-a}^{-b} f(u)(-du) = \int_{-b}^{-a} f(u)\, du = \int_{-b}^{-a} f(x)\, dx.$$

From the diagram, we see that the equality follows from the fact that we are reflecting the graph of f, and the limits of integration, about the y-axis.

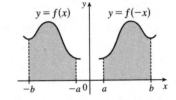

65. Let $u = 1 - x$. Then $x = 1 - u$ and $dx = -du$, so

$$\int_0^1 x^a (1-x)^b\, dx = \int_1^0 (1-u)^a\, u^b (-du) = \int_0^1 u^b (1-u)^a\, du = \int_0^1 x^b (1-x)^a\, dx.$$

5 Review

CONCEPT CHECK

1. (a) $\sum_{i=1}^n f(x_i^*)\, \Delta x_i$ is an expression for a Riemann sum of a function f.

x_i^* is a point in the ith subinterval $[x_{i-1}, x_i]$ and $\Delta x_i = x_i - x_{i-1}$ is the length of the ith subinterval.

(b) See Figure 3 in Section 5.2.

(c) In Section 5.2, see Figure 5 and the paragraph before it.

2. (a) See the definition of the definite integral in Definition 5.2.2. If f is continuous and the subintervals have equal lengths, then the expression for the definite integral simplifies to the one in Theorem 5.2.4.

(b) It's the area under the curve $y = f(x)$ from a to b.

(c) In Section 5.2, see Figure 6 and the paragraph before it.

3. See the statement of the Midpoint Rule on page 268.

4. (a) See the Evaluation Theorem on page 275.

(b) See the Net Change Theorem on page 279.

(c) $\int_{t_1}^{t_2} r(t)\, dt$ represents the change in the amount of water in the reservoir between time t_1 and time t_2.

5. (a) $\int f(x)\, dx$ is an antiderivative of f (or the family of functions $\{F \mid F' = f\}$). Any two such functions differ by a constant.

(b) The connection is given by the Evaluation Theorem: $\int_a^b f(x)\, dx = \Big[\int f(x)\, dx\Big]_a^b$ if f is continuous.

6. See the Fundamental Theorem of Calculus on page 288.

7. (a) $\int_{60}^{120} v(t)\, dt$ represents the change in position of the particle from $t = 60$ to $t = 120$ seconds.

(b) $\int_{60}^{120} |v(t)|\, dt$ represents the total distance traveled by the particle from $t = 60$ to 120 seconds.

(c) $\int_{60}^{120} a(t)\, dt$ represents the change in the velocity of the particle from $t = 60$ to $t = 120$ seconds.

8. (a) The average value of a function f on an interval $[a, b]$ is $f_{\text{ave}} = \dfrac{1}{b-a} \displaystyle\int_a^b f(x)\, dx$.

(b) The Mean Value Theorem for Integrals says that there is a number c at which the value of f is exactly equal to the average value of the function, that is, $f(c) = f_{\text{ave}}$. For a geometric interpretation of the Mean Value Theorem for Integrals, see Figure 10 in Section 5.4 and the discussion that accompanies it.

9. The precise version of this statement is given by the Fundamental Theorem of Calculus. See the statement of this theorem and the paragraph that follows it on page 288.

10. See the Substitution Rule (5.5.4). This says that it is permissible to operate with the dx after an integral sign as if it were a differential.

TRUE-FALSE QUIZ

1. True by Property 2 of the Integral in Section 5.2.

3. True by Property 3 of the Integral in Section 5.2.

5. False.   For example, let $f(x) = x^2$. Then $\int_0^1 \sqrt{x^2}\, dx = \int_0^1 x\, dx = \frac{1}{2}$, but $\sqrt{\int_0^1 x^2\, dx} = \sqrt{\frac{1}{3}} = \frac{1}{\sqrt{3}}$.

7. True by Comparison Property 7 of the Integral in Section 5.2.

9. True.   The integrand is an odd function that is continuous on $[-1, 1]$, so the result follows from Theorem 5.5.7(b).

11. False.   The function $f(x) = 1/x^4$ is not bounded on the interval $[-2, 1]$. It has an infinite discontinuity at $x = 0$, so it is not integrable on the interval. (If the integral were to exist, a positive value would be expected, by Comparison Property 6 of Integrals.)

13. True by FTC1.

EXERCISES

1. (a)

$$L_6 = \sum_{i=1}^{6} f(x_{i-1})\, \Delta x \quad [\Delta x = \tfrac{6-0}{6} = 1]$$

$$= f(x_0) \cdot 1 + f(x_1) \cdot 1 + f(x_2) \cdot 1 + f(x_3) \cdot 1 + f(x_4) \cdot 1 + f(x_5) \cdot 1$$

$$\approx 2 + 3.5 + 4 + 2 + (-1) + (-2.5) = 8$$

The Riemann sum represents the sum of the areas of the four rectangles above the x-axis minus the sum of the areas of the two rectangles below the x-axis.

(b)

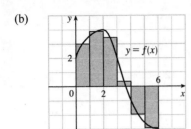

$$M_6 = \sum_{i=1}^{6} f(\overline{x}_i)\,\Delta x \quad [\Delta x = \tfrac{6-0}{6} = 1]$$

$$= f(\overline{x}_1)\cdot 1 + f(\overline{x}_2)\cdot 1 + f(\overline{x}_3)\cdot 1 + f(\overline{x}_4)\cdot 1 + f(\overline{x}_5)\cdot 1 + f(\overline{x}_6)\cdot 1$$

$$= f(0.5) + f(1.5) + f(2.5) + f(3.5) + f(4.5) + f(5.5)$$

$$\approx 3 + 3.9 + 3.4 + 0.3 + (-2) + (-2.9) = 5.7$$

The Riemann sum represents the sum of the areas of the four rectangles above the x-axis minus the sum of the areas of the two rectangles below the x-axis.

3. $\int_0^1 \left(x + \sqrt{1-x^2}\,\right) dx = \int_0^1 x\,dx + \int_0^1 \sqrt{1-x^2}\,dx = I_1 + I_2$.

I_1 can be interpreted as the area of the triangle shown in the figure and I_2 can be interpreted as the area of the quarter-circle.

Area $= \tfrac{1}{2}(1)(1) + \tfrac{1}{4}(\pi)(1)^2 = \tfrac{1}{2} + \tfrac{\pi}{4}$.

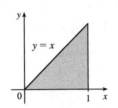

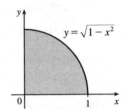

5. First note that either a or b must be the graph of $\int_0^x f(t)\,dt$, since $\int_0^0 f(t)\,dt = 0$, and $c(0) \neq 0$. Now notice that $b > 0$ when c is increasing, and that $c > 0$ when a is increasing. It follows that c is the graph of $f(x)$, b is the graph of $f'(x)$, and a is the graph of $\int_0^x f(t)\,dt$.

7. $\int_1^2 (8x^3 + 3x^2)\,dx = \left[8\cdot\tfrac{1}{4}x^4 + 3\cdot\tfrac{1}{3}x^3\right]_1^2 = \left[2x^4 + x^3\right]_1^2 = (2\cdot 2^4 + 2^3) - (2+1) = 40 - 3 = 37$

9. $\int_0^1 (1 - x^9)\,dx = \left[x - \tfrac{1}{10}x^{10}\right]_0^1 = \left(1 - \tfrac{1}{10}\right) - 0 = \tfrac{9}{10}$

11. $\displaystyle\int_1^9 \frac{\sqrt{u} - 2u^2}{u}\,du = \int_1^9 (u^{-1/2} - 2u)\,du = \left[2u^{1/2} - u^2\right]_1^9 = (6 - 81) - (2 - 1) = -76$

13. Let $u = y^2 + 1$, so $du = 2y\,dy$ and $y\,dy = \tfrac{1}{2}\,du$. When $y = 0$, $u = 1$; when $y = 1$, $u = 2$.

Thus, $\int_0^1 y(y^2+1)^5\,dy = \int_1^2 u^5\left(\tfrac{1}{2}\,du\right) = \tfrac{1}{2}\left[\tfrac{1}{6}u^6\right]_1^2 = \tfrac{1}{12}(64 - 1) = \tfrac{63}{12} = \tfrac{21}{4}$.

15. Let $u = v^3$, so $du = 3v^2\,dv$. When $v = 0$, $u = 0$; when $v = 1$, $u = 1$.

Thus, $\int_0^1 v^2 \cos(v^3)\,dv = \int_0^1 \cos u\left(\tfrac{1}{3}\,du\right) = \tfrac{1}{3}\left[\sin u\right]_0^1 = \tfrac{1}{3}(\sin 1 - 0) = \tfrac{1}{3}\sin 1$.

17. $\int_0^1 e^{\pi t}\,dt = \left[\tfrac{1}{\pi}e^{\pi t}\right]_0^1 = \tfrac{1}{\pi}(e^\pi - 1)$

19. $\displaystyle\int_2^4 \frac{1 + x - x^2}{x^2}\,dx = \int_2^4 \left(\frac{1}{x^2} + \frac{x}{x^2} - \frac{x^2}{x^2}\right) dx = \int_2^4 \left(x^{-2} + \frac{1}{x} - 1\right) dx = \left[-\frac{1}{x} + \ln|x| - x\right]_2^4$

$$= \left(-\tfrac{1}{4} + \ln 4 - 4\right) - \left(-\tfrac{1}{2} + \ln 2 - 2\right) = \ln 2 - \tfrac{7}{4}$$

21. Let $u = x^2 + 4x$. Then $du = (2x + 4)\,dx = 2(x+2)\,dx$, so

$$\int \frac{x+2}{\sqrt{x^2+4x}}\,dx = \int u^{-1/2}\left(\tfrac{1}{2}\,du\right) = \tfrac{1}{2}\cdot 2u^{1/2} + C = \sqrt{u} + C = \sqrt{x^2+4x} + C.$$

23. Let $u = \sin \pi t$. Then $du = \pi\cos \pi t\,dt$, so $\int \sin \pi t\,\cos \pi t\,dt = \int u\left(\tfrac{1}{\pi}\,du\right) = \tfrac{1}{\pi}\cdot\tfrac{1}{2}u^2 + C = \tfrac{1}{2\pi}(\sin \pi t)^2 + C$.

25. Let $u = \sqrt{x}$. Then $du = \dfrac{dx}{2\sqrt{x}}$, so $\displaystyle\int \frac{e^{\sqrt{x}}}{\sqrt{x}}\,dx = 2\int e^u\,du = 2e^u + C = 2e^{\sqrt{x}} + C$.

27. Let $u = \ln(\cos x)$. Then $du = \dfrac{-\sin x}{\cos x}\,dx = -\tan x\,dx$, so

$\int \tan x \ln(\cos x)\,dx = -\int u\,du = -\frac{1}{2}u^2 + C = -\frac{1}{2}[\ln(\cos x)]^2 + C.$

29. Let $u = 1 + x^4$. Then $du = 4x^3\,dx$, so $\displaystyle\int \dfrac{x^3}{1+x^4}\,dx = \dfrac{1}{4}\int \dfrac{1}{u}\,du = \frac{1}{4}\ln|u| + C = \frac{1}{4}\ln(1 + x^4) + C.$

31. Let $u = 1 + \sec\theta$. Then $du = \sec\theta\tan\theta\,d\theta$, so

$\displaystyle\int \dfrac{\sec\theta\tan\theta}{1+\sec\theta}\,d\theta = \int \dfrac{1}{1+\sec\theta}\,(\sec\theta\tan\theta\,d\theta) = \int \dfrac{1}{u}\,du = \ln|u| + C = \ln|1 + \sec\theta| + C.$

33. From the graph, it appears that the area under the curve $y = x\sqrt{x}$ between

$x = 0$ and $x = 4$ is somewhat less than half the area of an 8×4 rectangle,

so perhaps about 13 or 14. To find the exact value, we evaluate

$\int_0^4 x\sqrt{x}\,dx = \int_0^4 x^{3/2}\,dx = \left[\frac{2}{5}x^{5/2}\right]_0^4 = \frac{2}{5}(4)^{5/2} = \frac{64}{5} = 12.8.$

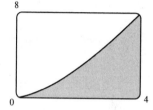

35. By FTC1, $F(x) = \int_1^x \sqrt{1+t^4}\,dt \quad\Rightarrow\quad F'(x) = \sqrt{1+x^4}.$

37. $y = \displaystyle\int_{\sqrt{x}}^x \dfrac{e^t}{t}\,dt = \int_{\sqrt{x}}^1 \dfrac{e^t}{t}\,dt + \int_1^x \dfrac{e^t}{t}\,dt = -\int_1^{\sqrt{x}} \dfrac{e^t}{t}\,dt + \int_1^x \dfrac{e^t}{t}\,dt \quad\Rightarrow$

$\dfrac{dy}{dx} = -\dfrac{d}{dx}\left(\int_1^{\sqrt{x}} \dfrac{e^t}{t}\,dt\right) + \dfrac{d}{dx}\left(\int_1^x \dfrac{e^t}{t}\,dt\right).$ Let $u = \sqrt{x}$. Then

$\dfrac{d}{dx}\int_1^{\sqrt{x}} \dfrac{e^t}{t}\,dt = \dfrac{d}{dx}\int_1^u \dfrac{e^t}{t}\,dt = \dfrac{d}{du}\left(\int_1^u \dfrac{e^t}{t}\,dt\right)\dfrac{du}{dx} = \dfrac{e^u}{u}\cdot\dfrac{1}{2\sqrt{x}} = \dfrac{e^{\sqrt{x}}}{\sqrt{x}}\cdot\dfrac{1}{2\sqrt{x}} = \dfrac{e^{\sqrt{x}}}{2x},$ so $\dfrac{dy}{dx} = -\dfrac{e^{\sqrt{x}}}{2x} + \dfrac{e^x}{x}.$

39. If $1 \le x \le 3$, then $\sqrt{1^2 + 3} \le \sqrt{x^2 + 3} \le \sqrt{3^2 + 3} \quad\Rightarrow\quad 2 \le \sqrt{x^2 + 3} \le 2\sqrt{3}$, so

$2(3 - 1) \le \int_1^3 \sqrt{x^2 + 3}\,dx \le 2\sqrt{3}(3 - 1);$ that is, $4 \le \int_1^3 \sqrt{x^2 + 3}\,dx \le 4\sqrt{3}.$

41. Let $f(x) = \sqrt{1 + x^3}$ on $[0, 1]$. The Midpoint Rule with $n = 5$ gives

$$\int_0^1 \sqrt{1 + x^3}\,dx \approx \tfrac{1}{5}[f(0.1) + f(0.3) + f(0.5) + f(0.7) + f(0.9)]$$

$$= \tfrac{1}{5}\left[\sqrt{1 + (0.1)^3} + \sqrt{1 + (0.3)^3} + \cdots + \sqrt{1 + (0.9)^3}\right] \approx 1.110$$

43. Note that $r(t) = b'(t)$, where $b(t) = $ the number of barrels of oil consumed up to time t. So, by the Net Change Theorem,

$\int_0^3 r(t)\,dt = b(3) - b(0)$ represents the number of barrels of oil consumed from Jan. 1, 2000, through Jan. 1, 2003.

45. We use the Midpoint Rule with $n = 6$ and $\Delta t = \dfrac{24 - 0}{6} = 4$. The increase in the bee population was

$$\int_0^{24} r(t)\,dt \approx M_6 = 4[r(2) + r(6) + r(10) + r(14) + r(18) + r(22)]$$

$$\approx 4[50 + 1000 + 7000 + 8550 + 1350 + 150] = 4(18{,}100) = 72{,}400$$

47. $\displaystyle\lim_{h\to 0} f_{\text{ave}} = \lim_{h\to 0} \dfrac{1}{(x + h) - x}\int_x^{x+h} f(t)\,dt = \lim_{h\to 0} \dfrac{F(x + h) - F(x)}{h},$ where $F(x) = \int_a^x f(t)\,dt.$ But we recognize this

limit as being $F'(x)$ by the definition of a derivative. Therefore, $\displaystyle\lim_{h\to 0} f_{\text{ave}} = F'(x) = f(x)$ by FTC1.

49. Area under the curve $y = \sinh cx$ between $x = 0$ and $x = 1$ is

equal to 1 $\Rightarrow$ $\int_0^1 \sinh cx\, dx = 1$ $\Rightarrow$ $\frac{1}{c}\left[\cosh cx\right]_0^1 = 1$ $\Rightarrow$

$\frac{1}{c}(\cosh c - 1) = 1$ $\Rightarrow$ $\cosh c - 1 = c$ $\Rightarrow$ $\cosh c = c + 1$.

From the graph, we get $c = 0$ and $c \approx 1.6161$, but $c = 0$ isn't a

solution for this problem since the curve $y = \sinh cx$ becomes $y = 0$

and the area under it is 0. Thus, $c \approx 1.6161$.

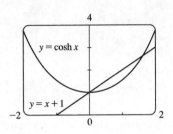

51. Using FTC1, we differentiate both sides of the given equation, $\int_0^x f(t)\, dt = xe^{2x} + \int_0^x e^{-t} f(t)\, dt$, and get

$$f(x) = e^{2x} + 2xe^{2x} + e^{-x} f(x) \quad \Rightarrow \quad f(x)(1 - e^{-x}) = e^{2x} + 2xe^{2x} \quad \Rightarrow \quad f(x) = \frac{e^{2x}(1 + 2x)}{1 - e^{-x}}.$$

53. Let $u = f(x)$ and $du = f'(x)\, dx$. So $2\int_a^b f(x)f'(x)\, dx = 2\int_{f(a)}^{f(b)} u\, du = \left[u^2\right]_{f(a)}^{f(b)} = [f(b)]^2 - [f(a)]^2$.

6 □ TECHNIQUES OF INTEGRATION

6.1 Integration by Parts

1. Let $u = \ln x$, $dv = x\,dx$ $\Rightarrow$ $du = dx/x$, $v = \frac{1}{2}x^2$. Then by Equation 2, $\int u\,dv = uv - \int v\,du$,

$\int x \ln x\,dx = \frac{1}{2}x^2 \ln x - \int \frac{1}{2}x^2\,(dx/x) = \frac{1}{2}x^2 \ln x - \frac{1}{2}\int x\,dx = \frac{1}{2}x^2 \ln x - \frac{1}{2} \cdot \frac{1}{2}x^2 + C = \frac{1}{2}x^2 \ln x - \frac{1}{4}x^2 + C.$

Note: A mnemonic device which is helpful for selecting u when using integration by parts is the LIATE principle of precedence for u:

$\underline{\text{L}}$ogarithmic

$\underline{\text{I}}$nverse trigonometric

$\underline{\text{A}}$lgebraic

$\underline{\text{T}}$rigonometric

$\underline{\text{E}}$xponential

If the integrand has several factors, then we try to choose among them a u which appears as high as possible on the list. For example, in $\int xe^{2x}\,dx$ the integrand is xe^{2x}, which is the product of an algebraic function (x) and an exponential function (e^{2x}). Since $\underline{\text{A}}$lgebraic appears before $\underline{\text{E}}$xponential, we choose $u = x$. Sometimes the integration turns out to be similar regardless of the selection of u and dv, but it is advisable to refer to LIATE when in doubt.

3. Let $u = x$, $dv = \cos 5x\,dx$ $\Rightarrow$ $du = dx$, $v = \frac{1}{5}\sin 5x$. Then by Equation 2,

$\int x \cos 5x\,dx = \frac{1}{5}x \sin 5x - \int \frac{1}{5}\sin 5x\,dx = \frac{1}{5}x \sin 5x + \frac{1}{25}\cos 5x + C.$

5. Let $u = r$, $dv = e^{r/2}\,dr$ $\Rightarrow$ $du = dr$, $v = 2e^{r/2}$. Then $\int re^{r/2}\,dr = 2re^{r/2} - \int 2e^{r/2}\,dr = 2re^{r/2} - 4e^{r/2} + C.$

7. Let $u = x^2$, $dv = \sin \pi x\,dx$ $\Rightarrow$ $du = 2x\,dx$ and $v = -\frac{1}{\pi}\cos \pi x$. Then

$I = \int x^2 \sin \pi x\,dx = -\frac{1}{\pi}x^2 \cos \pi x + \frac{2}{\pi}\int x \cos \pi x\,dx$ $(\star)$. Next let $U = x$, $dV = \cos \pi x\,dx$ $\Rightarrow$

$dU = dx$, $V = \frac{1}{\pi}\sin \pi x$, so $\int x \cos \pi x\,dx = \frac{1}{\pi}x \sin \pi x - \frac{1}{\pi}\int \sin \pi x\,dx = \frac{1}{\pi}x \sin \pi x + \frac{1}{\pi^2}\cos \pi x + C_1.$

Substituting for $\int x \cos \pi x\,dx$ in $(\star)$, we get

$I = -\frac{1}{\pi}x^2 \cos \pi x + \frac{2}{\pi}\left(\frac{1}{\pi}x \sin \pi x + \frac{1}{\pi^2}\cos \pi x + C_1\right) = -\frac{1}{\pi}x^2 \cos \pi x + \frac{2}{\pi^2}x \sin \pi x + \frac{2}{\pi^3}\cos \pi x + C$, where $C = \frac{2}{\pi}C_1.$

9. Let $u = \ln(2x + 1)$, $dv = dx$ $\Rightarrow$ $du = \dfrac{2}{2x + 1}\,dx$, $v = x$. Then

$$\int \ln(2x + 1)\,dx = x \ln(2x + 1) - \int \frac{2x}{2x + 1}\,dx = x \ln(2x + 1) - \int \frac{(2x + 1) - 1}{2x + 1}\,dx$$

$$= x \ln(2x + 1) - \int \left(1 - \frac{1}{2x + 1}\right)dx = x \ln(2x + 1) - x + \frac{1}{2}\ln(2x + 1) + C$$

$$= \frac{1}{2}(2x + 1)\ln(2x + 1) - x + C$$

11. Let $u = \arctan 4t$, $dv = dt$ $\Rightarrow$ $du = \dfrac{4}{1 + (4t)^2}\,dt = \dfrac{4}{1 + 16t^2}\,dt$, $v = t$. Then

$$\int \arctan 4t\,dt = t \arctan 4t - \int \frac{4t}{1 + 16t^2}\,dt = t \arctan 4t - \frac{1}{8}\int \frac{32t}{1 + 16t^2}\,dt = t \arctan 4t - \frac{1}{8}\ln(1 + 16t^2) + C.$$

13. First let $u = \sin 3\theta$, $dv = e^{2\theta}\, d\theta$ $\Rightarrow$ $du = 3\cos 3\theta\, d\theta$, $v = \frac{1}{2}e^{2\theta}$. Then

$I = \int e^{2\theta} \sin 3\theta\, d\theta = \frac{1}{2}e^{2\theta}\sin 3\theta - \frac{3}{2}\int e^{2\theta}\cos 3\theta\, d\theta$. Next let $U = \cos 3\theta$, $dV = e^{2\theta}\, d\theta$ $\Rightarrow$ $dU = -3\sin 3\theta\, d\theta$,

$V = \frac{1}{2}e^{2\theta}$ to get $\int e^{2\theta}\cos 3\theta\, d\theta = \frac{1}{2}e^{2\theta}\cos 3\theta + \frac{3}{2}\int e^{2\theta}\sin 3\theta\, d\theta$. Substituting in the previous formula gives

$I = \frac{1}{2}e^{2\theta}\sin 3\theta - \frac{3}{4}e^{2\theta}\cos 3\theta - \frac{9}{4}\int e^{2\theta}\sin 3\theta\, d\theta = \frac{1}{2}e^{2\theta}\sin 3\theta - \frac{3}{4}e^{2\theta}\cos 3\theta - \frac{9}{4}I$ $\Rightarrow$

$\frac{13}{4}I = \frac{1}{2}e^{2\theta}\sin 3\theta - \frac{3}{4}e^{2\theta}\cos 3\theta + C_1$. Hence, $I = \frac{1}{13}e^{2\theta}(2\sin 3\theta - 3\cos 3\theta) + C$, where $C = \frac{4}{13}C_1$.

15. Let $u = t$, $dv = \sin 3t\, dt$ $\Rightarrow$ $du = dt$, $v = -\frac{1}{3}\cos 3t$. Then

$\int_0^\pi t \sin 3t\, dt = \left[-\frac{1}{3}t\cos 3t\right]_0^\pi + \frac{1}{3}\int_0^\pi \cos 3t\, dt = \left(\frac{1}{3}\pi - 0\right) + \frac{1}{9}\left[\sin 3t\right]_0^\pi = \frac{\pi}{3}$.

17. Let $u = \ln x$, $dv = x^{-2}\, dx$ $\Rightarrow$ $du = \dfrac{1}{x}\, dx$, $v = -x^{-1}$. By (6),

$\displaystyle\int_1^2 \frac{\ln x}{x^2}\, dx = \left[-\frac{\ln x}{x}\right]_1^2 + \int_1^2 x^{-2}\, dx = -\frac{1}{2}\ln 2 + \ln 1 + \left[-\frac{1}{x}\right]_1^2 = -\frac{1}{2}\ln 2 + 0 - \frac{1}{2} + 1 = \frac{1}{2} - \frac{1}{2}\ln 2$.

19. Let $u = y$, $dv = \dfrac{dy}{e^{2y}} = e^{-2y}\, dy$ $\Rightarrow$ $du = dy$, $v = -\dfrac{1}{2}e^{-2y}$. Then

$\displaystyle\int_0^1 \frac{y}{e^{2y}}\, dy = \left[-\frac{1}{2}ye^{-2y}\right]_0^1 + \frac{1}{2}\int_0^1 e^{-2y}\, dy = \left(-\frac{1}{2}e^{-2} + 0\right) - \frac{1}{4}\left[e^{-2y}\right]_0^1 = -\frac{1}{2}e^{-2} - \frac{1}{4}e^{-2} + \frac{1}{4} = \frac{1}{4} - \frac{3}{4}e^{-2}$.

21. Let $u = \sin^{-1} x$, $dv = dx$ $\Rightarrow$ $du = \dfrac{dx}{\sqrt{1-x^2}}$, $v = x$. By (6),

$\displaystyle I = \int_0^{1/2} \sin^{-1} x\, dx = \left[x\sin^{-1} x\right]_0^{1/2} - \int_0^{1/2} \frac{x\, dx}{\sqrt{1-x^2}} = \frac{1}{2}\cdot\frac{\pi}{6} - \int_0^{1/2} \frac{x\, dx}{\sqrt{1-x^2}}$. To evaluate the last integral,

let $t = 1 - x^2$, so $dt = -2x\, dx$ and $x\, dx = -\frac{1}{2}\, dt$. When $x = 0$, $t = 1$; when $x = \frac{1}{2}$, $t = \frac{3}{4}$.

So $\displaystyle\int_0^{1/2} \frac{x\, dx}{\sqrt{1-x^2}} = \int_1^{3/4} \frac{1}{\sqrt{t}}\left(-\frac{1}{2}\, dt\right) = \frac{1}{2}\int_{3/4}^1 t^{-1/2}\, dt = \frac{1}{2}\left[2t^{1/2}\right]_{3/4}^1 = \sqrt{1} - \sqrt{\frac{3}{4}} = 1 - \frac{\sqrt{3}}{2}$.

Thus, $I = \frac{\pi}{12} - \left(1 - \frac{\sqrt{3}}{2}\right) = \frac{\pi}{12} - 1 + \frac{\sqrt{3}}{2} = \frac{1}{12}\left(\pi - 12 + 6\sqrt{3}\right)$.

23. Let $u = (\ln x)^2$, $dv = dx$ $\Rightarrow$ $du = \dfrac{2}{x}\ln x\, dx$, $v = x$. By (6), $I = \int_1^2 (\ln x)^2\, dx = \left[x(\ln x)^2\right]_1^2 - 2\int_1^2 \ln x\, dx$.

To evaluate the last integral, let $U = \ln x$, $dV = dx$ $\Rightarrow$ $dU = \dfrac{1}{x}\, dx$, $V = x$. Thus,

$$I = \left[x(\ln x)^2\right]_1^2 - 2\left(\left[x\ln x\right]_1^2 - \int_1^2 dx\right) = \left[x(\ln x)^2 - 2x\ln x + 2x\right]_1^2$$

$$= \left(2(\ln 2)^2 - 4\ln 2 + 4\right) - (0 - 0 + 2) = 2(\ln 2)^2 - 4\ln 2 + 2$$

25. Let $w = \sqrt{x}$, so that $x = w^2$ and $dx = 2w\, dw$. Thus, $\int \sin\sqrt{x}\, dx = \int 2w\sin w\, dw$. Now use parts with $u = 2w$,

$dv = \sin w\, dw$, $du = 2\, dw$, $v = -\cos w$ to get

$$\int 2w\sin w\, dw = -2w\cos w + \int 2\cos w\, dw = -2w\cos w + 2\sin w + C$$

$$= -2\sqrt{x}\cos\sqrt{x} + 2\sin\sqrt{x} + C = 2(\sin\sqrt{x} - \sqrt{x}\cos\sqrt{x}) + C$$

27. Let $x = \theta^2$, so that $dx = 2\theta\, d\theta$. Thus, $\int_{\sqrt{\pi/2}}^{\sqrt{\pi}} \theta^3 \cos(\theta^2)\, d\theta = \int_{\sqrt{\pi/2}}^{\sqrt{\pi}} \theta^2 \cos(\theta^2) \cdot \frac{1}{2}(2\theta\, d\theta) = \frac{1}{2}\int_{\pi/2}^{\pi} x \cos x\, dx$. Now use

parts with $u = x$, $dv = \cos x\, dx$, $du = dx$, $v = \sin x$ to get

$$\frac{1}{2}\int_{\pi/2}^{\pi} x \cos x\, dx = \frac{1}{2}\left(\left[x \sin x\right]_{\pi/2}^{\pi} - \int_{\pi/2}^{\pi} \sin x\, dx \right) = \frac{1}{2}\left[x \sin x + \cos x\right]_{\pi/2}^{\pi}$$

$$= \frac{1}{2}(\pi \sin \pi + \cos \pi) - \frac{1}{2}\left(\frac{\pi}{2} \sin \frac{\pi}{2} + \cos \frac{\pi}{2}\right) = \frac{1}{2}(\pi \cdot 0 - 1) - \frac{1}{2}\left(\frac{\pi}{2} \cdot 1 + 0\right) = -\frac{1}{2} - \frac{\pi}{4}$$

29. (a) Take $n = 2$ in Example 6 to get $\displaystyle\int \sin^2 x\, dx = -\frac{1}{2}\cos x \sin x + \frac{1}{2}\int 1\, dx = \frac{x}{2} - \frac{\sin 2x}{4} + C.$

(b) $\int \sin^4 x\, dx = -\frac{1}{4}\cos x \sin^3 x + \frac{3}{4}\int \sin^2 x\, dx = -\frac{1}{4}\cos x \sin^3 x + \frac{3}{8}x - \frac{3}{16}\sin 2x + C.$

31. (a) From Example 6, $\displaystyle\int \sin^n x\, dx = -\frac{1}{n}\cos x \sin^{n-1} x + \frac{n-1}{n}\int \sin^{n-2} x\, dx$. Using (6),

$$\int_0^{\pi/2} \sin^n x\, dx = \left[-\frac{\cos x \sin^{n-1} x}{n} \right]_0^{\pi/2} + \frac{n-1}{n}\int_0^{\pi/2} \sin^{n-2} x\, dx$$

$$= (0 - 0) + \frac{n-1}{n}\int_0^{\pi/2} \sin^{n-2} x\, dx = \frac{n-1}{n}\int_0^{\pi/2} \sin^{n-2} x\, dx$$

(b) Using $n = 3$ in part (a), we have $\int_0^{\pi/2} \sin^3 x\, dx = \frac{2}{3}\int_0^{\pi/2} \sin x\, dx = \left[-\frac{2}{3}\cos x\right]_0^{\pi/2} = \frac{2}{3}.$

Using $n = 5$ in part (a), we have $\int_0^{\pi/2} \sin^5 x\, dx = \frac{4}{5}\int_0^{\pi/2} \sin^3 x\, dx = \frac{4}{5} \cdot \frac{2}{3} = \frac{8}{15}.$

(c) The formula holds for $n = 1$ (that is, $2n + 1 = 3$) by (b). Assume it holds for some $k \geq 1$. Then

$$\int_0^{\pi/2} \sin^{2k+1} x\, dx = \frac{2 \cdot 4 \cdot 6 \cdot \cdots \cdot (2k)}{3 \cdot 5 \cdot 7 \cdot \cdots \cdot (2k+1)}. \text{ By Example 6,}$$

$$\int_0^{\pi/2} \sin^{2k+3} x\, dx = \frac{2k+2}{2k+3}\int_0^{\pi/2} \sin^{2k+1} x\, dx = \frac{2k+2}{2k+3} \cdot \frac{2 \cdot 4 \cdot 6 \cdot \cdots \cdot (2k)}{3 \cdot 5 \cdot 7 \cdot \cdots \cdot (2k+1)}$$

$$= \frac{2 \cdot 4 \cdot 6 \cdot \cdots \cdot (2k)[2\,(k+1)]}{3 \cdot 5 \cdot 7 \cdot \cdots \cdot (2k+1)[2\,(k+1)+1]},$$

so the formula holds for $n = k + 1$. By induction, the formula holds for all $n \geq 1$.

33. Let $u = (\ln x)^n$, $dv = dx$ $\Rightarrow$ $du = n(\ln x)^{n-1}(dx/x)$, $v = x$. Then

$\int (\ln x)^n\, dx = x(\ln x)^n - \int nx(\ln x)^{n-1}(dx/x) = x(\ln x)^n - n\int (\ln x)^{n-1}\, dx.$

35. Let $u = (x^2 + a^2)^n$, $dv = dx$ $\Rightarrow$ $du = n(x^2 + a^2)^{n-1} 2x\, dx$, $v = x$. Then

$$\int (x^2 + a^2)^n\, dx = x(x^2 + a^2)^n - 2n\int x^2(x^2 + a^2)^{n-1}\, dx$$

$$= x(x^2 + a^2)^n - 2n\left[\int (x^2 + a^2)^n\, dx - a^2\int (x^2 + a^2)^{n-1}\, dx\right] \quad [\text{since } x^2 = (x^2 + a^2) - a^2] \quad \Rightarrow$$

$$(2n + 1)\int (x^2 + a^2)^n\, dx = x(x^2 + a^2)^n + 2na^2\int (x^2 + a^2)^{n-1}\, dx, \text{ and}$$

$$\int (x^2 + a^2)^n\, dx = \frac{x(x^2 + a^2)^n}{2n + 1} + \frac{2na^2}{2n + 1}\int (x^2 + a^2)^{n-1}\, dx \quad [\text{provided } 2n + 1 \neq 0].$$

37. Take $n = 3$ in Exercise 33 to get $\int (\ln x)^3 \, dx = x \, (\ln x)^3 - 3 \int (\ln x)^2 \, dx = x(\ln x)^3 - 3x(\ln x)^2 + 6x \ln x - 6x + C$

[by Exercise 23].

Or: Instead of using Exercise 23, apply Exercise 33 again with $n = 2$.

39. The average value of $f(x) = x^2 \ln x$ on the interval $[1, 3]$ is $f_{\text{ave}} = \dfrac{1}{3-1} \displaystyle\int_1^3 x^2 \ln x \, dx = \tfrac{1}{2} I$.

Let $u = \ln x$, $dv = x^2 \, dx$ $\Rightarrow$ $du = (1/x) \, dx$, $v = \tfrac{1}{3} x^3$.

So $I = \left[\tfrac{1}{3} x^3 \ln x\right]_1^3 - \int_1^3 \tfrac{1}{3} x^2 dx = (9 \ln 3 - 0) - \left[\tfrac{1}{9} x^3\right]_1^3 = 9 \ln 3 - \left(3 - \tfrac{1}{9}\right) = 9 \ln 3 - \tfrac{26}{9}$.

Thus, $f_{\text{ave}} = \tfrac{1}{2} I = \tfrac{1}{2}\left(9 \ln 3 - \tfrac{26}{9}\right) = \tfrac{9}{2} \ln 3 - \tfrac{13}{9}$.

41. Since $v(t) > 0$ for all t, the desired distance is $s(t) = \int_0^t v(w) \, dw = \int_0^t w^2 e^{-w} \, dw$.

First let $u = w^2$, $dv = e^{-w} \, dw$ $\Rightarrow$ $du = 2w \, dw$, $v = -e^{-w}$. Then $s(t) = \left[-w^2 e^{-w}\right]_0^t + 2 \int_0^t we^{-w} \, dw$.

Next let $U = w$, $dV = e^{-w} \, dw$ $\Rightarrow$ $dU = dw$, $V = -e^{-w}$. Then

$$s(t) = -t^2 e^{-t} + 2\left(\left[-we^{-w}\right]_0^t + \int_0^t e^{-w} \, dw\right) = -t^2 e^{-t} + 2\left(-te^{-t} + 0 + \left[-e^{-w}\right]_0^t\right)$$

$$= -t^2 e^{-t} + 2(-te^{-t} - e^{-t} + 1) = -t^2 e^{-t} - 2te^{-t} - 2e^{-t} + 2 = 2 - e^{-t}(t^2 + 2t + 2) \text{ meters}$$

43. For $I = \int_1^4 x f''(x) \, dx$, let $u = x$, $dv = f''(x) \, dx$ $\Rightarrow$ $du = dx$, $v = f'(x)$. Then

$$I = \left[x f'(x)\right]_1^4 - \int_1^4 f'(x) \, dx = 4 f'(4) - 1 \cdot f'(1) - [f(4) - f(1)] = 4 \cdot 3 - 1 \cdot 5 - (7 - 2) = 12 - 5 - 5 = 2.$$

We used the fact that f'' is continuous to guarantee that I exists.

6.2   Trigonometric Integrals and Substitutions

The symbols $\overset{s}{=}$ and $\overset{c}{=}$ indicate the use of the substitutions $\{u = \sin x, du = \cos x \, dx\}$ and $\{u = \cos x, du = -\sin x \, dx\}$, respectively.

1. $\int \sin^3 x \cos^2 x \, dx = \int \sin^2 x \cos^2 x \sin x \, dx = \int (1 - \cos^2 x) \cos^2 x \sin x \, dx \overset{c}{=} \int (1 - u^2) u^2 \, (-du)$

$$= \int (u^2 - 1) u^2 \, du = \int (u^4 - u^2) \, du = \tfrac{1}{5} u^5 - \tfrac{1}{3} u^3 + C = \tfrac{1}{5} \cos^5 x - \tfrac{1}{3} \cos^3 x + C$$

3. $\int_{\pi/2}^{3\pi/4} \sin^5 x \cos^3 x \, dx = \int_{\pi/2}^{3\pi/4} \sin^5 x \cos^2 x \cos x \, dx = \int_{\pi/2}^{3\pi/4} \sin^5 x \, (1 - \sin^2 x) \cos x \, dx \overset{s}{=} \int_1^{\sqrt{2}/2} u^5 (1 - u^2) \, du$

$$= \int_1^{\sqrt{2}/2} (u^5 - u^7) \, du = \left[\tfrac{1}{6} u^6 - \tfrac{1}{8} u^8\right]_1^{\sqrt{2}/2} = \left(\tfrac{1/8}{6} - \tfrac{1/16}{8}\right) - \left(\tfrac{1}{6} - \tfrac{1}{8}\right) = -\tfrac{11}{384}$$

5. $\int_0^{\pi/2} \cos^2 \theta \, d\theta = \int_0^{\pi/2} \tfrac{1}{2}(1 + \cos 2\theta) \, d\theta$ [half-angle identity] $= \tfrac{1}{2}\left[\theta + \tfrac{1}{2} \sin 2\theta\right]_0^{\pi/2} = \tfrac{1}{2}\left[\left(\tfrac{\pi}{2} + 0\right) - (0 + 0)\right] = \tfrac{\pi}{4}$

7. $\int_0^\pi \sin^4(3t) \, dt = \int_0^\pi \left[\sin^2(3t)\right]^2 dt = \int_0^\pi \left[\tfrac{1}{2}(1 - \cos 6t)\right]^2 dt = \tfrac{1}{4} \int_0^\pi (1 - 2 \cos 6t + \cos^2 6t) \, dt$

$$= \tfrac{1}{4} \int_0^\pi \left[1 - 2 \cos 6t + \tfrac{1}{2}(1 + \cos 12t)\right] dt = \tfrac{1}{4} \int_0^\pi \left(\tfrac{3}{2} - 2 \cos 6t + \tfrac{1}{2} \cos 12t\right) dt$$

$$= \tfrac{1}{4}\left[\tfrac{3}{2} t - \tfrac{1}{3} \sin 6t + \tfrac{1}{24} \sin 12t\right]_0^\pi = \tfrac{1}{4}\left[\left(\tfrac{3\pi}{2} - 0 + 0\right) - (0 - 0 + 0)\right] = \tfrac{3\pi}{8}$$

9. $\int (1 + \cos \theta)^2 \, d\theta = \int (1 + 2 \cos \theta + \cos^2 \theta) \, d\theta = \theta + 2 \sin \theta + \tfrac{1}{2} \int (1 + \cos 2\theta) \, d\theta$

$$= \theta + 2 \sin \theta + \tfrac{1}{2} \theta + \tfrac{1}{4} \sin 2\theta + C = \tfrac{3}{2} \theta + 2 \sin \theta + \tfrac{1}{4} \sin 2\theta + C$$

11. $\int_0^{\pi/4} \sin^4 x \cos^2 x \, dx = \int_0^{\pi/4} \sin^2 x \, (\sin x \cos x)^2 \, dx = \int_0^{\pi/4} \frac{1}{2}(1 - \cos 2x)\left(\frac{1}{2}\sin 2x\right)^2 \, dx$

$\qquad = \frac{1}{8}\int_0^{\pi/4}(1 - \cos 2x)\sin^2 2x \, dx = \frac{1}{8}\int_0^{\pi/4}\sin^2 2x \, dx - \frac{1}{8}\int_0^{\pi/4}\sin^2 2x \cos 2x \, dx$

$\qquad = \frac{1}{16}\int_0^{\pi/4}(1 - \cos 4x)\, dx - \frac{1}{16}\left[\frac{1}{3}\sin^3 2x\right]_0^{\pi/4} = \frac{1}{16}\left[x - \frac{1}{4}\sin 4x - \frac{1}{3}\sin^3 2x\right]_0^{\pi/4}$

$\qquad = \frac{1}{16}\left(\frac{\pi}{4} - 0 - \frac{1}{3}\right) = \frac{1}{192}(3\pi - 4)$

13. $\displaystyle\int \cos^2 x \, \tan^3 x \, dx = \int \frac{\sin^3 x}{\cos x}\, dx \overset{c}{=} \int \frac{(1 - u^2)\,(-du)}{u} = \int \left[\frac{-1}{u} + u\right] du$

$\qquad = -\ln|u| + \frac{1}{2}u^2 + C = \frac{1}{2}\cos^2 x - \ln|\cos x| + C$

15. $\displaystyle\int \frac{1 - \sin x}{\cos x}\, dx = \int (\sec x - \tan x)\, dx = \ln|\sec x + \tan x| - \ln|\sec x| + C \qquad \begin{bmatrix} \text{by (1) and the boxed} \\ \text{formula above it} \end{bmatrix}$

$\qquad = \ln|(\sec x + \tan x)\cos x| + C = \ln|1 + \sin x| + C$

$\qquad = \ln(1 + \sin x) + C \quad \text{since } 1 + \sin x \geq 0$

$\textit{Or:} \displaystyle\int \frac{1 - \sin x}{\cos x}\, dx = \int \frac{1 - \sin x}{\cos x} \cdot \frac{1 + \sin x}{1 + \sin x}\, dx = \int \frac{(1 - \sin^2 x)dx}{\cos x\,(1 + \sin x)} = \int \frac{\cos x \, dx}{1 + \sin x}$

$\qquad = \int \frac{dw}{w} \qquad [\text{where } w = 1 + \sin x, \, dw = \cos x \, dx]$

$\qquad = \ln|w| + C = \ln|1 + \sin x| + C = \ln(1 + \sin x) + C$

17. Let $u = \tan x$, $du = \sec^2 x \, dx$. Then $\int \sec^2 x \, \tan x \, dx = \int u \, du = \frac{1}{2}u^2 + C = \frac{1}{2}\tan^2 x + C$.

$\textit{Or:}$ Let $v = \sec x$, $dv = \sec x \tan x \, dx$. Then $\int \sec^2 x \, \tan x \, dx = \int v \, dv = \frac{1}{2}v^2 + C = \frac{1}{2}\sec^2 x + C$.

19. $\int \tan^2 x \, dx = \int (\sec^2 x - 1)\, dx = \tan x - x + C$

21. $\int \sec^6 t \, dt = \int \sec^4 t \cdot \sec^2 t \, dt = \int (\tan^2 t + 1)^2 \sec^2 t \, dt = \int (u^2 + 1)^2 \, du \qquad [u = \tan t, \, du = \sec^2 t \, dt]$

$\qquad = \int (u^4 + 2u^2 + 1)\, du = \frac{1}{5}u^5 + \frac{2}{3}u^3 + u + C = \frac{1}{5}\tan^5 t + \frac{2}{3}\tan^3 t + \tan t + C$

23. $\int_0^{\pi/3} \tan^5 x \sec^4 x \, dx = \int_0^{\pi/3} \tan^5 x \,(\tan^2 x + 1)\sec^2 x \, dx = \int_0^{\sqrt{3}} u^5(u^2 + 1)\, du \qquad [u = \tan x, \, du = \sec^2 x \, dx]$

$\qquad = \int_0^{\sqrt{3}}(u^7 + u^5)\, du = \left[\frac{1}{8}u^8 + \frac{1}{6}u^6\right]_0^{\sqrt{3}} = \frac{81}{8} + \frac{27}{6} = \frac{81}{8} + \frac{9}{2} = \frac{81}{8} + \frac{36}{8} = \frac{117}{8}$

Alternate solution:

$\int_0^{\pi/3}\tan^5 x \sec^4 x \, dx = \int_0^{\pi/3}\tan^4 x \sec^3 x \sec x \tan x \, dx = \int_0^{\pi/3}(\sec^2 x - 1)^2 \sec^3 x \sec x \tan x \, dx$

$\qquad = \int_1^2 (u^2 - 1)^2 u^3 \, du \quad [u = \sec x, \, du = \sec x \tan x \, dx] \quad = \int_1^2(u^4 - 2u^2 + 1)u^3 \, du$

$\qquad = \int_1^2(u^7 - 2u^5 + u^3)\, du = \left[\frac{1}{8}u^8 - \frac{1}{3}u^6 + \frac{1}{4}u^4\right]_1^2 = \left(32 - \frac{64}{3} + 4\right) - \left(\frac{1}{8} - \frac{1}{3} + \frac{1}{4}\right) = \frac{117}{8}$

25. $\int \tan^3 x \sec x \, dx = \int \tan^2 x \sec x \tan x \, dx = \int (\sec^2 x - 1)\sec x \tan x \, dx$

$\qquad = \int (u^2 - 1)\, du \qquad [u = \sec x, \, du = \sec x \tan x \, dx]$

$\qquad = \frac{1}{3}u^3 - u + C = \frac{1}{3}\sec^3 x - \sec x + C$

27. $\int \tan^5 x \, dx = \int (\sec^2 x - 1)^2 \tan x \, dx = \int \sec^4 x \tan x \, dx - 2 \int \sec^2 x \tan x \, dx + \int \tan x \, dx$

$\qquad = \int \sec^3 x \sec x \tan x \, dx - 2 \int \tan x \sec^2 x \, dx + \int \tan x \, dx$

$\qquad = \frac{1}{4}\sec^4 x - \tan^2 x + \ln|\sec x| + C \qquad [\text{or } \frac{1}{4}\sec^4 x - \sec^2 x + \ln|\sec x| + C]$

29. $\int_{\pi/6}^{\pi/2} \cot^2 x \, dx = \int_{\pi/6}^{\pi/2} (\csc^2 x - 1) \, dx = \left[-\cot x - x \right]_{\pi/6}^{\pi/2} = \left(0 - \frac{\pi}{2} \right) - \left(-\sqrt{3} - \frac{\pi}{6} \right) = \sqrt{3} - \frac{\pi}{3}$

31. $\int \cot^3 \alpha \, \csc^3 \alpha \, d\alpha = \int \cot^2 \alpha \, \csc^2 \alpha \cdot \csc \alpha \, \cot \alpha \, d\alpha = \int (\csc^2 \alpha - 1) \csc^2 \alpha \cdot \csc \alpha \, \cot \alpha \, d\alpha$

$$= \int (u^2 - 1) u^2 \cdot (-du) \qquad [u = \csc \alpha, \, du = -\csc \alpha \, \cot \alpha \, d\alpha]$$

$$= \int (u^2 - u^4) \, du = \tfrac{1}{3} u^3 - \tfrac{1}{5} u^5 + C = \tfrac{1}{3} \csc^3 \alpha - \tfrac{1}{5} \csc^5 \alpha + C$$

33. $I = \int \csc x \, dx = \int \dfrac{\csc x \, (\csc x - \cot x)}{\csc x - \cot x} \, dx = \int \dfrac{-\csc x \, \cot x + \csc^2 x}{\csc x - \cot x} \, dx.$ Let $u = \csc x - \cot x \;\Rightarrow\;$

$du = (-\csc x \, \cot x + \csc^2 x) \, dx.$ Then $I = \int du/u = \ln |u| = \ln |\csc x - \cot x| + C.$

35. (a) $\tfrac{1}{2} [\cos(A - B) - \cos(A + B)] = \tfrac{1}{2} [(\cos A \cos B + \sin A \sin B) - (\cos A \cos B - \sin A \sin B)]$

$$= \tfrac{1}{2} (2 \sin A \sin B) = \sin A \sin B$$

(b) Use Equation 2(b): $\qquad \int \sin 5x \sin 2x \, dx = \int \tfrac{1}{2} [\cos(5x - 2x) - \cos(5x + 2x)] \, dx$

$$= \tfrac{1}{2} \int (\cos 3x - \cos 7x) \, dx = \tfrac{1}{6} \sin 3x - \tfrac{1}{14} \sin 7x + C$$

37. Let $x = 3 \sec \theta$, where $0 \le \theta < \frac{\pi}{2}$ or $\pi \le \theta < \frac{3\pi}{2}$.

Then $dx = 3 \sec \theta \, \tan \theta \, d\theta$ and

$$\sqrt{x^2 - 9} = \sqrt{9 \sec^2 \theta - 9} = \sqrt{9(\sec^2 \theta - 1)} = \sqrt{9 \tan^2 \theta}$$

$$= 3 |\tan \theta| = 3 \tan \theta \text{ for the relevant values of } \theta.$$

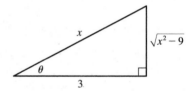

$$\int \dfrac{1}{x^2 \sqrt{x^2 - 9}} \, dx = \int \dfrac{1}{9 \sec^2 \theta \cdot 3 \tan \theta} \, 3 \sec \theta \, \tan \theta \, d\theta = \dfrac{1}{9} \int \cos \theta \, d\theta = \tfrac{1}{9} \sin \theta + C = \dfrac{1}{9} \dfrac{\sqrt{x^2 - 9}}{x} + C$$

Note that $-\sec(\theta + \pi) = \sec \theta$, so the figure is sufficient for the case $\pi \le \theta < \frac{3\pi}{2}$.

39. Let $x = 3 \tan \theta$, where $-\frac{\pi}{2} < \theta < \frac{\pi}{2}$. Then $dx = 3 \sec^2 \theta \, d\theta$ and

$$\sqrt{x^2 + 9} = \sqrt{9 \tan^2 \theta + 9} = \sqrt{9(\tan^2 \theta + 1)} = \sqrt{9 \sec^2 \theta}$$

$$= 3 |\sec \theta| = 3 \sec \theta \text{ for the relevant values of } \theta.$$

$$\int \dfrac{x^3}{\sqrt{x^2 + 9}} \, dx = \int \dfrac{3^3 \tan^3 \theta}{3 \sec \theta} \, 3 \sec^2 \theta \, d\theta = 3^3 \int \tan^3 \theta \, \sec \theta \, d\theta = 3^3 \int \tan^2 \theta \, \tan \theta \, \sec \theta \, d\theta$$

$$= 3^3 \int (\sec^2 \theta - 1) \tan \theta \, \sec \theta \, d\theta = 3^3 \int (u^2 - 1) du \qquad [u = \sec \theta, \, du = \sec \theta \, \tan \theta \, d\theta]$$

$$= 3^3 \left(\tfrac{1}{3} u^3 - u \right) + C = 3^3 \left(\tfrac{1}{3} \sec^3 \theta - \sec \theta \right) + C = 3^3 \left[\dfrac{1}{3} \dfrac{(x^2 + 9)^{3/2}}{3^3} - \dfrac{\sqrt{x^2 + 9}}{3} \right] + C$$

$$= \tfrac{1}{3} (x^2 + 9)^{3/2} - 9 \sqrt{x^2 + 9} + C \text{ or } \tfrac{1}{3} (x^2 - 18) \sqrt{x^2 + 9} + C$$

41. Let $t = \sec\theta$, so $dt = \sec\theta\tan\theta\,d\theta$, $t = \sqrt{2} \;\Rightarrow\; \theta = \frac{\pi}{4}$, and $t = 2 \;\Rightarrow\; \theta = \frac{\pi}{3}$. Then

$$\int_{\sqrt{2}}^{2} \frac{1}{t^3\sqrt{t^2-1}}\,dt = \int_{\pi/4}^{\pi/3} \frac{1}{\sec^3\theta\tan\theta}\sec\theta\tan\theta\,d\theta = \int_{\pi/4}^{\pi/3} \frac{1}{\sec^2\theta}\,d\theta$$

$$= \int_{\pi/4}^{\pi/3} \cos^2\theta\,d\theta = \int_{\pi/4}^{\pi/3} \tfrac{1}{2}(1+\cos2\theta)\,d\theta = \tfrac{1}{2}\big[\theta + \tfrac{1}{2}\sin2\theta\big]_{\pi/4}^{\pi/3}$$

$$= \tfrac{1}{2}\Big[\Big(\tfrac{\pi}{3} + \tfrac{1}{2}\tfrac{\sqrt{3}}{2}\Big) - \Big(\tfrac{\pi}{4} + \tfrac{1}{2}\cdot1\Big)\Big] = \tfrac{1}{2}\Big(\tfrac{\pi}{12} + \tfrac{\sqrt{3}}{4} - \tfrac{1}{2}\Big) = \tfrac{\pi}{24} + \tfrac{\sqrt{3}}{8} - \tfrac{1}{4}$$

43. Let $x = 5\sin\theta$, so $dx = 5\cos\theta\,d\theta$. Then

$$\int \frac{1}{x^2\sqrt{25-x^2}}\,dx = \int \frac{1}{5^2\sin^2\theta\cdot5\cos\theta}5\cos\theta\,d\theta = \frac{1}{25}\int\csc^2\theta\,d\theta$$

$$= -\frac{1}{25}\cot\theta + C = -\frac{1}{25}\frac{\sqrt{25-x^2}}{x} + C$$

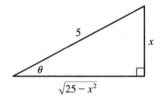

45. Let $x = 4\tan\theta$, where $-\frac{\pi}{2} < \theta < \frac{\pi}{2}$. Then $dx = 4\sec^2\theta\,d\theta$ and

$$\sqrt{x^2+16} = \sqrt{16\tan^2\theta+16} = \sqrt{16(\tan^2\theta+1)} = \sqrt{16\sec^2\theta}$$

$$= 4\,|\sec\theta| = 4\sec\theta \quad\text{for the relevant values of }\theta.$$

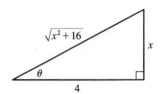

$$\int \frac{dx}{\sqrt{x^2+16}} = \int \frac{4\sec^2\theta\,d\theta}{4\sec\theta} = \int\sec\theta\,d\theta = \ln|\sec\theta+\tan\theta| + C_1 = \ln\left|\frac{\sqrt{x^2+16}}{4} + \frac{x}{4}\right| + C_1$$

$$= \ln\left|\sqrt{x^2+16}+x\right| - \ln|4| + C_1 = \ln\left(\sqrt{x^2+16}+x\right) + C, \text{ where } C = C_1 - \ln4.$$

(Since $\sqrt{x^2+16}+x > 0$, we don't need the absolute value.)

47. Let $2x = \sin\theta$, where $-\frac{\pi}{2} \le \theta \le \frac{\pi}{2}$.

Then $x = \frac{1}{2}\sin\theta$, $dx = \frac{1}{2}\cos\theta\,d\theta$, and $\sqrt{1-4x^2} = \sqrt{1-(2x)^2} = \cos\theta$.

$$\int\sqrt{1-4x^2}\,dx = \int\cos\theta\left(\tfrac{1}{2}\cos\theta\right)d\theta = \tfrac{1}{4}\int(1+\cos2\theta)\,d\theta$$

$$= \tfrac{1}{4}\left(\theta + \tfrac{1}{2}\sin2\theta\right) + C = \tfrac{1}{4}(\theta + \sin\theta\,\cos\theta) + C$$

$$= \tfrac{1}{4}\left[\sin^{-1}(2x) + 2x\sqrt{1-4x^2}\right] + C$$

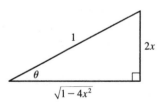

49. Let $x = 3\sec\theta$, where $0 \le \theta < \frac{\pi}{2}$ or $\pi \le \theta < \frac{3\pi}{2}$.

Then $dx = 3\sec\theta\tan\theta\,d\theta$ and $\sqrt{x^2-9} = 3\tan\theta$, so

$$\int \frac{\sqrt{x^2-9}}{x^3}\,dx = \int \frac{3\tan\theta}{27\sec^3\theta}3\sec\theta\tan\theta\,d\theta = \frac{1}{3}\int \frac{\tan^2\theta}{\sec^2\theta}\,d\theta$$

$$= \tfrac{1}{3}\int\sin^2\theta\,d\theta = \tfrac{1}{3}\int\tfrac{1}{2}(1-\cos2\theta)\,d\theta = \tfrac{1}{6}\theta - \tfrac{1}{12}\sin2\theta + C = \tfrac{1}{6}\theta - \tfrac{1}{6}\sin\theta\,\cos\theta + C$$

$$= \frac{1}{6}\sec^{-1}\left(\frac{x}{3}\right) - \frac{1}{6}\frac{\sqrt{x^2-9}}{x}\frac{3}{x} + C = \frac{1}{6}\sec^{-1}\left(\frac{x}{3}\right) - \frac{\sqrt{x^2-9}}{2x^2} + C$$

51. Let $x = a\sin\theta$, where $-\frac{\pi}{2} \le \theta \le \frac{\pi}{2}$. Then $dx = a\cos\theta\,d\theta$ and

$$\int \frac{x^2\,dx}{(a^2 - x^2)^{3/2}} = \int \frac{a^2\sin^2\theta\,a\cos\theta\,d\theta}{a^3\cos^3\theta} = \int \tan^2\theta\,d\theta = \int (\sec^2\theta - 1)\,d\theta$$

$$= \tan\theta - \theta + C = \frac{x}{\sqrt{a^2 - x^2}} - \sin^{-1}\left(\frac{x}{a}\right) + C$$

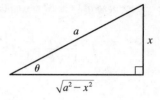

53. Let $u = x^2 - 7$, so $du = 2x\,dx$. Then $\int \frac{x}{\sqrt{x^2 - 7}}\,dx = \frac{1}{2}\int \frac{1}{\sqrt{u}}\,du = \frac{1}{2}\cdot 2\sqrt{u} + C = \sqrt{x^2 - 7} + C$.

55. Let $x = \tan\theta$, where $-\frac{\pi}{2} < \theta < \frac{\pi}{2}$. Then $dx = \sec^2\theta\,d\theta$ and $\sqrt{1 + x^2} = \sec\theta$, so

$$\int \frac{\sqrt{1 + x^2}}{x}\,dx = \int \frac{\sec\theta}{\tan\theta}\sec^2\theta\,d\theta = \int \frac{\sec\theta}{\tan\theta}(1 + \tan^2\theta)\,d\theta$$

$$= \int (\csc\theta + \sec\theta\tan\theta)\,d\theta$$

$$= \ln|\csc\theta - \cot\theta| + \sec\theta + C \qquad \text{[by Exercise 33]}$$

$$= \ln\left|\frac{\sqrt{1 + x^2}}{x} - \frac{1}{x}\right| + \frac{\sqrt{1 + x^2}}{1} + C = \ln\left|\frac{\sqrt{1 + x^2} - 1}{x}\right| + \sqrt{1 + x^2} + C$$

57. Let $u = x^2$, $du = 2x\,dx$. Then

$$\int x\sqrt{1 - x^4}\,dx = \int \sqrt{1 - u^2}\left(\tfrac{1}{2}\,du\right) = \tfrac{1}{2}\int \cos\theta\cdot\cos\theta\,d\theta \qquad \begin{bmatrix} \text{where } u = \sin\theta,\, du = \cos\theta\,d\theta \\ \text{and } \sqrt{1 - u^2} = \cos\theta \end{bmatrix}$$

$$= \tfrac{1}{2}\int \tfrac{1}{2}(1 + \cos 2\theta)\,d\theta = \tfrac{1}{4}\theta + \tfrac{1}{8}\sin 2\theta + C = \tfrac{1}{4}\theta + \tfrac{1}{4}\sin\theta\cos\theta + C$$

$$= \tfrac{1}{4}\sin^{-1}u + \tfrac{1}{4}u\sqrt{1 - u^2} + C = \tfrac{1}{4}\sin^{-1}(x^2) + \tfrac{1}{4}x^2\sqrt{1 - x^4} + C$$

59. $9x^2 + 6x - 8 = (3x + 1)^2 - 9$, so let $u = 3x + 1$, $du = 3dx$. Then $\int \frac{dx}{\sqrt{9x^2 + 6x - 8}} = \int \frac{\frac{1}{3}\,du}{\sqrt{u^2 - 9}}$.

Now let $u = 3\sec\theta$, where $0 \le \theta < \frac{\pi}{2}$ or $\pi \le \theta < \frac{3\pi}{2}$. Then $du = 3\sec\theta\tan\theta\,d\theta$ and $\sqrt{u^2 - 9} = 3\tan\theta$, so

$$\int \frac{\frac{1}{3}\,du}{\sqrt{u^2 - 9}} = \int \frac{\sec\theta\tan\theta\,d\theta}{3\tan\theta} = \tfrac{1}{3}\int \sec\theta\,d\theta = \tfrac{1}{3}\ln|\sec\theta + \tan\theta| + C_1 = \tfrac{1}{3}\ln\left|\frac{u + \sqrt{u^2 - 9}}{3}\right| + C_1$$

$$= \tfrac{1}{3}\ln|u + \sqrt{u^2 - 9}| + C = \tfrac{1}{3}\ln|3x + 1 + \sqrt{9x^2 + 6x - 8}| + C$$

61. $x^2 + 2x + 2 = (x + 1)^2 + 1$. Let $u = x + 1$, $du = dx$. Then

$$\int \frac{dx}{(x^2 + 2x + 2)^2} = \int \frac{du}{(u^2 + 1)^2} = \int \frac{\sec^2\theta\,d\theta}{\sec^4\theta} \qquad \begin{bmatrix} \text{where } u = \tan\theta,\, du = \sec^2\theta\,d\theta, \\ \text{and } u^2 + 1 = \sec^2\theta \end{bmatrix}$$

$$= \int \cos^2\theta\,d\theta = \tfrac{1}{2}\int (1 + \cos 2\theta)\,d\theta = \tfrac{1}{2}(\theta + \sin\theta\cos\theta) + C$$

$$= \frac{1}{2}\left[\tan^{-1}u + \frac{u}{1 + u^2}\right] + C = \frac{1}{2}\left[\tan^{-1}(x + 1) + \frac{x + 1}{x^2 + 2x + 2}\right] + C$$

63. $s = f(t) = \int_0^t \sin \omega u \cos^2 \omega u\, du$. Let $y = \cos \omega u \;\Rightarrow\; dy = -\omega \sin \omega u\, du$. Then

$$s = -\frac{1}{\omega} \int_1^{\cos \omega t} y^2\, dy = -\frac{1}{\omega}\left[\frac{1}{3}y^3\right]_1^{\cos \omega t} = \frac{1}{3\omega}(1 - \cos^3 \omega t).$$

65. The average value of $f(x) = \sqrt{x^2 - 1}/x$ on the interval $[1, 7]$ is

$$\frac{1}{7-1} \int_1^7 \frac{\sqrt{x^2 - 1}}{x}\, dx = \frac{1}{6} \int_0^\alpha \frac{\tan \theta}{\sec \theta} \cdot \sec \theta\, \tan \theta\, d\theta \qquad \left[\begin{array}{l}\text{where } x = \sec \theta,\, dx = \sec \theta\, \tan \theta\, d\theta, \\ \sqrt{x^2 - 1} = \tan \theta,\, \text{and } \alpha = \sec^{-1} 7\end{array}\right]$$

$$= \frac{1}{6} \int_0^\alpha \tan^2 \theta\, d\theta = \frac{1}{6} \int_0^\alpha (\sec^2 \theta - 1)\, d\theta = \frac{1}{6}\Big[\tan \theta - \theta\Big]_0^\alpha$$

$$= \frac{1}{6}(\tan \alpha - \alpha) = \frac{1}{6}\left(\sqrt{48} - \sec^{-1} 7\right)$$

67. Area of $\triangle POQ = \frac{1}{2}(r \cos \theta)(r \sin \theta) = \frac{1}{2} r^2 \sin \theta \cos \theta$. Area of region $PQR = \int_{r \cos \theta}^r \sqrt{r^2 - x^2}\, dx$. Let $x = r \cos u \;\Rightarrow$

$dx = -r \sin u\, du$ for $\theta \le u \le \frac{\pi}{2}$. Then we obtain

$$\int \sqrt{r^2 - x^2}\, dx = \int r \sin u\, (-r \sin u)\, du = -r^2 \int \sin^2 u\, du = -\frac{1}{2} r^2 (u - \sin u \cos u) + C$$

$$= -\frac{1}{2} r^2 \cos^{-1}(x/r) + \frac{1}{2} x \sqrt{r^2 - x^2} + C$$

so $\qquad$ area of region $PQR = \frac{1}{2}\Big[-r^2 \cos^{-1}(x/r) + x\sqrt{r^2 - x^2}\Big]_{r \cos \theta}^r$

$$= \frac{1}{2}\Big[0 - \left(-r^2 \theta + r \cos \theta\, r \sin \theta\right)\Big] = \frac{1}{2} r^2 \theta - \frac{1}{2} r^2 \sin \theta \cos \theta$$

and thus, (area of sector POR) = (area of $\triangle POQ$) + (area of region PQR) = $\frac{1}{2} r^2 \theta$.

6.3   Partial Fractions

1. (a) $\dfrac{2x}{(x + 3)(3x + 1)} = \dfrac{A}{x + 3} + \dfrac{B}{3x + 1}$

(b) $\dfrac{1}{x^3 + 2x^2 + x} = \dfrac{1}{x(x^2 + 2x + 1)} = \dfrac{1}{x(x + 1)^2} = \dfrac{A}{x} + \dfrac{B}{x + 1} + \dfrac{C}{(x + 1)^2}$

3. (a) $\dfrac{2}{x^2 + 3x - 4} = \dfrac{2}{(x + 4)(x - 1)} = \dfrac{A}{x + 4} + \dfrac{B}{x - 1}$

(b) $x^2 + x + 1$ is irreducible, so $\dfrac{x^2}{(x - 1)(x^2 + x + 1)} = \dfrac{A}{x - 1} + \dfrac{Bx + C}{x^2 + x + 1}$.

5. (a) $\dfrac{x^4}{x^4 - 1} = \dfrac{(x^4 - 1) + 1}{x^4 - 1} = 1 + \dfrac{1}{x^4 - 1}$ [or use long division] $= 1 + \dfrac{1}{(x^2 - 1)(x^2 + 1)}$

$$= 1 + \dfrac{1}{(x - 1)(x + 1)(x^2 + 1)} = 1 + \dfrac{A}{x - 1} + \dfrac{B}{x + 1} + \dfrac{Cx + D}{x^2 + 1}$$

(b) $\dfrac{t^4 + t^2 + 1}{(t^2 + 1)(t^2 + 4)^2} = \dfrac{At + B}{t^2 + 1} + \dfrac{Ct + D}{t^2 + 4} + \dfrac{Et + F}{(t^2 + 4)^2}$

7. $\displaystyle\int \dfrac{x}{x - 6}\, dx = \int \dfrac{(x - 6) + 6}{x - 6}\, dx = \int \left(1 + \dfrac{6}{x - 6}\right) dx = x + 6 \ln|x - 6| + C$

9. $\dfrac{x-9}{(x+5)(x-2)} = \dfrac{A}{x+5} + \dfrac{B}{x-2}$. Multiply both sides by $(x+5)(x-2)$ to get $x-9 = A(x-2) + B(x+5)$ $(\star)$, or

equivalently, $x-9 = (A+B)x - 2A + 5B$. Equating coefficients of x on each side of the equation gives us $1 = A+B$ **(1)**

and equating constants gives us $-9 = -2A + 5B$ **(2)**. Adding two times **(1)** to **(2)** gives us $-7 = 7B$ $\Leftrightarrow$ $B = -1$ and

hence, $A = 2$. [Alternatively, to find the coefficients A and B, we may use substitution as follows: substitute 2 for x in $(\star)$

to get $-7 = 7B$ $\Leftrightarrow$ $B = -1$, then substitute -5 for x in $(\star)$ to get $-14 = -7A$ $\Leftrightarrow$ $A = 2$.]

Thus, $\displaystyle\int \dfrac{x-9}{(x+5)(x-2)}\,dx = \int\left(\dfrac{2}{x+5} + \dfrac{-1}{x-2}\right)dx = 2\ln|x+5| - \ln|x-2| + C$.

To find the constants in problems involving partial fractions, we may use the coefficient comparison method or the substitution method (as in the solution
for Exercise 9) or a combination of both methods.

11. $\dfrac{1}{x^2-1} = \dfrac{1}{(x+1)(x-1)} = \dfrac{A}{x+1} + \dfrac{B}{x-1}$. Multiply both sides by $(x+1)(x-1)$ to get $1 = A(x-1) + B(x+1)$.

Substituting 1 for x gives $1 = 2B$ $\Leftrightarrow$ $B = \frac{1}{2}$. Substituting -1 for x gives $1 = -2A$ $\Leftrightarrow$ $A = -\frac{1}{2}$. Thus,

$$\int_2^3 \dfrac{1}{x^2-1}\,dx = \int_2^3\left(\dfrac{-1/2}{x+1} + \dfrac{1/2}{x-1}\right)dx = \left[-\tfrac{1}{2}\ln|x+1| + \tfrac{1}{2}\ln|x-1|\right]_2^3$$

$$= \left(-\tfrac{1}{2}\ln 4 + \tfrac{1}{2}\ln 2\right) - \left(-\tfrac{1}{2}\ln 3 + \tfrac{1}{2}\ln 1\right) = \tfrac{1}{2}(\ln 2 + \ln 3 - \ln 4) \quad \left[\text{or } \tfrac{1}{2}\ln\tfrac{3}{2}\right]$$

13. $\displaystyle\int \dfrac{ax}{x^2-bx}\,dx = \int \dfrac{ax}{x(x-b)}\,dx = \int \dfrac{a}{x-b}\,dx = a\ln|x-b| + C$

15. $\dfrac{2x+3}{(x+1)^2} = \dfrac{A}{x+1} + \dfrac{B}{(x+1)^2}$ $\Rightarrow$ $2x+3 = A(x+1) + B$. Take $x = -1$ to get $B = 1$, and equate coefficients

of x to get $A = 2$. Now

$$\int_0^1 \dfrac{2x+3}{(x+1)^2}\,dx = \int_0^1\left[\dfrac{2}{x+1} + \dfrac{1}{(x+1)^2}\right]dx = \left[2\ln(x+1) - \dfrac{1}{x+1}\right]_0^1$$

$$= 2\ln 2 - \tfrac{1}{2} - (2\ln 1 - 1) = 2\ln 2 + \tfrac{1}{2}$$

17. $\dfrac{4y^2-7y-12}{y(y+2)(y-3)} = \dfrac{A}{y} + \dfrac{B}{y+2} + \dfrac{C}{y-3}$ $\Rightarrow$ $4y^2-7y-12 = A(y+2)(y-3) + By(y-3) + Cy(y+2)$. Setting

$y = 0$ gives $-12 = -6A$, so $A = 2$. Setting $y = -2$ gives $18 = 10B$, so $B = \frac{9}{5}$. Setting $y = 3$ gives $3 = 15C$, so $C = \frac{1}{5}$.

Now

$$\int_1^2 \dfrac{4y^2-7y-12}{y(y+2)(y-3)}\,dy = \int_1^2\left(\dfrac{2}{y} + \dfrac{9/5}{y+2} + \dfrac{1/5}{y-3}\right)dy = \left[2\ln|y| + \tfrac{9}{5}\ln|y+2| + \tfrac{1}{5}\ln|y-3|\right]_1^2$$

$$= 2\ln 2 + \tfrac{9}{5}\ln 4 + \tfrac{1}{5}\ln 1 - 2\ln 1 - \tfrac{9}{5}\ln 3 - \tfrac{1}{5}\ln 2$$

$$= 2\ln 2 + \tfrac{18}{5}\ln 2 - \tfrac{1}{5}\ln 2 - \tfrac{9}{5}\ln 3 = \tfrac{27}{5}\ln 2 - \tfrac{9}{5}\ln 3 = \tfrac{9}{5}(3\ln 2 - \ln 3) = \tfrac{9}{5}\ln\tfrac{8}{3}$$

19. $\dfrac{1}{(x+5)^2(x-1)} = \dfrac{A}{x+5} + \dfrac{B}{(x+5)^2} + \dfrac{C}{x-1}$ $\Rightarrow$ $1 = A(x+5)(x-1) + B(x-1) + C(x+5)^2$.

Setting $x = -5$ gives $1 = -6B$, so $B = -\frac{1}{6}$. Setting $x = 1$ gives $1 = 36C$, so $C = \frac{1}{36}$. Setting $x = -2$ gives

$1 = A(3)(-3) + B(-3) + C(3^2) = -9A - 3B + 9C = -9A + \frac{1}{2} + \frac{1}{4} = -9A + \frac{3}{4}$, so $9A = -\frac{1}{4}$ and $A = -\frac{1}{36}$.

Now $\displaystyle\int \frac{1}{(x+5)^2(x-1)}\,dx = \int \left[\frac{-1/36}{x+5} - \frac{1/6}{(x+5)^2} + \frac{1/36}{x-1}\right]dx = -\frac{1}{36}\ln|x+5| + \frac{1}{6(x+5)} + \frac{1}{36}\ln|x-1| + C.$

21. $\dfrac{5x^2 + 3x - 2}{x^3 + 2x^2} = \dfrac{5x^2 + 3x - 2}{x^2(x+2)} = \dfrac{A}{x} + \dfrac{B}{x^2} + \dfrac{C}{x+2}.$ Multiply by $x^2(x+2)$ to get

$5x^2 + 3x - 2 = Ax(x+2) + B(x+2) + Cx^2.$ Set $x = -2$ to get $C = 3$, and take $x = 0$ to get $B = -1.$

Equating the coefficients of x^2 gives $5 = A + C \;\Rightarrow\; A = 2.$

So $\displaystyle\int \frac{5x^2 + 3x - 2}{x^3 + 2x^2}\,dx = \int\left(\frac{2}{x} - \frac{1}{x^2} + \frac{3}{x+2}\right)dx = 2\ln|x| + \frac{1}{x} + 3\ln|x+2| + C.$

23. $\dfrac{10}{(x-1)(x^2+9)} = \dfrac{A}{x-1} + \dfrac{Bx+C}{x^2+9}.$ Multiply both sides by $(x-1)(x^2+9)$ to get

$10 = A(x^2 + 9) + (Bx + C)(x - 1)$ $(*)$. Substituting 1 for x gives $10 = 10A \;\Leftrightarrow\; A = 1.$ Substituting 0 for x

gives $10 = 9A - C \;\Rightarrow\; C = 9(1) - 10 = -1.$ The coefficients of the x^2-terms in $(*)$ must be equal,

so $0 = A + B \;\Rightarrow\; B = -1.$ Thus,

$$\int \frac{10}{(x-1)(x^2+9)}\,dx = \int\left(\frac{1}{x-1} + \frac{-x-1}{x^2+9}\right)dx = \int\left(\frac{1}{x-1} - \frac{x}{x^2+9} - \frac{1}{x^2+9}\right)dx$$

$$= \ln|x-1| - \tfrac{1}{2}\ln(x^2 + 9) \;[\text{let } u = x^2 + 9] \;-\; \tfrac{1}{3}\tan^{-1}\left(\tfrac{x}{3}\right) \;[\text{Formula 10}] \;+\; C$$

25. $\dfrac{x^3 + x^2 + 2x + 1}{(x^2+1)(x^2+2)} = \dfrac{Ax+B}{x^2+1} + \dfrac{Cx+D}{x^2+2}.$ Multiply both sides by $(x^2+1)(x^2+2)$ to get

$x^3 + x^2 + 2x + 1 = (Ax + B)(x^2 + 2) + (Cx + D)(x^2 + 1) \;\Leftrightarrow\;$

$x^3 + x^2 + 2x + 1 = (Ax^3 + Bx^2 + 2Ax + 2B) + (Cx^3 + Dx^2 + Cx + D) \;\Leftrightarrow\;$

$x^3 + x^2 + 2x + 1 = (A+C)x^3 + (B+D)x^2 + (2A+C)x + (2B+D).$ Comparing coefficients gives us

the following system of equations:

$$A + C = 1 \quad \textbf{(1)} \qquad\qquad B + D = 1 \quad \textbf{(2)}$$
$$2A + C = 2 \quad \textbf{(3)} \qquad\qquad 2B + D = 1 \quad \textbf{(4)}$$

Subtracting equation **(1)** from equation **(3)** gives us $A = 1$, so $C = 0.$ Subtracting equation **(2)** from equation **(4)** gives us

$B = 0$, so $D = 1.$ Thus, $I = \displaystyle\int \frac{x^3 + x^2 + 2x + 1}{(x^2+1)(x^2+2)}\,dx = \int\left(\frac{x}{x^2+1} + \frac{1}{x^2+2}\right)dx.$ For $\displaystyle\int \frac{x}{x^2+1}\,dx$, let $u = x^2 + 1$

so $du = 2x\,dx$ and then $\displaystyle\int \frac{x}{x^2+1}\,dx = \frac{1}{2}\int\frac{1}{u}\,du = \frac{1}{2}\ln|u| + C = \frac{1}{2}\ln(x^2+1) + C.$ For $\displaystyle\int \frac{1}{x^2+2}\,dx$, use

Formula 10 with $a = \sqrt{2}.$ So $\displaystyle\int \frac{1}{x^2+2}\,dx = \int \frac{1}{x^2 + (\sqrt{2})^2}\,dx = \frac{1}{\sqrt{2}}\tan^{-1}\frac{x}{\sqrt{2}} + C.$

Thus, $I = \dfrac{1}{2}\ln(x^2 + 1) + \dfrac{1}{\sqrt{2}}\tan^{-1}\dfrac{x}{\sqrt{2}} + C.$

27. $\displaystyle\int \frac{x+4}{x^2+2x+5}\,dx = \int \frac{x+1}{x^2+2x+5}\,dx + \int \frac{3}{x^2+2x+5}\,dx = \frac{1}{2}\int \frac{(2x+2)\,dx}{x^2+2x+5} + \int \frac{3\,dx}{(x+1)^2+4}$

$\displaystyle = \frac{1}{2}\ln\left|x^2+2x+5\right| + 3\int \frac{2\,du}{4(u^2+1)}$ [where $x+1 = 2u$ and $dx = 2\,du$]

$\displaystyle = \frac{1}{2}\ln(x^2+2x+5) + \frac{3}{2}\tan^{-1}u + C = \frac{1}{2}\ln(x^2+2x+5) + \frac{3}{2}\tan^{-1}\left(\frac{x+1}{2}\right) + C$

29. $\displaystyle\frac{1}{x^3-1} = \frac{1}{(x-1)(x^2+x+1)} = \frac{A}{x-1} + \frac{Bx+C}{x^2+x+1}$ $\Rightarrow$ $1 = A(x^2+x+1) + (Bx+C)(x-1)$.

Take $x = 1$ to get $A = \frac{1}{3}$. Equating coefficients of x^2 and then comparing the constant terms, we get

$0 = \frac{1}{3} + B, 1 = \frac{1}{3} - C$, so $B = -\frac{1}{3}, C = -\frac{2}{3}$ $\Rightarrow$

$\displaystyle\int \frac{1}{x^3-1}\,dx = \int \frac{\frac{1}{3}}{x-1}\,dx + \int \frac{-\frac{1}{3}x - \frac{2}{3}}{x^2+x+1}\,dx = \frac{1}{3}\ln|x-1| - \frac{1}{3}\int \frac{x+2}{x^2+x+1}\,dx$

$\displaystyle = \frac{1}{3}\ln|x-1| - \frac{1}{3}\int \frac{x+1/2}{x^2+x+1}\,dx - \frac{1}{3}\int \frac{(3/2)\,dx}{(x+1/2)^2+3/4}$

$\displaystyle = \frac{1}{3}\ln|x-1| - \frac{1}{6}\ln(x^2+x+1) - \frac{1}{2}\left(\frac{2}{\sqrt{3}}\right)\tan^{-1}\left(\frac{x+\frac{1}{2}}{\sqrt{3}/2}\right) + K$

$\displaystyle = \frac{1}{3}\ln|x-1| - \frac{1}{6}\ln(x^2+x+1) - \frac{1}{\sqrt{3}}\tan^{-1}\left(\frac{1}{\sqrt{3}}(2x+1)\right) + K$

31. $\displaystyle\frac{1}{x^4-x^2} = \frac{1}{x^2(x-1)(x+1)} = \frac{A}{x} + \frac{B}{x^2} + \frac{C}{x-1} + \frac{D}{x+1}$. Multiply by $x^2(x-1)(x+1)$ to get

$1 = Ax(x-1)(x+1) + B(x-1)(x+1) + Cx^2(x+1) + Dx^2(x-1)$. Setting $x = 1$ gives $C = \frac{1}{2}$,

taking $x = -1$ gives $D = -\frac{1}{2}$. Equating the coefficients of x^3 gives $0 = A + C + D = A$. Finally, setting $x = 0$

yields $B = -1$. Now $\displaystyle\int \frac{dx}{x^4-x^2} = \int\left[\frac{-1}{x^2} + \frac{1/2}{x-1} - \frac{1/2}{x+1}\right]dx = \frac{1}{x} + \frac{1}{2}\ln\left|\frac{x-1}{x+1}\right| + C.$

33. $\displaystyle\int \frac{x-3}{(x^2+2x+4)^2}\,dx = \int \frac{x-3}{(x^2+2x+4)^2}\,dx = \int \frac{x-3}{[(x+1)^2+3]^2}\,dx = \int \frac{u-4}{(u^2+3)^2}\,du$ [with $u = x+1$]

$\displaystyle = \int \frac{u\,du}{(u^2+3)^2} - 4\int \frac{du}{(u^2+3)^2} = \frac{1}{2}\int \frac{dv}{v^2} - 4\int \frac{\sqrt{3}\sec^2\theta\,d\theta}{9\sec^4\theta}$ $\begin{bmatrix} v = u^2+3 \text{ in the first integral;} \\ u = \sqrt{3}\tan\theta \text{ in the second} \end{bmatrix}$

$\displaystyle = \frac{-1}{(2v)} - \frac{4\sqrt{3}}{9}\int \cos^2\theta\,d\theta = \frac{-1}{2(u^2+3)} - \frac{2\sqrt{3}}{9}(\theta + \sin\theta\cos\theta) + C$

$\displaystyle = \frac{-1}{2(x^2+2x+4)} - \frac{2\sqrt{3}}{9}\left[\tan^{-1}\left(\frac{x+1}{\sqrt{3}}\right) + \frac{\sqrt{3}(x+1)}{x^2+2x+4}\right] + C$

$\displaystyle = \frac{-1}{2(x^2+2x+4)} - \frac{2\sqrt{3}}{9}\tan^{-1}\left(\frac{x+1}{\sqrt{3}}\right) - \frac{2(x+1)}{3(x^2+2x+4)} + C$

35. Let $u = \sqrt{x}$, so $u^2 = x$ and $dx = 2u\,du$. Thus,

$$\int_9^{16} \frac{\sqrt{x}}{x-4}\,dx = \int_3^4 \frac{u}{u^2-4}\,2u\,du = 2\int_3^4 \frac{u^2}{u^2-4}\,du = 2\int_3^4 \left(1 + \frac{4}{u^2-4}\right)du \qquad \text{[by long division]}$$

$$= 2 + 8\int_3^4 \frac{du}{(u+2)(u-2)} \quad (\star)$$

Multiply $\dfrac{1}{(u+2)(u-2)} = \dfrac{A}{u+2} + \dfrac{B}{u-2}$ by $(u+2)(u-2)$ to get $1 = A(u-2) + B(u+2)$. Equating coefficients we

get $A + B = 0$ and $-2A + 2B = 1$. Solving gives us $B = \frac{1}{4}$ and $A = -\frac{1}{4}$, so $\dfrac{1}{(u+2)(u-2)} = \dfrac{-1/4}{u+2} + \dfrac{1/4}{u-2}$ and $(\star)$ is

$$2 + 8\int_3^4 \left(\frac{-1/4}{u+2} + \frac{1/4}{u-2}\right)du = 2 + 8\left[-\tfrac{1}{4}\ln|u+2| + \tfrac{1}{4}\ln|u-2|\right]_3^4 = 2 + \left[2\ln|u-2| - 2\ln|u+2|\right]_3^4$$

$$= 2 + 2\left[\ln\left|\frac{u-2}{u+2}\right|\right]_3^4 = 2 + 2\left(\ln\tfrac{2}{6} - \ln\tfrac{1}{5}\right) = 2 + 2\ln\tfrac{2/6}{1/5}$$

$$= 2 + 2\ln\tfrac{5}{3} \ \text{ or } \ 2 + \ln\left(\tfrac{5}{3}\right)^2 = 2 + \ln\tfrac{25}{9}$$

37. Let $u = \sqrt[3]{x^2+1}$. Then $x^2 = u^3 - 1$, $2x\,dx = 3u^2\,du \ \Rightarrow$

$$\int \frac{x^3\,dx}{\sqrt[3]{x^2+1}} = \int \frac{(u^3-1)\frac{3}{2}u^2\,du}{u} = \frac{3}{2}\int (u^4 - u)\,du = \frac{3}{10}u^5 - \frac{3}{4}u^2 + C = \frac{3}{10}(x^2+1)^{5/3} - \frac{3}{4}(x^2+1)^{2/3} + C.$$

39. Let $u = e^x$. Then $x = \ln u$, $dx = \dfrac{du}{u} \ \Rightarrow$

$$\int \frac{e^{2x}\,dx}{e^{2x} + 3e^x + 2} = \int \frac{u^2\,(du/u)}{u^2 + 3u + 2} = \int \frac{u\,du}{(u+1)(u+2)} = \int \left[\frac{-1}{u+1} + \frac{2}{u+2}\right]du$$

$$= 2\ln|u+2| - \ln|u+1| + C = \ln\left[\frac{(e^x+2)^2}{e^x+1}\right] + C$$

41. Let $u = \ln(x^2 - x + 2)$, $dv = dx$. Then $du = \dfrac{2x-1}{x^2-x+2}\,dx$, $v = x$, and (by integration by parts)

$$\int \ln(x^2 - x + 2)\,dx = x\ln(x^2 - x + 2) - \int \frac{2x^2 - x}{x^2 - x + 2}\,dx = x\ln(x^2 - x + 2) - \int \left(2 + \frac{x-4}{x^2-x+2}\right)dx$$

$$= x\ln(x^2 - x + 2) - 2x - \int \frac{\frac{1}{2}(2x-1)}{x^2 - x + 2}\,dx + \frac{7}{2}\int \frac{dx}{(x-\frac{1}{2})^2 + \frac{7}{4}}$$

$$= x\ln(x^2 - x + 2) - 2x - \frac{1}{2}\ln(x^2 - x + 2) + \frac{7}{2}\int \frac{\frac{\sqrt{7}}{2}\,du}{\frac{7}{4}(u^2+1)} \qquad \left[\begin{array}{l} \text{where } x - \frac{1}{2} = \frac{\sqrt{7}}{2}u, \\ \quad dx = \frac{\sqrt{7}}{2}\,du, \\ (x-\frac{1}{2})^2 + \frac{7}{4} = \frac{7}{4}(u^2+1) \end{array}\right]$$

$$= (x - \tfrac{1}{2})\ln(x^2 - x + 2) - 2x + \sqrt{7}\tan^{-1}u + C$$

$$= (x - \tfrac{1}{2})\ln(x^2 - x + 2) - 2x + \sqrt{7}\tan^{-1}\left(\frac{2x-1}{\sqrt{7}}\right) + C$$

43. $\dfrac{P+S}{P\left[(r-1)P-S\right]} = \dfrac{A}{P} + \dfrac{B}{(r-1)P-S}$ $\Rightarrow$ $P+S = A\left[(r-1)P-S\right]+BP = \left[(r-1)A+B\right]P - AS$ $\Rightarrow$

$(r-1)A+B=1, -A=1$ $\Rightarrow$ $A=-1, B=r.$ Now

$$t = \int \frac{P+S}{P\left[(r-1)P-S\right]}\,dP = \int\left[\frac{-1}{P}+\frac{r}{(r-1)P-S}\right]dP = -\int\frac{dP}{P}+\frac{r}{r-1}\int\frac{r-1}{(r-1)P-S}\,dP$$

so $t = -\ln P + \dfrac{r}{r-1}\ln|(r-1)P-S| + C.$ Here $r=0.10$ and $S=900$, so

$t = -\ln P + \frac{0.1}{-0.9}\ln|-0.9P-900| + C = -\ln P - \frac{1}{9}\ln\big(|-1|\,|0.9P+900|\big) = -\ln P - \frac{1}{9}\ln(0.9P+900) + C.$

When $t=0$, $P=10{,}000$, so $0 = -\ln 10{,}000 - \frac{1}{9}\ln(9900) + C.$ Thus, $C = \ln 10{,}000 + \frac{1}{9}\ln 9900$ $[\approx 10.2326]$,

so our equation becomes

$$t = \ln 10{,}000 - \ln P + \tfrac{1}{9}\ln 9900 - \tfrac{1}{9}\ln(0.9P+900) = \ln\frac{10{,}000}{P} + \frac{1}{9}\ln\frac{9900}{0.9P+900}$$

$$= \ln\frac{10{,}000}{P} + \frac{1}{9}\ln\frac{1100}{0.1P+100} = \ln\frac{10{,}000}{P} + \frac{1}{9}\ln\frac{11{,}000}{P+1000}$$

45. There are only finitely many values of x where $Q(x)=0$ (assuming that Q is not the zero polynomial). At all other values of x, $F(x)/Q(x) = G(x)/Q(x)$, so $F(x)=G(x)$. In other words, the values of F and G agree at all except perhaps finitely many values of x. By continuity of F and G, the polynomials F and G must agree at those values of x too.

More explicitly: If a is a value of x such that $Q(a)=0$, then $Q(x)\neq 0$ for all x sufficiently close to a. Thus,

$$F(a) = \lim_{x\to a} F(x) \qquad \text{[by continuity of } F]$$

$$= \lim_{x\to a} G(x) \qquad \text{[whenever } Q(x)\neq 0]$$

$$= G(a) \qquad \text{[by continuity of } G]$$

6.4 Integration with Tables and Computer Algebra Systems

Keep in mind that there are several ways to approach many of these exercises, and different methods can lead to different forms of the answer.

1. $\displaystyle\int_0^1 2x\cos^{-1}x\,dx \overset{91}{=} 2\left[\frac{2x^2-1}{4}\cos^{-1}x - \frac{x\sqrt{1-x^2}}{4}\right]_0^1 = 2\left[\left(\tfrac{1}{4}\cdot 0 - 0\right)-\left(-\tfrac{1}{4}\cdot\tfrac{\pi}{2}-0\right)\right] = 2\left(\tfrac{\pi}{8}\right) = \tfrac{\pi}{4}$

3. Let $u=\pi x$ $\Rightarrow$ $du = \pi\,dx$, so

$$\int \sec^3(\pi x)\,dx = \tfrac{1}{\pi}\int \sec^3 u\,du \overset{71}{=} \tfrac{1}{\pi}\left(\tfrac{1}{2}\sec u\tan u + \tfrac{1}{2}\ln|\sec u + \tan u|\right) + C$$

$$= \tfrac{1}{2\pi}\sec\pi x\tan\pi x + \tfrac{1}{2\pi}\ln|\sec\pi x + \tan\pi x| + C$$

5. Let $u=2x$ and $a=3$. Then $du=2\,dx$ and

$$\int\frac{dx}{x^2\sqrt{4x^2+9}} = \int\frac{\tfrac{1}{2}\,du}{\dfrac{u^2}{4}\sqrt{u^2+a^2}} = 2\int\frac{du}{u^2\sqrt{a^2+u^2}} \overset{28}{=} -2\frac{\sqrt{a^2+u^2}}{a^2u} + C$$

$$= -2\frac{\sqrt{4x^2+9}}{9\cdot 2x} + C = -\frac{\sqrt{4x^2+9}}{9x} + C$$

7. $\int x^3 \sin x \, dx \stackrel{84}{=} -x^3 \cos x + 3 \int x^2 \cos x \, dx$, $\int x^2 \cos x \, dx \stackrel{85}{=} x^2 \sin x - 2 \int x \sin x \, dx$, and

$\int x \sin x \, dx \stackrel{82}{=} \sin x - x \cos x + C$. Substituting, we get

$\int x^3 \sin x \, dx = -x^3 \cos x + 3 \left[x^2 \sin x - 2(\sin x - x \cos x) \right] + C = -x^3 \cos x + 3x^2 \sin x - 6 \sin x + 6x \cos x + C$.

So $\int_0^\pi x^3 \sin x \, dx = \left[-x^3 \cos x + 3x^2 \sin x - 6 \sin x + 6x \cos x \right]_0^\pi = \left(-\pi^3 \cdot -1 + 6\pi \cdot -1 \right) - (0) = \pi^3 - 6\pi$.

9. $\displaystyle\int \frac{\tan^3(1/z)}{z^2} \, dz \quad \begin{bmatrix} u = 1/z, \\ du = -dz/z^2 \end{bmatrix} = -\int \tan^3 u \, du \stackrel{69}{=} -\frac{1}{2} \tan^2 u - \ln|\cos u| + C = -\frac{1}{2} \tan^2 \left(\frac{1}{z} \right) - \ln\left|\cos\left(\frac{1}{z} \right)\right| + C$

11. Let $z = 6 + 4y - 4y^2 = 6 - (4y^2 - 4y + 1) + 1 = 7 - (2y - 1)^2$, $u = 2y - 1$, and $a = \sqrt{7}$. Then $z = a^2 - u^2$, $du = 2 \, dy$, and

$$\int y\sqrt{6 + 4y - 4y^2} \, dy = \int y\sqrt{z} \, dy = \int \tfrac{1}{2}(u + 1)\sqrt{a^2 - u^2} \, \tfrac{1}{2} \, du = \tfrac{1}{4} \int u\sqrt{a^2 - u^2} \, du + \tfrac{1}{4} \int \sqrt{a^2 - u^2} \, du$$

$$= \tfrac{1}{4} \int \sqrt{a^2 - u^2} \, du - \tfrac{1}{8} \int (-2u)\sqrt{a^2 - u^2} \, du$$

$$\stackrel{30}{=} \frac{u}{8}\sqrt{a^2 - u^2} + \frac{a^2}{8}\sin^{-1}\left(\frac{u}{a}\right) - \frac{1}{8}\int \sqrt{w} \, dw \qquad [w = a^2 - u^2, \, dw = -2u \, du]$$

$$= \frac{2y - 1}{8}\sqrt{6 + 4y - 4y^2} + \frac{7}{8}\sin^{-1}\left(\frac{2y - 1}{\sqrt{7}}\right) - \frac{1}{8} \cdot \frac{2}{3}w^{3/2} + C$$

$$= \frac{2y - 1}{8}\sqrt{6 + 4y - 4y^2} + \frac{7}{8}\sin^{-1}\left(\frac{2y - 1}{\sqrt{7}}\right) - \frac{1}{12}(6 + 4y - 4y^2)^{3/2} + C$$

This can be rewritten as

$$\sqrt{6 + 4y - 4y^2}\left[\frac{1}{8}(2y - 1) - \frac{1}{12}(6 + 4y - 4y^2)\right] + \frac{7}{8}\sin^{-1}\left(\frac{2y - 1}{\sqrt{7}}\right) + C$$

$$= \left(\frac{1}{3}y^2 - \frac{1}{12}y - \frac{5}{8}\right)\sqrt{6 + 4y - 4y^2} + \frac{7}{8}\sin^{-1}\left(\frac{2y - 1}{\sqrt{7}}\right) + C$$

$$= \frac{1}{24}(8y^2 - 2y - 15)\sqrt{6 + 4y - 4y^2} + \frac{7}{8}\sin^{-1}\left(\frac{2y - 1}{\sqrt{7}}\right) + C$$

13. Let $u = \sin x$. Then $du = \cos x \, dx$, so

$$\int \sin^2 x \cos x \ln(\sin x) \, dx = \int u^2 \ln u \, du \stackrel{101}{=} \frac{u^{2+1}}{(2+1)^2}\left[(2+1)\ln u - 1\right] + C$$

$$= \tfrac{1}{9}u^3(3\ln u - 1) + C = \tfrac{1}{9}\sin^3 x \left[3\ln(\sin x) - 1\right] + C$$

15. Let $u = e^x$ and $a = \sqrt{3}$. Then $du = e^x \, dx$ and

$$\int \frac{e^x}{3 - e^{2x}} \, dx = \int \frac{du}{a^2 - u^2} \stackrel{19}{=} \frac{1}{2a}\ln\left|\frac{u + a}{u - a}\right| + C = \frac{1}{2\sqrt{3}}\ln\left|\frac{e^x + \sqrt{3}}{e^x - \sqrt{3}}\right| + C.$$

17. $\displaystyle\int \frac{x^4 \, dx}{\sqrt{x^{10} - 2}} = \int \frac{x^4 \, dx}{\sqrt{(x^5)^2 - 2}} = \frac{1}{5}\int \frac{du}{\sqrt{u^2 - 2}} \qquad [u = x^5, \, du = 5x^4 \, dx]$

$$\stackrel{43}{=} \tfrac{1}{5}\ln\left|u + \sqrt{u^2 - 2}\right| + C = \tfrac{1}{5}\ln\left|x^5 + \sqrt{x^{10} - 2}\right| + C$$

19. Let $u = \ln x$ and $a = 2$. Then $du = \dfrac{dx}{x}$ and

$$\int \frac{\sqrt{4 + (\ln x)^2}}{x}\, dx = \int \sqrt{a^2 + u^2}\, du \overset{21}{=} \frac{u}{2}\sqrt{a^2 + u^2} + \frac{a^2}{2}\ln\left(u + \sqrt{a^2 + u^2}\right) + C$$

$$= \tfrac{1}{2}(\ln x)\sqrt{4 + (\ln x)^2} + 2\ln\left[\ln x + \sqrt{4 + (\ln x)^2}\right] + C$$

21. Let $u = e^x$. Then $x = \ln u$, $dx = du/u$, so

$$\int \sqrt{e^{2x} - 1}\, dx = \int \frac{\sqrt{u^2 - 1}}{u}\, du \overset{41}{=} \sqrt{u^2 - 1} - \cos^{-1}(1/u) + C = \sqrt{e^{2x} - 1} - \cos^{-1}(e^{-x}) + C.$$

23. (a) $\dfrac{d}{du}\left[\dfrac{1}{b^3}\left(a + bu - \dfrac{a^2}{a + bu} - 2a\ln|a + bu|\right) + C\right] = \dfrac{1}{b^3}\left[b + \dfrac{ba^2}{(a + bu)^2} - \dfrac{2ab}{(a + bu)}\right]$

$$= \frac{1}{b^3}\left[\frac{b(a + bu)^2 + ba^2 - (a + bu)2ab}{(a + bu)^2}\right]$$

$$= \frac{1}{b^3}\left[\frac{b^3 u^2}{(a + bu)^2}\right] = \frac{u^2}{(a + bu)^2}$$

(b) Let $t = a + bu \;\Rightarrow\; dt = b\, du$. Note that $u = \dfrac{t - a}{b}$ and $du = \dfrac{1}{b}\, dt$.

$$\int \frac{u^2\, du}{(a + bu)^2} = \frac{1}{b^3}\int \frac{(t - a)^2}{t^2}\, dt = \frac{1}{b^3}\int \frac{t^2 - 2at + a^2}{t^2}\, dt = \frac{1}{b^3}\int\left(1 - \frac{2a}{t} + \frac{a^2}{t^2}\right)dt$$

$$= \frac{1}{b^3}\left(t - 2a\ln|t| - \tfrac{a^2}{t}\right) + C = \frac{1}{b^3}\left(a + bu - \frac{a^2}{a + bu} - 2a\ln|a + bu|\right) + C$$

25. Maple, Mathematica and Derive all give $\int x^2\sqrt{5 - x^2}\, dx = -\tfrac{1}{4}x(5 - x^2)^{3/2} + \tfrac{5}{8}x\sqrt{5 - x^2} + \tfrac{25}{8}\sin^{-1}\left(\tfrac{1}{\sqrt{5}}x\right)$.

Using Formula 31, we get $\int x^2\sqrt{5 - x^2}\, dx = \tfrac{1}{8}x(2x^2 - 5)\sqrt{5 - x^2} + \tfrac{1}{8}(5^2)\sin^{-1}\left(\tfrac{1}{\sqrt{5}}x\right) + C$.

But $-\tfrac{1}{4}x(5 - x^2)^{3/2} + \tfrac{5}{8}x\sqrt{5 - x^2} = \tfrac{1}{8}x\sqrt{5 - x^2}\left[5 - 2(5 - x^2)\right] = \tfrac{1}{8}x(2x^2 - 5)\sqrt{5 - x^2}$, and the $\sin^{-1}$ terms

are the same in each expression, so the answers are equivalent.

27. Maple and Derive both give $\int \sin^3 x \cos^2 x\, dx = -\tfrac{1}{5}\sin^2 x \cos^3 x - \tfrac{2}{15}\cos^3 x$ (although Derive factors the expression), and

Mathematica gives $\int \sin^3 x \cos^2 x\, dx = -\tfrac{1}{8}\cos x - \tfrac{1}{48}\cos 3x + \tfrac{1}{80}\cos 5x$. We can use a CAS to show that both of these

expressions are equal to $-\tfrac{1}{3}\cos^3 x + \tfrac{1}{5}\cos^5 x$. Using Formula 86, we write

$$\int \sin^3 x \cos^2 x\, dx = -\tfrac{1}{5}\sin^2 x \cos^3 x + \tfrac{2}{5}\int \sin x \cos^2 x\, dx = -\tfrac{1}{5}\sin^2 x \cos^3 x + \tfrac{2}{5}\left(-\tfrac{1}{3}\cos^3 x\right) + C$$

$$= -\tfrac{1}{5}\sin^2 x \cos^3 x - \tfrac{2}{15}\cos^3 x + C$$

29. Maple gives $\int x\sqrt{1 + 2x}\, dx = \tfrac{1}{10}(1 + 2x)^{5/2} - \tfrac{1}{6}(1 + 2x)^{3/2}$, Mathematica gives $\sqrt{1 + 2x}\left(\tfrac{2}{5}x^2 + \tfrac{1}{15}x - \tfrac{1}{15}\right)$, and Derive

gives $\tfrac{1}{15}(1 + 2x)^{3/2}(3x - 1)$. The first two expressions can be simplified to Derive's result. If we use Formula 54, we get

$$\int x\sqrt{1 + 2x}\, dx = \frac{2}{15(2)^2}(3 \cdot 2x - 2 \cdot 1)(1 + 2x)^{3/2} + C = \tfrac{1}{30}(6x - 2)(1 + 2x)^{3/2} + C = \tfrac{1}{15}(3x - 1)(1 + 2x)^{3/2}.$$

31. Maple gives $\int \tan^5 x\, dx = \frac{1}{4} \tan^4 x - \frac{1}{2} \tan^2 x + \frac{1}{2} \ln(1 + \tan^2 x)$, Mathematica gives

$\int \tan^5 x\, dx = \frac{1}{4}[-1 - 2\cos(2x)] \sec^4 x - \ln(\cos x)$, and Derive gives $\int \tan^5 x\, dx = \frac{1}{4}\tan^4 x - \frac{1}{2}\tan^2 x - \ln(\cos x)$.

These expressions are equivalent, and none includes absolute value bars or a constant of integration. Note that Mathematica's

and Derive's expressions suggest that the integral is undefined where $\cos x < 0$, which is not the case.

Using Formula 75, $\int \tan^5 x\, dx = \frac{1}{5-1} \tan^{5-1} x - \int \tan^{5-2} x\, dx = \frac{1}{4} \tan^4 x - \int \tan^3 x\, dx$.

Using Formula 69, $\int \tan^3 x\, dx = \frac{1}{2} \tan^2 x + \ln|\cos x| + C$, so $\int \tan^5 x\, dx = \frac{1}{4} \tan^4 x - \frac{1}{2} \tan^2 x - \ln|\cos x| + C$.

33. Derive gives $I = \displaystyle\int 2^x \sqrt{4^x - 1}\, dx = \dfrac{2^{x-1}\sqrt{2^{2x} - 1}}{\ln 2} - \dfrac{\ln\left(\sqrt{2^{2x} - 1} + 2^x\right)}{2 \ln 2}$ immediately. Neither Maple nor Mathematica

is able to evaluate I in its given form. However, if we instead write I as $\int 2^x \sqrt{(2^x)^2 - 1}\, dx$, both systems give the same

answer as Derive (after minor simplification). Our trick works because the CAS now recognizes 2^x as a promising

substitution.

6.5 Approximate Integration

1. (a) $\Delta x = (b - a)/n = (4 - 0)/2 = 2$

$$L_2 = \sum_{i=1}^{2} f(x_{i-1}) \Delta x = f(x_0) \cdot 2 + f(x_1) \cdot 2 = 2\,[f(0) + f(2)] = 2(0.5 + 2.5) = 6$$

$$R_2 = \sum_{i=1}^{2} f(x_i) \Delta x = f(x_1) \cdot 2 + f(x_2) \cdot 2 = 2\,[f(2) + f(4)] = 2(2.5 + 3.5) = 12$$

$$M_2 = \sum_{i=1}^{2} f(\overline{x}_i)\Delta x = f(\overline{x}_1) \cdot 2 + f(\overline{x}_2) \cdot 2 = 2\,[f(1) + f(3)] \approx 2(1.6 + 3.2) = 9.6$$

(b)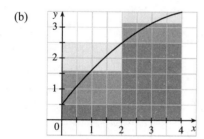

L_2 is an underestimate, since the area under the small rectangles is less than the area under the curve, and R_2 is an overestimate, since the area under the large rectangles is greater than the area under the curve. It appears that M_2 is an overestimate, though it is fairly close to I. See the solution to Exercise 37 for a proof of the fact that if f is concave down on $[a, b]$, then the Midpoint Rule is an overestimate of $\int_a^b f(x)\, dx$.

(c) $T_2 = \left(\frac{1}{2}\Delta x\right)[f(x_0) + 2f(x_1) + f(x_2)] = \frac{2}{2}[f(0) + 2f(2) + f(4)] = 0.5 + 2(2.5) + 3.5 = 9.$

This approximation is an underestimate, since the graph is concave down. Thus, $T_2 = 9 < I$. See the solution to

Exercise 37 for a general proof of this conclusion.

(d) For any n, we will have $L_n < T_n < I < M_n < R_n$.

3. $f(x) = \cos(x^2)$, $\Delta x = \frac{1-0}{4} = \frac{1}{4}$

(a) $T_4 = \frac{1}{4 \cdot 2} \left[f(0) + 2f\left(\frac{1}{4}\right) + 2f\left(\frac{2}{4}\right) + 2f\left(\frac{3}{4}\right) + f(1) \right] \approx 0.895759$

(b) $M_4 = \frac{1}{4} \left[f\left(\frac{1}{8}\right) + f\left(\frac{3}{8}\right) + f\left(\frac{5}{8}\right) + f\left(\frac{7}{8}\right) \right] \approx 0.908907$

The graph shows that f is concave down on $[0, 1]$. So T_4 is an

underestimate and M_4 is an overestimate. We can conclude that

$0.895759 < \int_0^1 \cos(x^2)\,dx < 0.908907$.

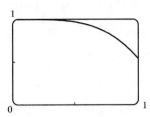

5. $f(x) = x^2 \sin x$, $\Delta x = \dfrac{b-a}{n} = \dfrac{\pi - 0}{8} = \dfrac{\pi}{8}$

(a) $M_8 = \frac{\pi}{8} \left[f\left(\frac{\pi}{16}\right) + f\left(\frac{3\pi}{16}\right) + f\left(\frac{5\pi}{16}\right) + \cdots + f\left(\frac{15\pi}{16}\right) \right] \approx 5.932957$

(b) $S_8 = \frac{\pi}{8 \cdot 3} \left[f(0) + 4f\left(\frac{\pi}{8}\right) + 2f\left(\frac{2\pi}{8}\right) + 4f\left(\frac{3\pi}{8}\right) + 2f\left(\frac{4\pi}{8}\right) + 4f\left(\frac{5\pi}{8}\right) + 2f\left(\frac{6\pi}{8}\right) + 4f\left(\frac{7\pi}{8}\right) + f(\pi) \right]$

≈ 5.869247

Actual: $\int_0^\pi x^2 \sin x\,dx \overset{84}{=} \left[-x^2 \cos x \right]_0^\pi + 2 \int_0^\pi x \cos x\,dx \overset{83}{=} \left[-\pi^2(-1) - 0 \right] + 2[\cos x + x \sin x]_0^\pi$

$= \pi^2 + 2[(-1+0) - (1+0)] = \pi^2 - 4 \approx 5.869604$

Errors: $E_M = \text{actual} - M_8 = \int_0^\pi x^2 \sin x\,dx - M_8 \approx -0.063353$

$E_S = \text{actual} - S_8 = \int_0^\pi x^2 \sin x\,dx - S_8 \approx 0.000357$

7. $f(x) = \sqrt[4]{1 + x^2}$, $\Delta x = \dfrac{2-0}{8} = \dfrac{1}{4}$

(a) $T_8 = \frac{1}{4 \cdot 2} \left[f(0) + 2f\left(\frac{1}{4}\right) + 2f\left(\frac{1}{2}\right) + \cdots + 2f\left(\frac{3}{2}\right) + 2f\left(\frac{7}{4}\right) + f(2) \right] \approx 2.413790$

(b) $M_8 = \frac{1}{4} \left[f\left(\frac{1}{8}\right) + f\left(\frac{3}{8}\right) + \cdots + f\left(\frac{13}{8}\right) + f\left(\frac{15}{8}\right) \right] \approx 2.411453$

(c) $S_8 = \frac{1}{4 \cdot 3} \left[f(0) + 4f\left(\frac{1}{4}\right) + 2f\left(\frac{1}{2}\right) + 4f\left(\frac{3}{4}\right) + 2f(1) + 4f\left(\frac{5}{4}\right) + 2f\left(\frac{3}{2}\right) + 4f\left(\frac{7}{4}\right) + f(2) \right] \approx 2.412232$

9. $f(x) = \dfrac{\ln x}{1 + x}$, $\Delta x = \dfrac{2-1}{10} = \dfrac{1}{10}$

(a) $T_{10} = \frac{1}{10 \cdot 2} [f(1) + 2f(1.1) + 2f(1.2) + \cdots + 2f(1.8) + 2f(1.9) + f(2)] \approx 0.146879$

(b) $M_{10} = \frac{1}{10} [f(1.05) + f(1.15) + \cdots + f(1.85) + f(1.95)] \approx 0.147391$

(c) $S_{10} = \frac{1}{10 \cdot 3} [f(1) + 4f(1.1) + 2f(1.2) + 4f(1.3) + 2f(1.4) + 4f(1.5) + 2f(1.6) + 4f(1.7)$

$\qquad + 2f(1.8) + 4f(1.9) + f(2)]$

≈ 0.147219

11. $f(t) = e^{\sqrt{t}} \sin t$, $\Delta t = \dfrac{4-0}{8} = \dfrac{1}{2}$

(a) $T_8 = \frac{1}{2 \cdot 2} \left[f(0) + 2f\left(\frac{1}{2}\right) + 2f(1) + 2f\left(\frac{3}{2}\right) + 2f(2) + 2f\left(\frac{5}{2}\right) + 2f(3) + 2f\left(\frac{7}{2}\right) + f(4) \right] \approx 4.513618$

(b) $M_8 = \frac{1}{2} \left[f\left(\frac{1}{4}\right) + f\left(\frac{3}{4}\right) + f\left(\frac{5}{4}\right) + f\left(\frac{7}{4}\right) + f\left(\frac{9}{4}\right) + f\left(\frac{11}{4}\right) + f\left(\frac{13}{4}\right) + f\left(\frac{15}{4}\right) \right] \approx 4.748256$

(c) $S_8 = \frac{1}{2 \cdot 3} \left[f(0) + 4f\left(\frac{1}{2}\right) + 2f(1) + 4f\left(\frac{3}{2}\right) + 2f(2) + 4f\left(\frac{5}{2}\right) + 2f(3) + 4f\left(\frac{7}{2}\right) + f(4) \right] \approx 4.675111$

13. $f(x) = \dfrac{\cos x}{x}, \Delta x = \dfrac{5-1}{8} = \dfrac{1}{2}$

(a) $T_8 = \frac{1}{2 \cdot 2} \left[f(1) + 2f\left(\frac{3}{2}\right) + 2f(2) + \cdots + 2f(4) + 2f\left(\frac{9}{2}\right) + f(5) \right] \approx -0.495333$

(b) $M_8 = \frac{1}{2} \left[f\left(\frac{5}{4}\right) + f\left(\frac{7}{4}\right) + f\left(\frac{9}{4}\right) + f\left(\frac{11}{4}\right) + f\left(\frac{13}{4}\right) + f\left(\frac{15}{4}\right) + f\left(\frac{17}{4}\right) + f\left(\frac{19}{4}\right) \right] \approx -0.543321$

(c) $S_8 = \frac{1}{2 \cdot 3} \left[f(1) + 4f\left(\frac{3}{2}\right) + 2f(2) + 4f\left(\frac{5}{2}\right) + 2f(3) + 4f\left(\frac{7}{2}\right) + 2f(4) + 4f\left(\frac{9}{2}\right) + f(5) \right] \approx -0.526123$

15. $f(y) = \dfrac{1}{1+y^5}, \Delta y = \dfrac{3-0}{6} = \dfrac{1}{2}$

(a) $T_6 = \frac{1}{2 \cdot 2} \left[f(0) + 2f\left(\frac{1}{2}\right) + 2f\left(\frac{2}{2}\right) + 2f\left(\frac{3}{2}\right) + 2f\left(\frac{4}{2}\right) + 2f\left(\frac{5}{2}\right) + f(3) \right] \approx 1.064275$

(b) $M_6 = \frac{1}{2} \left[f\left(\frac{1}{4}\right) + f\left(\frac{3}{4}\right) + f\left(\frac{5}{4}\right) + f\left(\frac{7}{4}\right) + f\left(\frac{9}{4}\right) + f\left(\frac{11}{4}\right) \right] \approx 1.067416$

(c) $S_6 = \frac{1}{2 \cdot 3} \left[f(0) + 4f\left(\frac{1}{2}\right) + 2f\left(\frac{2}{2}\right) + 4f\left(\frac{3}{2}\right) + 2f\left(\frac{4}{2}\right) + 4f\left(\frac{5}{2}\right) + f(3) \right] \approx 1.074915$

17. (a) $f(x) = e^{-x^2}, \Delta x = \dfrac{2-0}{10} = \dfrac{1}{5}$

$T_{10} = \frac{1}{5 \cdot 2} \{ f(0) + 2[f(0.2) + f(0.4) + \cdots + f(1.8)] + f(2) \} \approx 0.881839$

$M_{10} = \frac{1}{5} [f(0.1) + f(0.3) + f(0.5) + \cdots + f(1.7) + f(1.9)] \approx 0.882202$

(b) $f(x) = e^{-x^2}, f'(x) = -2xe^{-x^2}, f''(x) = (4x^2 - 2)e^{-x^2}, f'''(x) = 4x(3 - 2x^2)e^{-x^2}$.

$f'''(x) = 0 \iff x = 0$ or $x = \pm\sqrt{\frac{3}{2}}$. So to find the maximum value of $|f''(x)|$ on $[0, 2]$, we need only consider its

values at $x = 0, x = 2$, and $x = \sqrt{\frac{3}{2}}$. $|f''(0)| = 2, |f''(2)| \approx 0.2564$ and $\left| f''\left(\sqrt{\frac{3}{2}}\right) \right| \approx 0.8925$. Thus, taking $K = 2$,

$a = 0, b = 2$, and $n = 10$ in Theorem 3, we get $|E_T| \le 2 \cdot 2^3 / (12 \cdot 10^2) = \frac{1}{75} = 0.01\overline{3}$, and $|E_M| \le |E_T|/2 \le 0.00\overline{6}$.

(c) Take $K = 2$ [as in part (b)] in Theorem 3. $|E_T| \le \dfrac{K(b-a)^3}{12n^2} \le 10^{-5} \iff \dfrac{2(2-0)^3}{12n^2} \le 10^{-5} \iff \frac{3}{4}n^2 \ge 10^5 \iff$

$n \ge 365.1\ldots \iff n \ge 366$. Take $n = 366$ for T_n. For E_M, again take $K = 2$ in Theorem 3 to get $|E_M| \le 10^{-5} \iff$

$\frac{3}{2}n^2 \ge 10^5 \iff n \ge 258.2 \implies n \ge 259$. Take $n = 259$ for M_n.

19. (a) $T_{10} = \frac{1}{10 \cdot 2} \{ f(0) + 2[f(0.1) + f(0.2) + \cdots + f(0.9)] + f(1) \} \approx 1.71971349$

$S_{10} = \frac{1}{10 \cdot 3} [f(0) + 4f(0.1) + 2f(0.2) + 4f(0.3) + \cdots + 4f(0.9) + f(1)] \approx 1.71828278$

Since $I = \int_0^1 e^x dx = [e^x]_0^1 = e - 1 \approx 1.71828183$, $E_T = I - T_{10} \approx -0.00143166$ and

$E_S = I - S_{10} \approx -0.00000095$.

(b) $f(x) = e^x \implies f''(x) = e^x \le e$ for $0 \le x \le 1$. Taking $K = e, a = 0, b = 1$, and $n = 10$ in Theorem 3, we get

$|E_T| \le e(1)^3 / (12 \cdot 10^2) \approx 0.002265 > 0.00143166$ [actual $|E_T|$ from (a)]. $f^{(4)}(x) = e^x < e$ for $0 \le x \le 1$.

Using Theorem 4, we have $|E_S| \le e(1)^5 / (180 \cdot 10^4) \approx 0.0000015 > 0.00000095$ [actual $|E_S|$ from (a)].

We see that the actual errors are about two-thirds the size of the error estimates.

(c) From part (b), we take $K = e$ to get $|E_T| \leq \dfrac{K(b-a)^3}{12n^2} \leq 0.00001 \quad \Rightarrow \quad n^2 \geq \dfrac{e(1^3)}{12(0.00001)} \quad \Rightarrow \quad n \geq 150.5.$

Take $n = 151$ for T_n. Now $|E_M| \leq \dfrac{K(b-a)^3}{24n^2} \leq 0.00001 \quad \Rightarrow \quad n \geq 106.4.$ Take $n = 107$ for M_n.

Finally, $|E_S| \leq \dfrac{K(b-a)^5}{180n^4} \leq 0.00001 \quad \Rightarrow \quad n^4 \geq \dfrac{e(1^5)}{180(0.00001)} \quad \Rightarrow \quad n \geq 6.23.$

Take $n = 8$ for S_n (since n has to be even for Simpson's Rule).

21. (a) Using a CAS, we differentiate $f(x) = e^{\cos x}$ twice, and find that

$f''(x) = e^{\cos x}(\sin^2 x - \cos x)$. From the graph, we see that the

maximum value of $|f''(x)|$ occurs at the endpoints of the

interval $[0, 2\pi]$. Since $f''(0) = -e$, we can use $K = e$ or $K = 2.8$.

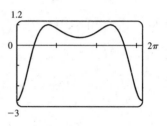

(b) A CAS gives $M_{10} \approx 7.954926518$. (In Maple, use `student[middlesum]`.)

(c) Using Theorem 3 for the Midpoint Rule, with $K = e$, we get $|E_M| \leq \dfrac{e(2\pi - 0)^3}{24 \cdot 10^2} \approx 0.280945995.$

With $K = 2.8$, we get $|E_M| \leq \dfrac{2.8(2\pi - 0)^3}{24 \cdot 10^2} = 0.289391916.$

(d) A CAS gives $I \approx 7.954926521$.

(e) The actual error is only about 3×10^{-9}, much less than the estimate in part (c).

(f) We use the CAS to differentiate twice more, and then graph

$f^{(4)}(x) = e^{\cos x}(\sin^4 x - 6\sin^2 x \cos x + 3 - 7\sin^2 x + \cos x).$

From the graph, we see that the maximum value of $\left|f^{(4)}(x)\right|$ occurs

at the endpoints of the interval $[0, 2\pi]$. Since $f^{(4)}(0) = 4e$, we can use

$K = 4e$ or $K = 10.9$.

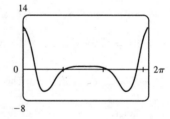

(g) A CAS gives $S_{10} \approx 7.953789422$. (In Maple, use `student[simpson]`.)

(h) Using Theorem 4 with $K = 4e$, we get $|E_S| \leq \dfrac{4e(2\pi - 0)^5}{180 \cdot 10^4} \approx 0.059153618.$

With $K = 10.9$, we get $|E_S| \leq \dfrac{10.9(2\pi - 0)^5}{180 \cdot 10^4} \approx 0.059299814.$

(i) The actual error is about $7.954926521 - 7.953789422 \approx 0.00114$. This is quite a bit smaller than the estimate in part (h), though the difference is not nearly as great as it was in the case of the Midpoint Rule.

(j) To ensure that $|E_S| \leq 0.0001$, we use Theorem 4: $|E_S| \leq \dfrac{4e(2\pi)^5}{180 \cdot n^4} \leq 0.0001 \quad \Rightarrow \quad \dfrac{4e(2\pi)^5}{180 \cdot 0.0001} \leq n^4 \quad \Rightarrow$

$n^4 \geq 5{,}915{,}362 \quad \Leftrightarrow \quad n \geq 49.3.$ So we must take $n \geq 50$ to ensure that $|I - S_n| \leq 0.0001.$ ($K = 10.9$ leads to the same value of n.)

23. $I = \int_0^1 x^3 dx = \left[\frac{1}{4}x^4\right]_0^1 = 0.25.$ $f(x) = x^3.$

$n = 4$: $L_4 = \frac{1}{4}\left[0^3 + \left(\frac{1}{4}\right)^3 + \left(\frac{2}{4}\right)^3 + \left(\frac{3}{4}\right)^3\right] = 0.140625$

$R_4 = \frac{1}{4}\left[\left(\frac{1}{4}\right)^3 + \left(\frac{2}{4}\right)^3 + \left(\frac{3}{4}\right)^3 + 1^3\right] = 0.390625$

$T_4 = \frac{1}{4 \cdot 2}\left[0^3 + 2\left(\frac{1}{4}\right)^3 + 2\left(\frac{2}{4}\right)^3 + 2\left(\frac{3}{4}\right)^3 + 1^3\right] = 0.265625,$

$M_4 = \frac{1}{4}\left[\left(\frac{1}{8}\right)^3 + \left(\frac{3}{8}\right)^3 + \left(\frac{5}{8}\right)^3 + \left(\frac{7}{8}\right)^3\right] = 0.2421875,$

$E_L = I - L_4 = \frac{1}{4} - 0.140625 = 0.109375,\ E_R = \frac{1}{4} - 0.390625 = -0.140625,$

$E_T = \frac{1}{4} - 0.265625 = -0.015625,\ E_M = \frac{1}{4} - 0.2421875 = 0.0078125$

$n = 8$: $L_8 = \frac{1}{8}\left[f(0) + f\left(\frac{1}{8}\right) + f\left(\frac{2}{8}\right) + \cdots + f\left(\frac{7}{8}\right)\right] \approx 0.191406$

$R_8 = \frac{1}{8}\left[f\left(\frac{1}{8}\right) + f\left(\frac{2}{8}\right) + \cdots + f\left(\frac{7}{8}\right) + f(1)\right] \approx 0.316406$

$T_8 = \frac{1}{8 \cdot 2}\left\{f(0) + 2\left[f\left(\frac{1}{8}\right) + f\left(\frac{2}{8}\right) + \cdots + f\left(\frac{7}{8}\right)\right] + f(1)\right\} \approx 0.253906$

$M_8 = \frac{1}{8}\left[f\left(\frac{1}{16}\right) + f\left(\frac{3}{16}\right) + \cdots + f\left(\frac{13}{16}\right) + f\left(\frac{15}{16}\right)\right] = 0.248047$

$E_L \approx \frac{1}{4} - 0.191406 \approx 0.058594,\ E_R \approx \frac{1}{4} - 0.316406 \approx -0.066406,$

$E_T \approx \frac{1}{4} - 0.253906 \approx -0.003906,\ E_M \approx \frac{1}{4} - 0.248047 \approx 0.001953.$

$n = 16$: $L_{16} = \frac{1}{16}\left[f(0) + f\left(\frac{1}{16}\right) + f\left(\frac{2}{16}\right) + \cdots + f\left(\frac{15}{16}\right)\right] \approx 0.219727$

$R_{16} = \frac{1}{16}\left[f\left(\frac{1}{16}\right) + f\left(\frac{2}{16}\right) + \cdots + f\left(\frac{15}{16}\right) + f(1)\right] \approx 0.282227$

$T_{16} = \frac{1}{16 \cdot 2}\left\{f(0) + 2\left[f\left(\frac{1}{16}\right) + f\left(\frac{2}{16}\right) + \cdots + f\left(\frac{15}{16}\right)\right] + f(1)\right\} \approx 0.250977$

$M_{16} = \frac{1}{16}\left[f\left(\frac{1}{32}\right) + f\left(\frac{3}{32}\right) + \cdots + f\left(\frac{31}{32}\right)\right] \approx 0.249512$

$E_L \approx \frac{1}{4} - 0.219727 \approx 0.030273,\ E_R \approx \frac{1}{4} - 0.282227 \approx -0.032227,$

$E_T \approx \frac{1}{4} - 0.250977 \approx -0.000977,\ E_M \approx \frac{1}{4} - 0.249512 \approx 0.000488.$

n	L_n	R_n	T_n	M_n
4	0.140625	0.390625	0.265625	0.242188
8	0.191406	0.316406	0.253906	0.248047
16	0.219727	0.282227	0.250977	0.249512

n	E_L	E_R	E_T	E_M
4	0.109375	−0.140625	−0.015625	0.007813
8	0.058594	−0.066406	−0.003906	0.001953
16	0.030273	−0.032227	−0.000977	0.000488

Observations:

1. E_L and E_R are always opposite in sign, as are E_T and E_M.

2. As n is doubled, E_L and E_R are decreased by about a factor of 2, and E_T and E_M are decreased by a factor of about 4.

3. The Midpoint approximation is about twice as accurate as the Trapezoidal approximation.

4. All the approximations become more accurate as the value of n increases.

5. The Midpoint and Trapezoidal approximations are much more accurate than the endpoint approximations.

25. $\Delta x = (4 - 0)/4 = 1$

(a) $T_4 = \frac{1}{2}[f(0) + 2f(1) + 2f(2) + 2f(3) + f(4)] \approx \frac{1}{2}[0 + 2(3) + 2(5) + 2(3) + 1] = 11.5$

(b) $M_4 = 1 \cdot [f(0.5) + f(1.5) + f(2.5) + f(3.5)] \approx 1 + 4.5 + 4.5 + 2 = 12$

(c) $S_4 = \frac{1}{3}[f(0) + 4f(1) + 2f(2) + 4f(3) + f(4)] \approx \frac{1}{3}[0 + 4(3) + 2(5) + 4(3) + 1] = 11.\overline{6}$

27. By the Net Change Theorem, the increase in velocity is equal to $\int_0^6 a(t)\,dt$. We use Simpson's Rule with $n = 6$ and

$\Delta t = (6 - 0)/6 = 1$ to estimate this integral:

$$\int_0^6 a(t)\,dt \approx S_6 = \frac{1}{3}[a(0) + 4a(1) + 2a(2) + 4a(3) + 2a(4) + 4a(5) + a(6)]$$

$$\approx \frac{1}{3}[0 + 4(0.5) + 2(4.1) + 4(9.8) + 2(12.9) + 4(9.5) + 0] = \frac{1}{3}(113.2) = 37.7\overline{3} \text{ ft/s}$$

29. By the Net Change Theorem, the energy used is equal to $\int_0^6 P(t)\,dt$. We use Simpson's Rule with $n = 12$ and

$\Delta t = (6 - 0)/12 = \frac{1}{2}$ to estimate this integral:

$$\int_0^6 P(t)\,dt \approx S_{12} = \frac{1/2}{3}[P(0) + 4P(0.5) + 2P(1) + 4P(1.5) + 2P(2) + 4P(2.5)$$

$$+ 2P(3) + 4P(3.5) + 2P(4) + 4P(4.5) + 2P(5) + 4P(5.5) + P(6)]$$

$$= \frac{1}{6}[1814 + 4(1735) + 2(1686) + 4(1646) + 2(1637) + 4(1609) + 2(1604)$$

$$+ 4(1611) + 2(1621) + 4(1666) + 2(1745) + 4(1886) + 2052]$$

$$= \frac{1}{6}(61{,}064) = 10{,}177.\overline{3} \text{ megawatt-hours}$$

31. (a) We are given the function values at the endpoints of 8 intervals of length 0.4, so we'll use the Midpoint Rule with

$n = 8/2 = 4$ and $\Delta x = (3.2 - 0)/4 = 0.8$.

$\int_0^{3.2} f(x)\,dx \approx M_4 = 0.8[f(0.4) + f(1.2) + f(2.0) + f(2.8)] = 0.8[6.5 + 6.4 + 7.6 + 8.8] = 0.8(29.3) = 23.44$

(b) $-4 \le f''(x) \le 1 \;\Rightarrow\; |f''(x)| \le 4$, so use $K = 4$, $a = 0$, $b = 3.2$, and $n = 4$ in Theorem 3.

So $|E_M| \le \dfrac{4(3.2 - 0)^3}{24(4)^2} = \dfrac{128}{375} = 0.341\overline{3}$.

33. $I(\theta) = \dfrac{N^2 \sin^2 k}{k^2}$, where $k = \dfrac{\pi N d \sin \theta}{\lambda}$, $N = 10{,}000$, $d = 10^{-4}$, and $\lambda = 632.8 \times 10^{-9}$.

So $I(\theta) = \dfrac{(10^4)^2 \sin^2 k}{k^2}$, where $k = \dfrac{\pi (10^4)(10^{-4}) \sin \theta}{632.8 \times 10^{-9}}$. Now $n = 10$ and $\Delta \theta = \dfrac{10^{-6} - (-10^{-6})}{10} = 2 \times 10^{-7}$,

so $M_{10} = 2 \times 10^{-7}[I(-0.0000009) + I(-0.0000007) + \cdots + I(0.0000009)] \approx 59.4$.

35. Consider the function f whose graph is shown.

The area $\int_0^2 f(x)\,dx$ is close to 2. The Trapezoidal Rule gives

$T_2 = \frac{2-0}{2 \cdot 2}[f(0) + 2f(1) + f(2)] = \frac{1}{2}[1 + 2 \cdot 1 + 1] = 2$.

The Midpoint Rule gives $M_2 = \frac{2-0}{2}[f(0.5) + f(1.5)] = 1[0 + 0] = 0$,

so the Trapezoidal Rule is more accurate.

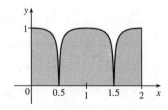

37. Since the Trapezoidal and Midpoint approximations on the interval $[a, b]$ are the sums of the Trapezoidal and Midpoint

approximations on the subintervals $[x_{i-1}, x_i]$, $i = 1, 2, \ldots, n$, we can focus our attention on one such interval. The condition

$f''(x) < 0$ for $a \le x \le b$ means that the graph of f is concave down as in Figure 5. In that figure, T_n is the area of the

trapezoid $AQRD$, $\int_a^b f(x)\, dx$ is the area of the region $AQPRD$, and M_n is the area of the trapezoid $ABCD$, so

$T_n < \int_a^b f(x)\, dx < M_n$. In general, the condition $f'' < 0$ implies that the graph of f on $[a, b]$ lies above the chord joining the

points $(a, f(a))$ and $(b, f(b))$. Thus, $\int_a^b f(x)\, dx > T_n$. Since M_n is the area under a tangent to the graph, and since $f'' < 0$

implies that the tangent lies above the graph, we also have $M_n > \int_a^b f(x)\, dx$. Thus, $T_n < \int_a^b f(x)\, dx < M_n$.

39. $T_n = \frac{1}{2}\,\Delta x\,[f(x_0) + 2f(x_1) + \cdots + 2f(x_{n-1}) + f(x_n)]$ and

$M_n = \Delta x\,[f(\overline{x}_1) + f(\overline{x}_2) + \cdots + f(\overline{x}_{n-1}) + f(\overline{x}_n)]$, where $\overline{x}_i = \frac{1}{2}(x_{i-1} + x_i)$. Now

$T_{2n} = \frac{1}{2}\left(\frac{1}{2}\Delta x\right)[f(x_0) + 2f(\overline{x}_1) + 2f(x_1) + 2f(\overline{x}_2) + 2f(x_2) + \cdots + 2f(\overline{x}_{n-1}) + 2f(x_{n-1}) + 2f(\overline{x}_n) + f(x_n)]$, so

$\frac{1}{2}(T_n + M_n) = \frac{1}{2}T_n + \frac{1}{2}M_n$

$= \frac{1}{4}\,\Delta x[f(x_0) + 2f(x_1) + \cdots + 2f(x_{n-1}) + f(x_n)]$

$\qquad + \frac{1}{4}\Delta x 2f(\overline{x}_1) + 2f(\overline{x}_2) + \cdots + 2f(\overline{x}_{n-1}) + 2f(\overline{x}_n)$

$= T_{2n}$

6.6   Improper Integrals

1. (a) Since $\int_1^\infty x^4 e^{-x^4}\, dx$ has an infinite interval of integration, it is an improper integral of Type I.

(b) Since $y = \sec x$ has an infinite discontinuity at $x = \frac{\pi}{2}$, $\int_0^{\pi/2} \sec x\, dx$ is a Type II improper integral.

(c) Since $y = \dfrac{x}{(x-2)(x-3)}$ has an infinite discontinuity at $x = 2$, $\displaystyle\int_0^2 \dfrac{x}{x^2 - 5x + 6}\, dx$ is a Type II improper integral.

(d) Since $\displaystyle\int_{-\infty}^0 \dfrac{1}{x^2 + 5}\, dx$ has an infinite interval of integration, it is an improper integral of Type I.

3. The area under the graph of $y = 1/x^3 = x^{-3}$ between $x = 1$ and $x = t$ is

$A(t) = \int_1^t x^{-3}\, dx = \left[-\frac{1}{2}x^{-2}\right]_1^t = -\frac{1}{2}t^{-2} - \left(-\frac{1}{2}\right) = \frac{1}{2} - 1/(2t^2)$. So the area for $1 \le x \le 10$ is

$A(10) = 0.5 - 0.005 = 0.495$, the area for $1 \le x \le 100$ is $A(100) = 0.5 - 0.00005 = 0.49995$, and the area for

$1 \le x \le 1000$ is $A(1000) = 0.5 - 0.0000005 = 0.4999995$. The total area under the curve for $x \ge 1$ is

$\displaystyle\lim_{t\to\infty} A(t) = \lim_{t\to\infty}\left[\frac{1}{2} - 1/(2t^2)\right] = \frac{1}{2}$.

5. $I = \displaystyle\int_1^\infty \dfrac{1}{(3x+1)^2}\, dx = \lim_{t\to\infty} \int_1^t \dfrac{1}{(3x+1)^2}\, dx$. Now

$\displaystyle\int \dfrac{1}{(3x+1)^2}\, dx = \dfrac{1}{3}\int \dfrac{1}{u^2}\, du \quad [u = 3x+1,\ du = 3\,dx] \quad = -\dfrac{1}{3u} + C = -\dfrac{1}{3(3x+1)} + C$,

so $I = \displaystyle\lim_{t\to\infty}\left[-\dfrac{1}{3(3x+1)}\right]_1^t = \lim_{t\to\infty}\left[-\dfrac{1}{3(3t+1)} + \dfrac{1}{12}\right] = 0 + \dfrac{1}{12} = \dfrac{1}{12}$.   Convergent

7. $\int_{-\infty}^{-1} \frac{1}{\sqrt{2-w}} \, dw = \lim_{t \to -\infty} \int_{t}^{-1} \frac{1}{\sqrt{2-w}} \, dw = \lim_{t \to -\infty} \left[-2\sqrt{2-w} \right]_{t}^{-1}$ $[u = 2 - w, \, du = -dw]$

$= \lim_{t \to -\infty} \left[-2\sqrt{3} + 2\sqrt{2-t} \right] = \infty.$ Divergent

9. $\int_{4}^{\infty} e^{-y/2} \, dy = \lim_{t \to \infty} \int_{4}^{t} e^{-y/2} \, dy = \lim_{t \to \infty} \left[-2e^{-y/2} \right]_{4}^{t} = \lim_{t \to \infty} \left(-2e^{-t/2} + 2e^{-2} \right) = 0 + 2e^{-2} = 2e^{-2}.$ Convergent

11. $\int_{2\pi}^{\infty} \sin \theta \, d\theta = \lim_{t \to \infty} \int_{2\pi}^{t} \sin \theta \, d\theta = \lim_{t \to \infty} \left[-\cos \theta \right]_{2\pi}^{t} = \lim_{t \to \infty} \left(-\cos t + 1 \right).$

This limit does not exist, so the integral is divergent.   Divergent

13. $\int_{-\infty}^{\infty} xe^{-x^2} \, dx = \int_{-\infty}^{0} xe^{-x^2} \, dx + \int_{0}^{\infty} xe^{-x^2} \, dx.$

$\int_{-\infty}^{0} xe^{-x^2} \, dx = \lim_{t \to -\infty} \left(-\frac{1}{2} \right) \left[e^{-x^2} \right]_{t}^{0} = \lim_{t \to -\infty} \left(-\frac{1}{2} \right) \left(1 - e^{-t^2} \right) = -\frac{1}{2} \cdot 1 = -\frac{1}{2}, \text{ and}$

$\int_{0}^{\infty} xe^{-x^2} \, dx = \lim_{t \to \infty} \left(-\frac{1}{2} \right) \left[e^{-x^2} \right]_{0}^{t} = \lim_{t \to \infty} \left(-\frac{1}{2} \right) \left(e^{-t^2} - 1 \right) = -\frac{1}{2} \cdot (-1) = \frac{1}{2}. \text{ Therefore,}$

$\int_{-\infty}^{\infty} xe^{-x^2} \, dx = -\frac{1}{2} + \frac{1}{2} = 0.$ Convergent

15. $\int_{0}^{\infty} se^{-5s} \, ds = \lim_{t \to \infty} \int_{0}^{t} se^{-5s} \, ds = \lim_{t \to \infty} \left[-\frac{1}{5} se^{-5s} - \frac{1}{25} e^{-5s} \right]_{0}^{t}$ $\begin{bmatrix} \text{by integration by} \\ \text{parts with } u = s \end{bmatrix}$

$= \lim_{t \to \infty} \left(-\frac{1}{5} te^{-5t} - \frac{1}{25} e^{-5t} + \frac{1}{25} \right) = 0 - 0 + \frac{1}{25}$ [by l'Hospital's Rule]

$= \frac{1}{25}$ Convergent

17. $\int_{1}^{\infty} \frac{\ln x}{x} \, dx = \lim_{t \to \infty} \left[\frac{(\ln x)^2}{2} \right]_{1}^{t}$ [by substitution with $u = \ln x, \, du = dx/x$] $= \lim_{t \to \infty} \frac{(\ln t)^2}{2} = \infty.$ Divergent

19. Integrate by parts with $u = \ln x, \, dv = dx/x^2 \Rightarrow du = dx/x, \, v = -1/x.$

$\int_{1}^{\infty} \frac{\ln x}{x^2} \, dx = \lim_{t \to \infty} \int_{1}^{t} \frac{\ln x}{x^2} \, dx = \lim_{t \to \infty} \left[-\frac{\ln x}{x} - \frac{1}{x} \right]_{1}^{t} = \lim_{t \to \infty} \left(-\frac{\ln t}{t} - \frac{1}{t} + 0 + 1 \right) = -0 - 0 + 0 + 1 = 1$

since $\lim_{t \to \infty} \frac{\ln t}{t} \overset{\text{H}}{=} \lim_{t \to \infty} \frac{1/t}{1} = 0.$ Convergent

21. $\int_{-\infty}^{\infty} \frac{x^2}{9 + x^6} \, dx = \int_{-\infty}^{0} \frac{x^2}{9 + x^6} \, dx + \int_{0}^{\infty} \frac{x^2}{9 + x^6} \, dx = 2 \int_{0}^{\infty} \frac{x^2}{9 + x^6} \, dx$ [since the integrand is even].

Now $\int \frac{x^2 \, dx}{9 + x^6} \begin{bmatrix} u = x^3 \\ du = 3x^2 dx \end{bmatrix} = \int \frac{\frac{1}{3} du}{9 + u^2} \begin{bmatrix} u = 3v \\ du = 3 \, dv \end{bmatrix} = \int \frac{\frac{1}{3}(3 \, dv)}{9 + 9v^2} = \frac{1}{9} \int \frac{dv}{1 + v^2}$

$= \frac{1}{9} \tan^{-1} v + C = \frac{1}{9} \tan^{-1} \left(\frac{u}{3} \right) + C = \frac{1}{9} \tan^{-1} \left(\frac{x^3}{3} \right) + C,$

so $2 \int_{0}^{\infty} \frac{x^2}{9 + x^6} \, dx = 2 \lim_{t \to \infty} \int_{0}^{t} \frac{x^2}{9 + x^6} \, dx = 2 \lim_{t \to \infty} \left[\frac{1}{9} \tan^{-1} \left(\frac{x^3}{3} \right) \right]_{0}^{t} = 2 \lim_{t \to \infty} \frac{1}{9} \tan^{-1} \left(\frac{t^3}{3} \right) = \frac{2}{9} \cdot \frac{\pi}{2} = \frac{\pi}{9}.$

Convergent

23. $\int_0^1 \dfrac{3}{x^5}\,dx = \lim\limits_{t\to 0^+}\int_t^1 3x^{-5}\,dx = \lim\limits_{t\to 0^+}\left[-\dfrac{3}{4x^4}\right]_t^1 = -\dfrac{3}{4}\lim\limits_{t\to 0^+}\left(1-\dfrac{1}{t^4}\right) = \infty.$ Divergent

25. $\int_{-2}^{14} \dfrac{dx}{\sqrt[4]{x+2}} = \lim\limits_{t\to -2^+}\int_t^{14}(x+2)^{-1/4}\,dx = \lim\limits_{t\to -2^+}\left[\dfrac{4}{3}(x+2)^{3/4}\right]_t^{14} = \dfrac{4}{3}\lim\limits_{t\to -2^+}\left[16^{3/4}-(t+2)^{3/4}\right]$

$= \dfrac{4}{3}(8-0) = \dfrac{32}{3}$ Convergent

27. There is an infinite discontinuity at $x = 1$. $\int_0^{33}(x-1)^{-1/5}\,dx = \int_0^1(x-1)^{-1/5}\,dx + \int_1^{33}(x-1)^{-1/5}\,dx.$

Here $\int_0^1(x-1)^{-1/5}\,dx = \lim\limits_{t\to 1^-}\int_0^t(x-1)^{-1/5}\,dx = \lim\limits_{t\to 1^-}\left[\dfrac{5}{4}(x-1)^{4/5}\right]_0^t = \lim\limits_{t\to 1^-}\left[\dfrac{5}{4}(t-1)^{4/5}-\dfrac{5}{4}\right] = -\dfrac{5}{4}$

and $\int_1^{33}(x-1)^{-1/5}\,dx = \lim\limits_{t\to 1^+}\int_t^{33}(x-1)^{-1/5}\,dx = \lim\limits_{t\to 1^+}\left[\dfrac{5}{4}(x-1)^{4/5}\right]_t^{33} = \lim\limits_{t\to 1^+}\left[\dfrac{5}{4}\cdot 16 - \dfrac{5}{4}(t-1)^{4/5}\right] = 20.$

Thus, $\int_0^{33}(x-1)^{-1/5}\,dx = -\dfrac{5}{4}+20 = \dfrac{75}{4}.$ Convergent

29. There is an infinite discontinuity at $x = 0$. $\int_{-1}^1 \dfrac{e^x}{e^x-1}\,dx = \int_{-1}^0 \dfrac{e^x}{e^x-1}\,dx + \int_0^1 \dfrac{e^x}{e^x-1}\,dx.$

$\int_{-1}^0 \dfrac{e^x}{e^x-1}\,dx = \lim\limits_{t\to 0^-}\int_{-1}^t \dfrac{e^x}{e^x-1}\,dx = \lim\limits_{t\to 0^-}\left[\ln|e^x-1|\right]_{-1}^t = \lim\limits_{t\to 0^-}\left[\ln|e^t-1| - \ln|e^{-1}-1|\right] = -\infty,$

so $\int_{-1}^1 \dfrac{e^x}{e^x-1}\,dx$ is divergent. The integral $\int_0^1 \dfrac{e^x}{e^x-1}\,dx$ also diverges since

$\int_0^1 \dfrac{e^x}{e^x-1}\,dx = \lim\limits_{t\to 0^+}\int_t^1 \dfrac{e^x}{e^x-1}\,dx = \lim\limits_{t\to 0^+}\left[\ln|e^x-1|\right]_t^1 = \lim\limits_{t\to 0^+}\left[\ln|e-1| - \ln|e^t-1|\right] = \infty.$ Divergent

31. $I = \int_0^2 z^2\ln z\,dz = \lim\limits_{t\to 0^+}\int_t^2 z^2\ln z\,dz \overset{101}{=} \lim\limits_{t\to 0^+}\left[\dfrac{z^3}{3^2}(3\ln z - 1)\right]_t^2 = \lim\limits_{t\to 0^+}\left[\dfrac{8}{9}(3\ln 2 - 1) - \dfrac{1}{9}t^3(3\ln t - 1)\right]$

$= \dfrac{8}{3}\ln 2 - \dfrac{8}{9} - \dfrac{1}{9}\lim\limits_{t\to 0^+}\left[t^3(3\ln t - 1)\right] = \dfrac{8}{3}\ln 2 - \dfrac{8}{9} - \dfrac{1}{9}L$

Now $L = \lim\limits_{t\to 0^+}\left[t^3(3\ln t - 1)\right] = \lim\limits_{t\to 0^+}\dfrac{3\ln t - 1}{t^{-3}} \overset{H}{=} \lim\limits_{t\to 0^+}\dfrac{3/t}{-3/t^4} = \lim\limits_{t\to 0^+}(-t^3) = 0.$

Thus, $L = 0$ and $I = \dfrac{8}{3}\ln 2 - \dfrac{8}{9}.$ Convergent

33.

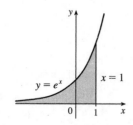

Area $= \int_{-\infty}^1 e^x\,dx = \lim\limits_{t\to -\infty}\left[e^x\right]_t^1 = e - \lim\limits_{t\to -\infty}e^t = e$

35.

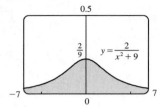

$$\text{Area} = \int_{-\infty}^{\infty} \frac{2}{x^2+9}\,dx = 2 \cdot 2 \int_{0}^{\infty} \frac{1}{x^2+9}\,dx$$

$$= 4 \lim_{t \to \infty} \int_{0}^{t} \frac{1}{x^2+9}\,dx = 4 \lim_{t \to \infty} \left[\frac{1}{3}\tan^{-1}\frac{x}{3} \right]_{0}^{t}$$

$$= \frac{4}{3} \lim_{t \to \infty} \left[\tan^{-1}\frac{t}{3} - 0 \right] = \frac{4}{3} \cdot \frac{\pi}{2} = \frac{2\pi}{3}$$

37.

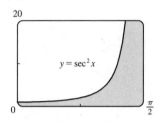

$$\text{Area} = \int_{0}^{\pi/2} \sec^2 x\,dx = \lim_{t \to (\pi/2)^-} \int_{0}^{t} \sec^2 x\,dx$$

$$= \lim_{t \to (\pi/2)^-} \left[\tan x\right]_{0}^{t} = \lim_{t \to (\pi/2)^-} (\tan t - 0)$$

$$= \infty$$

Infinite area

39. (a)

t	$\int_{1}^{t} g(x)\,dx$
2	0.447453
5	0.577101
10	0.621306
100	0.668479
1000	0.672957
10,000	0.673407

$$g(x) = \frac{\sin^2 x}{x^2}.$$

It appears that the integral is convergent.

(b) $-1 \le \sin x \le 1 \;\Rightarrow\; 0 \le \sin^2 x \le 1 \;\Rightarrow\; 0 \le \dfrac{\sin^2 x}{x^2} \le \dfrac{1}{x^2}$. Since $\displaystyle\int_{1}^{\infty} \frac{1}{x^2}\,dx$ is convergent

(Equation 2 with $p = 2 > 1$), $\displaystyle\int_{1}^{\infty} \frac{\sin^2 x}{x^2}\,dx$ is convergent by the Comparison Theorem.

(c)

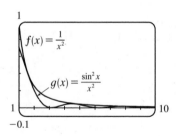

Since $\int_{1}^{\infty} f(x)\,dx$ is finite and the area under $g(x)$ is less than the area under $f(x)$ on any interval $[1, t]$, $\int_{1}^{\infty} g(x)\,dx$ must be finite; that is, the integral is convergent.

41. For $x \ge 1$, $\dfrac{\cos^2 x}{1+x^2} \le \dfrac{1}{1+x^2} < \dfrac{1}{x^2}$. $\displaystyle\int_{1}^{\infty} \frac{1}{x^2}\,dx$ is convergent by (2) with $p = 2 > 1$, so $\displaystyle\int_{1}^{\infty} \frac{\cos^2 x}{1+x^2}\,dx$ is convergent by the Comparison Theorem.

43. For $x \ge 1$, $x + e^{2x} > e^{2x} > 0 \;\Rightarrow\; \dfrac{1}{x+e^{2x}} \le \dfrac{1}{e^{2x}} = e^{-2x}$ on $[1, \infty)$.

$\int_{1}^{\infty} e^{-2x}\,dx = \lim_{t \to \infty} \left[-\frac{1}{2}e^{-2x} \right]_{1}^{t} = \lim_{t \to \infty} \left[-\frac{1}{2}e^{-2t} + \frac{1}{2}e^{-2} \right] = \frac{1}{2}e^{-2}$. Therefore, $\int_{1}^{\infty} e^{-2x}\,dx$ is convergent, and by the

Comparison Theorem, $\displaystyle\int_{1}^{\infty} \frac{dx}{x+e^{2x}}$ is also convergent.

45. $\dfrac{1}{x \sin x} \geq \dfrac{1}{x}$ on $\left(0, \frac{\pi}{2}\right]$ since $0 \leq \sin x \leq 1$. $\displaystyle\int_0^{\pi/2} \dfrac{dx}{x} = \lim_{t \to 0^+} \int_t^{\pi/2} \dfrac{dx}{x} = \lim_{t \to 0^+} \left[\ln x\right]_t^{\pi/2}.$

But $\ln t \to -\infty$ as $t \to 0^+$, so $\displaystyle\int_0^{\pi/2} \dfrac{dx}{x}$ is divergent, and by the Comparison Theorem, $\displaystyle\int_0^{\pi/2} \dfrac{dx}{x \sin x}$ is also divergent.

47. $\displaystyle\int_0^\infty \dfrac{dx}{\sqrt{x}\,(1+x)} = \int_0^1 \dfrac{dx}{\sqrt{x}\,(1+x)} + \int_1^\infty \dfrac{dx}{\sqrt{x}\,(1+x)} = \lim_{t \to 0^+} \int_t^1 \dfrac{dx}{\sqrt{x}\,(1+x)} + \lim_{t \to \infty} \int_1^t \dfrac{dx}{\sqrt{x}\,(1+x)}.$

Now $\displaystyle\int \dfrac{dx}{\sqrt{x}\,(1+x)} = \int \dfrac{2u\,du}{u(1+u^2)}$ $[u = \sqrt{x},\, x = u^2,\, dx = 2u\,du]$

$\qquad\qquad = 2\displaystyle\int \dfrac{du}{1+u^2} = 2\tan^{-1} u + C = 2\tan^{-1}\sqrt{x} + C,$

so $\displaystyle\int_0^\infty \dfrac{dx}{\sqrt{x}\,(1+x)} = \lim_{t \to 0^+} \left[2\tan^{-1}\sqrt{x}\right]_t^1 + \lim_{t \to \infty} \left[2\tan^{-1}\sqrt{x}\right]_1^t$

$\qquad\qquad = \displaystyle\lim_{t \to 0^+} \left[2\left(\tfrac{\pi}{4}\right) - 2\tan^{-1}\sqrt{t}\right] + \lim_{t \to \infty} \left[2\tan^{-1}\sqrt{t} - 2\left(\tfrac{\pi}{4}\right)\right] = \tfrac{\pi}{2} - 0 + 2\left(\tfrac{\pi}{2}\right) - \tfrac{\pi}{2} = \pi.$

49. If $p = 1$, then $\displaystyle\int_0^1 \dfrac{dx}{x^p} = \lim_{t \to 0^+} \int_t^1 \dfrac{dx}{x} = \lim_{t \to 0^+} \left[\ln x\right]_t^1 = \infty.$ Divergent.

If $p \neq 1$, then $\displaystyle\int_0^1 \dfrac{dx}{x^p} = \lim_{t \to 0^+} \int_t^1 \dfrac{dx}{x^p}$ $\left[\begin{matrix}\text{note that the integral is}\\ \text{not improper if } p < 0\end{matrix}\right]$ $= \displaystyle\lim_{t \to 0^+} \left[\dfrac{x^{-p+1}}{-p+1}\right]_t^1 = \lim_{t \to 0^+} \dfrac{1}{1-p}\left[1 - \dfrac{1}{t^{p-1}}\right]$

If $p > 1$, then $p - 1 > 0$, so $\dfrac{1}{t^{p-1}} \to \infty$ as $t \to 0^+$, and the integral diverges.

If $p < 1$, then $p - 1 < 0$, so $\dfrac{1}{t^{p-1}} \to 0$ as $t \to 0^+$ and $\displaystyle\int_0^1 \dfrac{dx}{x^p} = \dfrac{1}{1-p}\left[\lim_{t \to 0^+} \left(1 - t^{1-p}\right)\right] = \dfrac{1}{1-p}.$

Thus, the integral converges if and only if $p < 1$, and in that case its value is $\dfrac{1}{1-p}$.

51. (a) $I = \displaystyle\int_{-\infty}^\infty x\,dx = \int_{-\infty}^0 x\,dx + \int_0^\infty x\,dx$, and $\displaystyle\int_0^\infty x\,dx = \lim_{t \to \infty} \int_0^t x\,dx = \lim_{t \to \infty} \left[\tfrac{1}{2}x^2\right]_0^t = \lim_{t \to \infty} \left[\tfrac{1}{2}t^2 - 0\right] = \infty,$

so I is divergent.

(b) $\displaystyle\int_{-t}^t x\,dx = \left[\tfrac{1}{2}x^2\right]_{-t}^t = \tfrac{1}{2}t^2 - \tfrac{1}{2}t^2 = 0$, so $\displaystyle\lim_{t \to \infty} \int_{-t}^t x\,dx = 0$. Therefore, $\displaystyle\int_{-\infty}^\infty x\,dx \neq \lim_{t \to \infty} \int_{-t}^t x\,dx.$

53. We would expect a small percentage of bulbs to burn out in the first few hundred hours, most of the bulbs to burn out after close to 700 hours, and a few overachievers to burn on and on.

(a)

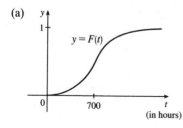

(b) $r(t) = F'(t)$ is the rate at which the fraction $F(t)$ of burnt-out bulbs increases as t increases. This could be interpreted as a fractional burnout rate.

(c) $\int_0^\infty r(t)\,dt = \lim_{x \to \infty} F(x) = 1$, since all of the bulbs will eventually burn out.

55. $I = \int_0^\infty te^{kt}\,dt = \lim_{s \to \infty}\left[\frac{1}{k^2}(kt-1)e^{kt}\right]_0^s$ [Formula 96, or parts] $= \lim_{s \to \infty}\left[\left(\frac{1}{k}se^{ks} - \frac{1}{k^2}e^{ks}\right) - \left(-\frac{1}{k^2}\right)\right]$.

Since $k < 0$ the first two terms approach 0 (you can verify that the first term does so with l'Hospital's Rule), so the limit is

equal to $1/k^2$. Thus, $M = -kI = -k\left(1/k^2\right) = -1/k = -1/(-0.000121) \approx 8264.5$ years.

57. $I = \int_a^\infty \frac{1}{x^2+1}\,dx = \lim_{t \to \infty}\int_a^t \frac{1}{x^2+1}\,dx = \lim_{t \to \infty}\left[\tan^{-1}x\right]_a^t = \lim_{t \to \infty}(\tan^{-1}t - \tan^{-1}a) = \frac{\pi}{2} - \tan^{-1}a.$

$I < 0.001 \;\Rightarrow\; \frac{\pi}{2} - \tan^{-1}a < 0.001 \;\Rightarrow\; \tan^{-1}a > \frac{\pi}{2} - 0.001 \;\Rightarrow\; a > \tan\left(\frac{\pi}{2} - 0.001\right) \approx 1000.$

59. We use integration by parts: Let $u = x$, $dv = xe^{-x^2}\,dx \;\Rightarrow\; du = dx$, $v = -\frac{1}{2}e^{-x^2}$. So

$\int_0^\infty x^2 e^{-x^2}\,dx = \lim_{t \to \infty}\left[-\frac{1}{2}xe^{-x^2}\right]_0^t + \frac{1}{2}\int_0^\infty e^{-x^2}\,dx = \lim_{t \to \infty}\left[-t\Big/\left(2e^{t^2}\right)\right] + \frac{1}{2}\int_0^\infty e^{-x^2}\,dx = \frac{1}{2}\int_0^\infty e^{-x^2}\,dx$

(The limit is 0 by l'Hospital's Rule.)

61. For the first part of the integral, let $x = 2\tan\theta \;\Rightarrow\; dx = 2\sec^2\theta\,d\theta.$

$\int \frac{1}{\sqrt{x^2+4}}\,dx = \int \frac{2\sec^2\theta}{2\sec\theta}\,d\theta = \int \sec\theta\,d\theta = \ln|\sec\theta + \tan\theta|.$

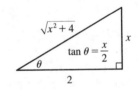

From the figure, $\tan\theta = \dfrac{x}{2}$, and $\sec\theta = \dfrac{\sqrt{x^2+4}}{2}$. So

$$I = \int_0^\infty \left(\frac{1}{\sqrt{x^2+4}} - \frac{C}{x+2}\right)dx = \lim_{t \to \infty}\left[\ln\left|\frac{\sqrt{x^2+4}}{2} + \frac{x}{2}\right| - C\ln|x+2|\right]_0^t$$

$$= \lim_{t \to \infty}\left[\ln\frac{\sqrt{t^2+4}+t}{2} - C\ln(t+2) - (\ln 1 - C\ln 2)\right]$$

$$= \lim_{t \to \infty}\left[\ln\left(\frac{\sqrt{t^2+4}+t}{2\,(t+2)^C}\right) + \ln 2^C\right] = \ln\left(\lim_{t \to \infty}\frac{t + \sqrt{t^2+4}}{(t+2)^C}\right) + \ln 2^{C-1}$$

Now $L = \lim_{t \to \infty}\dfrac{t + \sqrt{t^2+4}}{(t+2)^C} \overset{\text{H}}{=} \lim_{t \to \infty}\dfrac{1 + t/\sqrt{t^2+4}}{C\,(t+2)^{C-1}} = \dfrac{2}{C\,\lim\limits_{t \to \infty}(t+2)^{C-1}}.$

If $C < 1$, $L = \infty$ and I diverges.

If $C = 1$, $L = 2$ and I converges to $\ln 2 + \ln 2^0 = \ln 2.$

If $C > 1$, $L = 0$ and I diverges to $-\infty.$

6 Review

1. See Formula 6.1.1 or 6.1.2. We try to choose $u = f(x)$ to be a function that becomes simpler when differentiated (or at least not more complicated) as long as $dv = g'(x)\,dx$ can be readily integrated to give v.

2. See the margin note on page 312.

3. If $\sqrt{a^2 - x^2}$ occurs, try $x = a\sin\theta$; if $\sqrt{a^2 + x^2}$ occurs, try $x = a\tan\theta$, and if $\sqrt{x^2 - a^2}$ occurs, try $x = a\sec\theta$. See the Table of Trigonometric Substitutions on page 315.

4. See Equation 2 and Expressions 6, 8, and 10 in Section 6.3.

5. See the Midpoint Rule, the Trapezoidal Rule, and Simpson's Rule, as well as their associated error bounds, all in Section 6.6. We would expect the best estimate to be given by Simpson's Rule.

6. See Definitions 1(a), (b), and (c) in Section 6.6.

7. See Definitions 3(b), (a), and (c) in Section 6.6.

8. See the Comparison Theorem on page 351.

1. False. Since the numerator has a higher degree than the denominator, $\dfrac{x(x^2+4)}{x^2-4} = x + \dfrac{8x}{x^2-4} = x + \dfrac{A}{x+2} + \dfrac{B}{x-2}$.

3. False. It can be put in the form $\dfrac{A}{x} + \dfrac{B}{x^2} + \dfrac{C}{x-4}$.

5. False. This is an improper integral, since the denominator vanishes at $x = 1$.

$$\int_0^4 \frac{x}{x^2-1}\,dx = \int_0^1 \frac{x}{x^2-1}\,dx + \int_1^4 \frac{x}{x^2-1}\,dx \text{ and}$$

$$\int_0^1 \frac{x}{x^2-1}\,dx = \lim_{t\to 1^-} \int_0^t \frac{x}{x^2-1}\,dx = \lim_{t\to 1^-}\left[\tfrac{1}{2}\ln|x^2-1|\right]_0^t = \lim_{t\to 1^-} \tfrac{1}{2}\ln|t^2-1| = \infty$$

So the integral diverges.

7. False. See Exercise 51 in Section 6.6.

9. (a) True. See the end of Section 6.4.

(b) False. Examples include the functions $f(x) = e^{x^2}$, $g(x) = \sin(x^2)$, and $h(x) = \dfrac{\sin x}{x}$.

11. False. If $f(x) = 1/x$, then f is continuous and decreasing on $[1, \infty)$ with $\lim\limits_{x\to\infty} f(x) = 0$, but $\int_1^\infty f(x)\,dx$ is divergent.

13. False. Take $f(x) = 1$ for all x and $g(x) = -1$ for all x. Then $\int_a^\infty f(x)\,dx = \infty$ [divergent]

and $\int_a^\infty g(x)\,dx = -\infty$ [divergent], but $\int_a^\infty [f(x) + g(x)]\,dx = 0$ [convergent].

EXERCISES

1. $\int_0^5 \frac{x}{x+10}\,dx = \int_0^5 \left(1 - \frac{10}{x+10}\right)dx = \Big[x - 10\ln(x+10)\Big]_0^5 = 5 - 10\ln 15 + 10\ln 10 = 5 + 10\ln\frac{10}{15} = 5 + 10\ln\frac{2}{3}$

3. $\int_0^{\pi/2} \frac{\cos\theta}{1 + \sin\theta}\,d\theta = \Big[\ln(1 + \sin\theta)\Big]_0^{\pi/2} = \ln 2 - \ln 1 = \ln 2$

5. Let $u = \sec x$. Then $du = \sec x \tan x\,dx$, so

$$\int \tan^7 x \sec^3 x\,dx = \int \tan^6 x \sec^2 x \sec x \tan x\,dx = \int (u^2 - 1)^3 u^2\,du = \int (u^8 - 3u^6 + 3u^4 - u^2)\,du$$

$$= \tfrac{1}{9}u^9 - \tfrac{3}{7}u^7 + \tfrac{3}{5}u^5 - \tfrac{1}{3}u^3 + C = \tfrac{1}{9}\sec^9 x - \tfrac{3}{7}\sec^7 x + \tfrac{3}{5}\sec^5 x - \tfrac{1}{3}\sec^3 x + C$$

7. Let $u = \ln t$, $du = dt/t$. Then $\int \frac{\sin(\ln t)}{t}\,dt = \int \sin u\,du = -\cos u + C = -\cos(\ln t) + C$.

9. $\int_1^4 x^{3/2}\ln x\,dx \quad \begin{bmatrix} u = \ln x, & dv = x^{3/2}\,dx, \\ du = dx/x & v = \tfrac{2}{5}x^{5/2} \end{bmatrix} = \frac{2}{5}\Big[x^{5/2}\ln x\Big]_1^4 - \frac{2}{5}\int_1^4 x^{3/2}\,dx = \tfrac{2}{5}(32\ln 4 - \ln 1) - \frac{2}{5}\Big[\tfrac{2}{5}x^{5/2}\Big]_1^4$

$$= \tfrac{2}{5}(64\ln 2) - \tfrac{4}{25}(32 - 1) = \tfrac{128}{5}\ln 2 - \tfrac{124}{25} \quad \Big[\text{or } \tfrac{64}{5}\ln 4 - \tfrac{124}{25}\Big]$$

11. Let $x = \sec\theta$. Then

$$\int_1^2 \frac{\sqrt{x^2 - 1}}{x}\,dx = \int_0^{\pi/3} \frac{\tan\theta}{\sec\theta}\sec\theta\tan\theta\,d\theta = \int_0^{\pi/3}\tan^2\theta\,d\theta = \int_0^{\pi/3}(\sec^2\theta - 1)\,d\theta = \Big[\tan\theta - \theta\Big]_0^{\pi/3} = \sqrt{3} - \tfrac{\pi}{3}.$$

13. $\int \frac{dx}{x^3 + x} = \int \left(\frac{1}{x} - \frac{x}{x^2 + 1}\right)dx = \ln|x| - \tfrac{1}{2}\ln(x^2 + 1) + C$

15. $\int \sin^2\theta\cos^5\theta\,d\theta = \int \sin^2\theta\,(\cos^2\theta)^2\cos\theta\,d\theta = \int \sin^2\theta\,(1 - \sin^2\theta)^2\cos\theta\,d\theta$

$$= \int u^2\left(1 - u^2\right)^2\,du \quad [u = \sin\theta, du = \cos\theta\,d\theta] = \int u^2(1 - 2u^2 + u^4)\,du$$

$$= \int (u^2 - 2u^4 + u^6)\,du = \tfrac{1}{3}u^3 - \tfrac{2}{5}u^5 + \tfrac{1}{7}u^7 + C = \tfrac{1}{3}\sin^3\theta - \tfrac{2}{5}\sin^5\theta + \tfrac{1}{7}\sin^7\theta + C$$

17. Integrate by parts with $u = x$, $dv = \sec x \tan x\,dx \Rightarrow du = dx$, $v = \sec x$:

$$\int x \sec x \tan x\,dx = x\sec x - \int \sec x\,dx \overset{14}{=} x\sec x - \ln|\sec x + \tan x| + C.$$

19. $\int \frac{x + 1}{9x^2 + 6x + 5}\,dx = \int \frac{x + 1}{(9x^2 + 6x + 1) + 4}\,dx = \int \frac{x + 1}{(3x + 1)^2 + 4}\,dx \quad [u = 3x + 1, u = 3\,dx]$

$$= \int \frac{\left[\tfrac{1}{3}(u - 1)\right] + 1}{u^2 + 4}\left(\tfrac{1}{3}\,du\right) = \frac{1}{3}\cdot\frac{1}{3}\int \frac{(u - 1) + 3}{u^2 + 4}\,du$$

$$= \frac{1}{9}\int \frac{u}{u^2 + 4}\,du + \frac{1}{9}\int \frac{2}{u^2 + 2^2}\,du = \frac{1}{9}\cdot\frac{1}{2}\ln(u^2 + 4) + \frac{2}{9}\cdot\frac{1}{2}\tan^{-1}\left(\tfrac{1}{2}u\right) + C$$

$$= \tfrac{1}{18}\ln(9x^2 + 6x + 5) + \tfrac{1}{9}\tan^{-1}\left[\tfrac{1}{2}(3x + 1)\right] + C$$

21.

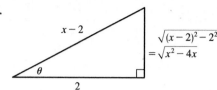

$$\int \frac{dx}{\sqrt{x^2 - 4x}} = \int \frac{dx}{\sqrt{(x^2 - 4x + 4) - 4}} = \int \frac{dx}{\sqrt{(x-2)^2 - 2^2}}$$

$$= \int \frac{2 \sec\theta \tan\theta \, d\theta}{2\tan\theta} \quad \begin{bmatrix} x - 2 = 2\sec\theta, \\ dx = 2\sec\theta \tan\theta \, d\theta \end{bmatrix}$$

$$= \int \sec\theta \, d\theta = \ln|\sec\theta + \tan\theta| + C_1$$

$$= \ln \left| \frac{x-2}{2} + \frac{\sqrt{x^2 - 4x}}{2} \right| + C_1$$

$$= \ln \left| x - 2 + \sqrt{x^2 - 4x} \right| + C, \text{ where } C = C_1 - \ln 2$$

23. Let $u = \cot 4x$. Then $du = -4\csc^2 4x \, dx \quad \Rightarrow$

$\int \csc^4 4x \, dx = \int (\cot^2 4x + 1)\csc^2 4x \, dx = \int (u^2 + 1)\left(-\frac{1}{4}\, du\right) = -\frac{1}{4}\left(\frac{1}{3}u^3 + u\right) + C = -\frac{1}{12}(\cot^3 4x + 3\cot 4x) + C.$

25. $\dfrac{3x^3 - x^2 + 6x - 4}{(x^2 + 1)(x^2 + 2)} = \dfrac{Ax + B}{x^2 + 1} + \dfrac{Cx + D}{x^2 + 2} \quad \Rightarrow \quad 3x^3 - x^2 + 6x - 4 = (Ax + B)(x^2 + 2) + (Cx + D)(x^2 + 1).$

Equating the coefficients gives $A + C = 3$, $B + D = -1$, $2A + C = 6$, and $2B + D = -4 \quad \Rightarrow$

$A = 3$, $C = 0$, $B = -3$, and $D = 2$. Now

$$\int \frac{3x^3 - x^2 + 6x - 4}{(x^2 + 1)(x^2 + 2)} \, dx = 3\int \frac{x - 1}{x^2 + 1} \, dx + 2\int \frac{dx}{x^2 + 2} = \tfrac{3}{2}\ln(x^2 + 1) - 3\tan^{-1} x + \sqrt{2}\tan^{-1}\left(\frac{1}{\sqrt{2}}x\right) + C.$$

27. $\int_0^{\pi/2} \cos^3 x \sin 2x \, dx = \int_0^{\pi/2} \cos^3 x \, (2\sin x \, \cos x) \, dx = \int_0^{\pi/2} 2\cos^4 x \, \sin x \, dx = \left[-\tfrac{2}{5}\cos^5 x\right]_0^{\pi/2} = \tfrac{2}{5}$

29. The product of an odd function and an even function is an odd function, so $f(x) = x^5 \sec x$ is an odd function.

By Theorem 5.5.7(b), $\int_{-1}^{1} x^5 \sec x \, dx = 0.$

31. Let $u = \sqrt{e^x - 1}$. Then $u^2 = e^x - 1$ and $2u\, du = e^x \, dx$. Also, $e^x + 8 = u^2 + 9$. Thus,

$$\int_0^{\ln 10} \frac{e^x \sqrt{e^x - 1}}{e^x + 8} \, dx = \int_0^3 \frac{u \cdot 2u \, du}{u^2 + 9} = 2\int_0^3 \frac{u^2}{u^2 + 9} \, du = 2\int_0^3 \left(1 - \frac{9}{u^2 + 9}\right) du = 2\left[u - \frac{9}{3}\tan^{-1}\left(\frac{u}{3}\right)\right]_0^3$$

$$= 2\left[(3 - 3\tan^{-1} 1) - 0\right] = 2\left(3 - 3\cdot\frac{\pi}{4}\right) = 6 - \frac{3\pi}{2}$$

33. Let $x = 2\sin\theta \quad \Rightarrow \quad (4 - x^2)^{3/2} = (2\cos\theta)^3$, $dx = 2\cos\theta \, d\theta$, so

$$\int \frac{x^2}{(4 - x^2)^{3/2}} \, dx = \int \frac{4\sin^2\theta}{8\cos^3\theta} \, 2\cos\theta \, d\theta = \int \tan^2\theta \, d\theta = \int (\sec^2\theta - 1)d\theta$$

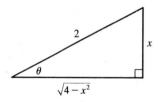

$$= \tan\theta - \theta + C = \frac{x}{\sqrt{4 - x^2}} - \sin^{-1}\left(\frac{x}{2}\right) + C$$

35. $\displaystyle\int \frac{1}{\sqrt{x + x^{3/2}}} \, dx = \int \frac{dx}{\sqrt{x\,(1 + \sqrt{x})}} = \int \frac{dx}{\sqrt{x}\sqrt{1 + \sqrt{x}}} \quad \begin{bmatrix} u = 1 + \sqrt{x}, \\ du = \dfrac{dx}{2\sqrt{x}} \end{bmatrix} = \int \frac{2\, du}{\sqrt{u}} = \int 2u^{-1/2} \, du$

$$= 4\sqrt{u} + C = 4\sqrt{1 + \sqrt{x}} + C$$

37. $\int (\cos x + \sin x)^2 \cos 2x \, dx = \int (\cos^2 x + 2 \sin x \, \cos x + \sin^2 x) \cos 2x \, dx = \int (1 + \sin 2x) \cos 2x \, dx$

$$= \int \cos 2x \, dx + \tfrac{1}{2} \int \sin 4x \, dx = \tfrac{1}{2} \sin 2x - \tfrac{1}{8} \cos 4x + C$$

Or: $\int (\cos x + \sin x)^2 \cos 2x \, dx = \int (\cos x + \sin x)^2 (\cos^2 x - \sin^2 x) \, dx$

$$= \int (\cos x + \sin x)^3 (\cos x - \sin x) \, dx = \tfrac{1}{4} (\cos x + \sin x)^4 + C_1$$

39. We'll integrate $I = \displaystyle\int \frac{xe^{2x}}{(1+2x)^2} \, dx$ by parts with $u = xe^{2x}$ and $dv = \dfrac{dx}{(1+2x)^2}$. Then $du = (x \cdot 2e^{2x} + e^{2x} \cdot 1) \, dx$

and $v = -\dfrac{1}{2} \cdot \dfrac{1}{1+2x}$, so

$$I = -\frac{1}{2} \cdot \frac{xe^{2x}}{1+2x} - \int \left[-\frac{1}{2} \cdot \frac{e^{2x}(2x+1)}{1+2x} \right] dx = -\frac{xe^{2x}}{4x+2} + \frac{1}{2} \cdot \frac{1}{2} e^{2x} + C = e^{2x} \left(\frac{1}{4} - \frac{x}{4x+2} \right) + C.$$

Thus, $\displaystyle\int_0^{1/2} \frac{xe^{2x}}{(1+2x)^2} \, dx = \left[e^{2x} \left(\frac{1}{4} - \frac{x}{4x+2} \right) \right]_0^{1/2} = e \left(\frac{1}{4} - \frac{1}{8} \right) - 1 \left(\frac{1}{4} - 0 \right) = \frac{1}{8} e - \frac{1}{4}.$

41. $\displaystyle\int_1^\infty \frac{1}{(2x+1)^3} \, dx = \lim_{t \to \infty} \int_1^t \frac{1}{(2x+1)^3} \, dx = \lim_{t \to \infty} \int_1^t \tfrac{1}{2}(2x+1)^{-3} 2 \, dx$

$$= \lim_{t \to \infty} \left[-\frac{1}{4(2x+1)^2} \right]_1^t = -\frac{1}{4} \lim_{t \to \infty} \left[\frac{1}{(2t+1)^2} - \frac{1}{9} \right] = -\frac{1}{4} \left(0 - \frac{1}{9} \right) = \frac{1}{36}$$

43. $\displaystyle\int \frac{dx}{x \ln x} \quad \begin{bmatrix} u = \ln x, \\ du = dx/x \end{bmatrix} = \int \frac{du}{u} = \ln |u| + C = \ln |\ln x| + C$, so

$$\int_2^\infty \frac{dx}{x \ln x} = \lim_{t \to \infty} \int_2^t \frac{dx}{x \ln x} = \lim_{t \to \infty} \Big[\ln |\ln x| \Big]_2^t = \lim_{t \to \infty} [\ln(\ln t) - \ln(\ln 2)] = \infty, \text{ so the integral is divergent.}$$

45. $\displaystyle\int_0^4 \frac{\ln x}{\sqrt{x}} \, dx = \lim_{t \to 0^+} \int_t^4 \frac{\ln x}{\sqrt{x}} \, dx \overset{\star}{=} \lim_{t \to 0^+} \left[2\sqrt{x} \ln x - 4\sqrt{x} \right]_t^4$

$$= \lim_{t \to 0^+} \left[(2 \cdot 2 \ln 4 - 4 \cdot 2) - (2\sqrt{t} \ln t - 4\sqrt{t}) \right] \overset{\star\star}{=} (4 \ln 4 - 8) - (0 - 0) = 4 \ln 4 - 8$$

$(\star)$ Let $u = \ln x$, $dv = \dfrac{1}{\sqrt{x}} \, dx \ \Rightarrow \ du = \dfrac{1}{x} \, dx$, $v = 2\sqrt{x}$. Then

$$\int \frac{\ln x}{\sqrt{x}} \, dx = 2\sqrt{x} \ln x - 2 \int \frac{dx}{\sqrt{x}} = 2\sqrt{x} \ln x - 4\sqrt{x} + C$$

$(\star\star)$ $\displaystyle\lim_{t \to 0^+} \left(2\sqrt{t} \ln t \right) = \lim_{t \to 0^+} \frac{2 \ln t}{t^{-1/2}} \overset{H}{=} \lim_{t \to 0^+} \frac{2/t}{-\frac{1}{2} t^{-3/2}} = \lim_{t \to 0^+} \left(-4\sqrt{t} \right) = 0$

47. $\displaystyle\int_0^3 \frac{dx}{x^2 - x - 2} = \int_0^3 \frac{dx}{(x+1)(x-2)} = \int_0^2 \frac{dx}{(x+1)(x-2)} + \int_2^3 \frac{dx}{(x+1)(x-2)}$, and

$$\int_2^3 \frac{dx}{x^2 - x - 2} = \lim_{t \to 2^+} \int_t^3 \left[\frac{-1/3}{x+1} + \frac{1/3}{x-2} \right] dx = \lim_{t \to 2^+} \left[\frac{1}{3} \ln \left| \frac{x-2}{x+1} \right| \right]_t^3 = \lim_{t \to 2^+} \left[\frac{1}{3} \ln \frac{1}{4} - \frac{1}{3} \ln \left| \frac{t-2}{t+1} \right| \right] = \infty,$$

so $\displaystyle\int_0^3 \frac{dx}{x^2 - x - 2}$ diverges.

49. Let $u = 2x + 1$. Then

$$\int_{-\infty}^{\infty} \frac{dx}{4x^2 + 4x + 5} = \int_{-\infty}^{\infty} \frac{\frac{1}{2}\,du}{u^2 + 4} = \frac{1}{2}\int_{-\infty}^{0} \frac{du}{u^2 + 4} + \frac{1}{2}\int_{0}^{\infty} \frac{du}{u^2 + 4}$$

$$= \frac{1}{2}\lim_{t\to-\infty}\left[\frac{1}{2}\tan^{-1}\left(\frac{1}{2}u\right)\right]_t^0 + \frac{1}{2}\lim_{t\to\infty}\left[\frac{1}{2}\tan^{-1}\left(\frac{1}{2}u\right)\right]_0^t = \frac{1}{4}\left[0 - \left(-\frac{\pi}{2}\right)\right] + \frac{1}{4}\left[\frac{\pi}{2} - 0\right] = \frac{\pi}{4}$$

51. $u = e^x \ \Rightarrow \ du = e^x\,dx$, so

$$\int e^x \sqrt{1 - e^{2x}}\,dx = \int \sqrt{1 - u^2}\,du \overset{30}{=} \frac{1}{2}u\sqrt{1 - u^2} + \frac{1}{2}\sin^{-1}u + C = \frac{1}{2}\left[e^x\sqrt{1 - e^{2x}} + \sin^{-1}(e^x)\right] + C.$$

53. $\displaystyle\int \sqrt{x^2 + x + 1}\,dx = \int \sqrt{x^2 + x + \frac{1}{4} + \frac{3}{4}}\,dx = \int \sqrt{\left(x + \frac{1}{2}\right)^2 + \frac{3}{4}}\,dx$

$$= \int \sqrt{u^2 + \left(\frac{\sqrt{3}}{2}\right)^2}\,du \qquad [u = x + \frac{1}{2},\, du = dx]$$

$$\overset{21}{=} \frac{1}{2}u\sqrt{u^2 + \frac{3}{4}} + \frac{3}{8}\ln\left(u + \sqrt{u^2 + \frac{3}{4}}\right) + C$$

$$= \frac{2x + 1}{4}\sqrt{x^2 + x + 1} + \frac{3}{8}\ln\left(x + \frac{1}{2} + \sqrt{x^2 + x + 1}\right) + C$$

55. For $n \geq 0$, $\int_0^\infty x^n\,dx = \lim_{t\to\infty}\left[x^{n+1}/(n+1)\right]_0^t = \infty$. For $n < 0$, $\int_0^\infty x^n\,dx = \int_0^1 x^n\,dx + \int_1^\infty x^n\,dx$. Both integrals are

improper. By (6.6.2), the second integral diverges if $-1 \leq n < 0$. By Exercise 6.6.49, the first integral diverges if $n \leq -1$.

Thus, $\int_0^\infty x^n\,dx$ is divergent for all values of n.

57. $f(x) = \sqrt{1 + x^4}$, $\ \Delta x = \dfrac{b - a}{n} = \dfrac{1 - 0}{10} = \dfrac{1}{10}$.

(a) $T_{10} = \frac{1}{10\cdot 2}\{f(0) + 2\left[f(0.1) + f(0.2) + \cdots + f(0.9)\right] + f(1)\} \approx 1.090608$

(b) $M_{10} = \frac{1}{10}\left[f\left(\frac{1}{20}\right) + f\left(\frac{3}{20}\right) + f\left(\frac{5}{20}\right) + \cdots + f\left(\frac{19}{20}\right)\right] \approx 1.088840$

(c) $S_{10} = \frac{1}{10\cdot 3}[f(0) + 4f(0.1) + 2f(0.2) + \cdots + 4f(0.9) + f(1)] \approx 1.089429$

f is concave upward, so the Trapezoidal Rule gives us an overestimate, the Midpoint Rule gives an underestimate, and we

cannot tell whether Simpson's Rule gives us an overestimate or an underestimate.

59. $f(x) = (1 + x^4)^{1/2}$, $f'(x) = \frac{1}{2}(1 + x^4)^{-1/2}(4x^3) = 2x^3(1 + x^4)^{-1/2}$, $f''(x) = (2x^6 + 6x^2)(1 + x^4)^{-3/2}$.

A graph of f'' on $[0, 1]$ shows that it has its maximum at $x = 1$, so $|f''(x)| \leq f''(1) = \sqrt{8}$ on $[0, 1]$. By taking $K = \sqrt{8}$, we

find that the error in Exercise 57(a) is bounded by $\dfrac{K(b - a)^3}{12n^2} = \dfrac{\sqrt{8}}{1200} \approx 0.0024$, and in (b) by about $\frac{1}{2}(0.0024) = 0.0012$.

Note: Another way to estimate K is to let $x = 1$ in the factor $2x^6 + 6x^2$ (maximizing the numerator) and let $x = 0$ in the

factor $(1 + x^4)^{-3/2}$ (minimizing the denominator). Doing so gives us $K = 8$ and errors of $0.00\overline{6}$ and $0.00\overline{3}$. Using $K = 8$ for

the Trapezoidal Rule, we have $|E_T| \leq \dfrac{K(b - a)^3}{12n^2} \leq 0.00001 \ \Leftrightarrow \ \dfrac{8(1 - 0)^3}{12n^2} \leq \dfrac{1}{100,000} \ \Leftrightarrow \ n^2 \geq \dfrac{800,000}{12} \ \Leftrightarrow$

$n \gtrsim 258.2$, so we should take $n = 259$. For the Midpoint Rule, $|E_M| \leq \dfrac{K(b - a)^3}{24n^2} \leq 0.00001 \ \Leftrightarrow \ n^2 \geq \dfrac{800,000}{24} \ \Leftrightarrow$

$n \gtrsim 182.6$, so we should take $n = 183$.

61. $\Delta t = \left(\frac{10}{60} - 0\right)/10 = \frac{1}{60}$.

Distance traveled $= \int_0^{10} v\, dt \approx S_{10}$

$$= \frac{1}{60 \cdot 3}[40 + 4(42) + 2(45) + 4(49) + 2(52) + 4(54) + 2(56) + 4(57) + 2(57) + 4(55) + 56]$$

$$= \frac{1}{180}(1544) = 8.5\overline{7} \text{ mi}$$

63. (a) $f(x) = \sin(\sin x)$. A CAS gives

$$f^{(4)}(x) = \sin(\sin x)\left[\cos^4 x + 7\cos^2 x - 3\right]$$
$$+ \cos(\sin x)\left[6\cos^2 x \, \sin x + \sin x\right]$$

From the graph, we see that $\left|f^{(4)}(x)\right| < 3.8$ for $x \in [0, \pi]$.

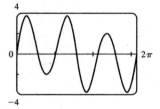

(b) We use Simpson's Rule with $f(x) = \sin(\sin x)$ and $\Delta x = \frac{\pi}{10}$:

$$\int_0^\pi f(x)\, dx \approx \frac{\pi}{10 \cdot 3}\left[f(0) + 4f\left(\frac{\pi}{10}\right) + 2f\left(\frac{2\pi}{10}\right) + \cdots + 4f\left(\frac{9\pi}{10}\right) + f(\pi)\right] \approx 1.786721$$

From part (a), we know that $\left|f^{(4)}(x)\right| < 3.8$ on $[0, \pi]$, so we use Theorem 6.5.4 with $K = 3.8$, and estimate the error

as $|E_S| \le \dfrac{3.8(\pi - 0)^5}{180(10)^4} \approx 0.000646$.

(c) If we want the error to be less than 0.00001, we must have $|E_S| \le \dfrac{3.8\pi^5}{180n^4} \le 0.00001$, so

$n^4 \ge \dfrac{3.8\pi^5}{180(0.00001)} \approx 646{,}041.6 \quad \Rightarrow \quad n \ge 28.35$. Since n must be even for Simpson's Rule, we must have $n \ge 30$ to

ensure the desired accuracy.

65. By the Fundamental Theorem of Calculus,

$$\int_0^\infty f'(x)\, dx = \lim_{t \to \infty} \int_0^t f'(x)\, dx = \lim_{t \to \infty}[f(t) - f(0)] = \lim_{t \to \infty} f(t) - f(0) = 0 - f(0) = -f(0)$$

7 □ APPLICATIONS OF INTEGRATION

7.1 Areas Between Curves

1. $A = \displaystyle\int_{x=0}^{x=4} (y_T - y_B)\,dx = \int_0^4 [(5x - x^2) - x]\,dx = \int_0^4 (4x - x^2)\,dx = \left[2x^2 - \tfrac{1}{3}x^3\right]_0^4 = \left(32 - \tfrac{64}{3}\right) - (0) = \tfrac{32}{3}$

3. $A = \displaystyle\int_{y=-1}^{y=1} (x_R - x_L)\,dy = \int_{-1}^1 [e^y - (y^2 - 2)]\,dy = \int_{-1}^1 (e^y - y^2 + 2)\,dy = \left[e^y - \tfrac{1}{3}y^3 + 2y\right]_{-1}^1$

$\qquad = \left(e^1 - \tfrac{1}{3} + 2\right) - \left(e^{-1} + \tfrac{1}{3} - 2\right) = e - \dfrac{1}{e} + \dfrac{10}{3}$

5. $A = \displaystyle\int_{-1}^2 [(9 - x^2) - (x + 1)]\,dx$

$\qquad = \displaystyle\int_{-1}^2 (8 - x - x^2)\,dx$

$\qquad = \left[8x - \dfrac{x^2}{2} - \dfrac{x^3}{3}\right]_{-1}^2$

$\qquad = \left(16 - 2 - \tfrac{8}{3}\right) - \left(-8 - \tfrac{1}{2} + \tfrac{1}{3}\right)$

$\qquad = 22 - 3 + \tfrac{1}{2} = \tfrac{39}{2}$

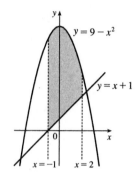 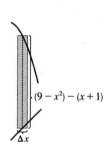

7. The curves intersect when $x = x^2 \;\Rightarrow\; x^2 - x = 0 \;\Leftrightarrow\; x(x - 1) = 0 \;\Leftrightarrow\; x = 0, 1$.

$\qquad A = \displaystyle\int_0^1 (x - x^2)\,dx = \left[\tfrac{1}{2}x^2 - \tfrac{1}{3}x^3\right]_0^1$

$\qquad\quad = \tfrac{1}{2} - \tfrac{1}{3} = \tfrac{1}{6}$

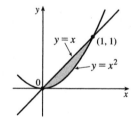 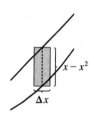

9. $12 - x^2 = x^2 - 6 \;\Leftrightarrow\; 2x^2 = 18 \;\Leftrightarrow\; x^2 = 9 \;\Leftrightarrow\; x = \pm 3$, so

$$A = \int_{-3}^3 [(12 - x^2) - (x^2 - 6)]\,dx = 2\int_0^3 (18 - 2x^2)\,dx \qquad \text{[by symmetry]}$$

$$= 2\left[18x - \tfrac{2}{3}x^3\right]_0^3 = 2[(54 - 18) - 0] = 2(36) = 72$$

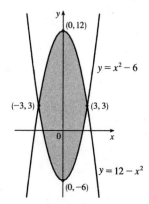 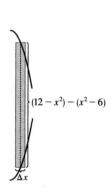

11. $2y^2 = 1 - y$ $\Leftrightarrow$ $2y^2 + y - 1 = 0$ $\Leftrightarrow$ $(2y - 1)(y + 1) = 0$ $\Leftrightarrow$ $y = \frac{1}{2}$ or -1, so $x = \frac{1}{2}$ or 2 and

$$A = \int_{-1}^{1/2} [(1 - y) - 2y^2] \, dy = \int_{-1}^{1/2} (1 - y - 2y^2) \, dy = \left[y - \frac{1}{2}y^2 - \frac{2}{3}y^3 \right]_{-1}^{1/2}$$

$$= \left(\frac{1}{2} - \frac{1}{8} - \frac{1}{12} \right) - \left(-1 - \frac{1}{2} + \frac{2}{3} \right) = \frac{7}{24} - \left(-\frac{5}{6} \right) = \frac{7}{24} + \frac{20}{24} = \frac{27}{24} = \frac{9}{8}$$

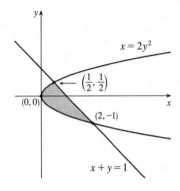

13. $2y^2 = 4 + y^2$ $\Leftrightarrow$ $y^2 = 4$ $\Leftrightarrow$ $y = \pm 2$, so

$$A = \int_{-2}^{2} [(4 + y^2) - 2y^2] \, dy = 2 \int_{0}^{2} (4 - y^2) \, dy \qquad \text{[by symmetry]}$$

$$= 2 \left[4y - \frac{1}{3}y^3 \right]_0^2 = 2 \left(8 - \frac{8}{3} \right) = \frac{32}{3}$$

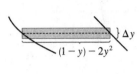
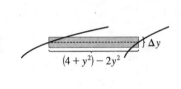

15. $1/x = x$ $\Leftrightarrow$ $1 = x^2$ $\Leftrightarrow$ $x = \pm 1$ and $1/x = \frac{1}{4}x$ $\Leftrightarrow$ $4 = x^2$ $\Leftrightarrow$ $x = \pm 2$, so for $x > 0$,

$$A = \int_0^1 \left(x - \frac{1}{4}x \right) dx + \int_1^2 \left(\frac{1}{x} - \frac{1}{4}x \right) dx = \int_0^1 \left(\frac{3}{4}x \right) dx + \int_1^2 \left(\frac{1}{x} - \frac{1}{4}x \right) dx$$

$$= \left[\frac{3}{8}x^2 \right]_0^1 + \left[\ln |x| - \frac{1}{8}x^2 \right]_1^2 = \frac{3}{8} + \left(\ln 2 - \frac{1}{2} \right) - \left(0 - \frac{1}{8} \right) = \ln 2$$

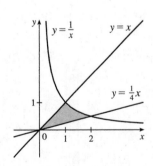

17.

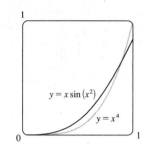

From the graph, we see that the curves intersect at $x = 0$ and $x = a \approx 0.896$, with $x \sin(x^2) > x^4$ on $(0, a)$. So the area A of the region bounded by the curves is

$$A = \int_0^a \left[x\sin(x^2) - x^4 \right] dx = \left[-\tfrac{1}{2}\cos(x^2) - \tfrac{1}{5}x^5 \right]_0^a$$

$$= -\tfrac{1}{2}\cos(a^2) - \tfrac{1}{5}a^5 + \tfrac{1}{2} \approx 0.037$$

19.

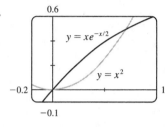

From the graph, we see that the curves intersect at $x = 0$ and $x = a \approx 0.70$, with $xe^{-x/2} > x^2$ on $(0, a)$. So the area A of the region bounded by the curves is

$$A = \int_0^a \left(xe^{-x/2} - x^2 \right) dx$$

$$= \left[4\left(-\tfrac{1}{2}x - 1\right)e^{-x/2} - \tfrac{1}{3}x^3 \right]_0^a \qquad \text{[Formula 96 with } a = -\tfrac{1}{2}\text{]}$$

$$\approx 0.08$$

21. $\cos x = \sin 2x = 2\sin x \cos x \quad \Leftrightarrow \quad 2\sin x \cos x - \cos x = 0 \quad \Leftrightarrow \quad \cos x\,(2\sin x - 1) = 0 \quad \Leftrightarrow$

$2\sin x = 1$ or $\cos x = 0 \quad \Leftrightarrow \quad x = \tfrac{\pi}{6}$ or $\tfrac{\pi}{2}$.

$A = \int_0^{\pi/6} (\cos x - \sin 2x)\, dx + \int_{\pi/6}^{\pi/2} (\sin 2x - \cos x)\, dx$

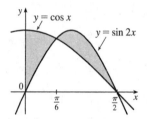

$= \left[\sin x + \tfrac{1}{2}\cos 2x \right]_0^{\pi/6} + \left[-\tfrac{1}{2}\cos 2x - \sin x \right]_{\pi/6}^{\pi/2}$

$= \left(\tfrac{1}{2} + \tfrac{1}{2}\cdot\tfrac{1}{2} \right) - \left(0 + \tfrac{1}{2}\cdot 1 \right) + \left[-\tfrac{1}{2}\cdot(-1) - 1 \right] - \left(-\tfrac{1}{2}\cdot\tfrac{1}{2} - \tfrac{1}{2} \right)$

$= \tfrac{3}{4} - \tfrac{1}{2} - \tfrac{1}{2} + \tfrac{3}{4} = \tfrac{1}{2}$

23. As in Example 4, we approximate the distance between the two cars after ten seconds using Simpson's Rule with $\Delta t = 1\,\text{s} = \frac{1}{3600}$ h.

$\text{distance}_{\text{Kelly}} - \text{distance}_{\text{Chris}} = \int_0^{10} v_K\, dt - \int_0^{10} v_C\, dt = \int_0^{10} (v_K - v_C)\, dt \approx S_{10}$

$= \frac{1}{3\cdot 3600}[(0 - 0) + 4(22 - 20) + 2(37 - 32) + 4(52 - 46) + 2(61 - 54) + 4(71 - 62)$

$\qquad + 2(80 - 69) + 4(86 - 75) + 2(93 - 81) + 4(98 - 86) + (102 - 90)]$

$= \frac{1}{10{,}800}(242) = \frac{121}{5400}\ \text{mi}$

So after 10 seconds, Kelly's car is about $\dfrac{121}{5400}\ \text{mi}\left(5280\dfrac{\text{ft}}{\text{mi}} \right) \approx 118\ \text{ft}$ ahead of Chris's.

25. If $x = $ distance from left end of pool and $w = w(x) = $ width at x, then Simpson's Rule with $n = 8$ and $\Delta x = 2$ gives

$\text{Area} = \int_0^{16} w\, dx \approx \tfrac{2}{3}[0 + 4(6.2) + 2(7.2) + 4(6.8) + 2(5.6) + 4(5.0) + 2(4.8) + 4(4.8) + 0] = \tfrac{2}{3}(126.4) \approx 84\ \text{m}^2.$

27. For $0 \le t \le 10$, $b(t) > d(t)$, so the area between the curves is given by

$$\int_0^{10} [b(t) - d(t)]\, dt = \int_0^{10} \left(2200e^{0.024t} - 1460e^{0.018t} \right) dt = \left[\frac{2200}{0.024}e^{0.024t} - \frac{1460}{0.018}e^{0.018t} \right]_0^{10}$$

$$= \left(\frac{275{,}000}{3}e^{0.24} - \frac{730{,}000}{9}e^{0.18} \right) - \left(\frac{275{,}000}{3} - \frac{730{,}000}{9} \right) \approx 8868\ \text{people}$$

This area represents the increase in population over a 10-year period.

29. Let the equation of the large circle be $x^2 + y^2 = R^2$. Then the equation of

the small circle is $x^2 + (y - b)^2 = r^2$, where $b = \sqrt{R^2 - r^2}$ is the distance

between the centers of the circles. The desired area is

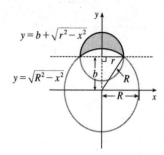

$A = \int_{-r}^{r} \left[\left(b + \sqrt{r^2 - x^2} \right) - \sqrt{R^2 - x^2} \right] dx$

$= 2 \int_0^r \left(b + \sqrt{r^2 - x^2} - \sqrt{R^2 - x^2} \right) dx$

$= 2 \int_0^r b \, dx + 2 \int_0^r \sqrt{r^2 - x^2} \, dx - 2 \int_0^r \sqrt{R^2 - x^2} \, dx$

The first integral is just $2br = 2r\sqrt{R^2 - r^2}$. The second integral represents the area of a quarter-circle of radius r, so its value

is $\frac{1}{4}\pi r^2$. To evaluate the other integral, note that

$$\int \sqrt{a^2 - x^2} \, dx = \int a^2 \cos^2 \theta \, d\theta \quad [x = a\sin\theta, \ dx = a\cos\theta \, d\theta] \quad = \left(\tfrac{1}{2}a^2\right)\int (1 + \cos 2\theta) \, d\theta$$

$$= \tfrac{1}{2}a^2\left(\theta + \tfrac{1}{2}\sin 2\theta\right) + C = \tfrac{1}{2}a^2(\theta + \sin\theta \, \cos\theta) + C$$

$$= \frac{a^2}{2}\arcsin\left(\frac{x}{a}\right) + \frac{a^2}{2}\left(\frac{x}{a}\right)\frac{\sqrt{a^2 - x^2}}{a} + C = \frac{a^2}{2}\arcsin\left(\frac{x}{a}\right) + \frac{x}{2}\sqrt{a^2 - x^2} + C$$

Thus, the desired area is

$$A = 2r\sqrt{R^2 - r^2} + 2\left(\tfrac{1}{4}\pi r^2\right) - \left[R^2 \arcsin(x/R) + x\sqrt{R^2 - x^2} \right]_0^r$$

$$= 2r\sqrt{R^2 - r^2} + \tfrac{1}{2}\pi r^2 - \left[R^2 \arcsin(r/R) + r\sqrt{R^2 - r^2} \right] = r\sqrt{R^2 - r^2} + \tfrac{\pi}{2}r^2 - R^2 \arcsin(r/R)$$

31. We first assume that $c > 0$, since c can be replaced by $-c$ in both equations without changing the graphs, and if $c = 0$ the

curves do not enclose a region. We see from the graph that the enclosed area A lies between $x = -c$ and $x = c$, and by

symmetry, it is equal to four times the area in the first quadrant. The enclosed area is

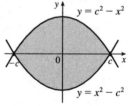

$A = 4 \int_0^c (c^2 - x^2) \, dx = 4\left[c^2 x - \tfrac{1}{3}x^3\right]_0^c = 4\left(c^3 - \tfrac{1}{3}c^3\right) = 4\left(\tfrac{2}{3}c^3\right) = \tfrac{8}{3}c^3$.

So $A = 576 \ \Leftrightarrow \ \tfrac{8}{3}c^3 = 576 \ \Leftrightarrow \ c^3 = 216 \ \Leftrightarrow \ c = \sqrt[3]{216} = 6$. Note

that $c = -6$ is another solution, since the graphs are the same.

33.

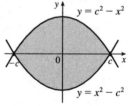

By the symmetry of the problem, we consider only the first quadrant, where

$y = x^2 \ \Rightarrow \ x = \sqrt{y}$. We are looking for a number b such that

$\int_0^b \sqrt{y} \, dy = \int_b^4 \sqrt{y} \, dy \ \Rightarrow \ \tfrac{2}{3}\left[y^{3/2}\right]_0^b = \tfrac{2}{3}\left[y^{3/2}\right]_b^4 \ \Rightarrow$

$b^{3/2} = 4^{3/2} - b^{3/2} \ \Rightarrow \ 2b^{3/2} = 8 \ \Rightarrow \ b^{3/2} = 4 \ \Rightarrow \ b = 4^{2/3} \approx 2.52$.

35. The area under the graph of f from 0 to t is equal to $\int_0^t f(x) \, dx$, so the requirement is that $\int_0^t f(x) \, dx = t^3$ for all t. We

differentiate both sides of this equation with respect to t (with the help of FTC1) to get $f(t) = 3t^2$. This function is positive

and continuous, as required.

37. The curve and the line will determine a region when they intersect at

two or more points. So we solve the equation $x/(x^2 + 1) = mx$ $\Rightarrow$

$x = x(mx^2 + m)$ $\Rightarrow$ $x(mx^2 + m) - x = 0$ $\Rightarrow$

$x(mx^2 + m - 1) = 0$ $\Rightarrow$ $x = 0$ or $mx^2 + m - 1 = 0$ $\Rightarrow$

$x = 0$ or $x^2 = \dfrac{1 - m}{m}$ $\Rightarrow$ $x = 0$ or $x = \pm\sqrt{\dfrac{1}{m} - 1}$.

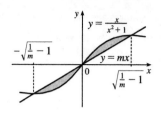

Note that if $m = 1$, this has only the solution $x = 0$, and no region is determined. But if $1/m - 1 > 0$ $\Leftrightarrow$ $1/m > 1$ $\Leftrightarrow$

$0 < m < 1$, then there are two solutions. [Another way of seeing this is to observe that the slope of the tangent to

$y = x/(x^2 + 1)$ at the origin is $y' = 1$ and therefore we must have $0 < m < 1$.] Note that we cannot just integrate between

the positive and negative roots, since the curve and the line cross at the origin. Since mx and $x/(x^2 + 1)$ are both odd

functions, the total area is twice the area between the curves on the interval $\left[0, \sqrt{1/m - 1}\,\right]$. So the total area enclosed is

$$2\int_0^{\sqrt{1/m-1}} \left[\frac{x}{x^2 + 1} - mx\right] dx = 2\left[\frac{1}{2}\ln(x^2 + 1) - \frac{1}{2}mx^2\right]_0^{\sqrt{1/m-1}}$$

$$= [\ln(1/m - 1 + 1) - m(1/m - 1)] - (\ln 1 - 0)$$

$$= \ln(1/m) - 1 + m = m - \ln m - 1$$

7.2   Volumes

1. A cross-section is a disk with radius $1/x$, so its area is

$A(x) = \pi(1/x)^2$.

$$V = \int_1^2 A(x)\,dx = \int_1^2 \pi\left(\frac{1}{x}\right)^2 dx$$

$$= \pi\int_1^2 \frac{1}{x^2}\,dx = \pi\left[-\frac{1}{x}\right]_1^2$$

$$= \pi\left[-\frac{1}{2} - (-1)\right] = \frac{\pi}{2}$$

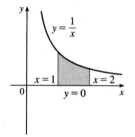 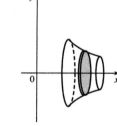

3. A cross-section is a disk with radius $2\sqrt{y}$, so its area is

$A(y) = \pi\left(2\sqrt{y}\right)^2$.

$$V = \int_0^9 A(y)\,dy = \int_0^9 \pi\left(2\sqrt{y}\right)^2 dy$$

$$= 4\pi\int_0^9 y\,dy = 4\pi\left[\frac{1}{2}y^2\right]_0^9$$

$$= 2\pi(81) = 162\pi$$

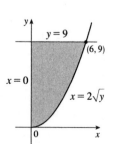 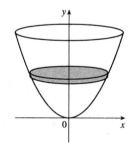

5. A cross-section is a washer (annulus) with inner radius x^3 and outer radius x, so its area is

$A(x) = \pi(x)^2 - \pi(x^3)^2 = \pi(x^2 - x^6)$.

$V = \int_0^1 A(x)\,dx = \int_0^1 \pi(x^2 - x^6)\,dx$

$= \pi\left[\frac{1}{3}x^3 - \frac{1}{7}x^7\right]_0^1 = \pi\left(\frac{1}{3} - \frac{1}{7}\right) = \frac{4\pi}{21}$

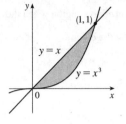
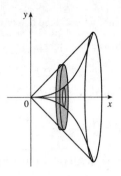

7. A cross-section is a washer with inner radius y^2 and outer radius $2y$, so its area is

$A(y) = \pi(2y)^2 - \pi(y^2)^2 = \pi(4y^2 - y^4)$.

$V = \int_0^2 A(y)\,dy = \pi\int_0^2 (4y^2 - y^4)\,dy$

$= \pi\left[\frac{4}{3}y^3 - \frac{1}{5}y^5\right]_0^2 = \pi\left(\frac{32}{3} - \frac{32}{5}\right) = \frac{64\pi}{15}$

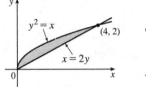
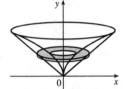

9. A cross-section is a washer with inner radius $1 - \sqrt{x}$ and outer radius $1 - x$, so its area is

$A(x) = \pi(1-x)^2 - \pi\left(1 - \sqrt{x}\right)^2$

$= \pi\left[(1 - 2x + x^2) - \left(1 - 2\sqrt{x} + x\right)\right]$

$= \pi\left(-3x + x^2 + 2\sqrt{x}\right)$

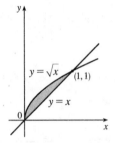
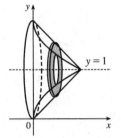

$V = \int_0^1 A(x)\,dx = \pi\int_0^1 \left(-3x + x^2 + 2\sqrt{x}\right) dx = \pi\left[-\frac{3}{2}x^2 + \frac{1}{3}x^3 + \frac{4}{3}x^{3/2}\right]_0^1 = \pi\left(-\frac{3}{2} + \frac{5}{3}\right) = \frac{\pi}{6}$

11. $y = x^2 \;\Rightarrow\; x = \sqrt{y}$ for $x \geq 0$. The outer radius is the distance from $x = -1$ to $x = \sqrt{y}$ and the inner radius is the distance from $x = -1$ to $x = y^2$.

$V = \int_0^1 \pi\left\{\left[\sqrt{y} - (-1)\right]^2 - \left[y^2 - (-1)\right]^2\right\} dy = \pi\int_0^1 \left[\left(\sqrt{y} + 1\right)^2 - \left(y^2 + 1\right)^2\right] dy$

$= \pi\int_0^1 \left(y + 2\sqrt{y} + 1 - y^4 - 2y^2 - 1\right) dy = \pi\int_0^1 \left(y + 2\sqrt{y} - y^4 - 2y^2\right) dy$

$= \pi\left[\frac{1}{2}y^2 + \frac{4}{3}y^{3/2} - \frac{1}{5}y^5 - \frac{2}{3}y^3\right]_0^1 = \pi\left(\frac{1}{2} + \frac{4}{3} - \frac{1}{5} - \frac{2}{3}\right) = \frac{29}{30}\pi$

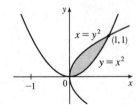
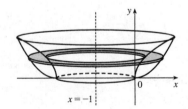

13. $y = \sqrt{x} \implies x = y^2$ and $y = x^3 \implies x = \sqrt[3]{y}$. A cross-section is a

washer with inner radius $1 - \sqrt[3]{y}$ and outer radius $1 - y^2$, so its area is

$A(y) = \pi(1 - y^2)^2 - \pi\left(1 - \sqrt[3]{y}\right)^2$.

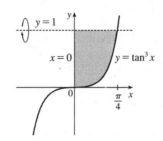

$V = \int_0^1 A(y)\, dy = \int_0^1 \left[\pi(1 - y^2)^2 - \pi\left(1 - \sqrt[3]{y}\right)^2\right] dy$

$= \pi \int_0^1 \left[(1 - 2y^2 + y^4) - (1 - 2y^{1/3} + y^{2/3})\right] dy$

$= \pi \int_0^1 (-2y^2 + y^4 + 2y^{1/3} - y^{2/3})\, dy = \pi\left[-\frac{2}{3}y^3 + \frac{1}{5}y^5 + \frac{3}{2}y^{4/3} - \frac{3}{5}y^{5/3}\right]_0^1 = \pi\left(-\frac{2}{3} + \frac{1}{5} + \frac{3}{2} - \frac{3}{5}\right) = \frac{13\pi}{30}$

15. $V = \pi \displaystyle\int_0^{\pi/4} (1 - \tan^3 x)^2\, dx$

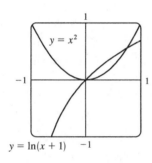

17.

$y = x^2$ and $y = \ln(x + 1)$ intersect at $x = 0$ and at $x = a \approx 0.747$.

$V = \pi \displaystyle\int_0^a \left\{[\ln(x + 1)]^2 - (x^2)^2\right\} dx \approx 0.132$

19. $V = \pi \displaystyle\int_0^{\pi} \left\{[\sin^2 x - (-1)]^2 - [0 - (-1)]^2\right\} dx$

$\overset{\text{CAS}}{=} \dfrac{11}{8}\pi^2$

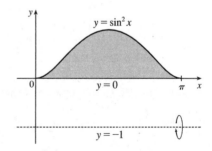

21. (a) $\pi \int_0^{\pi/2} \cos^2 x\, dx$ describes the volume of the solid obtained by rotating the region

$\mathcal{R} = \left\{(x, y) \mid 0 \le x \le \frac{\pi}{2}, 0 \le y \le \cos x\right\}$ of the xy-plane about the x-axis.

(b) $\pi \int_0^1 (y^4 - y^8)\, dy = \pi \int_0^1 \left[(y^2)^2 - (y^4)^2\right] dy$ describes the volume of the solid obtained by rotating the region

$\mathcal{R} = \left\{(x, y) \mid 0 \le y \le 1, y^4 \le x \le y^2\right\}$ of the xy-plane about the y-axis.

23. There are 10 subintervals over the 15-cm length, so we'll use $n = 10/2 = 5$ for the Midpoint Rule.

$V = \int_0^{15} A(x)\, dx \approx M_5 = \frac{15-0}{5}[A(1.5) + A(4.5) + A(7.5) + A(10.5) + A(13.5)]$

$= 3(18 + 79 + 106 + 128 + 39) = 3 \cdot 370 = 1110 \text{ cm}^3$

25. We'll form a right circular cone with height h and base radius r by revolving the line $y = \frac{r}{h}x$ about the x-axis.

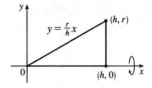

$$V = \pi \int_0^h \left(\frac{r}{h}x\right)^2 dx = \pi \int_0^h \frac{r^2}{h^2}x^2\,dx = \pi\frac{r^2}{h^2}\left[\frac{1}{3}x^3\right]_0^h$$

$$= \pi\frac{r^2}{h^2}\left(\frac{1}{3}h^3\right) = \frac{1}{3}\pi r^2 h$$

Another solution: Revolve $x = -\dfrac{r}{h}y + r$ about the y-axis.

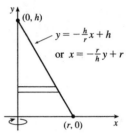

$$V = \pi \int_0^h \left(-\frac{r}{h}y + r\right)^2 dy \overset{*}{=} \pi \int_0^h \left[\frac{r^2}{h^2}y^2 - \frac{2r^2}{h}y + r^2\right] dy$$

$$= \pi\left[\frac{r^2}{3h^2}y^3 - \frac{r^2}{h}y^2 + r^2 y\right]_0^h = \pi\left(\tfrac{1}{3}r^2 h - r^2 h + r^2 h\right) = \tfrac{1}{3}\pi r^2 h$$

* Or use substitution with $u = r - \dfrac{r}{h}y$ and $du = -\dfrac{r}{h}\,dy$ to get

$$\pi \int_r^0 u^2\left(-\frac{h}{r}\,du\right) = -\pi\frac{h}{r}\left[\frac{1}{3}u^3\right]_r^0 = -\pi\frac{h}{r}\left(-\frac{1}{3}r^3\right) = \frac{1}{3}\pi r^2 h.$$

27. $x^2 + y^2 = r^2 \iff x^2 = r^2 - y^2$

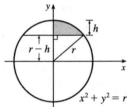

$$V = \pi \int_{r-h}^r (r^2 - y^2)\,dy = \pi\left[r^2 y - \frac{y^3}{3}\right]_{r-h}^r$$

$$= \pi\left\{\left[r^3 - \frac{r^3}{3}\right] - \left[r^2(r-h) - \frac{(r-h)^3}{3}\right]\right\}$$

$$= \pi\left\{\tfrac{2}{3}r^3 - \tfrac{1}{3}(r-h)[3r^2 - (r-h)^2]\right\}$$

$$= \tfrac{1}{3}\pi\left\{2r^3 - (r-h)[3r^2 - (r^2 - 2rh + h^2)]\right\}$$

$$= \tfrac{1}{3}\pi\left\{2r^3 - (r-h)[2r^2 + 2rh - h^2]\right\} = \tfrac{1}{3}\pi(2r^3 - 2r^3 - 2r^2 h + rh^2 + 2r^2 h + 2rh^2 - h^3)$$

$$= \tfrac{1}{3}\pi(3rh^2 - h^3) = \tfrac{1}{3}\pi h^2(3r - h), \text{ or, equivalently, } \pi h^2\left(r - \frac{h}{3}\right)$$

29. For a cross-section at height y, we see from similar triangles that $\dfrac{\alpha/2}{b/2} = \dfrac{h-y}{h}$, so

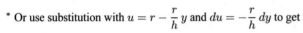

$\alpha = b\left(1 - \dfrac{y}{h}\right)$. Similarly, for cross-sections having $2b$ as their base and β replacing

α, $\beta = 2b\left(1 - \dfrac{y}{h}\right)$. So

$$V = \int_0^h A(y)\,dy = \int_0^h \left[b\left(1 - \frac{y}{h}\right)\right]\left[2b\left(1 - \frac{y}{h}\right)\right] dy = \int_0^h 2b^2\left(1 - \frac{y}{h}\right)^2 dy$$

$$= 2b^2 \int_0^h \left(1 - \frac{2y}{h} + \frac{y^2}{h^2}\right) dy = 2b^2\left[y - \frac{y^2}{h} + \frac{y^3}{3h^2}\right]_0^h = 2b^2\left[h - h + \tfrac{1}{3}h\right]$$

$$= \tfrac{2}{3}b^2 h \quad [= \tfrac{1}{3}Bh \text{ where } B \text{ is the area of the base, as with any pyramid.}]$$

31. A cross-section at height z is a triangle similar to the base, so we'll multiply the legs of the base triangle, 3 and 4, by a proportionality factor of $(5-z)/5$. Thus, the triangle at height z has area $A(z) = \frac{1}{2} \cdot 3\left(\frac{5-z}{5}\right) \cdot 4\left(\frac{5-z}{5}\right) = 6\left(1 - \frac{z}{5}\right)^2$, so

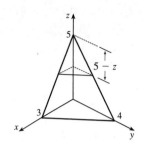

$$V = \int_0^5 A(z)\,dz = 6 \int_0^5 (1 - z/5)^2\,dz = 6 \int_1^0 u^2(-5\,du) \qquad \begin{bmatrix} u = 1 - z/5, \\ du = -\frac{1}{5}dz \end{bmatrix}$$

$$= -30\left[\tfrac{1}{3}u^3\right]_1^0 = -30\left(-\tfrac{1}{3}\right) = 10 \text{ cm}^3$$

33. If l is a leg of the isosceles right triangle and $2y$ is the hypotenuse, then $l^2 + l^2 = (2y)^2 \;\Rightarrow\; 2l^2 = 4y^2 \;\Rightarrow\; l^2 = 2y^2$.

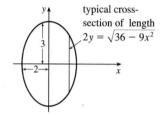

typical cross-section of length $2y = \sqrt{36 - 9x^2}$

$$V = \int_{-2}^2 A(x)\,dx = 2\int_0^2 A(x)\,dx = 2\int_0^2 \tfrac{1}{2}(l)(l)\,dx = 2\int_0^2 y^2\,dx$$

$$= 2\int_0^2 \tfrac{1}{4}(36 - 9x^2)\,dx = \tfrac{9}{2}(4 - x^2)\,dx = \tfrac{9}{2}\left[4x - \tfrac{1}{3}x^3\right]_0^2 = \tfrac{9}{2}\left(8 - \tfrac{8}{3}\right) = 24$$

35. The cross-section of the base corresponding to the coordinate y has length $2x = 2\sqrt{y}$. The square has area $A(y) = \left(2\sqrt{y}\right)^2 = 4y$, so

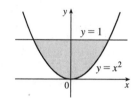

$$V = \int_0^1 A(y)\,dy = \int_0^1 4y\,dy = \left[2y^2\right]_0^1 = 2$$

37. A typical cross-section perpendicular to the y-axis in the base has length $\ell(y) = 3 - \frac{3}{2}y$. This length is the leg of an isosceles right triangle, so

$$A(y) = \tfrac{1}{2}\left[\ell(y)\right]^2 \quad \left[\tfrac{1}{2}bh \text{ with base } = \text{ height}\right] \;=\; \tfrac{1}{2}\left[3\left(1 - \tfrac{1}{2}y\right)\right]^2 = \tfrac{9}{2}\left(1 - \tfrac{1}{2}y\right)^2.$$

Thus,

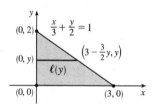

$$V = \int_0^2 A(y)\,dy = \tfrac{9}{2}\int_1^0 u^2(-2\,du) \qquad \begin{bmatrix} u = 1 - \tfrac{1}{2}y, \\ du = -\tfrac{1}{2}y \end{bmatrix} \;=\; -9\left[\tfrac{1}{3}u^3\right]_1^0 = -9\left(-\tfrac{1}{3}\right) = 3.$$

39. (a) The radius of the barrel is the same at each end by symmetry, since the function $y = R - cx^2$ is even. Since the barrel is obtained by rotating the graph of the function y about the x-axis, this radius is equal to the value of y at $x = \frac{1}{2}h$, which is $R - c\left(\frac{1}{2}h\right)^2 = R - d = r$.

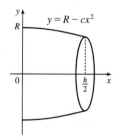

(b) The barrel is symmetric about the y-axis, so its volume is twice the volume of that part of the barrel for $x > 0$. Also, the barrel is a volume of rotation, so

$$V = 2\int_0^{h/2} \pi y^2\,dx = 2\pi \int_0^{h/2} (R - cx^2)^2\,dx$$

$$= 2\pi\left[R^2 x - \tfrac{2}{3}Rcx^3 + \tfrac{1}{5}c^2 x^5\right]_0^{h/2} = 2\pi\left(\tfrac{1}{2}R^2 h - \tfrac{1}{12}Rch^3 + \tfrac{1}{160}c^2 h^5\right)$$

Trying to make this look more like the expression we want, we rewrite it as $V = \frac{1}{3}\pi h\left[2R^2 + \left(R^2 - \frac{1}{2}Rch^2 + \frac{3}{80}c^2 h^4\right)\right]$.

But $R^2 - \frac{1}{2}Rch^2 + \frac{3}{80}c^2 h^4 = \left(R - \frac{1}{4}ch^2\right)^2 - \frac{1}{40}c^2 h^4 = (R - d)^2 - \frac{2}{5}\left(\frac{1}{4}ch^2\right)^2 = r^2 - \frac{2}{5}d^2$.

Substituting this back into V, we see that $V = \frac{1}{3}\pi h\left(2R^2 + r^2 - \frac{2}{5}d^2\right)$, as required.

41. (a) The torus is obtained by rotating the circle $(x - R)^2 + y^2 = r^2$ about

the y-axis. Solving for x, we see that the right half of the circle is given by

$x = R + \sqrt{r^2 - y^2} = f(y)$ and the left half by $x = R - \sqrt{r^2 - y^2} = g(y)$.

So

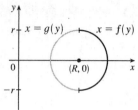

$$V = \pi \int_{-r}^{r} \left\{ [f(y)]^2 - [g(y)]^2 \right\} dy$$

$$= 2\pi \int_{0}^{r} \left[\left(R^2 + 2R\sqrt{r^2 - y^2} + r^2 - y^2 \right) - \left(R^2 - 2R\sqrt{r^2 - y^2} + r^2 - y^2 \right) \right] dy$$

$$= 2\pi \int_{0}^{r} 4R\sqrt{r^2 - y^2}\, dy = 8\pi R \int_{0}^{r} \sqrt{r^2 - y^2}\, dy$$

(b) Observe that the integral represents a quarter of the area of a circle with radius r, so

$$8\pi R \int_{0}^{r} \sqrt{r^2 - y^2}\, dy = 8\pi R \cdot \tfrac{1}{4}\pi r^2 = 2\pi^2 r^2 R$$

43. (a) Volume$(S_1) = \int_{0}^{h} A(z)\, dz =$ Volume(S_2) since the cross-sectional area $A(z)$ at height z is the same for both solids.

(b) By Cavalieri's Principle, the volume of the cylinder in the figure is the same as that of a right circular cylinder with radius r

and height h, that is, $\pi r^2 h$.

45. The volume is obtained by rotating the area common to two circles of

radius r, as shown. The volume of the right half is

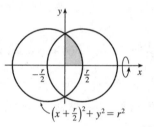

$$V_{\text{right}} = \pi \int_{0}^{r/2} y^2\, dx = \pi \int_{0}^{r/2} \left[r^2 - \left(\tfrac{1}{2}r + x \right)^2 \right] dx$$

$$= \pi \left[r^2 x - \tfrac{1}{3}\left(\tfrac{1}{2}r + x \right)^3 \right]_{0}^{r/2} = \pi \left[\left(\tfrac{1}{2}r^3 - \tfrac{1}{3}r^3 \right) - \left(0 - \tfrac{1}{24}r^3 \right) \right] = \tfrac{5}{24}\pi r^3$$

So by symmetry, the total volume is twice this, or $\tfrac{5}{12}\pi r^3$.

Another solution: We observe that the volume is the twice the volume of a cap of a sphere, so we can use the formula from

Exercise 27 with $h = \tfrac{1}{2}r$: $V = 2 \cdot \tfrac{1}{3}\pi h^2 (3r - h) = \tfrac{2}{3}\pi \left(\tfrac{1}{2}r \right)^2 \left(3r - \tfrac{1}{2}r \right) = \tfrac{5}{12}\pi r^3$.

47. Take the x-axis to be the axis of the cylindrical hole of

radius r. A quarter of the cross-section through y,

perpendicular to the y-axis, is the rectangle shown. Using the

Pythagorean Theorem twice, we see that the dimensions of

this rectangle are $x = \sqrt{R^2 - y^2}$ and $z = \sqrt{r^2 - y^2}$, so

$\tfrac{1}{4}A(y) = xz = \sqrt{r^2 - y^2}\sqrt{R^2 - y^2}$, and

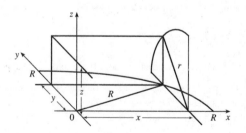

$$V = \int_{-r}^{r} A(y)\, dy = \int_{-r}^{r} 4\sqrt{r^2 - y^2}\sqrt{R^2 - y^2}\, dy = 8\int_{0}^{r} \sqrt{r^2 - y^2}\sqrt{R^2 - y^2}\, dy$$

7.3   Volumes by Cylindrical Shells

1. If we were to use the "washer" method, we would first have to locate the
local maximum point (a, b) of $y = x(x-1)^2$ using the methods of
Chapter 4. Then we would have to solve the equation $y = x(x-1)^2$
for x in terms of y to obtain the functions $x = g_1(y)$ and $x = g_2(y)$
shown in the first figure. This step would be difficult because it
involves the cubic formula. Finally we would find the volume using
$V = \pi \int_0^b \{[g_1(y)]^2 - [g_2(y)]^2\}\, dy$. Using shells, we find that a
typical approximating shell has radius x, so its circumference is $2\pi x$.
Its height is y, that is, $x(x-1)^2$. So the total volume is

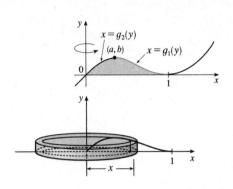

$$V = \int_0^1 2\pi x[x(x-1)^2]\, dx = 2\pi \int_0^1 (x^4 - 2x^3 + x^2)\, dx = 2\pi \left[\frac{x^5}{5} - 2\frac{x^4}{4} + \frac{x^3}{3}\right]_0^1 = \frac{\pi}{15}$$

3. $V = \displaystyle\int_1^2 2\pi x \cdot \frac{1}{x}\, dx = 2\pi \int_1^2 1\, dx$

$\quad = 2\pi \left[x\right]_1^2 = 2\pi(2-1) = 2\pi$

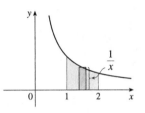

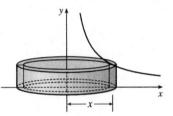

5. $V = \int_0^1 2\pi x e^{-x^2}\, dx$. Let $u = x^2$. Thus, $du = 2x\, dx$, so $V = \pi \int_0^1 e^{-u}\, du = \pi\left[-e^{-u}\right]_0^1 = \pi(1 - 1/e)$.

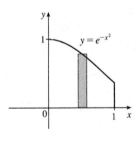

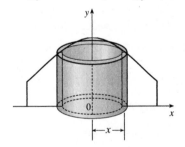

7. The curves intersect when $4(x-2)^2 = x^2 - 4x + 7 \iff 4x^2 - 16x + 16 = x^2 - 4x + 7 \iff$
$3x^2 - 12x + 9 = 0 \iff 3(x^2 - 4x + 3) = 0 \iff 3(x-1)(x-3) = 0$, so $x = 1$ or 3.

$$V = 2\pi \int_1^3 \{x[(x^2 - 4x + 7) - 4(x-2)^2]\}\, dx = 2\pi \int_1^3 [x(x^2 - 4x + 7 - 4x^2 + 16x - 16)]\, dx$$

$$= 2\pi \int_1^3 [x(-3x^2 + 12x - 9)]\, dx = 2\pi(-3) \int_1^3 (x^3 - 4x^2 + 3x)\, dx = -6\pi\left[\frac{1}{4}x^4 - \frac{4}{3}x^3 + \frac{3}{2}x^2\right]_1^3$$

$$= -6\pi\left[\left(\frac{81}{4} - 36 + \frac{27}{2}\right) - \left(\frac{1}{4} - \frac{4}{3} + \frac{3}{2}\right)\right] = -6\pi\left(20 - 36 + 12 + \frac{4}{3}\right) = -6\pi\left(-\frac{8}{3}\right) = 16\pi$$

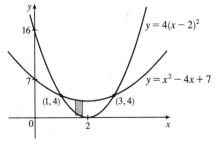

9. $V = \int_1^2 2\pi y (1 + y^2)\, dy = 2\pi \int_1^2 (y + y^3)\, dy = 2\pi \left[\frac{1}{2}y^2 + \frac{1}{4}y^4 \right]_1^2 = 2\pi \left[(2 + 4) - \left(\frac{1}{2} + \frac{1}{4} \right) \right] = 2\pi \left(\frac{21}{4} \right) = \frac{21\pi}{2}$

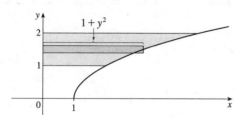
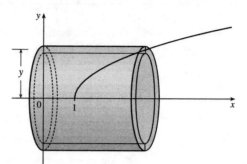

11. $V = 2\pi \int_0^8 [y(\sqrt[3]{y} - 0)]\, dy$

$\quad\quad = 2\pi \int_0^8 y^{4/3}\, dy = 2\pi \left[\frac{3}{7} y^{7/3} \right]_0^8$

$\quad\quad = \frac{6\pi}{7}(8^{7/3}) = \frac{6\pi}{7}(2^7) = \frac{768\pi}{7}$

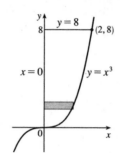
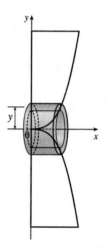

13. The curves intersect when $4x^2 = 6 - 2x \quad \Leftrightarrow \quad 2x^2 + x - 3 = 0 \quad \Leftrightarrow \quad (2x + 3)(x - 1) = 0 \quad \Leftrightarrow \quad x = -\frac{3}{2}$ or 1.

Solving the equations for x gives us $y = 4x^2 \quad \Rightarrow \quad x = \pm \frac{1}{2}\sqrt{y}$ and $2x + y = 6 \quad \Rightarrow \quad x = -\frac{1}{2}y + 3$.

$$V = 2\pi \int_0^4 \left\{ y \left[\left(\frac{1}{2}\sqrt{y} \right) - \left(-\frac{1}{2}\sqrt{y} \right) \right] \right\} dy + 2\pi \int_4^9 \left\{ y \left[\left(-\frac{1}{2}y + 3 \right) - \left(-\frac{1}{2}\sqrt{y} \right) \right] \right\} dy$$

$$= 2\pi \int_0^4 (y\sqrt{y})\, dy + 2\pi \int_4^9 \left(-\frac{1}{2}y^2 + 3y + \frac{1}{2}y^{3/2} \right) dy = 2\pi \left[\frac{2}{5}y^{5/2} \right]_0^4 + 2\pi \left[-\frac{1}{6}y^3 + \frac{3}{2}y^2 + \frac{1}{5}y^{5/2} \right]_4^9$$

$$= 2\pi \left(\frac{2}{5} \cdot 32 \right) + 2\pi \left[\left(-\frac{243}{2} + \frac{243}{2} + \frac{243}{5} \right) - \left(-\frac{32}{3} + 24 + \frac{32}{5} \right) \right] = \frac{128}{5}\pi + 2\pi \left(\frac{433}{15} \right) = \frac{1250}{15}\pi = \frac{250}{3}\pi$$

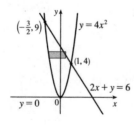
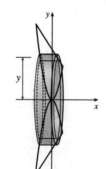

15. $V = \int_1^2 2\pi(x-1)x^2 \, dx = 2\pi \left[\frac{1}{4}x^4 - \frac{1}{3}x^3\right]_1^2$

$= 2\pi \left[\left(4 - \frac{8}{3}\right) - \left(\frac{1}{4} - \frac{1}{3}\right)\right] = \frac{17}{6}\pi$

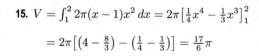

17. $V = \int_1^2 2\pi(4-x)x^2 \, dx = 2\pi \left[\frac{4}{3}x^3 - \frac{1}{4}x^4\right]_1^2$

$= 2\pi \left[\left(\frac{32}{3} - 4\right) - \left(\frac{4}{3} - \frac{1}{4}\right)\right] = \frac{67}{6}\pi$

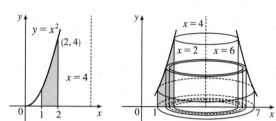

19. $V = \int_0^2 2\pi(3-y)(5-x) \, dy$

$= \int_0^2 2\pi(3-y)\left(5 - y^2 - 1\right) dy$

$= \int_0^2 2\pi\left(12 - 4y - 3y^2 + y^3\right) dy$

$= 2\pi\left[12y - 2y^2 - y^3 + \frac{1}{4}y^4\right]_0^2$

$= 2\pi(24 - 8 - 8 + 4) = 24\pi$

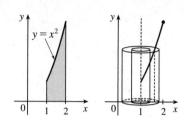

21. $V = \int_1^2 2\pi x \ln x \, dx$

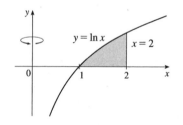

23. $V = \int_0^1 2\pi[x - (-1)]\left(\sin\frac{\pi}{2}x - x^4\right) dx$

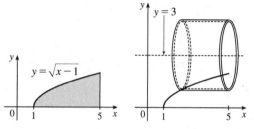

25. $V = \int_0^\pi 2\pi(4-y)\sqrt{\sin y}\, dy$

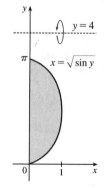

27. $\Delta x = \dfrac{\pi/4 - 0}{4} = \dfrac{\pi}{16}$.

$V = \int_0^{\pi/4} 2\pi x \tan x \, dx \approx 2\pi \cdot \frac{\pi}{16}\left(\frac{\pi}{32}\tan\frac{\pi}{32} + \frac{3\pi}{32}\tan\frac{3\pi}{32} + \frac{5\pi}{32}\tan\frac{5\pi}{32} + \frac{7\pi}{32}\tan\frac{7\pi}{32}\right) \approx 1.142$

29. $\int_0^3 2\pi x^5 \, dx = 2\pi \int_0^3 x(x^4) \, dx$. The solid is obtained by rotating the region $0 \le y \le x^4$, $0 \le x \le 3$ about the y-axis using cylindrical shells.

31. $\int_0^1 2\pi(3-y)(1-y^2) \, dy$. The solid is obtained by rotating the region bounded by (i) $x = 1 - y^2$, $x = 0$, and $y = 0$ or (ii) $x = y^2$, $x = 1$, and $y = 0$ about the line $y = 3$ using cylindrical shells.

33. Use disks:

$V = \int_{-2}^1 \pi(x^2 + x - 2)^2 \, dx = \pi \int_{-2}^1 (x^4 + 2x^3 - 3x^2 - 4x + 4) \, dx = \pi\left[\frac{1}{5}x^5 + \frac{1}{2}x^4 - x^3 - 2x^2 + 4x\right]_{-2}^1$

$= \pi\left[\left(\frac{1}{5} + \frac{1}{2} - 1 - 2 + 4\right) - \left(-\frac{32}{5} + 8 + 8 - 8 - 8\right)\right] = \pi\left(\frac{33}{5} + \frac{3}{2}\right) = \frac{81}{10}\pi$

35. Use shells:

$$V = \int_1^4 2\pi[x - (-1)][5 - (x + 4/x)]\, dx = 2\pi \int_1^4 (x + 1)(5 - x - 4/x)\, dx$$

$$= 2\pi \int_1^4 (5x - x^2 - 4 + 5 - x - 4/x)\, dx$$

$$= 2\pi \int_1^4 (-x^2 + 4x + 1 - 4/x)\, dx = 2\pi\left[-\tfrac{1}{3}x^3 + 2x^2 + x - 4\ln x\right]_1^4$$

$$= 2\pi\left[\left(-\tfrac{64}{3} + 32 + 4 - 4\ln 4\right) - \left(-\tfrac{1}{3} + 2 + 1 - 0\right)\right]$$

$$= 2\pi(12 - 4\ln 4) = 8\pi(3 - \ln 4)$$

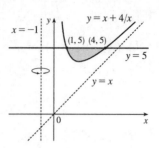

37. Use disks: $V = \pi \int_0^2 \left[\sqrt{1 - (y - 1)^2}\right]^2 dy = \pi \int_0^2 (2y - y^2)\, dy = \pi\left[y^2 - \tfrac{1}{3}y^3\right]_0^2 = \pi\left(4 - \tfrac{8}{3}\right) = \tfrac{4}{3}\pi$

39. $V = 2\int_0^r 2\pi x \sqrt{r^2 - x^2}\, dx = -2\pi \int_0^r (r^2 - x^2)^{1/2}(-2x)\, dx = \left[-2\pi \cdot \tfrac{2}{3}(r^2 - x^2)^{3/2}\right]_0^r = -\tfrac{4}{3}\pi(0 - r^3) = \tfrac{4}{3}\pi r^3$

41. $V = 2\pi \int_0^r x\left(-\dfrac{h}{r}x + h\right) dx = 2\pi h \int_0^r \left(-\dfrac{x^2}{r} + x\right) dx = 2\pi h\left[-\dfrac{x^3}{3r} + \dfrac{x^2}{2}\right]_0^r = 2\pi h\, \dfrac{r^2}{6} = \dfrac{\pi r^2 h}{3}$

43. Using the formula for volumes of rotation and the figure, we see that

Volume $= \int_0^d \pi b^2\, dy - \int_0^c \pi a^2\, dy - \int_c^d \pi\left[f^{-1}(y)\right]^2 dy = \pi b^2 d - \pi a^2 c - \int_c^d \pi\left[f^{-1}(y)\right]^2 dy$. Let $y = f(x)$,

which gives $dy = f'(x)\, dx$ and $f^{-1}(y) = x$, so that $V = \pi b^2 d - \pi a^2 c - \pi \int_a^b x^2 f'(x)\, dx$.

Now integrate by parts with $u = x^2$, and $dv = f'(x)\, dx \Rightarrow du = 2x\, dx, v = f(x)$, and

$\int_a^b x^2 f'(x)\, dx = \left[x^2 f(x)\right]_a^b - \int_a^b 2x f(x)\, dx = b^2 f(b) - a^2 f(a) - \int_a^b 2x f(x)\, dx$, but $f(a) = c$ and $f(b) = d \Rightarrow$

$V = \pi b^2 d - \pi a^2 c - \pi\left[b^2 d - a^2 c - \int_a^b 2x f(x)\, dx\right] = \int_a^b 2\pi x f(x)\, dx$.

7.4 Arc Length

1. $y = 2 - 3x \Rightarrow L = \int_{-2}^1 \sqrt{1 + (dy/dx)^2}\, dx = \int_{-2}^1 \sqrt{1 + (-3)^2}\, dx = \sqrt{10}\left[1 - (-2)\right] = 3\sqrt{10}$.

The arc length can be calculated using the distance formula, since the curve is a line segment, so

$L = $ [distance from $(-2, 8)$ to $(1, -1)$] $= \sqrt{[1 - (-2)]^2 + [(-1) - 8]^2} = \sqrt{90} = 3\sqrt{10}$.

3. $y = 1 + 6x^{3/2} \Rightarrow dy/dx = 9x^{1/2} \Rightarrow 1 + (dy/dx)^2 = 1 + 81x$. So

$L = \int_0^1 \sqrt{1 + 81x}\, dx = \int_1^{82} u^{1/2}\left(\tfrac{1}{81}\, du\right)$ [where $u = 1 + 81x$ and $du = 81\, dx$]

$= \tfrac{1}{81} \cdot \tfrac{2}{3}\left[u^{3/2}\right]_1^{82} = \tfrac{2}{243}\left(82\sqrt{82} - 1\right)$

5. $y = \dfrac{x^5}{6} + \dfrac{1}{10x^3} \Rightarrow \dfrac{dy}{dx} = \dfrac{5}{6}x^4 - \dfrac{3}{10}x^{-4} \Rightarrow$

$1 + (dy/dx)^2 = 1 + \tfrac{25}{36}x^8 - \tfrac{1}{2} + \tfrac{9}{100}x^{-8} = \tfrac{25}{36}x^8 + \tfrac{1}{2} + \tfrac{9}{100}x^{-8} = \left(\tfrac{5}{6}x^4 + \tfrac{3}{10}x^{-4}\right)^2$. So

$L = \int_1^2 \sqrt{\left(\tfrac{5}{6}x^4 + \tfrac{3}{10}x^{-4}\right)^2}\, dx = \int_1^2 \left(\tfrac{5}{6}x^4 + \tfrac{3}{10}x^{-4}\right) dx = \left[\tfrac{1}{6}x^5 - \tfrac{1}{10}x^{-3}\right]_1^2$

$= \left(\tfrac{32}{6} - \tfrac{1}{80}\right) - \left(\tfrac{1}{6} - \tfrac{1}{10}\right) = \tfrac{31}{6} + \tfrac{7}{80} = \tfrac{1261}{240}$

7. $x = \frac{1}{3}\sqrt{y}\,(y-3) = \frac{1}{3}y^{3/2} - y^{1/2} \;\Rightarrow\; dx/dy = \frac{1}{2}y^{1/2} - \frac{1}{2}y^{-1/2} \;\Rightarrow$

$1 + (dx/dy)^2 = 1 + \frac{1}{4}y - \frac{1}{2} + \frac{1}{4}y^{-1} = \frac{1}{4}y + \frac{1}{2} + \frac{1}{4}y^{-1} = \left(\frac{1}{2}y^{1/2} + \frac{1}{2}y^{-1/2}\right)^2.$ So

$L = \int_1^9 \left(\frac{1}{2}y^{1/2} + \frac{1}{2}y^{-1/2}\right) dy = \frac{1}{2}\left[\frac{2}{3}y^{3/2} + 2y^{1/2}\right]_1^9 = \frac{1}{2}\left[\left(\frac{2}{3}\cdot 27 + 2\cdot 3\right) - \left(\frac{2}{3}\cdot 1 + 2\cdot 1\right)\right]$

$\quad = \frac{1}{2}\left(24 - \frac{8}{3}\right) = \frac{1}{2}\left(\frac{64}{3}\right) = \frac{32}{3}$

9. $y = \ln(\sec x) \;\Rightarrow\; \dfrac{dy}{dx} = \dfrac{\sec x \tan x}{\sec x} = \tan x \;\Rightarrow\; 1 + \left(\dfrac{dy}{dx}\right)^2 = 1 + \tan^2 x = \sec^2 x,$ so

$L = \int_0^{\pi/4} \sqrt{\sec^2 x}\,dx = \int_0^{\pi/4} |\sec x|\,dx = \int_0^{\pi/4} \sec x\,dx = \Big[\ln(\sec x + \tan x)\Big]_0^{\pi/4}$

$\quad = \ln(\sqrt{2}+1) - \ln(1+0) = \ln(\sqrt{2}+1)$

11. $y = \cosh x \;\Rightarrow\; y' = \sinh x \;\Rightarrow\; 1 + (y')^2 = 1 + \sinh^2 x = \cosh^2 x.$

So $L = \int_0^1 \cosh x\,dx = [\sinh x]_0^1 = \sinh 1 = \frac{1}{2}(e - 1/e).$

13. $y = e^x \;\Rightarrow\; y' = e^x \;\Rightarrow\; 1 + (y')^2 = 1 + e^{2x}.$ So

$$L = \int_0^1 \sqrt{1 + e^{2x}}\,dx = \int_1^e \sqrt{1+u^2}\,\frac{du}{u} \qquad [u = e^x,\text{ so } x = \ln u,\, dx = du/u]$$

$$= \int_1^e \frac{\sqrt{1+u^2}}{u^2}\,u\,du = \int_{\sqrt{2}}^{\sqrt{1+e^2}} \frac{v}{v^2-1}\,v\,dv \qquad \left[v = \sqrt{1+u^2},\text{ so } v^2 = 1+u^2,\, v\,dv = u\,du\right]$$

$$= \int_{\sqrt{2}}^{\sqrt{1+e^2}} \left(1 + \frac{1/2}{v-1} - \frac{1/2}{v+1}\right) dv = \left[v + \frac{1}{2}\ln\frac{v-1}{v+1}\right]_{\sqrt{2}}^{\sqrt{1+e^2}}$$

$$= \sqrt{1+e^2} + \frac{1}{2}\ln\frac{\sqrt{1+e^2}-1}{\sqrt{1+e^2}+1} - \sqrt{2} - \frac{1}{2}\ln\frac{\sqrt{2}-1}{\sqrt{2}+1}$$

$$= \sqrt{1+e^2} - \sqrt{2} + \ln(\sqrt{1+e^2}-1) - 1 - \ln(\sqrt{2}-1)$$

Or: Use Formula 23 for $\int \left(\sqrt{1+u^2}/u\right) du$, or substitute $u = \tan\theta.$

15. $y = \cos x \;\Rightarrow\; dy/dx = -\sin x \;\Rightarrow\; 1 + (dy/dx)^2 = 1 + \sin^2 x.$ So $L = \int_0^{2\pi} \sqrt{1 + \sin^2 x}\,dx.$

17. $x = y + y^3 \;\Rightarrow\; dx/dy = 1 + 3y^2 \;\Rightarrow\; 1 + (dx/dy)^2 = 1 + (1 + 3y^2)^2 = 9y^4 + 6y^2 + 2.$

So $L = \int_1^4 \sqrt{9y^4 + 6y^2 + 2}\,dy.$

19. $y = xe^{-x} \;\Rightarrow\; dy/dx = e^{-x} - xe^{-x} = e^{-x}(1 - x) \;\Rightarrow\; 1 + (dy/dx)^2 = 1 + e^{-2x}(1-x)^2.$ Let

$f(x) = \sqrt{1 + (dy/dx)^2} = \sqrt{1 + e^{-2x}(1-x)^2}.$ Then $L = \int_0^5 f(x)\,dx.$ Since $n = 10,\ \Delta x = \frac{5-0}{10} = \frac{1}{2}.$ Now

$$L \approx S_{10} = \frac{1/2}{3}\left[f(0) + 4f\left(\tfrac{1}{2}\right) + 2f(1) + 4f\left(\tfrac{3}{2}\right) + 2f(2) + 4f\left(\tfrac{5}{2}\right) + 2f(3) + 4f\left(\tfrac{7}{2}\right) + 2f(4) + 4f\left(\tfrac{9}{2}\right) + f(5)\right]$$

$$\approx 5.115840$$

The value of the integral produced by a calculator is 5.113568 (to six decimal places).

21. $y = \sec x$ $\Rightarrow$ $dy/dx = \sec x \tan x$ $\Rightarrow$ $L = \int_0^{\pi/3} f(x)\,dx$, where $f(x) = \sqrt{1 + \sec^2 x \tan^2 x}$.

Since $n = 10$, $\Delta x = \dfrac{\pi/3 - 0}{10} = \dfrac{\pi}{30}$. Now

$$L \approx S_{10} = \frac{\pi/30}{3}\left[f(0) + 4f\left(\frac{\pi}{30}\right) + 2f\left(\frac{2\pi}{30}\right) + 4f\left(\frac{3\pi}{30}\right) + 2f\left(\frac{4\pi}{30}\right) + 4f\left(\frac{5\pi}{30}\right)\right.$$
$$\left. + 2f\left(\frac{6\pi}{30}\right) + 4f\left(\frac{7\pi}{30}\right) + 2f\left(\frac{8\pi}{30}\right) + 4f\left(\frac{9\pi}{30}\right) + f\left(\frac{\pi}{3}\right)\right] \approx 1.569619$$

The value of the integral produced by a calculator is 1.569259 (to six decimal places).

23. $x = \ln\left(1 - y^2\right)$ $\Rightarrow$ $\dfrac{dx}{dy} = \dfrac{-2y}{1 - y^2}$ $\Rightarrow$ $1 + \left(\dfrac{dx}{dy}\right)^2 = 1 + \dfrac{4y^2}{(1-y^2)^2} = \dfrac{\left(1+y^2\right)^2}{(1-y^2)^2}$. So

$$L = \int_0^{1/2} \sqrt{\frac{(1+y^2)^2}{(1-y^2)^2}}\,dy = \int_0^{1/2} \frac{1+y^2}{1-y^2}\,dy = \ln 3 - \tfrac{1}{2} \quad \text{[from a CAS]}$$

25. $y^{2/3} = 1 - x^{2/3}$ $\Rightarrow$ $y = (1 - x^{2/3})^{3/2}$ $\Rightarrow$

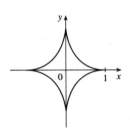

$\dfrac{dy}{dx} = \tfrac{3}{2}(1 - x^{2/3})^{1/2}\left(-\tfrac{2}{3}x^{-1/3}\right) = -x^{-1/3}(1-x^{2/3})^{1/2}$ $\Rightarrow$

$\left(\dfrac{dy}{dx}\right)^2 = x^{-2/3}(1 - x^{2/3}) = x^{-2/3} - 1$. Thus

$L = 4\int_0^1 \sqrt{1 + (x^{-2/3} - 1)}\,dx = 4\int_0^1 x^{-1/3}\,dx = 4\lim_{t\to 0^+}\left[\tfrac{3}{2}x^{2/3}\right]_t^1 = 6$.

27. $y = 2x^{3/2}$ $\Rightarrow$ $y' = 3x^{1/2}$ $\Rightarrow$ $1 + (y')^2 = 1 + 9x$. The arc length function with starting point $P_0(1, 2)$ is

$s(x) = \int_1^x \sqrt{1 + 9t}\,dt = \left[\tfrac{2}{27}(1 + 9t)^{3/2}\right]_1^x = \tfrac{2}{27}\left[(1 + 9x)^{3/2} - 10\sqrt{10}\right]$.

29. The prey hits the ground when $y = 0$ $\Leftrightarrow$ $180 - \tfrac{1}{45}x^2 = 0$ $\Leftrightarrow$ $x^2 = 45 \cdot 180$ $\Rightarrow$ $x = \sqrt{8100} = 90$, since x

must be positive. $y' = -\tfrac{2}{45}x$ $\Rightarrow$ $1 + (y')^2 = 1 + \tfrac{4}{45^2}x^2$, so the distance traveled by the prey is

$L = \int_0^{90} \sqrt{1 + \tfrac{4}{45^2}x^2}\,dx = \int_0^4 \sqrt{1 + u^2}\left(\tfrac{45}{2}\,du\right)$ $\left[u = \tfrac{2}{45}x,\ du = \tfrac{2}{45}\,dx\right]$

$\overset{21}{=} \tfrac{45}{2}\left[\tfrac{1}{2}u\sqrt{1 + u^2} + \tfrac{1}{2}\ln\left(u + \sqrt{1 + u^2}\right)\right]_0^4 = \tfrac{45}{2}\left[2\sqrt{17} + \tfrac{1}{2}\ln\left(4 + \sqrt{17}\right)\right] = 45\sqrt{17} + \tfrac{45}{4}\ln\left(4 + \sqrt{17}\right) \approx 209.1$ m

31. The sine wave has amplitude 1 and period 14, since it goes through two periods in a distance of 28 in., so its equation is

$y = 1\sin\left(\tfrac{2\pi}{14}x\right) = \sin\left(\tfrac{\pi}{7}x\right)$. The width w of the flat metal sheet needed to make the panel is the arc length of the sine curve

from $x = 0$ to $x = 28$. We set up the integral to evaluate w using the arc length formula with $\dfrac{dy}{dx} = \tfrac{\pi}{7}\cos\left(\tfrac{\pi}{7}x\right)$:

$L = \int_0^{28} \sqrt{1 + \left[\tfrac{\pi}{7}\cos\left(\tfrac{\pi}{7}x\right)\right]^2}\,dx = 2\int_0^{14} \sqrt{1 + \left[\tfrac{\pi}{7}\cos\left(\tfrac{\pi}{7}x\right)\right]^2}\,dx$. This integral would be very difficult to evaluate exactly,

so we use a CAS, and find that $L \approx 29.36$ inches.

7.5   Applications to Physics and Engineering

1. $W = \int_a^b f(x)\,dx = \int_0^9 \frac{10}{(1+x)^2}\,dx = 10\int_1^{10}\frac{1}{u^2}\,du \quad [u = 1+x,\ du = dx] \quad = 10\left[-\frac{1}{u}\right]_1^{10} = 10\left(-\frac{1}{10}+1\right) = 9$ ft-lb

3. The force function is given by $F(x)$ (in newtons) and the work (in joules) is the area under the curve, given by

$\int_0^8 F(x)\,dx = \int_0^4 F(x)\,dx + \int_4^8 F(x)\,dx = \frac{1}{2}(4)(30) + (4)(30) = 180$ J.

5. $10 = f(x) = kx = \frac{1}{3}k$ [4 inches $= \frac{1}{3}$ foot],   so $k = 30$ lb/ft and $f(x) = 30x$. Now 6 inches $= \frac{1}{2}$ foot, so

$W = \int_0^{1/2} 30x\,dx = \left[15x^2\right]_0^{1/2} = \frac{15}{4}$ ft-lb.

7. (a) If $\int_0^{0.12} kx\,dx = 2$ J, then $2 = \left[\frac{1}{2}kx^2\right]_0^{0.12} = \frac{1}{2}k(0.0144) = 0.0072k$ and $k = \frac{2}{0.0072} = \frac{2500}{9} \approx 277.78$ N/m.

Thus, the work needed to stretch the spring from 35 cm to 40 cm is

$\int_{0.05}^{0.10} \frac{2500}{9}x\,dx = \left[\frac{1250}{9}x^2\right]_{1/20}^{1/10} = \frac{1250}{9}\left(\frac{1}{100} - \frac{1}{400}\right) = \frac{25}{24} \approx 1.04$ J.

(b) $f(x) = kx$, so $30 = \frac{2500}{9}x$ and $x = \frac{270}{2500}$ m $= 10.8$ cm

In Exercises 9–16, n is the number of subintervals of length Δx, and x_i^* is a sample point in the ith subinterval $[x_{i-1}, x_i]$.

9. (a) The portion of the rope from x ft to $(x + \Delta x)$ ft below the top of the building weighs $\frac{1}{2}\Delta x$ lb and must be lifted x_i^* ft,

so its contribution to the total work is $\frac{1}{2}x_i^*\,\Delta x$ ft-lb. The total work is

$$W = \lim_{n\to\infty} \sum_{i=1}^n \frac{1}{2}x_i^*\,\Delta x = \int_0^{50} \frac{1}{2}x\,dx = \left[\frac{1}{4}x^2\right]_0^{50} = \frac{2500}{4} = 625 \text{ ft-lb}$$

Notice that the exact height of the building does not matter (as long as it is more than 50 ft).

(b) When half the rope is pulled to the top of the building, the work to lift the top half of the rope is

$W_1 = \int_0^{25} \frac{1}{2}x\,dx = \left[\frac{1}{4}x^2\right]_0^{25} = \frac{625}{4}$ ft-lb. The bottom half of the rope is lifted 25 ft and the work needed to accomplish

that is $W_2 = \int_{25}^{50} \frac{1}{2}\cdot 25\,dx = \frac{25}{2}[x]_{25}^{50} = \frac{625}{2}$ ft-lb. The total work done in pulling half the rope to the top of the building is

$W = W_1 + W_2 = \frac{625}{2} + \frac{625}{4} = \frac{3}{4}\cdot 625 = \frac{1875}{4}$ ft-lb.

11. The work needed to lift the cable is $\lim_{n\to\infty}\sum_{i=1}^n 2x_i^*\,\Delta x = \int_0^{500} 2x\,dx = \left[x^2\right]_0^{500} = 250{,}000$ ft-lb. The work needed to lift

the coal is 800 lb $\cdot\ 500$ ft $= 400{,}000$ ft-lb. Thus, the total work required is $250{,}000 + 400{,}000 = 650{,}000$ ft-lb.

13. At a height of x meters $(0 \le x \le 12)$, the mass of the rope is $(0.8 \text{ kg/m})(12 - x \text{ m}) = (9.6 - 0.8x)$ kg and the mass of the

water is $\left(\frac{36}{12} \text{ kg/m}\right)(12 - x \text{ m}) = (36 - 3x)$ kg. The mass of the bucket is 10 kg, so the total mass is

$(9.6 - 0.8x) + (36 - 3x) + 10 = (55.6 - 3.8x)$ kg, and hence, the total force is $9.8(55.6 - 3.8x)$ N. The work needed to lift

the bucket Δx m through the ith subinterval of $[0, 12]$ is $9.8(55.6 - 3.8x_i^*)\Delta x$, so the total work is

$$W = \lim_{n\to\infty} \sum_{i=1}^n 9.8(55.6 - 3.8x_i^*)\,\Delta x = \int_0^{12}(9.8)(55.6 - 3.8x)\,dx = 9.8\left[55.6x - 1.9x^2\right]_0^{12} = 9.8(393.6) \approx 3857 \text{ J}$$

15. A "slice" of water Δx m thick and lying at a depth of x_i^* m (where $0 \le x_i^* \le \frac{1}{2}$) has volume $(2 \times 1 \times \Delta x)$ m^3, a mass of

$2000\,\Delta x$ kg, weighs about $(9.8)(2000\,\Delta x) = 19{,}600\,\Delta x$ N, and thus requires about $19{,}600x_i^*\,\Delta x$ J of work for its removal.

So $W = \lim_{n\to\infty} \sum_{i=1}^n 19{,}600x_i^*\,\Delta x = \int_0^{1/2} 19{,}600x\,dx = \left[9800x^2\right]_0^{1/2} = 2450$ J.

17. (a) A rectangular "slice" of water Δx m thick and lying x m above the bottom has width x m and volume $8x\,\Delta x$ m^3.

It weighs about $(9.8 \times 1000)(8x\,\Delta x)$ N, and must be lifted $(5-x)$ m by the pump, so the work needed is about

$(9.8 \times 10^3)(5-x)(8x\,\Delta x)$ J. The total work required is

$$W \approx \int_0^3 (9.8 \times 10^3)(5-x)8x\,dx = (9.8 \times 10^3)\int_0^3 (40x - 8x^2)\,dx = (9.8 \times 10^3)\left[20x^2 - \tfrac{8}{3}x^3\right]_0^3$$

$$= (9.8 \times 10^3)(180 - 72) = (9.8 \times 10^3)(108) = 1058.4 \times 10^3 \approx 1.06 \times 10^6 \text{ J}$$

(b) If only 4.7×10^5 J of work is done, then only the water above a certain level

(call it h) will be pumped out. So we use the same formula as in par (a), except that

the work is fixed, and we are trying to find the lower limit of integration:

$$4.7 \times 10^5 \approx \int_h^3 (9.8 \times 10^3)(5-x)8x\,dx = (9.8 \times 10^3)\left[20x^2 - \tfrac{8}{3}x^3\right]_h^3 \quad \Leftrightarrow$$

$$\tfrac{4.7}{9.8} \times 10^2 \approx 48 = \left(20 \cdot 3^2 - \tfrac{8}{3}\cdot 3^3\right) - \left(20h^2 - \tfrac{8}{3}h^3\right) \quad \Leftrightarrow$$

$2h^3 - 15h^2 + 45 = 0$. To find the solution of this equation, we plot $2h^3 - 15h^2 + 45$ between $h = 0$ and $h = 3$.

We see that the equation is satisfied for $h \approx 2.0$. So the depth of water remaining in the tank is about 2.0 m.

19. $V = \pi r^2 x$, so V is a function of x and P can also be regarded as a function of x. If $V_1 = \pi r^2 x_1$ and $V_2 = \pi r^2 x_2$, then

$$W = \int_{x_1}^{x_2} F(x)\,dx = \int_{x_1}^{x_2} \pi r^2 P(V(x))\,dx = \int_{x_1}^{x_2} P(V(x))\,dV(x) \qquad [\text{Let } V(x) = \pi r^2 x, \text{ so } dV(x) = \pi r^2\,dx.]$$

$$= \int_{V_1}^{V_2} P(V)\,dV \quad \text{by the Substitution Rule.}$$

21. (a) $W = \displaystyle\int_a^b F(r)\,dr = \int_a^b G\frac{m_1 m_2}{r^2}\,dr = Gm_1 m_2\left[\frac{-1}{r}\right]_a^b = Gm_1 m_2\left(\frac{1}{a} - \frac{1}{b}\right)$

(b) By part (a), $W = GMm\left(\dfrac{1}{R} - \dfrac{1}{R + 1{,}000{,}000}\right)$ where $M =$ mass of the Earth in kg, $R =$ radius of the Earth in m,

and $m =$ mass of the satellite in kg. (Note that 1000 km $= 1{,}000{,}000$ m.) Thus,

$$W = \left(6.67 \times 10^{-11}\right)\left(5.98 \times 10^{24}\right)(1000) \times \left(\frac{1}{6.37 \times 10^6} - \frac{1}{7.37 \times 10^6}\right) \approx 8.50 \times 10^9 \text{ J}$$

23. Since an equation for the shape is $x^2 + y^2 = 10^2$ ($x \geq 0$), we have

$y = \sqrt{100 - x^2}$. Thus, the area of the ith strip is $2\sqrt{100 - (x_i^*)^2}\,\Delta x$

and the pressure on the strip is $\rho g x_i^*$, so the hydrostatic force on

the strip is $\rho g x_i^* \cdot 2\sqrt{100 - (x_i^*)^2}\,\Delta x$ and the total force on the

plate $\approx \displaystyle\sum_{i=1}^n \rho g x_i^* \cdot 2\sqrt{100 - (x_i^*)^2}\,\Delta x$. The total force

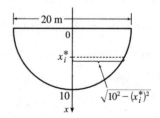

$$F = \lim_{n \to \infty} \sum_{i=1}^n \rho g x_i^* \cdot 2\sqrt{100 - (x_i^*)^2}\,\Delta x = \int_0^{10} 2\rho g x\sqrt{100 - x^2}\,dx$$

$$= -\rho g \int_0^{10} (100 - x^2)^{1/2}(-2x)\,dx = -\rho g\left[\tfrac{2}{3}(100 - x^2)^{3/2}\right]_0^{10} = -\tfrac{2}{3}\rho g(0 - 1000)$$

$$= \tfrac{2000}{3}\rho g \approx \tfrac{2000}{3} \cdot 1000 \cdot 9.8 \approx 6.5 \times 10^6 \text{ N} \qquad \left[\rho \approx 1000 \text{ kg/m}^3 \text{ and } g \approx 9.8 \text{ m/s}^2\right]$$

25. Using similar triangles, $\dfrac{4 \text{ ft wide}}{8 \text{ ft high}} = \dfrac{a \text{ ft wide}}{x_i^* \text{ ft high}}$, so $a = \frac{1}{2} x_i^*$ and the width

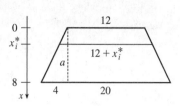

of the ith rectangular strip is $12 + 2a = 12 + x_i^*$. The area of the strip

is $(12 + x_i^*)\, \Delta x$. The pressure on the strip is δx_i^*.

$$F = \lim_{n \to \infty} \sum_{i=1}^{n} \delta x_i^*(12 + x_i^*)\, \Delta x = \int_0^8 \delta x \cdot (12 + x)\, dx = \delta \int_0^8 (12x + x^2)\, dx$$

$$= \delta \left[6x^2 + \frac{x^3}{3} \right]_0^8 = \delta\left(384 + \frac{512}{3} \right) = (62.5)\frac{1664}{3} \approx 3.47 \times 10^4 \text{ lb}$$

27. By similar triangles, $\dfrac{8}{4\sqrt{3}} = \dfrac{w_i}{x_i^*} \;\Rightarrow\; w_i = \dfrac{2x_i^*}{\sqrt{3}}$. The area of the ith

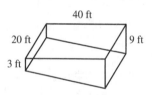

rectangular strip is $\dfrac{2x_i^*}{\sqrt{3}}\, \Delta x$ and the pressure on it is $\rho g \left(4\sqrt{3} - x_i^* \right)$.

$$F = \int_0^{4\sqrt{3}} \rho g \left(4\sqrt{3} - x \right) \frac{2x}{\sqrt{3}}\, dx = 8\rho g \int_0^{4\sqrt{3}} x\, dx - \frac{2\rho g}{\sqrt{3}} \int_0^{4\sqrt{3}} x^2\, dx$$

$$= 4\rho g \left[x^2 \right]_0^{4\sqrt{3}} - \frac{2\rho g}{3\sqrt{3}} \left[x^3 \right]_0^{4\sqrt{3}} = 192\rho g - \frac{2\rho g}{3\sqrt{3}} 64 \cdot 3\sqrt{3} = 192\rho g - 128\rho g = 64\rho g$$

$$\approx 64(840)(9.8) \approx 5.27 \times 10^5 \text{ N}$$

29. (a) The area of a strip is $20\, \Delta x$ and the pressure on it is δx_i.

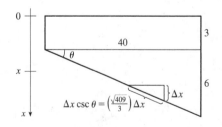

$$F = \int_0^3 \delta x 20\, dx = 20\delta \left[\tfrac{1}{2} x^2 \right]_0^3 = 20\delta \cdot \tfrac{9}{2} = 90\delta$$

$$= 90(62.5) = 5625 \text{ lb} \approx 5.63 \times 10^3 \text{ lb}$$

(b) $F = \int_0^9 \delta x 20\, dx = 20\delta \left[\tfrac{1}{2} x^2 \right]_0^9 = 20\delta \cdot \frac{81}{2} = 810\delta = 810(62.5) = 50{,}625 \text{ lb} \approx 5.06 \times 10^4 \text{ lb}$.

(c) For the first 3 ft, the length of the side is constant at 40 ft. For $3 < x \le 9$, we can use similar triangles to find the length a:

$$\frac{a}{40} = \frac{9 - x}{6} \;\Rightarrow\; a = 40 \cdot \frac{9 - x}{6}.$$

$$F = \int_0^3 \delta x 40\, dx + \int_3^9 \delta x (40)\frac{9-x}{6}\, dx = 40\delta \left[\tfrac{1}{2} x^2 \right]_0^3 + \frac{20}{3}\delta \int_3^9 \left(9x - x^2 \right) dx_3^9 = 180\delta + \frac{20}{3}\delta \left[\tfrac{9}{2} x^2 - \tfrac{1}{3} x^3 \right]_3^9$$

$$= 180\delta + \frac{20}{3}\delta \left[\left(\frac{729}{2} - 243 \right) - \left(\frac{81}{2} - 9 \right) \right] = 180\delta + 600\delta = 780\delta = 780(62.5) = 48{,}750 \text{ lb} \approx 4.88 \times 10^4 \text{ lb}$$

(d) For any right triangle with hypotenuse on the bottom,

$$\sin\theta = \frac{\Delta x}{\text{hypotenuse}} \;\Rightarrow\;$$

$$\text{hypotenuse} = \Delta x \csc\theta = \Delta x \frac{\sqrt{40^2 + 6^2}}{6} = \frac{\sqrt{409}}{3} \Delta x.$$

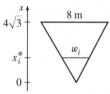

$$F = \int_3^9 \delta x 20 \frac{\sqrt{409}}{3}\, dx = \tfrac{1}{3} \left(20\sqrt{409} \right)\delta \left[\tfrac{1}{2} x^2 \right]_3^9$$

$$= \tfrac{1}{3} \cdot 10\sqrt{409}\, \delta(81 - 9) \approx 303{,}356 \text{ lb} \approx 3.03 \times 10^5 \text{ lb}$$

31. $F = \int_2^5 \rho g x \cdot w(x)\, dx$, where $w(x)$ is the width of the plate at depth x. Since $n = 6$, $\Delta x = \frac{5-2}{6} = \frac{1}{2}$, and

$$F \approx S_6 = \rho g \cdot \frac{1/2}{3}[2 \cdot w(2) + 4 \cdot 2.5 \cdot w(2.5) + 2 \cdot 3 \cdot w(3) + 4 \cdot 3.5 \cdot w(3.5)$$
$$+ 2 \cdot 4 \cdot w(4) + 4 \cdot 4.5 \cdot w(4.5) + 5 \cdot w(5)]$$
$$= \tfrac{1}{6}\rho g(2 \cdot 0 + 10 \cdot 0.8 + 6 \cdot 1.7 + 14 \cdot 2.4 + 8 \cdot 2.9 + 18 \cdot 3.3 + 5 \cdot 3.6)$$
$$= \tfrac{1}{6}(1000)(9.8)(152.4) \approx 2.5 \times 10^5 \text{ N}$$

33. $m = \sum_{i=1}^{3} m_i = 6 + 5 + 10 = 21.$

$$M_x = \sum_{i=1}^{3} m_i y_i = 6(5) + 5(-2) + 10(-1) = 10; \quad M_y = \sum_{i=1}^{3} m_i x_i = 6(1) + 5(3) + 10(-2) = 1.$$

$\overline{x} = \dfrac{M_y}{m} = \dfrac{1}{21}$ and $\overline{y} = \dfrac{M_x}{m} = \dfrac{10}{21}$, so the center of mass of the system is $\left(\frac{1}{21}, \frac{10}{21}\right)$.

35. Since the region in the figure is symmetric about the y-axis, we know
that $\overline{x} = 0$. The region is "bottom-heavy," so we know that $\overline{y} < 2$,
and we might guess that $\overline{y} = 1.5$.

$A = \int_{-2}^{2}(4 - x^2)\, dx = 2\int_0^2 (4 - x^2)\, dx = 2\left[4x - \frac{1}{3}x^3\right]_0^2$

$\quad = 2\left(8 - \frac{8}{3}\right) = \frac{32}{3}$

$\overline{x} = \frac{1}{A}\int_{-2}^{2} x(4 - x^2)\, dx = 0$ since $f(x) = x(4 - x^2)$ is an odd

function (or since the region is symmetric about the y-axis).

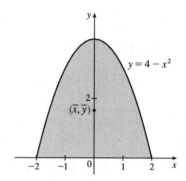

$$\overline{y} = \frac{1}{A}\int_{-2}^{2}\frac{1}{2}(4 - x^2)^2\, dx = \frac{3}{32} \cdot \frac{1}{2} \cdot 2\int_0^2 (16 - 8x^2 + x^4)\, dx = \frac{3}{32}\left[16x - \frac{8}{3}x^3 + \frac{1}{5}x^5\right]_0^2$$

$$= \frac{3}{32}\left(32 - \frac{64}{3} + \frac{32}{5}\right) = 3\left(1 - \frac{2}{3} + \frac{1}{5}\right) = 3\left(\frac{8}{15}\right) = \frac{8}{5}$$

Thus, the centroid is $(\overline{x}, \overline{y}) = \left(0, \frac{8}{5}\right)$.

37. The region in the figure is "right-heavy" and "bottom-heavy," so we know $\overline{x} > 0.5$
and $\overline{y} < 1$, and we might guess that $\overline{x} = 0.6$ and $\overline{y} = 0.9$.

$A = \int_0^1 e^x\, dx = \left[e^x\right]_0^1 = e - 1,$

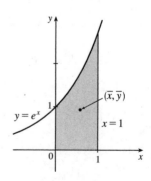

$\overline{x} = \dfrac{1}{A}\displaystyle\int_0^1 xe^x\, dx = \dfrac{1}{e-1}\left[xe^x - e^x\right]_0^1$ [by parts]

$\quad = \dfrac{1}{e-1}[0 - (-1)] = \dfrac{1}{e-1},$

$\overline{y} = \dfrac{1}{A}\displaystyle\int_0^1 \tfrac{1}{2}(e^x)^2\, dx = \dfrac{1}{e-1} \cdot \tfrac{1}{4}\left[e^{2x}\right]_0^1 = \dfrac{1}{4(e-1)}(e^2 - 1) = \dfrac{e+1}{4}.$

Thus, the centroid is $(\overline{x}, \overline{y}) = \left(\dfrac{1}{e-1}, \dfrac{e+1}{4}\right) \approx (0.58, 0.93).$

39. $A = \int_0^1 \left(\sqrt{x} - x\right) dx = \left[\frac{2}{3}x^{3/2} - \frac{1}{2}x^2\right]_0^1 = \frac{2}{3} - \frac{1}{2} = \frac{1}{6}.$

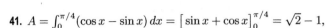

$\bar{x} = \frac{1}{A} \int_0^1 x(\sqrt{x} - x)\,dx = 6\int_0^1 (x^{3/2} - x^2)\,dx = 6\left[\frac{2}{5}x^{5/2} - \frac{1}{3}x^3\right]_0^1$

$\quad = 6\left(\frac{2}{5} - \frac{1}{3}\right) = 6\left(\frac{1}{15}\right) = \frac{2}{5},$

$\bar{y} = \frac{1}{A} \int_0^1 \frac{1}{2}\left[\left(\sqrt{x}\right)^2 - x^2\right] dx = 6\cdot\frac{1}{2}\int_0^1 (x - x^2)\,dx = 3\left[\frac{1}{2}x^2 - \frac{1}{3}x^3\right]_0^1$

$\quad = 3\left(\frac{1}{2} - \frac{1}{3}\right) = \frac{1}{2}.$

Thus, the centroid is $(\bar{x}, \bar{y}) = \left(\frac{2}{5}, \frac{1}{2}\right).$

41. $A = \int_0^{\pi/4}(\cos x - \sin x)\,dx = \left[\sin x + \cos x\right]_0^{\pi/4} = \sqrt{2} - 1,$

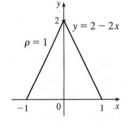

$\bar{x} = A^{-1}\int_0^{\pi/4} x(\cos x - \sin x)\,dx$

$\quad = A^{-1}\left[x(\sin x + \cos x) + \cos x - \sin x\right]_0^{\pi/4}\quad\text{[integration by parts]}$

$\quad = A^{-1}\left(\frac{\pi}{4}\sqrt{2} - 1\right) = \frac{\frac{1}{4}\pi\sqrt{2} - 1}{\sqrt{2} - 1},$

$\bar{y} = A^{-1}\int_0^{\pi/4} \frac{1}{2}(\cos^2 x - \sin^2 x)\,dx = \frac{1}{2A}\int_0^{\pi/4}\cos 2x\,dx = \frac{1}{4A}\left[\sin 2x\right]_0^{\pi/4} = \frac{1}{4A} = \frac{1}{4\left(\sqrt{2} - 1\right)}$

Thus, the centroid is $(\bar{x}, \bar{y}) = \left(\dfrac{\pi\sqrt{2} - 4}{4(\sqrt{2} - 1)}, \dfrac{1}{4(\sqrt{2} - 1)}\right) \approx (0.27, 0.60).$

43. By symmetry, $M_y = 0$ and $\bar{x} = 0$. $A = \frac{1}{2}bh = \frac{1}{2}\cdot 2\cdot 2 = 2.$

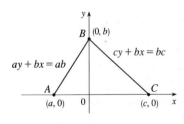

$M_x = \rho\int_{-1}^1 \frac{1}{2}(2 - 2x)^2\,dx = 2\rho\int_0^1 \frac{1}{2}(2 - 2x)^2\,dx$

$\quad = \left(2\cdot 1\cdot\frac{1}{2}\cdot 2^2\right)\int_0^1 (1 - x)^2\,dx = 4\int_1^0 u^2(-du)\quad [u = 1 - x, \, du = -dx]$

$\quad = -4\left[\frac{1}{3}u^3\right]_1^0 = -4\left(-\frac{1}{3}\right) = \frac{4}{3}$

$\bar{y} = \frac{1}{m}M_x = \frac{1}{\rho A}M_x = \frac{1}{1\cdot 2}\cdot\frac{4}{3} = \frac{2}{3}.$ Thus, the centroid is $(\bar{x}, \bar{y}) = \left(0, \frac{2}{3}\right).$

45. Choose x- and y-axes so that the base (one side of the triangle) lies along the x-axis with the other vertex along the positive y-axis as shown. From geometry, we know the medians intersect at a point $\frac{2}{3}$ of the way from each vertex (along the median) to the opposite side. The median from B goes to the midpoint $\left(\frac{1}{2}(a + c), 0\right)$ of side AC, so the point of intersection of the medians is $\left(\frac{2}{3}\cdot\frac{1}{2}(a + c), \frac{1}{3}b\right) = \left(\frac{1}{3}(a + c), \frac{1}{3}b\right).$

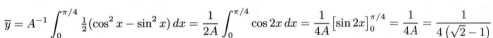

This can also be verified by finding the equations of two medians, and solving them simultaneously to find their point of intersection. Now let us compute the location of the centroid of the triangle. The area is $A = \frac{1}{2}(c - a)b.$

$\bar{x} = \frac{1}{A}\left[\int_a^0 x\cdot\frac{b}{a}(a - x)\,dx + \int_0^c x\cdot\frac{b}{c}(c - x)\,dx\right] = \frac{1}{A}\left[\frac{b}{a}\int_a^0 (ax - x^2)\,dx + \frac{b}{c}\int_0^c (cx - x^2)\,dx\right]$

$\quad = \frac{b}{Aa}\left[\frac{1}{2}ax^2 - \frac{1}{3}x^3\right]_a^0 + \frac{b}{Ac}\left[\frac{1}{2}cx^2 - \frac{1}{3}x^3\right]_0^c = \frac{b}{Aa}\left[-\frac{1}{2}a^3 + \frac{1}{3}a^3\right] + \frac{b}{Ac}\left[\frac{1}{2}c^3 - \frac{1}{3}c^3\right]$

$\quad = \frac{2}{a(c - a)}\cdot\frac{-a^3}{6} + \frac{2}{c(c - a)}\cdot\frac{c^3}{6} = \frac{1}{3(c - a)}(c^2 - a^2) = \frac{a + c}{3}$

and

$$\overline{y} = \frac{1}{A}\left[\int_a^0 \frac{1}{2}\left(\frac{b}{a}(a-x)\right)^2 dx + \int_0^c \frac{1}{2}\left(\frac{b}{c}(c-x)\right)^2 dx\right]$$

$$= \frac{1}{A}\left[\frac{b^2}{2a^2}\int_a^0 (a^2 - 2ax + x^2)\, dx + \frac{b^2}{2c^2}\int_0^c (c^2 - 2cx + x^2)\, dx\right]$$

$$= \frac{1}{A}\left[\frac{b^2}{2a^2}\left[a^2 x - ax^2 + \frac{1}{3}x^3\right]_a^0 + \frac{b^2}{2c^2}\left[c^2 x - cx^2 + \frac{1}{3}x^3\right]_0^c\right]$$

$$= \frac{1}{A}\left[\frac{b^2}{2a^2}\left(-a^3 + a^3 - \frac{1}{3}a^3\right) + \frac{b^2}{2c^2}\left(c^3 - c^3 + \frac{1}{3}c^3\right)\right] = \frac{1}{A}\left[\frac{b^2}{6}(-a+c)\right] = \frac{2}{(c-a)b}\cdot\frac{(c-a)b^2}{6} = \frac{b}{3}$$

Thus, the centroid is $(\overline{x}, \overline{y}) = \left(\dfrac{a+c}{3}, \dfrac{b}{3}\right)$, as claimed.

Remarks: Actually the computation of $\overline{y}$ is all that is needed. By considering each side of the triangle in turn to be the base, we see that the centroid is $\frac{1}{3}$ of the way from each side to the opposite vertex and must therefore be the intersection of the medians.

The computation of $\overline{y}$ in this problem (and many others) can be simplified by using horizontal rather than vertical approximating rectangles. If the length of a thin rectangle at coordinate y is $\ell(y)$, then its area is $\ell(y)\,\Delta y$, its mass is $\rho\ell(y)\,\Delta y$, and its moment about the x-axis is $\Delta M_x = \rho y\ell(y)\,\Delta y$. Thus,

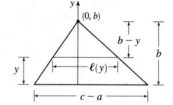

$$M_x = \int \rho y\ell(y)\, dy \text{ and } \overline{y} = \frac{\int \rho y\ell(y)\, dy}{\rho A} = \frac{1}{A}\int y\ell(y)\, dy.$$

In this problem, $\ell(y) = \dfrac{c-a}{b}(b-y)$ by similar triangles, so

$$\overline{y} = \frac{1}{A}\int_0^b \frac{c-a}{b} y(b-y)dy = \frac{2}{b^2}\int_0^b (by - y^2)\, dy = \frac{2}{b^2}\left[\frac{1}{2}by^2 - \frac{1}{3}y^3\right]_0^b = \frac{2}{b^2}\cdot\frac{b^3}{6} = \frac{b}{3}$$

Notice that only one integral is needed when this method is used.

47. Divide the lamina into two triangles and one rectangle with respective masses of 2, 2 and 4, so that the total mass is 8. Using the result of Exercise 45, the triangles have centroids $\left(-1, \frac{2}{3}\right)$ and $\left(1, \frac{2}{3}\right)$. The centroid of the rectangle (its center) is $\left(0, -\frac{1}{2}\right)$.

So, using Formulas 9 and 11, we have $\overline{y} = \dfrac{M_x}{m} = \dfrac{1}{m}\displaystyle\sum_{i=1}^3 m_i y_i = \frac{1}{8}\left[2\left(\frac{2}{3}\right) + 2\left(\frac{2}{3}\right) + 4\left(-\frac{1}{2}\right)\right] = \frac{1}{8}\left(\frac{2}{3}\right) = \frac{1}{12}$, and $\overline{x} = 0$,

since the lamina is symmetric about the line $x = 0$. Thus, the centroid is $(\overline{x}, \overline{y}) = \left(0, \frac{1}{12}\right)$.

49. A cone of height h and radius r can be generated by rotating a right triangle about one of its legs as shown. By Exercise 45, $\overline{x} = \frac{1}{3}r$, so by the Theorem of Pappus, the volume of the cone is

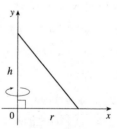

$$V = Ad = \left(\frac{1}{2}\cdot\text{base}\cdot\text{height}\right)\cdot(2\pi\overline{x}) = \frac{1}{2}rh\cdot 2\pi\left(\frac{1}{3}r\right) = \frac{1}{3}\pi r^2 h.$$

51. Suppose the region lies between two curves $y = f(x)$ and $y = g(x)$ where $f(x) \geq g(x)$, as illustrated in Figure 16. Choose points x_i with $a = x_0 < x_1 < \cdots < x_n = b$ and choose x_i^* to be the midpoint of the ith subinterval; that is, $x_i^* = \overline{x}_i = \frac{1}{2}(x_{i-1} + x_i)$. Then the centroid of the ith approximating rectangle R_i is its center $C_i = \left(\overline{x}_i, \frac{1}{2}[f(\overline{x}_i) + g(\overline{x}_i)]\right)$. Its area is $[f(\overline{x}_i) - g(\overline{x}_i)]\,\Delta x$, so its mass is $\rho[f(\overline{x}_i) - g(\overline{x}_i)]\,\Delta x$. Thus,

$M_y(R_i) = \rho[f(\overline{x}_i) - g(\overline{x}_i)]\,\Delta x \cdot \overline{x}_i = \rho\overline{x}_i[f(\overline{x}_i) - g(\overline{x}_i)]\,\Delta x$ and

$M_x(R_i) = \rho[f(\overline{x}_i) - g(\overline{x}_i)]\,\Delta x \cdot \frac{1}{2}[f(\overline{x}_i) + g(\overline{x}_i)] = \rho \cdot \frac{1}{2}\left[f(\overline{x}_i)^2 - g(\overline{x}_i)^2\right]\,\Delta x$. Summing over i and taking

the limit as $n \to \infty$, we get $M_y = \lim\limits_{n\to\infty} \sum_i \rho\overline{x}_i\,[f(\overline{x}_i) - g(\overline{x}_i)]\,\Delta x = \rho \int_a^b x[f(x) - g(x)]\,dx$ and

$M_x = \lim\limits_{n\to\infty} \sum_i \rho \cdot \frac{1}{2}\left[f(\overline{x}_i)^2 - g(\overline{x}_i)^2\right]\,\Delta x = \rho \int_a^b \frac{1}{2}[f(x)^2 - g(x)^2]\,dx$. Thus,

$$\overline{x} = \frac{M_y}{m} = \frac{M_y}{\rho A} = \frac{1}{A}\int_a^b x[f(x) - g(x)]\,dx \qquad \text{and} \qquad \overline{y} = \frac{M_x}{m} = \frac{M_x}{\rho A} = \frac{1}{A}\int_a^b \frac{1}{2}[f(x)^2 - g(x)^2]\,dx$$

7.6   Differential Equations

1. $\dfrac{dy}{dx} = \dfrac{y}{x} \ \Rightarrow \ \dfrac{dy}{y} = \dfrac{dx}{x} \ [y \neq 0] \ \Rightarrow \ \displaystyle\int \dfrac{dy}{y} = \int \dfrac{dx}{x} \ \Rightarrow \ \ln|y| = \ln|x| + C \ \Rightarrow$

$|y| = e^{\ln|x| + C} = e^{\ln|x|}e^C = e^C\,|x| \ \Rightarrow \ y = Kx$, where $K = \pm e^C$ is a constant. (In our derivation, K was nonzero, but

we can restore the excluded case $y = 0$ by allowing K to be zero.)

3. $(x^2 + 1)y' = xy \ \Rightarrow \ \dfrac{dy}{dx} = \dfrac{xy}{x^2 + 1} \ \Rightarrow \ \dfrac{dy}{y} = \dfrac{x\,dx}{x^2 + 1} \ [y \neq 0] \ \Rightarrow \ \displaystyle\int \dfrac{dy}{y} = \int \dfrac{x\,dx}{x^2 + 1} \ \Rightarrow$

$\ln|y| = \frac{1}{2}\ln(x^2 + 1) + C \ \ [u = x^2 + 1, du = 2x\,dx] \ = \ln(x^2 + 1)^{1/2} + \ln e^C = \ln(e^C\sqrt{x^2 + 1}) \ \Rightarrow$

$|y| = e^C\sqrt{x^2 + 1} \ \Rightarrow \ y = K\sqrt{x^2 + 1}$, where $K = \pm e^C$ is a constant. (In our derivation, K was nonzero, but we can

restore the excluded case $y = 0$ by allowing K to be zero.)

5. $(1 + \tan y)\,y' = x^2 + 1 \ \Rightarrow \ (1 + \tan y)\dfrac{dy}{dx} = x^2 + 1 \ \Rightarrow \ \left(1 + \dfrac{\sin y}{\cos y}\right)dy = (x^2 + 1)\,dx \ \Rightarrow$

$\displaystyle\int \left(1 - \dfrac{-\sin y}{\cos y}\right)dy = \int (x^2 + 1)\,dx \ \Rightarrow \ y - \ln|\cos y| = \frac{1}{3}x^3 + x + C$. Note: The left side is equivalent to

$y + \ln|\sec y|$.

7. $\dfrac{du}{dt} = 2 + 2u + t + tu \ \Rightarrow \ \dfrac{du}{dt} = (1 + u)(2 + t) \ \Rightarrow \ \displaystyle\int \dfrac{du}{1 + u} = \int (2 + t)\,dt \ \ [u \neq -1] \ \Rightarrow$

$\ln|1 + u| = \frac{1}{2}t^2 + 2t + C \ \Rightarrow \ |1 + u| = e^{t^2/2 + 2t + C} = Ke^{t^2/2 + 2t}$, where $K = e^C \ \Rightarrow \ 1 + u = \pm Ke^{t^2/2 + 2t} \ \Rightarrow$

$u = -1 \pm Ke^{t^2/2 + 2t}$ where $K > 0$. $\ u = -1$ is also a solution, so $u = -1 + Ae^{t^2/2 + 2t}$, where A is an arbitrary constant.

9. $\dfrac{du}{dt} = \dfrac{2t + \sec^2 t}{2u}, \ u(0) = -5. \ \ \int 2u\,du = \int (2t + \sec^2 t)\,dt \ \Rightarrow \ u^2 = t^2 + \tan t + C$, where

$[u(0)]^2 = 0^2 + \tan 0 + C \ \Rightarrow \ C = (-5)^2 = 25$. Therefore, $u^2 = t^2 + \tan t + 25$, so $u = \pm\sqrt{t^2 + \tan t + 25}$.

Since $u(0) = -5$, we must have $u = -\sqrt{t^2 + \tan t + 25}$.

11. $x\cos x = (2y + e^{3y})\,y' \ \Rightarrow \ x\cos x\,dx = (2y + e^{3y})\,dy \ \Rightarrow \ \int(2y + e^{3y})\,dy = \int x\cos x\,dx \ \Rightarrow$

$y^2 + \frac{1}{3}e^{3y} = x\sin x + \cos x + C$ [where the second integral is evaluated using integration by parts]. Now $y(0) = 0 \ \Rightarrow$

$0 + \frac{1}{3} = 0 + 1 + C \ \Rightarrow \ C = -\frac{2}{3}$. Thus, a solution is $y^2 + \frac{1}{3}e^{3y} = x\sin x + \cos x - \frac{2}{3}$. We cannot solve explicitly for y.

13. $y' \tan x = a + y, \quad 0 < x < \pi/2 \quad \Rightarrow \quad \dfrac{dy}{dx} = \dfrac{a+y}{\tan x} \quad \Rightarrow \quad \dfrac{dy}{a+y} = \cot x \, dx \quad [a+y \neq 0] \quad \Rightarrow$

$\displaystyle\int \dfrac{dy}{a+y} = \int \dfrac{\cos x}{\sin x} \, dx \quad \Rightarrow \quad \ln|a+y| = \ln|\sin x| + C \quad \Rightarrow \quad |a+y| = e^{\ln|\sin x|+C} = e^{\ln|\sin x|} \cdot e^{C} = e^{C} \, |\sin x| \quad \Rightarrow$

$a + y = K \sin x$, where $K = \pm e^{C}$. (In our derivation, K was nonzero, but we can restore the excluded case $y = -a$ by

allowing K to be zero.) $\quad y(\pi/3) = a \quad \Rightarrow \quad a + a = K \sin(\pi/3) \quad \Rightarrow \quad 2a = K \dfrac{\sqrt{3}}{2} \quad \Rightarrow \quad K = \dfrac{4a}{\sqrt{3}}.$

Thus, $a + y = \dfrac{4a}{\sqrt{3}} \sin x$ and so $y = \dfrac{4a}{\sqrt{3}} \sin x - a$.

15. $\dfrac{dy}{dx} = 4x^3 y, \; y(0) = 7. \quad \dfrac{dy}{y} = 4x^3 \, dx \quad [\text{if } y \neq 0] \quad \Rightarrow \quad \displaystyle\int \dfrac{dy}{y} = \int 4x^3 \, dx \quad \Rightarrow \quad \ln|y| = x^4 + C \quad \Rightarrow$

$e^{\ln|y|} = e^{x^4 + C} \quad \Rightarrow \quad |y| = e^{x^4} e^{C} \quad \Rightarrow \quad y = A e^{x^4}; \quad y(0) = 7 \quad \Rightarrow \quad A = 7 \quad \Rightarrow \quad y = 7 e^{x^4}.$

17. (a) $y' = 2x\sqrt{1 - y^2} \quad \Rightarrow \quad \dfrac{dy}{dx} = 2x\sqrt{1 - y^2} \quad \Rightarrow \quad \dfrac{dy}{\sqrt{1-y^2}} = 2x \, dx \quad \Rightarrow \quad \displaystyle\int \dfrac{dy}{\sqrt{1-y^2}} = \int 2x \, dx \quad \Rightarrow$

$\sin^{-1} y = x^2 + C$ for $-\dfrac{\pi}{2} \leq x^2 + C \leq \dfrac{\pi}{2}$.

(b) $y(0) = 0 \quad \Rightarrow \quad \sin^{-1} 0 = 0^2 + C \quad \Rightarrow \quad C = 0$, so $\sin^{-1} y = x^2$

and $y = \sin(x^2)$ for $-\sqrt{\pi/2} \leq x \leq \sqrt{\pi/2}$.

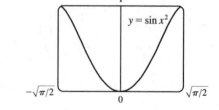

(c) For $\sqrt{1 - y^2}$ to be a real number, we must have $-1 \leq y \leq 1$;

that is, $-1 \leq y(0) \leq 1$. Thus, the initial-value problem

$y' = 2x\sqrt{1 - y^2}, \; y(0) = 2$ does *not* have a solution.

19. $\dfrac{dy}{dx} = \dfrac{\sin x}{\sin y}, \; y(0) = \dfrac{\pi}{2}.$ So $\int \sin y \, dy = \int \sin x \, dx \quad \Leftrightarrow$

$-\cos y = -\cos x + C \quad \Leftrightarrow \quad \cos y = \cos x - C.$ From the initial condition,

we need $\cos \dfrac{\pi}{2} = \cos 0 - C \quad \Rightarrow \quad 0 = 1 - C \quad \Rightarrow \quad C = 1$, so the solution is

$\cos y = \cos x - 1.$ Note that we cannot take $\cos^{-1}$ of both sides, since that would

unnecessarily restrict the solution to the case where $-1 \leq \cos x - 1 \quad \Leftrightarrow \quad 0 \leq \cos x$,

as $\cos^{-1}$ is defined only on $[-1, 1]$. Instead we plot the graph using Maple's

`plots[implicitplot]` or Mathematica's `Plot [Evaluate[···]]`.

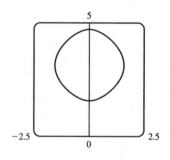

21. $y' = 2 - y$. The slopes at each point are independent of x, so the slopes are the same along each line parallel to the x-axis.
Thus, III is the direction field for this equation. Note that for $y = 2$, $y' = 0$.

23. $y' = x + y - 1 = 0$ on the line $y = -x + 1$. Direction field IV satisfies this condition. Notice also that on the line $y = -x$ we
have $y' = -1$, which is true in IV.

25. (a) $y(0) = 1$

(b) $y(0) = 2$

(c) $y(0) = -1$

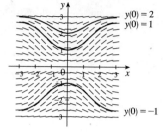

27.

x	y	$y' = 1 + y$
0	0	1
0	1	2
0	2	3
0	-3	-2
0	-2	-1

Note that for $y = -1$, $y' = 0$. The three solution curves sketched go through $(0, 0)$, $(0, -1)$, and $(0, -2)$.

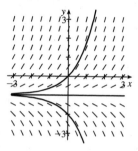

29.

x	y	$y' = y - 2x$
-2	-2	2
-2	2	6
2	2	-2
2	-2	-6

Note that $y' = 0$ for any point on the line $y = 2x$. The slopes are positive to the left of the line and negative to the right of the line. The solution curve in the graph passes through $(1, 0)$.

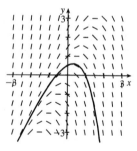

31.

x	y	$y' = y + xy$
0	± 2	± 2
1	± 2	± 4
-3	± 2	∓ 4

Note that $y' = y(x + 1) = 0$ for any point on $y = 0$ or on $x = -1$. The slopes are positive when the factors y and $x + 1$ have the same sign and negative when they have opposite signs. The solution curve in the graph passes through $(0, 1)$.

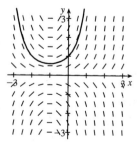

33. (a) $\dfrac{dP}{dt} = k(M - P)$ is always positive, so the level of performance P is increasing. As P gets close to M, dP/dt gets close

to 0; that is, the performance levels off, as explained in part (a).

(b) $\dfrac{dP}{dt} = k(M - P) \quad \Leftrightarrow \quad \displaystyle\int \frac{dP}{P - M} = \int (-k)\, dt \quad \Leftrightarrow \quad \ln|P - M| = -kt + C \quad \Leftrightarrow \quad |P - M| = e^{-kt + C} \quad \Leftrightarrow$

$P - M = Ae^{-kt} \quad [A = \pm e^{C}] \quad \Leftrightarrow \quad P = M + Ae^{-kt}$. If we assume that performance is at level 0 when $t = 0$, then

$P(0) = 0 \quad \Leftrightarrow \quad 0 = M + A \quad \Leftrightarrow \quad A = -M \quad \Leftrightarrow \quad P(t) = M - Me^{-kt}$. $\quad \displaystyle\lim_{t \to \infty} P(t) = M - M \cdot 0 = M$.

35. (a) $\dfrac{dC}{dt} = r - kC \quad \Rightarrow \quad \dfrac{dC}{dt} = -(kC - r) \quad \Rightarrow \quad \displaystyle\int \frac{dC}{kC - r} = \int -dt \quad \Rightarrow \quad (1/k)\ln|kC - r| = -t + M_1 \quad \Rightarrow$

$\ln|kC - r| = -kt + M_2 \quad \Rightarrow \quad |kC - r| = e^{-kt + M_2} \quad \Rightarrow \quad kC - r = M_3 e^{-kt} \quad \Rightarrow \quad kC = M_3 e^{-kt} + r \quad \Rightarrow$

$C(t) = M_4 e^{-kt} + r/k$. $\quad C(0) = C_0 \quad \Rightarrow \quad C_0 = M_4 + r/k \quad \Rightarrow \quad M_4 = C_0 - r/k \quad \Rightarrow$

$C(t) = (C_0 - r/k)e^{-kt} + r/k$.

(b) If $C_0 < r/k$, then $C_0 - r/k < 0$ and the formula for $C(t)$ shows that $C(t)$ increases and $\displaystyle\lim_{t \to \infty} C(t) = r/k$.

As t increases, the formula for $C(t)$ shows how the role of C_0 steadily diminishes as that of r/k increases.

37. The differential equation is a logistic equation with $k = 0.00008$, carrying capacity $M = 1000$, and initial population

$y_0 = P(0) = 100$. So Equation 8 gives the population at time t as

$$P(t) = \frac{100 \cdot 1000}{100 + (1000 - 100)e^{-0.08t}} = \frac{100{,}000}{100 + 900e^{-0.08t}} = \frac{1000}{1 + 9e^{-0.08t}}$$

So the population sizes when $t = 40$ and 80 are

$$P(40) = \frac{1000}{1 + 9e^{-3.2}} \approx 731.6 \qquad P(80) = \frac{1000}{1 + 9e^{-6.4}} \approx 985.3$$

The population reaches 900 when $\dfrac{1000}{1 + 9e^{-0.08t}} = 900$. Solving this equation for t, we get $1 + 9e^{-0.08t} = \frac{10}{9} \quad \Rightarrow$

$e^{-0.08t} = \frac{1}{81} \quad \Rightarrow \quad -0.08t = \ln\frac{1}{81} = -\ln 81 \quad \Rightarrow \quad t = \frac{\ln 81}{0.08} \approx 54.9$. So the population reaches 900 when t is

approximately 55.

39. (a) Our assumption is that $\dfrac{dy}{dt} = ky(1 - y)$, where y is the fraction of the population that has heard the rumor.

(b) The equation in part (a) is the logistic differential equation (7) with $M = 1$, so the solution is given by (8):

$y = \dfrac{y_0}{y_0 + (1 - y_0)e^{-kt}}$.

(c) Let t be the number of hours since 8 AM. Then $y_0 = y(0) = \frac{80}{1000} = 0.08$ and $y(4) = \frac{1}{2}$, so

$\dfrac{1}{2} = y(4) = \dfrac{0.08}{0.08 + 0.92e^{-4k}}$. Thus, $0.08 + 0.92e^{-4k} = 0.16$, $e^{-4k} = \frac{0.08}{0.92} = \frac{2}{23}$, and $e^{-k} = \left(\frac{2}{23}\right)^{1/4}$,

so $y = \dfrac{0.08}{0.08 + 0.92(2/23)^{t/4}} = \dfrac{2}{2 + 23(2/23)^{t/4}}$. Solving this equation for t, we get

$$2y + 23y\left(\frac{2}{23}\right)^{t/4} = 2 \;\Rightarrow\; \left(\frac{2}{23}\right)^{t/4} = \frac{2 - 2y}{23y} \;\Rightarrow\; \left(\frac{2}{23}\right)^{t/4} = \frac{2}{23}\cdot\frac{1-y}{y} \;\Rightarrow\; \left(\frac{2}{23}\right)^{t/4-1} = \frac{1-y}{y}.$$

It follows that $\dfrac{t}{4} - 1 = \dfrac{\ln[(1-y)/y]}{\ln\frac{2}{23}}$, so $t = 4\left[1 + \dfrac{\ln((1-y)/y)}{\ln\frac{2}{23}}\right]$.

When $y = 0.9$, $\dfrac{1-y}{y} = \frac{1}{9}$, so $t = 4\left(1 - \dfrac{\ln 9}{\ln\frac{2}{23}}\right) \approx 7.6$ h or 7 h 36 min. Thus, 90% of the population will have heard

the rumor by 3:36 PM.

41. (a) $\dfrac{dy}{dt} = ky(M - y) \;\Rightarrow$

$$\frac{d^2y}{dt^2} = ky\left(-\frac{dy}{dt}\right) + k(M - y)\frac{dy}{dt} = k\frac{dy}{dt}(M - 2y) = k[ky(M-y)](M - 2y) = k^2y(M-y)(M-2y)$$

(b) y grows fastest when y' has a maximum, that is, when $y'' = 0$. From part (a), $y'' = 0 \;\Leftrightarrow\; y = 0,\, y = M$, or $y = M/2$.
Since $0 < y < M$, we see that $y'' = 0 \;\Leftrightarrow\; y = M/2$.

43. (a) Let $y(t)$ be the amount of salt (in kg) after t minutes. Then $y(0) = 15$. The amount of liquid in the tank is 1000 L at all

times, so the concentration at time t (in minutes) is $y(t)/1000$ kg/L and $\dfrac{dy}{dt} = -\left[\dfrac{y(t)}{1000}\dfrac{\text{kg}}{\text{L}}\right]\left(10\,\dfrac{\text{L}}{\text{min}}\right) = -\dfrac{y(t)}{100}\dfrac{\text{kg}}{\text{min}}$.

$$\int \frac{dy}{y} = -\frac{1}{100}\int dt \;\Rightarrow\; \ln y = -\frac{t}{100} + C, \text{ and } y(0) = 15 \;\Rightarrow\; \ln 15 = C, \text{ so } \ln y = \ln 15 - \frac{t}{100}.$$

It follows that $\ln\left(\dfrac{y}{15}\right) = -\dfrac{t}{100}$ and $\dfrac{y}{15} = e^{-t/100}$, so $y = 15e^{-t/100}$ kg.

(b) After 20 minutes, $y = 15e^{-20/100} = 15e^{-0.2} \approx 12.3$ kg.

45. Let $y(t)$ be the amount of alcohol in the vat after t minutes. Then $y(0) = 0.04(500) = 20$ gal. The amount of beer in the vat
is 500 gallons at all times, so the percentage at time t (in minutes) is $y(t)/500 \times 100$, and the change in the amount of alcohol
with respect to time t is

$$\frac{dy}{dt} = \text{rate in} - \text{rate out} = 0.06\left(5\,\frac{\text{gal}}{\text{min}}\right) - \frac{y(t)}{500}\left(5\,\frac{\text{gal}}{\text{min}}\right) = 0.3 - \frac{y}{100} = \frac{30 - y}{100}\frac{\text{gal}}{\text{min}}$$

Hence, $\displaystyle\int \frac{dy}{30 - y} = \int \frac{dt}{100}$ and $-\ln|30 - y| = \frac{1}{100}t + C$. Because $y(0) = 20$, we have $-\ln 10 = C$, so

$-\ln|30 - y| = \frac{1}{100}t - \ln 10 \;\Rightarrow\; \ln|30 - y| = -t/100 + \ln 10 \;\Rightarrow\; \ln|30 - y| = \ln e^{-t/100} + \ln 10 \;\Rightarrow$

$\ln|30 - y| = \ln(10e^{-t/100}) \;\Rightarrow\; |30 - y| = 10e^{-t/100}$. Since y is continuous, $y(0) = 20$, and the right-hand side is

never zero, we deduce that $30 - y$ is always positive. Thus, $30 - y = 10e^{-t/100} \;\Rightarrow\; y = 30 - 10e^{-t/100}$.

The percentage of alcohol is $p(t) = y(t)/500 \times 100 = y(t)/5 = 6 - 2e^{-t/100}$. The percentage of alcohol after one hour

is $p(60) = 6 - 2e^{-60/100} \approx 4.9$.

47. Assume that the raindrop begins at rest, so that $v(0) = 0$. $dm/dt = km$ and $(mv)' = gm \;\Rightarrow\; mv' + vm' = gm \;\Rightarrow$

$$mv' + v(km) = gm \;\Rightarrow\; v' + vk = g \;\Rightarrow\; \frac{dv}{dt} = g - kv \;\Rightarrow\; \int \frac{dv}{g - kv} = \int dt \;\Rightarrow$$

$-(1/k)\ln|g - kv| = t + C$ ⟹ $\ln|g - kv| = -kt - kC$ ⟹ $g - kv = Ae^{-kt}$. $v(0) = 0$ ⟹ $A = g$. So

$kv = g - ge^{-kt}$ ⟹ $v = (g/k)(1 - e^{-kt})$. Since $k > 0$, as $t \to \infty$, $e^{-kt} \to 0$ and therefore, $\lim\limits_{t \to \infty} v(t) = g/k$.

49. (a) The rate of growth of the area is jointly proportional to $\sqrt{A(t)}$ and $M - A(t)$; that is, the rate is proportional to the

product of those two quantities. So for some constant k, $dA/dt = k\sqrt{A}\,(M - A)$. We are interested in the maximum of

the function dA/dt (when the tissue grows the fastest), so we differentiate, using the Chain Rule and then substituting for

dA/dt from the differential equation:

$$\frac{d}{dt}\left(\frac{dA}{dt}\right) = k\left[\sqrt{A}(-1)\frac{dA}{dt} + (M - A) \cdot \tfrac{1}{2}A^{-1/2}\frac{dA}{dt}\right] = \tfrac{1}{2}kA^{-1/2}\frac{dA}{dt}[-2A + (M - A)]$$

$$= \tfrac{1}{2}kA^{-1/2}\left[k\sqrt{A}(M - A)\right][M - 3A] = \tfrac{1}{2}k^2(M - A)(M - 3A)$$

This is 0 when $M - A = 0$ [this situation never actually occurs, since the graph of $A(t)$ is asymptotic to the line $y = M$,

as in the logistic model] and when $M - 3A = 0$ ⟺ $A(t) = M/3$. This represents a maximum by the First Derivative

Test, since $\dfrac{d}{dt}\left(\dfrac{dA}{dt}\right)$ goes from positive to negative when $A(t) = M/3$.

(b) From the CAS, we get $A(t) = M\left(\dfrac{Ce^{\sqrt{M}kt} - 1}{Ce^{\sqrt{M}kt} + 1}\right)^2$. To get C in terms of the initial area A_0 and the maximum area M,

we substitute $t = 0$ and $A = A_0 = A(0)$: $A_0 = M\left(\dfrac{C - 1}{C + 1}\right)^2$ ⟺ $(C + 1)\sqrt{A_0} = (C - 1)\sqrt{M}$ ⟺

$C\sqrt{A_0} + \sqrt{A_0} = C\sqrt{M} - \sqrt{M}$ ⟺ $\sqrt{M} + \sqrt{A_0} = C\sqrt{M} - C\sqrt{A_0}$ ⟺

$\sqrt{M} + \sqrt{A_0} = C\left(\sqrt{M} - \sqrt{A_0}\right)$ ⟺ $C = \dfrac{\sqrt{M} + \sqrt{A_0}}{\sqrt{M} - \sqrt{A_0}}$. (Notice that if $A_0 = 0$, then $C = 1$.)

7 Review

CONCEPT CHECK

1. (a) See Section 7.1, Figure 2 and Equations 7.1.1 and 7.1.2.

(b) Instead of using "top minus bottom" and integrating from left to right, we use "right minus left" and integrate from bottom

to top. See Figures 8 and 9 in Section 7.1.

2. The numerical value of the area represents the number of meters by which Sue is ahead of Kathy after 1 minute.

3. (a) See the discussion on pages 363–64.

(b) See the discussion between Examples 5 and 6 in Section 7.2. If the cross-section is a disk, find the radius in terms of x or y

and use $A = \pi(\text{radius})^2$. If the cross-section is a washer, find the inner radius r_{in} and outer radius r_{out} and use

$A = \pi\left(r_{out}^2\right) - \pi\left(r_{in}^2\right)$.

4. (a) $V = 2\pi r h\,\Delta r = $ (circumference) (height) (thickness)

 (b) For a typical shell, find the circumference and height in terms of x or y and calculate

 $V = \int_a^b$ (circumference) (height) $(dx$ or $dy)$, where a and b are the limits on x or y.

 (c) Sometimes slicing produces washers or disks whose radii are difficult (or impossible) to find explicitly. On other occasions, the cylindrical shell method leads to an easier integral than slicing does.

5. (a) The length of a curve is defined to be the limit of the lengths of the inscribed polygons, as described near Figure 3 in Section 7.4.

 (b) See Equation 7.4.2.

 (c) See Equation 7.4.4.

6. $\int_0^6 f(x)\,dx$ represents the amount of work done. Its units are newton-meters, or joules.

7. Let $c(x)$ be the cross-sectional length of the wall (measured parallel to the surface of the fluid) at depth x. Then the hydrostatic force against the wall is given by $F = \int_a^b \delta x c(x)\,dx$, where a and b are the lower and upper limits for x at points of the wall and δ is the weight density of the fluid.

8. (a) The center of mass is the point at which the plate balances horizontally.

 (b) See Equations 7.5.12.

9. If a plane region $\mathcal{R}$ that lies entirely on one side of a line ℓ in its plane is rotated about ℓ, then the volume of the resulting solid is the product of the area of $\mathcal{R}$ and the distance traveled by the centroid of $\mathcal{R}$.

10. (a) A differential equation is an equation that contains an unknown function and one or more of its derivatives.

 (b) The order of a differential equation is the order of the highest derivative that occurs in the equation.

 (c) An initial condition is a condition of the form $y(t_0) = y_0$.

11. See the paragraph preceding Example 6 in Section 7.6.

12. A separable equation is a first-order differential equation in which the expression for dy/dx can be factored as a function of x times a function of y, that is, $dy/dx = g(x)f(y)$. We can solve the equation by integrating both sides of the equation $dy/f(y) = g(x)\,dx$ and solving for y.

EXERCISES

1. $0 = x^2 - x - 6 = (x-3)(x+2) \iff x = 3$ or -2. So

$$A = \int_{-2}^{3} \left[0 - (x^2 - x - 6)\right] dx = \int_{-2}^{3} \left(-x^2 + x + 6\right) dx$$

$$= \left[-\tfrac{1}{3}x^3 + \tfrac{1}{2}x^2 + 6x\right]_{-2}^{3}$$

$$= \left(-9 + \tfrac{9}{2} + 18\right) - \left(\tfrac{8}{3} + 2 - 12\right)$$

$$= \tfrac{125}{6}$$

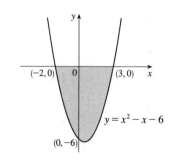

3. $A = \int_0^1 \left[(e^x - 1) - (x^2 - x) \right] dx$

$\qquad = \int_0^1 \left(e^x - 1 - x^2 + x \right) dx = \left[e^x - x - \tfrac{1}{3}x^3 + \tfrac{1}{2}x^2 \right]_0^1$

$\qquad = \left(e - 1 - \tfrac{1}{3} + \tfrac{1}{2} \right) - (1 - 0 - 0 + 0) = e - \tfrac{11}{6}$

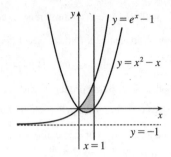

5. Using washers with inner radius x^2 and outer radius $2x$, we have

$\qquad V = \pi \int_0^2 \left[(2x)^2 - (x^2)^2 \right] dx = \pi \int_0^2 (4x^2 - x^4)\, dx$

$\qquad\qquad = \pi \left[\tfrac{4}{3}x^3 - \tfrac{1}{5}x^5 \right]_0^2 = \pi \left(\tfrac{32}{3} - \tfrac{32}{5} \right) = 32\pi \cdot \tfrac{2}{15}$

$\qquad\qquad = \tfrac{64\pi}{15}$

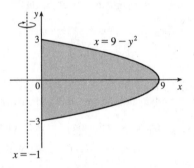

7. $V = \pi \int_{-3}^3 \left\{ [(9 - y^2) - (-1)]^2 - [0 - (-1)]^2 \right\} dy$

$\qquad = 2\pi \int_0^3 \left[(10 - y^2)^2 - 1 \right] dy$

$\qquad = 2\pi \int_0^3 (100 - 20y^2 + y^4 - 1)\, dy$

$\qquad = 2\pi \int_0^3 (99 - 20y^2 + y^4)\, dy = 2\pi \left[99y - \tfrac{20}{3}y^3 + \tfrac{1}{5}y^5 \right]_0^3$

$\qquad = 2\pi \left(297 - 180 + \tfrac{243}{5} \right) = \tfrac{1656\pi}{5}$

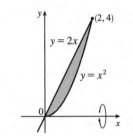

9. The graph of $x^2 - y^2 = a^2$ is a hyperbola with right and left branches. Solving for

$\quad y$ gives us $y^2 = x^2 - a^2 \;\Rightarrow\; y = \pm\sqrt{x^2 - a^2}$. We'll use shells and the height

$\quad$ of each shell is $\sqrt{x^2 - a^2} - \left(-\sqrt{x^2 - a^2} \right) = 2\sqrt{x^2 - a^2}$.

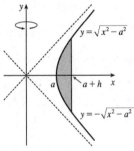

$\qquad$ The volume is $V = \int_a^{a+h} 2\pi x \cdot 2\sqrt{x^2 - a^2}\, dx$. To evaluate, let $u = x^2 - a^2$,

$\quad$ so $du = 2x\, dx$ and $x\, dx = \tfrac{1}{2}\, du$. When $x = a$, $u = 0$, and when $x = a + h$,

$\quad u = (a + h)^2 - a^2 = a^2 + 2ah + h^2 - a^2 = 2ah + h^2$.

$\qquad$ Thus, $V = 4\pi \int_0^{2ah+h^2} \sqrt{u} \left(\tfrac{1}{2}\, du \right) = 2\pi \left[\tfrac{2}{3}u^{3/2} \right]_0^{2ah+h^2} = \tfrac{4}{3}\pi (2ah + h^2)^{3/2}$.

11. $V = \int_0^1 \pi [(1 - x^3)^2 - (1 - x^2)^2]\, dx$

13. (a) A cross-section is a washer with inner radius x^2 and outer radius x.

$\qquad V = \int_0^1 \pi \left[(x)^2 - (x^2)^2 \right] dx = \int_0^1 \pi (x^2 - x^4)\, dx = \pi \left[\tfrac{1}{3}x^3 - \tfrac{1}{5}x^5 \right]_0^1 = \pi \left[\tfrac{1}{3} - \tfrac{1}{5} \right] = \tfrac{2\pi}{15}$

$\quad$ (b) A cross-section is a washer with inner radius y and outer radius $\sqrt{y}$.

$\qquad V = \int_0^1 \pi \left[\left(\sqrt{y} \right)^2 - y^2 \right] dy = \int_0^1 \pi (y - y^2)\, dy = \pi \left[\tfrac{1}{2}y^2 - \tfrac{1}{3}y^3 \right]_0^1 = \pi \left[\tfrac{1}{2} - \tfrac{1}{3} \right] = \tfrac{\pi}{6}$

(c) A cross-section is a washer with inner radius $2 - x$ and outer radius $2 - x^2$.

$$V = \int_0^1 \pi\left[(2-x^2)^2 - (2-x)^2\right] dx = \int_0^1 \pi(x^4 - 5x^2 + 4x)\, dx = \pi\left[\tfrac{1}{5}x^5 - \tfrac{5}{3}x^3 + 2x^2\right]_0^1 = \pi\left[\tfrac{1}{5} - \tfrac{5}{3} + 2\right] = \tfrac{8\pi}{15}$$

15. (a) Using the Midpoint Rule on $[0, 1]$ with $f(x) = \tan(x^2)$ and $n = 4$, we estimate

$$A = \int_0^1 \tan(x^2)\, dx \approx \tfrac{1}{4}\left[\tan\left(\left(\tfrac{1}{8}\right)^2\right) + \tan\left(\left(\tfrac{3}{8}\right)^2\right) + \tan\left(\left(\tfrac{5}{8}\right)^2\right) + \tan\left(\left(\tfrac{7}{8}\right)^2\right)\right] \approx \tfrac{1}{4}(1.53) \approx 0.38$$

(b) Using the Midpoint Rule on $[0, 1]$ with $f(x) = \pi \tan^2(x^2)$ (for disks) and $n = 4$, we estimate

$$V = \int_0^1 f(x)\, dx \approx \tfrac{1}{4}\pi\left[\tan^2\left(\left(\tfrac{1}{8}\right)^2\right) + \tan^2\left(\left(\tfrac{3}{8}\right)^2\right) + \tan^2\left(\left(\tfrac{5}{8}\right)^2\right) + \tan^2\left(\left(\tfrac{7}{8}\right)^2\right)\right] \approx \tfrac{\pi}{4}(1.114) \approx 0.87$$

17. The solid is obtained by rotating the region $\mathcal{R} = \left\{(x,y) \mid 0 \le x \le \tfrac{\pi}{2}, 0 \le y \le \cos x\right\}$ about the y-axis.

19. The solid is obtained by rotating the region $\mathcal{R} = \left\{(x,y) \mid 0 \le y \le 2, 0 \le x \le 4 - y^2\right\}$ about the x-axis.

21. Take the base to be the disk $x^2 + y^2 \le 9$. Then $V = \int_{-3}^3 A(x)\, dx$, where $A(x_0)$ is the area of the isosceles right triangle whose hypotenuse lies along the line $x = x_0$ in the xy-plane. The length of the hypotenuse is $2\sqrt{9 - x^2}$ and the length of each leg is $\sqrt{2}\sqrt{9 - x^2}$. $\quad A(x) = \tfrac{1}{2}\left(\sqrt{2}\sqrt{9 - x^2}\right)^2 = 9 - x^2$, so

$$V = 2\int_0^3 A(x)\, dx = 2\int_0^3 (9 - x^2)\, dx = 2\left[9x - \tfrac{1}{3}x^3\right]_0^3 = 2(27 - 9) = 36$$

23. Equilateral triangles with sides measuring $\tfrac{1}{4}x$ meters have height $\tfrac{1}{4}x \sin 60° = \tfrac{\sqrt{3}}{8}x$. Therefore,

$$A(x) = \tfrac{1}{2} \cdot \tfrac{1}{4}x \cdot \tfrac{\sqrt{3}}{8}x = \tfrac{\sqrt{3}}{64}x^2. \quad V = \int_0^{20} A(x)\, dx = \tfrac{\sqrt{3}}{64}\int_0^{20} x^2\, dx = \tfrac{\sqrt{3}}{64}\left[\tfrac{1}{3}x^3\right]_0^{20} = \tfrac{8000\sqrt{3}}{64 \cdot 3} = \tfrac{125\sqrt{3}}{3}\ \text{m}^3.$$

25. $y = \tfrac{1}{6}(x^2 + 4)^{3/2} \;\Rightarrow\; dy/dx = \tfrac{1}{4}(x^2 + 4)^{1/2}(2x) \;\Rightarrow$

$$1 + (dy/dx)^2 = 1 + \left[\tfrac{1}{2}x(x^2 + 4)^{1/2}\right]^2 = 1 + \tfrac{1}{4}x^2(x^2 + 4) = \tfrac{1}{4}x^4 + x^2 + 1 = \left(\tfrac{1}{2}x^2 + 1\right)^2.$$

Thus, $L = \int_0^3 \sqrt{\left(\tfrac{1}{2}x^2 + 1\right)^2}\, dx = \int_0^3 \left(\tfrac{1}{2}x^2 + 1\right) dx = \left[\tfrac{1}{6}x^3 + x\right]_0^3 = \tfrac{15}{2}.$

27. $y = e^{-x^2} \;\Rightarrow\; dy/dx = -2xe^{-x^2} \;\Rightarrow\; 1 + (dy/dx)^2 = 1 + 4x^2e^{-2x^2}$. Let $f(x) = \sqrt{1 + 4x^2e^{-2x^2}}$. Then

$$L = \int_0^3 f(x)\, dx \approx S_6 = \tfrac{(3 - 0)/6}{3}\left[f(0) + 4f(0.5) + 2f(1) + 4f(1.5) + 2f(2) + 4f(2.5) + f(3)\right] \approx 3.292287.$$

29. $f(x) = kx \;\Rightarrow\; 30\ \text{N} = k(15 - 12)\ \text{cm} \;\Rightarrow\; k = 10\ \text{N/cm} = 1000\ \text{N/m}. \quad 20\ \text{cm} - 12\ \text{cm} = 0.08\ \text{m} \;\Rightarrow$

$$W = \int_0^{0.08} kx\, dx = 1000\int_0^{0.08} x\, dx = 500\left[x^2\right]_0^{0.08} = 500(0.08)^2 = 3.2\ \text{N·m} = 3.2\ \text{J}.$$

31. (a) The parabola has equation $y = ax^2$ with vertex at the origin and passing through

$(4, 4)$. $\quad 4 = a \cdot 4^2 \;\Rightarrow\; a = \tfrac{1}{4} \;\Rightarrow\; y = \tfrac{1}{4}x^2 \;\Rightarrow\; x^2 = 4y \;\Rightarrow$

$x = 2\sqrt{y}$. Each circular disk has radius $2\sqrt{y}$ and is moved $4 - y$ ft.

$$W = \int_0^4 \pi\left(2\sqrt{y}\right)^2 62.5\,(4 - y)\, dy = 250\pi\int_0^4 y(4 - y)\, dy$$

$$= 250\pi\left[2y^2 - \tfrac{1}{3}y^3\right]_0^4 = 250\pi\left(32 - \tfrac{64}{3}\right) = \tfrac{8000\pi}{3} \approx 8378\ \text{ft-lb}$$

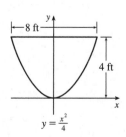

(b) In part (a) we knew the final water level (0) but not the amount of work done. Here we use the same equation, except with the work fixed, and the lower limit of integration (that is, the final water level — call it h) unknown: $W = 4000$ ⟺

$250\pi\left[2y^2 - \frac{1}{3}y^3\right]_h^4 = 4000$ ⟺ $\frac{16}{\pi} = \left[\left(32 - \frac{64}{3}\right) - \left(2h^2 - \frac{1}{3}h^3\right)\right]$ ⟺

$h^3 - 6h^2 + 32 - \frac{48}{\pi} = 0$. We graph the function $f(h) = h^3 - 6h^2 + 32 - \frac{48}{\pi}$

on the interval $[0, 4]$ to see where it is 0. From the graph, $f(h) = 0$ for $h \approx 2.1$.

So the depth of water remaining is about 2.1 ft.

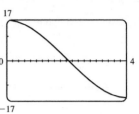

33. As in Example 5 of Section 7.5, $\dfrac{a}{2-x} = \dfrac{1}{2}$ ⟹ $2a = 2 - x$ and $w = 2(1.5 + a) = 3 + 2a = 3 + 2 - x = 5 - x$.

Thus, $F = \int_0^2 \rho g x(5 - x)\,dx = \rho g\left[\frac{5}{2}x^2 - \frac{1}{3}x^3\right]_0^2 = \rho g\left(10 - \frac{8}{3}\right) = \frac{22}{3}\delta$ $[\rho g = \delta]$ $\approx \frac{22}{3} \cdot 62.5 \approx 458$ lb.

35. $A = \int_{-2}^1 [(4 - x^2) - (x + 2)]\,dx = \int_{-2}^1 (2 - x - x^2)\,dx = \left[2x - \frac{1}{2}x^2 - \frac{1}{3}x^3\right]_{-2}^1$

$= \left(2 - \frac{1}{2} - \frac{1}{3}\right) - \left(-4 - 2 + \frac{8}{3}\right) = \frac{9}{2}$ ⟹

$\bar{x} = A^{-1}\int_{-2}^1 x(2 - x - x^2)\,dx = \frac{2}{9}\int_{-2}^1 (2x - x^2 - x^3)\,dx = \frac{2}{9}\left[x^2 - \frac{1}{3}x^3 - \frac{1}{4}x^4\right]_{-2}^1$

$= \frac{2}{9}\left[\left(1 - \frac{1}{3} - \frac{1}{4}\right) - \left(4 + \frac{8}{3} - 4\right)\right] = -\frac{1}{2}$

and $\bar{y} = A^{-1}\int_{-2}^1 \frac{1}{2}\left[(4 - x^2)^2 - (x + 2)^2\right]dx = \frac{1}{9}\int_{-2}^1 (x^4 - 9x^2 - 4x + 12)\,dx$

$= \frac{1}{9}\left[\frac{1}{5}x^5 - 3x^3 - 2x^2 + 12x\right]_{-2}^1 = \frac{1}{9}\left[\left(\frac{1}{5} - 3 - 2 + 12\right) - \left(-\frac{32}{5} + 24 - 8 - 24\right)\right] = \frac{12}{5}$

Thus, the centroid is $(\bar{x}, \bar{y}) = \left(-\frac{1}{2}, \frac{12}{5}\right)$.

37. The centroid of this circle, $(1, 0)$, travels a distance $2\pi(1)$ when the lamina is rotated about the y-axis. The area of the circle is $\pi(1)^2$. So by the Theorem of Pappus, $V = A(2\pi\bar{x}) = \pi(1)^2 2\pi(1) = 2\pi^2$.

39. $(3y^2 + 2y)\,y' = x\cos x$ ⟹ $(3y^2 + 2y)\,dy = (x\cos x)\,dx$ ⟹ $\int(3y^2 + 2y)\,dy = \int(x\cos x)\,dx$ ⟹

$y^3 + y^2 = \cos x + x\sin x + C$. For the last step, use integration by parts or Formula 83 in the Table of Integrals.

41. $\dfrac{dr}{dt} + 2tr = r$ ⟹ $\dfrac{dr}{dt} = r - 2tr = r(1 - 2t)$ ⟹ $\displaystyle\int \frac{dr}{r} = \int (1 - 2t)\,dt$ ⟹ $\ln|r| = t - t^2 + C$ ⟹

$|r| = e^{t - t^2 + C} = ke^{t - t^2}$. Since $r(0) = 5$, $5 = ke^0 = k$. Thus, $r(t) = 5e^{t - t^2}$.

43. (a)

We sketch the direction field and four solution curves, as shown.

Note that the slope $y' = x/y$ is not defined on the line $y = 0$.

(b) $y' = x/y$ ⟺ $y\,dy = x\,dx$ ⟺ $y^2 = x^2 + C$. For $C = 0$, this is the pair of lines $y = \pm x$. For $C \neq 0$, it is the hyperbola $x^2 - y^2 = -C$.

8 □ SERIES

8.1 Sequences

1. (a) A sequence is an ordered list of numbers. It can also be defined as a function whose domain is the set of positive integers.

   (b) The terms a_n approach 8 as n becomes large. In fact, we can make a_n as close to 8 as we like by taking n sufficiently large.

   (c) The terms a_n become large as n becomes large. In fact, we can make a_n as large as we like by taking n sufficiently large.

3. The first six terms of $a_n = \dfrac{n}{2n+1}$ are $\dfrac{1}{3}, \dfrac{2}{5}, \dfrac{3}{7}, \dfrac{4}{9}, \dfrac{5}{11}, \dfrac{6}{13}$. It appears that the sequence is approaching $\dfrac{1}{2}$.

$$\lim_{n\to\infty} \frac{n}{2n+1} = \lim_{n\to\infty} \frac{1}{2+1/n} = \frac{1}{2}$$

5. $\left\{1, -\frac{2}{3}, \frac{4}{9}, -\frac{8}{27}, \ldots\right\}$. Each term is $-\frac{2}{3}$ times the preceding one, so $a_n = \left(-\frac{2}{3}\right)^{n-1}$.

7. $\{2, 7, 12, 17, \ldots\}$. Each term is larger than the preceding one by 5, so $a_n = a_1 + d(n-1) = 2 + 5(n-1) = 5n - 3$.

9. $a_n = \dfrac{3+5n^2}{n+n^2} = \dfrac{(3+5n^2)/n^2}{(n+n^2)/n^2} = \dfrac{5+3/n^2}{1+1/n}$, so $a_n \to \dfrac{5+0}{1+0} = 5$ as $n \to \infty$.   Converges

11. $a_n = \dfrac{2^n}{3^{n+1}} = \dfrac{1}{3}\left(\dfrac{2}{3}\right)^n$, so $\displaystyle\lim_{n\to\infty} a_n = \frac{1}{3}\lim_{n\to\infty}\left(\frac{2}{3}\right)^n = \frac{1}{3}\cdot 0 = 0$ by (8) with $r = \frac{2}{3}$.   Converges

13. $a_n = \dfrac{(n+2)!}{n!} = \dfrac{(n+2)(n+1)n!}{n!} = (n+2)(n+1)$, so $a_n \to \infty$ as $n \to \infty$ and the sequence diverges.

15. $a_n = \dfrac{(-1)^{n-1}n}{n^2+1} = \dfrac{(-1)^{n-1}}{n+1/n}$, so $0 \le |a_n| = \dfrac{1}{n+1/n} \le \dfrac{1}{n} \to 0$ as $n \to \infty$, so $a_n \to 0$ by the Squeeze Theorem and

   Theorem 6.   Converges

17. $a_n = \dfrac{e^n + e^{-n}}{e^{2n}-1} \cdot \dfrac{e^{-n}}{e^{-n}} = \dfrac{1+e^{-2n}}{e^n - e^{-n}} \to \dfrac{1+0}{e^n - 0} \to 0$ as $n \to \infty$.   Converges

19. $a_n = n^2 e^{-n} = \dfrac{n^2}{e^n}$. Since $\displaystyle\lim_{x\to\infty}\frac{x^2}{e^x} \overset{\text{H}}{=} \lim_{x\to\infty}\frac{2x}{e^x} \overset{\text{H}}{=} \lim_{x\to\infty}\frac{2}{e^x} = 0$, it follows from Theorem 3 that $\displaystyle\lim_{n\to\infty} a_n = 0$.   Converges

21. $0 \le \dfrac{\cos^2 n}{2^n} \le \dfrac{1}{2^n}$ [since $0 \le \cos^2 n \le 1$],   so since $\displaystyle\lim_{n\to\infty}\frac{1}{2^n} = 0$, $\left\{\dfrac{\cos^2 n}{2^n}\right\}$ converges to 0 by the Squeeze Theorem.

23. $y = \left(1 + \dfrac{2}{x}\right)^x \;\Rightarrow\; \ln y = x \ln\left(1 + \dfrac{2}{x}\right)$, so

$$\lim_{x\to\infty}\ln y = \lim_{x\to\infty}\frac{\ln(1+2/x)}{1/x} \overset{\text{H}}{=} \lim_{x\to\infty}\frac{\left(\dfrac{1}{1+2/x}\right)\left(-\dfrac{2}{x^2}\right)}{-1/x^2} = \lim_{x\to\infty}\frac{2}{1+2/x} = 2 \;\Rightarrow$$

$$\lim_{x\to\infty}\left(1 + \frac{2}{x}\right)^x = \lim_{x\to\infty} e^{\ln y} = e^2, \text{ so by Theorem 2, } \lim_{n\to\infty}\left(1 + \frac{2}{n}\right)^n = e^2. \quad \text{Convergent}$$

25. $\{0, 1, 0, 0, 1, 0, 0, 0, 1, \ldots\}$ diverges since the sequence takes on only two values, 0 and 1, and never stays arbitrarily close to either one (or any other value) for n sufficiently large.

27. $a_n = \ln(2n^2 + 1) - \ln(n^2 + 1) = \ln\left(\dfrac{2n^2 + 1}{n^2 + 1}\right) = \ln\left(\dfrac{2 + 1/n^2}{1 + 1/n^2}\right) \to \ln 2$ as $n \to \infty$. Convergent

29. (a) $a_n = 1000(1.06)^n \;\Rightarrow\; a_1 = 1060, a_2 = 1123.60, a_3 = 1191.02, a_4 = 1262.48$, and $a_5 = 1338.23$.

(b) $\lim\limits_{n\to\infty} a_n = 1000 \lim\limits_{n\to\infty} (1.06)^n$, so the sequence diverges by (8) with $r = 1.06 > 1$.

31. Since $\{a_n\}$ is a decreasing sequence, $a_n > a_{n+1}$ for all $n \geq 1$. Because all of its terms lie between 5 and 8, $\{a_n\}$ is a bounded sequence. By the Monotonic Sequence Theorem, $\{a_n\}$ is convergent; that is, $\{a_n\}$ has a limit L. L must be less than 8 since $\{a_n\}$ is decreasing, so $5 \leq L < 8$.

33. $a_n = \dfrac{1}{2n + 3}$ is decreasing since $a_{n+1} = \dfrac{1}{2(n+1) + 3} = \dfrac{1}{2n + 5} < \dfrac{1}{2n + 3} = a_n$ for each $n \geq 1$. The sequence is

bounded since $0 < a_n \leq \frac{1}{5}$ for all $n \geq 1$. Note that $a_1 = \frac{1}{5}$.

35. $a_n = \cos(n\pi/2)$ is not monotonic. The first few terms are $0, -1, 0, 1, 0, -1, 0, 1, \ldots$. In fact, the sequence consists of the terms $0, -1, 0, 1$ repeated over and over again in that order. The sequence is bounded since $|a_n| \leq 1$ for all $n \geq 1$.

37. $a_1 = 2^{1/2}, a_2 = 2^{3/4}, a_3 = 2^{7/8}, \ldots$, so $a_n = 2^{(2^n - 1)/2^n} = 2^{1 - (1/2^n)}$. $\lim\limits_{n\to\infty} a_n = \lim\limits_{n\to\infty} 2^{1-(1/2^n)} = 2^1 = 2$.

Alternate solution: Let $L = \lim\limits_{n\to\infty} a_n$. (We could show the limit exists by showing that $\{a_n\}$ is bounded and increasing.)

Then L must satisfy $L = \sqrt{2 \cdot L} \;\Rightarrow\; L^2 = 2L \;\Rightarrow\; L(L - 2) = 0$. $L \neq 0$ since the sequence increases, so $L = 2$.

39. We show by induction that $\{a_n\}$ is increasing and bounded above by 3. Let P_n be the proposition that $a_{n+1} > a_n$ and $0 < a_n < 3$. Clearly P_1 is true. Assume that P_n is true.

Then $a_{n+1} > a_n \;\Rightarrow\; \dfrac{1}{a_{n+1}} < \dfrac{1}{a_n} \;\Rightarrow\; -\dfrac{1}{a_{n+1}} > -\dfrac{1}{a_n}$. Now $a_{n+2} = 3 - \dfrac{1}{a_{n+1}} > 3 - \dfrac{1}{a_n} = a_{n+1} \;\Leftrightarrow\; P_{n+1}$.

This proves that $\{a_n\}$ is increasing and bounded above by 3, so $1 = a_1 < a_n < 3$, that is, $\{a_n\}$ is bounded, and hence convergent by the Monotonic Sequence Theorem. If $L = \lim\limits_{n\to\infty} a_n$, then $\lim\limits_{n\to\infty} a_{n+1} = L$ also, so L must satisfy

$L = 3 - 1/L \;\Rightarrow\; L^2 - 3L + 1 = 0 \;\Rightarrow\; L = \frac{3 \pm \sqrt{5}}{2}$. But $L > 1$, so $L = \frac{3 + \sqrt{5}}{2}$.

41. (a) Let a_n be the number of rabbit pairs in the nth month. Clearly $a_1 = 1 = a_2$. In the nth month, each pair that is 2 or more months old (that is, a_{n-2} pairs) will produce a new pair to add to the a_{n-1} pairs already present. Thus, $a_n = a_{n-1} + a_{n-2}$, so that $\{a_n\} = \{f_n\}$, the Fibonacci sequence.

(b) $a_n = \dfrac{f_{n+1}}{f_n} \;\Rightarrow\; a_{n-1} = \dfrac{f_n}{f_{n-1}} = \dfrac{f_{n-1} + f_{n-2}}{f_{n-1}} = 1 + \dfrac{f_{n-2}}{f_{n-1}} = 1 + \dfrac{1}{f_{n-1}/f_{n-2}} = 1 + \dfrac{1}{a_{n-2}}$. If $L = \lim\limits_{n\to\infty} a_n$,

then $L = \lim\limits_{n\to\infty} a_{n-1}$ and $L = \lim\limits_{n\to\infty} a_{n-2}$, so L must satisfy $L = 1 + \dfrac{1}{L} \;\Rightarrow\; L^2 - L - 1 = 0 \;\Rightarrow\; L = \frac{1 + \sqrt{5}}{2}$

(since L must be positive).

43. $(0.8)^n < 0.000001 \;\Rightarrow\; \ln(0.8)^n < \ln(0.000001) \;\Rightarrow\; n \ln(0.8) < \ln(0.000001) \;\Rightarrow\; n > \dfrac{\ln(0.000001)}{\ln(0.8)} \;\Rightarrow\;$

$n > 61.9$, so n must be at least 62 to satisfy the given inequality.

45. If $\lim\limits_{n\to\infty} |a_n| = 0$ then $\lim\limits_{n\to\infty} -|a_n| = 0$, and since $-|a_n| \leq a_n \leq |a_n|$, we have that $\lim\limits_{n\to\infty} a_n = 0$ by the Squeeze Theorem.

47. (a) Suppose $\{p_n\}$ converges to p. Then $p_{n+1} = \dfrac{bp_n}{a + p_n}$ $\Rightarrow$ $\lim\limits_{n\to\infty} p_{n+1} = \dfrac{b \lim\limits_{n\to\infty} p_n}{a + \lim\limits_{n\to\infty} p_n}$ $\Rightarrow$ $p = \dfrac{bp}{a + p}$ $\Rightarrow$

$p^2 + ap = bp$ $\Rightarrow$ $p(p + a - b) = 0$ $\Rightarrow$ $p = 0$ or $p = b - a$.

(b) $p_{n+1} = \dfrac{bp_n}{a + p_n} = \dfrac{\left(\dfrac{b}{a}\right)p_n}{1 + \dfrac{p_n}{a}} < \left(\dfrac{b}{a}\right)p_n$ since $1 + \dfrac{p_n}{a} > 1$.

(c) By part (b), $p_1 < \left(\dfrac{b}{a}\right)p_0$, $p_2 < \left(\dfrac{b}{a}\right)p_1 < \left(\dfrac{b}{a}\right)^2 p_0$, $p_3 < \left(\dfrac{b}{a}\right)p_2 < \left(\dfrac{b}{a}\right)^3 p_0$, etc. In general, $p_n < \left(\dfrac{b}{a}\right)^n p_0$,

so $\lim\limits_{n\to\infty} p_n \le \lim\limits_{n\to\infty} \left(\dfrac{b}{a}\right)^n \cdot p_0 = 0$ since $b < a$. $\left[\text{By result 8, } \lim\limits_{n\to\infty} r^n = 0 \text{ if } -1 < r < 1. \text{ Here } r = \dfrac{b}{a} \in (0, 1).\right]$

(d) Let $a < b$. We first show, by induction, that if $p_0 < b - a$, then $p_n < b - a$ and $p_{n+1} > p_n$.

For $n = 0$, we have $p_1 - p_0 = \dfrac{bp_0}{a + p_0} - p_0 = \dfrac{p_0(b - a - p_0)}{a + p_0} > 0$ since $p_0 < b - a$. So $p_1 > p_0$.

Now we suppose the assertion is true for $n = k$, that is, $p_k < b - a$ and $p_{k+1} > p_k$. Then

$b - a - p_{k+1} = b - a - \dfrac{bp_k}{a + p_k} = \dfrac{a(b - a) + bp_k - ap_k - bp_k}{a + p_k} = \dfrac{a(b - a - p_k)}{a + p_k} > 0$ because $p_k < b - a$.

So $p_{k+1} < b - a$. And $p_{k+2} - p_{k+1} = \dfrac{bp_{k+1}}{a + p_{k+1}} - p_{k+1} = \dfrac{p_{k+1}(b - a - p_{k+1})}{a + p_{k+1}} > 0$ since $p_{k+1} < b - a$. Therefore,

$p_{k+2} > p_{k+1}$. Thus, the assertion is true for $n = k + 1$. It is therefore true for all n by mathematical induction. A similar

proof by induction shows that if $p_0 > b - a$, then $p_n > b - a$ and $\{p_n\}$ is decreasing.

In either case the sequence $\{p_n\}$ is bounded and monotonic, so it is convergent by the Monotonic Sequence Theorem.

It then follows from part (a) that $\lim\limits_{n\to\infty} p_n = b - a$.

8.2 Series

1. (a) A sequence is an ordered list of numbers whereas a series is the *sum* of a list of numbers.

(b) A series is convergent if the sequence of partial sums is a convergent sequence. A series is divergent if it is not convergent.

3. $5 - \dfrac{10}{3} + \dfrac{20}{9} - \dfrac{40}{27} + \cdots$ is a geometric series with $a = 5$ and $r = -\dfrac{2}{3}$. Since $|r| = \dfrac{2}{3} < 1$, the series converges to

$\dfrac{a}{1 - r} = \dfrac{5}{1 - (-2/3)} = \dfrac{5}{5/3} = 3$.

5. $\sum\limits_{n=1}^{\infty} 5\left(\dfrac{2}{3}\right)^{n-1}$ is a geometric series with $a = 5$ and $r = \dfrac{2}{3}$. Since $|r| = \dfrac{2}{3} < 1$, the series converges to

$\dfrac{a}{1 - r} = \dfrac{5}{1 - 2/3} = \dfrac{5}{1/3} = 15$.

7. $\sum\limits_{n=0}^{\infty} \dfrac{\pi^n}{3^{n+1}} = \dfrac{1}{3} \sum\limits_{n=0}^{\infty} \left(\dfrac{\pi}{3}\right)^n$ is a geometric series with ratio $r = \dfrac{\pi}{3}$. Since $|r| > 1$, the series diverges.

9. $\sum\limits_{n=1}^{\infty} \dfrac{1}{2n} = \dfrac{1}{2} \sum\limits_{n=1}^{\infty} \dfrac{1}{n}$ diverges since each of its partial sums is $\dfrac{1}{2}$ times the corresponding partial sum of the harmonic series

$\sum\limits_{n=1}^{\infty} \dfrac{1}{n}$, which diverges. $\left[\text{If } \sum\limits_{n=1}^{\infty} \dfrac{1}{2n} \text{ were to converge, then } \sum\limits_{n=1}^{\infty} \dfrac{1}{n} \text{ would also have to converge by Theorem 8(i).}\right]$

In general, constant multiples of divergent series are divergent.

11. $\sum\limits_{k=2}^{\infty} \dfrac{k^2}{k^2-1}$ diverges by the Test for Divergence since $\lim\limits_{k\to\infty} a_k = \lim\limits_{k\to\infty} \dfrac{k^2}{k^2-1} = 1 \neq 0$.

13. Converges.

$$\sum_{n=1}^{\infty} \frac{1+2^n}{3^n} = \sum_{n=1}^{\infty} \left(\frac{1}{3^n} + \frac{2^n}{3^n} \right) = \sum_{n=1}^{\infty} \left[\left(\frac{1}{3}\right)^n + \left(\frac{2}{3}\right)^n \right] \qquad \text{[sum of two convergent geometric series]}$$

$$= \frac{1/3}{1-1/3} + \frac{2/3}{1-2/3} = \frac{1}{2} + 2 = \frac{5}{2}$$

15. $\sum\limits_{n=1}^{\infty} \sqrt[n]{2} = 2 + \sqrt{2} + \sqrt[3]{2} + \sqrt[4]{2} + \cdots$ diverges by the Test for Divergence since

$$\lim_{n\to\infty} a_n = \lim_{n\to\infty} \sqrt[n]{2} = \lim_{n\to\infty} 2^{1/n} = 2^0 = 1 \neq 0.$$

17. $\lim\limits_{n\to\infty} a_n = \lim\limits_{n\to\infty} \arctan n = \frac{\pi}{2} \neq 0$, so the series diverges by the Test for Divergence.

19. Using partial fractions, the partial sums of the series $\sum\limits_{n=2}^{\infty} \dfrac{2}{n^2-1}$ are

$$s_n = \sum_{i=2}^{n} \frac{2}{(i-1)(i+1)} = \sum_{i=2}^{n} \left(\frac{1}{i-1} - \frac{1}{i+1} \right)$$

$$= \left(1 - \frac{1}{3}\right) + \left(\frac{1}{2} - \frac{1}{4}\right) + \left(\frac{1}{3} - \frac{1}{5}\right) + \cdots + \left(\frac{1}{n-3} - \frac{1}{n-1}\right) + \left(\frac{1}{n-2} - \frac{1}{n}\right)$$

This sum is a telescoping series and $s_n = 1 + \dfrac{1}{2} - \dfrac{1}{n-1} - \dfrac{1}{n}$.

Thus, $\sum\limits_{n=2}^{\infty} \dfrac{2}{n^2-1} = \lim\limits_{n\to\infty} s_n = \lim\limits_{n\to\infty} \left(1 + \dfrac{1}{2} - \dfrac{1}{n-1} - \dfrac{1}{n}\right) = \dfrac{3}{2}$.

21. For the series $\sum\limits_{n=1}^{\infty} \dfrac{3}{n(n+3)}$, $s_n = \sum\limits_{i=1}^{n} \dfrac{3}{i(i+3)} = \sum\limits_{i=1}^{n} \left(\dfrac{1}{i} - \dfrac{1}{i+3}\right)$ [using partial fractions]. The latter sum is

$$\left(1 - \tfrac{1}{4}\right) + \left(\tfrac{1}{2} - \tfrac{1}{5}\right) + \left(\tfrac{1}{3} - \tfrac{1}{6}\right) + \left(\tfrac{1}{4} - \tfrac{1}{7}\right) + \cdots + \left(\tfrac{1}{n-3} - \tfrac{1}{n}\right) + \left(\tfrac{1}{n-2} - \tfrac{1}{n+1}\right) + \left(\tfrac{1}{n-1} - \tfrac{1}{n+2}\right) + \left(\tfrac{1}{n} - \tfrac{1}{n+3}\right)$$

$$= 1 + \tfrac{1}{2} + \tfrac{1}{3} - \tfrac{1}{n+1} - \tfrac{1}{n+2} - \tfrac{1}{n+3} \quad \text{[telescoping series]}$$

Thus, $\sum\limits_{n=1}^{\infty} \dfrac{3}{n(n+3)} = \lim\limits_{n\to\infty} s_n = \lim\limits_{n\to\infty} \left(1 + \tfrac{1}{2} + \tfrac{1}{3} - \tfrac{1}{n+1} - \tfrac{1}{n+2} - \tfrac{1}{n+3}\right) = 1 + \tfrac{1}{2} + \tfrac{1}{3} = \tfrac{11}{6}$. Converges

23. $0.\overline{2} = \dfrac{2}{10} + \dfrac{2}{10^2} + \cdots$ is a geometric series with $a = \dfrac{2}{10}$ and $r = \dfrac{1}{10}$. It converges to $\dfrac{a}{1-r} = \dfrac{2/10}{1-1/10} = \dfrac{2}{9}$.

25. $3.\overline{417} = 3 + \dfrac{417}{10^3} + \dfrac{417}{10^6} + \cdots$. Now $\dfrac{417}{10^3} + \dfrac{417}{10^6} + \cdots$ is a geometric series with $a = \dfrac{417}{10^3}$ and $r = \dfrac{1}{10^3}$.

It converges to $\dfrac{a}{1-r} = \dfrac{417/10^3}{1-1/10^3} = \dfrac{417/10^3}{999/10^3} = \dfrac{417}{999}$. Thus, $3.\overline{417} = 3 + \dfrac{417}{999} = \dfrac{3414}{999} = \dfrac{1138}{333}$.

27. $\sum\limits_{n=1}^{\infty} \dfrac{x^n}{3^n} = \sum\limits_{n=1}^{\infty} \left(\dfrac{x}{3}\right)^n$ is a geometric series with $r = \dfrac{x}{3}$, so the series converges $\Leftrightarrow |r| < 1 \Leftrightarrow \dfrac{|x|}{3} < 1 \Leftrightarrow |x| < 3$;

that is, $-3 < x < 3$. In that case, the sum of the series is $\dfrac{a}{1-r} = \dfrac{x/3}{1-x/3} = \dfrac{x/3}{1-x/3} \cdot \dfrac{3}{3} = \dfrac{x}{3-x}$.

29. $\sum\limits_{n=0}^{\infty} \dfrac{\cos^n x}{2^n}$ is a geometric series with first term 1 and ratio $r = \dfrac{\cos x}{2}$, so it converges $\Leftrightarrow$ $|r| < 1$. But $|r| = \dfrac{|\cos x|}{2} \leq \dfrac{1}{2}$

for all x. Thus, the series converges for all real values of x and the sum of the series is $\dfrac{1}{1 - (\cos x)/2} = \dfrac{2}{2 - \cos x}$.

31. For $n = 1$, $a_1 = 0$ since $s_1 = 0$. For $n > 1$,

$$a_n = s_n - s_{n-1} = \frac{n-1}{n+1} - \frac{(n-1)-1}{(n-1)+1} = \frac{(n-1)n - (n+1)(n-2)}{(n+1)n} = \frac{2}{n(n+1)}$$

Also, $\sum\limits_{n=1}^{\infty} a_n = \lim\limits_{n\to\infty} s_n = \lim\limits_{n\to\infty} \dfrac{1 - 1/n}{1 + 1/n} = 1.$

33. (a) The first step in the chain occurs when the local government spends D dollars. The people who receive it spend a

fraction c of those D dollars, that is, Dc dollars. Those who receive the Dc dollars spend a fraction c of it, that is,

Dc^2 dollars. Continuing in this way, we see that the total spending after n transactions is

$$S_n = D + Dc + Dc^2 + \cdots + Dc^{n-1} = \frac{D(1 - c^n)}{1 - c} \text{ by (3)}.$$

(b) $\lim\limits_{n\to\infty} S_n = \lim\limits_{n\to\infty} \dfrac{D(1 - c^n)}{1 - c} = \dfrac{D}{1-c} \lim\limits_{n\to\infty} (1 - c^n) = \dfrac{D}{1-c}$ [since $0 < c < 1$ $\Rightarrow$ $\lim\limits_{n\to\infty} c^n = 0$]

$= \dfrac{D}{s}$ [since $c + s = 1$] $= kD$ [since $k = 1/s$]

If $c = 0.8$, then $s = 1 - c = 0.2$ and the multiplier is $k = 1/s = 5$.

35. $\sum\limits_{n=2}^{\infty} (1 + c)^{-n}$ is a geometric series with $a = (1 + c)^{-2}$ and $r = (1 + c)^{-1}$, so the series converges when

$\left|(1 + c)^{-1}\right| < 1$ $\Leftrightarrow$ $|1 + c| > 1$ $\Leftrightarrow$ $1 + c > 1$ or $1 + c < -1$ $\Leftrightarrow$ $c > 0$ or $c < -2$. We calculate the sum of the

series and set it equal to 2: $\dfrac{(1 + c)^{-2}}{1 - (1 + c)^{-1}} = 2$ $\Leftrightarrow$ $\left(\dfrac{1}{1+c}\right)^2 = 2 - 2\left(\dfrac{1}{1+c}\right)$ $\Leftrightarrow$ $1 = 2(1 + c)^2 - 2(1 + c)$ $\Leftrightarrow$

$2c^2 + 2c - 1 = 0$ $\Leftrightarrow$ $c = \dfrac{-2 \pm \sqrt{12}}{4} = \dfrac{\pm\sqrt{3} - 1}{2}$. However, the negative root is inadmissible because $-2 < \dfrac{-\sqrt{3} - 1}{2} < 0$.

So $c = \dfrac{\sqrt{3} - 1}{2}$.

37. Let d_n be the diameter of C_n. We draw lines from the centers of the C_i to the

center of D (or C), and using the Pythagorean Theorem, we can write

$1^2 + \left(1 - \tfrac{1}{2}d_1\right)^2 = \left(1 + \tfrac{1}{2}d_1\right)^2$ $\Leftrightarrow$

$1 = \left(1 + \tfrac{1}{2}d_1\right)^2 - \left(1 - \tfrac{1}{2}d_1\right)^2 = 2d_1$ [difference of squares] $\Rightarrow$ $d_1 = \tfrac{1}{2}$.

Similarly,

$1 = \left(1 + \tfrac{1}{2}d_2\right)^2 - \left(1 - d_1 - \tfrac{1}{2}d_2\right)^2$

$= 2d_2 + 2d_1 - d_1^2 - d_1 d_2 = (2 - d_1)(d_1 + d_2)$ $\Leftrightarrow$

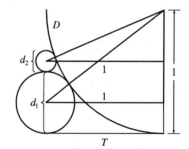

$d_2 = \dfrac{1}{2 - d_1} - d_1 = \dfrac{(1 - d_1)^2}{2 - d_1}$, $1 = \left(1 + \tfrac{1}{2}d_3\right)^2 - \left(1 - d_1 - d_2 - \tfrac{1}{2}d_3\right)^2$ $\Leftrightarrow$ $d_3 = \dfrac{[1 - (d_1 + d_2)]^2}{2 - (d_1 + d_2)}$, and in general,

$d_{n+1} = \dfrac{\left(1 - \sum_{i=1}^{n} d_i\right)^2}{2 - \sum_{i=1}^{n} d_i}$. If we actually calculate d_2 and d_3 from the formulas above, we find that they are $\dfrac{1}{6} = \dfrac{1}{2 \cdot 3}$ and

$\frac{1}{12} = \frac{1}{3 \cdot 4}$ respectively, so we suspect that in general, $d_n = \dfrac{1}{n(n+1)}$. To prove this, we use induction: Assume that for all

$k \le n$, $d_k = \dfrac{1}{k(k+1)} = \dfrac{1}{k} - \dfrac{1}{k+1}$. Then $\sum\limits_{i=1}^{n} d_i = 1 - \dfrac{1}{n+1} = \dfrac{n}{n+1}$ [telescoping sum]. Substituting this into our

formula for d_{n+1}, we get $d_{n+1} = \dfrac{\left[1 - \dfrac{n}{n+1}\right]^2}{2 - \left(\dfrac{n}{n+1}\right)} = \dfrac{\dfrac{1}{(n+1)^2}}{\dfrac{n+2}{n+1}} = \dfrac{1}{(n+1)(n+2)}$, and the induction is complete.

Now, we observe that the partial sums $\sum\limits_{i=1}^{n} d_i$ of the diameters of the circles approach 1 as $n \to \infty$; that is,

$$\sum_{n=1}^{\infty} a_n = \sum_{n=1}^{\infty} \frac{1}{n(n+1)} = 1,$$ which is what we wanted to prove.

39. The series $1 - 1 + 1 - 1 + 1 - 1 + \cdots$ diverges (geometric series with $r = -1$) so we cannot say that

$0 = 1 - 1 + 1 - 1 + 1 - 1 + \cdots$.

41. $\sum\limits_{n=1}^{\infty} ca_n = \lim\limits_{n \to \infty} \sum\limits_{i=1}^{n} ca_i = \lim\limits_{n \to \infty} c \sum\limits_{i=1}^{n} a_i = c \lim\limits_{n \to \infty} \sum\limits_{i=1}^{n} a_i = c \sum\limits_{n=1}^{\infty} a_n$, which exists by hypothesis.

43. Suppose on the contrary that $\sum(a_n + b_n)$ converges. Then $\sum(a_n + b_n)$ and $\sum a_n$ are convergent series. So by Theorem 8,

$\sum[(a_n + b_n) - a_n]$ would also be convergent. But $\sum[(a_n + b_n) - a_n] = \sum b_n$, a contradiction, since $\sum b_n$ is given to be

divergent.

45. The partial sums $\{s_n\}$ form an increasing sequence, since $s_n - s_{n-1} = a_n > 0$ for all n. Also, the sequence $\{s_n\}$ is bounded

since $s_n \le 1000$ for all n. So by the Monotonic Sequence Theorem, the sequence of partial sums converges, that is, the series

$\sum a_n$ is convergent.

47. (a) At the first step, only the interval $\left(\frac{1}{3}, \frac{2}{3}\right)$ (length $\frac{1}{3}$) is removed. At the second step, we remove the intervals $\left(\frac{1}{9}, \frac{2}{9}\right)$ and

$\left(\frac{7}{9}, \frac{8}{9}\right)$, which have a total length of $2 \cdot \left(\frac{1}{3}\right)^2$. At the third step, we remove 2^2 intervals, each of length $\left(\frac{1}{3}\right)^3$. In general, at

the nth step we remove 2^{n-1} intervals, each of length $\left(\frac{1}{3}\right)^n$, for a length of $2^{n-1} \cdot \left(\frac{1}{3}\right)^n = \frac{1}{3}\left(\frac{2}{3}\right)^{n-1}$. Thus, the total

length of all removed intervals is $\sum\limits_{n=1}^{\infty} \frac{1}{3}\left(\frac{2}{3}\right)^{n-1} = \frac{1/3}{1 - 2/3} = 1$ [geometric series with $a = \frac{1}{3}$ and $r = \frac{2}{3}$]. Notice that at

the nth step, the leftmost interval that is removed is $\left(\left(\frac{1}{3}\right)^n, \left(\frac{2}{3}\right)^n\right)$, so we never remove 0, and 0 is in the Cantor set. Also,

the rightmost interval removed is $\left(1 - \left(\frac{2}{3}\right)^n, 1 - \left(\frac{1}{3}\right)^n\right)$, so 1 is never removed. Some other numbers in the Cantor set

are $\frac{1}{3}$, $\frac{2}{3}$, $\frac{1}{9}$, $\frac{2}{9}$, $\frac{7}{9}$, and $\frac{8}{9}$.

(b) The area removed at the first step is $\frac{1}{9}$; at the second step, $8 \cdot \left(\frac{1}{9}\right)^2$; at the third step, $(8)^2 \cdot \left(\frac{1}{9}\right)^3$. In general, the area

removed at the nth step is $(8)^{n-1}\left(\frac{1}{9}\right)^n = \frac{1}{9}\left(\frac{8}{9}\right)^{n-1}$, so the total area of all removed squares is

$$\sum_{n=1}^{\infty} \frac{1}{9}\left(\frac{8}{9}\right)^{n-1} = \frac{\frac{1}{9}}{1 - \frac{8}{9}} = 1.$$

49. (a) For $\sum\limits_{n=1}^{\infty} \dfrac{n}{(n+1)!}$, $s_1 = \dfrac{1}{1 \cdot 2} = \dfrac{1}{2}$, $s_2 = \dfrac{1}{2} + \dfrac{2}{1 \cdot 2 \cdot 3} = \dfrac{5}{6}$, $s_3 = \dfrac{5}{6} + \dfrac{3}{1 \cdot 2 \cdot 3 \cdot 4} = \dfrac{23}{24}$,

$s_4 = \dfrac{23}{24} + \dfrac{4}{1 \cdot 2 \cdot 3 \cdot 4 \cdot 5} = \dfrac{119}{120}$. The denominators are $(n+1)!$, so a guess would be $s_n = \dfrac{(n+1)! - 1}{(n+1)!}$.

(b) For $n = 1$, $s_1 = \dfrac{1}{2} = \dfrac{2! - 1}{2!}$, so the formula holds for $n = 1$. Assume $s_k = \dfrac{(k+1)! - 1}{(k+1)!}$. Then

$$s_{k+1} = \dfrac{(k+1)! - 1}{(k+1)!} + \dfrac{k+1}{(k+2)!} = \dfrac{(k+1)! - 1}{(k+1)!} + \dfrac{k+1}{(k+1)!(k+2)}$$

$$= \dfrac{(k+2)! - (k+2) + k + 1}{(k+2)!} = \dfrac{(k+2)! - 1}{(k+2)!}$$

Thus, the formula is true for $n = k + 1$. So by induction, the guess is correct.

(c) $\lim\limits_{n \to \infty} s_n = \lim\limits_{n \to \infty} \dfrac{(n+1)! - 1}{(n+1)!} = \lim\limits_{n \to \infty} \left[1 - \dfrac{1}{(n+1)!} \right] = 1$ and so $\sum\limits_{n=1}^{\infty} \dfrac{n}{(n+1)!} = 1$.

8.3   The Integral and Comparison Tests

1. The picture shows that $a_2 = \dfrac{1}{2^{1.3}} < \displaystyle\int_1^2 \dfrac{1}{x^{1.3}}\, dx$,

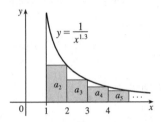

$a_3 = \dfrac{1}{3^{1.3}} < \displaystyle\int_2^3 \dfrac{1}{x^{1.3}}\, dx$, and so on, so $\sum\limits_{n=2}^{\infty} \dfrac{1}{n^{1.3}} < \displaystyle\int_1^{\infty} \dfrac{1}{x^{1.3}}\, dx$. The

integral converges by (6.6.2) with $p = 1.3 > 1$, so the series converges.

3. (a) We cannot say anything about $\sum a_n$. If $a_n > b_n$ for all n and $\sum b_n$ is convergent, then $\sum a_n$ could be convergent or divergent. (See the note after Example 4.)

(b) If $a_n < b_n$ for all n, then $\sum a_n$ is convergent. [This is part (i) of the Comparison Test.]

5. $\sum\limits_{n=1}^{\infty} n^b$ is a p-series with $p = -b$. $\sum\limits_{n=1}^{\infty} b^n$ is a geometric series. By (1), the p-series is convergent if $p > 1$. In this case,

$\sum\limits_{n=1}^{\infty} n^b = \sum\limits_{n=1}^{\infty} \left(1/n^{-b}\right)$, so $-b > 1$ $\Leftrightarrow$ $b < -1$ are the values for which the series converge. A geometric series

$\sum\limits_{n=1}^{\infty} ar^{n-1}$ converges if $|r| < 1$, so $\sum\limits_{n=1}^{\infty} b^n$ converges if $|b| < 1$ $\Leftrightarrow$ $-1 < b < 1$.

7. The function $f(x) = 1/x^4$ is continuous, positive, and decreasing on $[1, \infty)$, so the Integral Test applies.

$\displaystyle\int_1^{\infty} \dfrac{1}{x^4}\, dx = \lim\limits_{t \to \infty} \int_1^t x^{-4}\, dx = \lim\limits_{t \to \infty} \left[\dfrac{x^{-3}}{-3} \right]_1^t = \lim\limits_{t \to \infty} \left(-\dfrac{1}{3t^3} + \dfrac{1}{3} \right) = \dfrac{1}{3}$. Since this improper integral is convergent, the

series $\sum\limits_{n=1}^{\infty} \dfrac{1}{n^4}$ is also convergent by the Integral Test.

9. $\dfrac{1}{n^2 + n + 1} < \dfrac{1}{n^2}$ for all $n \geq 1$, so $\sum\limits_{n=1}^{\infty} \dfrac{1}{n^2 + n + 1}$ converges by comparison with $\sum\limits_{n=1}^{\infty} \dfrac{1}{n^2}$, which converges because it is a

p-series with $p = 2 > 1$.

11. $1 + \dfrac{1}{8} + \dfrac{1}{27} + \dfrac{1}{64} + \dfrac{1}{125} + \cdots = \displaystyle\sum_{n=1}^{\infty} \dfrac{1}{n^3}$. This is a p-series with $p = 3 > 1$, so it converges by (1).

13. $f(x) = xe^{-x}$ is continuous and positive on $[1, \infty)$. $f'(x) = -xe^{-x} + e^{-x} = e^{-x}(1 - x) < 0$ for $x > 1$, so f is

decreasing on $[1, \infty)$. Thus, the Integral Test applies.

$$\int_1^{\infty} xe^{-x}\,dx = \lim_{b \to \infty} \int_1^b xe^{-x}\,dx = \lim_{b \to \infty} \left[-xe^{-x} - e^{-x}\right]_1^b \qquad \text{[by parts]}$$

$$= \lim_{b \to \infty} \left[-be^{-b} - e^{-b} + e^{-1} + e^{-1}\right] = 2/e$$

since $\displaystyle\lim_{b \to \infty} be^{-b} = \lim_{b \to \infty} \left(b/e^b\right) \overset{\text{H}}{=} \lim_{b \to \infty} \left(1/e^b\right) = 0$ and $\displaystyle\lim_{b \to \infty} e^{-b} = 0$. Thus, $\displaystyle\sum_{n=1}^{\infty} ne^{-n}$ converges.

15. $f(x) = \dfrac{1}{x \ln x}$ is continuous and positive on $[2, \infty)$, and also decreasing since $f'(x) = -\dfrac{1 + \ln x}{x^2 (\ln x)^2} < 0$ for $x > 2$, so we can

use the Integral Test. $\displaystyle\int_2^{\infty} \dfrac{1}{x \ln x}\,dx = \lim_{t \to \infty} \left[\ln(\ln x)\right]_2^t = \lim_{t \to \infty} \left[\ln(\ln t) - \ln(\ln 2)\right] = \infty$, so the series $\displaystyle\sum_{n=2}^{\infty} \dfrac{1}{n \ln n}$ diverges.

17. $\dfrac{\cos^2 n}{n^2 + 1} \le \dfrac{1}{n^2 + 1} < \dfrac{1}{n^2}$, so the series $\displaystyle\sum_{n=1}^{\infty} \dfrac{\cos^2 n}{n^2 + 1}$ converges by comparison with the p-series $\displaystyle\sum_{n=1}^{\infty} \dfrac{1}{n^2}$ $[p = 2 > 1]$.

19. $\dfrac{n - 1}{n\,4^n}$ is positive for $n > 1$ and $\dfrac{n - 1}{n\,4^n} < \dfrac{n}{n\,4^n} = \dfrac{1}{4^n} = \left(\dfrac{1}{4}\right)^n$, so $\displaystyle\sum_{n=1}^{\infty} \dfrac{n - 1}{n\,4^n}$ converges by comparison with the convergent

geometric series $\displaystyle\sum_{n=1}^{\infty} \left(\dfrac{1}{4}\right)^n$.

21. Use the Limit Comparison Test with $a_n = \dfrac{1}{\sqrt{n^2 + 1}}$ and $b_n = \dfrac{1}{n}$:

$$\lim_{n \to \infty} \dfrac{a_n}{b_n} = \lim_{n \to \infty} \dfrac{n}{\sqrt{n^2 + 1}} = \lim_{n \to \infty} \dfrac{1}{\sqrt{1 + (1/n^2)}} = 1 > 0. \text{ Since the harmonic series } \sum_{n=1}^{\infty} \dfrac{1}{n} \text{ diverges,}$$

so does $\displaystyle\sum_{n=1}^{\infty} \dfrac{1}{\sqrt{n^2 + 1}}$.

23. $\dfrac{2 + (-1)^n}{n\sqrt{n}} \le \dfrac{3}{n\sqrt{n}}$, and $\displaystyle\sum_{n=1}^{\infty} \dfrac{3}{n\sqrt{n}}$ converges because it is a constant multiple of the convergent p-series $\displaystyle\sum_{n=1}^{\infty} \dfrac{1}{n\sqrt{n}}$

$\left[p = \frac{3}{2} > 1\right]$, so the given series converges by the Comparison Test.

25. Use the Limit Comparison Test with $a_n = \sin\left(\dfrac{1}{n}\right)$ and $b_n = \dfrac{1}{n}$. Then $\sum a_n$ and $\sum b_n$ are series with positive terms and

$$\lim_{n \to \infty} \dfrac{a_n}{b_n} = \lim_{n \to \infty} \dfrac{\sin(1/n)}{1/n} = \lim_{\theta \to 0} \dfrac{\sin \theta}{\theta} = 1 > 0. \text{ Since } \sum_{n=1}^{\infty} b_n \text{ is the divergent harmonic series, } \sum_{n=1}^{\infty} \sin(1/n) \text{ also}$$

diverges.

Note: We could also use l'Hospital's Rule to evaluate the limit:

$$\lim_{x \to \infty} \dfrac{\sin(1/x)}{1/x} \overset{\text{H}}{=} \lim_{x \to \infty} \dfrac{\cos(1/x) \cdot (-1/x^2)}{-1/x^2} = \lim_{x \to \infty} \cos \dfrac{1}{x} = \cos 0 = 1$$

27. We have already shown (in Exercise 15) that when $p = 1$ the series $\sum\limits_{n=2}^{\infty} \dfrac{1}{n(\ln n)^p}$ diverges, so assume that $p \neq 1$.

$f(x) = \dfrac{1}{x(\ln x)^p}$ is continuous and positive on $[2, \infty)$, and $f'(x) = -\dfrac{p + \ln x}{x^2(\ln x)^{p+1}} < 0$ if $x > e^{-p}$, so that f is eventually

decreasing and we can use the Integral Test.

$$\int_2^{\infty} \frac{1}{x(\ln x)^p}\, dx = \lim_{t \to \infty} \left[\frac{(\ln x)^{1-p}}{1-p}\right]_2^t \quad \text{[for } p \neq 1\text{]} = \lim_{t \to \infty} \left[\frac{(\ln t)^{1-p}}{1-p}\right] - \frac{(\ln 2)^{1-p}}{1-p}$$

This limit exists whenever $1 - p < 0 \;\Leftrightarrow\; p > 1$, so the series converges for $p > 1$.

29. (a)

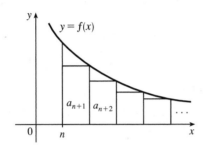   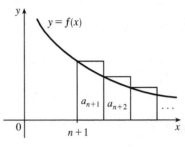

We use the same notation and ideas as in the Integral Test, assuming that f is decreasing on $[n, \infty)$. Comparing the areas

of the rectangles with the area under $y = f(x)$ for $x > n$ in the first figure, we see that

$$R_n = a_{n+1} + a_{n+2} + \cdots \leq \int_n^{\infty} f(x)\, dx$$

Similarly, we see from the second figure that

$$R_n = a_{n+1} + a_{n+2} + \cdots \geq \int_{n+1}^{\infty} f(x)\, dx$$

So we have proved that $\int_{n+1}^{\infty} f(x)\, dx \leq R_n \leq \int_n^{\infty} f(x)\, dx$.

(b) If we add s_n to each side of the inequalities in part (a), we get $s_n + \int_{n+1}^{\infty} f(x)\, dx \leq s \leq s_n + \int_n^{\infty} f(x)\, dx$

because $s_n + R_n = s$.

31. (a) $f(x) = \dfrac{1}{x^2}$ is positive and continuous and $f'(x) = -\dfrac{2}{x^3}$ is negative for $x > 0$, and so the Integral Test applies.

$$\sum_{n=1}^{\infty} \frac{1}{n^2} \approx s_{10} = \frac{1}{1^2} + \frac{1}{2^2} + \frac{1}{3^2} + \cdots + \frac{1}{10^2} \approx 1.549768. \text{ From Exercise 29(a) we have}$$

$$R_{10} \leq \int_{10}^{\infty} \frac{1}{x^2}\, dx = \lim_{t \to \infty}\left[\frac{-1}{x}\right]_{10}^t = \lim_{t \to \infty}\left(-\frac{1}{t} + \frac{1}{10}\right) = \frac{1}{10}, \text{ so the error is at most } 0.1.$$

(b) $s_{10} + \displaystyle\int_{11}^{\infty} \frac{1}{x^2}\, dx \leq s \leq s_{10} + \int_{10}^{\infty} \frac{1}{x^2}\, dx \;\Rightarrow\; s_{10} + \frac{1}{11} \leq s \leq s_{10} + \frac{1}{10} \;\Rightarrow$

$1.549768 + 0.090909 = 1.640677 \leq s \leq 1.549768 + 0.1 = 1.649768$, so we get $s \approx 1.64522$ (the average of 1.640677

and 1.649768) with error ≤ 0.005 (the maximum of $1.649768 - 1.64522$ and $1.64522 - 1.640677$, rounded up).

(c) $R_n \leq \displaystyle\int_n^{\infty} \frac{1}{x^2}\, dx = \frac{1}{n}.$ So $R_n < 0.001$ if $\dfrac{1}{n} < \dfrac{1}{1000} \;\Leftrightarrow\; n > 1000.$

33. (a) From the figure, $a_2 + a_3 + \cdots + a_n \le \int_1^n f(x)\,dx$, so with

$$f(x) = \frac{1}{x}, \frac{1}{2} + \frac{1}{3} + \frac{1}{4} + \cdots + \frac{1}{n} \le \int_1^n \frac{1}{x}\,dx = \ln n. \text{ Thus,}$$

$$s_n = 1 + \frac{1}{2} + \frac{1}{3} + \frac{1}{4} + \cdots + \frac{1}{n} \le 1 + \ln n.$$

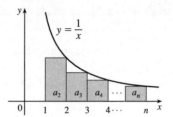

(b) By part (a), $s_{10^6} \le 1 + \ln 10^6 \approx 14.82 < 15$ and $s_{10^9} \le 1 + \ln 10^9 \approx 21.72 < 22$.

35. Since $\dfrac{d_n}{10^n} \le \dfrac{9}{10^n}$ for each n, and since $\displaystyle\sum_{n=1}^{\infty} \dfrac{9}{10^n}$ is a convergent geometric series $\left(|r| = \frac{1}{10} < 1\right)$, $0.d_1 d_2 d_3 \ldots = \displaystyle\sum_{n=1}^{\infty} \dfrac{d_n}{10^n}$

will always converge by the Comparison Test.

37. Yes. Since $\sum a_n$ is a convergent series with positive terms, $\displaystyle\lim_{n\to\infty} a_n = 0$ by Theorem 8.2.6, and $\sum b_n = \sum \sin(a_n)$ is a

series with positive terms (for large enough n). We have $\displaystyle\lim_{n\to\infty} \frac{b_n}{a_n} = \lim_{n\to\infty} \frac{\sin(a_n)}{a_n} = 1 > 0$ by Theorem 1.4.5. Thus, $\sum b_n$

is also convergent by the Limit Comparison Test.

39. (a) Since $\displaystyle\lim_{n\to\infty} \frac{a_n}{b_n} = \infty$, there is an integer N such that $\dfrac{a_n}{b_n} > 1$ whenever $n > N$. (Take $M = 1$ in Definition 8.1.5.)

Then $a_n > b_n$ whenever $n > N$ and since $\sum b_n$ is divergent, $\sum a_n$ is also divergent by the Comparison Test.

(b) (i) If $a_n = \dfrac{1}{\ln n}$ and $b_n = \dfrac{1}{n}$ for $n \ge 2$, then $\displaystyle\lim_{n\to\infty} \frac{a_n}{b_n} = \lim_{n\to\infty} \frac{n}{\ln n} = \lim_{x\to\infty} \frac{x}{\ln x} \overset{\text{H}}{=} \lim_{x\to\infty} \frac{1}{1/x} = \lim_{x\to\infty} x = \infty$, so by

part (a), $\displaystyle\sum_{n=2}^{\infty} \dfrac{1}{\ln n}$ is divergent.

(ii) If $a_n = \dfrac{\ln n}{n}$ and $b_n = \dfrac{1}{n}$, then $\displaystyle\sum_{n=1}^{\infty} b_n$ is the divergent harmonic series and $\displaystyle\lim_{n\to\infty} \frac{a_n}{b_n} = \lim_{n\to\infty} \ln n = \lim_{x\to\infty} \ln x = \infty$,

so $\displaystyle\sum_{n=1}^{\infty} a_n$ diverges by part (a).

41. Since $\sum a_n$ converges, $\displaystyle\lim_{n\to\infty} a_n = 0$, so there exists N such that $|a_n - 0| < 1$ for all $n > N \;\Rightarrow\; 0 \le a_n < 1$ for all

$n > N \;\Rightarrow\; 0 \le a_n^2 \le a_n$. Since $\sum a_n$ converges, so does $\sum a_n^2$ by the Comparison Test.

8.4 Other Convergence Tests

1. (a) An alternating series is a series whose terms are alternately positive and negative.

(b) An alternating series $\displaystyle\sum_{n=1}^{\infty} (-1)^{n-1} b_n$ converges if $0 < b_{n+1} \le b_n$ for all n and $\displaystyle\lim_{n\to\infty} b_n = 0$. (This is the Alternating

Series Test.)

(c) The error involved in using the partial sum s_n as an approximation to the total sum s is the remainder $R_n = s - s_n$ and the

size of the error is smaller than b_{n+1}; that is, $|R_n| \le b_{n+1}$. (This is the Alternating Series Estimation Theorem.)

3. $\dfrac{4}{7} - \dfrac{4}{8} + \dfrac{4}{9} - \dfrac{4}{10} + \dfrac{4}{11} - \cdots = \displaystyle\sum_{n=1}^{\infty} (-1)^{n-1} \dfrac{4}{n+6}$. Now $b_n = \dfrac{4}{n+6} > 0$, $\{b_n\}$ is decreasing, and $\displaystyle\lim_{n\to\infty} b_n = 0$, so the

series converges by the Alternating Series Test.

5. $b_n = \dfrac{1}{\sqrt{n}} > 0$, $\{b_n\}$ is decreasing, and $\lim\limits_{n\to\infty} b_n = 0$, so the series $\sum\limits_{n=1}^{\infty} \dfrac{(-1)^{n-1}}{\sqrt{n}}$ converges by the Alternating Series Test.

7. $\sum\limits_{n=1}^{\infty} a_n = \sum\limits_{n=1}^{\infty} (-1)^n \dfrac{3n-1}{2n+1} = \sum\limits_{n=1}^{\infty} (-1)^n b_n$. Now $\lim\limits_{n\to\infty} b_n = \lim\limits_{n\to\infty} \dfrac{3-1/n}{2+1/n} = \dfrac{3}{2} \neq 0$. Since $\lim\limits_{n\to\infty} a_n \neq 0$

(in fact the limit does not exist), the series diverges by the Test for Divergence.

9. The series $\sum\limits_{n=1}^{\infty} \dfrac{(-2)^n}{n!} = \sum\limits_{n=1}^{\infty} (-1)^n \dfrac{2^n}{n!}$ satisfies (i) of the Alternating Series Test because

$b_{n+1} = \dfrac{2^{n+1}}{(n+1)!} = \dfrac{2 \cdot 2^n}{(n+1)n!} = \dfrac{2}{n+1} \cdot \dfrac{2^n}{n!} = \dfrac{2}{n+1} \cdot b_n \leq b_n$ and (ii) $\lim\limits_{n\to\infty} \dfrac{2^n}{n!} = \dfrac{2}{n} \cdot \dfrac{2}{n-1} \cdot \ldots \cdot \dfrac{2}{2} \cdot \dfrac{2}{1} = 0$, so the

series is convergent. Now $b_7 = 2^7/7! \approx 0.025 > 0.01$ and $b_8 = 2^8/8! \approx 0.006 < 0.01$, so by the Alternating Series

Estimation Theorem, $n = 7$. (That is, since the 8th term is less than the desired error, we need to add the first 7 terms to get the

sum to the desired accuracy.)

11. The series $\sum\limits_{n=1}^{\infty} \dfrac{(-1)^{n+1}}{n^6}$ satisfies (i) of the Alternating Series Test because $\dfrac{1}{(n+1)^6} < \dfrac{1}{n^6}$ and (ii) $\lim\limits_{n\to\infty} \dfrac{1}{n^6} = 0$, so the

series is convergent. Now $b_5 = \dfrac{1}{5^6} = 0.000064 > 0.00005$ and $b_6 = \dfrac{1}{6^6} \approx 0.00002 < 0.00005$, so by the Alternating Series

Estimation Theorem, $n = 5$. (That is, since the 6th term is less than the desired error, we need to add the first 5 terms to get the

sum to the desired accuracy.)

13. $b_7 = \dfrac{1}{7^5} = \dfrac{1}{16{,}807} \approx 0.000\,059\,5$, so

$\sum\limits_{n=1}^{\infty} \dfrac{(-1)^{n+1}}{n^5} \approx s_6 = \sum\limits_{n=1}^{6} \dfrac{(-1)^{n+1}}{n^5} = 1 - \frac{1}{32} + \frac{1}{243} - \frac{1}{1024} + \frac{1}{3125} - \frac{1}{7776} \approx 0.972\,080$. Adding b_7 to s_6 does not change

the fourth decimal place of s_6, so the sum of the series, correct to four decimal places, is 0.9721.

15. $b_7 = \dfrac{7^2}{10^7} = 0.000\,004\,9$, so

$\sum\limits_{n=1}^{\infty} \dfrac{(-1)^{n-1} n^2}{10^n} \approx s_6 = \sum\limits_{n=1}^{6} \dfrac{(-1)^{n-1} n^2}{10^n} = \frac{1}{10} - \frac{4}{100} + \frac{9}{1000} - \frac{16}{10{,}000} + \frac{25}{100{,}000} - \frac{36}{1{,}000{,}000} = 0.067\,614$. Adding b_7 to s_6

does not change the fourth decimal place of s_6, so the sum of the series, correct to four decimal places, is 0.0676.

17. $\sum\limits_{n=1}^{\infty} \dfrac{(-1)^{n-1}}{n} = 1 - \frac{1}{2} + \frac{1}{3} - \frac{1}{4} + \cdots + \frac{1}{49} - \frac{1}{50} + \frac{1}{51} - \frac{1}{52} + \cdots$. The 50th partial sum of this series is an

underestimate, since $\sum\limits_{n=1}^{\infty} \dfrac{(-1)^{n-1}}{n} = s_{50} + \left(\dfrac{1}{51} - \dfrac{1}{52} \right) + \left(\dfrac{1}{53} - \dfrac{1}{54} \right) + \cdots$, and the terms in parentheses are all positive.

The result can be seen geometrically in Figure 1.

19. Using the Ratio Test, $\lim\limits_{n\to\infty} \left| \dfrac{a_{n+1}}{a_n} \right| = \lim\limits_{n\to\infty} \left| \dfrac{(-3)^{n+1}/(n+1)^3}{(-3)^n/n^3} \right| = \lim\limits_{n\to\infty} \left| \dfrac{(-3)n^3}{(n+1)^3} \right| = 3 \lim\limits_{n\to\infty} \left(\dfrac{n}{n+1} \right)^3 = 3 > 1$,

so the series diverges.

21. $\sum\limits_{n=0}^{\infty} \dfrac{(-10)^n}{n!}$. Using the Ratio Test, $\lim\limits_{n\to\infty} \left| \dfrac{a_{n+1}}{a_n} \right| = \lim\limits_{n\to\infty} \left| \dfrac{(-10)^{n+1}}{(n+1)!} \cdot \dfrac{n!}{(-10)^n} \right| = \lim\limits_{n\to\infty} \left| \dfrac{-10}{n+1} \right| = 0 < 1$, so the series is

absolutely convergent.

23. $\sum\limits_{n=1}^{\infty} \dfrac{(-1)^{n+1}}{\sqrt[4]{n}}$ converges by the Alternating Series Test, but $\sum\limits_{n=1}^{\infty} \dfrac{1}{\sqrt[4]{n}}$ is a divergent p-series $\left[p = \tfrac{1}{4} \leq 1 \right]$,

so the given series is conditionally convergent.

25. $\lim\limits_{n\to\infty} \left| \dfrac{a_{n+1}}{a_n} \right| = \lim\limits_{n\to\infty} \left[\dfrac{10^{n+1}}{(n+2)4^{2(n+1)+1}} \cdot \dfrac{(n+1)4^{2n+1}}{10^n} \right] = \lim\limits_{n\to\infty} \left[\dfrac{10^{n+1}}{(n+2)4^{2n+3}} \cdot \dfrac{(n+1)4^{2n+1}}{10^n} \right]$

$\qquad = \lim\limits_{n\to\infty} \left(\dfrac{10}{4^2} \cdot \dfrac{n+1}{n+2} \right) = \dfrac{5}{8} < 1$,

so the series is absolutely convergent by the Ratio Test. Since the terms of this series are positive, absolute convergence is the

same as convergence.

27. $\dfrac{|\cos(n\pi/3)|}{n!} \leq \dfrac{1}{n!}$ and $\sum\limits_{n=1}^{\infty} \dfrac{1}{n!}$ converges (use the Ratio Test), so the series $\sum\limits_{n=1}^{\infty} \dfrac{\cos(n\pi/3)}{n!}$ converges absolutely by the

Comparison Test.

29. $\left| \dfrac{(-1)^n \arctan n}{n^2} \right| < \dfrac{\pi/2}{n^2}$, so since $\sum\limits_{n=1}^{\infty} \dfrac{\pi/2}{n^2} = \dfrac{\pi}{2} \sum\limits_{n=1}^{\infty} \dfrac{1}{n^2}$ converges $\ [p = 2 > 1]$, the given series $\sum\limits_{n=1}^{\infty} \dfrac{(-1)^n \arctan n}{n^2}$

converges absolutely by the Comparison Test.

31. $\lim\limits_{n\to\infty} \sqrt[n]{|a_n|} = \lim\limits_{n\to\infty} \left(\dfrac{n^n}{3^{1+3n}} \right)^{1/n} = \lim\limits_{n\to\infty} \dfrac{n}{\sqrt[n]{3} \cdot 3^3} = \infty$, so the series $\sum\limits_{n=1}^{\infty} \dfrac{n^n}{3^{1+3n}}$ is divergent by the Root Test.

Or: $\lim\limits_{n\to\infty} \left| \dfrac{a_{n+1}}{a_n} \right| = \lim\limits_{n\to\infty} \left[\dfrac{(n+1)^{n+1}}{3^{4+3n}} \cdot \dfrac{3^{1+3n}}{n^n} \right] = \lim\limits_{n\to\infty} \left[\dfrac{1}{3^3} \cdot \left(\dfrac{n+1}{n} \right)^n (n+1) \right]$

$\qquad = \dfrac{1}{27} \lim\limits_{n\to\infty} \left(1 + \dfrac{1}{n} \right)^n \lim\limits_{n\to\infty} (n+1) = \tfrac{1}{27} e \lim\limits_{n\to\infty} (n+1) = \infty$,

so the series is divergent by the Ratio Test.

33. $\lim\limits_{n\to\infty} \sqrt[n]{|a_n|} = \lim\limits_{n\to\infty} \dfrac{n^2+1}{2n^2+1} = \lim\limits_{n\to\infty} \dfrac{1+1/n^2}{2+1/n^2} = \dfrac{1}{2} < 1$, so the series $\sum\limits_{n=1}^{\infty} \left(\dfrac{n^2+1}{2n^2+1} \right)^n$ is absolutely convergent by the

Root Test.

35. Use the Ratio Test with the series

$1 - \dfrac{1 \cdot 3}{3!} + \dfrac{1 \cdot 3 \cdot 5}{5!} - \dfrac{1 \cdot 3 \cdot 5 \cdot 7}{7!} + \cdots + (-1)^{n-1} \dfrac{1 \cdot 3 \cdot 5 \cdots (2n-1)}{(2n-1)!} + \cdots = \sum\limits_{n=1}^{\infty} (-1)^{n-1} \dfrac{1 \cdot 3 \cdot 5 \cdots (2n-1)}{(2n-1)!}$.

$\lim\limits_{n\to\infty} \left| \dfrac{a_{n+1}}{a_n} \right| = \lim\limits_{n\to\infty} \left| \dfrac{(-1)^n \cdot 1 \cdot 3 \cdot 5 \cdots (2n-1)[2(n+1)-1]}{[2(n+1)-1]!} \cdot \dfrac{(2n-1)!}{(-1)^{n-1} \cdot 1 \cdot 3 \cdot 5 \cdots (2n-1)} \right|$

$\qquad = \lim\limits_{n\to\infty} \left| \dfrac{(-1)(2n+1)(2n-1)!}{(2n+1)(2n)(2n-1)!} \right| = \lim\limits_{n\to\infty} \dfrac{1}{2n} = 0 < 1$,

so the given series is absolutely convergent and therefore convergent.

37. $\displaystyle\sum_{n=1}^{\infty} \frac{2 \cdot 4 \cdot 6 \cdot \cdots \cdot (2n)}{n!} = \sum_{n=1}^{\infty} \frac{(2 \cdot 1) \cdot (2 \cdot 2) \cdot (2 \cdot 3) \cdot \cdots \cdot (2 \cdot n)}{n!} = \sum_{n=1}^{\infty} \frac{2^n n!}{n!} = \sum_{n=1}^{\infty} 2^n$, which diverges by the Test for

Divergence since $\displaystyle\lim_{n\to\infty} 2^n = \infty$.

39. (a) $\displaystyle\lim_{n\to\infty} \left| \frac{1/(n+1)^3}{1/n^3} \right| = \lim_{n\to\infty} \frac{n^3}{(n+1)^3} = \lim_{n\to\infty} \frac{1}{(1+1/n)^3} = 1$.   Inconclusive

(b) $\displaystyle\lim_{n\to\infty} \left| \frac{(n+1)}{2^{n+1}} \cdot \frac{2^n}{n} \right| = \lim_{n\to\infty} \frac{n+1}{2n} = \lim_{n\to\infty} \left(\frac{1}{2} + \frac{1}{2n} \right) = \frac{1}{2}$.   Conclusive (convergent)

(c) $\displaystyle\lim_{n\to\infty} \left| \frac{(-3)^n}{\sqrt{n+1}} \cdot \frac{\sqrt{n}}{(-3)^{n-1}} \right| = 3 \lim_{n\to\infty} \sqrt{\frac{n}{n+1}} = 3 \lim_{n\to\infty} \sqrt{\frac{1}{1+1/n}} = 3$.   Conclusive (divergent)

(d) $\displaystyle\lim_{n\to\infty} \left| \frac{\sqrt{n+1}}{1+(n+1)^2} \cdot \frac{1+n^2}{\sqrt{n}} \right| = \lim_{n\to\infty} \left[\sqrt{1+\frac{1}{n}} \cdot \frac{1/n^2+1}{1/n^2+(1+1/n)^2} \right] = 1$.   Inconclusive

41. (a) $\displaystyle\lim_{n\to\infty} \left| \frac{a_{n+1}}{a_n} \right| = \lim_{n\to\infty} \left| \frac{x^{n+1}}{(n+1)!} \cdot \frac{n!}{x^n} \right| = \lim_{n\to\infty} \left| \frac{x}{n+1} \right| = |x| \lim_{n\to\infty} \frac{1}{n+1} = |x| \cdot 0 = 0 < 1$, so by the Ratio Test the

series $\displaystyle\sum_{n=0}^{\infty} \frac{x^n}{n!}$ converges for all x.

(b) Since the series of part (a) always converges, we must have $\displaystyle\lim_{n\to\infty} \frac{x^n}{n!} = 0$ by Theorem 8.2.6.

43. Following the hint, we get that $|a_n| < r^n$ for $n \geq N$, and so since the geometric series $\displaystyle\sum_{n=1}^{\infty} r^n$ converges $(0 < r < 1)$, the

series $\displaystyle\sum_{n=N}^{\infty} |a_n|$ converges as well by the Comparison Test, and hence so does $\displaystyle\sum_{n=1}^{\infty} |a_n|$, so $\displaystyle\sum_{n=1}^{\infty} a_n$ is absolutely convergent.

8.5   Power Series

1. A power series is a series of the form $\displaystyle\sum_{n=0}^{\infty} c_n x^n = c_0 + c_1 x + c_2 x^2 + c_3 x^3 + \cdots$, where x is a variable and the

c_n's are constants called the coefficients of the series. More generally, a series of the form

$$\sum_{n=0}^{\infty} c_n(x-a)^n = c_0 + c_1(x-a) + c_2(x-a)^2 + \cdots$$ is called a power series in $(x-a)$ or a power series centered at a or a

power series about a, where a is a constant.

3. If $a_n = \dfrac{x^n}{\sqrt{n}}$, then $\displaystyle\lim_{n\to\infty} \left| \frac{a_{n+1}}{a_n} \right| = \lim_{n\to\infty} \left| \frac{x^{n+1}}{\sqrt{n+1}} \cdot \frac{\sqrt{n}}{x^n} \right| = \lim_{n\to\infty} \left| \frac{x}{\sqrt{n+1}/\sqrt{n}} \right| = \lim_{n\to\infty} \frac{|x|}{\sqrt{1+1/n}} = |x|$.

By the Ratio Test, the series $\displaystyle\sum_{n=1}^{\infty} \frac{x^n}{\sqrt{n}}$ converges when $|x| < 1$, so the radius of convergence $R = 1$. Now we'll check the

endpoints, that is, $x = \pm 1$. When $x = 1$, the series $\displaystyle\sum_{n=1}^{\infty} \frac{1}{\sqrt{n}}$ diverges because it is a p-series with $p = \frac{1}{2} \leq 1$. When $x = -1$,

the series $\displaystyle\sum_{n=1}^{\infty} \frac{(-1)^n}{\sqrt{n}}$ converges by the Alternating Series Test. Thus, the interval of convergence is $I = [-1, 1)$.

5. If $a_n = \dfrac{(-1)^{n-1}x^n}{n^3}$, then

$$\lim_{n \to \infty} \left| \frac{a_{n+1}}{a_n} \right| = \lim_{n \to \infty} \left| \frac{(-1)^n x^{n+1}}{(n+1)^3} \cdot \frac{n^3}{(-1)^{n-1}x^n} \right| = \lim_{n \to \infty} \left| \frac{(-1)xn^3}{(n+1)^3} \right| = \lim_{n \to \infty} \left[\left(\frac{n}{n+1} \right)^3 |x| \right] = 1^3 \cdot |x| = |x|$$

converges by the Alternating Series Test. When $x = -1$, the series $\displaystyle\sum_{n=1}^{\infty} \frac{(-1)^{n-1}(-1)^n}{n^3} = -\sum_{n=1}^{\infty} \frac{1}{n^3}$ converges because it is

a constant multiple of a convergent p-series $[p = 3 > 1]$. Thus, the interval of convergence is $I = [-1, 1]$.

7. If $a_n = \dfrac{x^n}{n!}$, then $\displaystyle\lim_{n \to \infty} \left| \frac{a_{n+1}}{a_n} \right| = \lim_{n \to \infty} \left| \frac{x^{n+1}}{(n+1)!} \cdot \frac{n!}{x^n} \right| = \lim_{n \to \infty} \left| \frac{x}{n+1} \right| = |x| \lim_{n \to \infty} \frac{1}{n+1} = |x| \cdot 0 = 0 < 1$ for *all* real x.

So, by the Ratio Test, $R = \infty$, and $I = (-\infty, \infty)$.

9. $a_n = \dfrac{(-2)^n x^n}{\sqrt[4]{n}}$, so $\displaystyle\lim_{n \to \infty} \left| \frac{a_{n+1}}{a_n} \right| = \lim_{n \to \infty} \frac{2^{n+1}|x|^{n+1}}{\sqrt[4]{n+1}} \cdot \frac{\sqrt[4]{n}}{2^n |x|^n} = \lim_{n \to \infty} 2|x| \sqrt[4]{\frac{n}{n+1}} = 2|x|$, so by the Ratio Test, the

series converges when $2|x| < 1 \iff |x| < \frac{1}{2}$, so $R = \frac{1}{2}$. When $x = -\frac{1}{2}$, we get the divergent p-series $\displaystyle\sum_{n=1}^{\infty} \frac{1}{\sqrt[4]{n}}$

$\left[p = \frac{1}{4} \leq 1 \right]$. When $x = \frac{1}{2}$, we get the series $\displaystyle\sum_{n=1}^{\infty} \frac{(-1)^n}{\sqrt[4]{n}}$, which converges by the Alternating Series Test.

Thus, $I = \left(-\frac{1}{2}, \frac{1}{2} \right]$.

11. If $a_n = (-1)^n \dfrac{x^n}{4^n \ln n}$, then $\displaystyle\lim_{n \to \infty} \left| \frac{a_{n+1}}{a_n} \right| = \lim_{n \to \infty} \left| \frac{x^{n+1}}{4^{n+1}\ln(n+1)} \cdot \frac{4^n \ln n}{x^n} \right| = \frac{|x|}{4} \lim_{n \to \infty} \frac{\ln n}{\ln(n+1)} = \frac{|x|}{4} \cdot 1$

[by l'Hospital's Rule] $= \dfrac{|x|}{4}$. By the Ratio Test, the series converges when $\dfrac{|x|}{4} < 1 \iff |x| < 4$, so $R = 4$. When

$x = -4$, $\displaystyle\sum_{n=2}^{\infty} (-1)^n \frac{x^n}{4^n \ln n} = \sum_{n=2}^{\infty} \frac{[(-1)(-4)]^n}{4^n \ln n} = \sum_{n=2}^{\infty} \frac{1}{\ln n}$. Since $\ln n < n$ for $n \geq 2$, $\dfrac{1}{\ln n} > \dfrac{1}{n}$ and $\displaystyle\sum_{n=2}^{\infty} \frac{1}{n}$ is the

divergent harmonic series (without the $n = 1$ term), $\displaystyle\sum_{n=2}^{\infty} \frac{1}{\ln n}$ is divergent by the Comparison Test. When $x = 4$,

$\displaystyle\sum_{n=2}^{\infty} (-1)^n \frac{x^n}{4^n \ln n} = \sum_{n=2}^{\infty} (-1)^n \frac{1}{\ln n}$, which converges by the Alternating Series Test. Thus, $I = (-4, 4]$.

13. If $a_n = (-1)^n \dfrac{(x+2)^n}{n\,2^n}$, then $\displaystyle\lim_{n \to \infty} \left| \frac{a_{n+1}}{a_n} \right| = \lim_{n \to \infty} \left[\frac{|x+2|^{n+1}}{(n+1)2^{n+1}} \cdot \frac{n\,2^n}{|x+2|^n} \right] = \lim_{n \to \infty} \frac{n}{n+1} \cdot \frac{|x+2|}{2} = \frac{|x+2|}{2}$.

By the Ratio Test, the series converges when $\dfrac{|x+2|}{2} < 1 \iff |x+2| < 2$ [so $R = 2$] $\iff -2 < x+2 < 2 \iff$

$-4 < x < 0$. When $x = -4$, the series becomes $\displaystyle\sum_{n=1}^{\infty} (-1)^n \frac{(-2)^n}{n\,2^n} = \sum_{n=1}^{\infty} \frac{2^n}{n\,2^n} = \sum_{n=1}^{\infty} \frac{1}{n}$, which is the divergent harmonic

series. When $x = 0$, the series is $\displaystyle\sum_{n=1}^{\infty} \frac{(-1)^n}{n}$, the alternating harmonic series, which converges by the Alternating Series Test.

Thus, $I = (-4, 0]$.

15. $a_n = \dfrac{n}{b^n}(x-a)^n$, where $b > 0$.

$$\lim_{n\to\infty}\left|\frac{a_{n+1}}{a_n}\right| = \lim_{n\to\infty}\frac{(n+1)\,|x-a|^{n+1}}{b^{n+1}}\cdot\frac{b^n}{n\,|x-a|^n} = \lim_{n\to\infty}\left(1+\frac{1}{n}\right)\frac{|x-a|}{b} = \frac{|x-a|}{b}.$$

By the Ratio Test, the series converges when $\dfrac{|x-a|}{b} < 1 \;\Leftrightarrow\; |x-a| < b \;\; [\text{so } R = b] \;\Leftrightarrow\; -b < x - a < b \;\Leftrightarrow\;$

$a - b < x < a + b$. When $|x-a| = b$, $\lim\limits_{n\to\infty}|a_n| = \lim\limits_{n\to\infty} n = \infty$, so the series diverges. Thus, $I = (a-b, a+b)$.

17. If $a_n = n!(2x-1)^n$, then $\lim\limits_{n\to\infty}\left|\dfrac{a_{n+1}}{a_n}\right| = \lim\limits_{n\to\infty}\left|\dfrac{(n+1)!(2x-1)^{n+1}}{n!(2x-1)^n}\right| = \lim\limits_{n\to\infty}(n+1)\,|2x-1| \to \infty$ as $n \to \infty$ for

all $x \neq \frac{1}{2}$. Since the series diverges for all $x \neq \frac{1}{2}$, $R = 0$ and $I = \{\frac{1}{2}\}$.

19. (a) We are given that the power series $\sum\limits_{n=0}^{\infty} c_n x^n$ is convergent for $x = 4$. So by Theorem 3, it must converge for at least

$-4 < x \le 4$. In particular, it converges when $x = -2$; that is, $\sum\limits_{n=0}^{\infty} c_n(-2)^n$ is convergent.

(b) It does not follow that $\sum\limits_{n=0}^{\infty} c_n(-4)^n$ is necessarily convergent. [See the comments after Theorem 3 about convergence at

the endpoint of an interval. An example is $c_n = (-1)^n/(n4^n)$.]

21. If $a_n = \dfrac{(n!)^k}{(kn)!}x^n$, then

$$\lim_{n\to\infty}\left|\frac{a_{n+1}}{a_n}\right| = \lim_{n\to\infty}\frac{[(n+1)!]^k\,(kn)!}{(n!)^k\,[k(n+1)]!}\,|x| = \lim_{n\to\infty}\frac{(n+1)^k}{(kn+k)(kn+k-1)\cdots(kn+2)(kn+1)}\,|x|$$

$$= \lim_{n\to\infty}\left[\frac{(n+1)}{(kn+1)}\frac{(n+1)}{(kn+2)}\cdots\frac{(n+1)}{(kn+k)}\right]|x| = \lim_{n\to\infty}\left[\frac{n+1}{kn+1}\right]\lim_{n\to\infty}\left[\frac{n+1}{kn+2}\right]\cdots\lim_{n\to\infty}\left[\frac{n+1}{kn+k}\right]|x|$$

$$= \left(\frac{1}{k}\right)^k|x| < 1 \;\;\Leftrightarrow\;\; |x| < k^k \text{ for convergence, and the radius of convergence is } R = k^k.$$

23. (a) If $a_n = \dfrac{(-1)^n\,x^{2n+1}}{n!(n+1)!\,2^{2n+1}}$, then

$$\lim_{n\to\infty}\left|\frac{a_{n+1}}{a_n}\right| = \lim_{n\to\infty}\left|\frac{x^{2n+3}}{(n+1)!(n+2)!\,2^{2n+3}}\cdot\frac{n!(n+1)!\,2^{2n+1}}{x^{2n+1}}\right| = \left(\frac{x}{2}\right)^2\lim_{n\to\infty}\frac{1}{(n+1)(n+2)} = 0 \text{ for all } x.$$

So $J_1(x)$ converges for all x and its domain is $(-\infty, \infty)$.

(b), (c) The initial terms of $J_1(x)$ up to $n = 5$ are $a_0 = \dfrac{x}{2}$,

$a_1 = -\dfrac{x^3}{16}$, $a_2 = \dfrac{x^5}{384}$, $a_3 = -\dfrac{x^7}{18{,}432}$, $a_4 = \dfrac{x^9}{1{,}474{,}560}$,

and $a_5 = -\dfrac{x^{11}}{176{,}947{,}200}$. The partial sums seem to

approximate $J_1(x)$ well near the origin, but as $|x|$ increases,

we need to take a large number of terms to get a good

approximation.

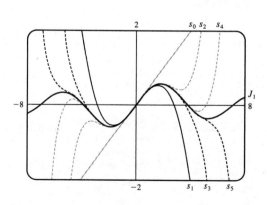

25. $s_{2n-1} = 1 + 2x + x^2 + 2x^3 + x^4 + 2x^5 + \cdots + x^{2n-2} + 2x^{2n-1}$

$$= 1(1+2x) + x^2(1+2x) + x^4(1+2x) + \cdots + x^{2n-2}(1+2x) = (1+2x)(1 + x^2 + x^4 + \cdots + x^{2n-2})$$

$$= (1+2x)\frac{1 - x^{2n}}{1 - x^2} \quad \text{[by (8.2.3)]} \quad \text{with } r = x^2] \quad \rightarrow \frac{1 + 2x}{1 - x^2} \text{ as } n \rightarrow \infty \quad \text{[by (8.2.4)], when } |x| < 1.$$

Also $s_{2n} = s_{2n-1} + x^{2n} \rightarrow \dfrac{1+2x}{1-x^2}$ since $x^{2n} \rightarrow 0$ for $|x| < 1$. Therefore, $s_n \rightarrow \dfrac{1+2x}{1-x^2}$ since s_{2n} and s_{2n-1} both

approach $\dfrac{1+2x}{1-x^2}$ as $n \rightarrow \infty$. Thus, the interval of convergence is $(-1,1)$ and $f(x) = \dfrac{1+2x}{1-x^2}$.

27. We use the Root Test on the series $\sum c_n x^n$. We need $\lim\limits_{n\to\infty} \sqrt[n]{|c_n x^n|} = |x| \lim\limits_{n\to\infty} \sqrt[n]{|c_n|} = c\,|x| < 1$ for convergence, or

$|x| < 1/c$, so $R = 1/c$.

29. For $2 < x < 3$, $\sum c_n x^n$ diverges and $\sum d_n x^n$ converges. By Exercise 8.2.43, $\sum (c_n + d_n)\,x^n$ diverges. Since both series

converge for $|x| < 2$, the radius of convergence of $\sum (c_n + d_n)\,x^n$ is 2.

8.6 Representing Functions as Power Series

1. If $f(x) = \sum\limits_{n=0}^{\infty} c_n x^n$ has radius of convergence 10, then $f'(x) = \sum\limits_{n=1}^{\infty} n c_n x^{n-1}$ also has radius of convergence 10 by

Theorem 2.

3. Our goal is to write the function in the form $\dfrac{1}{1-r}$, and then use Equation (1) to represent the function as a sum of a power

series. $f(x) = \dfrac{1}{1+x} = \dfrac{1}{1-(-x)} = \sum\limits_{n=0}^{\infty} (-x)^n = \sum\limits_{n=0}^{\infty} (-1)^n x^n$ with $|-x| < 1 \quad \Leftrightarrow \quad |x| < 1$, so $R = 1$ and $I = (-1, 1)$.

5. Replacing x with x^3 in (1) gives $f(x) = \dfrac{1}{1 - x^3} = \sum\limits_{n=0}^{\infty} (x^3)^n = \sum\limits_{n=0}^{\infty} x^{3n}$. The series converges when $|x^3| < 1 \quad \Leftrightarrow$

$|x|^3 < 1 \quad \Leftrightarrow \quad |x| < \sqrt[3]{1} \quad \Leftrightarrow \quad |x| < 1$. Thus, $R = 1$ and $I = (-1, 1)$.

7. $f(x) = \dfrac{1}{x-5} = -\dfrac{1}{5}\left(\dfrac{1}{1 - x/5}\right) = -\dfrac{1}{5} \sum\limits_{n=0}^{\infty} \left(\dfrac{x}{5}\right)^n$ or equivalently, $-\sum\limits_{n=0}^{\infty} \dfrac{1}{5^{n+1}} x^n$. The series converges when $\left|\dfrac{x}{5}\right| < 1$;

that is, when $|x| < 5$, so $I = (-5, 5)$.

9. $f(x) = \dfrac{x}{9 + x^2} = \dfrac{x}{9}\left[\dfrac{1}{1 + (x/3)^2}\right] = \dfrac{x}{9}\left[\dfrac{1}{1 - \{-(x/3)^2\}}\right] = \dfrac{x}{9} \sum\limits_{n=0}^{\infty} \left[-\left(\dfrac{x}{3}\right)^2\right]^n = \dfrac{x}{9} \sum\limits_{n=0}^{\infty} (-1)^n \dfrac{x^{2n}}{9^n}$

$$= \sum\limits_{n=0}^{\infty} (-1)^n \dfrac{x^{2n+1}}{9^{n+1}}$$

The geometric series $\sum\limits_{n=0}^{\infty} \left[-\left(\dfrac{x}{3}\right)^2\right]^n$ converges when $\left|-\left(\dfrac{x}{3}\right)^2\right| < 1 \quad \Leftrightarrow \quad \dfrac{|x^2|}{9} < 1 \quad \Leftrightarrow \quad |x|^2 < 9 \quad \Leftrightarrow \quad |x| < 3$,

so $R = 3$ and $I = (-3, 3)$.

11. $f(x) = \dfrac{3}{x^2 + x - 2} = \dfrac{3}{(x+2)(x-1)} = \dfrac{A}{x+2} + \dfrac{B}{x-1}$ $\Rightarrow$ $3 = A(x-1) + B(x+2)$. Taking $x = -2$, we get

$A = -1$. Taking $x = 1$, we get $B = 1$. Thus,

$$\dfrac{3}{x^2 + x - 2} = \dfrac{1}{x-1} - \dfrac{1}{x+2} = -\dfrac{1}{1-x} - \dfrac{1}{2}\dfrac{1}{1+x/2} = -\sum_{n=0}^{\infty} x^n - \dfrac{1}{2}\sum_{n=0}^{\infty}\left(-\dfrac{x}{2}\right)^n$$

$$= \sum_{n=0}^{\infty}\left[-1 - \dfrac{1}{2}\left(-\dfrac{1}{2}\right)^n\right] x^n = \sum_{n=0}^{\infty}\left[-1 + \left(-\dfrac{1}{2}\right)^{n+1}\right] x^n = \sum_{n=0}^{\infty}\left[\dfrac{(-1)^{n+1}}{2^{n+1}} - 1\right] x^n$$

We represented the given function as the sum of two geometric series; the first converges for $x \in (-1, 1)$ and the second

converges for $x \in (-2, 2)$. Thus, the sum converges for $x \in (-1, 1) = I$.

13. (a) $f(x) = \dfrac{1}{(1+x)^2} = \dfrac{d}{dx}\left(\dfrac{-1}{1+x}\right) = -\dfrac{d}{dx}\left[\sum_{n=0}^{\infty}(-1)^n x^n\right]$ [from Exercise 3]

$= \sum_{n=1}^{\infty}(-1)^{n+1}nx^{n-1}$ [from Theorem 2(i)] $= \sum_{n=0}^{\infty}(-1)^n(n+1)x^n$ with $R = 1$.

In the last step, note that we *decreased* the initial value of the summation variable n by 1, and then *increased* each

occurrence of n in the term by 1 [also note that $(-1)^{n+2} = (-1)^n$].

(b) $f(x) = \dfrac{1}{(1+x)^3} = -\dfrac{1}{2}\dfrac{d}{dx}\left[\dfrac{1}{(1+x)^2}\right] = -\dfrac{1}{2}\dfrac{d}{dx}\left[\sum_{n=0}^{\infty}(-1)^n(n+1)x^n\right]$ [from part (a)]

$= -\dfrac{1}{2}\sum_{n=1}^{\infty}(-1)^n(n+1)nx^{n-1} = \dfrac{1}{2}\sum_{n=0}^{\infty}(-1)^n(n+2)(n+1)x^n$ with $R = 1$.

(c) $f(x) = \dfrac{x^2}{(1+x)^3} = x^2 \cdot \dfrac{1}{(1+x)^3} = x^2 \cdot \dfrac{1}{2}\sum_{n=0}^{\infty}(-1)^n(n+2)(n+1)x^n$ [from part (b)]

$= \dfrac{1}{2}\sum_{n=0}^{\infty}(-1)^n(n+2)(n+1)x^{n+2}$

To write the power series with x^n rather than x^{n+2}, we will *decrease* each occurrence of n in the term by 2 and *increase*

the initial value of the summation variable by 2. This gives us $\dfrac{1}{2}\sum_{n=2}^{\infty}(-1)^n(n)(n-1)x^n$ with $R = 1$.

15. $f(x) = \ln(5-x) = -\displaystyle\int\dfrac{dx}{5-x} = -\dfrac{1}{5}\int\dfrac{dx}{1-x/5} = -\dfrac{1}{5}\int\left[\sum_{n=0}^{\infty}\left(\dfrac{x}{5}\right)^n\right]dx = C - \dfrac{1}{5}\sum_{n=0}^{\infty}\dfrac{x^{n+1}}{5^n(n+1)} = C - \sum_{n=1}^{\infty}\dfrac{x^n}{n\,5^n}.$

Putting $x = 0$, we get $C = \ln 5$ The series converges for $|x/5| < 1$ $\Leftrightarrow$ $|x| < 5$, so $R = 5$.

17. $\dfrac{1}{2-x} = \dfrac{1}{2(1-x/2)} = \dfrac{1}{2}\sum_{n=0}^{\infty}\left(\dfrac{x}{2}\right)^n = \sum_{n=0}^{\infty}\dfrac{1}{2^{n+1}}x^n$ for $\left|\dfrac{x}{2}\right| < 1$ $\Leftrightarrow$ $|x| < 2$. Now

$\dfrac{1}{(x-2)^2} = \dfrac{d}{dx}\left(\dfrac{1}{2-x}\right) = \dfrac{d}{dx}\left(\sum_{n=0}^{\infty}\dfrac{1}{2^{n+1}}x^n\right) = \sum_{n=1}^{\infty}\dfrac{n}{2^{n+1}}x^{n-1} = \sum_{n=0}^{\infty}\dfrac{n+1}{2^{n+2}}x^n$. So

$f(x) = \dfrac{x^3}{(x-2)^2} = x^3\sum_{n=0}^{\infty}\dfrac{n+1}{2^{n+2}}x^n = \sum_{n=0}^{\infty}\dfrac{n+1}{2^{n+2}}x^{n+3}$ or $\sum_{n=3}^{\infty}\dfrac{n-2}{2^{n-1}}x^n$ for $|x| < 2$. Thus, $R = 2$ and $I = (-2, 2)$.

19. $f(x) = \ln(3+x) = \int \frac{dx}{3+x} = \frac{1}{3}\int \frac{dx}{1+x/3} = \frac{1}{3}\int \frac{dx}{1-(-x/3)} = \frac{1}{3}\int \sum_{n=0}^{\infty}\left(-\frac{x}{3}\right)^n dx$

$$= C + \frac{1}{3}\sum_{n=0}^{\infty}\frac{(-1)^n}{(n+1)3^n}x^{n+1} = \ln 3 + \frac{1}{3}\sum_{n=1}^{\infty}\frac{(-1)^{n-1}}{n3^{n-1}}x^n \quad [C = f(0) = \ln 3] \quad = \ln 3 + \sum_{n=1}^{\infty}\frac{(-1)^{n-1}}{n\,3^n}x^n.$$

The series converges when $|-x/3| < 1 \Leftrightarrow |x| < 3$, so $R = 3$. The terms of the series are $a_0 = \ln 3$, $a_1 = \dfrac{x}{3}$,

$a_2 = -\dfrac{x^2}{18}$, $a_3 = \dfrac{x^3}{81}$, $a_4 = -\dfrac{x^4}{324}$, $a_5 = \dfrac{x^5}{1215}$, ...

As n increases, $s_n(x)$ approximates f better on the interval of convergence, which is $(-3, 3)$.

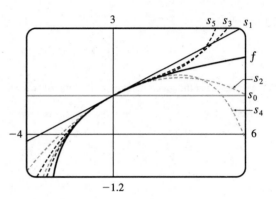

21. $f(x) = \ln\left(\dfrac{1+x}{1-x}\right) = \ln(1+x) - \ln(1-x) = \int \dfrac{dx}{1+x} + \int \dfrac{dx}{1-x} = \int \dfrac{dx}{1-(-x)} + \int \dfrac{dx}{1-x}$

$$= \int \left[\sum_{n=0}^{\infty}(-1)^n x^n + \sum_{n=0}^{\infty}x^n\right]dx = \int \left[(1-x+x^2-x^3+x^4-\cdots) + (1+x+x^2+x^3+x^4+\cdots)\right]dx$$

$$= \int (2+2x^2+2x^4+\cdots)\,dx = \int \sum_{n=0}^{\infty}2x^{2n}\,dx = C + \sum_{n=0}^{\infty}\frac{2x^{2n+1}}{2n+1}$$

But $f(0) = \ln\frac{1}{1} = 0$, so $C = 0$ and we have

$f(x) = \sum_{n=0}^{\infty}\dfrac{2x^{2n+1}}{2n+1}$ with $R = 1$. If $x = \pm 1$, then

$f(x) = \pm 2\sum_{n=0}^{\infty}\dfrac{1}{2n+1}$, which both diverge by the Limit

Comparison Test with $b_n = \dfrac{1}{n}$. As n increases, $s_n(x)$

approximates f better on the interval of convergence, which is $(-1, 1)$.

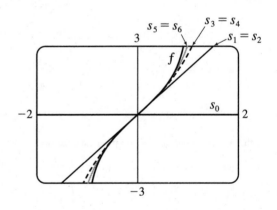

23. $\dfrac{t}{1-t^8} = t \cdot \dfrac{1}{1-t^8} = t\sum_{n=0}^{\infty}(t^8)^n = \sum_{n=0}^{\infty}t^{8n+1} \quad \Rightarrow \quad \int \dfrac{t}{1-t^8}\,dt = C + \sum_{n=0}^{\infty}\dfrac{t^{8n+2}}{8n+2}$. The series for $\dfrac{1}{1-t^8}$ converges

when $|t^8| < 1 \quad \Leftrightarrow \quad |t| < 1$, so $R = 1$ for that series and also the series for $t/(1-t^8)$. By Theorem 2, the series for

$\int \dfrac{t}{1-t^8}\,dt$ also has $R = 1$.

25. By Example 7, $\tan^{-1} x = \sum\limits_{n=0}^{\infty} (-1)^n \dfrac{x^{2n+1}}{2n+1}$ with $R = 1$, so

$$x - \tan^{-1} x = x - \left(x - \frac{x^3}{3} + \frac{x^5}{5} - \frac{x^7}{7} + \cdots \right) = \frac{x^3}{3} - \frac{x^5}{5} + \frac{x^7}{7} - \cdots = \sum_{n=1}^{\infty} (-1)^{n+1} \frac{x^{2n+1}}{2n+1} \text{ and}$$

$$\frac{x - \tan^{-1} x}{x^3} = \sum_{n=1}^{\infty} (-1)^{n+1} \frac{x^{2n-2}}{2n+1}, \text{ so}$$

$$\int \frac{x - \tan^{-1} x}{x^3} \, dx = C + \sum_{n=1}^{\infty} (-1)^{n+1} \frac{x^{2n-1}}{(2n+1)(2n-1)} = C + \sum_{n=1}^{\infty} (-1)^{n+1} \frac{x^{2n-1}}{4n^2 - 1}. \text{ By Theorem 2, } R = 1.$$

27. $\dfrac{1}{1 + x^5} = \dfrac{1}{1 - (-x^5)} = \sum\limits_{n=0}^{\infty} \left(-x^5 \right)^n = \sum\limits_{n=0}^{\infty} (-1)^n x^{5n} \quad \Rightarrow$

$$\int \frac{1}{1 + x^5} \, dx = \int \sum_{n=0}^{\infty} (-1)^n x^{5n} \, dx = C + \sum_{n=0}^{\infty} (-1)^n \frac{x^{5n+1}}{5n+1}. \text{ Thus,}$$

$$I = \int_0^{0.2} \frac{1}{1 + x^5} \, dx = \left[x - \frac{x^6}{6} + \frac{x^{11}}{11} - \cdots \right]_0^{0.2} = 0.2 - \frac{(0.2)^6}{6} + \frac{(0.2)^{11}}{11} - \cdots . \text{ The series is alternating, so if we use}$$

the first two terms, the error is at most $(0.2)^{11}/11 \approx 1.9 \times 10^{-9}$. So $I \approx 0.2 - (0.2)^6/6 \approx 0.199989$ to six decimal places.

29. We substitute $3x$ for x in Example 7, and find that

$$\int x \arctan(3x) \, dx = \int x \sum_{n=0}^{\infty} (-1)^n \frac{(3x)^{2n+1}}{2n+1} \, dx = \int \sum_{n=0}^{\infty} (-1)^n \frac{3^{2n+1} x^{2n+2}}{2n+1} \, dx$$

$$= C + \sum_{n=0}^{\infty} (-1)^n \frac{3^{2n+1} x^{2n+3}}{(2n+1)(2n+3)}$$

So

$$\int_0^{0.1} x \arctan(3x) \, dx = \left[\frac{3x^3}{1 \cdot 3} - \frac{3^3 x^5}{3 \cdot 5} + \frac{3^5 x^7}{5 \cdot 7} - \frac{3^7 x^9}{7 \cdot 9} + \cdots \right]_0^{0.1}$$

$$= \frac{1}{10^3} - \frac{9}{5 \times 10^5} + \frac{243}{35 \times 10^7} - \frac{2187}{63 \times 10^9} + \cdots .$$

The series is alternating, so if we use three terms, the error is at most $\dfrac{2187}{63 \times 10^9} \approx 3.5 \times 10^{-8}$. So

$$\int_0^{0.1} x \arctan(3x) \, dx \approx \frac{1}{10^3} - \frac{9}{5 \times 10^5} + \frac{243}{35 \times 10^7} \approx 0.000\,983 \text{ to six decimal places.}$$

31. Using the result of Example 6, $\ln(1 - x) = -\sum\limits_{n=1}^{\infty} \dfrac{x^n}{n}$, with $x = -0.1$, we have

$$\ln 1.1 = \ln[1 - (-0.1)] = 0.1 - \frac{0.01}{2} + \frac{0.001}{3} - \frac{0.0001}{4} + \frac{0.00001}{5} - \cdots . \text{ The series is alternating, so if we use only the}$$

first four terms, the error is at most $\dfrac{0.00001}{5} = 0.000\,002$. So $\ln 1.1 \approx 0.1 - \dfrac{0.01}{2} + \dfrac{0.001}{3} - \dfrac{0.0001}{4} \approx 0.09531$.

33. (a) $J_0(x) = \sum\limits_{n=0}^{\infty} \dfrac{(-1)^n x^{2n}}{2^{2n}(n!)^2}$, $J_0'(x) = \sum\limits_{n=1}^{\infty} \dfrac{(-1)^n 2nx^{2n-1}}{2^{2n}(n!)^2}$, and $J_0''(x) = \sum\limits_{n=1}^{\infty} \dfrac{(-1)^n 2n(2n-1)x^{2n-2}}{2^{2n}(n!)^2}$, so

$$x^2 J_0''(x) + x J_0'(x) + x^2 J_0(x) = \sum_{n=1}^{\infty} \frac{(-1)^n 2n(2n-1)x^{2n}}{2^{2n}(n!)^2} + \sum_{n=1}^{\infty} \frac{(-1)^n 2nx^{2n}}{2^{2n}(n!)^2} + \sum_{n=0}^{\infty} \frac{(-1)^n x^{2n+2}}{2^{2n}(n!)^2}$$

$$= \sum_{n=1}^{\infty} \frac{(-1)^n 2n(2n-1)x^{2n}}{2^{2n}(n!)^2} + \sum_{n=1}^{\infty} \frac{(-1)^n 2nx^{2n}}{2^{2n}(n!)^2} + \sum_{n=1}^{\infty} \frac{(-1)^{n-1} x^{2n}}{2^{2n-2}[(n-1)!]^2}$$

$$= \sum_{n=1}^{\infty} \frac{(-1)^n 2n(2n-1)x^{2n}}{2^{2n}(n!)^2} + \sum_{n=1}^{\infty} \frac{(-1)^n 2nx^{2n}}{2^{2n}(n!)^2} + \sum_{n=1}^{\infty} \frac{(-1)^n (-1)^{-1} 2^2 n^2 x^{2n}}{2^{2n}(n!)^2}$$

$$= \sum_{n=1}^{\infty} (-1)^n \left[\frac{2n(2n-1) + 2n - 2^2 n^2}{2^{2n}(n!)^2} \right] x^{2n}$$

$$= \sum_{n=1}^{\infty} (-1)^n \left[\frac{4n^2 - 2n + 2n - 4n^2}{2^{2n}(n!)^2} \right] x^{2n} = 0$$

(b) $\displaystyle\int_0^1 J_0(x)\,dx = \int_0^1 \left[\sum_{n=0}^{\infty} \frac{(-1)^n x^{2n}}{2^{2n}(n!)^2} \right] dx = \int_0^1 \left(1 - \frac{x^2}{4} + \frac{x^4}{64} - \frac{x^6}{2304} + \cdots \right) dx$

$$= \left[x - \frac{x^3}{3 \cdot 4} + \frac{x^5}{5 \cdot 64} - \frac{x^7}{7 \cdot 2304} + \cdots \right]_0^1 = 1 - \frac{1}{12} + \frac{1}{320} - \frac{1}{16{,}128} + \cdots$$

Since $\frac{1}{16{,}128} \approx 0.000062$, it follows from The Alternating Series Estimation Theorem that, correct to three decimal places,

$\int_0^1 J_0(x)\,dx \approx 1 - \frac{1}{12} + \frac{1}{320} \approx 0.920$.

35. (a) $f(x) = \sum\limits_{n=0}^{\infty} \dfrac{x^n}{n!} \;\Rightarrow\; f'(x) = \sum\limits_{n=1}^{\infty} \dfrac{nx^{n-1}}{n!} = \sum\limits_{n=1}^{\infty} \dfrac{x^{n-1}}{(n-1)!} = \sum\limits_{n=0}^{\infty} \dfrac{x^n}{n!} = f(x)$

(b) By Theorem 3.4.2, the only solution to the differential equation $df(x)/dx = f(x)$ is $f(x) = Ke^x$, but $f(0) = 1$,

so $K = 1$ and $f(x) = e^x$.

Or: We could solve the equation $df(x)/dx = f(x)$ as a separable differential equation.

37. If $a_n = \dfrac{x^n}{n^2}$, then by the Ratio Test, $\lim\limits_{n \to \infty} \left| \dfrac{a_{n+1}}{a_n} \right| = \lim\limits_{n \to \infty} \left| \dfrac{x^{n+1}}{(n+1)^2} \cdot \dfrac{n^2}{x^n} \right| = |x| \lim\limits_{n \to \infty} \left(\dfrac{n}{n+1} \right)^2 = |x| < 1$ for

convergence, so $R = 1$. When $x = \pm 1$, $\sum\limits_{n=1}^{\infty} \left| \dfrac{x^n}{n^2} \right| = \sum\limits_{n=1}^{\infty} \dfrac{1}{n^2}$ which is a convergent p-series ($p = 2 > 1$), so the interval of

convergence for f is $[-1, 1]$. By Theorem 2, the radii of convergence of f' and f'' are both 1, so we need only check the

endpoints. $f(x) = \sum\limits_{n=1}^{\infty} \dfrac{x^n}{n^2} \;\Rightarrow\; f'(x) = \sum\limits_{n=1}^{\infty} \dfrac{nx^{n-1}}{n^2} = \sum\limits_{n=0}^{\infty} \dfrac{x^n}{n+1}$, and this series diverges for $x = 1$ (harmonic series)

and converges for $x = -1$ (Alternating Series Test), so the interval of convergence is $[-1, 1)$. $f''(x) = \sum\limits_{n=1}^{\infty} \dfrac{nx^{n-1}}{n+1}$ diverges

at both 1 and -1 (Test for Divergence) since $\lim\limits_{n \to \infty} \dfrac{n}{n+1} = 1 \neq 0$, so its interval of convergence is $(-1, 1)$.

39. By Example 7, $\tan^{-1} x = \sum\limits_{n=0}^{\infty} (-1)^n \dfrac{x^{2n+1}}{2n+1}$ for $|x| < 1$. In particular, for $x = \dfrac{1}{\sqrt{3}}$,

we have $\dfrac{\pi}{6} = \tan^{-1}\left(\dfrac{1}{\sqrt{3}}\right) = \sum\limits_{n=0}^{\infty} (-1)^n \dfrac{\left(1/\sqrt{3}\right)^{2n+1}}{2n+1} = \sum\limits_{n=0}^{\infty} (-1)^n \left(\dfrac{1}{3}\right)^n \dfrac{1}{\sqrt{3}} \dfrac{1}{2n+1}$,

so $\pi = \dfrac{6}{\sqrt{3}} \sum\limits_{n=0}^{\infty} \dfrac{(-1)^n}{(2n+1)3^n} = 2\sqrt{3} \sum\limits_{n=0}^{\infty} \dfrac{(-1)^n}{(2n+1)3^n}$.

8.7 Taylor and Maclaurin Series

1. Using Theorem 5 with $\sum\limits_{n=0}^{\infty} b_n (x-5)^n$, $b_n = \dfrac{f^{(n)}(a)}{n!}$, so $b_8 = \dfrac{f^{(8)}(5)}{8!}$.

3. Since $f^{(n)}(0) = (n+1)!$, Equation 7 gives the Maclaurin series

$$\sum_{n=0}^{\infty} \dfrac{f^{(n)}(0)}{n!} x^n = \sum_{n=0}^{\infty} \dfrac{(n+1)!}{n!} x^n = \sum_{n=0}^{\infty} (n+1)x^n. \text{ Applying the Ratio Test with } a_n = (n+1)x^n \text{ gives us}$$

$\lim\limits_{n\to\infty} \left| \dfrac{a_{n+1}}{a_n} \right| = \lim\limits_{n\to\infty} \left| \dfrac{(n+2)x^{n+1}}{(n+1)x^n} \right| = |x| \lim\limits_{n\to\infty} \dfrac{n+2}{n+1} = |x| \cdot 1 = |x|$. For convergence, we must have $|x| < 1$, so the

radius of convergence $R = 1$.

5.

n	$f^{(n)}(x)$	$f^{(n)}(0)$
0	$\cos x$	1
1	$-\sin x$	0
2	$-\cos x$	-1
3	$\sin x$	0
4	$\cos x$	1
⋮	⋮	⋮

We use Equation 7 with $f(x) = \cos x$.

$$\cos x = f(0) + f'(0)x + \dfrac{f''(0)}{2!}x^2 + \dfrac{f^{(3)}(0)}{3!}x^3 + \dfrac{f^{(4)}(0)}{4!}x^4 + \cdots$$

$$= 1 - \dfrac{x^2}{2!} + \dfrac{x^4}{4!} - \cdots = \sum_{n=0}^{\infty} \dfrac{(-1)^n x^{2n}}{(2n)!}$$

If $a_n = \dfrac{(-1)^n x^{2n}}{(2n)!}$, then

$$\lim_{n\to\infty} \left| \dfrac{a_{n+1}}{a_n} \right| = \lim_{n\to\infty} \dfrac{x^{2n+2}}{(2n+2)!} \cdot \dfrac{(2n)!}{x^{2n}} = x^2 \lim_{n\to\infty} \dfrac{1}{(2n+2)(2n+1)}$$

$$= 0 < 1 \text{ for all } x.$$

So $R = \infty$ (Ratio Test).

7.

n	$f^{(n)}(x)$	$f^{(n)}(0)$
0	e^{5x}	1
1	$5e^{5x}$	5
2	$5^2 e^{5x}$	25
3	$5^3 e^{5x}$	125
4	$5^4 e^{5x}$	625
⋮	⋮	⋮

$$e^{5x} = \sum_{n=0}^{\infty} \dfrac{f^{(n)}(0)}{n!} x^n = \sum_{n=0}^{\infty} \dfrac{5^n}{n!} x^n$$

$$\lim_{n\to\infty} \left| \dfrac{a_{n+1}}{a_n} \right| = \lim_{n\to\infty} \left[\dfrac{5^{n+1} |x|^{n+1}}{(n+1)!} \cdot \dfrac{n!}{5^n |x|^n} \right] = \lim_{n\to\infty} \dfrac{5|x|}{n+1}$$

$$= 0 < 1 \text{ for all } x$$

So $R = \infty$.

9.

n	$f^{(n)}(x)$	$f^{(n)}(0)$
0	$\sinh x$	0
1	$\cosh x$	1
2	$\sinh x$	0
3	$\cosh x$	1
4	$\sinh x$	0
⋮	⋮	⋮

$$f^{(n)}(0) = \begin{cases} 0 & \text{if } n \text{ is even} \\ 1 & \text{if } n \text{ is odd} \end{cases} \quad \text{so } \sinh x = \sum_{n=0}^{\infty} \frac{x^{2n+1}}{(2n+1)!}.$$

Use the Ratio Test to find R. If $a_n = \dfrac{x^{2n+1}}{(2n+1)!}$, then

$$\lim_{n\to\infty}\left|\frac{a_{n+1}}{a_n}\right| = \lim_{n\to\infty}\left|\frac{x^{2n+3}}{(2n+3)!}\cdot\frac{(2n+1)!}{x^{2n+1}}\right| = x^2\cdot\lim_{n\to\infty}\frac{1}{(2n+3)(2n+2)}$$

$$= 0 < 1 \text{ for all } x$$

So $R = \infty$.

11.

n	$f^{(n)}(x)$	$f^{(n)}(2)$
0	$1 + x + x^2$	7
1	$1 + 2x$	5
2	2	2
3	0	0
4	0	0
⋮	⋮	⋮

$$f(x) = 7 + 5(x-2) + \frac{2}{2!}(x-2)^2 + \sum_{n=3}^{\infty}\frac{0}{n!}(x-2)^n$$

$$= 7 + 5(x-2) + (x-2)^2$$

Since $a_n = 0$ for large n, $R = \infty$.

13. Clearly, $f^{(n)}(x) = e^x$, so $f^{(n)}(3) = e^3$ and $e^x = \sum_{n=0}^{\infty}\dfrac{e^3}{n!}(x-3)^n$. If $a_n = \dfrac{e^3}{n!}(x-3)^n$, then

$$\lim_{n\to\infty}\left|\frac{a_{n+1}}{a_n}\right| = \lim_{n\to\infty}\left|\frac{e^3(x-3)^{n+1}}{(n+1)!}\cdot\frac{n!}{e^3(x-3)^n}\right| = \lim_{n\to\infty}\frac{|x-3|}{n+1} = 0 < 1 \text{ for all } x, \text{ so } R = \infty.$$

15.

n	$f^{(n)}(x)$	$f^{(n)}(\pi)$
0	$\cos x$	-1
1	$-\sin x$	0
2	$-\cos x$	1
3	$\sin x$	0
4	$\cos x$	-1
⋮	⋮	⋮

$$\cos x = \sum_{k=0}^{\infty}\frac{f^{(k)}(\pi)}{k!}(x-\pi)^k = -1 + \frac{(x-\pi)^2}{2!} - \frac{(x-\pi)^4}{4!} + \frac{(x-\pi)^6}{6!} - \cdots$$

$$= \sum_{n=0}^{\infty}(-1)^{n+1}\frac{(x-\pi)^{2n}}{(2n)!}$$

$$\lim_{n\to\infty}\left|\frac{a_{n+1}}{a_n}\right| = \lim_{n\to\infty}\left[\frac{|x-\pi|^{2n+2}}{(2n+2)!}\cdot\frac{(2n)!}{|x-\pi|^{2n}}\right] = \lim_{n\to\infty}\frac{|x-\pi|^2}{(2n+2)(2n+1)}$$

$$= 0 < 1 \text{ for all } x$$

So $R = \infty$.

17.

n	$f^{(n)}(x)$	$f^{(n)}(9)$
0	$x^{-1/2}$	$\frac{1}{3}$
1	$-\frac{1}{2}x^{-3/2}$	$-\frac{1}{2}\cdot\frac{1}{3^3}$
2	$\frac{3}{4}x^{-5/2}$	$-\frac{1}{2}\cdot\left(-\frac{3}{2}\right)\cdot\frac{1}{3^5}$
3	$-\frac{15}{8}x^{-7/2}$	$-\frac{1}{2}\cdot\left(-\frac{3}{2}\right)\cdot\left(-\frac{5}{2}\right)\cdot\frac{1}{3^7}$
$\vdots$	$\vdots$	$\vdots$

$$\frac{1}{\sqrt{x}} = \frac{1}{3} - \frac{1}{2\cdot 3^3}(x-9) + \frac{3}{2^2\cdot 3^5}\frac{(x-9)^2}{2!} - \frac{3\cdot 5}{2^3\cdot 3^7}\frac{(x-9)^3}{3!} + \cdots$$

$$= \frac{1}{3} + \sum_{n=0}^{\infty}(-1)^n\frac{1\cdot 3\cdot 5\cdot\;\cdots\;\cdot(2n-1)}{2^n\cdot 3^{2n+1}\cdot n!}(x-9)^n.$$

$$\lim_{n\to\infty}\left|\frac{a_{n+1}}{a_n}\right| = \lim_{n\to\infty}\left[\frac{1\cdot 3\cdot 5\cdot\;\cdots\;\cdot(2n-1)[2(n+1)-1]\,|x-9|^{n+1}}{2^{n+1}\cdot 3^{[2(n+1)+1]}\cdot(n+1)!}\cdot\frac{2^n\cdot 3^{2n+1}\cdot n!}{1\cdot 3\cdot 5\cdot\;\cdots\;\cdot(2n-1)\,|x-9|^n}\right]$$

$$= \lim_{n\to\infty}\left[\frac{(2n+1)\,|x-9|}{2\cdot 3^2(n+1)}\right] = \frac{1}{9}|x-9| < 1 \ \text{ for convergence, so } |x-9| < 9 \text{ and } R = 9.$$

19. If $f(x) = \cos x$, then by Taylor's Formula $R_n(x) = \dfrac{f^{(n+1)}(z)}{(n+1)!}x^{n+1}$, where $0 < |z| < |x|$. But $f^{(n+1)}(z) = \pm\sin z$ or

$\pm\cos z$. In each case, $\left|f^{(n+1)}(z)\right| \le 1$, so $|R_n(x)| \le \dfrac{1}{(n+1)!}|x|^{n+1}$. Thus, $|R_n(x)| \to 0$ as $n\to\infty$ by Equation 11.

So $\lim_{n\to\infty} R_n(x) = 0$ and, by Theorem 8, the series in Exercise 5 represents $\cos x$ for all x.

21. If $f(x) = \sinh x$, then by Taylor's Formula $R_n(x) = \dfrac{f^{(n+1)}(z)}{(n+1)!}x^{n+1}$, where $0 < |z| < |x|$. But for all n,

$\left|f^{(n+1)}(z)\right| \le \cosh z \le \cosh x$ (because all derivatives are either sinh or cosh, $|\sinh z| < |\cosh z|$ for all z, and

$|z| < |x| \Rightarrow \cosh z < \cosh x$). So $|R_n(x)| \le \dfrac{\cosh x}{(n+1)!}x^{n+1} \to 0$ as $n\to\infty$ by Equation 11. So by Theorem 8, the series

represents $\sinh x$ for all x.

23. The general binomial series in (18) is

$$(1+x)^k = \sum_{n=0}^{\infty}\binom{k}{n}x^n = 1 + kx + \frac{k(k-1)}{2!}x^2 + \frac{k(k-1)(k-2)}{3!}x^3 + \cdots.$$

$$(1+x)^{1/2} = \sum_{n=0}^{\infty}\binom{\frac{1}{2}}{n}x^n = 1 + \left(\tfrac{1}{2}\right)x + \frac{\left(\frac{1}{2}\right)\left(-\frac{1}{2}\right)}{2!}x^2 + \frac{\left(\frac{1}{2}\right)\left(-\frac{1}{2}\right)\left(-\frac{3}{2}\right)}{3!}x^3 + \cdots$$

$$= 1 + \frac{x}{2} - \frac{x^2}{2^2\cdot 2!} + \frac{1\cdot 3\cdot x^3}{2^3\cdot 3!} - \frac{1\cdot 3\cdot 5\cdot x^4}{2^4\cdot 4!} + \cdots$$

$$= 1 + \frac{x}{2} + \sum_{n=2}^{\infty}\frac{(-1)^{n-1}\,1\cdot 3\cdot 5\cdot\;\cdots\;\cdot(2n-3)x^n}{2^n\cdot n!} \ \text{ for } |x| < 1, \quad \text{so } R = 1.$$

25. $\dfrac{1}{(2+x)^3} = \dfrac{1}{[2(1+x/2)]^3} = \dfrac{1}{8}\left(1+\dfrac{x}{2}\right)^{-3} = \dfrac{1}{8}\sum\limits_{n=0}^{\infty}\binom{-3}{n}\left(\dfrac{x}{2}\right)^{n}$. The binomial coefficient is

$$\binom{-3}{n} = \frac{(-3)(-4)(-5)\cdots\cdots(-3-n+1)}{n!} = \frac{(-3)(-4)(-5)\cdots\cdots[-(n+2)]}{n!}$$

$$= \frac{(-1)^n \cdot 2\cdot 3\cdot 4\cdot 5\cdots\cdots(n+1)(n+2)}{2\cdot n!} = \frac{(-1)^n(n+1)(n+2)}{2}$$

Thus, $\dfrac{1}{(2+x)^3} = \dfrac{1}{8}\sum\limits_{n=0}^{\infty}\dfrac{(-1)^n(n+1)(n+2)}{2}\dfrac{x^n}{2^n} = \sum\limits_{n=0}^{\infty}\dfrac{(-1)^n(n+1)(n+2)x^n}{2^{n+4}}$ for $\left|\dfrac{x}{2}\right| < 1 \iff |x| < 2$, so $R = 2$.

27. $\cos x = \sum\limits_{n=0}^{\infty}(-1)^n\dfrac{x^{2n}}{(2n)!} \quad\Rightarrow\quad f(x) = \cos(\pi x) = \sum\limits_{n=0}^{\infty}\dfrac{(-1)^n(\pi x)^{2n}}{(2n)!} = \sum\limits_{n=0}^{\infty}\dfrac{(-1)^n\pi^{2n}x^{2n}}{(2n)!}, \quad R = \infty$

29. $\tan^{-1}x = \sum\limits_{n=0}^{\infty}(-1)^n\dfrac{x^{2n+1}}{2n+1} \quad\Rightarrow\quad f(x) = x\tan^{-1}x = x\sum\limits_{n=0}^{\infty}(-1)^n\dfrac{x^{2n+1}}{2n+1} = \sum\limits_{n=0}^{\infty}(-1)^n\dfrac{x^{2n+2}}{2n+1}, \quad R = 1$

31. $e^x = \sum\limits_{n=0}^{\infty}\dfrac{x^n}{n!} \quad\Rightarrow\quad f(x) = x^2 e^{-x} = x^2\sum\limits_{n=0}^{\infty}\dfrac{(-x)^n}{n!} = \sum\limits_{n=0}^{\infty}\dfrac{(-1)^n x^{n+2}}{n!}, \quad R = \infty$

33. We must write the binomial in the form $(1+$ expression$)$, so we'll factor out a 4.

$$\frac{x}{\sqrt{4+x^2}} = \frac{x}{\sqrt{4(1+x^2/4)}} = \frac{x}{2\sqrt{1+x^2/4}} = \frac{x}{2}\left(1+\frac{x^2}{4}\right)^{-1/2} = \frac{x}{2}\sum_{n=0}^{\infty}\binom{-\frac{1}{2}}{n}\left(\frac{x^2}{4}\right)^{n}$$

$$= \frac{x}{2}\left[1 + \left(-\tfrac{1}{2}\right)\frac{x^2}{4} + \frac{\left(-\frac{1}{2}\right)\left(-\frac{3}{2}\right)}{2!}\left(\frac{x^2}{4}\right)^2 + \frac{\left(-\frac{1}{2}\right)\left(-\frac{3}{2}\right)\left(-\frac{5}{2}\right)}{3!}\left(\frac{x^2}{4}\right)^3 + \cdots\right]$$

$$= \frac{x}{2} + \frac{x}{2}\sum_{n=1}^{\infty}(-1)^n\frac{1\cdot 3\cdot 5\cdots\cdots(2n-1)}{2^n\cdot 4^n\cdot n!}x^{2n}$$

$$= \frac{x}{2} + \sum_{n=1}^{\infty}(-1)^n\frac{1\cdot 3\cdot 5\cdots\cdots(2n-1)}{n!\,2^{3n+1}}x^{2n+1} \text{ and } \frac{x^2}{4} < 1 \iff \frac{|x|}{2} < 1 \iff |x| < 2, \quad \text{so } R = 2.$$

35. $\sin^2 x = \dfrac{1}{2}(1-\cos 2x) = \dfrac{1}{2}\left[1 - \sum\limits_{n=0}^{\infty}\dfrac{(-1)^n(2x)^{2n}}{(2n)!}\right] = \dfrac{1}{2}\left[1 - 1 - \sum\limits_{n=1}^{\infty}\dfrac{(-1)^n(2x)^{2n}}{(2n)!}\right]$

$$= \sum_{n=1}^{\infty}\frac{(-1)^{n+1}2^{2n-1}x^{2n}}{(2n)!}, \quad R = \infty$$

37. $\cos x = \sum\limits_{n=0}^{\infty}(-1)^n\dfrac{x^{2n}}{(2n)!} \quad\Rightarrow$

$$f(x) = \cos(x^2) = \sum_{n=0}^{\infty}\frac{(-1)^n\left(x^2\right)^{2n}}{(2n)!} = \sum_{n=0}^{\infty}\frac{(-1)^n x^{4n}}{(2n)!}, \quad R = \infty$$

Notice that, as n increases, $T_n(x)$ becomes a better approximation

to $f(x)$.

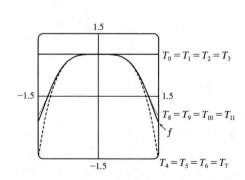

39. $e^x = \sum\limits_{n=0}^{\infty} \dfrac{x^n}{n!}$, so $e^{-0.2} = \sum\limits_{n=0}^{\infty} \dfrac{(-0.2)^n}{n!} = 1 - 0.2 + \dfrac{1}{2!}(0.2)^2 - \dfrac{1}{3!}(0.2)^3 + \dfrac{1}{4!}(0.2)^4 - \dfrac{1}{5!}(0.2)^5 + \dfrac{1}{6!}(0.2)^6 - \cdots$.

But $\dfrac{1}{6!}(0.2)^6 = 8.\overline{8} \times 10^{-8}$, so by the Alternating Series Estimation Theorem, $e^{-0.2} \approx \sum\limits_{n=0}^{5} \dfrac{(-0.2)^n}{n!} \approx 0.81873$, correct to

five decimal places.

41. (a) $1/\sqrt{1-x^2} = \left[1 + \left(-x^2\right)\right]^{-1/2} = 1 + \left(-\tfrac{1}{2}\right)\left(-x^2\right) + \dfrac{\left(-\tfrac{1}{2}\right)\left(-\tfrac{3}{2}\right)}{2!}\left(-x^2\right)^2 + \dfrac{\left(-\tfrac{1}{2}\right)\left(-\tfrac{3}{2}\right)\left(-\tfrac{5}{2}\right)}{3!}\left(-x^2\right)^3 + \cdots$

$$= 1 + \sum\limits_{n=1}^{\infty} \dfrac{1 \cdot 3 \cdot 5 \cdots \cdots (2n-1)}{2^n \cdot n!} x^{2n}$$

(b) $\sin^{-1} x = \displaystyle\int \dfrac{1}{\sqrt{1-x^2}}\, dx = C + x + \sum\limits_{n=1}^{\infty} \dfrac{1 \cdot 3 \cdot 5 \cdots \cdots (2n-1)}{(2n+1)2^n \cdot n!} x^{2n+1}$

$$= x + \sum\limits_{n=1}^{\infty} \dfrac{1 \cdot 3 \cdot 5 \cdots \cdots (2n-1)}{(2n+1)2^n \cdot n!} x^{2n+1} \quad \text{since } 0 = \sin^{-1} 0 = C.$$

43. $\cos x \overset{(17)}{=} \sum\limits_{n=0}^{\infty} (-1)^n \dfrac{x^{2n}}{(2n)!} \quad \Rightarrow \quad \cos(x^3) = \sum\limits_{n=0}^{\infty} (-1)^n \dfrac{(x^3)^{2n}}{(2n)!} = \sum\limits_{n=0}^{\infty} (-1)^n \dfrac{x^{6n}}{(2n)!} \quad \Rightarrow$

$x \cos(x^3) = \sum\limits_{n=0}^{\infty} (-1)^n \dfrac{x^{6n+1}}{(2n)!} \quad \Rightarrow \quad \displaystyle\int x \cos(x^3)\, dx = C + \sum\limits_{n=0}^{\infty} (-1)^n \dfrac{x^{6n+2}}{(6n+2)(2n)!}$, with $R = \infty$.

45. Using the series from Exercise 23 and substituting x^3 for x, we get

$$\int \sqrt{x^3 + 1}\, dx = \int \left[1 + \dfrac{x^3}{2} + \sum\limits_{n=2}^{\infty} \dfrac{(-1)^{n-1} 1 \cdot 3 \cdot 5 \cdots \cdots (2n-3)}{2^n n!} x^{3n}\right] dx$$

$$= C + x + \dfrac{x^4}{8} + \sum\limits_{n=2}^{\infty} \dfrac{(-1)^{n-1} 1 \cdot 3 \cdot 5 \cdots \cdots (2n-3)}{2^n n!(3n+1)} x^{3n+1}$$

47. By Exercise 43, $\displaystyle\int x \cos(x^3)\, dx = C + \sum\limits_{n=0}^{\infty} (-1)^n \dfrac{x^{6n+2}}{(6n+2)(2n)!}$, so

$$\int_0^1 x \cos(x^3)\, dx = \left[\sum\limits_{n=0}^{\infty} (-1)^n \dfrac{x^{6n+2}}{(6n+2)(2n)!}\right]_0^1 = \sum\limits_{n=0}^{\infty} \dfrac{(-1)^n}{(6n+2)(2n)!} = \dfrac{1}{2} - \dfrac{1}{8 \cdot 2!} + \dfrac{1}{14 \cdot 4!} - \dfrac{1}{20 \cdot 6!} + \cdots, \text{ but}$$

$\dfrac{1}{20 \cdot 6!} = \dfrac{1}{14{,}400} \approx 0.000\,069$, so $\displaystyle\int_0^1 x \cos(x^3)\, dx \approx \dfrac{1}{2} - \dfrac{1}{16} + \dfrac{1}{336} \approx 0.440$ (correct to three decimal places) by the

Alternating Series Estimation Theorem.

49. We first find a series representation for $f(x) = (1+x)^{-1/2}$, and

then substitute.

$\dfrac{1}{\sqrt{1+x}} = 1 - \dfrac{x}{2} + \dfrac{3}{4}\left(\dfrac{x^2}{2!}\right) - \dfrac{15}{8}\left(\dfrac{x^3}{3!}\right) + \cdots \quad \Rightarrow$

$\dfrac{1}{\sqrt{1+x^3}} = 1 - \dfrac{1}{2}x^3 + \dfrac{3}{8}x^6 - \dfrac{5}{16}x^9 + \cdots \quad \Rightarrow$

n	$f^{(n)}(x)$	$f^{(n)}(0)$
0	$(1+x)^{-1/2}$	1
1	$-\tfrac{1}{2}(1+x)^{-3/2}$	$-\tfrac{1}{2}$
2	$\tfrac{3}{4}(1+x)^{-5/2}$	$\tfrac{3}{4}$
3	$-\tfrac{15}{8}(1+x)^{-7/2}$	$-\tfrac{15}{8}$
$\vdots$	$\vdots$	$\vdots$

$$\int_0^{0.1} \frac{dx}{\sqrt{1+x^3}} = \left[x - \frac{1}{8}x^4 + \frac{3}{56}x^7 - \frac{1}{32}x^{10} + \cdots \right]_0^{0.1} \approx (0.1) - \frac{1}{8}(0.1)^4, \text{ by the Alternating Series Estimation}$$

Theorem, since $\frac{3}{56}(0.1)^7 \approx 0.000\,000\,005\,4 < 10^{-8}$, which is the maximum desired error. Therefore,

$$\int_0^{0.1} \frac{dx}{\sqrt{1+x^3}} \approx 0.099\,987\,50.$$

51. $\lim\limits_{x \to 0} \dfrac{x - \tan^{-1} x}{x^3} = \lim\limits_{x \to 0} \dfrac{x - \left(x - \frac{1}{3}x^3 + \frac{1}{5}x^5 - \frac{1}{7}x^7 + \cdots\right)}{x^3} = \lim\limits_{x \to 0} \dfrac{\frac{1}{3}x^3 - \frac{1}{5}x^5 + \frac{1}{7}x^7 - \cdots}{x^3}$

$$= \lim\limits_{x \to 0} \left(\frac{1}{3} - \frac{1}{5}x^2 + \frac{1}{7}x^4 - \cdots \right) = \frac{1}{3}$$

since power series are continuous functions.

53. $\lim\limits_{x \to 0} \dfrac{\sin x - x + \frac{1}{6}x^3}{x^5} = \lim\limits_{x \to 0} \dfrac{\left(x - \frac{1}{3!}x^3 + \frac{1}{5!}x^5 - \frac{1}{7!}x^7 + \cdots\right) - x + \frac{1}{6}x^3}{x^5}$

$$= \lim\limits_{x \to 0} \frac{\frac{1}{5!}x^5 - \frac{1}{7!}x^7 + \cdots}{x^5} = \lim\limits_{x \to 0} \left(\frac{1}{5!} - \frac{x^2}{7!} + \frac{x^4}{9!} - \cdots \right) = \frac{1}{5!} = \frac{1}{120}$$

since power series are continuous functions.

55. As in Example 9(a), we have $e^{-x^2} = 1 - \dfrac{x^2}{1!} + \dfrac{x^4}{2!} - \dfrac{x^6}{3!} + \cdots$ and we know that $\cos x = 1 - \dfrac{x^2}{2!} + \dfrac{x^4}{4!} - \cdots$ from

Equation 17. Therefore, $e^{-x^2} \cos x = \left(1 - x^2 + \frac{1}{2}x^4 - \cdots\right)\left(1 - \frac{1}{2}x^2 + \frac{1}{24}x^4 - \cdots\right)$. Writing only the terms with

degree ≤ 4, we get $e^{-x^2} \cos x = 1 - \frac{1}{2}x^2 + \frac{1}{24}x^4 - x^2 + \frac{1}{2}x^4 + \frac{1}{2}x^4 + \cdots = 1 - \frac{3}{2}x^2 + \frac{25}{24}x^4 + \cdots$.

57. $\dfrac{x}{\sin x} \overset{(16)}{=} \dfrac{x}{x - \frac{1}{6}x^3 + \frac{1}{120}x^5 - \cdots}.$

$$
\begin{array}{r}
1 + \frac{1}{6}x^2 + \frac{7}{360}x^4 + \cdots \\
x - \frac{1}{6}x^3 + \frac{1}{120}x^5 - \cdots \overline{\smash{\big)}\ x } \\
x - \frac{1}{6}x^3 + \frac{1}{120}x^5 - \cdots \\
\hline
\frac{1}{6}x^3 - \frac{1}{120}x^5 + \cdots \\
\frac{1}{6}x^3 - \frac{1}{36}x^5 + \cdots \\
\hline
\frac{7}{360}x^5 + \cdots \\
\frac{7}{360}x^5 + \cdots \\
\hline
\cdots
\end{array}
$$

From the long division,

$$\frac{x}{\sin x} = 1 + \frac{1}{6}x^2 + \frac{7}{360}x^4 + \cdots.$$

59. $\sum\limits_{n=0}^{\infty} (-1)^n \dfrac{x^{4n}}{n!} = \sum\limits_{n=0}^{\infty} \dfrac{\left(-x^4\right)^n}{n!} = e^{-x^4}$, by (12).

61. $\sum\limits_{n=0}^{\infty} \dfrac{(-1)^n \pi^{2n+1}}{4^{2n+1}(2n+1)!} = \sum\limits_{n=0}^{\infty} \dfrac{(-1)^n \left(\frac{\pi}{4}\right)^{2n+1}}{(2n+1)!} = \sin\frac{\pi}{4} = \frac{1}{\sqrt{2}}$, by (16).

63. $3 + \dfrac{9}{2!} + \dfrac{27}{3!} + \dfrac{81}{4!} + \cdots = \dfrac{3^1}{1!} + \dfrac{3^2}{2!} + \dfrac{3^3}{3!} + \dfrac{3^4}{4!} + \cdots = \sum\limits_{n=1}^{\infty} \dfrac{3^n}{n!} = \sum\limits_{n=0}^{\infty} \dfrac{3^n}{n!} - 1 = e^3 - 1$, by (12).

65. (a) $[1+(-x)]^{-2} = 1 + (-2)(-x) + \dfrac{(-2)(-3)}{2!}(-x)^2 + \dfrac{(-2)(-3)(-4)}{3!}(-x)^3 + \cdots$

$$= 1 + 2x + 3x^2 + 4x^3 + \cdots = \sum_{n=0}^{\infty}(n+1)x^n,$$

so $\dfrac{x}{(1-x)^2} = x\sum_{n=0}^{\infty}(n+1)x^n = \sum_{n=0}^{\infty}(n+1)x^{n+1} = \sum_{n=1}^{\infty}nx^n.$

(b) With $x = \frac{1}{2}$ in part (a), we have $\sum_{n=1}^{\infty}n\left(\frac{1}{2}\right)^n = \sum_{n=1}^{\infty}\dfrac{n}{2^n} = \dfrac{\frac{1}{2}}{\left(1-\frac{1}{2}\right)^2} = \dfrac{\frac{1}{2}}{\frac{1}{4}} = 2.$

67. (a) $(1+x^2)^{1/2} = 1 + \left(\frac{1}{2}\right)x^2 + \dfrac{\left(\frac{1}{2}\right)\left(-\frac{1}{2}\right)}{2!}(x^2)^2 + \dfrac{\left(\frac{1}{2}\right)\left(-\frac{1}{2}\right)\left(-\frac{3}{2}\right)}{3!}(x^2)^3 + \cdots$

$$= 1 + \dfrac{x^2}{2} + \sum_{n=2}^{\infty}\dfrac{(-1)^{n-1}\,1\cdot3\cdot5\cdots(2n-3)}{2^n\cdot n!}x^{2n}$$

(b) The coefficient of x^{10} (corresponding to $n=5$) in the above Maclaurin series is $\dfrac{f^{(10)}(0)}{10!}$, so

$$\dfrac{f^{(10)}(0)}{10!} = \dfrac{(-1)^4\cdot1\cdot3\cdot5\cdot7}{2^5\cdot5!} \;\Rightarrow\; f^{(10)}(0) = 10!\left(\dfrac{1\cdot3\cdot5\cdot7}{2^5\cdot5!}\right) = 99{,}225.$$

69. (a) $g(x) = \sum_{n=0}^{\infty}\binom{k}{n}x^n \;\Rightarrow\; g'(x) = \sum_{n=1}^{\infty}\binom{k}{n}nx^{n-1}$, so

$$(1+x)g'(x) = (1+x)\sum_{n=1}^{\infty}\binom{k}{n}nx^{n-1} = \sum_{n=1}^{\infty}\binom{k}{n}nx^{n-1} + \sum_{n=1}^{\infty}\binom{k}{n}nx^n$$

$$= \sum_{n=0}^{\infty}\binom{k}{n+1}(n+1)x^n + \sum_{n=0}^{\infty}\binom{k}{n}nx^n \quad \left[\begin{array}{c}\text{Replace }n\text{ with }n+1\\\text{in the first series}\end{array}\right]$$

$$= \sum_{n=0}^{\infty}(n+1)\dfrac{k(k-1)(k-2)\cdots(k-n+1)(k-n)}{(n+1)!}x^n$$

$$+ \sum_{n=0}^{\infty}\left[(n)\dfrac{k(k-1)(k-2)\cdots(k-n+1)}{n!}\right]x^n$$

$$= \sum_{n=0}^{\infty}\dfrac{(n+1)k(k-1)(k-2)\cdots(k-n+1)}{(n+1)!}[(k-n)+n]\,x^n$$

$$= k\sum_{n=0}^{\infty}\dfrac{k(k-1)(k-2)\cdots(k-n+1)}{n!}x^n = k\sum_{n=0}^{\infty}\binom{k}{n}x^n = kg(x)$$

Thus, $g'(x) = \dfrac{kg(x)}{1+x}.$

(b) $h(x) = (1+x)^{-k}g(x) \;\Rightarrow$

$$h'(x) = -k(1+x)^{-k-1}g(x) + (1+x)^{-k}g'(x) \qquad \text{[Product Rule]}$$

$$= -k(1+x)^{-k-1}g(x) + (1+x)^{-k}\dfrac{kg(x)}{1+x} \qquad \text{[from part (a)]}$$

$$= -k(1+x)^{-k-1}g(x) + k(1+x)^{-k-1}g(x) = 0$$

(c) From part (b) we see that $h(x)$ must be constant for $x \in (-1,1)$, so $h(x) = h(0) = 1$ for $x \in (-1,1)$.

Thus, $h(x) = 1 = (1+x)^{-k}g(x) \;\Leftrightarrow\; g(x) = (1+x)^k$ for $x \in (-1,1)$.

8.8 Applications of Taylor Polynomials

1. (a)

n	$f^{(n)}(x)$	$f^{(n)}(0)$	$T_n(x)$
0	$\cos x$	1	1
1	$-\sin x$	0	1
2	$-\cos x$	-1	$1 - \frac{1}{2}x^2$
3	$\sin x$	0	$1 - \frac{1}{2}x^2$
4	$\cos x$	1	$1 - \frac{1}{2}x^2 + \frac{1}{24}x^4$
5	$-\sin x$	0	$1 - \frac{1}{2}x^2 + \frac{1}{24}x^4$
6	$-\cos x$	-1	$1 - \frac{1}{2}x^2 + \frac{1}{24}x^4 - \frac{1}{720}x^6$

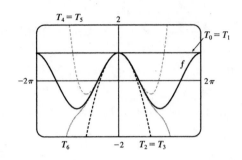

(b)

x	f	$T_0 = T_1$	$T_2 = T_3$	$T_4 = T_5$	T_6
$\frac{\pi}{4}$	0.7071	1	0.6916	0.7074	0.7071
$\frac{\pi}{2}$	0	1	-0.2337	0.0200	-0.0009
π	-1	1	-3.9348	0.1239	-1.2114

(c) As n increases, $T_n(x)$ is a good approximation to $f(x)$ on a larger and larger interval.

3.

n	$f^{(n)}(x)$	$f^{(n)}\left(\frac{\pi}{6}\right)$
0	$\sin x$	$\frac{1}{2}$
1	$\cos x$	$\frac{\sqrt{3}}{2}$
2	$-\sin x$	$-\frac{1}{2}$
3	$-\cos x$	$-\frac{\sqrt{3}}{2}$

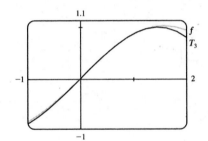

$$T_3(x) = \sum_{n=0}^{3} \frac{f^{(n)}\left(\frac{\pi}{6}\right)}{n!}\left(x - \frac{\pi}{6}\right)^n = \frac{1}{2} + \frac{\sqrt{3}}{2}\left(x - \frac{\pi}{6}\right) - \frac{1}{4}\left(x - \frac{\pi}{6}\right)^2 - \frac{\sqrt{3}}{12}\left(x - \frac{\pi}{6}\right)^3$$

5.

n	$f^{(n)}(x)$	$f^{(n)}(0)$
0	$\arcsin x$	0
1	$1/\sqrt{1 - x^2}$	1
2	$x/(1 - x^2)^{3/2}$	0
3	$(2x^2 + 1)/(1 - x^2)^{5/2}$	1

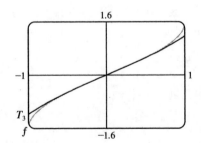

$$T_3(x) = \sum_{n=0}^{3} \frac{f^{(n)}(0)}{n!}x^n = x + \frac{x^3}{6}$$

7.

n	$f^{(n)}(x)$	$f^{(n)}(0)$
0	xe^{-2x}	0
1	$(1-2x)e^{-2x}$	1
2	$4(x-1)e^{-2x}$	-4
3	$4(3-2x)e^{-2x}$	12

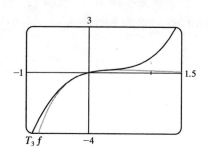

$T_3 f$

$$T_3(x) = \sum_{n=0}^{3} \frac{f^{(n)}(0)}{n!}x^n = \tfrac{0}{1}\cdot 1 + \tfrac{1}{1}x^1 + \tfrac{-4}{2}x^2 + \tfrac{12}{6}x^3 = x - 2x^2 + 2x^3$$

9. $f(x) = (1+x)^{1/2}$ $\qquad\qquad f(0) = 1$

$f'(x) = \tfrac{1}{2}(1+x)^{-1/2}$ $\qquad f'(0) = \tfrac{1}{2}$

$f''(x) = -\tfrac{1}{4}(1+x)^{-3/2}$

(a) $(1+x)^{1/2} \approx T_1(x) = 1 + \tfrac{1}{2}x$

(b) By Taylor's Formula, the remainder is $R_1(x) = \dfrac{f''(z)}{2!}x^2 = -\dfrac{1}{8(1+z)^{3/2}}x^2$, where z lies between 0 and x. Now

$0 \le x \le 0.1 \;\Rightarrow\; 0 \le x^2 \le 0.01$, and $0 < z < 0.1 \;\Rightarrow\; 1 < 1+z < 1.1$ so $|R_1(x)| < \tfrac{0.01}{8\cdot 1} = 0.00125$.

(c) From the graph of $|R_1(x)| = \left|\sqrt{1+x} - \left(1+\tfrac{1}{2}x\right)\right|$, it seems that

the error is at most 0.0013 on $(0, 0.1)$.

0.0015

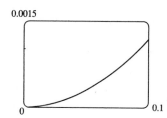

0 $\qquad\qquad$ 0.1

11.

n	$f^{(n)}(x)$	$f^{(n)}(0)$
0	$\tan x$	0
1	$\sec^2 x$	1
2	$2\sec^2 x \tan x$	0
3	$4\sec^2 x \tan^2 x + 2\sec^4 x$	2
4	$8\sec^2 x \tan^3 x + 16\sec^4 x \tan x$	

(a) $\tan x \approx T_3(x) = x + \tfrac{1}{3}x^3$

(b) The remainder is $R_3(x) = \dfrac{f^{(4)}(z)}{4!}x^4 = \dfrac{8\sec^2 z \tan^3 z + 16\sec^4 z \tan z}{4!}x^4 = \dfrac{\sec^2 z \tan^3 z + 2\sec^4 z \tan z}{3}x^4$

where z lies between 0 and x. Now $0 \le x^4 \le (\pi/6)^4$ and $0 < z < \pi/6 \;\Rightarrow\; \sec^2 z < \tfrac{4}{3}$ and $\tan z < \sqrt{3}/3$, so

$$|R_3(x)| \le \dfrac{\tfrac{4}{3}\left(\tfrac{1}{\sqrt{3}}\right)^3 + 2\left(\tfrac{2}{\sqrt{3}}\right)^4\left(\tfrac{1}{\sqrt{3}}\right)}{3}\left(\tfrac{\pi}{6}\right)^4 = \dfrac{4\sqrt{3}}{9}\left(\tfrac{\pi}{6}\right)^4 \approx 0.057859 < 0.06$$

(c) From the graph, it seems that the error is less than 0.006 on $(0, \pi)$.

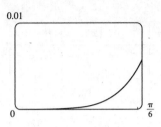

13.

n	$f^{(n)}(x)$	$f^{(n)}(0)$
0	e^{x^2}	1
1	$e^{x^2}(2x)$	0
2	$e^{x^2}(2 + 4x^2)$	2
3	$e^{x^2}(12x + 8x^3)$	0
4	$e^{x^2}(12 + 48x^2 + 16x^4)$	

(a) $f(x) = e^{x^2} \approx T_3(x) = 1 + \frac{2}{2!}x^2 = 1 + x^2$

(b) $R_3(x) = \dfrac{f^{(4)}(z)}{4!}x^4 = \dfrac{e^{z^2}(3 + 12z^2 + 4z^4)}{6}$, where z lies between 0 and x. Now $0 \le x \le 0.1 \quad \Rightarrow$

$|R_3(x)| \le \dfrac{e^{0.01}(3 + 0.12 + 0.0004)}{6}(0.1)^4 < 0.00006$.

(c) From the graph of $|R_3(x)| = \left| e^{x^2} - (1 + x^2) \right|$, it appears that the

error is less than 0.000051 on $[0, 0.1]$.

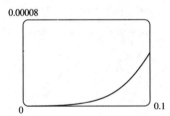

15. $f(x) = x^{3/4}$ $\qquad\qquad$ $f(16) = 8$

$\quad f'(x) = \frac{3}{4}x^{-1/4}$ $\qquad\qquad$ $f'(16) = \frac{3}{8}$

$\quad f''(x) = -\frac{3}{16}x^{-5/4}$ $\qquad\quad$ $f''(16) = -\frac{3}{512}$

$\quad f'''(x) = \frac{15}{64}x^{-9/4}$ $\qquad\quad$ $f'''(16) = \frac{15}{32,768}$

$\quad f^{(4)}(x) = -\frac{135}{256}x^{-13/4}$

(a) $x^{3/4} \approx T_3(x) = 8 + \frac{3}{8}(x - 16) - \frac{3}{1024}(x - 16)^2 + \frac{5}{65,536}(x - 16)^3$

(b) $R_3(x) \le \dfrac{-6z^{-4}}{4!}(x - 4)^4$, where z lies between x and 4. Now $3 \le x \le 5 \quad \Rightarrow \quad |x - 4| \le 1$ and $z > 3 \quad \Rightarrow$

$\dfrac{1}{z^4} < \dfrac{1}{3^4}$, so $|R_3(x)| < \dfrac{6}{4!\,3^4} 3 \le x \le 5 \quad \Rightarrow \quad |x - 4| \le 1$ and $z > 3 \quad \Rightarrow \quad \dfrac{1}{z^4} < \dfrac{1}{3^4}$, so

$|R_3(x)| < \dfrac{6}{4!\,3^4} = \dfrac{1}{324} \approx 0.0031$.

(c) It appears that the error is less than 3×10^{-6} on $(15, 17)$.

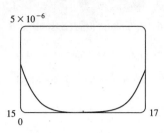

17. From Exercise 3, $\sin x = \frac{1}{2} + \frac{\sqrt{3}}{2}\left(x - \frac{\pi}{6}\right) - \frac{1}{4}\left(x - \frac{\pi}{6}\right)^2 - \frac{\sqrt{3}}{12}\left(x - \frac{\pi}{6}\right)^3 + R_3(x)$, where $|R_3(x)| \le \frac{1}{4!}\left|x - \frac{\pi}{6}\right|^4$ because

$\left|f^{(4)}(z)\right| = |\sin z| \le 1$. Now $x = 35° = (30° + 5°) = \left(\frac{\pi}{6} + \frac{\pi}{36}\right)$ radians, so the error is $\left|R_3\left(\frac{\pi}{36}\right)\right| \le \frac{\left(\frac{\pi}{36}\right)^4}{4!} < 0.000003$.

Therefore, to five decimal places, $\sin 35° \approx \frac{1}{2} + \frac{\sqrt{3}}{2}\left(\frac{\pi}{36}\right) - \frac{1}{4}\left(\frac{\pi}{36}\right)^2 - \frac{\sqrt{3}}{12}\left(\frac{\pi}{36}\right)^3 \approx 0.57358$.

19. All derivatives of e^x are e^x, so the remainder term is $R_n(x) = \frac{e^z}{(n+1)!} x^{n+1}$, where $0 < z < 0.1$. So we want

$R_n(0.1) \le \frac{e^{0.1}}{(n+1)!} (0.1)^{n+1} < 0.00001$, and we find that $n = 3$ satisfies this inequality. [In fact $R_3(0.1) < 0.0000046$.]

21. $\sin x = x - \frac{1}{3!}x^3 + \frac{1}{5!}x^5 - \cdots$. By the Alternating Series

Estimation Theorem, the error in the approximation

$\sin x = x - \frac{1}{3!}x^3$ is less than $\left|\frac{1}{5!}x^5\right| < 0.01$ ⇔

$\left|x^5\right| < 120(0.01)$ ⇔ $|x| < (1.2)^{1/5} \approx 1.037$. The curves

$y = x - \frac{1}{6}x^3$ and $y = \sin x - 0.01$ intersect at $x \approx 1.043$, so

the graph confirms our estimate. Since both the sine function

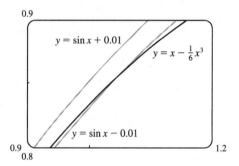

and the given approximation are odd functions, we need to check the estimate only for $x > 0$. Thus, the desired range of

values for x is $-1.037 < x < 1.037$.

23. Let $s(t)$ be the position function of the car, and for convenience set $s(0) = 0$. The velocity of the car is $v(t) = s'(t)$ and the

acceleration is $a(t) = s''(t)$, so the second degree Taylor polynomial is $T_2(t) = s(0) + v(0)t + \frac{a(0)}{2}t^2 = 20t + t^2$. We

estimate the distance traveled during the next second to be $s(1) \approx T_2(1) = 20 + 1 = 21$ m. The function $T_2(t)$ would not be

accurate over a full minute, since the car could not possibly maintain an acceleration of 2 m/s^2 for that long (if it did, its final

speed would be $140 \text{ m/s} \approx 313 \text{ mi/h}$!)

25. $E = \frac{q}{D^2} - \frac{q}{(D+d)^2} = \frac{q}{D^2} - \frac{q}{D^2(1 + d/D)^2} = \frac{q}{D^2}\left[1 - \left(1 + \frac{d}{D}\right)^{-2}\right]$.

We use the Binomial Series to expand $(1 + d/D)^{-2}$:

$E = \frac{q}{D^2}\left[1 - \left(1 - 2\left(\frac{d}{D}\right) + \frac{2 \cdot 3}{2!}\left(\frac{d}{D}\right)^2 - \frac{2 \cdot 3 \cdot 4}{3!}\left(\frac{d}{D}\right)^3 + \cdots\right)\right] = \frac{q}{D^2}\left[2\left(\frac{d}{D}\right) - 3\left(\frac{d}{D}\right)^2 + 4\left(\frac{d}{D}\right)^3 - \cdots\right]$

$\approx \frac{q}{D^2} \cdot 2\left(\frac{d}{D}\right) = 2qd \cdot \frac{1}{D^3}$

when D is much larger than d; that is, when P is far away from the dipole.

27. (a) L is the length of the arc subtended by the angle θ, so $L = R\theta$ $\Rightarrow$

$\theta = L/R$. Now $\sec\theta = (R + C)/R$ $\Rightarrow$ $R\sec\theta = R + C$ $\Rightarrow$

$C = R\sec\theta - R = R\sec(L/R) - R$.

(b) If $f(x) = \sec x$, then $f'(x) = \sec x\ \tan x$,

$f''(x) = \sec^3 x + \sec x\ \tan^2 x$, $f'''(x) = 5\sec^3 x\ \tan x + \sec x\ \tan^3 x$,

$f^{(4)}(x) = 5\sec^5 x + 18\sec^3 x\ \tan^2 x + \sec x\ \tan^4 x$. So $f(0) = 1$, $f'(0) = 0$, $f''(0) = 1$, $f'''(0) = 0$,

$f^{(4)}(0) = 5$, and $\sec x \approx T_4(x) = 1 + \frac{1}{2}x^2 + \frac{5}{24}x^4$. By part (a),

$$C \approx R\left[1 + \frac{1}{2}\left(\frac{L}{R}\right)^2 + \frac{5}{24}\left(\frac{L}{R}\right)^4\right] - R = R + \frac{1}{2}R\cdot\frac{L^2}{R^2} + \frac{5}{24}R\cdot\frac{L^4}{R^4} - R = \frac{L^2}{2R} + \frac{5L^4}{24R^3}.$$

(c) Taking $L = 100$ km and $R = 6370$ km, the formula in part (a) says that

$C = R\sec(L/R) - R = 6370\sec(100/6370) - 6370 \approx 0.785\,009\,965\,44$ km. The formula in part (b) says that

$$C \approx \frac{L^2}{2R} + \frac{5L^4}{24R^3} = \frac{100^2}{2\cdot 6370} + \frac{5\cdot 100^4}{24\cdot 6370^3} \approx 0.785\,009\,957\,36\text{ km.}$$

The difference between these two results is only $0.000\,000\,008\,08$ km, or $0.000\,008\,08$ m!

29. Using Taylor's Formula with $n = 1$, $a = x_n$, $x = r$, we get $f(r) = f(x_n) + f'(x_n)(r - x_n) + R_1(x)$, where

$R_1(x) = \frac{1}{2}f''(z)(r - x_n)^2$ and lies betwen x_n and r. But r is a root, so $f(r) = 0$ and Taylor's Formula becomes

$0 = f(x_n) + f'(x_n)(r - x_n) + \frac{1}{2}f''(z)(r - x_n)^2$. Taking the first two terms to the left side and dividing by $f'(x_n)$, we have

$x_n - r - \dfrac{f(x_n)}{f'(x_n)} = \dfrac{1}{2}\dfrac{f''(z)}{f'(x_n)}\,|x_n - r|^2$. By the formula for Newton's Method, we have

$|x_{n+1} - r| = \left|x_n - \dfrac{f(x_n)}{f'(x_n)} - r\right| = \dfrac{1}{2}\dfrac{|f''(z)|}{|f'(x_n)|}\,|x_n - r|^2 \le \dfrac{M}{2K}\,|x_n - r|^2$ since $|f''(z)| \le M$ and $|f'(x_n)| \ge K$.

8 Review

CONCEPT CHECK

1. (a) See Definition 8.1.1.

(b) See Definition 8.2.2.

(c) The terms of the sequence $\{a_n\}$ approach 3 as n becomes large.

(d) By adding sufficiently many terms of the series, we can make the partial sums as close to 3 as we like.

2. (a) A sequence $\{a_n\}$ is bounded if there are numbers m and M such that $m \le a_n \le M$ for all $n \ge 1$.

(b) A sequence is monotonic if it is either increasing or decreasing.

(c) By Theorem 8.1.11, every bounded, monotonic sequence is convergent.

3. (a) See (4) in Section 8.2.

(b) The p-series $\displaystyle\sum_{n=1}^{\infty}\frac{1}{n^p}$ is convergent if $p > 1$.

4. If $\sum a_n = 3$, then $\lim\limits_{n\to\infty} a_n = 0$ and $\lim\limits_{n\to\infty} s_n = 3$.

5. (a) See the Test for Divergence on page 425.

(b) See the Integral Test on page 431.

(c) See the Comparison Test on page 433.

(d) See the Limit Comparison Test on page 434.

(e) See the Alternating Series Test on page 438.

(f) See the Ratio Test on page 443.

(g) See the Root Test on page 445.

6. (a) A series $\sum a_n$ is called *absolutely convergent* if the series of absolute values $\sum |a_n|$ is convergent.

(b) If a series $\sum a_n$ is absolutely convergent, then it is convergent.

(c) A series $\sum a_n$ is called *conditionally convergent* if it is convergent but not absolutely convergent.

7. By adding terms until you reach the desired accuracy given by the Alternating Series Estimation Theorem on page 440.

8. (a) $\sum\limits_{n=0}^{\infty} c_n(x-a)^n$

(b) Given the power series $\sum\limits_{n=0}^{\infty} c_n(x-a)^n$, the radius of convergence is:

(i) 0 if the series converges only when $x = a$

(ii) ∞ if the series converges for all x, or

(iii) a positive number R such that the series converges if $|x-a| < R$ and diverges if $|x-a| > R$.

(c) The interval of convergence of a power series is the interval that consists of all values of x for which the series converges. Corresponding to the cases in part (b), the interval of convergence is: (i) the single point $\{a\}$, (ii) all real numbers, that is, the real number line $(-\infty, \infty)$, or (iii) an interval with endpoints $a - R$ and $a + R$ which can contain neither, either, or both of the endpoints. In this case, we must test the series for convergence at each endpoint to determine the interval of convergence.

9. (a), (b) See Theorem 8.6.2.

10. (a) $T_n(x) = \sum\limits_{i=0}^{n} \dfrac{f^{(i)}(a)}{i!}(x-a)^i$

(b) $\sum\limits_{n=0}^{\infty} \dfrac{f^{(n)}(a)}{n!}(x-a)^n$

(c) $\sum\limits_{n=0}^{\infty} \dfrac{f^{(n)}(0)}{n!}x^n$ [$a = 0$ in part (b)]

(d) See Theorem 8.7.8.

(e) See Taylor's Formula.

11. (a) – (e) See the table on page 466.

12. See the Binomial Series (8.7.18) for the expansion. The radius of convergence for the binomial series is 1.

TRUE-FALSE QUIZ

1. False.   See Note 2 on page 425.

3. True.   If $\lim_{n\to\infty} a_n = L$, then given any $\varepsilon > 0$, we can find a positive integer N such that $|a_n - L| < \varepsilon$ whenever $n > N$.

If $n > N$, then $2n + 1 > N$ and $|a_{2n+1} - L| < \varepsilon$. Thus, $\lim_{n\to\infty} a_{2n+1} = L$.

5. False.   For example, take $c_n = (-1)^n/(n6^n)$.

7. False, since $\lim_{n\to\infty}\left|\dfrac{a_{n+1}}{a_n}\right| = \lim_{n\to\infty}\left|\dfrac{1}{(n+1)^3}\cdot\dfrac{n^3}{1}\right| = \lim_{n\to\infty}\left|\dfrac{n^3}{(n+1)^3}\cdot\dfrac{1/n^3}{1/n^3}\right| = \lim_{n\to\infty}\dfrac{1}{(1+1/n)^3} = 1$.

9. False.   See the note after Example 4 in Section 8.3.

11. True.   See (8) in Section 8.1.

13. True.   By Theorem 8.7.5 the coefficient of x^3 is $\dfrac{f'''(0)}{3!} = \dfrac{1}{3} \;\Rightarrow\; f'''(0) = 2$.

Or: Use Theorem 8.6.2 to differentiate f three times.

15. False.   For example, let $a_n = b_n = (-1)^n$. Then $\{a_n\}$ and $\{b_n\}$ are divergent, but $a_n b_n = 1$, so $\{a_n b_n\}$ is convergent.

17. True by Theorem 8.4.1. $\left[\sum (-1)^n\, a_n \text{ is absolutely convergent and hence convergent.}\right]$

EXERCISES

1. $\left\{\dfrac{2+n^3}{1+2n^3}\right\}$ converges since $\lim_{n\to\infty}\dfrac{2+n^3}{1+2n^3} = \lim_{n\to\infty}\dfrac{2/n^3+1}{1/n^3+2} = \dfrac{1}{2}$.

3. $\lim_{n\to\infty} a_n = \lim_{n\to\infty}\dfrac{n^3}{1+n^2} = \lim_{n\to\infty}\dfrac{n}{1/n^2+1} = \infty$, so the sequence diverges.

5. $|a_n| = \left|\dfrac{n\sin n}{n^2+1}\right| \le \dfrac{n}{n^2+1} < \dfrac{1}{n}$, so $|a_n| \to 0$ as $n \to \infty$. Thus, $\lim_{n\to\infty} a_n = 0$. The sequence $\{a_n\}$ is convergent.

7. $\left\{\left(1+\dfrac{3}{n}\right)^{4n}\right\}$ is convergent. Let $y = \left(1+\dfrac{3}{x}\right)^{4x}$. Then

$$\lim_{x\to\infty}\ln y = \lim_{x\to\infty} 4x\ln(1+3/x) = \lim_{x\to\infty}\dfrac{\ln(1+3/x)}{1/(4x)} \overset{\text{H}}{=} \lim_{x\to\infty}\dfrac{\dfrac{1}{1+3/x}\left(-\dfrac{3}{x^2}\right)}{-1/(4x^2)} = \lim_{x\to\infty}\dfrac{12}{1+3/x} = 12$$

so $\lim_{x\to\infty} y = \lim_{n\to\infty}\left(1+\dfrac{3}{n}\right)^{4n} = e^{12}$.

9. $\dfrac{n}{n^3+1} < \dfrac{n}{n^3} = \dfrac{1}{n^2}$, so $\sum_{n=1}^{\infty}\dfrac{n}{n^3+1}$ converges by the Comparison Test with the convergent p-series $\sum_{n=1}^{\infty}\dfrac{1}{n^2}$ $[\,p=2>1\,]$.

11. $\lim\limits_{n\to\infty}\left|\dfrac{a_{n+1}}{a_n}\right| = \lim\limits_{n\to\infty}\left[\dfrac{(n+1)^3}{5^{n+1}}\cdot\dfrac{5^n}{n^3}\right] = \lim\limits_{n\to\infty}\left(1+\dfrac{1}{n}\right)^3\cdot\dfrac{1}{5} = \dfrac{1}{5} < 1$, so $\sum\limits_{n=1}^{\infty}\dfrac{n^3}{5^n}$ converges by the Ratio Test.

13. Let $f(x) = \dfrac{1}{x\sqrt{\ln x}}$. Then f is continuous, positive, and decreasing on $[2,\infty)$, so the Integral Test applies.

$$\int_2^{\infty} f(x)\,dx = \lim_{t\to\infty}\int_2^t \dfrac{1}{x\sqrt{\ln x}}\,dx \quad \begin{bmatrix} u=\ln x, \\ du=\frac{1}{x}\,dx \end{bmatrix} = \lim_{t\to\infty}\int_{\ln 2}^{\ln t} u^{-1/2}\,du$$

$$= \lim_{t\to\infty}\left[2\sqrt{u}\right]_{\ln 2}^{\ln t} = \lim_{t\to\infty}\left(2\sqrt{\ln t} - 2\sqrt{\ln 2}\right) = \infty, \text{ so the series } \sum_{n=2}^{\infty}\dfrac{1}{n\sqrt{\ln n}} \text{ diverges.}$$

15. $|a_n| = \left|\dfrac{\cos 3n}{1+(1.2)^n}\right| \leq \dfrac{1}{1+(1.2)^n} < \dfrac{1}{(1.2)^n} = \left(\dfrac{5}{6}\right)^n$, so $\sum\limits_{n=1}^{\infty}|a_n|$ converges by comparison with the convergent geometric

series $\sum\limits_{n=1}^{\infty}\left(\dfrac{5}{6}\right)^n$ $\left[r=\frac{5}{6}<1\right]$. It follows that $\sum\limits_{n=1}^{\infty} a_n$ converges (by Theorem 1 in Section 8.4).

17. $\lim\limits_{n\to\infty}\left|\dfrac{a_{n+1}}{a_n}\right| = \lim\limits_{n\to\infty}\dfrac{1\cdot3\cdot5\cdots\cdots(2n-1)(2n+1)}{5^{n+1}\,(n+1)!}\cdot\dfrac{5^n n!}{1\cdot3\cdot5\cdots\cdots(2n-1)} = \lim\limits_{n\to\infty}\dfrac{2n+1}{5(n+1)} = \dfrac{2}{5} < 1$, so the series

converges by the Ratio Test.

19. $b_n = \dfrac{\sqrt{n}}{n+1} > 0$, $\{b_n\}$ is decreasing, and $\lim\limits_{n\to\infty} b_n = 0$, so the series $\sum\limits_{n=1}^{\infty}(-1)^{n-1}\dfrac{\sqrt{n}}{n+1}$ converges by the Alternating Series

Test.

21. Consider the series of absolute values: $\sum\limits_{n=1}^{\infty} n^{-1/3}$ is a p-series with $p=\frac{1}{3}\leq 1$ and is therefore divergent. But if we apply the

Alternating Series Test, we see that $b_n = \dfrac{1}{\sqrt[3]{n}} > 0$, $\{b_n\}$ is decreasing, and $\lim\limits_{n\to\infty} b_n = 0$, so the series $\sum\limits_{n=1}^{\infty}(-1)^{n-1}n^{-1/3}$

converges. Thus, $\sum\limits_{n=1}^{\infty}(-1)^{n-1}n^{-1/3}$ is conditionally convergent.

23. $\left|\dfrac{a_{n+1}}{a_n}\right| = \left|\dfrac{(-1)^{n+1}\,(n+2)\,3^{n+1}}{2^{2n+3}}\cdot\dfrac{2^{2n+1}}{(-1)^n\,(n+1)\,3^n}\right| = \dfrac{n+2}{n+1}\cdot\dfrac{3}{4} = \dfrac{1+(2/n)}{1+(1/n)}\cdot\dfrac{3}{4} \to \dfrac{3}{4} < 1$ as $n\to\infty$, so by the

Ratio Test, $\sum\limits_{n=1}^{\infty}\dfrac{(-1)^n\,(n+1)\,3^n}{2^{2n+1}}$ is absolutely convergent.

25. $\dfrac{2^{2n+1}}{5^n} = \dfrac{2^{2n}\cdot2^1}{5^n} = \dfrac{(2^2)^n\cdot2}{5^n} = 2\left(\dfrac{4}{5}\right)^n$, so $\sum\limits_{n=1}^{\infty}\dfrac{2^{2n+1}}{5^n} = 2\sum\limits_{n=1}^{\infty}\left(\dfrac{4}{5}\right)^n$ is a geometric series with $a=\dfrac{8}{5}$ and $r=\dfrac{4}{5}$.

Since $|r| = \dfrac{4}{5} < 1$, the series converges to $\dfrac{a}{1-r} = \dfrac{8/5}{1-4/5} = \dfrac{8/5}{1/5} = 8$.

27. $\sum\limits_{n=1}^{\infty}\left[\tan^{-1}(n+1) - \tan^{-1} n\right] = \lim\limits_{n\to\infty} s_n$

$$= \lim_{n\to\infty}\left[(\tan^{-1}2 - \tan^{-1}1) + (\tan^{-1}3 - \tan^{-1}2) + \cdots\right.$$
$$+ \left.(\tan^{-1}(n+1) - \tan^{-1} n)\right]$$
$$= \lim_{n\to\infty}\left[\tan^{-1}(n+1) - \tan^{-1}1\right] = \dfrac{\pi}{2} - \dfrac{\pi}{4} = \dfrac{\pi}{4}$$

29. $1 - e + \dfrac{e^2}{2!} - \dfrac{e^3}{3!} + \dfrac{e^4}{4!} - \cdots = \sum\limits_{n=0}^{\infty} (-1)^n \dfrac{e^n}{n!} = \sum\limits_{n=0}^{\infty} \dfrac{(-e)^n}{n!} = e^{-e}$ since $e^x = \sum\limits_{n=0}^{\infty} \dfrac{x^n}{n!}$ for all x.

31. $\cosh x = \dfrac{1}{2}(e^x + e^{-x}) = \dfrac{1}{2}\left(\sum\limits_{n=0}^{\infty} \dfrac{x^n}{n!} + \sum\limits_{n=0}^{\infty} \dfrac{(-x)^n}{n!} \right)$

$$= \dfrac{1}{2}\left[\left(1 + x + \dfrac{x^2}{2!} + \dfrac{x^3}{3!} + \dfrac{x^4}{4!} + \cdots \right) + \left(1 - x + \dfrac{x^2}{2!} - \dfrac{x^3}{3!} + \dfrac{x^4}{4!} - \cdots \right) \right]$$

$$= \dfrac{1}{2}\left(2 + 2 \cdot \dfrac{x^2}{2!} + 2 \cdot \dfrac{x^4}{4!} + \cdots \right) = 1 + \dfrac{1}{2}x^2 + \sum\limits_{n=2}^{\infty} \dfrac{x^{2n}}{(2n)!}$$

$$\geq 1 + \dfrac{1}{2}x^2 \quad \text{for all } x$$

33. $\sum\limits_{n=1}^{\infty} \dfrac{(-1)^{n+1}}{n^5} = 1 - \dfrac{1}{32} + \dfrac{1}{243} - \dfrac{1}{1024} + \dfrac{1}{3125} - \dfrac{1}{7776} + \dfrac{1}{16,807} - \dfrac{1}{32,768} + \cdots .$

Since $b_8 = \dfrac{1}{8^5} = \dfrac{1}{32,768} < 0.000031$, $\sum\limits_{n=1}^{\infty} \dfrac{(-1)^{n+1}}{n^5} \approx \sum\limits_{n=1}^{7} \dfrac{(-1)^{n+1}}{n^5} \approx 0.9721.$

35. Use the Limit Comparison Test. $\lim\limits_{n\to\infty} \left| \dfrac{\left(\frac{n+1}{n}\right)a_n}{a_n} \right| = \lim\limits_{n\to\infty} \dfrac{n+1}{n} = \lim\limits_{n\to\infty} \left(1 + \dfrac{1}{n} \right) = 1 > 0.$

Since $\sum |a_n|$ is convergent, so is $\sum \left| \left(\dfrac{n+1}{n} \right) a_n \right|$, by the Limit Comparison Test.

37. $\lim\limits_{n\to\infty} \left| \dfrac{a_{n+1}}{a_n} \right| = \lim\limits_{n\to\infty} \left[\dfrac{|x+2|^{n+1}}{(n+1)\,4^{n+1}} \cdot \dfrac{n\,4^n}{|x+2|^n} \right] = \lim\limits_{n\to\infty} \left[\dfrac{n}{n+1}\dfrac{|x+2|}{4} \right] = \dfrac{|x+2|}{4} < 1 \iff |x+2| < 4$, so $R = 4$.

$|x+2| < 4 \iff -4 < x+2 < 4 \iff -6 < x < 2$. If $x = -6$, then the series $\sum\limits_{n=1}^{\infty} \dfrac{(x+2)^n}{n4^n}$ becomes

$\sum\limits_{n=1}^{\infty} \dfrac{(-4)^n}{n4^n} = \sum\limits_{n=1}^{\infty} \dfrac{(-1)^n}{n}$, the alternating harmonic series, which converges by the Alternating Series Test. When $x = 2$, the

series becomes the harmonic series $\sum\limits_{n=1}^{\infty} \dfrac{1}{n}$, which diverges. Thus, $I = [-6, 2)$.

39. $\lim\limits_{n\to\infty} \left| \dfrac{a_{n+1}}{a_n} \right| = \lim\limits_{n\to\infty} \left| \dfrac{2^{n+1}(x-3)^{n+1}}{\sqrt{n+4}} \cdot \dfrac{\sqrt{n+3}}{2^n(x-3)^n} \right| = 2\,|x-3| \lim\limits_{n\to\infty} \sqrt{\dfrac{n+3}{n+4}} = 2\,|x-3| < 1 \iff |x-3| < \dfrac{1}{2}$, so

$R = \dfrac{1}{2}$. $|x-3| < \dfrac{1}{2} \iff -\dfrac{1}{2} < x-3 < \dfrac{1}{2} \iff \dfrac{5}{2} < x < \dfrac{7}{2}$. For $x = \dfrac{7}{2}$, the series $\sum\limits_{n=1}^{\infty} \dfrac{2^n(x-3)^n}{\sqrt{n+3}}$ becomes

$\sum\limits_{n=0}^{\infty} \dfrac{1}{\sqrt{n+3}} = \sum\limits_{n=3}^{\infty} \dfrac{1}{n^{1/2}}$, which diverges ($p = \dfrac{1}{2} \leq 1$), but for $x = \dfrac{5}{2}$, we get $\sum\limits_{n=0}^{\infty} \dfrac{(-1)^n}{\sqrt{n+3}}$, which is a convergent

alternating series, so $I = \left[\dfrac{5}{2}, \dfrac{7}{2} \right)$.

41.

n	$f^{(n)}(x)$	$f^{(n)}\left(\frac{\pi}{6}\right)$
0	$\sin x$	$\frac{1}{2}$
1	$\cos x$	$\frac{\sqrt{3}}{2}$
2	$-\sin x$	$-\frac{1}{2}$
3	$-\cos x$	$-\frac{\sqrt{3}}{2}$
4	$\sin x$	$\frac{1}{2}$
$\vdots$	$\vdots$	$\vdots$

$$\sin x = f\left(\frac{\pi}{6}\right) + f'\left(\frac{\pi}{6}\right)\left(x - \frac{\pi}{6}\right) + \frac{f''\left(\frac{\pi}{6}\right)}{2!}\left(x - \frac{\pi}{6}\right)^2 + \frac{f^{(3)}\left(\frac{\pi}{6}\right)}{3!}\left(x - \frac{\pi}{6}\right)^3 + \frac{f^{(4)}\left(\frac{\pi}{6}\right)}{4!}\left(x - \frac{\pi}{6}\right)^4 +$$

$$= \frac{1}{2}\left[1 - \frac{1}{2!}\left(x - \frac{\pi}{6}\right)^2 + \frac{1}{4!}\left(x - \frac{\pi}{6}\right)^4 - \cdots\right] + \frac{\sqrt{3}}{2}\left[\left(x - \frac{\pi}{6}\right) - \frac{1}{3!}\left(x - \frac{\pi}{6}\right)^3 + \cdots\right]$$

$$= \frac{1}{2}\sum_{n=0}^{\infty}(-1)^n\frac{1}{(2n)!}\left(x - \frac{\pi}{6}\right)^{2n} + \frac{\sqrt{3}}{2}\sum_{n=0}^{\infty}(-1)^n\frac{1}{(2n+1)!}\left(x - \frac{\pi}{6}\right)^{2n+1}$$

43. $\dfrac{1}{1+x} = \dfrac{1}{1-(-x)} = \displaystyle\sum_{n=0}^{\infty}(-x)^n = \sum_{n=0}^{\infty}(-1)^n x^n$ for $|x| < 1$ $\Rightarrow$ $\dfrac{x^2}{1+x} = \displaystyle\sum_{n=0}^{\infty}(-1)^n x^{n+2}$ with $R = 1$.

45. $\dfrac{1}{1-x} = \displaystyle\sum_{n=0}^{\infty}x^n$ for $|x| < 1$ $\Rightarrow$ $\ln(1-x) = -\displaystyle\int\frac{dx}{1-x} = -\int\sum_{n=0}^{\infty}x^n\,dx = C - \sum_{n=0}^{\infty}\frac{x^{n+1}}{n+1}$.

$\ln(1-0) = C - 0$ $\Rightarrow$ $C = 0$ $\Rightarrow$ $\ln(1-x) = -\displaystyle\sum_{n=0}^{\infty}\frac{x^{n+1}}{n+1} = \sum_{n=1}^{\infty}\frac{-x^n}{n}$ with $R = 1$.

47. $\sin x = \displaystyle\sum_{n=0}^{\infty}\frac{(-1)^n x^{2n+1}}{(2n+1)!}$ $\Rightarrow$ $\sin(x^4) = \displaystyle\sum_{n=0}^{\infty}\frac{(-1)^n (x^4)^{2n+1}}{(2n+1)!} = \sum_{n=0}^{\infty}\frac{(-1)^n x^{8n+4}}{(2n+1)!}$ for all x, so the radius of

convergence is ∞.

49. $f(x) = \dfrac{1}{\sqrt[4]{16-x}} = \dfrac{1}{\sqrt[4]{16(1-x/16)}} = \dfrac{1}{\sqrt[4]{16}\left(1 - \frac{1}{16}x\right)^{1/4}} = \frac{1}{2}\left(1 - \frac{1}{16}x\right)^{-1/4}$

$$= \frac{1}{2}\left[1 + \left(-\frac{1}{4}\right)\left(-\frac{x}{16}\right) + \frac{\left(-\frac{1}{4}\right)\left(-\frac{5}{4}\right)}{2!}\left(-\frac{x}{16}\right)^2 + \frac{\left(-\frac{1}{4}\right)\left(-\frac{5}{4}\right)\left(-\frac{9}{4}\right)}{3!}\left(-\frac{x}{16}\right)^3 + \cdots\right]$$

$$= \frac{1}{2} + \sum_{n=1}^{\infty}\frac{1\cdot 5\cdot 9\cdots(4n-3)}{2\cdot 4^n\cdot n!\cdot 16^n}x^n = \frac{1}{2} + \sum_{n=1}^{\infty}\frac{1\cdot 5\cdot 9\cdots(4n-3)}{2^{6n+1}\,n!}x^n$$

for $\left|-\dfrac{x}{16}\right| < 1$ $\Leftrightarrow$ $|x| < 16$, so $R = 16$.

51. $e^x = \displaystyle\sum_{n=0}^{\infty}\frac{x^n}{n!}$, so $\dfrac{e^x}{x} = \frac{1}{x}\displaystyle\sum_{n=0}^{\infty}\frac{x^n}{n!} = \sum_{n=0}^{\infty}\frac{x^{n-1}}{n!} = x^{-1} + \sum_{n=1}^{\infty}\frac{x^{n-1}}{n!} = \frac{1}{x} + \sum_{n=1}^{\infty}\frac{x^{n-1}}{n!}$ and

$\displaystyle\int\frac{e^x}{x}\,dx = C + \ln|x| + \sum_{n=1}^{\infty}\frac{x^n}{n\cdot n!}$.

53. (a)

n	$f^{(n)}(x)$	$f^{(n)}(1)$
0	$x^{1/2}$	1
1	$\frac{1}{2}x^{-1/2}$	$\frac{1}{2}$
2	$-\frac{1}{4}x^{-3/2}$	$-\frac{1}{4}$
3	$\frac{3}{8}x^{-5/2}$	$\frac{3}{8}$
4	$-\frac{15}{16}x^{-7/2}$	$-\frac{15}{16}$
⋮	⋮	⋮

$$\sqrt{x} \approx T_3(x) = 1 + \frac{1/2}{1!}(x-1) - \frac{1/4}{2!}(x-1)^2 + \frac{3/8}{3!}(x-1)^3$$

$$= 1 + \tfrac{1}{2}(x-1) - \tfrac{1}{8}(x-1)^2 + \tfrac{1}{16}(x-1)^3$$

(b) 1.5

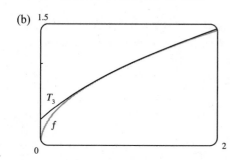

(c) By Taylor's Formula, $R_3(x) = \dfrac{f^{(4)}(z)}{4!}(x-1)^4 = -\dfrac{5(x-1)^4}{128\,z^{7/2}}$, with z

between x and 1. If $0.9 \le x \le 1.1$, then $0 \le |x-1| \le 0.1$

and $z^{7/2} > (0.9)^{7/2}$ so $|R_3(x)| < \dfrac{5(0.1)^4}{128(0.9)^{7/2}} < 0.000006$.

(d) 5×10^{-6}

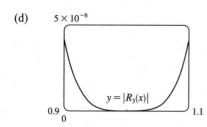

From the graph of $|R_3(x)| = |\sqrt{x} - T_3(x)|$, it appears

that the error is less than 5×10^{-6} on $[0.9, 1.1]$.

55. $\sin x = \displaystyle\sum_{n=0}^{\infty} (-1)^n \frac{x^{2n+1}}{(2n+1)!} = x - \frac{x^3}{3!} + \frac{x^5}{5!} - \frac{x^7}{7!} + \cdots$, so $\sin x - x = -\frac{x^3}{3!} + \frac{x^5}{5!} - \frac{x^7}{7!} + \cdots$ and

$\dfrac{\sin x - x}{x^3} = -\dfrac{1}{3!} + \dfrac{x^2}{5!} - \dfrac{x^4}{7!} + \cdots$. Thus, $\displaystyle\lim_{x \to 0} \dfrac{\sin x - x}{x^3} = \lim_{x \to 0} \left(-\dfrac{1}{6} + \dfrac{x^2}{120} - \dfrac{x^4}{5040} + \cdots \right) = -\dfrac{1}{6}$.

57. $f(x) = \displaystyle\sum_{n=0}^{\infty} c_n x^n \Rightarrow f(-x) = \sum_{n=0}^{\infty} c_n(-x)^n = \sum_{n=0}^{\infty} (-1)^n c_n x^n$

(a) If f is an odd function, then $f(-x) = -f(x) \Rightarrow \displaystyle\sum_{n=0}^{\infty} (-1)^n c_n x^n = \sum_{n=0}^{\infty} -c_n x^n$. The coefficients of any power series

are uniquely determined (by Theorem 8.7.5), so $(-1)^n c_n = -c_n$. If n is even, then $(-1)^n = 1$, so $c_n = -c_n \Rightarrow$

$2c_n = 0 \Rightarrow c_n = 0$. Thus, all even coefficients are 0, that is, $c_0 = c_2 = c_4 = \cdots = 0$.

(b) If f is even, then $f(-x) = f(x) \Rightarrow \displaystyle\sum_{n=0}^{\infty} (-1)^n c_n x^n = \sum_{n=0}^{\infty} c_n x^n \Rightarrow (-1)^n c_n = c_n$. If n is odd, then

$(-1)^n = -1$, so $-c_n = c_n \Rightarrow 2c_n = 0 \Rightarrow c_n = 0$. Thus, all odd coefficients are 0, that is,

$c_1 = c_3 = c_5 = \cdots = 0$.

9 ☐ PARAMETRIC EQUATIONS AND POLAR COORDINATES

9.1 Parametric Curves

1. $x = 1 + \sqrt{t}$, $y = t^2 - 4t$, $0 \le t \le 5$

t	0	1	2	3	4	5
x	1	2	$1+\sqrt{2}$	$1+\sqrt{3}$	3	$1+\sqrt{5}$
			2.41	2.73		3.24
y	0	-3	-4	-3	0	5

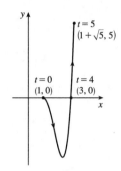

3. $x = 5\sin t$, $y = t^2$, $-\pi \le t \le \pi$

t	$-\pi$	$-\pi/2$	0	$\pi/2$	π
x	0	-5	0	5	0
y	π^2	$\pi^2/4$	0	$\pi^2/4$	π^2
	9.87	2.47		2.47	9.87

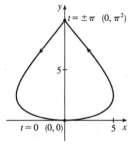

5. $x = 3t - 5$, $\quad y = 2t + 1$

(a)

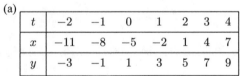

t	-2	-1	0	1	2	3	4
x	-11	-8	-5	-2	1	4	7
y	-3	-1	1	3	5	7	9

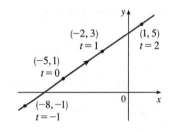

(b) $x = 3t - 5 \;\Rightarrow\; 3t = x + 5 \;\Rightarrow\; t = \frac{1}{3}(x+5) \;\Rightarrow$

$y = 2 \cdot \frac{1}{3}(x+5) + 1$, so $y = \frac{2}{3}x + \frac{13}{3}$.

7. $x = \sqrt{t}$, $y = 1 - t$

(a)

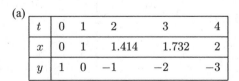

t	0	1	2	3	4
x	0	1	1.414	1.732	2
y	1	0	-1	-2	-3

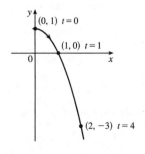

(b) $x = \sqrt{t} \;\Rightarrow\; t = x^2 \;\Rightarrow\; y = 1 - t = 1 - x^2$.

Since $t \ge 0$, $x \ge 0$.

9. (a) $x = \sin\theta$, $y = \cos\theta$, $0 \le \theta \le \pi$.

(b)

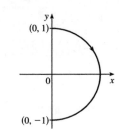

$x^2 + y^2 = \sin^2\theta + \cos^2\theta = 1$.

Since $0 \le \theta \le \pi$, we have $\sin\theta \ge 0$, so $x \ge 0$. Thus, the curve is the right

half of the circle $x^2 + y^2 = 1$.

11. (a) $x = \sin t$, $y = \csc t$, $0 < t < \frac{\pi}{2}$.

(b)

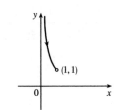

$y = \csc t = \dfrac{1}{\sin t} = \dfrac{1}{x}$. For $0 < t < \frac{\pi}{2}$, we have $0 < x < 1$ and $y > 1$.

Thus, the curve is the portion of the hyperbola $y = 1/x$ with $y > 1$.

13. (a) $x = e^{2t} \;\Rightarrow\; 2t = \ln x \;\Rightarrow\; t = \frac{1}{2}\ln x$.

(b)

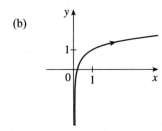

$y = t + 1 = \frac{1}{2}\ln x + 1$.

15. $x = 3 + 2\cos t$, $y = 1 + 2\sin t$, $\pi/2 \le t \le 3\pi/2$. By Example 4 with $r = 2$, $h = 3$, and $k = 1$, the motion of the particle

takes place on a circle centered at $(3, 1)$ with a radius of 2. As t goes from $\frac{\pi}{2}$ to $\frac{3\pi}{2}$, the particle starts at the point $(3, 3)$ and

moves counterclockwise to $(3, -1)$ [one-half of a circle].

17. $x = 5\sin t$, $y = 2\cos t \;\Rightarrow\; \sin t = \dfrac{x}{5}$, $\cos t = \dfrac{y}{2}$. $\sin^2 t + \cos^2 t = 1 \;\Rightarrow\; \left(\dfrac{x}{5}\right)^2 + \left(\dfrac{y}{2}\right)^2 = 1$. The motion of the

particle takes place on an ellipse centered at $(0, 0)$. As t goes from $-\pi$ to 5π, the particle starts at the point $(0, -2)$ and moves

clockwise around the ellipse 3 times.

19. When $t = -1$, $(x, y) = (0, -1)$. As t increases to 0, x decreases to -1 and y

increases to 0. As t increases from 0 to 1, x increases to 0 and y increases to 1. As

t increases beyond 1, both x and y increase. For $t < -1$, x is positive and

decreasing and y is negative and increasing. We could achieve greater accuracy by

estimating x- and y-values for selected values of t from the given graphs and

plotting the corresponding points.

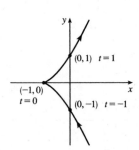

21. When $t = 0$ we see that $x = 0$ and $y = 0$, so the curve starts at the origin. As t

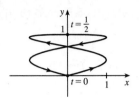

increases from 0 to $\frac{1}{2}$, the graphs show that y increases from 0 to 1 while x

increases from 0 to 1, decreases to 0 and to -1, then increases back to 0, so we

arrive at the point $(0, 1)$. Similarly, as t increases from $\frac{1}{2}$ to 1, y decreases from 1

to 0 while x repeats its pattern, and we arrive back at the origin. We could

achieve greater accuracy by estimating x- and y-values for selected values of t from the given graphs and plotting the

corresponding points.

23. As in Example 6, we let $y = t$ and $x = t - 3t^3 + t^5$ and use a t-interval

of $[-2\pi, 2\pi]$.

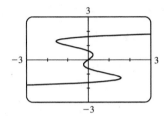

25. (a) $x = x_1 + (x_2 - x_1)t$, $y = y_1 + (y_2 - y_1)t$, $0 \le t \le 1$.   Clearly the curve passes through $P_1(x_1, y_1)$ when $t = 0$ and

through $P_2(x_2, y_2)$ when $t = 1$. For $0 < t < 1$, x is strictly between x_1 and x_2 and y is strictly between y_1 and y_2. For

every value of t, x and y satisfy the relation $y - y_1 = \dfrac{y_2 - y_1}{x_2 - x_1}(x - x_1)$, which is the equation of the line through

$P_1(x_1, y_1)$ and $P_2(x_2, y_2)$.

   Finally, any point (x, y) on that line satisfies $\dfrac{y - y_1}{y_2 - y_1} = \dfrac{x - x_1}{x_2 - x_1}$; if we call that common value t, then the given

parametric equations yield the point (x, y); and any (x, y) on the line between $P_1(x_1, y_1)$ and $P_2(x_2, y_2)$ yields a value of

t in $[0, 1]$. So the given parametric equations exactly specify the line segment from $P_1(x_1, y_1)$ to $P_2(x_2, y_2)$.

(b) $x = -2 + [3 - (-2)]t = -2 + 5t$ and $y = 7 + (-1 - 7)t = 7 - 8t$ for $0 \le t \le 1$.

27. (a) To get a clockwise orientation, we could change the equations to $x = 2\cos t$, $y = 1 - 2\sin t$, $0 \le t \le 2\pi$.

(b) To get three times around in the counterclockwise direction, we use the original equations $x = 2\cos t$, $y = 1 + 2\sin t$

   with the domain expanded to $0 \le t \le 6\pi$.

(c) To start at $(0, 3)$ using the original equations, we must have $x_1 = 0$; that is, $2\cos t = 0$. Hence, $t = \frac{\pi}{2}$. So we use

   $x = 2\cos t$, $y = 1 + 2\sin t$, $\frac{\pi}{2} \le t \le \frac{3\pi}{2}$.

   Alternatively, if we want t to start at 0, we could change the equations of the curve. For example, we could use

   $x = -2\sin t$, $y = 1 + 2\cos t$, $0 \le t \le \pi$.

29. *Big circle:* It's centered at $(2, 2)$ with a radius of 2, so by Example 4, parametric equations are

$$x = 2 + 2\cos t, \qquad y = 2 + 2\sin t, \qquad 0 \le t \le 2\pi$$

Small circles: They are centered at $(1, 3)$ and $(3, 3)$ with a radius of 0.1. By Example 4, parametric equations are

$$\begin{aligned}
\textit{(left)} \qquad & x = 1 + 0.1\cos t, & y = 3 + 0.1\sin t, & \qquad 0 \le t \le 2\pi \\
\text{and} \qquad \textit{(right)} \qquad & x = 3 + 0.1\cos t, & y = 3 + 0.1\sin t, & \qquad 0 \le t \le 2\pi
\end{aligned}$$

Semicircle: It's the lower half of a circle centered at $(2, 2)$ with radius 1. By Example 4, parametric equations are

$$x = 2 + 1\cos t, \qquad y = 2 + 1\sin t, \qquad \pi \le t \le 2\pi$$

To get all four graphs on the same screen with a typical graphing calculator, we need to change the last t-interval to $[0, 2\pi]$ in order to match the others. We can do this by changing t to $0.5t$. This change gives us the upper half. There are several ways to get the lower half—one is to change the "+" to a "−" in the y-assignment, giving us

$$x = 2 + 1\cos(0.5t), \qquad y = 2 - 1\sin(0.5t), \qquad 0 \le t \le 2\pi$$

31. (a) $x = t^3 \implies t = x^{1/3}$, so $y = t^2 = x^{2/3}$.

We get the entire curve $y = x^{2/3}$ traversed in a left to right direction.

(b) $x = t^6 \implies t = x^{1/6}$, so $y = t^4 = x^{4/6} = x^{2/3}$.

Since $x = t^6 \ge 0$, we only get the right half of the curve $y = x^{2/3}$.

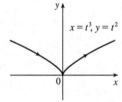

$x = t^3, \ y = t^2$

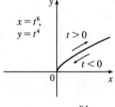

$x = t^6,$
$y = t^4$
$t > 0$
$t < 0$

(c) $x = e^{-3t} = (e^{-t})^3$ [so $e^{-t} = x^{1/3}$],

$y = e^{-2t} = (e^{-t})^2 = (x^{1/3})^2 = x^{2/3}$.

If $t < 0$, then x and y are both larger than 1. If $t > 0$, then x and y are between 0 and 1. Since $x > 0$ and $y > 0$, the curve never quite reaches the origin.

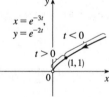

$x = e^{-3t},$
$y = e^{-2t}$
$t < 0$
$t > 0$
$(1, 1)$

33. The case $\frac{\pi}{2} < \theta < \pi$ is illustrated. C has coordinates $(r\theta, r)$ as in Example 6, and Q has coordinates $(r\theta, r + r\cos(\pi - \theta)) = (r\theta, r(1 - \cos\theta))$ [since $\cos(\pi - \alpha) = \cos\pi\cos\alpha + \sin\pi\sin\alpha = -\cos\alpha$], so P has coordinates $(r\theta - r\sin(\pi - \theta), r(1 - \cos\theta)) = (r(\theta - \sin\theta), r(1 - \cos\theta))$ [since $\sin(\pi - \alpha) = \sin\pi\cos\alpha - \cos\pi\sin\alpha = \sin\alpha$]. Again we have the parametric equations $x = r(\theta - \sin\theta)$, $y = r(1 - \cos\theta)$.

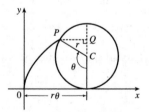

35. It is apparent that $x = |OQ|$ and $y = |QP| = |ST|$. From the diagram,

$x = |OQ| = a\cos\theta$ and $y = |ST| = b\sin\theta$. Thus, the parametric equations are

$x = a\cos\theta$ and $y = b\sin\theta$. To eliminate θ we rearrange: $\sin\theta = y/b \;\;\Rightarrow$

$\sin^2\theta = (y/b)^2$ and $\cos\theta = x/a \;\;\Rightarrow\;\; \cos^2\theta = (x/a)^2$. Adding the two

equations: $\sin^2\theta + \cos^2\theta = 1 = x^2/a^2 + y^2/b^2$. Thus, we have an ellipse.

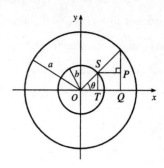

37. (a)

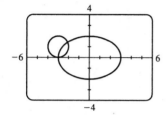

There are 2 points of intersection:

$(-3, 0)$ and approximately $(-2.1, 1.4)$.

(b) A collision point occurs when $x_1 = x_2$ and $y_1 = y_2$ for the same t. So solve the equations:

$$3\sin t = -3 + \cos t \quad \textbf{(1)}$$
$$2\cos t = 1 + \sin t \quad \textbf{(2)}$$

From **(2)**, $\sin t = 2\cos t - 1$. Substituting into **(1)**, we get $3(2\cos t - 1) = -3 + \cos t \;\;\Rightarrow\;\; 5\cos t = 0 \;\;(\star) \;\;\Rightarrow$

$\cos t = 0 \;\;\Rightarrow\;\; t = \frac{\pi}{2}$ or $\frac{3\pi}{2}$. We check that $t = \frac{3\pi}{2}$ satisfies **(1)** and **(2)** but $t = \frac{\pi}{2}$ does not. So the only collision point

occurs when $t = \frac{3\pi}{2}$, and this gives the point $(-3, 0)$. [We could check our work by graphing x_1 and x_2 together as

functions of t and, on another plot, y_1 and y_2 as functions of t. If we do so, we see that the only value of t for which *both*

pairs of graphs intersect is $t = \frac{3\pi}{2}$.]

(c) The circle is centered at $(3, 1)$ instead of $(-3, 1)$. There are still 2 intersection points: $(3, 0)$ and $(2.1, 1.4)$, but there are

no collision points, since $(\star)$ in part (b) becomes $5\cos t = 6 \;\;\Rightarrow\;\; \cos t = \frac{6}{5} > 1$.

39. $x = t^2$, $y = t^3 - ct$. We use a graphing device to produce the graphs for various values of c with $-\pi \le t \le \pi$. Note that all

the members of the family are symmetric about the x-axis. For $c < 0$, the graph does not cross itself, but for $c = 0$ it has a

cusp at $(0, 0)$ and for $c > 0$ the graph crosses itself at $x = c$, so the loop grows larger as c increases.

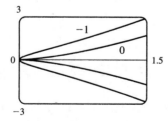

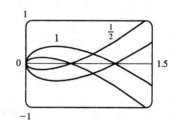

41. Note that all the Lissajous figures are symmetric about the x-axis. The parameters a and b simply stretch the graph in the x-and y-directions respectively. For $a = b = n = 1$ the graph is simply a circle with radius 1. For $n = 2$ the graph crosses itself at the origin and there are loops above and below the x-axis. In general, the figures have $n - 1$ points of intersection, all of which are on the y-axis, and a total of n closed loops.

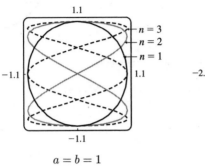

$a = b = 1$

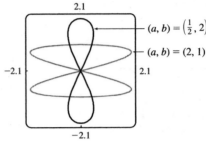

$n = 2$

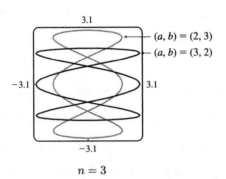

$n = 3$

9.2 Calculus with Parametric Curves

1. $x = t - t^3$, $y = 2 - 5t$ $\Rightarrow$ $\dfrac{dy}{dt} = -5$, $\dfrac{dx}{dt} = 1 - 3t^2$, and $\dfrac{dy}{dx} = \dfrac{dy/dt}{dx/dt} = \dfrac{-5}{1 - 3t^2}$ or $\dfrac{5}{3t^2 - 1}$.

3. $x = t^4 + 1$, $y = t^3 + t$; $t = -1$. $\dfrac{dy}{dt} = 3t^2 + 1$, $\dfrac{dx}{dt} = 4t^3$, and $\dfrac{dy}{dx} = \dfrac{dy/dt}{dx/dt} = \dfrac{3t^2 + 1}{4t^3}$. When $t = -1$,

$(x, y) = (2, -2)$ and $dy/dx = \frac{4}{-4} = -1$, so an equation of the tangent to the curve at the point corresponding to $t = -1$ is

$y - (-2) = (-1)(x - 2)$, or $y = -x$.

5. $x = e^{\sqrt{t}}$, $y = t - \ln t^2$; $t = 1$. $\dfrac{dy}{dt} = 1 - \dfrac{2t}{t^2} = 1 - \dfrac{2}{t}$, $\dfrac{dx}{dt} = \dfrac{e^{\sqrt{t}}}{2\sqrt{t}}$, and $\dfrac{dy}{dx} = \dfrac{dy/dt}{dx/dt} = \dfrac{1 - 2/t}{e^{\sqrt{t}}/(2\sqrt{t})} \cdot \dfrac{2t}{2t} = \dfrac{2t - 4}{\sqrt{t}\,e^{\sqrt{t}}}$.

When $t = 1$, $(x, y) = (e, 1)$ and $\dfrac{dy}{dx} = -\dfrac{2}{e}$, so an equation of the tangent line is $y - 1 = -\frac{2}{e}(x - e)$, or $y = -\frac{2}{e}x + 3$.

7. (a) $x = e^t$, $y = (t-1)^2$; $(1,1)$. $\dfrac{dy}{dt} = 2(t-1)$, $\dfrac{dx}{dt} = e^t$, and $\dfrac{dy}{dx} = \dfrac{dy/dt}{dx/dt} = \dfrac{2(t-1)}{e^t}$.

At $(1,1)$, $t = 0$ and $\dfrac{dy}{dx} = -2$, so an equation of the tangent is $y - 1 = -2(x-1)$, or $y = -2x + 3$.

(b) $x = e^t$ ⟹ $t = \ln x$, so $y = (t-1)^2 = (\ln x - 1)^2$ and $\dfrac{dy}{dx} = 2(\ln x - 1)\left(\dfrac{1}{x}\right)$. When $x = 1$,

$\dfrac{dy}{dx} = 2(-1)(1) = -2$, so an equation of the tangent is $y = -2x + 3$, as in part (a).

9. $x = 4 + t^2$, $y = t^2 + t^3$ ⟹ $\dfrac{dy}{dx} = \dfrac{dy/dt}{dx/dt} = \dfrac{2t + 3t^2}{2t} = 1 + \dfrac{3}{2}t$ ⟹

$\dfrac{d^2y}{dx^2} = \dfrac{d}{dx}\left(\dfrac{dy}{dx}\right) = \dfrac{d(dy/dx)/dt}{dx/dt} = \dfrac{(d/dt)(1 + \frac{3}{2}t)}{2t} = \dfrac{3/2}{2t} = \dfrac{3}{4t}$. The curve is CU when $\dfrac{d^2y}{dx^2} > 0$, that is, when $t > 0$.

11. $x = t - e^t$, $y = t + e^{-t}$ ⟹

$\dfrac{dy}{dx} = \dfrac{dy/dt}{dx/dt} = \dfrac{1 - e^{-t}}{1 - e^t} = \dfrac{1 - \dfrac{1}{e^t}}{1 - e^t} = \dfrac{\dfrac{e^t - 1}{e^t}}{1 - e^t} = -e^{-t}$ ⟹ $\dfrac{d^2y}{dx^2} = \dfrac{\dfrac{d}{dt}\left(\dfrac{dy}{dx}\right)}{dx/dt} = \dfrac{\dfrac{d}{dt}(-e^{-t})}{dx/dt} = \dfrac{e^{-t}}{1 - e^t}$.

The curve is CU when $e^t < 1$ [since $e^{-t} > 0$] ⟹ $t < 0$.

13. $x = 10 - t^2$, $y = t^3 - 12t$.

$dy/dt = 3t^2 - 12 = 3(t+2)(t-2)$, so $dy/dt = 0$ ⟺

$t = \pm 2$ ⟺ $(x, y) = (6, \mp 16)$. $dx/dt = -2t$, so

$dx/dt = 0$ ⟺ $t = 0$ ⟺ $(x, y) = (10, 0)$. The curve has

horizontal tangents at $(6, \pm 16)$ and a vertical tangent at $(10, 0)$.

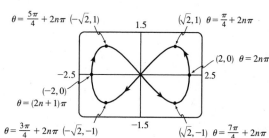

15. $x = 2\cos\theta$, $y = \sin 2\theta$.

$dy/d\theta = 2\cos 2\theta$, so $dy/d\theta = 0$ ⟺ $2\theta = \dfrac{\pi}{2} + n\pi$

(n an integer) ⟺ $\theta = \dfrac{\pi}{4} + \dfrac{\pi}{2}n$ ⟺

$(x, y) = (\pm\sqrt{2}, \pm 1)$. Also, $dx/d\theta = -2\sin\theta$, so

$dx/d\theta = 0$ ⟺ $\theta = n\pi$ ⟺ $(x, y) = (\pm 2, 0)$. The

curve has horizontal tangents at $(\pm\sqrt{2}, \pm 1)$ (four points),

and vertical tangents at $(\pm 2, 0)$.

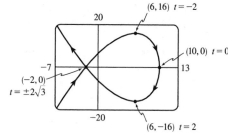

17. From the graph, it appears that the leftmost point on the curve $x = t^4 - t^2$,

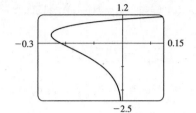

$y = t + \ln t$ is about $(-0.25, 0.36)$. To find the exact coordinates, we

find the value of t for which the graph has a vertical tangent, that is,

$0 = dx/dt = 4t^3 - 2t \iff 2t(2t^2 - 1) = 0 \iff$

$2t(\sqrt{2}\,t + 1)(\sqrt{2}\,t - 1) = 0 \iff t = 0$ or $\pm\frac{1}{\sqrt{2}}$. The negative and 0 roots are

inadmissible since $y(t)$ is only defined for $t > 0$, so the leftmost point must be

$$\left(x\!\left(\frac{1}{\sqrt{2}}\right), y\!\left(\frac{1}{\sqrt{2}}\right)\right) = \left(\left(\frac{1}{\sqrt{2}}\right)^4 - \left(\frac{1}{\sqrt{2}}\right)^2, \frac{1}{\sqrt{2}} + \ln\frac{1}{\sqrt{2}}\right) = \left(-\frac{1}{4}, \frac{1}{\sqrt{2}} - \frac{1}{2}\ln 2\right)$$

19. We graph the curve $x = t^4 - 2t^3 - 2t^2$,

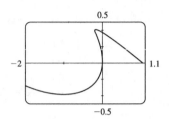

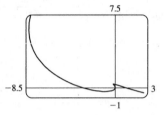

$y = t^3 - t$ in the viewing rectangle $[-2, 1.1]$

by $[-0.5, 0.5]$. This rectangle corresponds

approximately to $t \in [-1, 0.8]$. We estimate

that the curve has horizontal tangents at about

$(-1, -0.4)$ and $(-0.17, 0.39)$ and vertical tangents at about $(0, 0)$ and $(-0.19, 0.37)$. We calculate

$\dfrac{dy}{dx} = \dfrac{dy/dt}{dx/dt} = \dfrac{3t^2 - 1}{4t^3 - 6t^2 - 4t}$. The horizontal tangents occur when $dy/dt = 3t^2 - 1 = 0 \iff t = \pm\frac{1}{\sqrt{3}}$, so both

horizontal tangents are shown in our graph. The vertical tangents occur when $dx/dt = 2t(2t^2 - 3t - 2) = 0 \iff$

$2t(2t + 1)(t - 2) = 0 \iff t = 0, -\frac{1}{2}$ or 2. It seems that we have missed one vertical tangent, and indeed if we plot the

curve on the t-interval $[-1.2, 2.2]$ we see that there is another vertical tangent at $(-8, 6)$.

21. $x = \cos t$, $y = \sin t \cos t$. $\dfrac{dx}{dt} = -\sin t$, $\dfrac{dy}{dt} = -\sin^2 t + \cos^2 t = \cos 2t$.

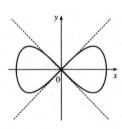

$(x, y) = (0, 0) \iff \cos t = 0 \iff t$ is an odd multiple of $\dfrac{\pi}{2}$. When

$t = \dfrac{\pi}{2}, \dfrac{dx}{dt} = -1$ and $\dfrac{dy}{dt} = -1$, so $\dfrac{dy}{dx} = 1$. When $t = \dfrac{3\pi}{2}, \dfrac{dx}{dt} = 1$ and

$\dfrac{dy}{dt} = -1$. So $\dfrac{dy}{dx} = -1$. Thus, $y = x$ and $y = -x$ are both tangent to the

curve at $(0, 0)$.

23. (a) $x = r\theta - d\sin\theta$, $y = r - d\cos\theta$. $\dfrac{dx}{d\theta} = r - d\cos\theta$, $\dfrac{dy}{d\theta} = d\sin\theta$. So $\dfrac{dy}{dx} = \dfrac{d\sin\theta}{r - d\cos\theta}$.

(b) If $0 < d < r$, then $|d\cos\theta| \le d < r$, so $r - d\cos\theta \ge r - d > 0$. This shows that $dx/d\theta$ never vanishes, so the trochoid

can have no vertical tangent if $d < r$.

25. The line with parametric equations $x = -7t$, $y = 12t - 5$ is $y = 12\left(-\frac{1}{7}x\right) - 5$, which has slope $-\frac{12}{7}$.

The curve $x = t^3 + 4t$, $y = 6t^2$ has slope $\dfrac{dy}{dx} = \dfrac{dy/dt}{dx/dt} = \dfrac{12t}{3t^2 + 4}$. This equals $-\frac{12}{7}$ $\Leftrightarrow$ $3t^2 + 4 = -7t$ $\Leftrightarrow$

$(3t + 4)(t + 1) = 0$ $\Leftrightarrow$ $t = -1$ or $t = -\frac{4}{3}$ $\Leftrightarrow$ $(x, y) = (-5, 6)$ or $\left(-\frac{208}{27}, \frac{32}{3}\right)$.

27. By symmetry of the ellipse about the x- and y-axes,

$$A = 4 \int_0^a y\, dx = 4 \int_{\pi/2}^0 b \sin\theta\, (-a \sin\theta)\, d\theta = 4ab \int_0^{\pi/2} \sin^2\theta\, d\theta = 4ab \int_0^{\pi/2} \tfrac{1}{2}(1 - \cos 2\theta)\, d\theta$$

$$= 2ab\left[\theta - \tfrac{1}{2}\sin 2\theta\right]_0^{\pi/2} = 2ab\left(\tfrac{\pi}{2}\right) = \pi ab$$

29. $A = \int_0^1 (y - 1)\, dx = \int_{\pi/2}^0 (e^t - 1)(-\sin t)\, dt = \int_0^{\pi/2} (e^t \sin t - \sin t)\, dt \overset{98}{=} \left[\tfrac{1}{2}e^t(\sin t - \cos t) + \cos t\right]_0^{\pi/2}$

$= \tfrac{1}{2}(e^{\pi/2} - 1)$

31. $A = \int_0^{2\pi r} y\, dx = \int_0^{2\pi} (r - d\cos\theta)(r - d\cos\theta)\, d\theta = \int_0^{2\pi} (r^2 - 2dr\cos\theta + d^2\cos^2\theta)\, d\theta$

$= \left[r^2\theta - 2dr\sin\theta + \tfrac{1}{2}d^2\left(\theta + \tfrac{1}{2}\sin 2\theta\right)\right]_0^{2\pi} = 2\pi r^2 + \pi d^2$

33. $x = t - t^2$, $y = \frac{4}{3}t^{3/2}$, $1 \le t \le 2$. $dx/dt = 1 - 2t$ and $dy/dt = 2t^{1/2}$, so

$(dx/dt)^2 + (dy/dt)^2 = (1 - 2t)^2 + (2t^{1/2})^2 = 1 - 4t + 4t^2 + 4t = 1 + 4t^2$. Thus,

$L = \int_a^b \sqrt{(dx/dt)^2 + (dy/dt)^2}\, dt = \int_1^2 \sqrt{1 + 4t^2}\, dt$.

35. $x = t + \cos t$, $y = t - \sin t$, $0 \le t \le 2\pi$. $dx/dt = 1 - \sin t$ and $dy/dt = 1 - \cos t$, so

$(dx/dt)^2 + (dy/dt)^2 = (1 - \sin t)^2 + (1 - \cos t)^2 = (1 - 2\sin t + \sin^2 t) + (1 - 2\cos t + \cos^2 t)$

$$= 3 - 2\sin t - 2\cos t$$

Thus, $L = \int_a^b \sqrt{(dx/dt)^2 + (dy/dt)^2}\, dt = \int_0^{2\pi} \sqrt{3 - 2\sin t - 2\cos t}\, dt$.

37.

$x = 1 + 3t^2$, $y = 4 + 2t^3$, $0 \le t \le 1$.

$dx/dt = 6t$ and $dy/dt = 6t^2$, so $(dx/dt)^2 + (dy/dt)^2 = 36t^2 + 36t^4$.

Thus, $\quad L = \int_0^1 \sqrt{36t^2 + 36t^4}\, dt$

$\qquad = \int_0^1 6t\sqrt{1 + t^2}\, dt = 6\int_1^2 \sqrt{u}\left(\tfrac{1}{2}du\right)$ $[u = 1 + t^2, du = 2t\, dt]$

$\qquad = 3\left[\tfrac{2}{3}u^{3/2}\right]_1^2 = 2(2^{3/2} - 1) = 2(2\sqrt{2} - 1)$

39. $x = \dfrac{t}{1+t}$, $y = \ln(1 + t)$, $0 \le t \le 2$. $\dfrac{dx}{dt} = \dfrac{(1+t)\cdot 1 - t\cdot 1}{(1+t)^2} = \dfrac{1}{(1+t)^2}$ and $\dfrac{dy}{dt} = \dfrac{1}{1+t}$, so

$\left(\dfrac{dx}{dt}\right)^2 + \left(\dfrac{dy}{dt}\right)^2 = \dfrac{1}{(1+t)^4} + \dfrac{1}{(1+t)^2} = \dfrac{1}{(1+t)^4}\left[1 + (1+t)^2\right] = \dfrac{t^2 + 2t + 2}{(1+t)^4}$. Thus,

$L = \int_0^2 \dfrac{\sqrt{t^2 + 2t + 2}}{(1+t)^2}\, dt = \int_1^3 \dfrac{\sqrt{u^2 + 1}}{u^2}\, du$ $[u = t + 1, du = dt]$ $\overset{24}{=} \left[-\dfrac{\sqrt{u^2 + 1}}{u} + \ln\left(u + \sqrt{u^2 + 1}\right)\right]_1^3$

$= -\dfrac{\sqrt{10}}{3} + \ln\left(3 + \sqrt{10}\right) + \sqrt{2} - \ln\left(1 + \sqrt{2}\right)$

41. $x = e^t \cos t, \ y = e^t \sin t, \ 0 \le t \le \pi$.

$$\left(\frac{dx}{dt}\right)^2 + \left(\frac{dy}{dt}\right)^2 = \left[e^t(\cos t - \sin t)\right]^2 + \left[e^t(\sin t + \cos t)\right]^2$$

$$= (e^t)^2 (\cos^2 t - 2\cos t \sin t + \sin^2 t)$$

$$+ (e^t)^2 (\sin^2 t + 2\sin t \cos t + \cos^2 t)$$

$$= e^{2t}(2\cos^2 t + 2\sin^2 t) = 2e^{2t}$$

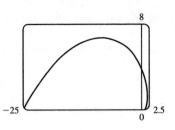

Thus, $L = \int_0^\pi \sqrt{2e^{2t}} \, dt = \int_0^\pi \sqrt{2}\, e^t \, dt = \sqrt{2}\left[e^t\right]_0^\pi = \sqrt{2}(e^\pi - 1)$.

43.

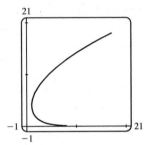

$x = e^t - t, \ y = 4e^{t/2}, \ -8 \le t \le 3$.

$$(dx/dt)^2 + (dy/dt)^2 = (e^t - 1)^2 + (2e^{t/2})^2 = e^{2t} - 2e^t + 1 + 4e^t$$

$$= e^{2t} + 2e^t + 1 = (e^t + 1)^2$$

$$L = \int_{-8}^3 \sqrt{(e^t + 1)^2} \, dt = \int_{-8}^3 (e^t + 1) \, dt = \left[e^t + t\right]_{-8}^{3t}$$

$$= (e^3 + 3) - (e^{-8} - 8) = e^3 - e^{-8} + 11$$

45. $x = t - e^t, \ y = t + e^t, \ -6 \le t \le 6$.

$$\left(\frac{dx}{dt}\right)^2 + \left(\frac{dy}{dt}\right)^2 = (1 - e^t)^2 + (1 + e^t)^2 = (1 - 2e^t + e^{2t}) + (1 + 2e^t + e^{2t}) = 2 + 2e^{2t}, \text{ so } L = \int_{-6}^6 \sqrt{2 + 2e^{2t}} \, dt.$$

Set $f(t) = \sqrt{2 + 2e^{2t}}$. Then by Simpson's Rule with $n = 6$ and $\Delta t = \frac{6 - (-6)}{6} = 2$, we get

$$L \approx \frac{2}{3}[f(-6) + 4f(-4) + 2f(-2) + 4f(0) + 2f(2) + 4f(4) + f(6)] \approx 612.3053.$$

47. $x = \sin^2 t, \ y = \cos^2 t, \ 0 \le t \le 3\pi$.

$$(dx/dt)^2 + (dy/dt)^2 = (2\sin t \cos t)^2 + (-2\cos t \sin t)^2 = 8\sin^2 t \cos^2 t = 2\sin^2 2t \ \Rightarrow$$

$$\text{Distance} = \int_0^{3\pi} \sqrt{2}\, |\sin 2t| \, dt = 6\sqrt{2} \int_0^{\pi/2} \sin 2t \, dt \quad \text{[by symmetry]} \ = -3\sqrt{2}\left[\cos 2t\right]_0^{\pi/2}$$

$$= -3\sqrt{2}(-1 - 1) = 6\sqrt{2}$$

The full curve is traversed as t goes from 0 to $\frac{\pi}{2}$, because the curve is the segment of $x + y = 1$ that lies in the first quadrant

(since $x, y \ge 0$), and this segment is completely traversed as t goes from 0 to $\frac{\pi}{2}$. Thus, $L = \int_0^{\pi/2} \sin 2t \, dt = \sqrt{2}$, as above.

49. $x = a\sin\theta, \ y = b\cos\theta, \ 0 \le \theta \le 2\pi$.

$$(dx/d\theta)^2 + (dy/d\theta)^2 = (a\cos\theta)^2 + (-b\sin\theta)^2 = a^2\cos^2\theta + b^2\sin^2\theta = a^2(1 - \sin^2\theta) + b^2\sin^2\theta$$

$$= a^2 - (a^2 - b^2)\sin^2\theta = a^2 - c^2\sin^2\theta = a^2\left(1 - \frac{c^2}{a^2}\sin^2\theta\right) = a^2(1 - e^2\sin^2\theta)$$

So $L = 4\int_0^{\pi/2} \sqrt{a^2(1 - e^2\sin^2\theta)} \, d\theta$ [by symmetry] $= 4a\int_0^{\pi/2} \sqrt{1 - e^2\sin^2\theta} \, d\theta$.

51. (a) Notice that $0 \le t \le 2\pi$ does not give the complete curve because
$x(0) \ne x(2\pi)$. In fact, we must take $t \in [0, 4\pi]$ in order to obtain the
complete curve, since the first term in each of the parametric
equations has period 2π and the second has period $\frac{2\pi}{11/2} = \frac{4\pi}{11}$, and
the least common integer multiple of these two numbers is 4π.

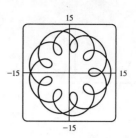

(b) We use the CAS to find the derivatives dx/dt and dy/dt, and then use Formula 1 to find the arc length. Recent versions of
Maple express the integral $\int_0^{4\pi} \sqrt{(dx/dt)^2 + (dy/dt)^2}\, dt$ as $88E\left(2\sqrt{2}\,i\right)$, where $E(x)$ is the elliptic integral

$$\int_0^1 \frac{\sqrt{1 - x^2 t^2}}{\sqrt{1 - t^2}}\, dt$$ and i is the imaginary number $\sqrt{-1}$. Some earlier versions of Maple (as well as Mathematica) cannot

do the integral exactly, so we use the command

`evalf(Int(sqrt(diff(x,t)^2+diff(y,t)^2),t=0..4*Pi));` to estimate the length, and find that the arc
length is approximately 294.03. Derive's `Para_arc_length` function in the utility file `Int_apps` simplifies the

integral to $11 \int_0^{4\pi} \sqrt{-4\cos t\, \cos\left(\frac{11t}{2}\right) - 4\sin t\, \sin\left(\frac{11t}{2}\right) + 5}\, dt$.

53. The coordinates of T are $(r\cos\theta, r\sin\theta)$. Since TP was
unwound from arc TA, TP has length $r\theta$. Also
$\angle PTQ = \angle PTR - \angle QTR = \frac{1}{2}\pi - \theta$, so P has coordinates
$x = r\cos\theta + r\theta\cos\left(\frac{1}{2}\pi - \theta\right) = r(\cos\theta + \theta\sin\theta)$,
$y = r\sin\theta - r\theta\sin\left(\frac{1}{2}\pi - \theta\right) = r(\sin\theta - \theta\cos\theta)$.

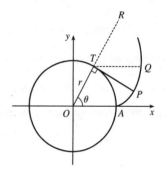

9.3 Polar Coordinates

1. (a) By adding 2π to $\frac{\pi}{2}$, we obtain the
point $\left(1, \frac{5\pi}{2}\right)$. The direction
opposite $\frac{\pi}{2}$ is $\frac{3\pi}{2}$, so $\left(-1, \frac{3\pi}{2}\right)$ is a
point that satisfies the $r < 0$
requirement.

(b) $\left(-2, \frac{\pi}{4}\right)$

(c) $(3, 2)$

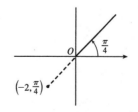

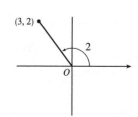

$\left(1, \frac{\pi}{2}\right)$

$\left(2, \frac{5\pi}{4}\right), \left(-2, \frac{9\pi}{4}\right)$

$(3, 2 + 2\pi), (-3, 2 + \pi)$

3. (a)

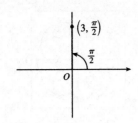

$x = 3 \cos \frac{\pi}{2} = 3(0) = 0$ and

$y = 3 \sin \frac{\pi}{2} = 3(1) = 3$ give us

the Cartesian coordinates $(0, 3)$.

(b)

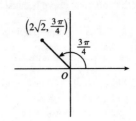

$x = 2\sqrt{2} \cos \frac{3\pi}{4}$

$= 2\sqrt{2}\left(-\frac{1}{\sqrt{2}}\right) = -2$ and

$y = 2\sqrt{2} \sin \frac{3\pi}{4} = 2\sqrt{2}\left(\frac{1}{\sqrt{2}}\right) = 2$

give us $(-2, 2)$.

(c)

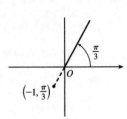

$x = -1 \cos \frac{\pi}{3} = -\frac{1}{2}$ and

$y = -1 \sin \frac{\pi}{3} = -\frac{\sqrt{3}}{2}$ give

us $\left(-\frac{1}{2}, -\frac{\sqrt{3}}{2}\right)$.

5. (a) $x = 1$ and $y = 1$ $\Rightarrow$ $r = \sqrt{1^2 + 1^2} = \sqrt{2}$ and $\theta = \tan^{-1}\left(\frac{1}{1}\right) = \frac{\pi}{4}$. Since $(1, 1)$ is in the first quadrant, the polar

coordinates are (i) $\left(\sqrt{2}, \frac{\pi}{4}\right)$ and (ii) $\left(-\sqrt{2}, \frac{5\pi}{4}\right)$.

(b) $x = 2\sqrt{3}$ and $y = -2$ $\Rightarrow$ $r = \sqrt{\left(2\sqrt{3}\right)^2 + (-2)^2} = \sqrt{12 + 4} = \sqrt{16} = 4$ and

$\theta = \tan^{-1}\left(-\frac{2}{2\sqrt{3}}\right) = \tan^{-1}\left(-\frac{1}{\sqrt{3}}\right) = -\frac{\pi}{6}$. Since $\left(2\sqrt{3}, -2\right)$ is in the fourth quadrant and $0 \le \theta \le 2\pi$, the polar

coordinates are (i) $\left(4, \frac{11\pi}{6}\right)$ and (ii) $\left(-4, \frac{5\pi}{6}\right)$.

7. The curves $r = 1$ and $r = 2$ represent circles with center O and radii 1 and 2. The region in the plane satisfying $1 \le r \le 2$

consists of both circles and the shaded region between them in the figure.

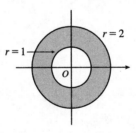

9. The region satisfying $0 \le r < 4$ and

$-\pi/2 \le \theta < \pi/6$ does not include the circle $r = 4$

nor the line $\theta = \frac{\pi}{6}$.

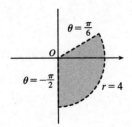

11. $2 < r < 3$, $\quad \frac{5\pi}{3} \le \theta \le \frac{7\pi}{3}$

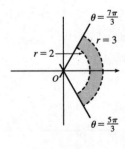

13. $r = 3\sin\theta \;\Rightarrow\; r^2 = 3r\sin\theta \;\Leftrightarrow\; x^2 + y^2 = 3y \;\Leftrightarrow\; x^2 + \left(y - \frac{3}{2}\right)^2 = \left(\frac{3}{2}\right)^2$, a circle of radius $\frac{3}{2}$ centered at $\left(0, \frac{3}{2}\right)$.

The first two equations are actually equivalent since $r^2 = 3r\sin\theta \;\Rightarrow\; r(r - 3\sin\theta) = 0 \;\Rightarrow\; r = 0$ or $r = 3\sin\theta$. But

$r = 3\sin\theta$ gives the point $r = 0$ (the pole) when $\theta = 0$. Thus, the single equation $r = 3\sin\theta$ is equivalent to the compound

condition $(r = 0$ or $r = 3\sin\theta)$.

15. $r = \csc\theta \;\Leftrightarrow\; r = \dfrac{1}{\sin\theta} \;\Leftrightarrow\; r\sin\theta = 1 \;\Leftrightarrow\; y = 1$, a horizontal line 1 unit above the x-axis.

17. $x = -y^2 \;\Leftrightarrow\; r\cos\theta = -r^2\sin^2\theta \;\Leftrightarrow\; \cos\theta = -r\sin^2\theta \;\Leftrightarrow\; r = -\dfrac{\cos\theta}{\sin^2\theta} = -\cot\theta\,\csc\theta$.

19. $x^2 + y^2 = 2cx \;\Leftrightarrow\; r^2 = 2cr\cos\theta \;\Leftrightarrow\; r^2 - 2cr\cos\theta = 0 \;\Leftrightarrow\; r(r - 2c\cos\theta) = 0 \;\Leftrightarrow\; r = 0$ or $r = 2c\cos\theta$.

$r = 0$ is included in $r = 2c\cos\theta$ when $\theta = \frac{\pi}{2} + n\pi$, so the curve is represented by the single equation $r = 2c\cos\theta$.

21. (a) The description leads immediately to the polar equation $\theta = \frac{\pi}{6}$, and the Cartesian equation $\tan\theta = y/x \;\Rightarrow\;$

$y = \left(\tan\frac{\pi}{6}\right)x = \frac{1}{\sqrt{3}}x$ is slightly more difficult to derive.

(b) The easier description here is the Cartesian equation $x = 3$.

23. $\theta = -\pi/6$

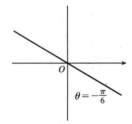

25. $r = \sin\theta \;\Leftrightarrow\; r^2 = r\sin\theta \;\Leftrightarrow\; x^2 + y^2 = y \;\Leftrightarrow\;$

$x^2 + \left(y - \frac{1}{2}\right)^2 = \left(\frac{1}{2}\right)^2$.

The reasoning here is the same as in Exercise 13.

This is a circle of radius $\frac{1}{2}$ centered at $\left(0, \frac{1}{2}\right)$.

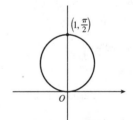

27. $r = 2(1 - \sin\theta)$. This curve is a

cardioid.

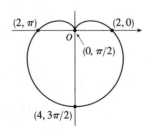

29. $r = \theta, \;\; \theta \geq 0$

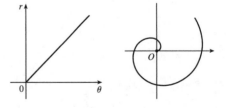

31. $r = \sin 2\theta$

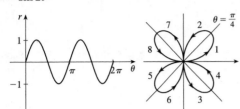

33. $r = 2\cos 4\theta$

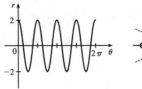

35. $r^2 = 4\cos 2\theta$

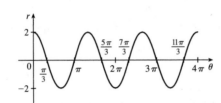 

37. $r = 2\cos\left(\frac{3}{2}\theta\right)$

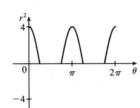 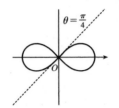

39. $r = 1 + 2\cos 2\theta$

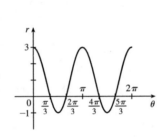 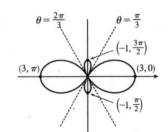

41. For $\theta = 0$, π, and 2π, r has its minimum value of about 0.5. For $\theta = \frac{\pi}{2}$ and $\frac{3\pi}{2}$, r attains its maximum value of 2. We see that the graph has a similar shape for $0 \le \theta \le \pi$ and $\pi \le \theta \le 2\pi$.

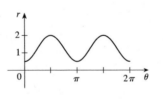 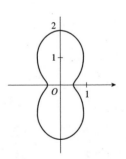

43. $x = r\cos\theta = (4 + 2\sec\theta)\cos\theta = 4\cos\theta + 2$. Now, $r \to \infty$ $\Rightarrow$ $(4 + 2\sec\theta) \to \infty$ $\Rightarrow$ $\theta \to \left(\frac{\pi}{2}\right)^-$ or $\theta \to \left(\frac{3\pi}{2}\right)^+$ (since we need only consider $0 \le \theta < 2\pi$), so $\lim\limits_{r\to\infty} x = \lim\limits_{\theta\to\pi/2^-}(4\cos\theta + 2) = 2$. Also,

$r \to -\infty$ $\Rightarrow$ $(4 + 2\sec\theta) \to -\infty$ $\Rightarrow$ $\theta \to \left(\frac{\pi}{2}\right)^+$ or $\theta \to \left(\frac{3\pi}{2}\right)^-$,

so $\lim\limits_{r\to-\infty} x = \lim\limits_{\theta\to\pi/2^+}(4\cos\theta + 2) = 2$. Therefore, $\lim\limits_{r\to\pm\infty} x = 2$ $\Rightarrow$

$x = 2$ is a vertical asymptote.

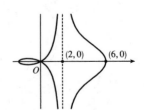

45. To show that $x = 1$ is an asymptote we must prove $\lim\limits_{r \to \pm\infty} x = 1$.

$x = r\cos\theta = (\sin\theta\,\tan\theta)\cos\theta = \sin^2\theta$. Now, $r \to \infty \;\Rightarrow\; \sin\theta\,\tan\theta \to \infty \;\Rightarrow$

$\theta \to \left(\frac{\pi}{2}\right)^-$, so $\lim\limits_{r \to \infty} x = \lim\limits_{\theta \to \pi/2^-} \sin^2\theta = 1$. Also, $r \to -\infty \;\Rightarrow\; \sin\theta\,\tan\theta \to -\infty \;\Rightarrow$

$\theta \to \left(\frac{\pi}{2}\right)^+$, so $\lim\limits_{r \to -\infty} x = \lim\limits_{\theta \to \pi/2^+} \sin^2\theta = 1$. Therefore, $\lim\limits_{r \to \pm\infty} x = 1 \;\Rightarrow\; x = 1$ is a

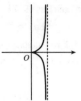

vertical asymptote. Also notice that $x = \sin^2\theta \geq 0$ for all θ, and $x = \sin^2\theta \leq 1$ for all θ. And $x \neq 1$, since the curve is not

defined at odd multiples of $\frac{\pi}{2}$. Therefore, the curve lies entirely within the vertical strip $0 \leq x < 1$.

47. $r = 2\sin\theta \;\Rightarrow\; x = r\cos\theta = 2\sin\theta\,\cos\theta = \sin 2\theta,\; y = r\sin\theta = 2\sin^2\theta \;\Rightarrow$

$\dfrac{dy}{dx} = \dfrac{dy/d\theta}{dx/d\theta} = \dfrac{2\cdot 2\sin\theta\,\cos\theta}{\cos 2\theta \cdot 2} = \dfrac{\sin 2\theta}{\cos 2\theta} = \tan 2\theta$. When $\theta = \dfrac{\pi}{6}$, $\dfrac{dy}{dx} = \tan\!\left(2\cdot\dfrac{\pi}{6}\right) = \tan\dfrac{\pi}{3} = \sqrt{3}$.

49. $r = 1/\theta \;\Rightarrow\; x = r\cos\theta = (\cos\theta)/\theta,\; y = r\sin\theta = (\sin\theta)/\theta \;\Rightarrow$

$$\dfrac{dy}{dx} = \dfrac{dy/d\theta}{dx/d\theta} = \dfrac{\sin\theta(-1/\theta^2) + (1/\theta)\cos\theta}{\cos\theta(-1/\theta^2) - (1/\theta)\sin\theta}\cdot\dfrac{\theta^2}{\theta^2} = \dfrac{-\sin\theta + \theta\cos\theta}{-\cos\theta - \theta\sin\theta}$$

When $\theta = \pi$, $\dfrac{dy}{dx} = \dfrac{-0 + \pi(-1)}{-(-1) - \pi(0)} = \dfrac{-\pi}{1} = -\pi$.

51. $r = 3\cos\theta \;\Rightarrow\; x = r\cos\theta = 3\cos\theta\,\cos\theta,\; y = r\sin\theta = 3\cos\theta\,\sin\theta \;\Rightarrow$

$dy/d\theta = -3\sin^2\theta + 3\cos^2\theta = 3\cos 2\theta = 0 \;\Rightarrow\; 2\theta = \frac{\pi}{2}$ or $\frac{3\pi}{2} \;\Leftrightarrow\; \theta = \frac{\pi}{4}$ or $\frac{3\pi}{4}$. So the tangent is horizontal

at $\left(\frac{3}{\sqrt{2}}, \frac{\pi}{4}\right)$ and $\left(-\frac{3}{\sqrt{2}}, \frac{3\pi}{4}\right)$ $\left[\text{same as }\left(\frac{3}{\sqrt{2}}, -\frac{\pi}{4}\right)\right]$. $dx/d\theta = -6\sin\theta\,\cos\theta = -3\sin 2\theta = 0 \;\Rightarrow\; 2\theta = 0$ or $\pi \;\Leftrightarrow$

$\theta = 0$ or $\frac{\pi}{2}$. So the tangent is vertical at $(3, 0)$ and $\left(0, \frac{\pi}{2}\right)$.

53. $r = 1 + \cos\theta \;\Rightarrow\; x = r\cos\theta = \cos\theta(1 + \cos\theta),\; y = r\sin\theta = \sin\theta(1 + \cos\theta) \;\Rightarrow$

$dy/d\theta = (1 + \cos\theta)\cos\theta - \sin^2\theta = 2\cos^2\theta + \cos\theta - 1 = (2\cos\theta - 1)(\cos\theta + 1) = 0 \;\Rightarrow$

$\cos\theta = \frac{1}{2}$ or $-1 \;\Rightarrow\; \theta = \frac{\pi}{3}, \pi,$ or $\frac{5\pi}{3} \;\Rightarrow\;$ horizontal tangent at $\left(\frac{3}{2}, \frac{\pi}{3}\right),\; (0, \pi)$ [the pole], and $\left(\frac{3}{2}, \frac{5\pi}{3}\right)$.

$dx/d\theta = -(1 + \cos\theta)\sin\theta - \cos\theta\,\sin\theta = -\sin\theta(1 + 2\cos\theta) = 0 \;\Rightarrow\; \sin\theta = 0$ or $\cos\theta = -\frac{1}{2} \;\Rightarrow$

$\theta = 0, \pi, \frac{2\pi}{3},$ or $\frac{4\pi}{3} \;\Rightarrow\;$ vertical tangent at $(2, 0),\; \left(\frac{1}{2}, \frac{2\pi}{3}\right),$ and $\left(\frac{1}{2}, \frac{4\pi}{3}\right)$. Note that the tangent is horizontal, not vertical

when $\theta = \pi$, since $\lim\limits_{\theta \to \pi} \dfrac{dy/d\theta}{dx/d\theta} = 0$.

55. $r = a\sin\theta + b\cos\theta \;\Rightarrow\; r^2 = ar\sin\theta + br\cos\theta \;\Rightarrow\; x^2 + y^2 = ay + bx \;\Rightarrow$

$x^2 - bx + \left(\frac{1}{2}b\right)^2 + y^2 - ay + \left(\frac{1}{2}a\right)^2 = \left(\frac{1}{2}b\right)^2 + \left(\frac{1}{2}a\right)^2 \;\Rightarrow\; \left(x - \frac{1}{2}b\right)^2 + \left(y - \frac{1}{2}a\right)^2 = \frac{1}{4}(a^2 + b^2)$, and this is a circle

with center $\left(\frac{1}{2}b, \frac{1}{2}a\right)$ and radius $\frac{1}{2}\sqrt{a^2 + b^2}$.

Note for Exercises 69–74: Maple is able to plot polar curves using the `polarplot` command, or using the `coords=polar` option in a regular `plot` command. In Mathematica, use `PolarPlot`. In Derive, change to `Polar` under `Options` `State`. If your graphing device cannot plot polar equations, you must convert to parametric equations. For example, in Exercise 69, $x = r \cos\theta = [1 + 2\sin(\theta/2)]\cos\theta$, $y = r\sin\theta = [1 + 2\sin(\theta/2)]\sin\theta$.

57. $r = e^{\sin\theta} - 2\cos(4\theta)$.

The parameter interval is $[0, 2\pi]$.

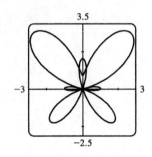

59. $r = 2 - 5\sin(\theta/6)$.

The parameter interval is $[-6\pi, 6\pi]$.

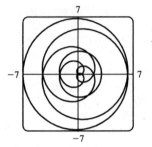

61.

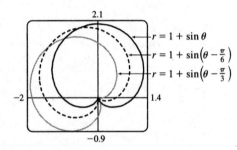

It appears that the graph of $r = 1 + \sin\left(\theta - \frac{\pi}{6}\right)$ is the same shape as the graph of $r = 1 + \sin\theta$, but rotated counter-clockwise about the origin by $\frac{\pi}{6}$. Similarly, the graph of $r = 1 + \sin\left(\theta - \frac{\pi}{3}\right)$ is rotated by $\frac{\pi}{3}$. In general, the graph of $r = f(\theta - \alpha)$ is the same shape as that of $r = f(\theta)$, but rotated counterclockwise through α about the origin. That is, for any point (r_0, θ_0) on the curve $r = f(\theta)$, the point $(r_0, \theta_0 + \alpha)$ is on the curve $r = f(\theta - \alpha)$, since $r_0 = f(\theta_0) = f((\theta_0 + \alpha) - \alpha)$.

63. (a) $r = \sin n\theta$. From the graphs, it seems that when n is even, the number of loops in the curve (called a rose) is $2n$, and when n is odd, the number of loops is simply n. This is because in the case of n odd, every point on the graph is traversed twice, due to the fact that

$$r(\theta + \pi) = \sin[n(\theta + \pi)] = \sin n\theta \cos n\pi + \cos n\theta \sin n\pi = \begin{cases} \sin n\theta & \text{if } n \text{ is even} \\ -\sin n\theta & \text{if } n \text{ is odd} \end{cases}$$

$n = 2$

$n = 3$

$n = 4 \qquad n = 5$

(b) The graph of $r = |\sin n\theta|$ has $2n$ loops whether n is odd or even, since $r(\theta + \pi) = r(\theta)$.

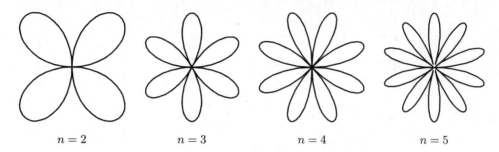

$n = 2$　　　　　$n = 3$　　　　　$n = 4$　　　　　$n = 5$

65. $r = \dfrac{1 - a\cos\theta}{1 + a\cos\theta}$. We start with $a = 0$, since in this case the curve is simply the circle $r = 1$.

As a increases, the graph moves to the left, and its right side becomes flattened. As a increases through about 0.4, the right side seems to grow a dimple, which upon closer investigation (with narrower θ-ranges) seems to appear at $a \approx 0.42$ (the actual value is $\sqrt{2} - 1$). As $a \to 1$, this dimple becomes more pronounced, and the curve begins to stretch out horizontally, until at $a = 1$ the denominator vanishes at $\theta = \pi$, and the dimple becomes an actual cusp. For $a > 1$ we must choose our parameter interval carefully, since $r \to \infty$ as $1 + a\cos\theta \to 0 \iff \theta \to \pm\cos^{-1}(-1/a)$. As a increases from 1, the curve splits into two parts. The left part has a loop, which grows larger as a increases, and the right part grows broader vertically, and its left tip develops a dimple when $a \approx 2.42$ (actually, $\sqrt{2} + 1$). As a increases, the dimple grows more and more pronounced. If $a < 0$, we get the same graph as we do for the corresponding positive a-value, but with a rotation through π about the pole, as happened when c was replaced with $-c$ in Exercise 64.

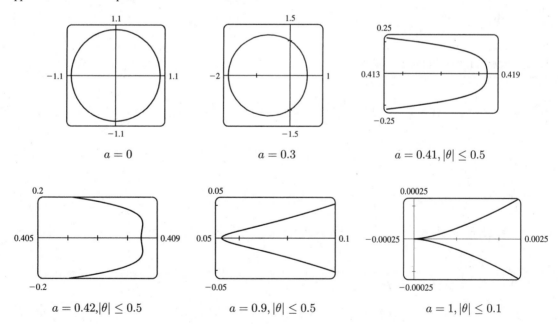

$a = 0$　　　　　$a = 0.3$　　　　　$a = 0.41, |\theta| \le 0.5$

$a = 0.42, |\theta| \le 0.5$　　　　　$a = 0.9, |\theta| \le 0.5$　　　　　$a = 1, |\theta| \le 0.1$

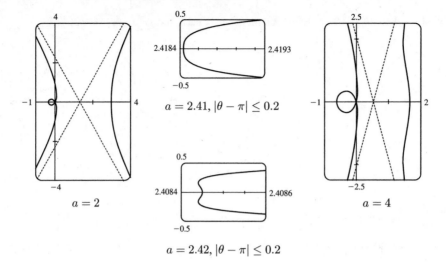

$a = 2.41, |\theta - \pi| \le 0.2$

$a = 2$

$a = 2.42, |\theta - \pi| \le 0.2$

$a = 4$

67.
$$\tan \psi = \tan(\phi - \theta) = \frac{\tan \phi - \tan \theta}{1 + \tan \phi \tan \theta} = \frac{\dfrac{dy}{dx} - \tan \theta}{1 + \dfrac{dy}{dx} \tan \theta} = \frac{\dfrac{dy/d\theta}{dx/d\theta} - \tan \theta}{1 + \dfrac{dy/d\theta}{dx/d\theta} \tan \theta} = \frac{\dfrac{dy}{d\theta} - \dfrac{dx}{d\theta} \tan \theta}{\dfrac{dx}{d\theta} + \dfrac{dy}{d\theta} \tan \theta}$$

$$= \frac{\left(\dfrac{dr}{d\theta} \sin \theta + r \cos \theta \right) - \tan \theta \left(\dfrac{dr}{d\theta} \cos \theta - r \sin \theta \right)}{\left(\dfrac{dr}{d\theta} \cos \theta - r \sin \theta \right) + \tan \theta \left(\dfrac{dr}{d\theta} \sin \theta + r \cos \theta \right)} = \frac{r \cos \theta + r \cdot \dfrac{\sin^2 \theta}{\cos \theta}}{\dfrac{dr}{d\theta} \cos \theta + \dfrac{dr}{d\theta} \cdot \dfrac{\sin^2 \theta}{\cos \theta}}$$

$$= \frac{r \cos^2 \theta + r \sin^2 \theta}{\dfrac{dr}{d\theta} \cos^2 \theta + \dfrac{dr}{d\theta} \sin^2 \theta} = \frac{r}{dr/d\theta}$$

9.4 Areas and Lengths in Polar Coordinates

1. $r = \sqrt{\theta}, \ 0 \le \theta \le \frac{\pi}{4}.$ $A = \int_0^{\pi/4} \frac{1}{2} r^2 \, d\theta = \int_0^{\pi/4} \frac{1}{2} \left(\sqrt{\theta} \right)^2 d\theta = \int_0^{\pi/4} \frac{1}{2} \theta \, d\theta = \left[\frac{1}{4} \theta^2 \right]_0^{\pi/4} = \frac{1}{64} \pi^2$

3. $r = \sin \theta, \ \frac{\pi}{3} \le \theta \le \frac{2\pi}{3}.$

$$A = \int_{\pi/3}^{2\pi/3} \frac{1}{2} \sin^2 \theta \, d\theta = \frac{1}{4} \int_{\pi/3}^{2\pi/3} (1 - \cos 2\theta) \, d\theta = \frac{1}{4} \left[\theta - \frac{1}{2} \sin 2\theta \right]_{\pi/3}^{2\pi/3}$$

$$= \frac{1}{4} \left[\frac{2\pi}{3} - \frac{1}{2} \sin \frac{4\pi}{3} - \frac{\pi}{3} + \frac{1}{2} \sin \frac{2\pi}{3} \right] = \frac{1}{4} \left[\frac{2\pi}{3} - \frac{1}{2} \left(-\frac{\sqrt{3}}{2} \right) - \frac{\pi}{3} + \frac{1}{2} \left(\frac{\sqrt{3}}{2} \right) \right] = \frac{1}{4} \left(\frac{\pi}{3} + \frac{\sqrt{3}}{2} \right) = \frac{\pi}{12} + \frac{\sqrt{3}}{8}$$

5. $r = \theta \ , 0 \le \theta \le \pi.$ $A = \int_0^\pi \frac{1}{2} \theta^2 d\theta = \left[\frac{1}{6} \theta^3 \right]_0^\pi = \frac{1}{6} \pi^3$

7. $r = 4 + 3\sin\theta$, $-\frac{\pi}{2} \le \theta \le \frac{\pi}{2}$.

$$A = \int_{-\pi/2}^{\pi/2} \tfrac{1}{2}(4 + 3\sin\theta)^2 \, d\theta = \tfrac{1}{2}\int_{-\pi/2}^{\pi/2}\left(16 + 24\sin\theta + 9\sin^2\theta\right)d\theta$$

$$= \tfrac{1}{2}\int_{-\pi/2}^{\pi/2}\left(16 + 9\sin^2\theta\right)d\theta \quad \text{[by Theorem 5.5.7(b)]}$$

$$= \tfrac{1}{2}\cdot 2 \int_{0}^{\pi/2}\left[16 + 9\cdot\tfrac{1}{2}(1 - \cos 2\theta)\right]d\theta \quad \text{[by Theorem 5.5.7(a)]}$$

$$= \int_{0}^{\pi/2}\left(\tfrac{41}{2} - \tfrac{9}{2}\cos 2\theta\right)d\theta = \left[\tfrac{41}{2}\theta - \tfrac{9}{4}\sin 2\theta\right]_0^{\pi/2} = \left(\tfrac{41\pi}{4} - 0\right) - (0 - 0) = \tfrac{41\pi}{4}$$

9. The curve $r^2 = 4\cos 2\theta$ goes through the pole when $\theta = \pi/4$,

so we'll find the area for $0 \le \theta \le \pi/4$ and multiply it by 4.

$$A = 4\int_0^{\pi/4} \tfrac{1}{2}r^2 \, d\theta = 2\int_0^{\pi/4}(4\cos 2\theta)\,d\theta = 8\int_0^{\pi/4}\cos 2\theta \, d\theta$$

$$= 4\left[\sin 2\theta\right]_0^{\pi/4} = 4(1 - 0) = 4$$

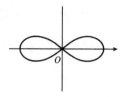

11. One-sixth of the area lies above the polar axis and is bounded

by the curve $r = 2\cos 3\theta$ for $\theta = 0$ to $\theta = \pi/6$.

$$A = 6\int_0^{\pi/6} \tfrac{1}{2}(2\cos 3\theta)^2 \, d\theta = 12\int_0^{\pi/6}\cos^2 3\theta \, d\theta$$

$$= \tfrac{12}{2}\int_0^{\pi/6}(1 + \cos 6\theta)d\theta$$

$$= 6\left[\theta + \tfrac{1}{6}\sin 6\theta\right]_0^{\pi/6} = 6\left(\tfrac{\pi}{6}\right) = \pi$$

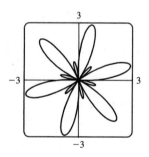

$r = 2\cos 3\theta$

$\theta = \frac{\pi}{6}$

13. $A = \int_0^{2\pi} \tfrac{1}{2}(1 + 2\sin 6\theta)^2 \, d\theta = \tfrac{1}{2}\int_0^{2\pi}(1 + 4\sin 6\theta + 4\sin^2 6\theta)d\theta$

$$= \tfrac{1}{2}\int_0^{2\pi}\left[1 + 4\sin 6\theta + 4\cdot\tfrac{1}{2}(1 - \cos 12\theta)\right]d\theta$$

$$= \tfrac{1}{2}\int_0^{2\pi}(3 + 4\sin 6\theta - 2\cos 12\theta)\,d\theta$$

$$= \tfrac{1}{2}\left[3\theta - \tfrac{2}{3}\cos 6\theta - \tfrac{1}{6}\sin 12\theta\right]_0^{2\pi}$$

$$= \tfrac{1}{2}\left[\left(6\pi - \tfrac{2}{3} - 0\right) - \left(0 - \tfrac{2}{3} - 0\right)\right] = 3\pi.$$

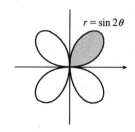

15. The shaded loop is traced out from $\theta = 0$ to $\theta = \pi/2$.

$$A = \int_0^{\pi/2} \tfrac{1}{2}r^2 \, d\theta = \tfrac{1}{2}\int_0^{\pi/2}\sin^2 2\theta \, d\theta$$

$$= \tfrac{1}{2}\int_0^{\pi/2}\tfrac{1}{2}(1 - \cos 4\theta)\,d\theta$$

$$= \tfrac{1}{4}\left[\theta - \tfrac{1}{4}\sin 4\theta\right]_0^{\pi/2} = \tfrac{1}{4}\left(\tfrac{\pi}{2}\right) = \tfrac{\pi}{8}$$

$r = \sin 2\theta$

17.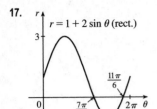

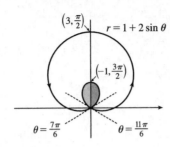

This is a limaçon, with inner loop traced out between $\theta = \frac{7\pi}{6}$ and $\frac{11\pi}{6}$ [found by solving $r = 0$].

$A = 2 \int_{7\pi/6}^{3\pi/2} \frac{1}{2}(1 + 2\sin\theta)^2 \, d\theta = \int_{7\pi/6}^{3\pi/2} (1 + 4\sin\theta + 4\sin^2\theta) \, d\theta = \int_{7\pi/6}^{3\pi/2} \left[1 + 4\sin\theta + 4 \cdot \frac{1}{2}(1 - \cos 2\theta)\right] d\theta$

$= \left[\theta - 4\cos\theta + 2\theta - \sin 2\theta\right]_{7\pi/6}^{3\pi/2} = \left(\frac{9\pi}{2}\right) - \left(\frac{7\pi}{2} + 2\sqrt{3} - \frac{\sqrt{3}}{2}\right) = \pi - \frac{3\sqrt{3}}{2}$

19. $4\sin\theta = 2 \iff \sin\theta = \frac{1}{2} \iff \theta = \frac{\pi}{6}$ or $\frac{5\pi}{6}$ (for $0 \le \theta \le 2\pi$). We'll subtract the unshaded area from the shaded area for $\pi/6 \le \theta \le \pi/2$ and double that value.

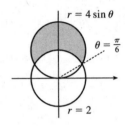

$A = 2 \int_{\pi/6}^{\pi/2} \frac{1}{2}(4\sin\theta)^2 \, d\theta - 2 \int_{\pi/6}^{\pi/2} \frac{1}{2}(2)^2 \, d\theta = 2 \int_{\pi/6}^{\pi/2} \frac{1}{2}\left[(4\sin\theta)^2 - 2^2\right] d\theta$

$= \int_{\pi/6}^{\pi/2} (16\sin^2\theta - 4) \, d\theta = \int_{\pi/6}^{\pi/2} [8(1 - \cos 2\theta) - 4] \, d\theta = \int_{\pi/6}^{\pi/2} (4 - 8\cos 2\theta) \, d\theta$

$= \left[4\theta - 4\sin 2\theta\right]_{\pi/6}^{\pi/2} = (2\pi - 0) - \left(\frac{2\pi}{3} - 4 \cdot \frac{\sqrt{3}}{2}\right) = \frac{4}{3}\pi + 2\sqrt{3}$

21. $3\cos\theta = 1 + \cos\theta \iff \cos\theta = \frac{1}{2} \Rightarrow \theta = \frac{\pi}{3}$ or $-\frac{\pi}{3}$.

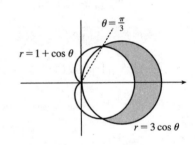

$A = 2 \int_0^{\pi/3} \frac{1}{2}\left[(3\cos\theta)^2 - (1 + \cos\theta)^2\right] d\theta$

$= \int_0^{\pi/3} (8\cos^2\theta - 2\cos\theta - 1) \, d\theta$

$= \int_0^{\pi/3} [4(1 + \cos 2\theta) - 2\cos\theta - 1] \, d\theta$

$= \int_0^{\pi/3} (3 + 4\cos 2\theta - 2\cos\theta) \, d\theta = \left[3\theta + 2\sin 2\theta - 2\sin\theta\right]_0^{\pi/3}$

$= \pi + \sqrt{3} - \sqrt{3} = \pi$

23. $A = 2 \int_0^{\pi/4} \frac{1}{2}\sin^2\theta \, d\theta = \int_0^{\pi/4} \frac{1}{2}(1 - \cos 2\theta) \, d\theta$

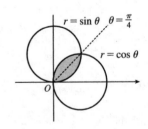

$= \frac{1}{2}\left[\theta - \frac{1}{2}\sin 2\theta\right]_0^{\pi/4} = \frac{1}{2}\left[\left(\frac{\pi}{4} - \frac{1}{2} \cdot 1\right) - (0 - 0)\right]$

$= \frac{1}{8}\pi - \frac{1}{4}$

25. $\sin 2\theta = \cos 2\theta \Rightarrow \dfrac{\sin 2\theta}{\cos 2\theta} = 1 \Rightarrow \tan 2\theta = 1 \Rightarrow$

$2\theta = \frac{\pi}{4} \Rightarrow \theta = \frac{\pi}{8} \Rightarrow$

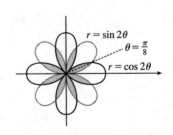

$A = 8 \cdot 2 \int_0^{\pi/8} \frac{1}{2}\sin^2 2\theta \, d\theta = 8 \int_0^{\pi/8} \frac{1}{2}(1 - \cos 4\theta) \, d$

$= 4\left[\theta - \frac{1}{4}\sin 4\theta\right]_0^{\pi/8} = 4\left(\frac{\pi}{8} - \frac{1}{4} \cdot 1\right) = \frac{1}{2}\pi - 1$

27. The darker shaded region (from $\theta = 0$ to $\theta = 2\pi/3$) represents $\frac{1}{2}$ of the desired

area plus $\frac{1}{2}$ of the area of the inner loop. From this area, we'll subtract $\frac{1}{2}$ of the

area of the inner loop (the lighter shaded region from $\theta = 2\pi/3$ to $\theta = \pi$), and

then double that difference to obtain the desired area.

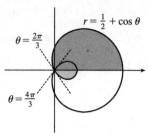

$r = \frac{1}{2} + \cos\theta$

$\theta = \frac{2\pi}{3}$

$\theta = \frac{4\pi}{3}$

$A = 2\left[\int_0^{2\pi/3} \frac{1}{2}\left(\frac{1}{2} + \cos\theta\right)^2 d\theta - \int_{2\pi/3}^\pi \frac{1}{2}\left(\frac{1}{2} + \cos\theta\right)^2 d\theta\right]$

$= \int_0^{2\pi/3}\left(\frac{1}{4} + \cos\theta + \cos^2\theta\right) d\theta - \int_{2\pi/3}^\pi\left(\frac{1}{4} + \cos\theta + \cos^2\theta\right) d\theta$

$= \int_0^{2\pi/3}\left[\frac{1}{4} + \cos\theta + \frac{1}{2}(1 + \cos 2\theta)\right] d\theta - \int_{2\pi/3}^\pi\left[\frac{1}{4} + \cos\theta + \frac{1}{2}(1 + \cos 2\theta)\right] d\theta$

$= \left[\frac{\theta}{4} + \sin\theta + \frac{\theta}{2} + \frac{\sin 2\theta}{4}\right]_0^{2\pi/3} - \left[\frac{\theta}{4} + \sin\theta + \frac{\theta}{2} + \frac{\sin 2\theta}{4}\right]_{2\pi/3}^\pi$

$= \left(\frac{\pi}{6} + \frac{\sqrt{3}}{2} + \frac{\pi}{3} - \frac{\sqrt{3}}{8}\right) - \left(\frac{\pi}{4} + \frac{\pi}{2}\right) + \left(\frac{\pi}{6} + \frac{\sqrt{3}}{2} + \frac{\pi}{3} - \frac{\sqrt{3}}{8}\right) = \frac{\pi}{4} + \frac{3}{4}\sqrt{3} = \frac{1}{4}\left(\pi + 3\sqrt{3}\right)$

29. The curves intersect at the pole since $\left(0, \frac{\pi}{2}\right)$ satisfies

$r = \cos\theta$ and $(0, 0)$ satisfies $r = 1 - \cos\theta$. Now

$\cos\theta = 1 - \cos\theta \;\Rightarrow\; 2\cos\theta = 1 \;\Rightarrow$

$\cos\theta = \frac{1}{2} \;\Rightarrow\; \theta = \frac{\pi}{3}$ or $\frac{5\pi}{3} \;\Rightarrow$

the other intersection points are $\left(\frac{1}{2}, \frac{\pi}{3}\right)$ and $\left(\frac{1}{2}, \frac{5\pi}{3}\right)$.

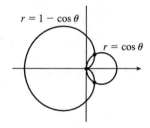

$r = 1 - \cos\theta$

$r = \cos\theta$

31. The pole is a point of intersection.

$\sin\theta = \sin 2\theta = 2\sin\theta\cos\theta \;\Leftrightarrow$

$\sin\theta\,(1 - 2\cos\theta) = 0 \;\Leftrightarrow$

$\sin\theta = 0$ or $\cos\theta = \frac{1}{2} \;\Rightarrow$

$\theta = 0, \pi, \frac{\pi}{3}, -\frac{\pi}{3} \;\Rightarrow\; \left(\frac{\sqrt{3}}{2}, \frac{\pi}{3}\right)$ and $\left(\frac{\sqrt{3}}{2}, \frac{2\pi}{3}\right)$

(by symmetry) are the other intersection points.

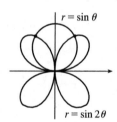

$r = \sin\theta$

$r = \sin 2\theta$

33. $L = \int_a^b \sqrt{r^2 + (dr/d\theta)^2}\, d\theta = \int_0^{\pi/3} \sqrt{(3\sin\theta)^2 + (3\cos\theta)^2}\, d\theta = \int_0^{\pi/3} \sqrt{9(\sin^2\theta + \cos^2\theta)}\, d\theta$

$= 3\int_0^{\pi/3} d\theta = 3[\theta]_0^{\pi/3} = 3\left(\frac{\pi}{3}\right) = \pi.$

As a check, note that the circumference of a circle with radius $\frac{3}{2}$ is $2\pi\left(\frac{3}{2}\right) = 3\pi$, and since $\theta = 0$ to $\pi = \frac{\pi}{3}$ traces out $\frac{1}{3}$ of the

circle (from $\theta = 0$ to $\theta = \pi$), $\frac{1}{3}(3\pi) = \pi.$

35. $L = \int_a^b \sqrt{r^2 + (dr/d\theta)^2}\, d\theta = \int_0^{2\pi} \sqrt{\left(\theta^2\right)^2 + (2\theta)^2}\, d\theta = \int_0^{2\pi} \sqrt{\theta^4 + 4\theta^2}\, d\theta$

$= \int_0^{2\pi} \sqrt{\theta^2\left(\theta^2 + 4\right)}\, d\theta = \int_0^{2\pi} \theta\sqrt{\theta^2 + 4}\, d\theta$

Now let $u = \theta^2 + 4$, so that $du = 2\theta\, d\theta \;\left[\theta\, d\theta = \frac{1}{2}\, du\right]$ and

$\int_0^{2\pi} \theta\sqrt{\theta^2 + 4}\, d\theta = \int_4^{4\pi^2 + 4} \frac{1}{2}\sqrt{u}\, du = \frac{1}{2} \cdot \frac{2}{3}\left[u^{3/2}\right]_4^{4(\pi^2 + 1)} = \frac{1}{3}\left[4^{3/2}(\pi^2 + 1)^{3/2} - 4^{3/2}\right] = \frac{8}{3}\left[(\pi^2 + 1)^{3/2} - 1\right]$

37. The curve $r = 3 \sin 2\theta$ is completely traced with $0 \le \theta \le 2\pi$.

$r^2 + (dr/d\theta)^2 = (3 \sin 2\theta)^2 + (6 \cos 2\theta)^2 \quad \Rightarrow$

$L = \int_0^{2\pi} \sqrt{9 \sin^2 2\theta + 36 \cos^2 2\theta} \, d\theta \approx 29.0653$

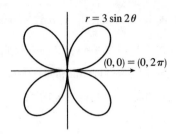

9.5 Conic Sections in Polar Coordinates

1. The directrix $y = 6$ is above the focus at the origin, so we use the form with "$+ e \sin \theta$" in the denominator. (See Theorem 8

and Figure 8.) $r = \dfrac{ed}{1 + e \sin \theta} = \dfrac{\frac{7}{4} \cdot 6}{1 + \frac{7}{4} \sin \theta} = \dfrac{42}{4 + 7 \sin \theta}$

3. The directrix $x = -5$ is to the left of the focus at the origin, so we use the form with "$- e \cos \theta$" in the denominator.

$r = \dfrac{ed}{1 - e \cos \theta} = \dfrac{\frac{3}{4} \cdot 5}{1 - \frac{3}{4} \cos \theta} = \dfrac{15}{4 - 3 \cos \theta}$

5. The vertex $(4, 3\pi/2)$ is 4 units below the focus at the origin, so the directrix is 8 units below the focus ($d = 8$), and we use the

form with "$-e \sin \theta$" in the denominator.

$e = 1$ for a parabola, so an equation is $r = \dfrac{ed}{1 - e \sin \theta} = \dfrac{1(8)}{1 - 1 \sin \theta} = \dfrac{8}{1 - \sin \theta}$.

7. The directrix $r = 4 \sec \theta$ (equivalent to $r \cos \theta = 4$ or $x = 4$) is to the right of the focus at the origin, so we will use the form

with "$+e \cos \theta$" in the denominator. The distance from the focus to the directrix is $d = 4$, so an equation is

$r = \dfrac{ed}{1 + e \cos \theta} = \dfrac{0.5(4)}{1 + 0.5 \cos \theta} \cdot \dfrac{2}{2} = \dfrac{4}{2 + \cos \theta}$.

9. $r = \dfrac{1}{1 + \sin \theta} = \dfrac{ed}{1 + e \sin \theta}$, where $d = e = 1$.

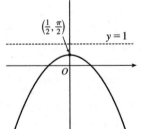

(a) Eccentricity $= e = 1$

(b) Since $e = 1$, the conic is a parabola.

(c) Since "$+e \sin \theta$" appears in the denominator, the directrix is above the focus

at the origin. $d = |Fl| = 1$, so an equation of the directrix is $y = 1$.

(d) The vertex is at $\left(\frac{1}{2}, \frac{\pi}{2}\right)$, midway between the focus and the directrix.

11. $r = \dfrac{12}{4 - \sin\theta} \cdot \dfrac{1/4}{1/4} = \dfrac{3}{1 - \frac{1}{4}\sin\theta}$, where $e = \frac{1}{4}$ and $ed = 3$ $\Rightarrow$ $d = 12$.

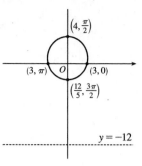

(a) Eccentricity $= e = \frac{1}{4}$

(b) Since $e = \frac{1}{4} < 1$, the conic is an ellipse.

(c) Since "$-e\sin\theta$" appears in the denominator, the directrix is below the focus at the

origin. $d = |Fl| = 12$, so an equation of the directrix is $y = -12$.

(d) The vertices are $\left(4, \frac{\pi}{2}\right)$ and $\left(\frac{12}{5}, \frac{3\pi}{2}\right)$, so the center is midway between them,

that is, $\left(\frac{4}{5}, \frac{\pi}{2}\right)$.

13. $r = \dfrac{9}{6 + 2\cos\theta} \cdot \dfrac{1/6}{1/6} = \dfrac{3/2}{1 + \frac{1}{3}\cos\theta}$, where $e = \frac{1}{3}$ and $ed = \frac{3}{2}$ $\Rightarrow$ $d = \frac{9}{2}$.

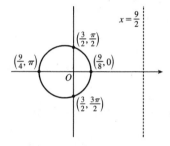

(a) Eccentricity $= e = \frac{1}{3}$

(b) Since $e = \frac{1}{3} < 1$, the conic is an ellipse.

(c) Since "$+e\cos\theta$" appears in the denominator, the directrix is to the right of the

focus at the origin. $d = |Fl| = \frac{9}{2}$, so an equation of the directrix is $x = \frac{9}{2}$.

(d) The vertices are $\left(\frac{9}{8}, 0\right)$ and $\left(\frac{9}{4}, \pi\right)$, so the center is midway between them, that is,

$\left(\frac{9}{16}, \pi\right)$.

15. $r = \dfrac{3}{4 - 8\cos\theta} \cdot \dfrac{1/4}{1/4} = \dfrac{3/4}{1 - 2\cos\theta}$, where $e = 2$ and $ed = \frac{3}{4}$ $\Rightarrow$ $d = \frac{3}{8}$.

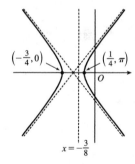

(a) Eccentricity $= e = 2$

(b) Since $e = 2 > 1$, the conic is a hyperbola.

(c) Since "$-e\cos\theta$" appears in the denominator, the directrix is to the left of the focus

at the origin. $d = |Fl| = \frac{3}{8}$, so an equation of the directrix is $x = -\frac{3}{8}$.

(d) The vertices are $\left(-\frac{3}{4}, 0\right)$ and $\left(\frac{1}{4}, \pi\right)$, so the center is midway between them,

that is, $\left(\frac{1}{2}, \pi\right)$.

17. For $e < 1$ the curve is an ellipse. It is nearly circular when e is close to 0. As e

increases, the graph is stretched out to the right, and grows larger (that is, its

right-hand focus moves to the right while its left-hand focus remains at the origin.)

At $e = 1$, the curve becomes a parabola with focus at the origin.

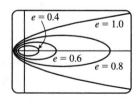

19. $|PF| = e|Pl| \Rightarrow r = e[d - r\cos(\pi - \theta)] = e(d + r\cos\theta) \Rightarrow$

$r(1 - e\cos\theta) = ed \Rightarrow r = \dfrac{ed}{1 - e\cos\theta}$

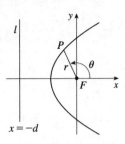

21. $|PF| = e|Pl| \Rightarrow r = e[d - r\sin(\theta - \pi)] = e(d + r\sin\theta) \Rightarrow$

$r(1 - e\sin\theta) = ed \Rightarrow r = \dfrac{ed}{1 - e\sin\theta}$

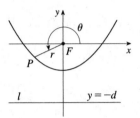

23. (a) If the directrix is $x = -d$, then $r = \dfrac{ed}{1 - e\cos\theta}$ [see Figure 8(b)], and, from (6), $a^2 = \dfrac{e^2 d^2}{(1 - e^2)^2} \Rightarrow$

$ed = a(1 - e^2)$. Therefore, $r = \dfrac{a(1 - e^2)}{1 - e\cos\theta}$.

(b) $e = 0.017$ and the major axis $= 2a = 2.99 \times 10^8 \Rightarrow a = 1.495 \times 10^8$.

Therefore $r = \dfrac{1.495 \times 10^8\,[1 - (0.017)^2]}{1 - 0.017\cos\theta} \approx \dfrac{1.49 \times 10^8}{1 - 0.017\cos\theta}$.

25. Here $2a = $ length of major axis $= 36.18$ AU $\Rightarrow a = 18.09$ AU and $e = 0.97$. By Exercise 23(a), the equation of the orbit

is $r = \dfrac{18.09\,[1 - (0.97)^2]}{1 - 0.97\cos\theta} \approx \dfrac{1.07}{1 - 0.97\cos\theta}$. By Exercise 24(a), the maximum distance from the comet to the sun is

$18.09(1 + 0.97) \approx 35.64$ AU or about 3.314 billion miles.

27. The minimum distance is at perihelion, where $4.6 \times 10^7 = r = a(1 - e) = a(1 - 0.206) = a(0.794) \Rightarrow$

$a = 4.6 \times 10^7/0.794$. So the maximum distance, which is at aphelion, is

$r = a(1 + e) = (4.6 \times 10^7/0.794)(1.206) \approx 7.0 \times 10^7$ km.

29. From Exercise 27, we have $e = 0.206$ and $a(1 - e) = 4.6 \times 10^7$ km. Thus, $a = 4.6 \times 10^7/0.794$. From Exercise 23, we can

write the equation of Mercury's orbit as $r = a\dfrac{1 - e^2}{1 - e\cos\theta}$. So since $\dfrac{dr}{d\theta} = \dfrac{-a(1 - e^2)e\sin\theta}{(1 - e\cos\theta)^2} \Rightarrow$

$r^2 + \left(\dfrac{dr}{d\theta}\right)^2 = \dfrac{a^2(1 - e^2)^2}{(1 - e\cos\theta)^2} + \dfrac{a^2(1 - e^2)^2 e^2\sin^2\theta}{(1 - e\cos\theta)^4} = \dfrac{a^2(1 - e^2)^2}{(1 - e\cos\theta)^4}(1 - 2e\cos\theta + e^2)$

the length of the orbit is $L = \displaystyle\int_0^{2\pi} \sqrt{r^2 + (dr/d\theta)^2}\,d\theta = a(1 - e^2)\int_0^{2\pi} \dfrac{\sqrt{1 + e^2 - 2e\cos\theta}}{(1 - e\cos\theta)^2}\,d\theta \approx 3.6 \times 10^8$ km.

This seems reasonable, since Mercury's orbit is nearly circular, and the circumference of a circle of radius a

is $2\pi a \approx 3.6 \times 10^8$ km.

9 Review

1. (a) A parametric curve is a set of points of the form $(x, y) = (f(t), g(t))$, where f and g are continuous functions of a variable t.

   (b) Sketching a parametric curve, like sketching the graph of a function, is difficult to do in general. We can plot points on the curve by finding $f(t)$ and $g(t)$ for various values of t, either by hand or with a calculator or computer. Sometimes, when f and g are given by formulas, we can eliminate t from the equations $x = f(t)$ and $y = g(t)$ to get a Cartesian equation relating x and y. It may be easier to graph that equation than to work with the original formulas for x and y in terms of t.

2. (a) You can find $\dfrac{dy}{dx}$ as a function of t by calculating $\dfrac{dy}{dx} = \dfrac{dy/dt}{dx/dt}$ (if $dx/dt \neq 0$).

   (b) Calculate the area as $\int_a^b y\, dx = \int_\alpha^\beta g(t) f'(t)\, dt$ [or $\int_\beta^\alpha g(t) f'(t)\, dt$ if the leftmost point is $(f(\beta), g(\beta))$ rather than $(f(\alpha), g(\alpha))$].

3. $L = \int_\alpha^\beta \sqrt{(dx/dt)^2 + (dy/dt)^2}\, dt = \int_\alpha^\beta \sqrt{[f'(t)]^2 + [g'(t)]^2}\, dt$

4. (a) See Figure 5 in Section 9.3.

   (b) $x = r\cos\theta,\ y = r\sin\theta$

   (c) To find a polar representation (r, θ) with $r \geq 0$ and $0 \leq \theta < 2\pi$, first calculate $r = \sqrt{x^2 + y^2}$. Then θ is specified by $\cos\theta = x/r$ and $\sin\theta = y/r$.]ifnum1=1

5. (a) Calculate $\dfrac{dy}{dx} = \dfrac{\dfrac{dy}{d\theta}}{\dfrac{dx}{d\theta}} = \dfrac{\dfrac{d}{d\theta}(y)}{\dfrac{d}{d\theta}(x)} = \dfrac{\dfrac{d}{d\theta}(r\sin\theta)}{\dfrac{d}{d\theta}(r\cos\theta)} = \dfrac{\left(\dfrac{dr}{d\theta}\right)\sin\theta + r\cos\theta}{\left(\dfrac{dr}{d\theta}\right)\cos\theta - r\sin\theta}$, where $r = f(\theta)$.

   (b) Calculate $A = \int_a^b \frac{1}{2}r^2\, d\theta = \int_a^b \frac{1}{2}[f(\theta)]^2\, d\theta$.

   (c) $L = \int_a^b \sqrt{(dx/d\theta)^2 + (dy/d\theta)^2}\, d\theta = \int_a^b \sqrt{r^2 + (dr/d\theta)^2}\, d\theta = \int_a^b \sqrt{[f(\theta)]^2 + [f'(\theta)]^2}\, d\theta$

6. (a) If a conic section has focus F and corresponding directrix l, then the eccentricity e is the fixed ratio $|PF|\,/\,|Pl|$ for points P of the conic section.

   (b) $e < 1$ for an ellipse; $e > 1$ for a hyperbola; $e = 1$ for a parabola.

   (c) $x = d$: $r = \dfrac{ed}{1 + e\cos\theta}$; $\quad x = -d$: $r = \dfrac{ed}{1 - e\cos\theta}$; $\quad y = d$: $r = \dfrac{ed}{1 + e\sin\theta}$; $\quad y = -d$: $r = \dfrac{ed}{1 - e\sin\theta}$.

TRUE-FALSE QUIZ

1. False. Consider the curve defined by $x = f(t) = (t-1)^3$ and $y = g(t) = (t-1)^2$. Then $g'(t) = 2(t-1)$, so $g'(1) = 0$, but its graph has a *vertical* tangent when $t = 1$. *Note:* The statement is true if $f'(1) \neq 0$ when $g'(1) = 0$.

3. False. For example, if $f(t) = \cos t$ and $g(t) = \sin t$ for $0 \leq t \leq 4\pi$, then the curve is a circle of radius 1, hence its length is 2π, but $\int_0^{4\pi} \sqrt{[f'(t)]^2 + [g'(t)]^2} \, dt = \int_0^{4\pi} \sqrt{(-\sin t)^2 + (\cos t)^2} \, dt = \int_0^{4\pi} 1 \, dt = 4\pi$, since as t increases from 0 to 4π, the circle is traversed twice.

5. True. The curve $r = 1 - \sin 2\theta$ is unchanged if we rotate it through $180°$ about O because

$$1 - \sin 2(\theta + \pi) = 1 - \sin(2\theta + 2\pi) = 1 - \sin 2\theta.$$ So it's unchanged if we replace r by $-r$. In other words, it's the same curve as $r = -(1 - \sin 2\theta) = \sin 2\theta - 1$.

7. False. The first pair of equations yields the portion of the parabola $y = x^2$ with $x \geq 0$, whereas the second pair of equations traces out the whole parabola $y = x^2$.

EXERCISES

1. $x = t^2 + 4t$, $y = 2 - t$, $-4 \leq t \leq 1$. $t = 2 - y$, so

$x = (2-y)^2 + 4(2-y) = 4 - 4y + y^2 + 8 - 4y = y^2 - 8y + 12$ ⟺

$x + 4 = y^2 - 8y + 16 = (y-4)^2$. This is part of a parabola with

vertex $(-4, 4)$, opening to the right.

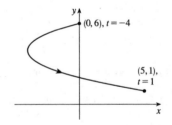

3. $x = \tan \theta$, $y = \cot \theta$. $y = 1/\tan \theta = 1/x$. The whole curve is traced

out as θ ranges over the open interval $\left(-\frac{\pi}{2}, \frac{\pi}{2}\right)$ [or any open interval of the

form $\left(-\frac{\pi}{2} + n\pi, \frac{\pi}{2} + n\pi\right)$, where n is an integer].

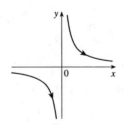

5. Three different sets of parametric equations for the curve $y = \sqrt{x}$ are

(i) $x = t$, $y = \sqrt{t}$, $t \geq 0$

(ii) $x = t^4$, $y = t^2$

(iii) $x = \tan^2 t$, $y = \tan t$, $0 \leq t < \pi/2$

There are many other sets of equations that also give this curve.

7. $r = 1 - \cos\theta$. This cardioid is symmetric about the polar axis.

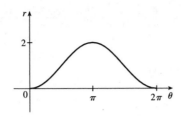

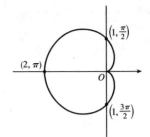

9. $r = 1 + \cos 2\theta$. The curve is symmetric about the pole and both the horizontal and vertical axes.

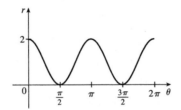

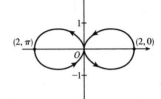

11. $r^2 = \sec 2\theta \quad\Rightarrow$

$r^2 \cos 2\theta = 1 \quad\Rightarrow$

$r^2(\cos^2\theta - \sin^2\theta) = 1 \quad\Rightarrow$

$r^2 \cos^2\theta - r^2 \sin^2\theta = 1 \quad\Rightarrow$

$x^2 - y^2 = 1$, a hyperbola

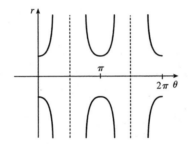

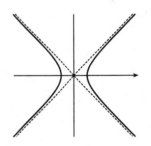

13. $r = \dfrac{1}{1 + \cos\theta} \quad\Rightarrow\quad e = 1 \quad\Rightarrow\quad$ parabola; $d = 1 \quad\Rightarrow\quad$ directrix $x = 1$

and vertex $\left(\frac{1}{2}, 0\right)$; y-intercepts are $\left(1, \frac{\pi}{2}\right)$ and $\left(1, \frac{3\pi}{2}\right)$.

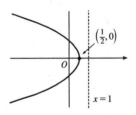

15. $x + y = 2 \quad\Leftrightarrow\quad r\cos\theta + r\sin\theta = 2 \quad\Leftrightarrow\quad r(\cos\theta + \sin\theta) = 2 \quad\Leftrightarrow\quad r = \dfrac{2}{\cos\theta + \sin\theta}$

17. $r = (\sin\theta)/\theta$. As $\theta \to \pm\infty$, $r \to 0$. As $\theta \to 0$, $r \to 1$. In the first figure, there are an infinite number of x-intercepts at $x = \pi n$, n a nonzero integer. These correspond to pole points in the second figure.

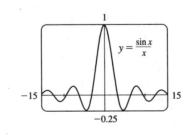

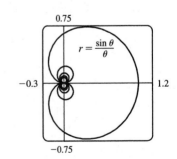

19. $x = \ln t$, $y = 1 + t^2$; $t = 1$. $\quad \dfrac{dy}{dt} = 2t$ and $\dfrac{dx}{dt} = \dfrac{1}{t}$, so $\dfrac{dy}{dx} = \dfrac{dy/dt}{dx/dt} = \dfrac{2t}{1/t} = 2t^2$.

When $t = 1$, $(x, y) = (0, 2)$ and $dy/dx = 2$.

21. $r = e^{-\theta} \quad \Rightarrow \quad y = r \sin \theta = e^{-\theta} \sin \theta$ and $x = r \cos \theta = e^{-\theta} \cos \theta \quad \Rightarrow$

$$\frac{dy}{dx} = \frac{dy/d\theta}{dx/d\theta} = \frac{\frac{dr}{d\theta} \sin \theta + r \cos \theta}{\frac{dr}{d\theta} \cos \theta - r \sin \theta} = \frac{-e^{-\theta} \sin \theta + e^{-\theta} \cos \theta}{-e^{-\theta} \cos \theta - e^{-\theta} \sin \theta} \cdot \frac{-e^{\theta}}{-e^{\theta}} = \frac{\sin \theta - \cos \theta}{\cos \theta + \sin \theta}.$$

When $\theta = \pi$, $\dfrac{dy}{dx} = \dfrac{0 - (-1)}{-1 + 0} = \dfrac{1}{-1} = -1$.

23. $x = t \cos t$, $y = t \sin t$. $\quad \dfrac{dy}{dx} = \dfrac{dy/dt}{dx/dt} = \dfrac{t \cos t + \sin t}{-t \sin t + \cos t}$. $\dfrac{d^2 y}{dx^2} = \dfrac{\frac{d}{dt}\left(\frac{dy}{dx}\right)}{dx/dt}$, where

$$\frac{d}{dt}\left(\frac{dy}{dx}\right) = \frac{(-t \sin t + \cos t)(-t \sin t + 2 \cos t) - (t \cos t + \sin t)(-t \cos t - 2 \sin t)}{(-t \sin t + \cos t)^2} = \frac{t^2 + 2}{(-t \sin t + \cos t)^2} \quad \Rightarrow$$

$$\frac{d^2 y}{dx^2} = \frac{t^2 + 2}{(-t \sin t + \cos t)^3}.$$

25. We graph the curve $x = t^3 - 3t$, $y = t^2 + t + 1$ for $-2.2 \le t \le 1.2$.

By zooming in or using a cursor, we find that the lowest point is about

$(1.4, 0.75)$. To find the exact values, we find the t-value at which

$dy/dt = 2t + 1 = 0 \quad \Leftrightarrow \quad t = -\tfrac{1}{2} \quad \Leftrightarrow \quad (x, y) = \left(\tfrac{11}{8}, \tfrac{3}{4}\right)$.

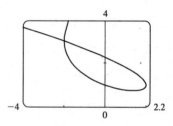

27. $x = 2a \cos t - a \cos 2t \quad \Rightarrow \quad \dfrac{dx}{dt} = -2a \sin t + 2a \sin 2t = 2a \sin t(2 \cos t - 1) = 0 \quad \Leftrightarrow \quad \sin t = 0$ or $\cos t = \tfrac{1}{2} \quad \Rightarrow$

$t = 0$, $\tfrac{\pi}{3}$, π, or $\tfrac{5\pi}{3}$.

$y = 2a \sin t - a \sin 2t \quad \Rightarrow \quad \dfrac{dy}{dt} = 2a \cos t - 2a \cos 2t = 2a(1 + \cos t - 2 \cos^2 t) = 2a(1 - \cos t)(1 + 2 \cos t) = 0 \quad \Rightarrow$

$t = 0$, $\tfrac{2\pi}{3}$, or $\tfrac{4\pi}{3}$.

Thus, the graph has vertical tangents where $t = \tfrac{\pi}{3}$, π and $\tfrac{5\pi}{3}$, and horizontal tangents where $t = \tfrac{2\pi}{3}$ and $\tfrac{4\pi}{3}$. To determine

what the slope is where $t = 0$, we use l'Hospital's Rule to evaluate $\lim\limits_{t \to 0} \dfrac{dy/dt}{dx/dt} = 0$, so there is a horizontal tangent there.

t	x	y
0	a	0
$\tfrac{\pi}{3}$	$\tfrac{3}{2}a$	$\tfrac{\sqrt{3}}{2}a$
$\tfrac{2\pi}{3}$	$-\tfrac{1}{2}a$	$\tfrac{3\sqrt{3}}{2}a$
π	$-3a$	0
$\tfrac{4\pi}{3}$	$-\tfrac{1}{2}a$	$-\tfrac{3\sqrt{3}}{2}a$
$\tfrac{5\pi}{3}$	$\tfrac{3}{2}a$	$-\tfrac{\sqrt{3}}{2}a$

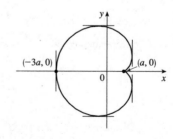

29. The curve $r^2 = 9 \cos 5\theta$ has 10 "petals." For instance, for $-\frac{\pi}{10} \leq \theta \leq \frac{\pi}{10}$, there are two petals, one with $r > 0$ and one

with $r < 0$. $A = 10 \int_{-\pi/10}^{\pi/10} \frac{1}{2}r^2 \, d\theta = 5 \int_{-\pi/10}^{\pi/10} 9 \cos 5\theta \, d\theta = 5 \cdot 9 \cdot 2 \int_0^{\pi/10} \cos 5\theta \, d\theta = 18 \left[\sin 5\theta \right]_0^{\pi/10} = 18.$

31. The curves intersect when $4 \cos \theta = 2 \quad \Rightarrow \quad \cos \theta = \frac{1}{2} \quad \Rightarrow$

$\theta = \pm \frac{\pi}{3}$ for $-\pi \leq \theta \leq \pi$. The points of intersection are

$\left(2, \frac{\pi}{3}\right)$ and $\left(2, -\frac{\pi}{3}\right)$.

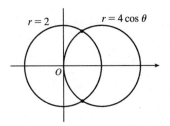

33. The curves intersect where $2 \sin \theta = \sin \theta + \cos \theta \quad \Rightarrow$

$\sin \theta = \cos \theta \quad \Rightarrow \theta = \frac{\pi}{4}$, and also at the origin (at which $\theta = \frac{3\pi}{4}$

on the second curve).

$A = \int_0^{\pi/4} \frac{1}{2}(2 \sin \theta)^2 \, d\theta + \int_{\pi/4}^{3\pi/4} \frac{1}{2}(\sin \theta + \cos \theta)^2 \, d\theta$

$= \int_0^{\pi/4}(1 - \cos 2\theta)s d\theta + \frac{1}{2}\int_{\pi/4}^{3\pi/4}(1 + \sin 2\theta) \, d\theta$

$= \left[\theta - \frac{1}{2}\sin 2\theta\right]_0^{\pi/4} + \left[\frac{1}{2}\theta - \frac{1}{4}\cos 2\theta\right]_{\pi/4}^{3\pi/4} = \frac{1}{2}(\pi - 1)$

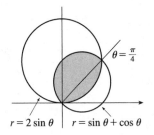

35. $x = 3t^2, \; y = 2t^3.$

$$L = \int_0^2 \sqrt{(dx/dt)^2 + (dy/dt)^2} \, dt = \int_0^2 \sqrt{(6t)^2 + (6t^2)^2} \, dt$$

$$= \int_0^2 \sqrt{36t^2 + 36t^4} \, dt = \int_0^2 \sqrt{36t^2} \sqrt{1 + t^2} \, dt$$

$$= \int_0^2 6\,|t|\, \sqrt{1 + t^2} \, dt = 6 \int_0^2 t \sqrt{1 + t^2} \, dt = 6 \int_1^5 u^{1/2} \left(\tfrac{1}{2} du\right) \qquad \left[u = 1 + t^2, \, du = 2t \, dt\right]$$

$$= 6 \cdot \tfrac{1}{2} \cdot \tfrac{2}{3}\left[u^{3/2}\right]_1^5 = 2(5^{3/2} - 1) = 2\left(5\sqrt{5} - 1\right)$$

37. $L = \int_\pi^{2\pi} \sqrt{r^2 + (dr/d\theta)^2} \, d\theta = \int_\pi^{2\pi} \sqrt{(1/\theta)^2 + \left(-1/\theta^2\right)^2} \, d\theta = \int_\pi^{2\pi} \dfrac{\sqrt{\theta^2 + 1}}{\theta^2} \, d\theta$

$\overset{24}{=} \left[-\dfrac{\sqrt{\theta^2 + 1}}{\theta} + \ln\left(\theta + \sqrt{\theta^2 + 1}\right)\right]_\pi^{2\pi} = \dfrac{\sqrt{\pi^2 + 1}}{\pi} - \dfrac{\sqrt{4\pi^2 + 1}}{2\pi} + \ln\left(\dfrac{2\pi + \sqrt{4\pi^2 + 1}}{\pi + \sqrt{\pi^2 + 1}}\right)$

$= \dfrac{2\sqrt{\pi^2 + 1} - \sqrt{4\pi^2 + 1}}{2\pi} + \ln\left(\dfrac{2\pi + \sqrt{4\pi^2 + 1}}{\pi + \sqrt{\pi^2 + 1}}\right)$

39. For all c except -1, the curve is asymptotic to the line $x = 1$. For $c < -1$,

the curve bulges to the right near $y = 0$. As c increases, the bulge becomes

smaller, until at $c = -1$ the curve is the straight line $x = 1$. As c continues

to increase, the curve bulges to the left, until at $c = 0$ there is a cusp at the

origin. For $c > 0$, there is a loop to the left of the origin, whose size and

roundness increase as c increases. Note that the x-intercept of the curve is

always $-c$.

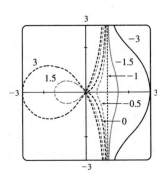

41. Directrix $x = 4$ $\Rightarrow$ $d = 4$, so $e = \frac{1}{3}$ $\Rightarrow$ $r = \dfrac{ed}{1 + e\cos\theta} = \dfrac{4}{3 + \cos\theta}$.

43. In polar coordinates, an equation for the circle is $r = 2a\sin\theta$. Thus, the coordinates of Q are $x = r\cos\theta = 2a\sin\theta\,\cos\theta$

and $y = r\sin\theta = 2a\sin^2\theta$. The coordinates of R are $x = 2a\cot\theta$ and $y = 2a$. Since P is the midpoint of QR, we use the

midpoint formula to get $x = a(\sin\theta\,\cos\theta + \cot\theta)$ and $y = a(1 + \sin^2\theta)$.

10 □ VECTORS AND THE GEOMETRY OF SPACE

10.1 Three-Dimensional Coordinate Systems

1. We start at the origin, which has coordinates $(0, 0, 0)$. First we move 4 units along the positive x-axis, affecting only the x-coordinate, bringing us to the point $(4, 0, 0)$. We then move 3 units straight downward, in the negative z-direction. Thus only the z-coordinate is affected, and we arrive at $(4, 0, -3)$.

3. The distance from a point to the xz-plane is the absolute value of the y-coordinate of the point. $Q(-5, -1, 4)$ has the y-coordinate with the smallest absolute value, so Q is the point closest to the xz-plane. $R(0, 3, 8)$ must lie in the yz-plane since the distance from R to the yz-plane, given by the x-coordinate of R, is 0.

5. The equation $x + y = 2$ represents the set of all points in $\mathbb{R}^3$ whose x- and y-coordinates have a sum of 2, or equivalently where $y = 2 - x$. This is the set $\{(x, 2 - x, z) \mid x \in \mathbb{R}, z \in \mathbb{R}\}$ which is a vertical plane that intersects the xy-plane in the line $y = 2 - x$, $z = 0$.

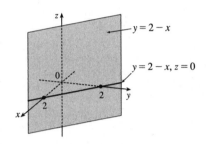

7. (a) We can find the lengths of the sides of the triangle by using the distance formula between pairs of vertices:

$$|PQ| = \sqrt{(7 - 3)^2 + [0 - (-2)]^2 + [1 - (-3)]^2} = \sqrt{16 + 4 + 16} = 6$$
$$|QR| = \sqrt{(1 - 7)^2 + (2 - 0)^2 + (1 - 1)^2} = \sqrt{36 + 4 + 0} = \sqrt{40} = 2\sqrt{10}$$
$$|RP| = \sqrt{(3 - 1)^2 + (-2 - 2)^2 + (-3 - 1)^2} = \sqrt{4 + 16 + 16} = 6$$

The longest side is QR, but the Pythagorean Theorem is not satisfied: $|PQ|^2 + |RP|^2 \neq |QR|^2$. Thus PQR is not a right triangle. PQR is isosceles, as two sides have the same length.

(b) Compute the lengths of the sides of the triangle by using the distance formula between pairs of vertices:

$$|PQ| = \sqrt{(4 - 2)^2 + [1 - (-1)]^2 + (1 - 0)^2} = \sqrt{4 + 4 + 1} = 3$$
$$|QR| = \sqrt{(4 - 4)^2 + (-5 - 1)^2 + (4 - 1)^2} = \sqrt{0 + 36 + 9} = \sqrt{45} = 3\sqrt{5}$$
$$|RP| = \sqrt{(2 - 4)^2 + [-1 - (-5)]^2 + (0 - 4)^2} = \sqrt{4 + 16 + 16} = 6$$

Since the Pythagorean Theorem is satisfied by $|PQ|^2 + |RP|^2 = |QR|^2$, PQR is a right triangle. PQR is not isosceles, as no two sides have the same length.

9. (a) First we find the distances between points:

$$|AB| = \sqrt{(3-2)^2 + (7-4)^2 + (-2-2)^2} = \sqrt{26}$$

$$|BC| = \sqrt{(1-3)^2 + (3-7)^2 + [3-(-2)]^2} = \sqrt{45} = 3\sqrt{5}$$

$$|AC| = \sqrt{(1-2)^2 + (3-4)^2 + (3-2)^2} = \sqrt{3}$$

In order for the points to lie on a straight line, the sum of the two shortest distances must be equal to the longest distance.

Since $\sqrt{26} + \sqrt{3} \neq 3\sqrt{5}$, the three points do not lie on a straight line.

(b) First we find the distances between points:

$$|DE| = \sqrt{(1-0)^2 + [-2-(-5)]^2 + (4-5)^2} = \sqrt{11}$$

$$|EF| = \sqrt{(3-1)^2 + [4-(-2)]^2 + (2-4)^2} = \sqrt{44} = 2\sqrt{11}$$

$$|DF| = \sqrt{(3-0)^2 + [4-(-5)]^2 + (2-5)^2} = \sqrt{99} = 3\sqrt{11}$$

Since $|DE| + |EF| = |DF|$, the three points lie on a straight line.

11. The radius of the sphere is the distance between $(4, 3, -1)$ and $(3, 8, 1)$: $r = \sqrt{(3-4)^2 + (8-3)^2 + [1-(-1)]^2} = \sqrt{30}$.

Thus, an equation of the sphere is $(x-3)^2 + (y-8)^2 + (z-1)^2 = 30$.

13. Completing squares in the equation $x^2 + y^2 + z^2 - 6x + 4y - 2z = 11$ gives

$(x^2 - 6x + 9) + (y^2 + 4y + 4) + (z^2 - 2z + 1) = 11 + 9 + 4 + 1 \quad \Rightarrow \quad (x-3)^2 + (y+2)^2 + (z-1)^2 = 25$, which we

recognize as an equation of a sphere with center $(3, -2, 1)$ and radius 5.

15. Completing squares in the equation gives $\left(x^2 - x + \frac{1}{4}\right) + \left(y^2 - y + \frac{1}{4}\right) + \left(z^2 - z + \frac{1}{4}\right) = \frac{1}{4} + \frac{1}{4} + \frac{1}{4} \quad \Rightarrow$

$\left(x - \frac{1}{2}\right)^2 + \left(y - \frac{1}{2}\right)^2 + \left(z - \frac{1}{2}\right)^2 = \frac{3}{4}$ which we recognize as an equation of a sphere with center $\left(\frac{1}{2}, \frac{1}{2}, \frac{1}{2}\right)$ and

radius $\sqrt{\frac{3}{4}} = \frac{\sqrt{3}}{2}$.

17. (a) If the midpoint of the line segment from $P_1(x_1, y_1, z_1)$ to $P_2(x_2, y_2, z_2)$ is $Q = \left(\dfrac{x_1 + x_2}{2}, \dfrac{y_1 + y_2}{2}, \dfrac{z_1 + z_2}{2}\right)$,

then the distances $|P_1Q|$ and $|QP_2|$ are equal, and each is half of $|P_1P_2|$. We verify that this is the case:

$$|P_1P_2| = \sqrt{(x_2 - x_1)^2 + (y_2 - y_1)^2 + (z_2 - z_1)^2}$$

$$|P_1Q| = \sqrt{\left[\tfrac{1}{2}(x_1 + x_2) - x_1\right]^2 + \left[\tfrac{1}{2}(y_1 + y_2) - y_1\right]^2 + \left[\tfrac{1}{2}(z_1 + z_2) - z_1\right]^2}$$

$$= \sqrt{\left(\tfrac{1}{2}x_2 - \tfrac{1}{2}x_1\right)^2 + \left(\tfrac{1}{2}y_2 - \tfrac{1}{2}y_1\right)^2 + \left(\tfrac{1}{2}z_2 - \tfrac{1}{2}z_1\right)^2}$$

$$= \sqrt{\left(\tfrac{1}{2}\right)^2 \left[(x_2 - x_1)^2 + (y_2 - y_1)^2 + (z_2 - z_1)^2\right]} = \tfrac{1}{2}\sqrt{(x_2 - x_1)^2 + (y_2 - y_1)^2 + (z_2 - z_1)^2}$$

$$= \tfrac{1}{2}|P_1P_2|$$

$$|QP_2| = \sqrt{\left[x_2 - \tfrac{1}{2}(x_1 + x_2)\right]^2 + \left[y_2 - \tfrac{1}{2}(y_1 + y_2)\right]^2 + \left[z_2 - \tfrac{1}{2}(z_1 + z_2)\right]^2}$$

$$= \sqrt{\left(\tfrac{1}{2}x_2 - \tfrac{1}{2}x_1\right)^2 + \left(\tfrac{1}{2}y_2 - \tfrac{1}{2}y_1\right)^2 + \left(\tfrac{1}{2}z_2 - \tfrac{1}{2}z_1\right)^2} = \tfrac{1}{2}\sqrt{(x_2 - x_1)^2 + (y_2 - y_1)^2 + (z_2 - z_1)^2}$$

$$= \sqrt{\left(\tfrac{1}{2}\right)^2 \left[(x_2 - x_1)^2 + (y_2 - y_1)^2 + (z_2 - z_1)^2\right]} = \tfrac{1}{2}|P_1 P_2|$$

So Q is indeed the midpoint of $P_1 P_2$.

(b) By part (a), the midpoints of sides AB, BC and CA are $P_1\left(-\tfrac{1}{2}, 1, 4\right)$, $P_2\left(1, \tfrac{1}{2}, 5\right)$ and $P_3\left(\tfrac{5}{2}, \tfrac{3}{2}, 4\right)$. (Recall that a median of a triangle is a line segment from a vertex to the midpoint of the opposite side.) Then the lengths of the medians are:

$$|AP_2| = \sqrt{0^2 + \left(\tfrac{1}{2} - 2\right)^2 + (5 - 3)^2} = \sqrt{\tfrac{9}{4} + 4} = \sqrt{\tfrac{25}{4}} = \tfrac{5}{2}$$

$$|BP_3| = \sqrt{\left(\tfrac{5}{2} + 2\right)^2 + \left(\tfrac{3}{2}\right)^2 + (4 - 5)^2} = \sqrt{\tfrac{81}{4} + \tfrac{9}{4} + 1} = \sqrt{\tfrac{94}{4}} = \tfrac{1}{2}\sqrt{94}$$

$$|CP_1| = \sqrt{\left(-\tfrac{1}{2} - 4\right)^2 + (1 - 1)^2 + (4 - 5)^2} = \sqrt{\tfrac{81}{4} + 1} = \tfrac{1}{2}\sqrt{85}$$

19. (a) Since the sphere touches the xy-plane, its radius is the distance from its center, $(2, -3, 6)$, to the xy-plane, namely 6. Therefore $r = 6$ and an equation of the sphere is $(x - 2)^2 + (y + 3)^2 + (z - 6)^2 = 6^2 = 36$.

(b) The radius of this sphere is the distance from its center $(2, -3, 6)$ to the yz-plane, which is 2. Therefore, an equation is

$$(x - 2)^2 + (y + 3)^2 + (z - 6)^2 = 4.$$

(c) Here the radius is the distance from the center $(2, -3, 6)$ to the xz-plane, which is 3. Therefore, an equation is

$$(x - 2)^2 + (y + 3)^2 + (z - 6)^2 = 9.$$

21. The equation $y = -4$ represents a plane parallel to the xz-plane and 4 units to the left of it.

23. The inequality $x > 3$ represents a half-space consisting of all points in front of the plane $x = 3$.

25. The inequality $0 \le z \le 6$ represents all points on or between the horizontal planes $z = 0$ (the xy-plane) and $z = 6$.

27. The inequality $x^2 + y^2 + z^2 \le 3$ is equivalent to $\sqrt{x^2 + y^2 + z^2} \le \sqrt{3}$, so the region consists of those points whose distance from the origin is at most $\sqrt{3}$. This is the set of all points on or inside the sphere with radius $\sqrt{3}$ and center $(0, 0, 0)$.

29. Here $x^2 + z^2 \le 9$ or equivalently $\sqrt{x^2 + z^2} \le 3$ which describes the set of all points in $\mathbb{R}^3$ whose distance from the y-axis is at most 3. Thus, the inequality represents the region consisting of all points on or inside a circular cylinder of radius 3 with axis the y-axis.

31. This describes all points with negative y-coordinates, that is, $y < 0$.

33. This describes a region all of whose points have a distance to the origin which is greater than r, but smaller than R. So inequalities describing the region are $r < \sqrt{x^2 + y^2 + z^2} < R$, or $r^2 < x^2 + y^2 + z^2 < R^2$.

35. We need to find a set of points $\{P(x, y, z) \mid |AP| = |BP|\}$.

$$\sqrt{(x+1)^2 + (y-5)^2 + (z-3)^2} = \sqrt{(x-6)^2 + (y-2)^2 + (z+2)^2} \quad \Rightarrow$$

$$(x+1)^2 + (y-5) + (z-3)^2 = (x-6)^2 + (y-2)^2 + (z+2)^2 \quad \Rightarrow$$

$$x^2 + 2x + 1 + y^2 - 10y + 25 + z^2 - 6z + 9 = x^2 - 12x + 36 + y^2 - 4y + 4 + z^2 + 4z + 4 \quad \Rightarrow \quad 14x - 6y - 10z = 9.$$

Thus the set of points is a plane perpendicular to the line segment joining A and B (since this plane must contain the perpendicular bisector of the line segment AB).

10.2 Vectors

1. Vectors are equal when they share the same length and direction (but not necessarily location). Using the symmetry of the parallelogram as a guide, we see that $\overrightarrow{AB} = \overrightarrow{DC}, \overrightarrow{DA} = \overrightarrow{CB}, \overrightarrow{DE} = \overrightarrow{EB}$, and $\overrightarrow{EA} = \overrightarrow{CE}$.

3. (a)

(b)
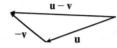

(c)

(d)
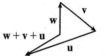

5. $\mathbf{a} = \langle -2 - 2, 1 - 3 \rangle = \langle -4, -2 \rangle$

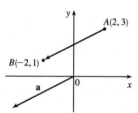

7. $\mathbf{a} = \langle 2 - 0, 3 - 3, -1 - 1 \rangle = \langle 2, 0, -2 \rangle$

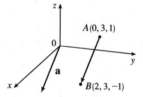

9. $\langle 3, -1 \rangle + \langle -2, 4 \rangle = \langle 3 + (-2), -1 + 4 \rangle$

$\qquad = \langle 1, 3 \rangle$

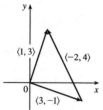

11. $\langle 0, 1, 2 \rangle + \langle 0, 0, -3 \rangle = \langle 0 + 0, 1 + 0, 2 + (-3) \rangle$

$\qquad = \langle 0, 1, -1 \rangle$

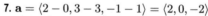

13. $\mathbf{a} + \mathbf{b} = \langle 5 + (-3), -12 + (-6) \rangle = \langle 2, -18 \rangle$

$2\mathbf{a} + 3\mathbf{b} = \langle 10, -24 \rangle + \langle -9, -18 \rangle = \langle 1, -42 \rangle$

$|\mathbf{a}| = \sqrt{5^2 + (-12)^2} = \sqrt{169} = 13$

$|\mathbf{a} - \mathbf{b}| = |\langle 5 - (-3), -12 - (-6) \rangle| = |\langle 8, -6 \rangle| = \sqrt{8^2 + (-6)^2} = \sqrt{100} = 10$

15. $\mathbf{a} + \mathbf{b} = (\mathbf{i} + 2\mathbf{j} - 3\mathbf{k}) + (-2\mathbf{i} - \mathbf{j} + 5\mathbf{k}) = -\mathbf{i} + \mathbf{j} + 2\mathbf{k}$

$2\mathbf{a} + 3\mathbf{b} = 2(\mathbf{i} + 2\mathbf{j} - 3\mathbf{k}) + 3(-2\mathbf{i} - \mathbf{j} + 5\mathbf{k}) = 2\mathbf{i} + 4\mathbf{j} - 6\mathbf{k} - 6\mathbf{i} - 3\mathbf{j} + 15\mathbf{k} = -4\mathbf{i} + \mathbf{j} + 9\mathbf{k}$

$|\mathbf{a}| = \sqrt{1^2 + 2^2 + (-3)^2} = \sqrt{14}$

$|\mathbf{a} - \mathbf{b}| = |(\mathbf{i} + 2\mathbf{j} - 3\mathbf{k}) - (-2\mathbf{i} - \mathbf{j} + 5\mathbf{k})| = |3\mathbf{i} + 3\mathbf{j} - 8\mathbf{k}| = \sqrt{3^2 + 3^2 + (-8)^2} = \sqrt{82}$

17. The vector $8\mathbf{i} - \mathbf{j} + 4\mathbf{k}$ has length $|8\mathbf{i} - \mathbf{j} + 4\mathbf{k}| = \sqrt{8^2 + (-1)^2 + 4^2} = \sqrt{81} = 9$, so by Equation 4 the unit vector with

the same direction is $\frac{1}{9}(8\mathbf{i} - \mathbf{j} + 4\mathbf{k}) = \frac{8}{9}\mathbf{i} - \frac{1}{9}\mathbf{j} + \frac{4}{9}\mathbf{k}$.

19. From the figure, we see that the x-component of $\mathbf{v}$ is

$v_1 = |\mathbf{v}|\cos(\pi/3) = 4 \cdot \frac{1}{2} = 2$ and the y-component is

$v_2 = |\mathbf{v}|\sin(\pi/3) = 4 \cdot \frac{\sqrt{3}}{2} = 2\sqrt{3}$. Thus

$\mathbf{v} = \langle v_1, v_2 \rangle = \langle 2, 2\sqrt{3} \rangle$.

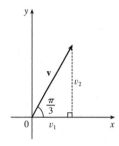

21. $|\mathbf{F}_1| = 10$ lb and $|\mathbf{F}_2| = 12$ lb.

$\mathbf{F}_1 = -|\mathbf{F}_1|\cos 45°\mathbf{i} + |\mathbf{F}_1|\sin 45°\mathbf{j} = -10\cos 45°\mathbf{i} + 10\sin 45°\mathbf{j} = -5\sqrt{2}\mathbf{i} + 5\sqrt{2}\mathbf{j}$

$\mathbf{F}_2 = |\mathbf{F}_2|\cos 30°\mathbf{i} + |\mathbf{F}_2|\sin 30°\mathbf{j} = 12\cos 30°\mathbf{i} + 12\sin 30°\mathbf{j} = 6\sqrt{3}\mathbf{i} + 6\mathbf{j}$

$\mathbf{F} = \mathbf{F}_1 + \mathbf{F}_2 = (6\sqrt{3} - 5\sqrt{2})\mathbf{i} + (6 + 5\sqrt{2})\mathbf{j} \approx 3.32\mathbf{i} + 13.07\mathbf{j}$

$|\mathbf{F}| \approx \sqrt{(3.32)^2 + (13.07)^2} \approx 13.5$ lb. $\tan\theta = \dfrac{6 + 5\sqrt{2}}{6\sqrt{3} - 5\sqrt{2}} \;\Rightarrow\; \theta = \tan^{-1}\dfrac{6 + 5\sqrt{2}}{6\sqrt{3} - 5\sqrt{2}} \approx 76°.$

23. With respect to the water's surface, the woman's velocity is the vector sum of the velocity of the ship with respect to the water,

and the woman's velocity with respect to the ship. If we let north be the positive y-direction, then

$\mathbf{v} = \langle 0, 22 \rangle + \langle -3, 0 \rangle = \langle -3, 22 \rangle$. The woman's speed is $|\mathbf{v}| = \sqrt{9 + 484} \approx 22.2$ mi/h. The vector $\mathbf{v}$ makes an angle θ

with the east, where $\theta = \tan^{-1}\left(\frac{22}{-3}\right) \approx 98°$. Therefore, the woman's direction is about $\mathrm{N}(98 - 90)°\mathrm{W} = \mathrm{N}8°\mathrm{W}$.

25. Let $\mathbf{T}_1$ and $\mathbf{T}_2$ represent the tension vectors in each side of the

clothesline as shown in the figure. $\mathbf{T}_1$ and $\mathbf{T}_2$ have equal vertical

components and opposite horizontal components, so we can write

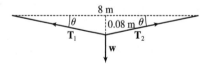

$\mathbf{T}_1 = -a\mathbf{i} + b\mathbf{j}$ and $\mathbf{T}_2 = a\mathbf{i} + b\mathbf{j}$ $[a, b > 0]$. By similar triangles, $\dfrac{b}{a} = \dfrac{0.08}{4}$ $\Rightarrow$ $a = 50b$. The force due to gravity

acting on the shirt has magnitude $0.8g \approx (0.8)(9.8) = 7.84$ N, hence we have $\mathbf{w} = -7.84\mathbf{j}$. The resultant $\mathbf{T}_1 + \mathbf{T}_2$

of the tensile forces counterbalances $\mathbf{w}$, so $\mathbf{T}_1 + \mathbf{T}_2 = -\mathbf{w}$ $\Rightarrow$ $(-a\mathbf{i} + b\mathbf{j}) + (a\mathbf{i} + b\mathbf{j}) = 7.84\mathbf{j}$ $\Rightarrow$

$(-50b\mathbf{i} + b\mathbf{j}) + (50b\mathbf{i} + b\mathbf{j}) = 2b\mathbf{j} = 7.84\mathbf{j}$ $\Rightarrow$ $b = \frac{7.84}{2} = 3.92$ and $a = 50b = 196$. Thus the tensions are

$\mathbf{T}_1 = -a\mathbf{i} + b\mathbf{j} = -196\mathbf{i} + 3.92\mathbf{j}$ and $\mathbf{T}_2 = a\mathbf{i} + b\mathbf{j} = 196\mathbf{i} + 3.92\mathbf{j}$.

Alternatively, we can find the value of θ and proceed as in Example 7.

27. (a), (b)

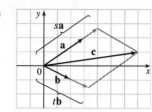

(c) From the sketch, we estimate that $s \approx 1.3$ and $t \approx 1.6$.

(d) $\mathbf{c} = s\,\mathbf{a} + t\,\mathbf{b}$ $\Leftrightarrow$ $7 = 3s + 2t$ and $1 = 2s - t$.

Solving these equations gives $s = \frac{9}{7}$ and $t = \frac{11}{7}$.

29. $|\mathbf{r} - \mathbf{r}_0|$ is the distance between the points (x, y, z) and (x_0, y_0, z_0), so the set of points is a sphere with radius 1 and center (x_0, y_0, z_0).

Alternate method: $|\mathbf{r} - \mathbf{r}_0| = 1$ $\Leftrightarrow$ $\sqrt{(x - x_0)^2 + (y - y_0)^2 + (z - z_0)^2} = 1$ $\Leftrightarrow$

$(x - x_0)^2 + (y - y_0)^2 + (z - z_0)^2 = 1$, which is the equation of a sphere with radius 1 and center (x_0, y_0, z_0).

31. $\mathbf{a} + (\mathbf{b} + \mathbf{c}) = \langle a_1, a_2 \rangle + (\langle b_1, b_2 \rangle + \langle c_1, c_2 \rangle) = \langle a_1, a_2 \rangle + \langle b_1 + c_1, b_2 + c_2 \rangle$

$$= \langle a_1 + b_1 + c_1, a_2 + b_2 + c_2 \rangle = \langle (a_1 + b_1) + c_1, (a_2 + b_2) + c_2 \rangle$$

$$= \langle a_1 + b_1, a_2 + b_2 \rangle + \langle c_1, c_2 \rangle = (\langle a_1, a_2 \rangle + \langle b_1, b_2 \rangle) + \langle c_1, c_2 \rangle$$

$$= (\mathbf{a} + \mathbf{b}) + \mathbf{c}$$

33. Consider triangle ABC, where D and E are the midpoints of AB and BC. We know that $\overrightarrow{AB} + \overrightarrow{BC} = \overrightarrow{AC}$ **(1)** and $\overrightarrow{DB} + \overrightarrow{BE} = \overrightarrow{DE}$ **(2)**. However, $\overrightarrow{DB} = \frac{1}{2}\overrightarrow{AB}$, and $\overrightarrow{BE} = \frac{1}{2}\overrightarrow{BC}$. Substituting these expressions for $\overrightarrow{DB}$ and $\overrightarrow{BE}$ into **(2)** gives $\frac{1}{2}\overrightarrow{AB} + \frac{1}{2}\overrightarrow{BC} = \overrightarrow{DE}$. Comparing this with **(1)** gives $\overrightarrow{DE} = \frac{1}{2}\overrightarrow{AC}$. Therefore $\overrightarrow{AC}$ and $\overrightarrow{DE}$ are parallel and $\left|\overrightarrow{DE}\right| = \frac{1}{2}\left|\overrightarrow{AC}\right|$.

10.3 The Dot Product

1. (a) $\mathbf{a} \cdot \mathbf{b}$ is a scalar, and the dot product is defined only for vectors, so $(\mathbf{a} \cdot \mathbf{b}) \cdot \mathbf{c}$ has no meaning.

(b) $(\mathbf{a} \cdot \mathbf{b})\,\mathbf{c}$ is a scalar multiple of a vector, so it does have meaning.

(c) Both $|\mathbf{a}|$ and $\mathbf{b} \cdot \mathbf{c}$ are scalars, so $|\mathbf{a}|\,(\mathbf{b} \cdot \mathbf{c})$ is an ordinary product of real numbers, and has meaning.

(d) Both $\mathbf{a}$ and $\mathbf{b} + \mathbf{c}$ are vectors, so the dot product $\mathbf{a} \cdot (\mathbf{b} + \mathbf{c})$ has meaning.

(e) $\mathbf{a} \cdot \mathbf{b}$ is a scalar, but $\mathbf{c}$ is a vector, and so the two quantities cannot be added and $\mathbf{a} \cdot \mathbf{b} + \mathbf{c}$ has no meaning.

(f) $|\mathbf{a}|$ is a scalar, and the dot product is defined only for vectors, so $|\mathbf{a}| \cdot (\mathbf{b} + \mathbf{c})$ has no meaning.

3. $\mathbf{a} \cdot \mathbf{b} = |\mathbf{a}|\,|\mathbf{b}| \cos\theta = (6)(5) \cos\frac{2\pi}{3} = 30\left(-\frac{1}{2}\right) = -15$

5. $\mathbf{a} \cdot \mathbf{b} = \langle 4, 1, \frac{1}{4} \rangle \cdot \langle 6, -3, -8 \rangle = (4)(6) + (1)(-3) + \left(\frac{1}{4}\right)(-8) = 19$

7. $\mathbf{a} \cdot \mathbf{b} = (\mathbf{i} - 2\mathbf{j} + 3\mathbf{k}) \cdot (5\mathbf{i} + 9\mathbf{k}) = (1)(5) + (-2)(0) + (3)(9) = 32$

9. $\mathbf{u}$, $\mathbf{v}$, and $\mathbf{w}$ are all unit vectors, so the triangle is an equilateral triangle. Thus the angle between $\mathbf{u}$ and $\mathbf{v}$ is $60°$ and $\mathbf{u} \cdot \mathbf{v} = |\mathbf{u}|\,|\mathbf{v}| \cos 60° = (1)(1)\left(\frac{1}{2}\right) = \frac{1}{2}$. If $\mathbf{w}$ is moved so it has the same initial point as $\mathbf{u}$, we can see that the angle between them is $120°$ and we have $\mathbf{u} \cdot \mathbf{w} = |\mathbf{u}|\,|\mathbf{w}| \cos 120° = (1)(1)\left(-\frac{1}{2}\right) = -\frac{1}{2}$.

11. (a) $\mathbf{i} \cdot \mathbf{j} = \langle 1, 0, 0 \rangle \cdot \langle 0, 1, 0 \rangle = (1)(0) + (0)(1) + (0)(0) = 0$. Similarly, $\mathbf{j} \cdot \mathbf{k} = (0)(0) + (1)(0) + (0)(1) = 0$ and

$\mathbf{k} \cdot \mathbf{i} = (0)(1) + (0)(0) + (1)(0) = 0$.

Another method: Because $\mathbf{i}$, $\mathbf{j}$, and $\mathbf{k}$ are mutually perpendicular, the cosine factor in each dot product is $\cos \frac{\pi}{2} = 0$.

(b) By Property 1 of the dot product, $\mathbf{i} \cdot \mathbf{i} = |\mathbf{i}|^2 = 1^2 = 1$ since $\mathbf{i}$ is a unit vector. Similarly, $\mathbf{j} \cdot \mathbf{j} = |\mathbf{j}|^2 = 1$ and

$\mathbf{k} \cdot \mathbf{k} = |\mathbf{k}|^2 = 1$.

13. $|\mathbf{a}| = \sqrt{(-8)^2 + 6^2} = 10$, $|\mathbf{b}| = \sqrt{\left(\sqrt{7}\right)^2 + 3^2} = 4$, and $\mathbf{a} \cdot \mathbf{b} = (-8)\left(\sqrt{7}\right) + (6)(3) = 18 - 8\sqrt{7}$. From the definition

of the dot product, we have $\cos\theta = \dfrac{\mathbf{a} \cdot \mathbf{b}}{|\mathbf{a}|\,|\mathbf{b}|} = \dfrac{18 - 8\sqrt{7}}{10 \cdot 4} = \dfrac{9 - 4\sqrt{7}}{20}$. So the angle between $\mathbf{a}$ and $\mathbf{b}$ is

$\theta = \cos^{-1}\left(\dfrac{9 - 4\sqrt{7}}{20}\right) \approx 95°$.

15. $|\mathbf{a}| = \sqrt{0^2 + 1^2 + 1^2} = \sqrt{2}$, $|\mathbf{b}| = \sqrt{1^2 + 2^2 + (-3)^2} = \sqrt{14}$, and $\mathbf{a} \cdot \mathbf{b} = (0)(1) + (1)(2) + (1)(-3) = -1$.

Then $\cos\theta = \dfrac{\mathbf{a} \cdot \mathbf{b}}{|\mathbf{a}|\,|\mathbf{b}|} = \dfrac{-1}{\sqrt{2} \cdot \sqrt{14}} = \dfrac{-1}{2\sqrt{7}}$ and $\theta = \cos^{-1}\left(-\dfrac{1}{2\sqrt{7}}\right) \approx 101°$.

17. (a) $\mathbf{a} \cdot \mathbf{b} = (-5)(6) + (3)(-8) + (7)(2) = -40 \neq 0$, so $\mathbf{a}$ and $\mathbf{b}$ are not orthogonal. Also, since $\mathbf{a}$ is not a scalar multiple

of $\mathbf{b}$, $\mathbf{a}$ and $\mathbf{b}$ are not parallel.

(b) $\mathbf{a} \cdot \mathbf{b} = (4)(-3) + (6)(2) = 0$, so $\mathbf{a}$ and $\mathbf{b}$ are orthogonal (and not parallel).

(c) $\mathbf{a} \cdot \mathbf{b} = (-1)(3) + (2)(4) + (5)(-1) = 0$, so $\mathbf{a}$ and $\mathbf{b}$ are orthogonal (and not parallel).

(d) Because $\mathbf{a} = -\frac{2}{3}\,\mathbf{b}$, $\mathbf{a}$ and $\mathbf{b}$ are parallel.

19. $\overrightarrow{QP} = \langle -1, -3, 2 \rangle$, $\overrightarrow{QR} = \langle 4, -2, -1 \rangle$, and $\overrightarrow{QP} \cdot \overrightarrow{QR} = -4 + 6 - 2 = 0$. Thus $\overrightarrow{QP}$ and $\overrightarrow{QR}$ are orthogonal, so the angle of

the triangle at vertex Q is a right angle.

21. Let $\mathbf{a} = a_1\,\mathbf{i} + a_2\,\mathbf{j} + a_3\,\mathbf{k}$ be a vector orthogonal to both $\mathbf{i} + \mathbf{j}$ and $\mathbf{i} + \mathbf{k}$. Then $\mathbf{a} \cdot (\mathbf{i} + \mathbf{j}) = 0 \iff a_1 + a_2 = 0$ and

$\mathbf{a} \cdot (\mathbf{i} + \mathbf{k}) = 0 \iff a_1 + a_3 = 0$, so $a_1 = -a_2 = -a_3$. Furthermore $\mathbf{a}$ is to be a unit vector, so $1 = a_1^2 + a_2^2 + a_3^2 = 3a_1^2$

implies $a_1 = \pm\frac{1}{\sqrt{3}}$. Thus $\mathbf{a} = \frac{1}{\sqrt{3}}\,\mathbf{i} - \frac{1}{\sqrt{3}}\,\mathbf{j} - \frac{1}{\sqrt{3}}\,\mathbf{k}$ and $\mathbf{a} = -\frac{1}{\sqrt{3}}\,\mathbf{i} + \frac{1}{\sqrt{3}}\,\mathbf{j} + \frac{1}{\sqrt{3}}\,\mathbf{k}$ are two such unit vectors.

23. $|\mathbf{a}| = \sqrt{3^2 + (-4)^2} = 5$. The scalar projection of $\mathbf{b}$ onto $\mathbf{a}$ is $\text{comp}_{\mathbf{a}}\,\mathbf{b} = \dfrac{\mathbf{a} \cdot \mathbf{b}}{|\mathbf{a}|} = \dfrac{3 \cdot 5 + (-4) \cdot 0}{5} = 3$ and the vector

projection of $\mathbf{b}$ onto $\mathbf{a}$ is $\text{proj}_{\mathbf{a}}\,\mathbf{b} = \left(\dfrac{\mathbf{a} \cdot \mathbf{b}}{|\mathbf{a}|}\right)\dfrac{\mathbf{a}}{|\mathbf{a}|} = 3 \cdot \frac{1}{5}\langle 3, -4 \rangle = \left\langle \frac{9}{5}, -\frac{12}{5} \right\rangle$.

25. $|\mathbf{a}| = \sqrt{9 + 36 + 4} = 7$ so the scalar projection of $\mathbf{b}$ onto $\mathbf{a}$ is $\text{comp}_{\mathbf{a}}\mathbf{b} = \dfrac{\mathbf{a} \cdot \mathbf{b}}{|\mathbf{a}|} = \frac{1}{7}(3 + 12 - 6) = \frac{9}{7}$. The vector

projection of $\mathbf{b}$ onto $\mathbf{a}$ is $\text{proj}_{\mathbf{a}}\mathbf{b} = \dfrac{9}{7}\dfrac{\mathbf{a}}{|\mathbf{a}|} = \frac{9}{7} \cdot \frac{1}{7}\langle 3, 6, -2 \rangle = \frac{9}{49}\langle 3, 6, -2 \rangle = \left\langle \frac{27}{49}, \frac{54}{49}, -\frac{18}{49} \right\rangle$.

27. $(\text{orth}_a \, \mathbf{b}) \cdot \mathbf{a} = (\mathbf{b} - \text{proj}_a \, \mathbf{b}) \cdot \mathbf{a} = \mathbf{b} \cdot \mathbf{a} - (\text{proj}_a \, \mathbf{b}) \cdot \mathbf{a} = \mathbf{b} \cdot \mathbf{a} - \dfrac{\mathbf{a} \cdot \mathbf{b}}{|\mathbf{a}|^2} \, \mathbf{a} \cdot \mathbf{a} = \mathbf{b} \cdot \mathbf{a} - \dfrac{\mathbf{a} \cdot \mathbf{b}}{|\mathbf{a}|^2} \, |\mathbf{a}|^2 = \mathbf{b} \cdot \mathbf{a} - \mathbf{a} \cdot \mathbf{b} = 0.$

So they are orthogonal by (7).

29. $\text{comp}_a \, \mathbf{b} = \dfrac{\mathbf{a} \cdot \mathbf{b}}{|\mathbf{a}|} = 2 \iff \mathbf{a} \cdot \mathbf{b} = 2\,|\mathbf{a}| = 2\sqrt{10}.$ If $\mathbf{b} = \langle b_1, b_2, b_3 \rangle$, then we need $3b_1 + 0b_2 - 1b_3 = 2\sqrt{10}.$

One possible solution is obtained by taking $b_1 = 0, b_2 = 0, b_3 = -2\sqrt{10}.$ In general, $\mathbf{b} = \langle s, t, 3s - 2\sqrt{10}\,\rangle,\ s, t \in \mathbb{R}.$

31. Here $\mathbf{D} = (4 - 2)\,\mathbf{i} + (9 - 3)\,\mathbf{j} + (15 - 0)\,\mathbf{k} = 2\,\mathbf{i} + 6\,\mathbf{j} + 15\,\mathbf{k}$ so $W = \mathbf{F} \cdot \mathbf{D} = 20 + 108 - 90 = 38$ J.

33. $W = |\mathbf{F}|\,|\mathbf{D}| \cos\theta = (25)(10) \cos 20° \approx 235$ ft-lb

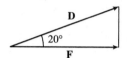

35. First note that $\mathbf{n} = \langle a, b \rangle$ is perpendicular to the line, because if $Q_1 = (a_1, b_1)$ and $Q_2 = (a_2, b_2)$ lie on the line, then

$\mathbf{n} \cdot \overrightarrow{Q_1 Q_2} = aa_2 - aa_1 + bb_2 - bb_1 = 0$, since $aa_2 + bb_2 = -c = aa_1 + bb_1$ from the equation of the line.

Let $P_2 = (x_2, y_2)$ lie on the line. Then the distance from P_1 to the line is the absolute value of the scalar projection

of $\overrightarrow{P_1 P_2}$ onto $\mathbf{n}$. $\text{comp}_\mathbf{n}\left(\overrightarrow{P_1 P_2}\right) = \dfrac{|\mathbf{n} \cdot \langle x_2 - x_1, y_2 - y_1 \rangle|}{|\mathbf{n}|} = \dfrac{|ax_2 - ax_1 + by_2 - by_1|}{\sqrt{a^2 + b^2}} = \dfrac{|ax_1 + by_1 + c|}{\sqrt{a^2 + b^2}}$

since $ax_2 + by_2 = -c$. The required distance is $\dfrac{|3 \cdot -2 + -4 \cdot 3 + 5|}{\sqrt{3^2 + 4^2}} = \dfrac{13}{5}.$

37. For convenience, consider the unit cube positioned so that its back left corner is at the origin, and its edges lie along the

coordinate axes. The diagonal of the cube that begins at the origin and ends at $(1, 1, 1)$ has vector representation $\langle 1, 1, 1 \rangle$.

The angle θ between this vector and the vector of the edge which also begins at the origin and runs along the x-axis [that is,

$\langle 1, 0, 0 \rangle$] is given by $\cos\theta = \dfrac{\langle 1, 1, 1 \rangle \cdot \langle 1, 0, 0 \rangle}{|\langle 1, 1, 1 \rangle| \, |\langle 1, 0, 0 \rangle|} = \dfrac{1}{\sqrt{3}} \Rightarrow \theta = \cos^{-1}\left(\dfrac{1}{\sqrt{3}}\right) \approx 55°.$

39. Consider the H—C—H combination consisting of the sole carbon atom and the two hydrogen atoms that are at $(1, 0, 0)$ and

$(0, 1, 0)$ (or any H—C—H combination, for that matter). Vector representations of the line segments emanating from the

carbon atom and extending to these two hydrogen atoms are $\langle 1 - \frac{1}{2}, 0 - \frac{1}{2}, 0 - \frac{1}{2} \rangle = \langle \frac{1}{2}, -\frac{1}{2}, -\frac{1}{2} \rangle$ and

$\langle 0 - \frac{1}{2}, 1 - \frac{1}{2}, 0 - \frac{1}{2} \rangle = \langle -\frac{1}{2}, \frac{1}{2}, -\frac{1}{2} \rangle.$ The bond angle, θ, is therefore given by

$\cos\theta = \dfrac{\langle \frac{1}{2}, -\frac{1}{2}, -\frac{1}{2} \rangle \cdot \langle -\frac{1}{2}, \frac{1}{2}, -\frac{1}{2} \rangle}{|\langle \frac{1}{2}, -\frac{1}{2}, -\frac{1}{2} \rangle| \, |\langle -\frac{1}{2}, \frac{1}{2}, -\frac{1}{2} \rangle|} = \dfrac{-\frac{1}{4} - \frac{1}{4} + \frac{1}{4}}{\sqrt{\frac{3}{4}}\sqrt{\frac{3}{4}}} = -\dfrac{1}{3} \Rightarrow \theta = \cos^{-1}\left(-\frac{1}{3}\right) \approx 109.5°.$

41. Let $\mathbf{a} = \langle a_1, a_2, a_3 \rangle$ and $= \langle b_1, b_2, b_3 \rangle$.

Property 2: $\mathbf{a} \cdot \mathbf{b} = \langle a_1, a_2, a_3 \rangle \cdot \langle b_1, b_2, b_3 \rangle = a_1 b_1 + a_2 b_2 + a_3 b_3$

$\qquad\qquad = b_1 a_1 + b_2 a_2 + b_3 a_3 = \langle b_1, b_2, b_3 \rangle \cdot \langle a_1, a_2, a_3 \rangle = \mathbf{b} \cdot \mathbf{a}$

Property 4: $(c\,\mathbf{a}) \cdot \mathbf{b} = \langle ca_1, ca_2, ca_3 \rangle \cdot \langle b_1, b_2, b_3 \rangle = (ca_1)b_1 + (ca_2)b_2 + (ca_3)b_3$

$\qquad\qquad\qquad = c\,(a_1 b_1 + a_2 b_2 + a_3 b_3) = c\,(\mathbf{a} \cdot \mathbf{b}) = a_1(cb_1) + a_2(cb_2) + a_3(cb_3)$

$\qquad\qquad\qquad = \langle a_1, a_2, a_3 \rangle \cdot \langle cb_1, cb_2, cb_3 \rangle = \mathbf{a} \cdot (c\,\mathbf{b})$

Property 5: $\mathbf{0} \cdot \mathbf{a} = \langle 0, 0, 0 \rangle \cdot \langle a_1, a_2, a_3 \rangle = (0)(a_1) + (0)(a_2) + (0)(a_3) = 0$

43. $|\mathbf{a} \cdot \mathbf{b}| = \big|\,|\mathbf{a}|\,|\mathbf{b}| \cos\theta\,\big| = |\mathbf{a}|\,|\mathbf{b}|\,|\cos\theta|$. Since $|\cos\theta| \le 1$, $|\mathbf{a} \cdot \mathbf{b}| = |\mathbf{a}|\,|\mathbf{b}|\,|\cos\theta| \le |\mathbf{a}|\,|\mathbf{b}|$.

Note: We have equality in the case of $\cos\theta = \pm 1$, so $\theta = 0$ or $\theta = \pi$, thus equality when $\mathbf{a}$ and $\mathbf{b}$ are parallel.

45. (a)

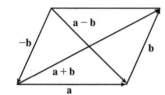

The Parallelogram Law states that the sum of the squares of the lengths of the diagonals of a parallelogram equals the sum of the squares of its (four) sides.

(b) $|\mathbf{a} + \mathbf{b}|^2 = (\mathbf{a} + \mathbf{b}) \cdot (\mathbf{a} + \mathbf{b}) = |\mathbf{a}|^2 + 2(\mathbf{a} \cdot \mathbf{b}) + |\mathbf{b}|^2$ and $|\mathbf{a} - \mathbf{b}|^2 = (\mathbf{a} - \mathbf{b}) \cdot (\mathbf{a} - \mathbf{b}) = |\mathbf{a}|^2 - 2(\mathbf{a} \cdot \mathbf{b}) + |\mathbf{b}|^2$.

Adding these two equations gives $|\mathbf{a} + \mathbf{b}|^2 + |\mathbf{a} - \mathbf{b}|^2 = 2\,|\mathbf{a}|^2 + 2\,|\mathbf{b}|^2$.

10.4 The Cross Product

1. $\mathbf{a} \times \mathbf{b} = \begin{vmatrix} \mathbf{i} & \mathbf{j} & \mathbf{k} \\ 1 & 2 & 0 \\ 0 & 3 & 1 \end{vmatrix} = \begin{vmatrix} 2 & 0 \\ 3 & 1 \end{vmatrix} \mathbf{i} - \begin{vmatrix} 1 & 0 \\ 0 & 1 \end{vmatrix} \mathbf{j} + \begin{vmatrix} 1 & 2 \\ 0 & 3 \end{vmatrix} \mathbf{k} = (2 - 0)\,\mathbf{i} - (1 - 0)\,\mathbf{j} + (3 - 0)\,\mathbf{k} = 2\,\mathbf{i} - \mathbf{j} + 3\,\mathbf{k}$

Now $(\mathbf{a} \times \mathbf{b}) \cdot \mathbf{a} = \langle 2, -1, 3 \rangle \cdot \langle 1, 2, 0 \rangle = 2 - 2 + 0 = 0$ and $(\mathbf{a} \times \mathbf{b}) \cdot \mathbf{b} = \langle 2, -1, 3 \rangle \cdot \langle 0, 3, 1 \rangle = 0 - 3 + 3 = 0$,

so $\mathbf{a} \times \mathbf{b}$ is orthogonal to both $\mathbf{a}$ and $\mathbf{b}$.

3. $\mathbf{a} \times \mathbf{b} = \begin{vmatrix} \mathbf{i} & \mathbf{j} & \mathbf{k} \\ 2 & 1 & -1 \\ 0 & 1 & 2 \end{vmatrix} = \begin{vmatrix} 1 & -1 \\ 1 & 2 \end{vmatrix} \mathbf{i} - \begin{vmatrix} 2 & -1 \\ 0 & 2 \end{vmatrix} \mathbf{j} + \begin{vmatrix} 2 & 1 \\ 0 & 1 \end{vmatrix} \mathbf{k}$

$\qquad\qquad = [2 - (-1)]\,\mathbf{i} - (4 - 0)\,\mathbf{j} + (2 - 0)\,\mathbf{k} = 3\,\mathbf{i} - 4\,\mathbf{j} + 2\,\mathbf{k}$

Now $(\mathbf{a} \times \mathbf{b}) \cdot \mathbf{a} = (3\,\mathbf{i} - 4\,\mathbf{j} + 2\,\mathbf{k}) \cdot (2\,\mathbf{i} + \mathbf{j} - \mathbf{k}) = 6 - 4 - 2 = 0$ and

$(\mathbf{a} \times \mathbf{b}) \cdot \mathbf{b} = (3\,\mathbf{i} - 4\,\mathbf{j} + 2\,\mathbf{k}) \cdot (\mathbf{j} + 2\,\mathbf{k}) = 0 - 4 + 4 = 0$, so $\mathbf{a} \times \mathbf{b}$ is orthogonal to both $\mathbf{a}$ and $\mathbf{b}$.

5. $\mathbf{a} \times \mathbf{b} = \begin{vmatrix} \mathbf{i} & \mathbf{j} & \mathbf{k} \\ 3 & 2 & 4 \\ 1 & -2 & -3 \end{vmatrix} = \begin{vmatrix} 2 & 4 \\ -2 & -3 \end{vmatrix} \mathbf{i} - \begin{vmatrix} 3 & 4 \\ 1 & -3 \end{vmatrix} \mathbf{j} + \begin{vmatrix} 3 & 2 \\ 1 & -2 \end{vmatrix} \mathbf{k}$

$\qquad = [-6 - (-8)]\,\mathbf{i} - (-9 - 4)\,\mathbf{j} + (-6 - 2)\,\mathbf{k} = 2\,\mathbf{i} + 13\,\mathbf{j} - 8\,\mathbf{k}$

Since $(\mathbf{a} \times \mathbf{b}) \cdot \mathbf{a} = (2\,\mathbf{i} + 13\,\mathbf{j} - 8\,\mathbf{k}) \cdot (3\,\mathbf{i} + 2\,\mathbf{j} + 4\,\mathbf{k}) = 6 + 26 - 32 = 0$, $\mathbf{a} \times \mathbf{b}$ is orthogonal to $\mathbf{a}$.

Since $(\mathbf{a} \times \mathbf{b}) \cdot \mathbf{b} = (2\,\mathbf{i} + 13\,\mathbf{j} - 8\,\mathbf{k}) \cdot (\mathbf{i} - 2\,\mathbf{j} - 3\,\mathbf{k}) = 2 - 26 + 24 = 0$, $\mathbf{a} \times \mathbf{b}$ is orthogonal to $\mathbf{b}$.

7. $\mathbf{a} \times \mathbf{b} = \begin{vmatrix} \mathbf{i} & \mathbf{j} & \mathbf{k} \\ t & t^2 & t^3 \\ 1 & 2t & 3t^2 \end{vmatrix} = \begin{vmatrix} t^2 & t^3 \\ 2t & 3t^2 \end{vmatrix} \mathbf{i} - \begin{vmatrix} t & t^3 \\ 1 & 3t^2 \end{vmatrix} \mathbf{j} + \begin{vmatrix} t & t^2 \\ 1 & 2t \end{vmatrix} \mathbf{k}$

$\qquad = (3t^4 - 2t^4)\,\mathbf{i} - (3t^3 - t^3)\,\mathbf{j} + (2t^2 - t^2)\,\mathbf{k} = t^4\,\mathbf{i} - 2t^3\,\mathbf{j} + t^2\,\mathbf{k}$

Since $(\mathbf{a} \times \mathbf{b}) \cdot \mathbf{a} = \langle t^4, -2t^3, t^2 \rangle \cdot \langle t, t^2, t^3 \rangle = t^5 - 2t^5 + t^5 = 0$, $\mathbf{a} \times \mathbf{b}$ is orthogonal to $\mathbf{a}$.

Since $(\mathbf{a} \times \mathbf{b}) \cdot \mathbf{b} = \langle t^4, -2t^3, t^2 \rangle \cdot \langle 1, 2t, 3t^2 \rangle = t^4 - 4t^4 + 3t^4 = 0$, $\mathbf{a} \times \mathbf{b}$ is orthogonal to $\mathbf{b}$.

9. (a) Since $\mathbf{b} \times \mathbf{c}$ is a vector, the dot product $\mathbf{a} \cdot (\mathbf{b} \times \mathbf{c})$ is meaningful and is a scalar.

(b) $\mathbf{b} \cdot \mathbf{c}$ is a scalar, so $\mathbf{a} \times (\mathbf{b} \cdot \mathbf{c})$ is meaningless, as the cross product is defined only for two *vectors*.

(c) Since $\mathbf{b} \times \mathbf{c}$ is a vector, the cross product $\mathbf{a} \times (\mathbf{b} \times \mathbf{c})$ is meaningful and results in another vector.

(d) $\mathbf{a} \cdot \mathbf{b}$ is a scalar, so the cross product $(\mathbf{a} \cdot \mathbf{b}) \times \mathbf{c}$ is meaningless.

(e) Since $(\mathbf{a} \cdot \mathbf{b})$ and $(\mathbf{c} \cdot \mathbf{d})$ are both scalars, the cross product $(\mathbf{a} \cdot \mathbf{b}) \times (\mathbf{c} \cdot \mathbf{d})$ is meaningless.

(f) $\mathbf{a} \times \mathbf{b}$ and $\mathbf{c} \times \mathbf{d}$ are both vectors, so the dot product $(\mathbf{a} \times \mathbf{b}) \cdot (\mathbf{c} \times \mathbf{d})$ is meaningful and is a scalar.

11. If we sketch $\mathbf{u}$ and $\mathbf{v}$ starting from the same initial point, we see that the angle between them is $30°$, so $|\mathbf{u} \times \mathbf{v}| = |\mathbf{u}|\,|\mathbf{v}| \sin 30° = (6)(8)\left(\frac{1}{2}\right) = 24$. By the right-hand rule, $\mathbf{u} \times \mathbf{v}$ is directed into the page.

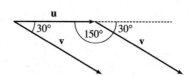

13. $\mathbf{a} \times \mathbf{b} = \begin{vmatrix} \mathbf{i} & \mathbf{j} & \mathbf{k} \\ 1 & 2 & 1 \\ 0 & 1 & 3 \end{vmatrix} = \begin{vmatrix} 2 & 1 \\ 1 & 3 \end{vmatrix} \mathbf{i} - \begin{vmatrix} 1 & 1 \\ 0 & 3 \end{vmatrix} \mathbf{j} + \begin{vmatrix} 1 & 2 \\ 0 & 1 \end{vmatrix} \mathbf{k} = (6 - 1)\,\mathbf{i} - (3 - 0)\,\mathbf{j} + (1 - 0)\,\mathbf{k} = 5\,\mathbf{i} - 3\,\mathbf{j} + \mathbf{k}$

$\mathbf{b} \times \mathbf{a} = \begin{vmatrix} \mathbf{i} & \mathbf{j} & \mathbf{k} \\ 0 & 1 & 3 \\ 1 & 2 & 1 \end{vmatrix} = \begin{vmatrix} 1 & 3 \\ 2 & 1 \end{vmatrix} \mathbf{i} - \begin{vmatrix} 0 & 3 \\ 1 & 1 \end{vmatrix} \mathbf{j} + \begin{vmatrix} 0 & 1 \\ 1 & 2 \end{vmatrix} \mathbf{k} = (1 - 6)\,\mathbf{i} - (0 - 3)\,\mathbf{j} + (0 - 1)\,\mathbf{k} = -5\,\mathbf{i} + 3\,\mathbf{j} - \mathbf{k}$

Notice $\mathbf{a} \times \mathbf{b} = -\mathbf{b} \times \mathbf{a}$ here, as we know is always true by Theorem 8.

15. We know that the cross product of two vectors is orthogonal to both. So we calculate

$$\langle 2, 0, -3 \rangle \times \langle -1, 4, 2 \rangle = \begin{vmatrix} \mathbf{i} & \mathbf{j} & \mathbf{k} \\ 2 & 0 & -3 \\ -1 & 4 & 2 \end{vmatrix} = \begin{vmatrix} 0 & -3 \\ 4 & 2 \end{vmatrix} \mathbf{i} - \begin{vmatrix} 2 & -3 \\ -1 & 2 \end{vmatrix} \mathbf{j} + \begin{vmatrix} 2 & 0 \\ -1 & 4 \end{vmatrix} \mathbf{k} = 12\,\mathbf{i} - \mathbf{j} + 8\,\mathbf{k}$$

So two unit vectors orthogonal to both are $\pm \dfrac{\langle 12, -1, 8 \rangle}{\sqrt{144 + 1 + 64}} = \pm \dfrac{\langle 12, -1, 8 \rangle}{\sqrt{209}}$, that is, $\left\langle \dfrac{12}{\sqrt{209}}, -\dfrac{1}{\sqrt{209}}, \dfrac{8}{\sqrt{209}} \right\rangle$

and $\left\langle -\dfrac{12}{\sqrt{209}}, \dfrac{1}{\sqrt{209}}, -\dfrac{8}{\sqrt{209}} \right\rangle$.

17. Let $\mathbf{a} = \langle a_1, a_2, a_3 \rangle$. Then

$$\mathbf{0} \times \mathbf{a} = \begin{vmatrix} \mathbf{i} & \mathbf{j} & \mathbf{k} \\ 0 & 0 & 0 \\ a_1 & a_2 & a_3 \end{vmatrix} = \begin{vmatrix} 0 & 0 \\ a_2 & a_3 \end{vmatrix} \mathbf{i} - \begin{vmatrix} 0 & 0 \\ a_1 & a_3 \end{vmatrix} \mathbf{j} + \begin{vmatrix} 0 & 0 \\ a_1 & a_2 \end{vmatrix} \mathbf{k} = \mathbf{0},$$

$$\mathbf{a} \times \mathbf{0} = \begin{vmatrix} \mathbf{i} & \mathbf{j} & \mathbf{k} \\ a_1 & a_2 & a_3 \\ 0 & 0 & 0 \end{vmatrix} = \begin{vmatrix} a_2 & a_3 \\ 0 & 0 \end{vmatrix} \mathbf{i} - \begin{vmatrix} a_1 & a_3 \\ 0 & 0 \end{vmatrix} \mathbf{j} + \begin{vmatrix} a_1 & a_2 \\ 0 & 0 \end{vmatrix} \mathbf{k} = \mathbf{0}.$$

19. $\mathbf{a} \times \mathbf{b} = \langle a_2 b_3 - a_3 b_2, a_3 b_1 - a_1 b_3, a_1 b_2 - a_2 b_1 \rangle$

$= \langle (-1)(b_2 a_3 - b_3 a_2), (-1)(b_3 a_1 - b_1 a_3), (-1)(b_1 a_2 - b_2 a_1) \rangle$

$= - \langle b_2 a_3 - b_3 a_2, b_3 a_1 - b_1 a_3, b_1 a_2 - b_2 a_1 \rangle = -\mathbf{b} \times \mathbf{a}$

21. $\mathbf{a} \times (\mathbf{b} + \mathbf{c}) = \mathbf{a} \times \langle b_1 + c_1, b_2 + c_2, b_3 + c_3 \rangle$

$= \langle a_2(b_3 + c_3) - a_3(b_2 + c_2), a_3(b_1 + c_1) - a_1(b_3 + c_3), a_1(b_2 + c_2) - a_2(b_1 + c_1) \rangle$

$= \langle a_2 b_3 + a_2 c_3 - a_3 b_2 - a_3 c_2, a_3 b_1 + a_3 c_1 - a_1 b_3 - a_1 c_3, a_1 b_2 + a_1 c_2 - a_2 b_1 - a_2 c_1 \rangle$

$= \langle (a_2 b_3 - a_3 b_2) + (a_2 c_3 - a_3 c_2), (a_3 b_1 - a_1 b_3) + (a_3 c_1 - a_1 c_3), (a_1 b_2 - a_2 b_1) + (a_1 c_2 - a_2 c_1) \rangle$

$= \langle a_2 b_3 - a_3 b_2, a_3 b_1 - a_1 b_3, a_1 b_2 - a_2 b_1 \rangle + \langle a_2 c_3 - a_3 c_2, a_3 c_1 - a_1 c_3, a_1 c_2 - a_2 c_1 \rangle$

$= (\mathbf{a} \times \mathbf{b}) + (\mathbf{a} \times \mathbf{c})$

23. By plotting the vertices, we can see that the parallelogram is determined by the

vectors $\overrightarrow{AB} = \langle 2, 3 \rangle$ and $\overrightarrow{AD} = \langle 4, -2 \rangle$. We know that the area of the parallelogram determined by two vectors is equal to the length of the cross product of these vectors.

In order to compute the cross product, we consider the vector $\overrightarrow{AB}$ as the three-

dimensional vector $\langle 2, 3, 0 \rangle$ (and similarly for $\overrightarrow{AD}$), and then the area of

parallelogram $ABCD$ is

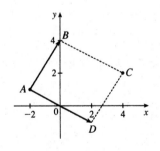

$$\left| \overrightarrow{AB} \times \overrightarrow{AD} \right| = \begin{Vmatrix} \mathbf{i} & \mathbf{j} & \mathbf{k} \\ 2 & 3 & 0 \\ 4 & -2 & 0 \end{Vmatrix} = |(0)\,\mathbf{i} - (0)\,\mathbf{j} + (-4 - 12)\,\mathbf{k}| = |-16\,\mathbf{k}| = 16$$

25. (a) Because the plane through P, Q, and R contains the vectors $\overrightarrow{PQ}$ and $\overrightarrow{PR}$, a vector orthogonal to both of these vectors

(such as their cross product) is also orthogonal to the plane. Here $\overrightarrow{PQ} = \langle -1, 2, 0 \rangle$ and $\overrightarrow{PR} = \langle -1, 0, 3 \rangle$, so

$$\overrightarrow{PQ} \times \overrightarrow{PR} = \langle (2)(3) - (0)(0), (0)(-1) - (-1)(3), (-1)(0) - (2)(-1) \rangle = \langle 6, 3, 2 \rangle$$

Therefore, $\langle 6, 3, 2 \rangle$ (or any scalar multiple thereof) is orthogonal to the plane through P, Q, and R.

(b) Note that the area of the triangle determined by P, Q, and R is equal to half of the area of the parallelogram determined by

the three points. From part (a), the area of the parallelogram is $\left| \overrightarrow{PQ} \times \overrightarrow{PR} \right| = |\langle 6, 3, 2 \rangle| = \sqrt{36 + 9 + 4} = 7$, so the area

of the triangle is $\frac{1}{2}(7) = \frac{7}{2}$.

27. (a) $\overrightarrow{PQ} = \langle 4, 3, -2 \rangle$ and $\overrightarrow{PR} = \langle 5, 5, 1 \rangle$, so a vector orthogonal to the plane through P, Q, and R is

$$\overrightarrow{PQ} \times \overrightarrow{PR} = \langle (3)(1) - (-2)(5), (-2)(5) - (4)(1), (4)(5) - (3)(5) \rangle = \langle 13, -14, 5 \rangle \text{ (or any scalar mutiple thereof)}.$$

(b) The area of the parallelogram determined by $\overrightarrow{PQ}$ and $\overrightarrow{PR}$ is

$$\left| \overrightarrow{PQ} \times \overrightarrow{PR} \right| = |\langle 13, -14, 5 \rangle| = \sqrt{13^2 + (-14)^2 + 5^2} = \sqrt{390}, \text{ so the area of triangle } PQR \text{ is } \frac{1}{2}\sqrt{390}.$$

29. We know that the volume of the parallelepiped determined by $\mathbf{a}$, $\mathbf{b}$, and $\mathbf{c}$ is the magnitude of their scalar triple product, which

$$\text{is } \mathbf{a} \cdot (\mathbf{b} \times \mathbf{c}) = \begin{vmatrix} 6 & 3 & -1 \\ 0 & 1 & 2 \\ 4 & -2 & 5 \end{vmatrix} = 6 \begin{vmatrix} 1 & 2 \\ -2 & 5 \end{vmatrix} - 3 \begin{vmatrix} 0 & 2 \\ 4 & 5 \end{vmatrix} + (-1) \begin{vmatrix} 0 & 1 \\ 4 & -2 \end{vmatrix} = 6(5+4) - 3(0-8) - (0-4) = 82.$$

Thus the volume of the parallelepiped is 82 cubic units.

31. $\mathbf{a} = \overrightarrow{PQ} = \langle 2, 1, 1 \rangle$, $\mathbf{b} = \overrightarrow{PR} = \langle 1, -1, 2 \rangle$, and $\mathbf{c} = \overrightarrow{PS} = \langle 0, -2, 3 \rangle$.

$$\mathbf{a} \cdot (\mathbf{b} \times \mathbf{c}) = \begin{vmatrix} 2 & 1 & 1 \\ 1 & -1 & 2 \\ 0 & -2 & 3 \end{vmatrix} = 2 \begin{vmatrix} -1 & 2 \\ -2 & 3 \end{vmatrix} - 1 \begin{vmatrix} 1 & 2 \\ 0 & 3 \end{vmatrix} + 1 \begin{vmatrix} 1 & -1 \\ 0 & -2 \end{vmatrix} = 2 - 3 - 2 = -3,$$

so the volume of the parallelepiped is 3 cubic units.

33. $\mathbf{u} \cdot (\mathbf{v} \times \mathbf{w}) = \begin{vmatrix} 1 & 5 & -2 \\ 3 & -1 & 0 \\ 5 & 9 & -4 \end{vmatrix} = 1 \begin{vmatrix} -1 & 0 \\ 9 & -4 \end{vmatrix} - 5 \begin{vmatrix} 3 & 0 \\ 5 & -4 \end{vmatrix} + (-2) \begin{vmatrix} 3 & -1 \\ 5 & 9 \end{vmatrix} = 4 + 60 - 64 = 0$, which says that the volume

of the parallelepiped determined by $\mathbf{u}$, $\mathbf{v}$ and $\mathbf{w}$ is 0, and thus these three vectors are coplanar.

35. The magnitude of the torque is $|\boldsymbol{\tau}| = |\mathbf{r} \times \mathbf{F}| = |\mathbf{r}| |\mathbf{F}| \sin \theta = (0.18 \text{ m})(60 \text{ N}) \sin(70 + 10)° = 10.8 \sin 80° \approx 10.6 \text{ N·m}.$

37. Using the notation of (1), $\mathbf{r} = \langle 0, 0.3, 0 \rangle$ and $\mathbf{F}$ has direction $\langle 0, 3, -4 \rangle$. The angle θ between them can be determined by

$$\cos\theta = \frac{\langle 0, 0.3, 0 \rangle \cdot \langle 0, 3, -4 \rangle}{|\langle 0, 0.3, 0 \rangle| \, |\langle 0, 3, -4 \rangle|} \;\Rightarrow\; \cos\theta = \frac{0.9}{(0.3)(5)} \;\Rightarrow\; \cos\theta = 0.6 \;\Rightarrow\; \theta \approx 53.1°. \text{ Then } |\boldsymbol{\tau}| = |\mathbf{r}|\,|\mathbf{F}|\sin\theta \;\Rightarrow\;$$

$100 = 0.3\,|\mathbf{F}|\sin 53.1° \;\Rightarrow\; |\mathbf{F}| \approx 417\text{ N}.$

39. (a)

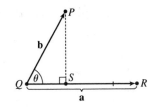

The distance between a point and a line is the length of the perpendicular from the point to the line, here $\left|\overrightarrow{PS}\right| = d$. But referring to triangle PQS,

$d = \left|\overrightarrow{PS}\right| = \left|\overrightarrow{QP}\right|\sin\theta = |\mathbf{b}|\sin\theta$. But θ is the angle between $\overrightarrow{QP} = \mathbf{b}$ and

$\overrightarrow{QR} = \mathbf{a}$. Thus by the definition of the cross product, $\sin\theta = \dfrac{|\mathbf{a} \times \mathbf{b}|}{|\mathbf{a}|\,|\mathbf{b}|}$ and so

$$d = |\mathbf{b}|\sin\theta = \frac{|\mathbf{b}|\,|\mathbf{a} \times \mathbf{b}|}{|\mathbf{a}|\,|\mathbf{b}|} = \frac{|\mathbf{a} \times \mathbf{b}|}{|\mathbf{a}|}.$$

(b) $\mathbf{a} = \overrightarrow{QR} = \langle -1, -2, -1 \rangle$ and $\mathbf{b} = \overrightarrow{QP} = \langle 1, -5, -7 \rangle$. Then

$\mathbf{a} \times \mathbf{b} = \langle (-2)(-7) - (-1)(-5), (-1)(1) - (-1)(-7), (-1)(-5) - (-2)(1) \rangle = \langle 9, -8, 7 \rangle.$

Thus the distance is $d = \dfrac{|\mathbf{a} \times \mathbf{b}|}{|\mathbf{a}|} = \dfrac{1}{\sqrt{6}}\sqrt{81 + 64 + 49} = \sqrt{\dfrac{194}{6}} = \sqrt{\dfrac{97}{3}}.$

41. $(\mathbf{a} - \mathbf{b}) \times (\mathbf{a} + \mathbf{b}) = (\mathbf{a} - \mathbf{b}) \times \mathbf{a} + (\mathbf{a} - \mathbf{b}) \times \mathbf{b}$ by Property 3 of Theorem 8

$\qquad = \mathbf{a} \times \mathbf{a} + (-\mathbf{b}) \times \mathbf{a} + \mathbf{a} \times \mathbf{b} + (-\mathbf{b}) \times \mathbf{b}$ by Property 4 of Theorem 8

$\qquad = (\mathbf{a} \times \mathbf{a}) - (\mathbf{b} \times \mathbf{a}) + (\mathbf{a} \times \mathbf{b}) - (\mathbf{b} \times \mathbf{b})$ by Property 2 of Theorem 8 (with $c = -1$)

$\qquad = \mathbf{0} - (\mathbf{b} \times \mathbf{a}) + (\mathbf{a} \times \mathbf{b}) - \mathbf{0}$ by Example 2

$\qquad = (\mathbf{a} \times \mathbf{b}) + (\mathbf{a} \times \mathbf{b})$ by Property 1 of Theorem 8

$\qquad = 2(\mathbf{a} \times \mathbf{b})$

43. $\mathbf{a} \times (\mathbf{b} \times \mathbf{c}) + \mathbf{b} \times (\mathbf{c} \times \mathbf{a}) + \mathbf{c} \times (\mathbf{a} \times \mathbf{b})$

$\qquad = [(\mathbf{a} \cdot \mathbf{c})\mathbf{b} - (\mathbf{a} \cdot \mathbf{b})\mathbf{c}] + [(\mathbf{b} \cdot \mathbf{a})\mathbf{c} - (\mathbf{b} \cdot \mathbf{c})\mathbf{a}] + [(\mathbf{c} \cdot \mathbf{b})\mathbf{a} - (\mathbf{c} \cdot \mathbf{a})\mathbf{b}]$ by Exercise 42

$\qquad = (\mathbf{a} \cdot \mathbf{c})\mathbf{b} - (\mathbf{a} \cdot \mathbf{b})\mathbf{c} + (\mathbf{a} \cdot \mathbf{b})\mathbf{c} - (\mathbf{b} \cdot \mathbf{c})\mathbf{a} + (\mathbf{b} \cdot \mathbf{c})\mathbf{a} - (\mathbf{a} \cdot \mathbf{c})\mathbf{b} = \mathbf{0}$

45. (a) No. If $\mathbf{a} \cdot \mathbf{b} = \mathbf{a} \cdot \mathbf{c}$, then $\mathbf{a} \cdot (\mathbf{b} - \mathbf{c}) = 0$, so $\mathbf{a}$ is perpendicular to $\mathbf{b} - \mathbf{c}$, which can happen if $\mathbf{b} \neq \mathbf{c}$. For example,

let $\mathbf{a} = \langle 1, 1, 1 \rangle$, $\mathbf{b} = \langle 1, 0, 0 \rangle$ and $\mathbf{c} = \langle 0, 1, 0 \rangle$.

(b) No. If $\mathbf{a} \times \mathbf{b} = \mathbf{a} \times \mathbf{c}$ then $\mathbf{a} \times (\mathbf{b} - \mathbf{c}) = \mathbf{0}$, which implies that $\mathbf{a}$ is parallel to $\mathbf{b} - \mathbf{c}$, which of course can happen

if $\mathbf{b} \neq \mathbf{c}$.

(c) Yes. Since $\mathbf{a} \cdot \mathbf{c} = \mathbf{a} \cdot \mathbf{b}$, $\mathbf{a}$ is perpendicular to $\mathbf{b} - \mathbf{c}$, by part (a). From part (b), $\mathbf{a}$ is also parallel to $\mathbf{b} - \mathbf{c}$. Thus since

$\mathbf{a} \neq \mathbf{0}$ but is both parallel and perpendicular to $\mathbf{b} - \mathbf{c}$, we have $\mathbf{b} - \mathbf{c} = \mathbf{0}$, so $\mathbf{b} = \mathbf{c}$.

10.5 Equations of Lines and Planes

1. (a) True; each of the first two lines has a direction vector parallel to the direction vector of the third line, so these vectors are each scalar multiples of the third direction vector. Then the first two direction vectors are also scalar multiples of each other, so these vectors, and hence the two lines, are parallel.

   (b) False; for example, the x- and y-axes are both perpendicular to the z-axis, yet the x- and y-axes are not parallel.

   (c) True; each of the first two planes has a normal vector parallel to the normal vector of the third plane, so these two normal vectors are parallel to each other and the planes are parallel.

   (d) False; for example, the xy- and yz-planes are not parallel, yet they are both perpendicular to the xz-plane.

   (e) False; the x- and y-axes are not parallel, yet they are both parallel to the plane $z = 1$.

   (f) True; if each line is perpendicular to a plane, then the lines' direction vectors are both parallel to a normal vector for the plane. Thus, the direction vectors are parallel to each other and the lines are parallel.

   (g) False; the planes $y = 1$ and $z = 1$ are not parallel, yet they are both parallel to the x-axis.

   (h) True; if each plane is perpendicular to a line, then any normal vector for each plane is parallel to a direction vector for the line. Thus, the normal vectors are parallel to each other and the planes are parallel.

   (i) True; see Figure 9 and the accompanying discussion.

   (j) False; they can be skew, as in Example 3.

   (k) True. Consider any normal vector for the plane and any direction vector for the line. If the normal vector is perpendicular to the direction vector, the line and plane are parallel. Otherwise, the vectors meet at an angle θ, $0° \leq \theta < 90°$, and the line will intersect the plane at an angle $90° - \theta$.

3. For this line, we have $\mathbf{r}_0 = -2\mathbf{i} + 4\mathbf{j} + 10\mathbf{k}$ and $\mathbf{v} = 3\mathbf{i} + \mathbf{j} - 8\mathbf{k}$, so a vector equation is

   $\mathbf{r} = \mathbf{r}_0 + t\mathbf{v} = (-2\mathbf{i} + 4\mathbf{j} + 10\mathbf{k}) + t(3\mathbf{i} + \mathbf{j} - 8\mathbf{k}) = (-2 + 3t)\mathbf{i} + (4 + t)\mathbf{j} + (10 - 8t)\mathbf{k}$ and parametric equations are $x = -2 + 3t$, $y = 4 + t$, $z = 10 - 8t$.

5. A line perpendicular to the given plane has the same direction as a normal vector to the plane, such as

   $\mathbf{n} = \langle 1, 3, 1 \rangle$. So $\mathbf{r}_0 = \mathbf{i} + 6\mathbf{k}$, and we can take $\mathbf{v} = \mathbf{i} + 3\mathbf{j} + \mathbf{k}$. Then a vector equation is

   $\mathbf{r} = (\mathbf{i} + 6\mathbf{k}) + t(\mathbf{i} + 3\mathbf{j} + \mathbf{k}) = (1 + t)\mathbf{i} + 3t\mathbf{j} + (6 + t)\mathbf{k}$, and parametric equations are $x = 1 + t$, $y = 3t$, $z = 6 + t$.

7. $\mathbf{v} = \langle 2 - 0, 1 - \frac{1}{2}, -3 - 1 \rangle = \langle 2, \frac{1}{2}, -4 \rangle$, and letting $P_0 = (2, 1, -3)$, parametric equations are $x = 2 + 2t$, $y = 1 + \frac{1}{2}t$,

   $z = -3 - 4t$, while symmetric equations are $\dfrac{x - 2}{2} = \dfrac{y - 1}{1/2} = \dfrac{z + 3}{-4}$ or $\dfrac{x - 2}{2} = 2y - 2 = \dfrac{z + 3}{-4}$.

9. The line has direction $\mathbf{v} = \langle 1, 2, 1 \rangle$. Letting $P_0 = (1, -1, 1)$, parametric equations are $x = 1 + t$, $y = -1 + 2t$, $z = 1 + t$

   and symmetric equations are $x - 1 = \dfrac{y + 1}{2} = z - 1$.

11. Direction vectors of the lines are $\mathbf{v}_1 = \langle -2 - (-4), 0 - (-6), -3 - 1 \rangle = \langle 2, 6, -4 \rangle$ and

$\mathbf{v}_2 = \langle 5 - 10, 3 - 18, 14 - 4 \rangle = \langle -5, -15, 10 \rangle$, and since $\mathbf{v}_2 = -\frac{5}{2}\mathbf{v}_1$, the direction vectors and thus the lines are parallel.

13. (a) A direction vector of the line with parametric equations $x = 1 + 2t$, $y = 3t$, $z = 5 - 7t$ is $\mathbf{v} = \langle 2, 3, -7 \rangle$ and the desired

parallel line must also have $\mathbf{v}$ as a direction vector. Here $P_0 = (0, 2, -1)$, so symmetric equations for the line are

$$\frac{x}{2} = \frac{y-2}{3} = \frac{z+1}{-7}.$$

(b) The line intersects the xy-plane when $z = 0$, so we need $\frac{x}{2} = \frac{y-2}{3} = \frac{1}{-7}$ or $x = -\frac{2}{7}$, $y = \frac{11}{7}$. Thus the point of

intersection with the xy-plane is $\left(-\frac{2}{7}, \frac{11}{7}, 0\right)$. Similarly for the yz-plane, we need $x = 0$ $\Leftrightarrow$ $0 = \frac{y-2}{3} = \frac{z+1}{-7}$ $\Leftrightarrow$

$y = 2$, $z = -1$. Thus the line intersects the yz-plane at $(0, 2, -1)$. For the xz-plane, we need $y = 0$ $\Leftrightarrow$

$\frac{x}{2} = -\frac{2}{3} = \frac{z+1}{-7}$ $\Leftrightarrow$ $x = -\frac{4}{3}$, $z = \frac{11}{3}$. So the line intersects the xz-plane at $\left(-\frac{4}{3}, 0, \frac{11}{3}\right)$.

15. From Equation 4, the line segment from $\mathbf{r}_0 = 2\mathbf{i} - \mathbf{j} + 4\mathbf{k}$ to $\mathbf{r}_1 = 4\mathbf{i} + 6\mathbf{j} + \mathbf{k}$ is

$\mathbf{r}(t) = (1-t)\mathbf{r}_0 + t\mathbf{r}_1 = (1-t)(2\mathbf{i} - \mathbf{j} + 4\mathbf{k}) + t(4\mathbf{i} + 6\mathbf{j} + \mathbf{k}) = (2\mathbf{i} - \mathbf{j} + 4\mathbf{k}) + t(2\mathbf{i} + 7\mathbf{j} - 3\mathbf{k})$, $0 \le t \le 1$.

17. Since the direction vectors are $\mathbf{v}_1 = \langle -6, 9, -3 \rangle$ and $\mathbf{v}_2 = \langle 2, -3, 1 \rangle$, we have $\mathbf{v}_1 = -3\mathbf{v}_2$ so the lines are parallel.

19. Since the direction vectors $\langle 1, 2, 3 \rangle$ and $\langle -4, -3, 2 \rangle$ are not scalar multiples of each other, the lines are not parallel, so we

check to see if the lines intersect. The parametric equations of the lines are L_1: $x = t$, $y = 1 + 2t$, $z = 2 + 3t$ and L_2:

$x = 3 - 4s$, $y = 2 - 3s$, $z = 1 + 2s$. For the lines to intersect, we must be able to find one value of t and one value of s that

produce the same point from the respective parametric equations. Thus we need to satisfy the following three equations:

$t = 3 - 4s$, $1 + 2t = 2 - 3s$, $2 + 3t = 1 + 2s$. Solving the first two equations we get $t = -1$, $s = 1$ and checking, we see

that these values don't satisfy the third equation. Thus the lines aren't parallel and don't intersect, so they must be skew lines.

21. Since the plane is perpendicular to the vector $\langle -2, 1, 5 \rangle$, we can take $\langle -2, 1, 5 \rangle$ as a normal vector to the plane.

$(6, 3, 2)$ is a point on the plane, so setting $a = -2$, $b = 1$, $c = 5$ and $x_0 = 6$, $y_0 = 3$, $z_0 = 2$ in Equation 6 gives

$-2(x - 6) + 1(y - 3) + 5(z - 2) = 0$ or $-2x + y + 5z = 1$ to be an equation of the plane.

23. Since the two planes are parallel, they will have the same normal vectors. So we can take $\mathbf{n} = \langle 2, -1, 3 \rangle$, and an equation of

the plane is $2(x - 0) - 1(y - 0) + 3(z - 0) = 0$ or $2x - y + 3z = 0$.

25. Here the vectors $\mathbf{a} = \langle 1 - 0, 0 - 1, 1 - 1 \rangle = \langle 1, -1, 0 \rangle$ and $\mathbf{b} = \langle 1 - 0, 1 - 1, 0 - 1 \rangle = \langle 1, 0, -1 \rangle$ lie in the plane, so

$\mathbf{a} \times \mathbf{b}$ is a normal vector to the plane. Thus, we can take $\mathbf{n} = \mathbf{a} \times \mathbf{b} = \langle 1 - 0, 0 + 1, 0 + 1 \rangle = \langle 1, 1, 1 \rangle$. If P_0 is the point

$(0, 1, 1)$, an equation of the plane is $1(x - 0) + 1(y - 1) + 1(z - 1) = 0$ or $x + y + z = 2$.

27. If we first find two nonparallel vectors in the plane, their cross product will be a normal vector to the plane. Since the given line lies in the plane, its direction vector $\mathbf{a} = \langle -2, 5, 4 \rangle$ is one vector in the plane. We can verify that the given point $(6, 0, -2)$ does not lie on this line, so to find another nonparallel vector $\mathbf{b}$ which lies in the plane, we can pick any point on the line and find a vector connecting the points. If we put $t = 0$, we see that $(4, 3, 7)$ is on the line, so

$\mathbf{b} = \langle 6 - 4, 0 - 3, -2 - 7 \rangle = \langle 2, -3, -9 \rangle$ and $\mathbf{n} = \mathbf{a} \times \mathbf{b} = \langle -45 + 12, 8 - 18, 6 - 10 \rangle = \langle -33, -10, -4 \rangle$. Thus, an equation of the plane is $-33(x - 6) - 10(y - 0) - 4[z - (-2)] = 0$ or $33x + 10y + 4z = 190$.

29. A direction vector for the line of intersection is $\mathbf{a} = \mathbf{n}_1 \times \mathbf{n}_2 = \langle 1, 1, -1 \rangle \times \langle 2, -1, 3 \rangle = \langle 2, -5, -3 \rangle$, and $\mathbf{a}$ is parallel to the desired plane. Another vector parallel to the plane is the vector connecting any point on the line of intersection to the given point $(-1, 2, 1)$ in the plane. Setting $x = 0$, the equations of the planes reduce to $y - z = 2$ and $-y + 3z = 1$ with simultaneous solution $y = \frac{7}{2}$ and $z = \frac{3}{2}$. So a point on the line is $\left(0, \frac{7}{2}, \frac{3}{2} \right)$ and another vector parallel to the plane is $\left\langle -1, -\frac{3}{2}, -\frac{1}{2} \right\rangle$. Then a normal vector to the plane is $\mathbf{n} = \langle 2, -5, -3 \rangle \times \left\langle -1, -\frac{3}{2}, -\frac{1}{2} \right\rangle = \langle -2, 4, -8 \rangle$ and an equation of the plane is $-2(x + 1) + 4(y - 2) - 8(z - 1) = 0$ or $x - 2y + 4z = -1$.

31. Substitute the parametric equations of the line into the equation of the plane: $(3 - t) - (2 + t) + 2(5t) = 9 \Rightarrow 8t = 8 \Rightarrow t = 1$. Therefore, the point of intersection of the line and the plane is given by $x = 3 - 1 = 2$, $y = 2 + 1 = 3$, and $z = 5(1) = 5$, that is, the point $(2, 3, 5)$.

33. Normal vectors for the planes are $\mathbf{n}_1 = \langle 1, 1, 1 \rangle$ and $\mathbf{n}_2 = \langle 1, -1, 1 \rangle$. The normals are not parallel, so neither are the planes. Furthermore, $\mathbf{n}_1 \cdot \mathbf{n}_2 = 1 - 1 + 1 = 1 \neq 0$, so the planes aren't perpendicular. The angle between them is given by

$$\cos \theta = \frac{\mathbf{n}_1 \cdot \mathbf{n}_2}{|\mathbf{n}_1|\,|\mathbf{n}_2|} = \frac{1}{\sqrt{3}\sqrt{3}} = \frac{1}{3} \Rightarrow \theta = \cos^{-1}\left(\frac{1}{3} \right) \approx 70.5°.$$

35. The normals are $\mathbf{n}_1 = \langle 1, -4, 2 \rangle$ and $\mathbf{n}_2 = \langle 2, -8, 4 \rangle$. Since $\mathbf{n}_2 = 2\mathbf{n}_1$, the normals (and thus the planes) are parallel.

37. (a) To find a point on the line of intersection, set one of the variables equal to a constant, say $z = 0$. (This will only work if the line of intersection crosses the xy-plane; otherwise, try setting x or y equal to 0.) Then the equations of the planes reduce to $x + y = 2$ and $3x - 4y = 6$. Solving these two equations gives $x = 2$, $y = 0$. So a point on the line of intersection is $(2, 0, 0)$. The direction of the line is $\mathbf{v} = \mathbf{n}_1 \times \mathbf{n}_2 = \langle 5 - 4, -3 - 5, -4 - 3 \rangle = \langle 1, -8, -7 \rangle$, and symmetric equations for the line are $x - 2 = \dfrac{y}{-8} = \dfrac{z}{-7}$.

(b) The angle between the planes satisfies $\cos \theta = \dfrac{\mathbf{n}_1 \cdot \mathbf{n}_2}{|\mathbf{n}_1|\,|\mathbf{n}_2|} = \dfrac{3 - 4 - 5}{\sqrt{3}\sqrt{50}} = -\dfrac{\sqrt{6}}{5}$. Therefore $\theta = \cos^{-1}\left(-\frac{\sqrt{6}}{5} \right) \approx 119°$

(or $61°$).

39. The plane contains the points $(a, 0, 0)$, $(0, b, 0)$ and $(0, 0, c)$. Thus the vectors $\mathbf{a} = \langle -a, b, 0 \rangle$ and $\mathbf{b} = \langle -a, 0, c \rangle$ lie in the

plane, and $\mathbf{n} = \mathbf{a} \times \mathbf{b} = \langle bc - 0, 0 + ac, 0 + ab \rangle = \langle bc, ac, ab \rangle$ is a normal vector to the plane. The equation of the plane is

therefore $bcx + acy + abz = abc + 0 + 0$ or $bcx + acy + abz = abc$. Notice that if $a \neq 0$, $b \neq 0$ and $c \neq 0$ then we can

rewrite the equation as $\dfrac{x}{a} + \dfrac{y}{b} + \dfrac{z}{c} = 1$. This is a good equation to remember!

41. Two vectors which are perpendicular to the required line are the normal of the given plane, $\langle 1, 1, 1 \rangle$, and a direction vector for

the given line, $\langle 1, -1, 2 \rangle$. So a direction vector for the required line is $\langle 1, 1, 1 \rangle \times \langle 1, -1, 2 \rangle = \langle 3, -1, -2 \rangle$. Thus L is given

by $\langle x, y, z \rangle = \langle 0, 1, 2 \rangle + t\langle 3, -1, -2 \rangle$, or in parametric form, $x = 3t$, $y = 1 - t$, $z = 2 - 2t$.

43. Let P_i have normal vector $\mathbf{n}_i$. Then $\mathbf{n}_1 = \langle 4, -2, 6 \rangle$, $\mathbf{n}_2 = \langle 4, -2, -2 \rangle$, $\mathbf{n}_3 = \langle -6, 3, -9 \rangle$, $\mathbf{n}_4 = \langle 2, -1, -1 \rangle$. Now

$\mathbf{n}_1 = -\frac{2}{3}\mathbf{n}_3$, so $\mathbf{n}_1$ and $\mathbf{n}_3$ are parallel, and hence P_1 and P_3 are parallel; similarly P_2 and P_4 are parallel because $\mathbf{n}_2 = 2\mathbf{n}_4$.

However, $\mathbf{n}_1$ and $\mathbf{n}_2$ are not parallel. $\left(0, 0, \frac{1}{2}\right)$ lies on P_1, but not on P_3, so they are not the same plane, but both P_2 and P_4

contain the point $(0, 0, -3)$, so these two planes are identical.

45. Let $Q = (2, 2, 0)$ and $R = (3, -1, 5)$, points on the line corresponding to $t = 0$ and $t = 1$.

Let $P = (1, 2, 3)$. Then $\mathbf{a} = \overrightarrow{QR} = \langle 1, -3, 5 \rangle$, $\mathbf{b} = \overrightarrow{QP} = \langle -1, 0, 3 \rangle$. The distance is

$$d = \frac{|\mathbf{a} \times \mathbf{b}|}{|\mathbf{a}|} = \frac{|\langle 1, -3, 5 \rangle \times \langle -1, 0, 3 \rangle|}{|\langle 1, -3, 5 \rangle|} = \frac{|\langle -9, -8, -3 \rangle|}{|\langle 1, -3, 5 \rangle|} = \frac{\sqrt{9^2 + 8^2 + 3^2}}{\sqrt{1^2 + 3^2 + 5^2}} = \frac{\sqrt{154}}{\sqrt{35}} = \sqrt{\frac{22}{5}}.$$

47. By Equation 9, the distance is $D = \dfrac{1}{\sqrt{1 + 4 + 4}} \left[(1)(2) + (-2)(8) + (-2)(5) - 1 \right] = \dfrac{25}{3}$.

49. Put $y = z = 0$ in the equation of the first plane to get the point $(-1, 0, 0)$ on the plane. Because the planes are parallel, the

distance D between them is the distance from $(-1, 0, 0)$ to the second plane. By Equation 9,

$$D = \frac{|3(-1) + 6(0) - 3(0) - 4|}{\sqrt{3^2 + 6^2 + (-3)^2}} = \frac{7}{3\sqrt{6}} \text{ or } \frac{7\sqrt{6}}{18}.$$

51. The distance between two parallel planes is the same as the distance between a point on one of the planes and the other plane.

Let $P_0 = (x_0, y_0, z_0)$ be a point on the plane given by $ax + by + cz + d_1 = 0$. Then $ax_0 + by_0 + cz_0 + d_1 = 0$ and the

distance between P_0 and the plane given by $ax + by + cz + d_2 = 0$ is, from Equation 9,

$$D = \frac{|ax_0 + by_0 + cz_0 + d_2|}{\sqrt{a^2 + b^2 + c^2}} = \frac{|-d_1 + d_2|}{\sqrt{a^2 + b^2 + c^2}} = \frac{|d_1 - d_2|}{\sqrt{a^2 + b^2 + c^2}}.$$

53. $L_1: x = y = z$ $\Rightarrow$ $x = y$ **(1)**. $L_2: x + 1 = y/2 = z/3$ $\Rightarrow$ $x + 1 = y/2$ **(2)**.   The solution of **(1)** and **(2)** is

$x = y = -2$. However, when $x = -2$, $x = z$ $\Rightarrow$ $z = -2$, but $x + 1 = z/3$ $\Rightarrow$ $z = -3$, a contradiction. Hence the

lines do not intersect. For L_1, $\mathbf{v}_1 = \langle 1, 1, 1 \rangle$, and for L_2, $\mathbf{v}_2 = \langle 1, 2, 3 \rangle$, so the lines are not parallel. Thus the lines are skew

lines. If two lines are skew, they can be viewed as lying in two parallel planes and so the distance between the skew lines

would be the same as the distance between these parallel planes. The common normal vector to the planes must be

perpendicular to both $\langle 1, 1, 1 \rangle$ and $\langle 1, 2, 3 \rangle$, the direction vectors of the two lines. So set

$\mathbf{n} = \langle 1, 1, 1 \rangle \times \langle 1, 2, 3 \rangle = \langle 3 - 2, -3 + 1, 2 - 1 \rangle = \langle 1, -2, 1 \rangle$. From above, we know that $(-2, -2, -2)$ and $(-2, -2, -3)$

are points of L_1 and L_2 respectively. So in the notation of Equation 8, $1(-2) - 2(-2) + 1(-2) + d_1 = 0$ $\Rightarrow$ $d_1 = 0$ and

$1(-2) - 2(-2) + 1(-3) + d_2 = 0$ $\Rightarrow$ $d_2 = 1$.

By Exercise 51, the distance between these two skew lines is $D = \dfrac{|0 - 1|}{\sqrt{1 + 4 + 1}} = \dfrac{1}{\sqrt{6}}$.

Alternate solution (without reference to planes): A vector which is perpendicular to both of the lines is

$\mathbf{n} = \langle 1, 1, 1 \rangle \times \langle 1, 2, 3 \rangle = \langle 1, -2, 1 \rangle$. Pick any point on each of the lines, say $(-2, -2, -2)$ and $(-2, -2, -3)$, and form the

vector $\mathbf{b} = \langle 0, 0, 1 \rangle$ connecting the two points. The distance between the two skew lines is the absolute value of the scalar

projection of $\mathbf{b}$ along $\mathbf{n}$, that is, $D = \dfrac{|\mathbf{n} \cdot \mathbf{b}|}{|\mathbf{n}|} = \dfrac{|1 \cdot 0 - 2 \cdot 0 + 1 \cdot 1|}{\sqrt{1 + 4 + 1}} = \dfrac{1}{\sqrt{6}}$.

55. If $a \neq 0$, then $ax + by + cz + d = 0$ $\Rightarrow$ $a(x + d/a) + b(y - 0) + c(z - 0) = 0$ which by **(7)** is the scalar equation of the

plane through the point $(-d/a, 0, 0)$ with normal vector $\langle a, b, c \rangle$. Similarly, if $b \neq 0$ (or if $c \neq 0$) the equation of the plane can

be rewritten as $a(x - 0) + b(y + d/b) + c(z - 0) = 0$ [or as $a(x - 0) + b(y - 0) + c(z + d/c) = 0$] which by **(7)** is the

scalar equation of a plane through the point $(0, -d/b, 0)$ [or the point $(0, 0, -d/c)$] with normal vector $\langle a, b, c \rangle$.

10.6 Cylinders and Quadric Surfaces

1. (a) In $\mathbb{R}^2$, the equation $y = x^2$ represents a parabola.

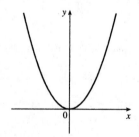

(b) In $\mathbb{R}^3$, the equation $y = x^2$ doesn't involve z, so any horizontal plane with equation $z = k$ intersects the graph in a curve with equation $y = x^2$. Thus, the surface is a parabolic cylinder, made up of infinitely many shifted copies of the same parabola. The rulings are parallel to the z-axis.

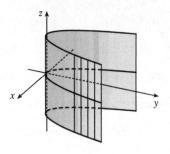

(c) In $\mathbb{R}^3$, the equation $z = y^2$ also represents a parabolic cylinder. Since x doesn't appear, the graph is formed by moving the parabola $z = y^2$ in the direction of the x-axis. Thus, the rulings of the cylinder are parallel to the x-axis.

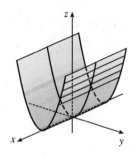

3. Since x is missing from the equation, the vertical traces $y^2 + 4z^2 = 4$, $x = k$, are copies of the same ellipse in the plane $x = k$. Thus, the surface $y^2 + 4z^2 = 4$ is an elliptic cylinder with rulings parallel to the x-axis.

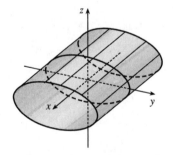

5. Since z is missing, each horizontal trace $x = y^2$, $z = k$, is a copy of the same parabola in the plane $z = k$. Thus, the surface $x - y^2 = 0$ is a parabolic cylinder with rulings parallel to the z-axis.

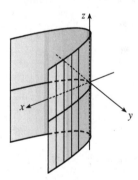

7. Since y is missing, each vertical trace $z = \cos x$, $y = k$ is a copy of a cosine curve in the plane $y = k$. Thus, the surface $z = \cos x$ is a cylindrical surface with rulings parallel to the y-axis.

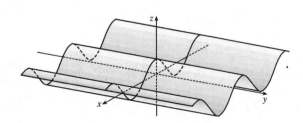

9. (a) The traces of $x^2 + y^2 - z^2 = 1$ in $x = k$ are $y^2 - z^2 = 1 - k^2$, a family of hyperbolas. (Note that the hyperbolas are

oriented differently for $-1 < k < 1$ than for $k < -1$ or $k > 1$.) The traces in $y = k$ are $x^2 - z^2 = 1 - k^2$, a similar

family of hyperbolas. The traces in $z = k$ are $x^2 + y^2 = 1 + k^2$, a family of circles. For $k = 0$, the trace in the

xy-plane, the circle is of radius 1. As $|k|$ increases, so does the radius of the circle. This behavior, combined with the

hyperbolic vertical traces, gives the graph of the hyperboloid of one sheet in Table 1.

(b) The shape of the surface is unchanged, but the hyperboloid is

rotated so that its axis is the y-axis. Traces in $y = k$ are circles,

while traces in $x = k$ and $z = k$ are hyperbolas.

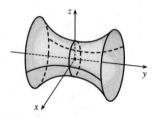

(c) Completing the square in y gives $x^2 + (y + 1)^2 - z^2 = 1$. The

surface is a hyperboloid identical to the one in part (a) but shifted

one unit in the negative y-direction.

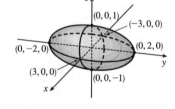

11. Traces: $x = k$, $9y^2 + 36z^2 = 36 - 4k^2$, an ellipse for $|k| < 3$;

$y = k$, $4x^2 + 36z^2 = 36 - 9k^2$, an ellipse for $|k| < 2$; $z = k$,

$4x^2 + 9y^2 = 36(1 - k^2)$, an ellipse for $|k| < 1$. Thus the surface is

an ellipsoid with center at the origin and axes along the x-, y- and

z-axes.

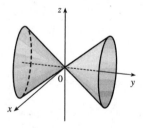

13. Traces: $x = k$, $y^2 = k^2 + z^2$ or $y^2 - z^2 = k^2$, a hyperbola for $k \neq 0$

and two intersecting lines for $k = 0$; $y = k$, $x^2 + z^2 = k^2$, a circle for

$k \neq 0$; $z = k$, $y^2 = x^2 + k^2$ or $y^2 - x^2 = k^2$, a hyperbola for $k \neq 0$

and two intersecting lines for $k = 0$. Thus the surface is a cone (right

circular) with axis the y-axis and vertex the origin.

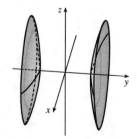

15. Traces: $x = k$, $4y^2 - z^2 = 4 + k^2$, a hyperbola; $y = k$,

$x^2 + z^2 = 4k^2 - 4$, a circle for $|k| > 1$; $z = k$, $4y^2 - x^2 = 4 + k^2$,

a hyperbola. Thus the surface is a hyperboloid of two sheets with

axis the y-axis.

17. Traces: $x = k$, $k^2 + 4z^2 - y = 0$ or $y - k^2 = 4z^2$, a parabola;

$y = k$, $x^2 + 4z^2 = k$, an ellipse for $k > 0$; $z = k$, $x^2 + 4k^2 - y = 0$

or $y - 4k^2 = x^2$, a parabola. Thus the surface is an elliptic paraboloid

with axis the y-axis and vertex the origin.

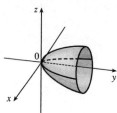
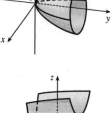

19. Traces: $x = k$, $y = z^2 - k^2$, parabolas; $y = k$, $k = z^2 - x^2$,

hyperbolas (note the hyperbolas are oriented differently for $k > 0$ than

for $k < 0$); $z = k$, $y = k^2 - x^2$, parabolas. Thus $\dfrac{y}{1} = \dfrac{z^2}{1^2} - \dfrac{x^2}{1^2}$ is a

hyperbolic paraboloid.

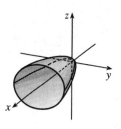

21. $z^2 = 4x^2 + 9y^2 + 36$ or $-4x^2 - 9y^2 + z^2 = 36$ or

$-\dfrac{x^2}{9} - \dfrac{y^2}{4} + \dfrac{z^2}{36} = 1$ represents a hyperboloid of two

sheets with axis the z-axis.

23. $x = 2y^2 + 3z^2$ or $x = \dfrac{y^2}{1/2} + \dfrac{z^2}{1/3}$ or $\dfrac{x}{6} = \dfrac{y^2}{3} + \dfrac{z^2}{2}$
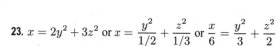

represents an elliptic paraboloid with vertex $(0, 0, 0)$ and

axis the x-axis.

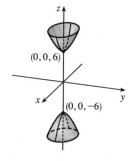

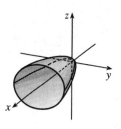

25. Completing squares in y and z gives

$4x^2 + (y - 2)^2 + 4(z - 3)^2 = 4$ or

$x^2 + \dfrac{(y - 2)^2}{4} + (z - 3)^2 = 1$, an ellipsoid with

center $(0, 2, 3)$.

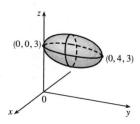

27. Completing squares in all three variables gives

$(x - 2)^2 - (y + 1)^2 + (z - 1)^2 = 0$ or

$(y + 1)^2 = (x - 2)^2 + (z - 1)^2$, a circular cone with

center $(2, -1, 1)$ and axis the horizontal line $x = 2$,

$z = 1$.

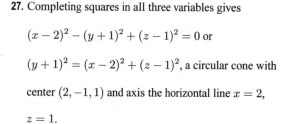
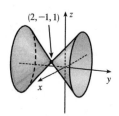

29.

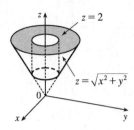

31. Let $P = (x, y, z)$ be an arbitrary point equidistant from $(-1, 0, 0)$ and the plane $x = 1$. Then the distance from P to

$(-1, 0, 0)$ is $\sqrt{(x+1)^2 + y^2 + z^2}$ and the distance from P to the plane $x = 1$ is $|x - 1| / \sqrt{1^2} = |x - 1|$

(by Equation 10.5.7). So $|x - 1| = \sqrt{(x+1)^2 + y^2 + z^2}$ $\Leftrightarrow$ $(x - 1)^2 = (x + 1)^2 + y^2 + z^2$ $\Leftrightarrow$

$x^2 - 2x + 1 = x^2 + 2x + 1 + y^2 + z^2$ $\Leftrightarrow$ $-4x = y^2 + z^2$. Thus the collection of all such points P is a circular

paraboloid with vertex at the origin, axis the x-axis, which opens in the negative direction.

33.

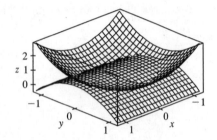

The curve of intersection looks like a bent ellipse. The projection

of this curve onto the xy-plane is the set of points $(x, y, 0)$ which

satisfy $x^2 + y^2 = 1 - y^2$ $\Leftrightarrow$ $x^2 + 2y^2 = 1$ $\Leftrightarrow$

$x^2 + \dfrac{y^2}{\left(1/\sqrt{2}\right)^2} = 1$. This is an equation of an ellipse.

10.7 Vector Functions and Space Curves

1. The component functions t^2, $\sqrt{t - 1}$, and $\sqrt{5 - t}$ are all defined when $t - 1 \geq 0$ $\Rightarrow$ $t \geq 1$ and $5 - t \geq 0$ $\Rightarrow$ $t \leq 5$,

so the domain of $\mathbf{r}(t)$ is $[1, 5]$.

3. $\lim\limits_{t \to 0^+} \cos t = \cos 0 = 1$, $\lim\limits_{t \to 0^+} \sin t = \sin 0 = 0$, $\lim\limits_{t \to 0^+} t \ln t = \lim\limits_{t \to 0^+} \dfrac{\ln t}{1/t} = \lim\limits_{t \to 0^+} \dfrac{1/t}{-1/t^2} = \lim\limits_{t \to 0^+} -t = 0$

[by l'Hospital's Rule]. Thus $\lim\limits_{t \to 0^+} \langle \cos t, \sin t, t \ln t \rangle = \left\langle \lim\limits_{t \to 0^+} \cos t, \lim\limits_{t \to 0^+} \sin t, \lim\limits_{t \to 0^+} t \ln t \right\rangle = \langle 1, 0, 0 \rangle$.

5. The corresponding parametric equations for this curve are $x = \sin t$, $y = t$.

We can make a table of values, or we can eliminate the parameter: $t = y$ $\Rightarrow$

$x = \sin y$, with $y \in \mathbb{R}$. By comparing different values of t, we find the direction in

which t increases as indicated in the graph.

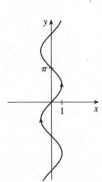

7. The corresponding parametric equations are $x = t$, $y = \cos 2t$, $z = \sin 2t$.

Note that $y^2 + z^2 = \cos^2 2t + \sin^2 2t = 1$, so the curve lies on the circular

cylinder $y^2 + z^2 = 1$. Since $x = t$, the curve is a helix.

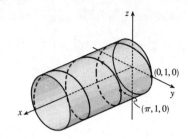

9. The corresponding parametric equations are $x = 1$, $y = \cos t$, $z = 2\sin t$.

Eliminating the parameter in y and z gives $y^2 + (z/2)^2 = \cos^2 t + \sin^2 t = 1$

or $y^2 + z^2/4 = 1$. Since $x = 1$, the curve is an ellipse centered at $(1, 0, 0)$ in

the plane $x = 1$.

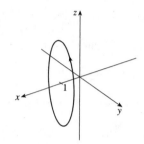

11. The parametric equations are $x = t^2$, $y = t^4$, $z = t^6$. These are positive

for $t \neq 0$ and 0 when $t = 0$. So the curve lies entirely in the first quadrant.

The projection of the graph onto the xy-plane is $y = x^2$, $y > 0$, a half parabola.

On the xz-plane $z = x^3$, $z > 0$, a half cubic, and the yz-plane, $y^3 = z^2$.

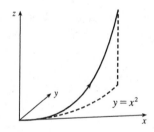

13. Taking $\mathbf{r}_0 = \langle 0, 0, 0 \rangle$ and $\mathbf{r}_1 = \langle 1, 2, 3 \rangle$, we have from Equation 10.5.4

$\mathbf{r}(t) = (1-t)\,\mathbf{r}_0 + t\,\mathbf{r}_1 = (1-t)\,\langle 0, 0, 0 \rangle + t\,\langle 1, 2, 3 \rangle,\ 0 \le t \le 1$ or $\mathbf{r}(t) = \langle t, 2t, 3t \rangle,\ 0 \le t \le 1$.

Parametric equations are $x = t$, $y = 2t$, $z = 3t$, $0 \le t \le 1$.

15. Taking $\mathbf{r}_0 = \langle 1, -1, 2 \rangle$ and $\mathbf{r}_1 = \langle 4, 1, 7 \rangle$, we have

$\mathbf{r}(t) = (1-t)\,\mathbf{r}_0 + t\,\mathbf{r}_1 = (1-t)\,\langle 1, -1, 2 \rangle + t\,\langle 4, 1, 7 \rangle,\ 0 \le t \le 1$ or $\mathbf{r}(t) = \langle 1 + 3t, -1 + 2t, 2 + 5t \rangle,\ 0 \le t \le 1$.

Parametric equations are $x = 1 + 3t$, $y = -1 + 2t$, $z = 2 + 5t$, $0 \le t \le 1$.

17. $x = \cos 4t$, $y = t$, $z = \sin 4t$.   At any point (x, y, z) on the curve, $x^2 + z^2 = \cos^2 4t + \sin^2 4t = 1$. So the curve lies on a

circular cylinder with axis the y-axis. Since $y = t$, this is a helix. So the graph is VI.

19. $x = t$, $y = 1/(1 + t^2)$, $z = t^2$.   Note that y and z are positive for all t. The curve passes through $(0, 1, 0)$ when $t = 0$.

As $t \to \infty$, $(x, y, z) \to (\infty, 0, \infty)$, and as $t \to -\infty$, $(x, y, z) \to (-\infty, 0, \infty)$. So the graph is IV.

21. $x = \cos t$, $y = \sin t$, $z = \sin 5t$. $x^2 + y^2 = \cos^2 t + \sin^2 t = 1$, so the curve lies on a circular cylinder with axis the

z-axis. Each of x, y and z is periodic, and at $t = 0$ and $t = 2\pi$ the curve passes through the same point, so the curve repeats

itself and the graph is V.

23. If $x = t \cos t$, $y = t \sin t$, $z = t$, then

$x^2 + y^2 = t^2 \cos^2 t + t^2 \sin^2 t = t^2 = z^2$, so the curve lies on the

cone $z^2 = x^2 + y^2$. Since $z = t$, the curve is a spiral on this cone.

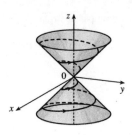

25. Parametric equations for the curve are $x = t$, $y = 0$, $z = 2t - t^2$. Substituting into the equation of the paraboloid

gives $2t - t^2 = t^2$ $\Rightarrow$ $2t = 2t^2$ $\Rightarrow$ $t = 0, 1$. Since $\mathbf{r}(0) = \mathbf{0}$ and $\mathbf{r}(1) = \mathbf{i} + \mathbf{k}$, the points of intersection

are $(0, 0, 0)$ and $(1, 0, 1)$.

27. If $t = -1$, then $x = 1$, $y = 4$, $z = 0$, so the curve passes through the point $(1, 4, 0)$. If $t = 3$, then $x = 9$, $y = -8$, $z = 28$,

so the curve passes through the point $(9, -8, 28)$. For the point $(4, 7, -6)$ to be on the curve, we require $y = 1 - 3t = 7$ $\Rightarrow$

$t = -2$. But then $z = 1 + (-2)^3 = -7 \neq -6$, so $(4, 7, -6)$ is not on the curve.

29. Both equations are solved for z, so we can substitute to eliminate z: $\sqrt{x^2 + y^2} = 1 + y$ $\Rightarrow$ $x^2 + y^2 = 1 + 2y + y^2$ $\Rightarrow$

$x^2 = 1 + 2y$ $\Rightarrow$ $y = \frac{1}{2}(x^2 - 1)$. We can form parametric equations for the curve C of intersection by choosing a

parameter $x = t$, then $y = \frac{1}{2}(t^2 - 1)$ and $z = 1 + y = 1 + \frac{1}{2}(t^2 - 1) = \frac{1}{2}(t^2 + 1)$. Thus a vector function representing C

is $\mathbf{r}(t) = t\,\mathbf{i} + \frac{1}{2}(t^2 - 1)\,\mathbf{j} + \frac{1}{2}(t^2 + 1)\,\mathbf{k}$.

31.

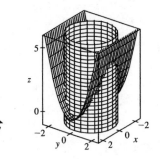

The projection of the curve C of intersection onto the

xy-plane is the circle $x^2 + y^2 = 4$, $z = 0$. Then we can write

$x = 2 \cos t$, $y = 2 \sin t$, $0 \leq t \leq 2\pi$. Since C also lies on

the surface $z = x^2$, we have $z = x^2 = (2 \cos t)^2 = 4 \cos^2 t$.

Then parametric equations for C are $x = 2 \cos t$, $y = 2 \sin t$,

$z = 4 \cos^2 t$, $0 \leq t \leq 2\pi$.

33. (a), (c)

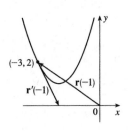

35. (a), (c)

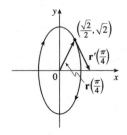

37. (a), (c)

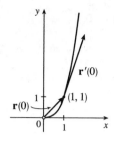

39. $\mathbf{r}'(t) = \left\langle \dfrac{d}{dt}\,[t^2],\, \dfrac{d}{dt}\,[1-t],\, \dfrac{d}{dt}\,[\sqrt{t}\,] \right\rangle = \left\langle 2t,\, -1,\, \dfrac{1}{2\sqrt{t}} \right\rangle$

41. $\mathbf{r}(t) = e^{t^2}\,\mathbf{i} - \mathbf{j} + \ln(1+3t)\,\mathbf{k}$ $\Rightarrow$ $\mathbf{r}'(t) = 2te^{t^2}\,\mathbf{i} + \dfrac{3}{1+3t}\,\mathbf{k}$

43. $\mathbf{r}'(t) = \mathbf{0} + \mathbf{b} + 2t\,\mathbf{c} = \mathbf{b} + 2t\,\mathbf{c}$ by Formulas 1 and 3 of Theorem 5.

45. $\mathbf{r}'(t) = -\sin t\,\mathbf{i} + 3\,\mathbf{j} + 4\cos 2t\,\mathbf{k}$ $\Rightarrow$ $\mathbf{r}'(0) = 3\,\mathbf{j} + 4\,\mathbf{k}$. Thus

$$\mathbf{T}(0) = \dfrac{\mathbf{r}'(0)}{|\mathbf{r}'(0)|} = \dfrac{1}{\sqrt{0^2 + 3^2 + 4^2}}\,(3\,\mathbf{j} + 4\,\mathbf{k}) = \tfrac{1}{5}(3\,\mathbf{j} + 4\,\mathbf{k}) = \tfrac{3}{5}\,\mathbf{j} + \tfrac{4}{5}\,\mathbf{k}.$$

47. $\mathbf{r}(t) = \langle t, t^2, t^3 \rangle$ $\Rightarrow$ $\mathbf{r}'(t) = \langle 1, 2t, 3t^2 \rangle$. Then $\mathbf{r}'(1) = \langle 1, 2, 3 \rangle$ and $|\mathbf{r}'(1)| = \sqrt{1^2 + 2^2 + 3^2} = \sqrt{14}$, so

$$\mathbf{T}(1) = \dfrac{\mathbf{r}'(1)}{|\mathbf{r}'(1)|} = \tfrac{1}{\sqrt{14}}\,\langle 1, 2, 3 \rangle = \left\langle \tfrac{1}{\sqrt{14}}, \tfrac{2}{\sqrt{14}}, \tfrac{3}{\sqrt{14}} \right\rangle. \quad \mathbf{r}''(t) = \langle 0, 2, 6t \rangle,\ \text{so}$$

$$\mathbf{r}'(t) \times \mathbf{r}''(t) = \begin{vmatrix} \mathbf{i} & \mathbf{j} & \mathbf{k} \\ 1 & 2t & 3t^2 \\ 0 & 2 & 6t \end{vmatrix} = \begin{vmatrix} 2t & 3t^2 \\ 2 & 6t \end{vmatrix}\mathbf{i} - \begin{vmatrix} 1 & 3t^2 \\ 0 & 6t \end{vmatrix}\mathbf{j} + \begin{vmatrix} 1 & 2t \\ 0 & 2 \end{vmatrix}\mathbf{k}$$

$$= (12t^2 - 6t^2)\,\mathbf{i} - (6t - 0)\,\mathbf{j} + (2 - 0)\,\mathbf{k} = \langle 6t^2, -6t, 2 \rangle.$$

49. The vector equation for the curve is $\mathbf{r}(t) = \langle t^5, t^4, t^3 \rangle$, so $\mathbf{r}'(t) = \langle 5t^4, 4t^3, 3t^2 \rangle$. The point $(1, 1, 1)$ corresponds to $t = 1$, so

the tangent vector there is $\mathbf{r}'(1) = \langle 5, 4, 3 \rangle$. Thus, the tangent line goes through the point $(1, 1, 1)$ and is parallel to the vector

$\langle 5, 4, 3 \rangle$. Parametric equations are $x = 1 + 5t$, $y = 1 + 4t$, $z = 1 + 3t$.

51. The vector equation for the curve is $\mathbf{r}(t) = \langle e^{-t}\cos t, e^{-t}\sin t, e^{-t} \rangle$, so

$$\mathbf{r}'(t) = \langle e^{-t}(-\sin t) + (\cos t)(-e^{-t}),\ e^{-t}\cos t + (\sin t)(-e^{-t}),\ (-e^{-t}) \rangle$$

$$= \langle -e^{-t}(\cos t + \sin t),\ e^{-t}(\cos t - \sin t),\ -e^{-t} \rangle$$

The point $(1, 0, 1)$ corresponds to $t = 0$, so the tangent vector there is

$\mathbf{r}'(0) = \langle -e^{0}(\cos 0 + \sin 0),\ e^{0}(\cos 0 - \sin 0),\ -e^{0} \rangle = \langle -1, 1, -1 \rangle$. Thus, the tangent line is parallel to the vector

$\langle -1, 1, -1 \rangle$ and parametric equations are $x = 1 + (-1)t = 1 - t$, $y = 0 + 1 \cdot t = t$, $z = 1 + (-1)t = 1 - t$.

53. (a) $\mathbf{r}(t) = \langle t^3, t^4, t^5 \rangle$ $\Rightarrow$ $\mathbf{r}'(t) = \langle 3t^2, 4t^3, 5t^4 \rangle$, and since $\mathbf{r}'(0) = \langle 0, 0, 0 \rangle = \mathbf{0}$, the curve is not smooth.

   (b) $\mathbf{r}(t) = \langle t^3 + t, t^4, t^5 \rangle$ $\Rightarrow$ $\mathbf{r}'(t) = \langle 3t^2 + 1, 4t^3, 5t^4 \rangle$. $\mathbf{r}'(t)$ is continuous since its component functions are

   continuous. Also, $\mathbf{r}'(t) \neq \mathbf{0}$, as the y- and z-components are 0 only for $t = 0$, but $\mathbf{r}'(0) = \langle 1, 0, 0 \rangle \neq \mathbf{0}$. Thus, the curve is

   smooth.

   (c) $\mathbf{r}(t) = \langle \cos^3 t, \sin^3 t \rangle$ $\Rightarrow$ $\mathbf{r}'(t) = \langle -3 \cos^2 t \sin t, 3 \sin^2 t \cos t \rangle$. Since

   $\mathbf{r}'(0) = \langle -3 \cos^2 0 \sin 0, 3 \sin^2 0 \cos 0 \rangle = \langle 0, 0 \rangle = \mathbf{0}$, the curve is not smooth.

55. The angle of intersection of the two curves is the angle between the two tangent vectors to the curves at the point of

   intersection. Since $\mathbf{r}_1'(t) = \langle 1, 2t, 3t^2 \rangle$ and $t = 0$ at $(0, 0, 0)$, $\mathbf{r}_1'(0) = \langle 1, 0, 0 \rangle$ is a tangent vector to $\mathbf{r}_1$ at $(0, 0, 0)$. Similarly,

   $\mathbf{r}_2'(t) = \langle \cos t, 2 \cos 2t, 1 \rangle$ and since $\mathbf{r}_2(0) = \langle 0, 0, 0 \rangle$, $\mathbf{r}_2'(0) = \langle 1, 2, 1 \rangle$ is a tangent vector to $\mathbf{r}_2$ at $(0, 0, 0)$. If θ is the angle

   between these two tangent vectors, then $\cos \theta = \frac{1}{\sqrt{1}\sqrt{6}} \langle 1, 0, 0 \rangle \cdot \langle 1, 2, 1 \rangle = \frac{1}{\sqrt{6}}$ and $\theta = \cos^{-1}\left(\frac{1}{\sqrt{6}}\right) \approx 66°$.

57. $\int_0^1 (16t^3 \, \mathbf{i} - 9t^2 \, \mathbf{j} + 25t^4 \, \mathbf{k}) \, dt = \left(\int_0^1 16t^3 \, dt\right) \mathbf{i} - \left(\int_0^1 9t^2 \, dt\right) \mathbf{j} + \left(\int_0^1 25t^4 \, dt\right) \mathbf{k}$

   $= [4t^4]_0^1 \, \mathbf{i} - [3t^3]_0^1 \, \mathbf{j} + [5t^5]_0^1 \, \mathbf{k} = 4\mathbf{i} - 3\mathbf{j} + 5\mathbf{k}$

59. $\int_0^{\pi/2} (3 \sin^2 t \cos t \, \mathbf{i} + 3 \sin t \cos^2 t \, \mathbf{j} + 2 \sin t \cos t \, \mathbf{k}) \, dt$

   $= \left(\int_0^{\pi/2} 3 \sin^2 t \cos t \, dt\right) \mathbf{i} + \left(\int_0^{\pi/2} 3 \sin t \cos^2 t \, dt\right) \mathbf{j} + \left(\int_0^{\pi/2} 2 \sin t \cos t \, dt\right) \mathbf{k}$

   $= [\sin^3 t]_0^{\pi/2} \, \mathbf{i} + [-\cos^3 t]_0^{\pi/2} \, \mathbf{j} + [\sin^2 t]_0^{\pi/2} \, \mathbf{k} = (1 - 0) \mathbf{i} + (0 + 1) \mathbf{j} + (1 - 0) \mathbf{k} = \mathbf{i} + \mathbf{j} + \mathbf{k}$

61. $\int (e^t \, \mathbf{i} + 2t \, \mathbf{j} + \ln t \, \mathbf{k}) \, dt = \left(\int e^t \, dt\right) \mathbf{i} + \left(\int 2t \, dt\right) \mathbf{j} + \left(\int \ln t \, dt\right) \mathbf{k}$

   $= e^t \, \mathbf{i} + t^2 \, \mathbf{j} + (t \ln t - t) \mathbf{k} + \mathbf{C}$, where $\mathbf{C}$ is a vector constant of integration.

63. $\mathbf{r}'(t) = 2t \, \mathbf{i} + 3t^2 \, \mathbf{j} + \sqrt{t} \, \mathbf{k}$ $\Rightarrow$ $\mathbf{r}(t) = t^2 \, \mathbf{i} + t^3 \, \mathbf{j} + \frac{2}{3} t^{3/2} \, \mathbf{k} + \mathbf{C}$, where $\mathbf{C}$ is a constant vector.

   But $\mathbf{i} + \mathbf{j} = \mathbf{r}(1) = \mathbf{i} + \mathbf{j} + \frac{2}{3} \mathbf{k} + \mathbf{C}$. Thus $\mathbf{C} = -\frac{2}{3} \mathbf{k}$ and $\mathbf{r}(t) = t^2 \, \mathbf{i} + t^3 \, \mathbf{j} + \left(\frac{2}{3} t^{3/2} - \frac{2}{3}\right) \mathbf{k}$.

65. For the particles to collide, we require $\mathbf{r}_1(t) = \mathbf{r}_2(t)$ $\Leftrightarrow$ $\langle t^2, 7t - 12, t^2 \rangle = \langle 4t - 3, t^2, 5t - 6 \rangle$. Equating components

   gives $t^2 = 4t - 3$, $7t - 12 = t^2$, and $t^2 = 5t - 6$. From the first equation, $t^2 - 4t + 3 = 0$ $\Leftrightarrow$ $(t - 3)(t - 1) = 0$ so

   $t = 1$ or $t = 3$. $t = 1$ does not satisfy the other two equations, but $t = 3$ does. The particles collide when $t = 3$, at the

   point $(9, 9, 9)$.

67. (a) $\lim_{t \to a} \mathbf{u}(t) + \lim_{t \to a} \mathbf{v}(t) = \langle \lim_{t \to a} u_1(t), \lim_{t \to a} u_2(t), \lim_{t \to a} u_3(t) \rangle + \langle \lim_{t \to a} v_1(t), \lim_{t \to a} v_2(t), \lim_{t \to a} v_3(t) \rangle$ and the limits of these

   component functions must each exist since the vector functions both possess limits as $t \to a$. Then adding the two vectors and using the addition property of limits for real-valued functions, we have that

$$\lim_{t\to a} \mathbf{u}(t) + \lim_{t\to a} \mathbf{v}(t) = \left\langle \lim_{t\to a} u_1(t) + \lim_{t\to a} v_1(t), \lim_{t\to a} u_2(t) + \lim_{t\to a} v_2(t), \lim_{t\to a} u_3(t) + \lim_{t\to a} v_3(t) \right\rangle$$

$$= \left\langle \lim_{t\to a} [u_1(t) + v_1(t)], \lim_{t\to a} [u_2(t) + v_2(t)], \lim_{t\to a} [u_3(t) + v_3(t)] \right\rangle$$

$$= \lim_{t\to a} \langle u_1(t) + v_1(t), u_2(t) + v_2(t), u_3(t) + v_3(t) \rangle \qquad \text{[using (1) backward]}$$

$$= \lim_{t\to a} [\mathbf{u}(t) + \mathbf{v}(t)]$$

(b) $\displaystyle \lim_{t\to a} c\mathbf{u}(t) = \lim_{t\to a} \langle cu_1(t), cu_2(t), cu_3(t) \rangle = \left\langle \lim_{t\to a} cu_1(t), \lim_{t\to a} cu_2(t), \lim_{t\to a} cu_3(t) \right\rangle$

$$= \left\langle c\lim_{t\to a} u_1(t), c\lim_{t\to a} u_2(t), c\lim_{t\to a} u_3(t) \right\rangle = c\left\langle \lim_{t\to a} u_1(t), \lim_{t\to a} u_2(t), \lim_{t\to a} u_3(t) \right\rangle$$

$$= c\lim_{t\to a} \langle u_1(t), u_2(t), u_3(t) \rangle = c\lim_{t\to a} \mathbf{u}(t)$$

(c) $\displaystyle \lim_{t\to a} \mathbf{u}(t) \cdot \lim_{t\to a} \mathbf{v}(t) = \left\langle \lim_{t\to a} u_1(t), \lim_{t\to a} u_2(t), \lim_{t\to a} u_3(t) \right\rangle \cdot \left\langle \lim_{t\to a} v_1(t), \lim_{t\to a} v_2(t), \lim_{t\to a} v_3(t) \right\rangle$

$$= \left[\lim_{t\to a} u_1(t)\right]\left[\lim_{t\to a} v_1(t)\right] + \left[\lim_{t\to a} u_2(t)\right]\left[\lim_{t\to a} v_2(t)\right] + \left[\lim_{t\to a} u_3(t)\right]\left[\lim_{t\to a} v_3(t)\right]$$

$$= \lim_{t\to a} u_1(t)v_1(t) + \lim_{t\to a} u_2(t)v_2(t) + \lim_{t\to a} u_3(t)v_3(t)$$

$$= \lim_{t\to a} [u_1(t)v_1(t) + u_2(t)v_2(t) + u_3(t)v_3(t)] = \lim_{t\to a} [\mathbf{u}(t) \cdot \mathbf{v}(t)]$$

(d) $\displaystyle \lim_{t\to a} \mathbf{u}(t) \times \lim_{t\to a} \mathbf{v}(t) = \left\langle \lim_{t\to a} u_1(t), \lim_{t\to a} u_2(t), \lim_{t\to a} u_3(t) \right\rangle \times \left\langle \lim_{t\to a} v_1(t), \lim_{t\to a} v_2(t), \lim_{t\to a} v_3(t) \right\rangle$

$$= \left\langle \left[\lim_{t\to a} u_2(t)\right]\left[\lim_{t\to a} v_3(t)\right] - \left[\lim_{t\to a} u_3(t)\right]\left[\lim_{t\to a} v_2(t)\right], \right.$$

$$\left[\lim_{t\to a} u_3(t)\right]\left[\lim_{t\to a} v_1(t)\right] - \left[\lim_{t\to a} u_1(t)\right]\left[\lim_{t\to a} v_3(t)\right],$$

$$\left. \left[\lim_{t\to a} u_1(t)\right]\left[\lim_{t\to a} v_2(t)\right] - \left[\lim_{t\to a} u_2(t)\right]\left[\lim_{t\to a} v_1(t)\right] \right\rangle$$

$$= \left\langle \lim_{t\to a} [u_2(t)v_3(t) - u_3(t)v_2(t)], \lim_{t\to a} [u_3(t)v_1(t) - u_1(t)v_3(t)], \right.$$

$$\left. \lim_{t\to a} [u_1(t)v_2(t) - u_2(t)v_1(t)] \right\rangle$$

$$= \lim_{t\to a} \langle u_2(t)v_3(t) - u_3(t)v_2(t), u_3(t)v_1(t) - u_1(t)v_3(t), u_1(t)v_2(t) - u_2(t)v_1(t) \rangle$$

$$= \lim_{t\to a} [\mathbf{u}(t) \times \mathbf{v}(t)]$$

For Exercises 69–71, let $\mathbf{u}(t) = \langle u_1(t), u_2(t), u_3(t) \rangle$ and $\mathbf{v}(t) = \langle v_1(t), v_2(t), v_3(t) \rangle$. In each of these exercises, the procedure is to apply Theorem 3 so that the corresponding properties of derivatives of real-valued functions can be used.

69. $\displaystyle \frac{d}{dt} [\mathbf{u}(t) + \mathbf{v}(t)] = \frac{d}{dt} \langle u_1(t) + v_1(t), u_2(t) + v_2(t), u_3(t) + v_3(t) \rangle$

$$= \left\langle \frac{d}{dt} [u_1(t) + v_1(t)], \frac{d}{dt} [u_2(t) + v_2(t)], \frac{d}{dt} [u_3(t) + v_3(t)] \right\rangle$$

$$= \langle u_1'(t) + v_1'(t), u_2'(t) + v_2'(t), u_3'(t) + v_3'(t) \rangle$$

$$= \langle u_1'(t), u_2'(t), u_3'(t) \rangle + \langle v_1'(t), v_2'(t), v_3'(t) \rangle = \mathbf{u}'(t) + \mathbf{v}'(t)$$

71. $\dfrac{d}{dt}\left[\mathbf{u}(t)\times\mathbf{v}(t)\right] = \dfrac{d}{dt}\langle u_2(t)v_3(t)-u_3(t)v_2(t),\, u_3(t)v_1(t)-u_1(t)v_3(t),\, u_1(t)v_2(t)-u_2(t)v_1(t)\rangle$

$$= \langle u_2'v_3(t)+u_2(t)v_3'(t)-u_3'(t)v_2(t)-u_3(t)v_2'(t),$$

$$u_3'(t)v_1(t)+u_3(t)v_1'\,(t)-u_1'(t)v_3(t)-u_1(t)v_3'(t),$$

$$u_1'(t)v_2(t)+u_1(t)v_2'(t)-u_2'(t)v_1(t)-u_2(t)v_1'(t)\rangle$$

$$= \langle u_2'(t)v_3(t)-u_3'(t)v_2\,(t)\,,\, u_3'(t)v_1(t)-u_1'(t)v_3(t),\, u_1'(t)v_2(t)-u_2'(t)v_1(t)\rangle$$

$$+ \langle u_2(t)v_3'(t)-u_3(t)v_2'(t),\, u_3(t)v_1'\,(t)-u_1(t)v_3'(t),\, u_1(t)v_2'(t)-u_2(t)v_1'(t)\rangle$$

$$= \mathbf{u}'(t)\times\mathbf{v}(t)+\mathbf{u}(t)\times\mathbf{v}'(t)$$

Alternate solution: Let $\mathbf{r}(t)=\mathbf{u}(t)\times\mathbf{v}(t)$. Then

$$\mathbf{r}(t+h)-\mathbf{r}(t) = [\mathbf{u}(t+h)\times\mathbf{v}(t+h)]-[\mathbf{u}(t)\times\mathbf{v}(t)]$$

$$= [\mathbf{u}(t+h)\times\mathbf{v}(t+h)]-[\mathbf{u}(t)\times\mathbf{v}(t)]+[\mathbf{u}(t+h)\times\mathbf{v}(t)]-[\mathbf{u}(t+h)\times\mathbf{v}(t)]$$

$$= \mathbf{u}(t+h)\times[\mathbf{v}(t+h)-\mathbf{v}(t)]+[\mathbf{u}(t+h)-\mathbf{u}(t)]\times\mathbf{v}(t)$$

(Be careful of the order of the cross product.) Dividing through by h and taking the limit as $h\to 0$ we have

$$\mathbf{r}'(t)=\lim_{h\to 0}\dfrac{\mathbf{u}(t+h)\times[\mathbf{v}(t+h)-\mathbf{v}(t)]}{h}+\lim_{h\to 0}\dfrac{[\mathbf{u}(t+h)-\mathbf{u}(t)]\times\mathbf{v}(t)}{h}=\mathbf{u}(t)\times\mathbf{v}'(t)+\mathbf{u}'(t)\times\mathbf{v}(t)$$

by Exercise 10.7.67(a) and Definition 3.

73. $\dfrac{d}{dt}\left[\mathbf{u}(t)\cdot\mathbf{v}(t)\right]=\mathbf{u}'(t)\cdot\mathbf{v}(t)+\mathbf{u}(t)\cdot\mathbf{v}'(t)$ [by Formula 4 of Theorem 5]

$$= (-4t\,\mathbf{j}+9t^2\,\mathbf{k})\cdot(t\,\mathbf{i}+\cos t\,\mathbf{j}+\sin t\,\mathbf{k})+(\mathbf{i}-2t^2\,\mathbf{j}+3t^3\,\mathbf{k})\cdot(\mathbf{i}-\sin t\,\mathbf{j}+\cos t\,\mathbf{k})$$

$$= -4t\cos t+9t^2\sin t+1+2t^2\sin t+3t^3\cos t$$

$$= 1-4t\cos t+11t^2\sin t+3t^3\cos t$$

75. $\dfrac{d}{dt}\left[\mathbf{r}(t)\times\mathbf{r}'(t)\right]=\mathbf{r}'(t)\times\mathbf{r}'(t)+\mathbf{r}(t)\times\mathbf{r}''(t)$ by Formula 5 of Theorem 5. But $\mathbf{r}'(t)\times\mathbf{r}'(t)=\mathbf{0}$ [by Example 2 in

Section 10.4]. Thus, $\dfrac{d}{dt}\left[\mathbf{r}(t)\times\mathbf{r}'(t)\right]=\mathbf{r}(t)\times\mathbf{r}''(t)$.

77. $\dfrac{d}{dt}\,|\mathbf{r}(t)|=\dfrac{d}{dt}\left[\mathbf{r}(t)\cdot\mathbf{r}(t)\right]^{1/2}=\tfrac{1}{2}[\mathbf{r}(t)\cdot\mathbf{r}(t)]^{-1/2}\,[2\mathbf{r}(t)\cdot\mathbf{r}'(t)]=\dfrac{1}{|\mathbf{r}(t)|}\,\mathbf{r}(t)\cdot\mathbf{r}'(t)$

79. Since $\mathbf{u}(t)=\mathbf{r}(t)\cdot[\mathbf{r}'(t)\times\mathbf{r}''(t)]$,

$$\mathbf{u}'(t) = \mathbf{r}'(t)\cdot[\mathbf{r}'(t)\times\mathbf{r}''(t)]+\mathbf{r}(t)\cdot\dfrac{d}{dt}\,[\mathbf{r}'(t)\times\mathbf{r}''(t)]$$

$$= 0+\mathbf{r}(t)\cdot[\mathbf{r}''(t)\times\mathbf{r}''(t)+\mathbf{r}'(t)\times\mathbf{r}'''(t)] \qquad [\text{since } \mathbf{r}'(t)\perp\mathbf{r}'(t)\times\mathbf{r}''(t)]$$

$$= \mathbf{r}(t)\cdot[\mathbf{r}'(t)\times\mathbf{r}'''(t)] \qquad\qquad\qquad\qquad [\text{since } \mathbf{r}''(t)\times\mathbf{r}''(t)=\mathbf{0}]$$

10.8 Arc Length and Curvature

1. $\mathbf{r}'(t) = \langle 2\cos t, 5, -2\sin t \rangle \;\Rightarrow\; |\mathbf{r}'(t)| = \sqrt{(2\cos t)^2 + 5^2 + (-2\sin t)^2} = \sqrt{29}$.

Then using Formula 3, we have $L = \int_{-10}^{10} |\mathbf{r}'(t)|\,dt = \int_{-10}^{10} \sqrt{29}\,dt = \sqrt{29}\,t\Big]_{-10}^{10} = 20\sqrt{29}$.

3. $\mathbf{r}'(t) = 2t\,\mathbf{j} + 3t^2\,\mathbf{k} \;\Rightarrow\; |\mathbf{r}'(t)| = \sqrt{4t^2 + 9t^4} = t\sqrt{4 + 9t^2}$ [since $t \geq 0$].

Then $L = \int_0^1 |\mathbf{r}'(t)|\,dt = \int_0^1 t\sqrt{4 + 9t^2}\,dt = \frac{1}{18} \cdot \frac{2}{3}(4 + 9t^2)^{3/2}\Big]_0^1 = \frac{1}{27}(13^{3/2} - 4^{3/2}) = \frac{1}{27}(13^{3/2} - 8)$.

5. The point $(2, 4, 8)$ corresponds to $t = 2$, so by Equation 2, $L = \int_0^2 \sqrt{(1)^2 + (2t)^2 + (3t^2)^2}\,dt$. If $f(t) = \sqrt{1 + 4t^2 + 9t^4}$,

then Simpson's Rule gives $L \approx \dfrac{2 - 0}{10 \cdot 3}\,[f(0) + 4f(0.2) + 2f(0.4) + \cdots + 4f(1.8) + f(2)] \approx 9.5706$.

7. $\mathbf{r}'(t) = 2\,\mathbf{i} - 3\,\mathbf{j} + 4\,\mathbf{k}$ and $\frac{ds}{dt} = |\mathbf{r}'(t)| = \sqrt{4 + 9 + 16} = \sqrt{29}$. Then $s = s(t) = \int_0^t |\mathbf{r}'(u)|\,du = \int_0^t \sqrt{29}\,du = \sqrt{29}\,t$.

Therefore, $t = \frac{1}{\sqrt{29}}s$, and substituting for t in the original equation, we have

$$\mathbf{r}(t(s)) = \frac{2}{\sqrt{29}}s\,\mathbf{i} + \left(1 - \frac{3}{\sqrt{29}}s\right)\mathbf{j} + \left(5 + \frac{4}{\sqrt{29}}s\right)\mathbf{k}.$$

9. Here $\mathbf{r}(t) = \langle 3\sin t, 4t, 3\cos t \rangle$, so $\mathbf{r}'(t) = \langle 3\cos t, 4, -3\sin t \rangle$ and $|\mathbf{r}'(t)| = \sqrt{9\cos^2 t + 16 + 9\sin^2 t} = \sqrt{25} = 5$.

The point $(0, 0, 3)$ corresponds to $t = 0$, so the arc length function beginning at $(0, 0, 3)$ and measuring in the positive

direction is given by $s(t) = \int_0^t |\mathbf{r}'(u)|\,du = \int_0^t 5\,du = 5t$. $s(t) = 5 \;\Rightarrow\; 5t = 5 \;\Rightarrow\; t = 1$, thus your location after

moving 5 units along the curve is $(3\sin 1, 4, 3\cos 1)$.

11. (a) $\mathbf{r}'(t) = \langle 2\cos t, 5, -2\sin t \rangle \;\Rightarrow\; |\mathbf{r}'(t)| = \sqrt{4\cos^2 t + 25 + 4\sin^2 t} = \sqrt{29}$. Then

$$\mathbf{T}(t) = \frac{\mathbf{r}'(t)}{|\mathbf{r}'(t)|} = \frac{1}{\sqrt{29}}\langle 2\cos t, 5, -2\sin t \rangle \text{ or } \left\langle \frac{2}{\sqrt{29}}\cos t, \frac{5}{\sqrt{29}}, -\frac{2}{\sqrt{29}}\sin t \right\rangle.$$

$$\mathbf{T}'(t) = \frac{1}{\sqrt{29}}\langle -2\sin t, 0, -2\cos t \rangle \;\Rightarrow\; |\mathbf{T}'(t)| = \frac{1}{\sqrt{29}}\sqrt{4\sin^2 t + 0 + 4\cos^2 t} = \frac{2}{\sqrt{29}}. \text{ Thus}$$

$$\mathbf{N}(t) = \frac{\mathbf{T}'(t)}{|\mathbf{T}'(t)|} = \frac{1/\sqrt{29}}{2/\sqrt{29}}\langle -2\sin t, 0, -2\cos t \rangle = \langle -\sin t, 0, -\cos t \rangle.$$

(b) $\kappa(t) = \dfrac{|\mathbf{T}'(t)|}{|\mathbf{r}'(t)|} = \dfrac{2/\sqrt{29}}{\sqrt{29}} = \dfrac{2}{29}$.

13. (a) $\mathbf{r}'(t) = \langle \sqrt{2}, e^t, -e^{-t} \rangle \;\Rightarrow\; |\mathbf{r}'(t)| = \sqrt{2 + e^{2t} + e^{-2t}} = \sqrt{(e^t + e^{-t})^2} = e^t + e^{-t}$. Then

$$\mathbf{T}(t) = \frac{\mathbf{r}'(t)}{|\mathbf{r}'(t)|} = \frac{1}{e^t + e^{-t}}\langle \sqrt{2}, e^t, -e^{-t} \rangle = \frac{1}{e^{2t} + 1}\langle \sqrt{2}e^t, e^{2t}, -1 \rangle \quad \left[\text{after multiplying by } \frac{e^t}{e^t}\right] \quad \text{and}$$

$$\mathbf{T}'(t) = \frac{1}{e^{2t} + 1}\langle \sqrt{2}e^t, 2e^{2t}, 0 \rangle - \frac{2e^{2t}}{(e^{2t} + 1)^2}\langle \sqrt{2}e^t, e^{2t}, -1 \rangle$$

$$= \frac{1}{(e^{2t} + 1)^2}\left[(e^{2t} + 1)\langle \sqrt{2}e^t, 2e^{2t}, 0 \rangle - 2e^{2t}\langle \sqrt{2}e^t, e^{2t}, -1 \rangle\right] = \frac{1}{(e^{2t} + 1)^2}\langle \sqrt{2}e^t\left(1 - e^{2t}\right), 2e^{2t}, 2e^{2t} \rangle$$

Then

$$|\mathbf{T}'(t)| = \frac{1}{(e^{2t}+1)^2}\sqrt{2e^{2t}(1-2e^{2t}+e^{4t})+4e^{4t}+4e^{4t}} = \frac{1}{(e^{2t}+1)^2}\sqrt{2e^{2t}(1+2e^{2t}+e^{4t})}$$

$$= \frac{1}{(e^{2t}+1)^2}\sqrt{2e^{2t}(1+e^{2t})^2} = \frac{\sqrt{2}e^t(1+e^{2t})}{(e^{2t}+1)^2} = \frac{\sqrt{2}e^t}{e^{2t}+1}$$

Therefore

$$\mathbf{N}(t) = \frac{\mathbf{T}'(t)}{|\mathbf{T}'(t)|} = \frac{e^{2t}+1}{\sqrt{2}e^t}\frac{1}{(e^{2t}+1)^2}\left\langle\sqrt{2}e^t(1-e^{2t}), 2e^{2t}, 2e^{2t}\right\rangle$$

$$= \frac{1}{\sqrt{2}e^t(e^{2t}+1)}\left\langle\sqrt{2}e^t(1-e^{2t}), 2e^{2t}, 2e^{2t}\right\rangle = \frac{1}{e^{2t}+1}\left\langle 1-e^{2t}, \sqrt{2}e^t, \sqrt{2}e^t\right\rangle$$

(b) $\kappa(t) = \dfrac{|\mathbf{T}'(t)|}{|\mathbf{r}'(t)|} = \dfrac{\sqrt{2}e^t}{e^{2t}+1}\cdot\dfrac{1}{e^t+e^{-t}} = \dfrac{\sqrt{2}e^t}{e^{3t}+2e^t+e^{-t}} = \dfrac{\sqrt{2}e^{2t}}{e^{4t}+2e^{2t}+1} = \dfrac{\sqrt{2}e^{2t}}{(e^{2t}+1)^2}.$

15. $\mathbf{r}'(t) = 2t\,\mathbf{i}+\mathbf{k}, \quad \mathbf{r}''(t) = 2\,\mathbf{i}, \quad |\mathbf{r}'(t)| = \sqrt{(2t)^2+0^2+1^2} = \sqrt{4t^2+1}, \quad \mathbf{r}'(t)\times\mathbf{r}''(t) = 2\,\mathbf{j}, \quad |\mathbf{r}'(t)\times\mathbf{r}''(t)| = 2.$

Then $\kappa(t) = \dfrac{|\mathbf{r}'(t)\times\mathbf{r}''(t)|}{|\mathbf{r}'(t)|^3} = \dfrac{2}{\left(\sqrt{4t^2+1}\right)^3} = \dfrac{2}{(4t^2+1)^{3/2}}.$

17. $\mathbf{r}'(t) = 3\,\mathbf{i}+4\cos t\,\mathbf{j}-4\sin t\,\mathbf{k}, \quad \mathbf{r}''(t) = -4\sin t\,\mathbf{j}-4\cos t\,\mathbf{k}, \quad |\mathbf{r}'(t)| = \sqrt{9+16\cos^2 t+16\sin^2 t} = \sqrt{9+16} = 5,$

$\mathbf{r}'(t)\times\mathbf{r}''(t) = -16\,\mathbf{i}+12\cos t\,\mathbf{j}-12\sin t\,\mathbf{k}, \quad |\mathbf{r}'(t)\times\mathbf{r}''(t)| = \sqrt{256+144\cos^2 t+144\sin^2 t} = \sqrt{400} = 20.$

Then $\kappa(t) = \dfrac{|\mathbf{r}'(t)\times\mathbf{r}''(t)|}{|\mathbf{r}'(t)|^3} = \dfrac{20}{5^3} = \dfrac{4}{25}.$

19. $\mathbf{r}'(t) = \left\langle 1, 2t, 3t^2\right\rangle.$ The point $(1,1,1)$ corresponds to $t = 1$, and $\mathbf{r}'(1) = \langle 1, 2, 3\rangle \;\Rightarrow\; |\mathbf{r}'(1)| = \sqrt{1+4+9} = \sqrt{14}.$

$\mathbf{r}''(t) = \langle 0, 2, 6t\rangle \;\Rightarrow\; \mathbf{r}''(1) = \langle 0, 2, 6\rangle. \quad \mathbf{r}'(1)\times\mathbf{r}''(1) = \langle 6, -6, 2\rangle,$ so $|\mathbf{r}'(1)\times\mathbf{r}''(1)| = \sqrt{36+36+4} = \sqrt{76}.$

Then $\kappa(1) = \dfrac{|\mathbf{r}'(1)\times\mathbf{r}''(1)|}{|\mathbf{r}'(1)|^3} = \dfrac{\sqrt{76}}{\sqrt{14}^3} = \dfrac{1}{7}\sqrt{\dfrac{19}{14}}.$

21. $f(x) = xe^x, \quad f'(x) = xe^x+e^x, \quad f''(x) = xe^x+2e^x,$

$$\kappa(x) = \frac{|f''(x)|}{[1+(f'(x))^2]^{3/2}} = \frac{|xe^x+2e^x|}{[1+(xe^x+e^x)^2]^{3/2}} = \frac{|x+2|\,e^x}{[1+(xe^x+e^x)^2]^{3/2}}$$

23. $f(x) = 4x^{5/2}, \quad f'(x) = 10x^{3/2}, \quad f''(x) = 15x^{1/2},$

$$\kappa(x) = \frac{|f''(x)|}{[1+(f'(x))^2]^{3/2}} = \frac{\left|15x^{1/2}\right|}{[1+(10x^{3/2})^2]^{3/2}} = \frac{15\sqrt{x}}{(1+100x^3)^{3/2}}$$

25. Since $y' = y'' = e^x$, the curvature is $\kappa(x) = \dfrac{|y''(x)|}{[1+(y'(x))^2]^{3/2}} = \dfrac{e^x}{(1+e^{2x})^{3/2}} = e^x(1+e^{2x})^{-3/2}.$

To find the maximum curvature, we first find the critical numbers of $\kappa(x)$:

$$\kappa'(x) = e^x(1+e^{2x})^{-3/2}+e^x\left(-\tfrac{3}{2}\right)(1+e^{2x})^{-5/2}(2e^{2x}) = e^x\frac{1+e^{2x}-3e^{2x}}{(1+e^{2x})^{5/2}} = e^x\frac{1-2e^{2x}}{(1+e^{2x})^{5/2}}.$$

$\kappa'(x) = 0$ when $1-2e^{2x} = 0$, so $e^{2x} = \tfrac{1}{2}$ or $x = -\tfrac{1}{2}\ln 2$. And since $1-2e^{2x} > 0$ for $x < -\tfrac{1}{2}\ln 2$ and

$1 - 2e^{2x} < 0$ for $x > -\frac{1}{2}\ln 2$, the maximum curvature is attained at the point $\left(-\frac{1}{2}\ln 2, e^{(-\ln 2)/2}\right) = \left(-\frac{1}{2}\ln 2, \frac{1}{\sqrt{2}}\right)$.

Since $\lim\limits_{x\to\infty} e^x(1 + e^{2x})^{-3/2} = 0$, $\kappa(x)$ approaches 0 as $x \to \infty$.

27. (a) C appears to be changing direction more quickly at P than Q, so we would expect the curvature to be greater at P.

(b) First we sketch approximate osculating circles at P and Q. Using the

axes scale as a guide, we measure the radius of the osculating circle

at P to be approximately 0.8 units, thus $\rho = \dfrac{1}{\kappa}$ $\Rightarrow$

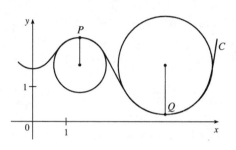

$\kappa = \dfrac{1}{\rho} \approx \dfrac{1}{0.8} \approx 1.3$. Similarly, we estimate the radius of the

osculating circle at Q to be 1.4 units, so $\kappa = \dfrac{1}{\rho} \approx \dfrac{1}{1.4} \approx 0.7$.

29. $y = x^{-2}$ $\Rightarrow$ $y' = -2x^{-3}$, $y'' = 6x^{-4}$, and

$$\kappa(x) = \frac{|y''|}{\left[1 + (y')^2\right]^{3/2}} = \frac{\left|6x^{-4}\right|}{\left[1 + (-2x^{-3})^2\right]^{3/2}} = \frac{6}{x^4\left(1 + 4x^{-6}\right)^{3/2}}.$$

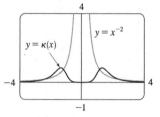

The appearance of the two humps in this graph is perhaps a little surprising, but it is

explained by the fact that $y = x^{-2}$ increases asymptotically at the origin from both

directions, and so its graph has very little bend there. (Note that $\kappa(0)$ is undefined.)

31. Notice that the curve b has two inflection points at which the graph appears almost straight. We would expect the curvature to

be 0 or nearly 0 at these values, but the curve a isn't near 0 there. Thus, a must be the graph of $y = f(x)$ rather than the graph

of curvature, and b is the graph of $y = \kappa(x)$.

33. $x = e^t \cos t$ $\Rightarrow$ $\dot{x} = e^t(\cos t - \sin t)$ $\Rightarrow$ $\ddot{x} = e^t(-\sin t - \cos t) + e^t(\cos t - \sin t) = -2e^t \sin t$,

$y = e^t \sin t$ $\Rightarrow$ $\dot{y} = e^t(\cos t + \sin t)$ $\Rightarrow$ $\ddot{y} = e^t(-\sin t + \cos t) + e^t(\cos t + \sin t) = 2e^t \cos t$. Then

$$\kappa(t) = \frac{|\dot{x}\ddot{y} - \dot{y}\ddot{x}|}{[\dot{x}^2 + \dot{y}^2]^{3/2}} = \frac{\left|e^t(\cos t - \sin t)(2e^t \cos t) - e^t(\cos t + \sin t)(-2e^t \sin t)\right|}{\left([e^t(\cos t - \sin t)]^2 + [e^t(\cos t + \sin t)]^2\right)^{3/2}}$$

$$= \frac{\left|2e^{2t}(\cos^2 t - \sin t \cos t + \sin t \cos t + \sin^2 t)\right|}{\left[e^{2t}(\cos^2 t - 2\cos t \sin t + \sin^2 t + \cos^2 t + 2\cos t \sin t + \sin^2 t)\right]^{3/2}} = \frac{\left|2e^{2t}(1)\right|}{[e^{2t}(1+1)]^{3/2}} = \frac{2e^{2t}}{e^{3t}(2)^{3/2}} = \frac{1}{\sqrt{2}\,e^t}$$

35. $(1, \frac{2}{3}, 1)$ corresponds to $t = 1$. $\mathbf{T}(t) = \dfrac{\mathbf{r}'(t)}{|\mathbf{r}'(t)|} = \dfrac{\langle 2t, 2t^2, 1\rangle}{\sqrt{4t^2 + 4t^4 + 1}} = \dfrac{\langle 2t, 2t^2, 1\rangle}{2t^2 + 1}$, so $\mathbf{T}(1) = \langle \frac{2}{3}, \frac{2}{3}, \frac{1}{3}\rangle$.

$\mathbf{T}'(t) = -4t(2t^2 + 1)^{-2}\langle 2t, 2t^2, 1\rangle + (2t^2 + 1)^{-1}\langle 2, 4t, 0\rangle$ [by Formula 3 of Theorem 10.7.5]

$\qquad = (2t^2 + 1)^{-2}\langle -8t^2 + 4t^2 + 2, -8t^3 + 8t^3 + 4t, -4t\rangle = 2(2t^2 + 1)^{-2}\langle 1 - 2t^2, 2t, -2t\rangle$

$\mathbf{N}(t) = \dfrac{\mathbf{T}'(t)}{|\mathbf{T}'(t)|} = \dfrac{2(2t^2 + 1)^{-2}\langle 1 - 2t^2, 2t, -2t\rangle}{2(2t^2 + 1)^{-2}\sqrt{(1 - 2t^2)^2 + (2t)^2 + (-2t)^2}} = \dfrac{\langle 1 - 2t^2, 2t, -2t\rangle}{\sqrt{1 - 4t^2 + 4t^4 + 8t^2}} = \dfrac{\langle 1 - 2t^2, 2t, -2t\rangle}{1 + 2t^2}$

$\mathbf{N}(1) = \langle -\frac{1}{3}, \frac{2}{3}, -\frac{2}{3}\rangle$ and $\mathbf{B}(1) = \mathbf{T}(1) \times \mathbf{N}(1) = \langle -\frac{4}{9} - \frac{2}{9}, -\left(-\frac{4}{9} + \frac{1}{9}\right), \frac{4}{9} + \frac{2}{9}\rangle = \langle -\frac{2}{3}, \frac{1}{3}, \frac{2}{3}\rangle$.

37. $(0, \pi, -2)$ corresponds to $t = \pi$. $\mathbf{r}(t) = \langle 2\sin 3t, t, 2\cos 3t \rangle$ $\Rightarrow$

$$\mathbf{T}(t) = \frac{\mathbf{r}'(t)}{|\mathbf{r}'(t)|} = \frac{\langle 6\cos 3t, 1, -6\sin 3t \rangle}{\sqrt{36\cos^2 3t + 1 + 36\sin^2 3t}} = \tfrac{1}{\sqrt{37}} \langle 6\cos 3t, 1, -6\sin 3t \rangle.$$

$\mathbf{T}(\pi) = \tfrac{1}{\sqrt{37}} \langle -6, 1, 0 \rangle$ is a normal vector for the normal plane, and so $\langle -6, 1, 0 \rangle$ is also normal. Thus an equation for the

plane is $-6(x - 0) + 1(y - \pi) + 0(z + 2) = 0$ or $y - 6x = \pi$.

$$\mathbf{T}'(t) = \tfrac{1}{\sqrt{37}} \langle -18\sin 3t, 0, -18\cos 3t \rangle \Rightarrow |\mathbf{T}'(t)| = \frac{\sqrt{18^2\sin^2 3t + 18^2\cos^2 3t}}{\sqrt{37}} = \frac{18}{\sqrt{37}} \Rightarrow$$

$$\mathbf{N}(t) = \frac{\mathbf{T}'(t)}{|\mathbf{T}'(t)|} = \langle -\sin 3t, 0, -\cos 3t \rangle. \text{ So } \mathbf{N}(\pi) = \langle 0, 0, 1 \rangle \text{ and } \mathbf{B}(\pi) = \tfrac{1}{\sqrt{37}} \langle -6, 1, 0 \rangle \times \langle 0, 0, 1 \rangle = \tfrac{1}{\sqrt{37}} \langle 1, 6, 0 \rangle.$$

Since $\mathbf{B}(\pi)$ is a normal to the osculating plane, so is $\langle 1, 6, 0 \rangle$.

An equation for the plane is $1(x - 0) + 6(y - \pi) + 0(z + 2) = 0$ or $x + 6y = 6\pi$.

39. The ellipse is given by the parametric equations $x = 2\cos t$, $y = 3\sin t$, so using the result from Exercise 32,

$$\kappa(t) = \frac{|\dot{x}\ddot{y} - \ddot{x}\dot{y}|}{[\dot{x}^2 + \dot{y}^2]^{3/2}} = \frac{|(-2\sin t)(-3\sin t) - (3\cos t)(-2\cos t)|}{(4\sin^2 t + 9\cos^2 t)^{3/2}} = \frac{6}{(4\sin^2 t + 9\cos^2 t)^{3/2}}.$$

At $(2, 0)$, $t = 0$. Now $\kappa(0) = \tfrac{6}{27} = \tfrac{2}{9}$, so the radius of the osculating circle is

$1/\kappa(0) = \tfrac{9}{2}$ and its center is $\left(-\tfrac{5}{2}, 0\right)$. Its equation is therefore $\left(x + \tfrac{5}{2}\right)^2 + y^2 = \tfrac{81}{4}$.

At $(0, 3)$, $t = \tfrac{\pi}{2}$, and $\kappa\left(\tfrac{\pi}{2}\right) = \tfrac{6}{8} = \tfrac{3}{4}$. So the radius of the osculating circle is $\tfrac{4}{3}$ and

its center is $\left(0, \tfrac{5}{3}\right)$. Hence its equation is $x^2 + \left(y - \tfrac{5}{3}\right)^2 = \tfrac{16}{9}$.

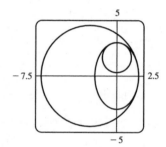

41. The tangent vector is normal to the normal plane, and the vector $\langle 6, 6, -8 \rangle$ is normal to the given plane.

But $\mathbf{T}(t) \parallel \mathbf{r}'(t)$ and $\langle 6, 6, -8 \rangle \parallel \langle 3, 3, -4 \rangle$, so we need to find t such that $\mathbf{r}'(t) \parallel \langle 3, 3, -4 \rangle$.

$\mathbf{r}(t) = \langle t^3, 3t, t^4 \rangle \Rightarrow \mathbf{r}'(t) = \langle 3t^2, 3, 4t^3 \rangle \parallel \langle 3, 3, -4 \rangle$ when $t = -1$. So the planes are parallel at the point $(-1, -3, 1)$.

43. $\kappa = \left| \dfrac{d\mathbf{T}}{ds} \right| = \left| \dfrac{d\mathbf{T}/dt}{ds/dt} \right| = \dfrac{|d\mathbf{T}/dt|}{ds/dt}$ and $\mathbf{N} = \dfrac{d\mathbf{T}/dt}{|d\mathbf{T}/dt|}$, so $\kappa\mathbf{N} = \dfrac{\dfrac{d\mathbf{T}}{dt} \dfrac{d\mathbf{T}}{dt}}{\dfrac{d\mathbf{T}}{dt} \dfrac{ds}{dt}} = \dfrac{d\mathbf{T}/dt}{ds/dt} = \dfrac{d\mathbf{T}}{ds}$ by the Chain Rule.

45. (a) $|\mathbf{B}| = 1 \Rightarrow \mathbf{B} \cdot \mathbf{B} = 1 \Rightarrow \dfrac{d}{ds}(\mathbf{B} \cdot \mathbf{B}) = 0 \Rightarrow 2\dfrac{d\mathbf{B}}{ds} \cdot \mathbf{B} = 0 \Rightarrow \dfrac{d\mathbf{B}}{ds} \perp \mathbf{B}$

(b) $\mathbf{B} = \mathbf{T} \times \mathbf{N} \Rightarrow$

$$\dfrac{d\mathbf{B}}{ds} = \dfrac{d}{ds}(\mathbf{T} \times \mathbf{N}) = \dfrac{d}{dt}(\mathbf{T} \times \mathbf{N}) \dfrac{1}{ds/dt} = \dfrac{d}{dt}(\mathbf{T} \times \mathbf{N}) \dfrac{1}{|\mathbf{r}'(t)|} = [(\mathbf{T}' \times \mathbf{N}) + (\mathbf{T} \times \mathbf{N}')] \dfrac{1}{|\mathbf{r}'(t)|}$$

$$= \left[\left(\mathbf{T}' \times \dfrac{\mathbf{T}'}{|\mathbf{T}'|} \right) + (\mathbf{T} \times \mathbf{N}') \right] \dfrac{1}{|\mathbf{r}'(t)|} = \dfrac{\mathbf{T} \times \mathbf{N}'}{|\mathbf{r}'(t)|} \Rightarrow \dfrac{d\mathbf{B}}{ds} \perp \mathbf{T}$$

(c) $\mathbf{B} = \mathbf{T} \times \mathbf{N}$ $\Rightarrow$ $\mathbf{T} \perp \mathbf{N}, \mathbf{B} \perp \mathbf{T}$ and $\mathbf{B} \perp \mathbf{N}$. So $\mathbf{B}, \mathbf{T}$ and $\mathbf{N}$ form an orthogonal set of vectors in the three-dimensional space $\mathbb{R}^3$. From parts (a) and (b), $d\mathbf{B}/ds$ is perpendicular to both $\mathbf{B}$ and $\mathbf{T}$, so $d\mathbf{B}/ds$ is parallel to $\mathbf{N}$. Therefore, $d\mathbf{B}/ds = -\tau(s)\mathbf{N}$, where $\tau(s)$ is a scalar.

(d) Since $\mathbf{B} = \mathbf{T} \times \mathbf{N}, \mathbf{T} \perp \mathbf{N}$ and both $\mathbf{T}$ and $\mathbf{N}$ are unit vectors, $\mathbf{B}$ is a unit vector mutually perpendicular to both $\mathbf{T}$ and $\mathbf{N}$. For a plane curve, $\mathbf{T}$ and $\mathbf{N}$ always lie in the plane of the curve, so that $\mathbf{B}$ is a constant unit vector always perpendicular to the plane. Thus $d\mathbf{B}/ds = \mathbf{0}$, but $d\mathbf{B}/ds = -\tau(s)\mathbf{N}$ and $\mathbf{N} \neq \mathbf{0}$, so $\tau(s) = 0$.

47. (a) $\mathbf{r}' = s'\,\mathbf{T}$ $\Rightarrow$ $\mathbf{r}'' = s''\,\mathbf{T} + s'\,\mathbf{T}' = s''\,\mathbf{T} + s'\,\dfrac{d\mathbf{T}}{ds}\,s' = s''\,\mathbf{T} + \kappa(s')^2\,\mathbf{N}$ by the first Serret-Frenet formula.

(b) Using part (a), we have

$$\mathbf{r}' \times \mathbf{r}'' = (s'\,\mathbf{T}) \times [s''\,\mathbf{T} + \kappa(s')^2\,\mathbf{N}]$$
$$= [(s'\,\mathbf{T}) \times (s''\,\mathbf{T})] + [(s'\mathbf{T}) \times (\kappa(s')^2\,\mathbf{N})] \qquad \text{[by Property 3 of the cross product]}$$
$$= (s's'')(\mathbf{T} \times \mathbf{T}) + \kappa(s')^3(\mathbf{T} \times \mathbf{N}) = \mathbf{0} + \kappa(s')^3\,\mathbf{B} = \kappa(s')^3\,\mathbf{B}$$

(c) Using part (a), we have

$$\mathbf{r}''' = [s''\,\mathbf{T} + \kappa(s')^2\,\mathbf{N}]' = s'''\,\mathbf{T} + s''\,\mathbf{T}' + \kappa'(s')^2\,\mathbf{N} + 2\kappa s's''\,\mathbf{N} + \kappa(s')^2\,\mathbf{N}'$$
$$= s'''\,\mathbf{T} + s''\,\dfrac{d\mathbf{T}}{ds}\,s' + \kappa'(s')^2\,\mathbf{N} + 2\kappa s's''\,\mathbf{N} + \kappa(s')^2\,\dfrac{d\mathbf{N}}{ds}\,s'$$
$$= s'''\,\mathbf{T} + s''s'\kappa\,\mathbf{N} + \kappa'(s')^2\,\mathbf{N} + 2\kappa s's''\,\mathbf{N} + \kappa(s')^3(-\kappa\,\mathbf{T} + \tau\,\mathbf{B}) \qquad \text{[by the second formula]}$$
$$= [s''' - \kappa^2(s')^3]\,\mathbf{T} + [3\kappa s's'' + \kappa'(s')^2]\,\mathbf{N} + \kappa\tau(s')^3\,\mathbf{B}$$

(d) Using parts (b) and (c) and the facts that $\mathbf{B} \cdot \mathbf{T} = 0, \mathbf{B} \cdot \mathbf{N} = 0$, and $\mathbf{B} \cdot \mathbf{B} = 1$, we get

$$\frac{(\mathbf{r}' \times \mathbf{r}'') \cdot \mathbf{r}'''}{|\mathbf{r}' \times \mathbf{r}''|^2} = \frac{\kappa(s')^3\,\mathbf{B} \cdot \left\{[s''' - \kappa^2(s')^3]\,\mathbf{T} + [3\kappa s's'' + \kappa'(s')^2]\,\mathbf{N} + \kappa\tau(s')^3\,\mathbf{B}\right\}}{|\kappa(s')^3\,\mathbf{B}|^2} = \frac{\kappa(s')^3\kappa\tau(s')^3}{[\kappa(s')^3]^2} = \tau.$$

49. For one helix, the vector equation is $\mathbf{r}(t) = \langle 10\cos t, 10\sin t, 34t/(2\pi) \rangle$ (measuring in angstroms), because the radius of each helix is 10 angstroms, and z increases by 34 angstroms for each increase of 2π in t. Using the arc length formula, letting t go from 0 to $2.9 \times 10^8 \times 2\pi$, we find the approximate length of each helix to be

$$L = \int_0^{2.9 \times 10^8 \times 2\pi} |\mathbf{r}'(t)|\, dt = \int_0^{2.9 \times 10^8 \times 2\pi} \sqrt{(-10\sin t)^2 + (10\cos t)^2 + \left(\tfrac{34}{2\pi}\right)^2}\, dt = \left. \sqrt{100 + \left(\tfrac{34}{2\pi}\right)^2}\, \right]_{}^{2.9 \times 10^8 \times 2\pi}$$

$$= 2.9 \times 10^8 \times 2\pi \sqrt{100 + \left(\tfrac{34}{2\pi}\right)^2} \approx 2.07 \times 10^{10}\ \text{Å} \text{— more than two meters!}$$

10.9   Motion in Space: Velocity and Acceleration

1. $\mathbf{r}(t) = \langle -\tfrac{1}{2}t^2, t \rangle$ $\Rightarrow$

$\mathbf{v}(t) = \mathbf{r}'(t) = \langle -t, 1 \rangle$

$\mathbf{a}(t) = \mathbf{r}''(t) = \langle -1, 0 \rangle$

$|\mathbf{v}(t)| = \sqrt{t^2 + 1}$

At $t = 2$:

$\mathbf{v}(2) = \langle -2, 1 \rangle$

$\mathbf{a}(2) = \langle -1, 0 \rangle$

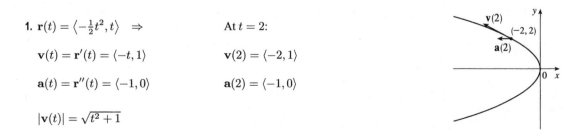

3. $(t) = 3\cos t\,\mathbf{i} + 2\sin t\,\mathbf{j} \implies$ At $t = \pi/3$:

$\mathbf{v}(t) = -3\sin t\,\mathbf{i} + 2\cos t\,\mathbf{j}$ $\mathbf{v}\left(\frac{\pi}{3}\right) = -\frac{3\sqrt{3}}{2}\mathbf{i} + \mathbf{j}$

$\mathbf{a}(t) = -3\cos t\,\mathbf{i} - 2\sin t\,\mathbf{j}$ $\mathbf{a}\left(\frac{\pi}{3}\right) = -\frac{3}{2}\mathbf{i} - \sqrt{3}\mathbf{j}$

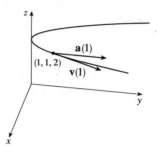

$|\mathbf{v}(t)| = \sqrt{9\sin^2 t + 4\cos^2 t} = \sqrt{4 + 5\sin^2 t}$

Notice that $x^2/9 + y^2/4 = \sin^2 t + \cos^2 t = 1$, so the path is an ellipse.

5. $\mathbf{r}(t) = t\,\mathbf{i} + t^2\,\mathbf{j} + 2\,\mathbf{k} \implies$ At $t = 1$:

$\mathbf{v}(t) = \mathbf{i} + 2t\,\mathbf{j}$ $\mathbf{v}(1) = \mathbf{i} + 2\,\mathbf{j}$

$\mathbf{a}(t) = 2\,\mathbf{j}$ $\mathbf{a}(1) = 2\,\mathbf{j}$

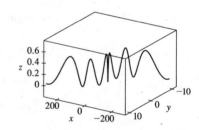

$|\mathbf{v}(t)| = \sqrt{1 + 4t^2}$

Here $x = t$, $y = t^2 \implies y = x^2$ and $z = 2$, so the path of the particle is a parabola in the plane $z = 2$.

7. $\mathbf{r}(t) = \langle t^2 + 1, t^3, t^2 - 1 \rangle \implies \mathbf{v}(t) = \mathbf{r}'(t) = \langle 2t, 3t^2, 2t \rangle$, $\mathbf{a}(t) = \mathbf{v}'(t) = \langle 2, 6t, 2 \rangle$,

$|\mathbf{v}(t)| = \sqrt{(2t)^2 + (3t^2)^2 + (2t)^2} = \sqrt{9t^4 + 8t^2} = |t|\sqrt{9t^2 + 8}$.

9. $\mathbf{r}(t) = \sqrt{2}\,t\,\mathbf{i} + e^t\,\mathbf{j} + e^{-t}\,\mathbf{k} \implies \mathbf{v}(t) = \mathbf{r}'(t) = \sqrt{2}\,\mathbf{i} + e^t\,\mathbf{j} - e^{-t}\,\mathbf{k}$, $\mathbf{a}(t) = \mathbf{v}'(t) = e^t\,\mathbf{j} + e^{-t}\,\mathbf{k}$,

$|\mathbf{v}(t)| = \sqrt{2 + e^{2t} + e^{-2t}} = \sqrt{(e^t + e^{-t})^2} = e^t + e^{-t}$.

11. $\mathbf{a}(t) = \mathbf{i} + 2\mathbf{j} \implies \mathbf{v}(t) = \int \mathbf{a}(t)\,dt = \int (\mathbf{i} + 2\mathbf{j})\,dt = t\,\mathbf{i} + 2t\,\mathbf{j} + \mathbf{C}$ and $\mathbf{k} = \mathbf{v}(0) = \mathbf{C}$,

so $\mathbf{C} = \mathbf{k}$ and $\mathbf{v}(t) = t\,\mathbf{i} + 2t\,\mathbf{j} + \mathbf{k}$. $\mathbf{r}(t) = \int \mathbf{v}(t)\,dt = \int (t\,\mathbf{i} + 2t\,\mathbf{j} + \mathbf{k})\,dt = \frac{1}{2}t^2\,\mathbf{i} + t^2\,\mathbf{j} + t\,\mathbf{k} + \mathbf{D}$.

But $\mathbf{i} = \mathbf{r}(0) = \mathbf{D}$, so $\mathbf{D} = \mathbf{i}$ and $\mathbf{r}(t) = \left(\frac{1}{2}t^2 + 1\right)\mathbf{i} + t^2\,\mathbf{j} + t\,\mathbf{k}$.

13. (a) $\mathbf{a}(t) = 2t\,\mathbf{i} + \sin t\,\mathbf{j} + \cos 2t\,\mathbf{k} \implies$ (b)

$\mathbf{v}(t) = \int (2t\,\mathbf{i} + \sin t\,\mathbf{j} + \cos 2t\,\mathbf{k})\,dt = t^2\,\mathbf{i} - \cos t\,\mathbf{j} + \frac{1}{2}\sin 2t\,\mathbf{k} + \mathbf{C}$

and $\mathbf{i} = \mathbf{v}(0) = -\mathbf{j} + \mathbf{C}$, so $\mathbf{C} = \mathbf{i} + \mathbf{j}$

and $\mathbf{v}(t) = (t^2 + 1)\mathbf{i} + (1 - \cos t)\mathbf{j} + \frac{1}{2}\sin 2t\,\mathbf{k}$.

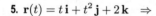

$\mathbf{r}(t) = \int [(t^2 + 1)\mathbf{i} + (1 - \cos t)\mathbf{j} + \frac{1}{2}\sin 2t\,\mathbf{k}]\,dt$

$= \left(\frac{1}{3}t^3 + t\right)\mathbf{i} + (t - \sin t)\mathbf{j} - \frac{1}{4}\cos 2t\,\mathbf{k} + \mathbf{D}$

But $\mathbf{j} = \mathbf{r}(0) = -\frac{1}{4}\mathbf{k} + \mathbf{D}$, so $\mathbf{D} = \mathbf{j} + \frac{1}{4}\mathbf{k}$ and $\mathbf{r}(t) = \left(\frac{1}{3}t^3 + t\right)\mathbf{i} + (t - \sin t + 1)\mathbf{j} + \left(\frac{1}{4} - \frac{1}{4}\cos 2t\right)\mathbf{k}$.

15. $\mathbf{r}(t) = \langle t^2, 5t, t^2 - 16t \rangle \implies \mathbf{v}(t) = \langle 2t, 5, 2t - 16 \rangle$, $|\mathbf{v}(t)| = \sqrt{4t^2 + 25 + 4t^2 - 64t + 256} = \sqrt{8t^2 - 64t + 281}$

and $\frac{d}{dt}|\mathbf{v}(t)| = \frac{1}{2}(8t^2 - 64t + 281)^{-1/2}(16t - 64)$. This is zero if and only if the numerator is zero, that is,

$16t - 64 = 0$ or $t = 4$. Since $\frac{d}{dt}|\mathbf{v}(t)| < 0$ for $t < 4$ and $\frac{d}{dt}|\mathbf{v}(t)| > 0$ for $t > 4$, the minimum speed of $\sqrt{153}$ is attained

at $t = 4$ units of time.

17. $|\mathbf{F}(t)| = 20$ N in the direction of the positive z-axis, so $\mathbf{F}(t) = 20\,\mathbf{k}$. Also $m = 4$ kg, $\mathbf{r}(0) = \mathbf{0}$ and $\mathbf{v}(0) = \mathbf{i} - \mathbf{j}$.

Since $20\mathbf{k} = \mathbf{F}(t) = 4\,\mathbf{a}(t)$, $\mathbf{a}(t) = 5\,\mathbf{k}$. Then $\mathbf{v}(t) = 5t\,\mathbf{k} + \mathbf{c}_1$ where $\mathbf{c}_1 = \mathbf{i} - \mathbf{j}$ so $\mathbf{v}(t) = \mathbf{i} - \mathbf{j} + 5t\,\mathbf{k}$ and the

speed is $|\mathbf{v}(t)| = \sqrt{1 + 1 + 25t^2} = \sqrt{25t^2 + 2}$. Also $\mathbf{r}(t) = t\,\mathbf{i} - t\,\mathbf{j} + \frac{5}{2}t^2\,\mathbf{k} + \mathbf{c}_2$ and $\mathbf{0} = \mathbf{r}(0)$, so $\mathbf{c}_2 = \mathbf{0}$

and $\mathbf{r}(t) = t\,\mathbf{i} - t\,\mathbf{j} + \frac{5}{2}t^2\,\mathbf{k}$.

19. $|\mathbf{v}(0)| = 500$ m/s and since the angle of elevation is $30°$, the direction of the velocity is $\frac{1}{2}\left(\sqrt{3}\,\mathbf{i} + \mathbf{j}\right)$. Thus

$\mathbf{v}(0) = 250\left(\sqrt{3}\,\mathbf{i} + \mathbf{j}\right)$ and if we set up the axes so the projectile starts at the origin, then $\mathbf{r}(0) = \mathbf{0}$. Ignoring air resistance, the

only force is that due to gravity, so $\mathbf{F}(t) = -mg\,\mathbf{j}$ where $g \approx 9.8$ m/s^2. Thus $\mathbf{a}(t) = -g\,\mathbf{j}$ and $\mathbf{v}(t) = -gt\,\mathbf{j} + \mathbf{c}_1$. But

$250\left(\sqrt{3}\,\mathbf{i} + \mathbf{j}\right) = \mathbf{v}(0) = \mathbf{c}_1$, so $\mathbf{v}(t) = 250\sqrt{3}\,\mathbf{i} + (250 - gt)\,\mathbf{j}$ and $\mathbf{r}(t) = 250\sqrt{3}\,t\,\mathbf{i} + \left(250t - \frac{1}{2}gt^2\right)\,\mathbf{j} + \mathbf{c}_2$ where

$\mathbf{0} = \mathbf{r}(0) = \mathbf{c}_2$. Thus $\mathbf{r}(t) = 250\sqrt{3}\,t\,\mathbf{i} + \left(250t - \frac{1}{2}gt^2\right)\,\mathbf{j}$.

(a) Setting $250t - \frac{1}{2}gt^2 = 0$ gives $t = 0$ or $t = \frac{500}{g} \approx 51.0$ s. So the range is $250\sqrt{3} \cdot \frac{500}{g} \approx 22$ km.

(b) $0 = \dfrac{d}{dt}\left(250t - \frac{1}{2}gt^2\right) = 250 - gt$ implies that the maximum height is attained when $t = 250/g \approx 25.5$ s.

  Thus, the maximum height is $(250)(250/g) - g(250/g)^2\frac{1}{2} = (250)^2/(2g) \approx 3.2$ km.

(c) From part (a), impact occurs at $t = 500/g \approx 51.0$. Thus, the velocity at impact is

  $\mathbf{v}(500/g) = 250\sqrt{3}\,\mathbf{i} + [250 - g(500/g)]\,\mathbf{j} = 250\sqrt{3}\,\mathbf{i} - 250\,\mathbf{j}$ and the speed is $|\mathbf{v}(500/g)| = 250\sqrt{3+1} = 500$ m/s.

21. As in Example 5, $\mathbf{r}(t) = (v_0 \cos 45°)t\,\mathbf{i} + \left[(v_0 \sin 45°)t - \frac{1}{2}gt^2\right]\,\mathbf{j} = \frac{1}{2}\left[v_0\sqrt{2}\,t\,\mathbf{i} + \left(v_0\sqrt{2}\,t - gt^2\right)\,\mathbf{j}\right]$. Then the ball

lands at $t = \dfrac{v_0\sqrt{2}}{g}$ s. Now since it lands 90 m away, $90 = \frac{1}{2}v_0\sqrt{2}\,\dfrac{v_0\sqrt{2}}{g}$ or $v_0^2 = 90g$ and the initial velocity

is $v_0 = \sqrt{90g} \approx 30$ m/s.

23. Let α be the angle of elevation. Then $v_0 = 150$ m/s and from Example 5, the horizontal distance traveled by the projectile is

$d = \dfrac{v_0^2 \sin 2\alpha}{g}$. Thus $\dfrac{150^2 \sin 2\alpha}{g} = 800 \;\Rightarrow\; \sin 2\alpha = \dfrac{800g}{150^2} \approx 0.3484 \;\Rightarrow\; 2\alpha \approx 20.4°$ or $180 - 20.4 = 159.6°$.

Two angles of elevation then are $\alpha \approx 10.2°$ and $\alpha \approx 79.8°$.

25. Place the catapult at the origin and assume the catapult is 100 meters from the city, so the city lies between $(100, 0)$

and $(600, 0)$. The initial speed is $v_0 = 80$ m/s and let θ be the angle the catapult is set at. As in Example 5, the trajectory of

the catapulted rock is given by $\mathbf{r}(t) = (80 \cos\theta)t\,\mathbf{i} + \left[(80\sin\theta)t - 4.9t^2\right]\,\mathbf{j}$. The top of the near city wall is at $(100, 15)$,

which the rock will hit when $(80\cos\theta)\,t = 100 \;\Rightarrow\; t = \dfrac{5}{4\cos\theta}$ and $(80\sin\theta)t - 4.9t^2 = 15 \;\Rightarrow$

$80\sin\theta \cdot \dfrac{5}{4\cos\theta} - 4.9\left(\dfrac{5}{4\cos\theta}\right)^2 = 15 \;\Rightarrow\; 100\tan\theta - 7.65625\sec^2\theta = 15$. Replacing $\sec^2\theta$ with $\tan^2\theta + 1$ gives

$7.65625\tan^2\theta - 100\tan\theta + 22.62625 = 0$. Using the quadratic formula, we have $\tan\theta \approx 0.230324, 12.8309 \;\Rightarrow$

$\theta \approx 13.0°, 85.5°$. So for $13.0° < \theta < 85.5°$, the rock will land beyond the near city wall. The base of the far wall is located at $(600, 0)$ which the rock hits if $(80 \cos \theta)t = 600 \Rightarrow t = \dfrac{15}{2 \cos \theta}$ and $(80 \sin \theta)t - 4.9t^2 = 0 \Rightarrow$

$80 \sin \theta \cdot \dfrac{15}{2 \cos \theta} - 4.9 \left(\dfrac{15}{2 \cos \theta} \right)^2 = 0 \Rightarrow 600 \tan \theta - 275.625 \sec^2 \theta = 0 \Rightarrow$

$275.625 \tan^2 \theta - 600 \tan \theta + 275.625 = 0$. Solutions are $\tan \theta \approx 0.658678, 1.51819 \Rightarrow \theta \approx 33.4°, 56.6°$. Thus the

rock lands beyond the enclosed city ground for $33.4° < \theta < 56.6°$, and the angles that allow the rock to land on city ground

are $13.0° < \theta < 33.4°, 56.6° < \theta < 85.5°$. If you consider that the rock can hit the far wall and bounce back into the city, we

calculate the angles that cause the rock to hit the top of the wall at $(600, 15)$: $(80 \cos \theta)t = 600 \Rightarrow t = \dfrac{15}{2 \cos \theta}$ and

$(80 \sin \theta)t - 4.9t^2 = 15 \Rightarrow 600 \tan \theta - 275.625 \sec^2 \theta = 15 \Rightarrow 275.625 \tan^2 \theta - 600 \tan \theta + 290.625 = 0$.

Solutions are $\tan \theta \approx 0.727506, 1.44936 \Rightarrow \theta \approx 36.0°, 55.4°$, so the catapult should be set with angle θ where

$13.0° < \theta < 36.0°, 55.4° < \theta < 85.5°$.

27. (a) After t seconds, the boat will be $5t$ meters west of point A. The velocity

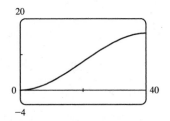

of the water at that location is $\frac{3}{400}(5t)(40 - 5t)\mathbf{j}$. The velocity of the

boat in still water is $5\mathbf{i}$, so the resultant velocity of the boat is

$\mathbf{v}(t) = 5\mathbf{i} + \frac{3}{400}(5t)(40 - 5t)\mathbf{j} = 5\mathbf{i} + \left(\frac{3}{2}t - \frac{3}{16}t^2 \right)\mathbf{j}$. Integrating, we obtain

$\mathbf{r}(t) = 5t\,\mathbf{i} + \left(\frac{3}{4}t^2 - \frac{1}{16}t^3 \right)\mathbf{j} + \mathbf{C}$. If we place the origin at A (and consider $\mathbf{j}$

to coincide with the northern direction) then $\mathbf{r}(0) = \mathbf{0} \Rightarrow \mathbf{C} = \mathbf{0}$ and we have $\mathbf{r}(t) = 5t\,\mathbf{i} + \left(\frac{3}{4}t^2 - \frac{1}{16}t^3 \right)\mathbf{j}$. The boat

reaches the east bank after 8 s, and it is located at $\mathbf{r}(8) = 5(8)\mathbf{i} + \left(\frac{3}{4}(8)^2 - \frac{1}{16}(8)^3 \right)\mathbf{j} = 40\,\mathbf{i} + 16\,\mathbf{j}$. Thus the boat is 16 m

downstream.

(b) Let α be the angle north of east that the boat heads. Then the velocity of the boat in still water is given by

$5(\cos \alpha)\mathbf{i} + 5(\sin \alpha)\mathbf{j}$. At t seconds, the boat is $5(\cos \alpha)t$ meters from the west bank, at which point the velocity

of the water is $\frac{3}{400}[5(\cos \alpha)t][40 - 5(\cos \alpha)t]\,\mathbf{j}$. The resultant velocity of the boat is given by

$\mathbf{v}(t) = 5(\cos \alpha)\mathbf{i} + \left[5 \sin \alpha + \frac{3}{400}(5t \cos \alpha)(40 - 5t \cos \alpha) \right]\mathbf{j} = (5 \cos \alpha)\mathbf{i} + \left(5 \sin \alpha + \frac{3}{2}t \cos \alpha - \frac{3}{16}t^2 \cos^2 \alpha \right)\mathbf{j}$.

Integrating, $\mathbf{r}(t) = (5t \cos \alpha)\mathbf{i} + \left(5t \sin \alpha + \frac{3}{4}t^2 \cos \alpha - \frac{1}{16}t^3 \cos^2 \alpha \right)\mathbf{j}$ (where we have again placed

the origin at A). The boat will reach the east bank when $5t \cos \alpha = 40 \Rightarrow t = \dfrac{40}{5 \cos \alpha} = \dfrac{8}{\cos \alpha}$.

In order to land at point $B(40, 0)$ we need $5t \sin \alpha + \frac{3}{4}t^2 \cos \alpha - \frac{1}{16}t^3 \cos^2 \alpha = 0 \Rightarrow$

$5 \left(\dfrac{8}{\cos \alpha} \right) \sin \alpha + \frac{3}{4} \left(\dfrac{8}{\cos \alpha} \right)^2 \cos \alpha - \frac{1}{16} \left(\dfrac{8}{\cos \alpha} \right)^3 \cos^2 \alpha = 0 \Rightarrow \dfrac{1}{\cos \alpha}(40 \sin \alpha + 48 - 32) = 0 \Rightarrow$

$40 \sin \alpha + 16 = 0 \Rightarrow \sin \alpha = -\frac{2}{5}$. Thus $\alpha = \sin^{-1}\left(-\frac{2}{5} \right) \approx -23.6°$, so the boat should head 23.6° south of

east (upstream). The path does seem realistic. The boat initially heads

upstream to counteract the effect of the current. Near the center of the river,

the current is stronger and the boat is pushed downstream. When the boat

nears the eastern bank, the current is slower and the boat is able to progress

upstream to arrive at point B.

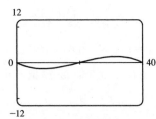

29. $\mathbf{r}(t) = \cos t\,\mathbf{i} + \sin t\,\mathbf{j} + t\,\mathbf{k}$ $\Rightarrow$ $\mathbf{r}'(t) = -\sin t\,\mathbf{i} + \cos t\,\mathbf{j} + \mathbf{k}$, $|\mathbf{r}'(t)| = \sqrt{\sin^2 t + \cos^2 t + 1} = \sqrt{2}$,

$\mathbf{r}''(t) = -\cos t\,\mathbf{i} - \sin t\,\mathbf{j}$, $\mathbf{r}'(t) \times \mathbf{r}''(t) = \sin t\,\mathbf{i} - \cos t\,\mathbf{j} + \mathbf{k}$.

Then $a_T = \dfrac{\mathbf{r}'(t)\cdot\mathbf{r}''(t)}{|\mathbf{r}'(t)|} = \dfrac{\sin t\,\cos t - \sin t\,\cos t}{\sqrt{2}} = 0$ and $a_N = \dfrac{|\mathbf{r}'(t)\times\mathbf{r}''(t)|}{|\mathbf{r}'(t)|} = \dfrac{\sqrt{\sin^2 t + \cos^2 t + 1}}{\sqrt{2}} = \dfrac{\sqrt{2}}{\sqrt{2}} = 1.$

31. $\mathbf{r}(t) = e^t\,\mathbf{i} + \sqrt{2}\,t\,\mathbf{j} + e^{-t}\,\mathbf{k}$ $\Rightarrow$ $\mathbf{r}'(t) = e^t\,\mathbf{i} + \sqrt{2}\,\mathbf{j} - e^{-t}\,\mathbf{k}$, $|\mathbf{r}(t)| = \sqrt{e^{2t} + 2 + e^{-2t}} = \sqrt{(e^t + e^{-t})^2} = e^t + e^{-t}$,

$\mathbf{r}''(t) = e^t\,\mathbf{i} + e^{-t}\,\mathbf{k}$. Then $a_T = \dfrac{e^{2t} - e^{-2t}}{e^t + e^{-t}} = \dfrac{(e^t + e^{-t})(e^t - e^{-t})}{e^t + e^{-t}} = e^t - e^{-t} = 2\sinh t$

and $a_N = \dfrac{|\sqrt{2}\,e^{-t}\,\mathbf{i} - 2\,\mathbf{j} - \sqrt{2}\,e^t\,\mathbf{k}|}{e^t + e^{-t}} = \dfrac{\sqrt{2(e^{-2t} + 2 + e^{2t})}}{e^t + e^{-t}} = \sqrt{2}\,\dfrac{e^t + e^{-t}}{e^t + e^{-t}} = \sqrt{2}.$

33. If the engines are turned off at time t, then the spacecraft will continue to travel in the direction of $\mathbf{v}(t)$, so we need a t such

that for some scalar $s > 0$, $\mathbf{r}(t) + s\,\mathbf{v}(t) = \langle 6, 4, 9\rangle$. $\mathbf{v}(t) = \mathbf{r}'(t) = \mathbf{i} + \dfrac{1}{t}\,\mathbf{j} + \dfrac{8t}{(t^2 + 1)^2}\,\mathbf{k}$ $\Rightarrow$

$\mathbf{r}(t) + s\,\mathbf{v}(t) = \left\langle 3 + t + s,\, 2 + \ln t + \dfrac{s}{t},\, 7 - \dfrac{4}{t^2 + 1} + \dfrac{8st}{(t^2 + 1)^2}\right\rangle$ $\Rightarrow$ $3 + t + s = 6$ $\Rightarrow$ $s = 3 - t$,

so $7 - \dfrac{4}{t^2 + 1} + \dfrac{8(3 - t)t}{(t^2 + 1)^2} = 9$ $\Leftrightarrow$ $\dfrac{24t - 12t^2 - 4}{(t^2 + 1)^2} = 2$ $\Leftrightarrow$ $t^4 + 8t^2 - 12t + 3 = 0.$

It is easily seen that $t = 1$ is a root of this polynomial. Also $2 + \ln 1 + \dfrac{3 - 1}{1} = 4$, so $t = 1$ is the desired solution.

10 Review

CONCEPT CHECK

1. A scalar is a real number, while a vector is a quantity that has both a real-valued magnitude and a direction.

2. To add two vectors geometrically, we can use either the Triangle Law or the Parallelogram Law, as illustrated in Figures 3 and 4 in Section 10.2. (See also the definition of vector addition on page 523.) Algebraically, we add the corresponding components of the vectors.

3. For $c > 0$, $c\,\mathbf{a}$ is a vector with the same direction as $\mathbf{a}$ and length c times the length of $\mathbf{a}$. If $c < 0$, $c\,\mathbf{a}$ points in the opposite direction as $\mathbf{a}$ and has length $|c|$ times the length of $\mathbf{a}$. (See Figures 7 and 15 in Section 10.2.) Algebraically, to find $c\,\mathbf{a}$ we multiply each component of $\mathbf{a}$ by c.

4. See (1) in Section 10.2.

5. See Theorem 10.3.3 and Definition 10.3.1.

6. The dot product can be used to determine the work done moving an object given the force and displacement vectors. The dot product can also be used to find the angle between two vectors and the scalar projection of one vector onto another. In particular, the dot product can determine if two vectors are orthogonal.

7. See the boxed equations on page 534 as well as Figures 3 and 4 and the accompanying discussion on pages 533–34.

8. See Theorem 10.4.6 and the preceding discussion; use either (1) or (4) in Section 10.4.

9. The cross product can be used to determine torque if the force and position vectors are known. In addition, the cross product can be used to create a vector orthogonal to two given vectors as well as to determine if two vectors are parallel. The cross product can also be used to find the area of a parallelogram determined by two vectors.

10. (a) The area of the parallelogram determined by $\mathbf{a}$ and $\mathbf{b}$ is the length of the cross product: $|\mathbf{a} \times \mathbf{b}|$.

    (b) The volume of the parallelepiped determined by $\mathbf{a}$, $\mathbf{b}$, and $\mathbf{c}$ is the magnitude of their scalar triple product: $|\mathbf{a} \cdot (\mathbf{b} \times \mathbf{c})|$.

11. If an equation of the plane is known, it can be written as $ax + by + cz + d = 0$. A normal vector, which is perpendicular to the plane, is $\langle a, b, c \rangle$ (or any scalar multiple of $\langle a, b, c \rangle$). If an equation is not known, we can use points on the plane to find two non-parallel vectors which lie in the plane. The cross product of these vectors is a vector perpendicular to the plane.

12. The angle between two intersecting planes is defined as the acute angle between their normal vectors. We can find this angle using Corollary 10.3.6.

13. See (1), (2), and (3) in Section 10.5.

14. See (5), (6), and (7) in Section 10.5.

15. (a) Two (nonzero) vectors are parallel if and only if one is a scalar multiple of the other. In addition, two nonzero vectors are parallel if and only if their cross product is $\mathbf{0}$.

    (b) Two vectors are perpendicular if and only if their dot product is 0.

    (c) Two planes are parallel if and only if their normal vectors are parallel.

16. (a) Determine the vectors $\overrightarrow{PQ} = \langle a_1, a_2, a_3 \rangle$ and $\overrightarrow{PR} = \langle b_1, b_2, b_3 \rangle$. If there is a scalar t such that $\langle a_1, a_2, a_3 \rangle = t \langle b_1, b_2, b_3 \rangle$, then the vectors are parallel and the points must all lie on the same line.

    Alternatively, if $\overrightarrow{PQ} \times \overrightarrow{PR} = \mathbf{0}$, then $\overrightarrow{PQ}$ and $\overrightarrow{PR}$ are parallel, so P, Q, and R are collinear.

    Thirdly, an algebraic method is to determine an equation of the line joining two of the points, and then check whether or not the third point satisfies this equation.

    (b) Find the vectors $\overrightarrow{PQ} = \mathbf{a}$, $\overrightarrow{PR} = \mathbf{b}$, $\overrightarrow{PS} = \mathbf{c}$. $\mathbf{a} \times \mathbf{b}$ is normal to the plane formed by P, Q and R, and so S lies on this plane if $\mathbf{a} \times \mathbf{b}$ and $\mathbf{c}$ are orthogonal, that is, if $(\mathbf{a} \times \mathbf{b}) \cdot \mathbf{c} = 0$. (Or use the reasoning in Example 5 in Section 10.4.) Alternatively, find an equation for the plane determined by three of the points and check whether or not the fourth point satisfies this equation.

17. (a) See Exercise 10.4.39.

    (b) See Example 7 in Section 10.5.

18. The traces of a surface are the curves of intersection of the surface with planes parallel to the coordinate planes. We can find the trace in the plane $x = k$ (parallel to the yz-plane) by setting $x = k$ and determining the curve represented by the resulting equation. Traces in the planes $y = k$ (parallel to the xz-plane) and $z = k$ (parallel to the xy-plane) are found similarly.

19. See Table 1 in Section 10.6.

20. A vector function is a function whose domain is a set of real numbers and whose range is a set of vectors. To find the derivative or integral, we can differentiate or integrate each component of the vector function.

21. The tip of the moving vector $\mathbf{r}(t)$ of a continuous vector function traces out a space curve.

22. (a) A curve represented by the vector function $\mathbf{r}(t)$ is smooth if $\mathbf{r}'(t)$ is continuous and $\mathbf{r}'(t) \neq \mathbf{0}$ on its parametric domain (except possibly at the endpoints).

(b) The tangent vector to a smooth curve at a point P with position vector $\mathbf{r}(t)$ is the vector $\mathbf{r}'(t)$. The tangent line at P is the line through P parallel to the tangent vector $\mathbf{r}'(t)$. The unit tangent vector is $\mathbf{T}(t) = \dfrac{\mathbf{r}'(t)}{|\mathbf{r}'(t)|}$.

23. (a)–(f) See Theorem 10.7.5.

24. Use Formula 10.8.2, or equivalently, 10.8.3.

25. (a) The curvature of a curve is $\kappa = \left| \dfrac{d\mathbf{T}}{ds} \right|$ where $\mathbf{T}$ is the unit tangent vector.

(b) $\kappa(t) = \left| \dfrac{\mathbf{T}'(t)}{\mathbf{r}'(t)} \right|$ **(c)** $\kappa(t) = \dfrac{|\mathbf{r}'(t) \times \mathbf{r}''(t)|}{|\mathbf{r}'(t)|^3}$ **(d)** $\kappa(x) = \dfrac{|f''(x)|}{[1 + (f'(x))^2]^{3/2}}$

26. The unit normal vector: $\mathbf{N}(t) = \dfrac{\mathbf{T}'(t)}{|\mathbf{T}'(t)|}$.   The binormal vector: $\mathbf{B}(t) = \mathbf{T}(t) \times \mathbf{N}(t)$.

27. (a) If $\mathbf{r}(t)$ is the position vector of the particle on the space curve, the velocity $\mathbf{v}(t) = \mathbf{r}'(t)$, the speed is given by $|\mathbf{v}(t)|$, and the acceleration $\mathbf{a}(t) = \mathbf{v}'(t) = \mathbf{r}''(t)$.

(b) $\mathbf{a} = a_T \mathbf{T} + a_N \mathbf{N}$ where $a_T = v'$ and $a_N = \kappa v^2$.

28. See the statement of Kepler's Laws on page 584.

TRUE-FALSE QUIZ

1. True, by Property 2 of the dot product. (See page 531.)

3. True. If θ is the angle between $\mathbf{u}$ and $\mathbf{v}$, then by Theorem 10.4.6, $|\mathbf{u} \times \mathbf{v}| = |\mathbf{u}| \, |\mathbf{v}| \sin\theta = |\mathbf{v}| \, |\mathbf{u}| \sin\theta = |\mathbf{v} \times \mathbf{u}|$.

(Or, by Theorem 10.4.8, $|\mathbf{u} \times \mathbf{v}| = |-\mathbf{v} \times \mathbf{u}| = |-1| \, |\mathbf{v} \times \mathbf{u}| = |\mathbf{v} \times \mathbf{u}|$.)

5. Property 2 of the cross product tells us that this is true.

7. This is true by Theorem 10.4.8 #5.

9. This is true because $\mathbf{u} \times \mathbf{v}$ is orthogonal to $\mathbf{u}$ (see Theorem 10.4.5), and the dot product of two orthogonal vectors is 0.

11. If $|\mathbf{u}| = 1$, $|\mathbf{v}| = 1$ and θ is the angle between these two vectors (so $0 \le \theta \le \pi$), then by Theorem 10.4.6,

$|\mathbf{u} \times \mathbf{v}| = |\mathbf{u}|\,|\mathbf{v}| \sin \theta = \sin \theta$, which is equal to 1 if and only if $\theta = \frac{\pi}{2}$ (that is, if and only if the two vectors are orthogonal).

Therefore, the assertion that the cross product of two unit vectors is a unit vector is false.

13. This is false. In $\mathbb{R}^2$, $x^2 + y^2 = 1$ represents a circle, but $\{(x, y, z) \mid x^2 + y^2 = 1\}$ represents a *three-dimensional surface*,

namely, a circular cylinder with axis the z-axis.

15. False. For example, $\mathbf{i} \cdot \mathbf{j} = 0$ but $\mathbf{i} \ne \mathbf{0}$ and $\mathbf{j} \ne \mathbf{0}$.

17. True. If we reparametrize the curve by replacing $u = t^3$, we have $\mathbf{r}(u) = u\,\mathbf{i} + 2u\,\mathbf{j} + 3u\,\mathbf{k}$, which is a line through the origin

with direction vector $\mathbf{i} + 2\,\mathbf{j} + 3\,\mathbf{k}$.

19. False. $\mathbf{r}'(t) = \langle -\sin t, 2t, 4t^3 \rangle$, and since $\mathbf{r}'(0) = \langle 0, 0, 0 \rangle = \mathbf{0}$, the curve is not smooth.

21. False. By Formula 5 of Theorem 10.7.5, $\dfrac{d}{dt} [\mathbf{u}(t) \times \mathbf{v}(t)] = \mathbf{u}'(t) \times \mathbf{v}(t) + \mathbf{u}(t) \times \mathbf{v}'(t)$.

23. False. κ is the magnitude of the rate of change of the unit tangent vector $\mathbf{T}$ with respect to arc length s, not with respect to t.

EXERCISES

1. (a) The radius of the sphere is the distance between the points $(-1, 2, 1)$ and $(6, -2, 3)$, namely,

$\sqrt{[6 - (-1)]^2 + (-2 - 2)^2 + (3 - 1)^2} = \sqrt{69}$. By the formula for an equation of a sphere (see page 520), an equation

of the sphere with center $(-1, 2, 1)$ and radius $\sqrt{69}$ is $(x + 1)^2 + (y - 2)^2 + (z - 1)^2 = 69$.

(b) The intersection of this sphere with the yz-plane is the set of points on the sphere whose x-coordinate is 0. Putting $x = 0$

into the equation, we have $(y - 2)^2 + (z - 1)^2 = 68$, $x = 0$ which represents a circle in the yz-plane with center $(0, 2, 1)$

and radius $\sqrt{68}$.

(c) Completing squares gives $(x - 4)^2 + (y + 1)^2 + (z + 3)^2 = -1 + 16 + 1 + 9 = 25$. Thus the sphere is centered at

$(4, -1, -3)$ and has radius 5.

3. $\mathbf{u} \cdot \mathbf{v} = |\mathbf{u}|\,|\mathbf{v}| \cos 45° = (2)(3)\frac{\sqrt{2}}{2} = 3\sqrt{2}$. $|\mathbf{u} \times \mathbf{v}| = |\mathbf{u}|\,|\mathbf{v}| \sin 45° = (2)(3)\frac{\sqrt{2}}{2} = 3\sqrt{2}$.

By the right-hand rule, $\mathbf{u} \times \mathbf{v}$ is directed out of the page.

5. For the two vectors to be orthogonal, we need $\langle 3, 2, x \rangle \cdot \langle 2x, 4, x \rangle = 0 \quad \Leftrightarrow \quad (3)(2x) + (2)(4) + (x)(x) = 0 \quad \Leftrightarrow$

$x^2 + 6x + 8 = 0 \quad \Leftrightarrow \quad (x+2)(x+4) = 0 \quad \Leftrightarrow \quad x = -2 \text{ or } x = -4.$

7. (a) $(\mathbf{u} \times \mathbf{v}) \cdot \mathbf{w} = \mathbf{u} \cdot (\mathbf{v} \times \mathbf{w}) = 2$

(b) $\mathbf{u} \cdot (\mathbf{w} \times \mathbf{v}) = \mathbf{u} \cdot [-(\mathbf{v} \times \mathbf{w})] = -\mathbf{u} \cdot (\mathbf{v} \times \mathbf{w}) = -2$

(c) $\mathbf{v} \cdot (\mathbf{u} \times \mathbf{w}) = (\mathbf{v} \times \mathbf{u}) \cdot \mathbf{w} = -(\mathbf{u} \times \mathbf{v}) \cdot \mathbf{w} = -2$

(d) $(\mathbf{u} \times \mathbf{v}) \cdot \mathbf{v} = \mathbf{u} \cdot (\mathbf{v} \times \mathbf{v}) = \mathbf{u} \cdot \mathbf{0} = 0$

9. For simplicity, consider a unit cube positioned with its back left corner at the origin. Vector representations of the diagonals

joining the points $(0,0,0)$ to $(1,1,1)$ and $(1,0,0)$ to $(0,1,1)$ are $\langle 1,1,1 \rangle$ and $\langle -1,1,1 \rangle$. Let θ be the angle between these

two vectors. $\langle 1,1,1 \rangle \cdot \langle -1,1,1 \rangle = -1+1+1 = 1 = |\langle 1,1,1 \rangle| \, |\langle -1,1,1 \rangle| \cos \theta = 3 \cos \theta \quad \Rightarrow \quad \cos \theta = \frac{1}{3} \quad \Rightarrow$

$\theta = \cos^{-1}\left(\frac{1}{3}\right) \approx 71°.$

11. $\overrightarrow{AB} = \langle 1,0,-1 \rangle$, $\overrightarrow{AC} = \langle 0,4,3 \rangle$, so

(a) a vector perpendicular to the plane is $\overrightarrow{AB} \times \overrightarrow{AC} = \langle 0+4, -(3+0), 4-0 \rangle = \langle 4,-3,4 \rangle.$

(b) $\frac{1}{2} \left| \overrightarrow{AB} \times \overrightarrow{AC} \right| = \frac{1}{2}\sqrt{16+9+16} = \frac{\sqrt{41}}{2}.$

13. Let F_1 be the magnitude of the force directed $20°$ away from the direction of shore, and let F_2 be the magnitude of the other

force. Separating these forces into components parallel to the direction of the resultant force and perpendicular to it gives

$F_1 \cos 20° + F_2 \cos 30° = 255$ **(1)**, and $F_1 \sin 20° - F_2 \sin 30° = 0 \quad \Rightarrow \quad F_1 = F_2 \dfrac{\sin 30°}{\sin 20°}$ **(2)**. Substituting **(2)**

into **(1)** gives $F_2(\sin 30° \cot 20° + \cos 30°) = 255 \quad \Rightarrow \quad F_2 \approx 114$ N. Substituting this into **(2)** gives $F_1 \approx 166$ N.

15. The line has direction $\mathbf{v} = \langle -3,2,3 \rangle$. Letting $P_0 = (4,-1,2)$, parametric equations are

$x = 4 - 3t, \ y = -1 + 2t, \ z = 2 + 3t.$

17. A direction vector for the line is a normal vector for the plane, $\mathbf{n} = \langle 2,-1,5 \rangle$, and parametric equations for the line are

$x = -2 + 2t, \ y = 2 - t, \ z = 4 + 5t.$

19. Here the vectors $\mathbf{a} = \langle 4-3, 0-(-1), 2-1 \rangle = \langle 1,1,1 \rangle$ and $\mathbf{b} = \langle 6-3, 3-(-1), 1-1 \rangle = \langle 3,4,0 \rangle$ lie in the plane,

so $\mathbf{n} = \mathbf{a} \times \mathbf{b} = \langle -4,3,1 \rangle$ is a normal vector to the plane and an equation of the plane is

$-4(x-3) + 3(y-(-1)) + 1(z-1) = 0$ or $-4x + 3y + z = -14.$

21. Substitution of the parametric equations into the equation of the plane gives $2x - y + z = 2(2-t) - (1+3t) + 4t = 2 \quad \Rightarrow$

$-t + 3 = 2 \quad \Rightarrow \quad t = 1$. When $t = 1$, the parametric equations give $x = 2 - 1 = 1$, $y = 1 + 3 = 4$ and $z = 4$. Therefore,

the point of intersection is $(1,4,4).$

23. Since the direction vectors $\langle 2, 3, 4 \rangle$ and $\langle 6, -1, 2 \rangle$ aren't parallel, neither are the lines. For the lines to intersect, the three

equations $1 + 2t = -1 + 6s$, $2 + 3t = 3 - s$, $3 + 4t = -5 + 2s$ must be satisfied simultaneously. Solving the first two

equations gives $t = \frac{1}{5}$, $s = \frac{2}{5}$ and checking we see these values don't satisfy the third equation. Thus the lines aren't parallel

and they don't intersect, so they must be skew.

25. By Exercise 10.5.51, $D = \dfrac{|2 - 24|}{\sqrt{26}} = \dfrac{22}{\sqrt{26}}$.

27. The equation $x = z$ represents a plane perpendicular to

the xz-plane and intersecting the xz-plane in the line

$x = z$, $y = 0$.

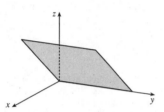

29. The equation $x^2 = y^2 + 4z^2$ represents a (right elliptical)

cone with vertex at the origin and axis the x-axis.

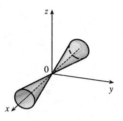

31. An equivalent equation is $-x^2 + \dfrac{y^2}{4} - z^2 = 1$, a

hyperboloid of two sheets with axis the y-axis. For

$|y| > 2$, traces parallel to the xz-plane are circles.

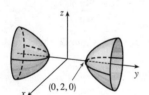

33. Completing the square in y gives

$$4x^2 + 4(y-1)^2 + z^2 = 4 \text{ or } x^2 + (y-1)^2 + \dfrac{z^2}{4} = 1,$$

an ellipsoid centered at $(0, 1, 0)$.

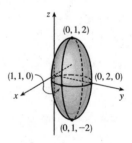

35. $4x^2 + y^2 = 16 \iff \dfrac{x^2}{4} + \dfrac{y^2}{16} = 1$. The equation of the ellipsoid is $\dfrac{x^2}{4} + \dfrac{y^2}{16} + \dfrac{z^2}{c^2} = 1$, since the horizontal trace in the

plane $z = 0$ must be the original ellipse. The traces of the ellipsoid in the yz-plane must be circles since the surface is obtained

by rotation about the x-axis. Therefore, $c^2 = 16$ and the equation of the ellipsoid is $\dfrac{x^2}{4} + \dfrac{y^2}{16} + \dfrac{z^2}{16} = 1 \iff$

$4x^2 + y^2 + z^2 = 16$.

37. (a) The corresponding parametric equations for the curve are $x = t$,

$y = \cos \pi t$, $z = \sin \pi t$. Since $y^2 + z^2 = 1$, the curve is contained in a

circular cylinder with axis the x-axis. Since $x = t$, the curve is a helix.

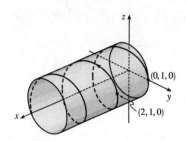

(b) $\mathbf{r}(t) = t\,\mathbf{i} + \cos \pi t\,\mathbf{j} + \sin \pi t\,\mathbf{k} \quad \Rightarrow$

$\mathbf{r}'(t) = \mathbf{i} - \pi \sin \pi t\,\mathbf{j} + \pi \cos \pi t\,\mathbf{k} \quad \Rightarrow$

$\mathbf{r}''(t) = -\pi^2 \cos \pi t\,\mathbf{j} - \pi^2 \sin \pi t\,\mathbf{k}$

39. The projection of the curve C of intersection onto the xy-plane is the circle $x^2 + y^2 = 16$, $z = 0$. So we can write

$x = 4\cos t$, $y = 4\sin t$, $0 \le t \le 2\pi$. From the equation of the plane, we have $z = 5 - x = 5 - 4\cos t$, so parametric

equations for C are $x = 4\cos t$, $y = 4\sin t$, $z = 5 - 4\cos t$, $0 \le t \le 2\pi$, and the corresponding vector function is

$\mathbf{r}(t) = 4\cos t\,\mathbf{i} + 4\sin t\,\mathbf{j} + (5 - 4\cos t)\,\mathbf{k}$, $0 \le t \le 2\pi$.

41. $\int_0^1 (t^2\,\mathbf{i} + t\cos \pi t\,\mathbf{j} + \sin \pi t\,\mathbf{k})\,dt = \left(\int_0^1 t^2\,dt\right)\mathbf{i} + \left(\int_0^1 t\cos \pi t\,dt\right)\mathbf{j} + \left(\int_0^1 \sin \pi t\,dt\right)\mathbf{k}$

$$= \left[\tfrac{1}{3}t^3\right]_0^1 \mathbf{i} + \left(\tfrac{t}{\pi}\sin \pi t\big]_0^1 - \int_0^1 \tfrac{1}{\pi}\sin \pi t\,dt\right)\mathbf{j} + \left[-\tfrac{1}{\pi}\cos \pi t\right]_0^1 \mathbf{k}$$

$$= \tfrac{1}{3}\mathbf{i} + \left[\tfrac{1}{\pi^2}\cos \pi t\right]_0^1 \mathbf{j} + \tfrac{2}{\pi}\mathbf{k} = \tfrac{1}{3}\mathbf{i} - \tfrac{2}{\pi^2}\mathbf{j} + \tfrac{2}{\pi}\mathbf{k}$$

where we integrated by parts in the y-component.

43. $\mathbf{r}(t) = \langle t^2, t^3, t^4 \rangle \quad \Rightarrow \quad \mathbf{r}'(t) = \langle 2t, 3t^2, 4t^3 \rangle \quad \Rightarrow \quad |\mathbf{r}'(t)| = \sqrt{4t^2 + 9t^4 + 16t^6}$ and

$L = \int_0^3 |\mathbf{r}'(t)|\,dt = \int_0^3 \sqrt{4t^2 + 9t^4 + 16t^6}\,dt$. Using Simpson's Rule with $f(t) = \sqrt{4t^2 + 9t^4 + 16t^6}$ and $n = 6$ we

have $\Delta t = \frac{3-0}{6} = \frac{1}{2}$ and

$$L \approx \tfrac{\Delta t}{3}\left[f(0) + 4f\left(\tfrac{1}{2}\right) + 2f(1) + 4f\left(\tfrac{3}{2}\right) + 2f(2) + 4f\left(\tfrac{5}{2}\right) + f(3)\right]$$

$$= \tfrac{1}{6}\left[\sqrt{0+0+0} + 4\cdot\sqrt{4\left(\tfrac{1}{2}\right)^2 + 9\left(\tfrac{1}{2}\right)^4 + 16\left(\tfrac{1}{2}\right)^6} + 2\cdot\sqrt{4(1)^2 + 9(1)^4 + 16(1)^6}\right.$$

$$+ 4\cdot\sqrt{4\left(\tfrac{3}{2}\right)^2 + 9\left(\tfrac{3}{2}\right)^4 + 16\left(\tfrac{3}{2}\right)^6} + 2\cdot\sqrt{4(2)^2 + 9(2)^4 + 16(2)^6}$$

$$\left.+ 4\cdot\sqrt{4\left(\tfrac{5}{2}\right)^2 + 9\left(\tfrac{5}{2}\right)^4 + 16\left(\tfrac{5}{2}\right)^6} + \sqrt{4(3)^2 + 9(3)^4 + 16(3)^6}\right]$$

$$\approx 86.631$$

45. The angle of intersection of the two curves, θ, is the angle between their respective tangents at the point of intersection.

For both curves the point $(1, 0, 0)$ occurs when $t = 0$.

$\mathbf{r}_1'(t) = -\sin t\,\mathbf{i} + \cos t\,\mathbf{j} + \mathbf{k} \quad \Rightarrow \quad \mathbf{r}_1'(0) = \mathbf{j} + \mathbf{k}$ and $\mathbf{r}_2'(t) = \mathbf{i} + 2t\,\mathbf{j} + 3t^2\,\mathbf{k} \quad \Rightarrow \quad \mathbf{r}_2'(0) = \mathbf{i}$.

$\mathbf{r}_1'(0) \cdot \mathbf{r}_2'(0) = (\mathbf{j} + \mathbf{k}) \cdot \mathbf{i} = 0$.   Therefore, the curves intersect in a right angle, that is, $\theta = \frac{\pi}{2}$.

47. (a) $\mathbf{T}(t) = \dfrac{\mathbf{r}'(t)}{|\mathbf{r}'(t)|} = \dfrac{\langle t^2, t, 1 \rangle}{|\langle t^2, t, 1 \rangle|} = \dfrac{\langle t^2, t, 1 \rangle}{\sqrt{t^4 + t^2 + 1}}$

(b) $\mathbf{T}'(t) = -\frac{1}{2}(t^4 + t^2 + 1)^{-3/2}(4t^3 + 2t)\langle t^2, t, 1\rangle + (t^4 + t^2 + 1)^{-1/2}\langle 2t, 1, 0\rangle$

$\qquad = \dfrac{-2t^3 - t}{(t^4 + t^2 + 1)^{3/2}}\langle t^2, t, 1\rangle + \dfrac{1}{(t^4 + t^2 + 1)^{1/2}}\langle 2t, 1, 0\rangle$

$\qquad = \dfrac{\langle -2t^5 - t^3, -2t^4 - t^2, -2t^3 - t\rangle + \langle 2t^5 + 2t^3 + 2t, t^4 + t^2 + 1, 0\rangle}{(t^4 + t^2 + 1)^{3/2}} = \dfrac{\langle 2t, -t^4 + 1, -2t^3 - t\rangle}{(t^4 + t^2 + 1)^{3/2}}$

$|\mathbf{T}'(t)| = \dfrac{\sqrt{4t^2 + t^8 - 2t^4 + 1 + 4t^6 + 4t^4 + t^2}}{(t^4 + t^2 + 1)^{3/2}} = \dfrac{\sqrt{t^8 + 4t^6 + 2t^4 + 5t^2}}{(t^4 + t^2 + 1)^{3/2}}$ and $\mathbf{N}(t) = \dfrac{\langle 2t, 1 - t^4, -2t^3 - t\rangle}{\sqrt{t^8 + 4t^6 + 2t^4 + 5t^2}}$.

(c) $\kappa(t) = \dfrac{|\mathbf{T}'(t)|}{|\mathbf{r}'(t)|} = \dfrac{\sqrt{t^8 + 4t^6 + 2t^4 + 5t^2}}{(t^4 + t^2 + 1)^2}$

49. $y' = 4x^3$, $y'' = 12x^2$ and $\kappa(x) = \dfrac{|y''|}{[1 + (y')^2]^{3/2}} = \dfrac{|12x^2|}{(1 + 16x^6)^{3/2}}$, so $\kappa(1) = \dfrac{12}{17^{3/2}}$.

51. $\mathbf{r}(t) = t\ln t\,\mathbf{i} + t\mathbf{j} + e^{-t}\mathbf{k}$, $\mathbf{v}(t) = \mathbf{r}'(t) = (1 + \ln t)\mathbf{i} + \mathbf{j} - e^{-t}\mathbf{k}$,

$|\mathbf{v}(t)| = \sqrt{(1 + \ln t)^2 + 1^2 + (-e^{-t})^2} = \sqrt{2 + 2\ln t + (\ln t)^2 + e^{-2t}}$, $\mathbf{a}(t) = \mathbf{v}'(t) = \frac{1}{t}\mathbf{i} + e^{-t}\mathbf{k}$

53. We set up the axes so that the shot leaves the athlete's hand 7 ft above the origin. Then we are given $\mathbf{r}(0) = 7\mathbf{j}$,

$|\mathbf{v}(0)| = 43$ ft/s, and $\mathbf{v}(0)$ has direction given by a 45° angle of elevation. Then a unit vector in the direction of $\mathbf{v}(0)$ is

$\frac{1}{\sqrt{2}}(\mathbf{i} + \mathbf{j})$ $\Rightarrow$ $\mathbf{v}(0) = \frac{43}{\sqrt{2}}(\mathbf{i} + \mathbf{j})$. Assuming air resistance is negligible, the only external force is due to gravity, so as in

Example 10.4.5 we have $\mathbf{a} = -g\mathbf{j}$ where here $g \approx 32$ ft/s². Since $\mathbf{v}'(t) = \mathbf{a}(t)$, we integrate, giving $\mathbf{v}(t) = -gt\mathbf{j} + \mathbf{C}$

where $\mathbf{C} = \mathbf{v}(0) = \frac{43}{\sqrt{2}}(\mathbf{i} + \mathbf{j})$ $\Rightarrow$ $\mathbf{v}(t) = \frac{43}{\sqrt{2}}\mathbf{i} + \left(\frac{43}{\sqrt{2}} - gt\right)\mathbf{j}$. Since $\mathbf{r}'(t) = \mathbf{v}(t)$ we integrate again, so

$\mathbf{r}(t) = \frac{43}{\sqrt{2}}t\mathbf{i} + \left(\frac{43}{\sqrt{2}}t - \frac{1}{2}gt^2\right)\mathbf{j} + \mathbf{D}$. But $\mathbf{D} = \mathbf{r}(0) = 7\mathbf{j}$ $\Rightarrow$ $\mathbf{r}(t) = \frac{43}{\sqrt{2}}t\mathbf{i} + \left(\frac{43}{\sqrt{2}}t - \frac{1}{2}gt^2 + 7\right)\mathbf{j}$.

(a) At 2 seconds, the shot is at $\mathbf{r}(2) = \frac{43}{\sqrt{2}}(2)\mathbf{i} + \left(\frac{43}{\sqrt{2}}(2) - \frac{1}{2}g(2)^2 + 7\right)\mathbf{j} \approx 60.8\,\mathbf{i} + 3.8\,\mathbf{j}$, so the shot is about 3.8 ft above
the ground, at a horizontal distance of 60.8 ft from the athlete.

(b) The shot reaches its maximum height when the vertical component of velocity is 0: $\frac{43}{\sqrt{2}} - gt = 0$ $\Rightarrow$

$t = \dfrac{43}{\sqrt{2}\,g} \approx 0.95$ s. Then $\mathbf{r}(0.95) \approx 28.9\,\mathbf{i} + 21.4\,\mathbf{j}$, so the maximum height is approximately 21.4 ft.

(c) The shot hits the ground when the vertical component of $\mathbf{r}(t)$ is 0, so $\frac{43}{\sqrt{2}}t - \frac{1}{2}gt^2 + 7 = 0$ $\Rightarrow$

$-16t^2 + \frac{43}{\sqrt{2}}t + 7 = 0$ $\Rightarrow$ $t \approx 2.11$ s. $\mathbf{r}(2.11) \approx 64.2\,\mathbf{i} - 0.08\,\mathbf{j}$, thus the shot lands approximately 64.2 ft from the

athlete.

55. By the Fundamental Theorem of Calculus, $\mathbf{r}'(t) = \langle \sin(\frac{1}{2}\pi t^2), \cos(\frac{1}{2}\pi t^2)\rangle$, $|\mathbf{r}'(t)| = 1$ and so $\mathbf{T}(t) = \mathbf{r}'(t)$.

Thus $\mathbf{T}'(t) = \pi t\langle \cos(\frac{1}{2}\pi t^2), -\sin(\frac{1}{2}\pi t^2)\rangle$ and the curvature is $\kappa = |\mathbf{T}'(t)| = \sqrt{(\pi t)^2(1)} = \pi\,|t|$.

11 □ PARTIAL DERIVATIVES

11.1 Functions of Several Variables

1. (a) $f(2,0) = 2^2 e^{3(2)(0)} = 4(1) = 4$

(b) Since both x^2 and the exponential function are defined everywhere, $x^2 e^{3xy}$ is defined for all choices of values for x and y. Thus the domain of f is $\mathbb{R}^2$.

(c) Because the range of $g(x,y) = 3xy$ is $\mathbb{R}$, and the range of e^x is $(0, \infty)$, the range of $e^{g(x,y)} = e^{3xy}$ is $(0, \infty)$. The range of x^2 is $[0, \infty)$, so the range of the product $x^2 e^{3xy}$ is $[0, \infty)$.

3. (a) $f(2,-1,6) = e^{\sqrt{6 - 2^2 - (-1)^2}} = e^{\sqrt{1}} = e.$

(b) $e^{\sqrt{z - x^2 - y^2}}$ is defined when $z - x^2 - y^2 \geq 0 \;\;\Rightarrow\;\; z \geq x^2 + y^2$. Thus the domain of f is $\{(x,y,z) \mid z \geq x^2 + y^2\}$.

(c) Since $\sqrt{z - x^2 - y^2} \geq 0$, we have $e^{\sqrt{z - x^2 - y^2}} \geq 1$. Thus the range of f is $[1, \infty)$.

5. $\sqrt{x+y}$ is defined only when $x + y \geq 0$, or $y \geq -x$. So the domain of f is $\{(x,y) \mid y \geq -x\}$.

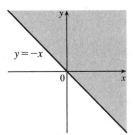

7. $\ln(9 - x^2 - 9y^2)$ is defined only when $9 - x^2 - 9y^2 > 0$, or $\frac{1}{9}x^2 + y^2 < 1$. So the domain of f is $\{(x,y) \mid \frac{1}{9}x^2 + y^2 < 1\}$, the interior of an ellipse.

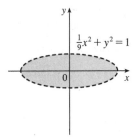

9. $\sqrt{y - x^2}$ is defined only when $y - x^2 \geq 0$, or $y \geq x^2$. In addition, f is not defined if $1 - x^2 = 0 \;\;\Rightarrow\;\; x = \pm 1$. Thus the domain of f is $\{(x,y) \mid y \geq x^2, x \neq \pm 1\}$.

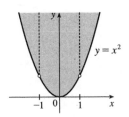

11. We need $1 - x^2 - y^2 - z^2 \geq 0$ or $x^2 + y^2 + z^2 \leq 1$, so $D = \{(x,y,z) \mid x^2 + y^2 + z^2 \leq 1\}$ (the points inside or on the sphere of radius 1, center the origin).

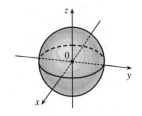

13. $z = 6 - 3x - 2y$ or $3x + 2y + z = 6$, a plane with intercepts 2, 3, and 6.

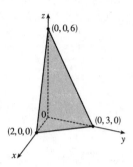

15. $z = y^2 + 1$, a parabolic cylinder

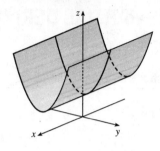

17. $z = 4x^2 + y^2 + 1$, an elliptic paraboloid with vertex at $(0, 0, 1)$.

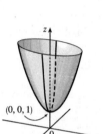

19. $z = \sqrt{x^2 + y^2}$ so $x^2 + y^2 = z^2$ and $z \geq 0$, the top half of a right circular cone.

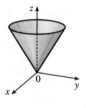

21. The point $(-3, 3)$ lies between the level curves with z-values 50 and 60. Since the point is a little closer to the level curve with $z = 60$, we estimate that $f(-3, 3) \approx 56$. The point $(3, -2)$ appears to be just about halfway between the level curves with z-values 30 and 40, so we estimate $f(3, -2) \approx 35$. The graph rises as we approach the origin, gradually from above, steeply from below.

23. Near A, the level curves are very close together, indicating that the terrain is quite steep. At B, the level curves are much farther apart, so we would expect the terrain to be much less steep than near A, perhaps almost flat.

25. The level curves are $(y - 2x)^2 = k$ or $y = 2x \pm \sqrt{k}$, $k \geq 0$, a family of pairs of parallel lines.

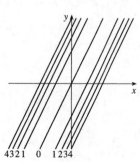

27. The level curves are $y - \ln x = k$ or $y = \ln x + k$.

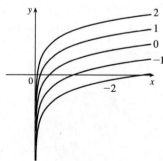

29. The level curves are $ye^x = k$ or $y = ke^{-x}$, a family of exponential curves.

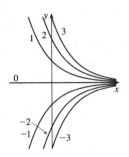

31. The level curves are $\sqrt{y^2 - x^2} = k$ or $y^2 - x^2 = k^2$, $k \geq 0$. When $k = 0$ the level curve is the pair of lines $y = \pm x$. For $k > 0$, the level curves are hyperbolas with axis the y-axis.

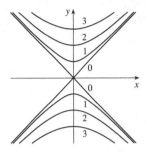

33. The contour map consists of the level curves $k = x^2 + 9y^2$, a family of ellipses with major axis the x-axis. (Or, if $k = 0$, the origin.)

The graph of $f(x, y)$ is the surface $z = x^2 + 9y^2$, an elliptic paraboloid.

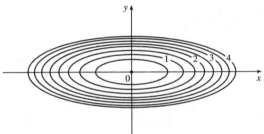

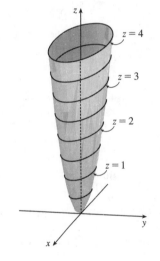

If we visualize lifting each ellipse $k = x^2 + 9y^2$ of the contour map to the plane $z = k$, we have horizontal traces that indicate the shape of the graph of f.

35. The isothermals are given by $k = 100/(1 + x^2 + 2y^2)$ or $x^2 + 2y^2 = (100 - k)/k$ $[0 < k \leq 100]$, a family of ellipses.

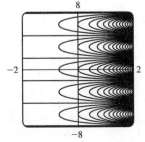

37. $f(x, y) = e^x \cos y$

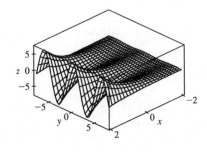

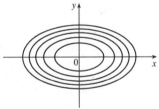

Traces parallel to the yz-plane (such as the left-front trace in the first graph above) are cosine curves. The amplitudes of these curves decrease as x decreases.

39. $f(x, y) = xy^2 - x^3$

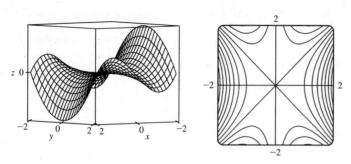

The traces parallel to the yz-plane (such as the left-front trace in the graph above) are parabolas; those parallel to the xz-plane (such as the right-front trace) are cubic curves. The surface is called a monkey saddle because a monkey sitting on the surface near the origin has places for both legs and tail to rest.

41. (a) C   (b) II

Reasons: This function is periodic in both x and y, and the function is the same when x is interchanged with y, so its graph is symmetric about the plane $y = x$. In addition, the function is 0 along the x- and y-axes. These conditions are satisfied only by C and II.

43. (a) F   (b) I

Reasons: This function is periodic in both x and y but is constant along the lines $y = x + k$, a condition satisfied only by F and I.

45. (a) B   (b) VI

Reasons: This function is 0 along the lines $x = \pm 1$ and $y = \pm 1$. The only contour map in which this could occur is VI. Also note that the trace in the xz-plane is the parabola $z = 1 - x^2$ and the trace in the yz-plane is the parabola $z = 1 - y^2$, so the graph is B.

47. $k = x + 3y + 5z$ is a family of parallel planes with normal vector $\langle 1, 3, 5 \rangle$.

49. $k = x^2 - y^2 + z^2$ are the equations of the level surfaces. For $k = 0$, the surface is a right circular cone with vertex the origin and axis the y-axis. For $k > 0$, we have a family of hyperboloids of one sheet with axis the y-axis. For $k < 0$, we have a family of hyperboloids of two sheets with axis the y-axis.

51. (a) The graph of g is the graph of f shifted upward 2 units.

(b) The graph of g is the graph of f stretched vertically by a factor of 2.

(c) The graph of g is the graph of f reflected about the xy-plane.

(d) The graph of $g(x, y) = -f(x, y) + 2$ is the graph of f reflected about the xy-plane and then shifted upward 2 units.

53. $f(x, y) = e^{cx^2 + y^2}$. First, if $c = 0$, the graph is the cylindrical surface $z = e^{y^2}$ (whose level curves are parallel lines). When $c > 0$, the vertical trace above the y-axis remains fixed while the sides of the surface in the x-direction "curl" upward, giving the graph a shape resembling an elliptic paraboloid. The level curves of the surface are ellipses centered at the origin.

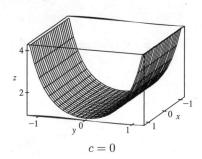

$c = 0$

For $0 < c < 1$, the ellipses have major axis the x-axis and the eccentricity increases as $c \to 0$.

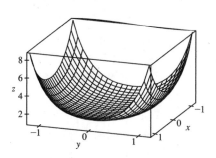 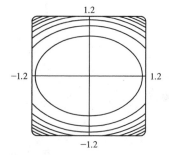

$c = 0.5$ (level curves in increments of 1)

For $c = 1$ the level curves are circles centered at the origin.

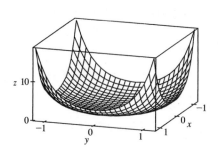 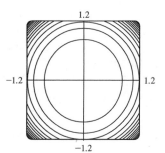

$c = 1$ (level curves in increments of 1)

When $c > 1$, the level curves are ellipses with major axis the y-axis, and the eccentricity increases as c increases.

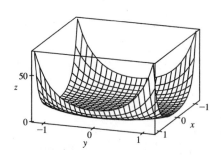 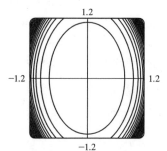

$c = 2$ (level curves in increments of 4)

For values of $c < 0$, the sides of the surface in the x-direction curl downward and approach the xy-plane (while the vertical trace $x = 0$ remains fixed), giving a saddle-shaped appearance to the graph near the point $(0, 0, 1)$. The level curves consist of a family of hyperbolas. As c decreases, the surface becomes flatter in the x-direction and the surface's approach to the curve in

the trace $x = 0$ becomes steeper, as the graphs demonstrate.

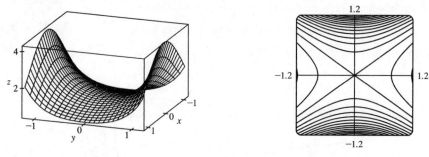

$c = -0.5$ (level curves in increments of 0.25)

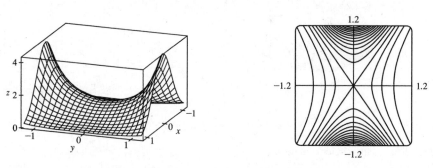

$c = -2$ (level curves in increments of 0.25)

11.2 Limits and Continuity

1. In general, we can't say anything about $f(3, 1)$! $\displaystyle\lim_{(x,y)\to(3,1)} f(x, y) = 6$ means that the values of $f(x, y)$ approach 6 as (x, y)

approaches, but is not equal to, $(3, 1)$. If f is continuous, we know that $\displaystyle\lim_{(x,y)\to(a,b)} f(x, y) = f(a, b)$, so

$$\lim_{(x,y)\to(3,1)} f(x, y) = f(3, 1) = 6.$$

3. $f(x, y) = x^5 + 4x^3 y - 5xy^2$ is a polynomial, and hence continuous, so

$$\lim_{(x,y)\to(5,-2)} f(x, y) = f(5, -2) = 5^5 + 4(5)^3(-2) - 5(5)(-2)^2 = 2025.$$

5. $f(x, y) = y^4/(x^4 + 3y^4)$.   First approach $(0, 0)$ along the x-axis. Then $f(x, 0) = 0/x^4 = 0$ for $x \neq 0$, so $f(x, y) \to 0$.

Now approach $(0, 0)$ along the y-axis. Then for $y \neq 0$, $f(0, y) = y^4/3y^4 = 1/3$, so $f(x, y) \to 1/3$. Since f has two different

limits along two different lines, the limit does not exist.

7. $f(x, y) = (xy \cos y)/(3x^2 + y^2)$.   On the x-axis, $f(x, 0) = 0$ for $x \neq 0$, so $f(x, y) \to 0$ as $(x, y) \to (0, 0)$ along the

x-axis. Approaching $(0, 0)$ along the line $y = x$, $f(x, x) = (x^2 \cos x)/4x^2 = \frac{1}{4} \cos x$ for $x \neq 0$, so $f(x, y) \to \frac{1}{4}$ along this

line. Thus the limit does not exist.

9. $f(x, y) = \dfrac{xy}{\sqrt{x^2 + y^2}}$. We can see that the limit along any line through $(0,0)$ is 0, as well as along other paths through

$(0, 0)$ such as $x = y^2$ and $y = x^2$. So we suspect that the limit exists and equals 0; we use the Squeeze Theorem to prove our

assertion. $0 \leq \left| \dfrac{xy}{\sqrt{x^2 + y^2}} \right| \leq |x|$ since $|y| \leq \sqrt{x^2 + y^2}$, and $|x| \to 0$ as $(x, y) \to (0, 0)$. So $\displaystyle \lim_{(x,y)\to(0,0)} f(x, y) = 0$.

11. Let $f(x, y) = 2x^2 y/(x^4 + y^2)$. Then $f(x, 0) = 0$ for $x \neq 0$, so $f(x, y) \to 0$ as $(x, y) \to (0, 0)$ along the x-axis. But

$f(x, x^2) = \dfrac{2x^4}{2x^4} = 1$ for $x \neq 0$, so $f(x, y) \to 1$ as $(x, y) \to (0, 0)$ along the parabola $y = x^2$. Thus the limit doesn't exist.

13. $\displaystyle \lim_{(x,y)\to(0,0)} \dfrac{x^2 + y^2}{\sqrt{x^2 + y^2 + 1} - 1} = \lim_{(x,y)\to(0,0)} \dfrac{x^2 + y^2}{\sqrt{x^2 + y^2 + 1} - 1} \cdot \dfrac{\sqrt{x^2 + y^2 + 1} + 1}{\sqrt{x^2 + y^2 + 1} + 1}$

$\qquad = \displaystyle \lim_{(x,y)\to(0,0)} \dfrac{(x^2 + y^2)\left(\sqrt{x^2 + y^2 + 1} + 1\right)}{x^2 + y^2} = \lim_{(x,y)\to(0,0)} \left(\sqrt{x^2 + y^2 + 1} + 1\right) = 2$

15. $f(x, y, z) = \dfrac{xy + yz^2 + xz^2}{x^2 + y^2 + z^4}$. Then $f(x, 0, 0) = 0/x^2 = 0$ for $x \neq 0$, so as $(x, y, z) \to (0, 0, 0)$ along the x-axis,

$f(x, y, z) \to 0$. But $f(x, x, 0) = x^2/(2x^2) = \frac{1}{2}$ for $x \neq 0$, so as $(x, y, z) \to (0, 0, 0)$ along the line $y = x$, $z = 0$,

$f(x, y, z) \to \frac{1}{2}$. Thus the limit doesn't exist.

17.

From the ridges on the graph, we see that as $(x, y) \to (0, 0)$ along the

lines under the two ridges, $f(x, y)$ approaches different values. So the

limit does not exist.

19. $h(x, y) = g(f(x, y)) = (2x + 3y - 6)^2 + \sqrt{2x + 3y - 6}$. Since f is a polynomial, it is continuous on $\mathbb{R}^2$ and g is

continuous on its domain $\{t \mid t \geq 0\}$. Thus h is continuous on its domain.

$D = \{(x, y) \mid 2x + 3y - 6 \geq 0\} = \{(x, y) \mid y \geq -\frac{2}{3}x + 2\}$, which consists of all points on or above the line $y = -\frac{2}{3}x + 2$.

21. The functions $\sin(xy)$ and $e^x - y^2$ are continuous everywhere, so $F(x, y) = \dfrac{\sin(xy)}{e^x - y^2}$ is continuous except where

$e^x - y^2 = 0 \;\Rightarrow\; y^2 = e^x \;\Rightarrow\; y = \pm\sqrt{e^x} = \pm e^{\frac{1}{2}x}$. Thus F is continuous on its domain $\{(x, y) \mid y \neq \pm e^{x/2}\}$.

23. $G(x, y) = \ln(x^2 + y^2 - 4) = g(f(x, y))$ where $f(x, y) = x^2 + y^2 - 4$, continuous on $\mathbb{R}^2$, and $g(t) = \ln t$, continuous on

its domain $\{t \mid t > 0\}$. Thus G is continuous on its domain $\{(x, y) \mid x^2 + y^2 - 4 > 0\} = \{(x, y) \mid x^2 + y^2 > 4\}$, the

exterior of the circle $x^2 + y^2 = 4$.

25. $\sqrt{y}$ is continuous on its domain $\{y \mid y \geq 0\}$ and $x^2 - y^2 + z^2$ is continuous everywhere, so $f(x, y, z) = \dfrac{\sqrt{y}}{x^2 - y^2 + z^2}$ is

continuous for $y \geq 0$ and $x^2 - y^2 + z^2 \neq 0 \;\Rightarrow\; y^2 \neq x^2 + z^2$, that is, $\{(x, y, z) \mid y \geq 0, y \neq \sqrt{x^2 + z^2}\}$.

27. $f(x, y) = \begin{cases} \dfrac{x^2 y^3}{2x^2 + y^2} & \text{if } (x, y) \neq (0, 0) \\ 1 & \text{if } (x, y) = (0, 0) \end{cases}$ The first piece of f is a rational function defined everywhere except at the

origin, so f is continuous on $\mathbb{R}^2$ except possibly at the origin. Since $x^2 \leq 2x^2 + y^2$, we have $\left| x^2 y^3 / (2x^2 + y^2) \right| \leq |y^3|$. We

know that $\left| y^3 \right| \to 0$ as $(x, y) \to (0, 0)$. So, by the Squeeze Theorem, $\displaystyle\lim_{(x,y) \to (0,0)} f(x, y) = \lim_{(x,y) \to (0,0)} \dfrac{x^2 y^3}{2x^2 + y^2} = 0$. But

$f(0, 0) = 1$, so f is discontinuous at $(0, 0)$. Therefore, f is continuous on the set $\{(x, y) \mid (x, y) \neq (0, 0)\}$.

29. $\displaystyle\lim_{(x,y) \to (0,0)} \dfrac{x^3 + y^3}{x^2 + y^2} = \lim_{r \to 0^+} \dfrac{(r \cos \theta)^3 + (r \sin \theta)^3}{r^2} = \lim_{r \to 0^+} (r \cos^3 \theta + r \sin^3 \theta) = 0$

31. Since $|\mathbf{x} - \mathbf{a}|^2 = |\mathbf{x}|^2 + |\mathbf{a}|^2 - 2 |\mathbf{x}| |\mathbf{a}| \cos \theta \geq |\mathbf{x}|^2 + |\mathbf{a}|^2 - 2 |\mathbf{x}| |\mathbf{a}| = (|\mathbf{x}| - |\mathbf{a}|)^2$, we have $\big| |\mathbf{x}| - |\mathbf{a}| \big| \leq |\mathbf{x} - \mathbf{a}|$. Let

$\epsilon > 0$ be given and set $\delta = \epsilon$. Then whenever $0 < |\mathbf{x} - \mathbf{a}| < \delta$, $\big| |\mathbf{x}| - |\mathbf{a}| \big| \leq |\mathbf{x} - \mathbf{a}| < \delta = \epsilon$. Hence $\lim_{\mathbf{x} \to \mathbf{a}} |\mathbf{x}| = |\mathbf{a}|$ and

$f(\mathbf{x}) = |\mathbf{x}|$ is continuous on $\mathbb{R}^n$.

11.3   Partial Derivatives

1. (a) $\partial T / \partial x$ represents the rate of change of T when we fix y and t and consider T as a function of the single variable x, which

describes how quickly the temperature changes when longitude changes but latitude and time are constant. $\partial T / \partial y$

represents the rate of change of T when we fix x and t and consider T as a function of y, which describes how quickly the

temperature changes when latitude changes but longitude and time are constant. $\partial T / \partial t$ represents the rate of change of T

when we fix x and y and consider T as a function of t, which describes how quickly the temperature changes over time for

a constant longitude and latitude.

(b) $f_x(158, 21, 9)$ represents the rate of change of temperature at longitude $158° \text{W}$, latitude $21° \text{N}$ at 9:00 AM when only

longitude varies. Since the air is warmer to the west than to the east, increasing longitude results in an increased air

temperature, so we would expect $f_x(158, 21, 9)$ to be positive. $f_y(158, 21, 9)$ represents the rate of change of temperature

at the same time and location when only latitude varies. Since the air is warmer to the south and cooler to the north,

increasing latitude results in a decreased air temperature, so we would expect $f_y(158, 21, 9)$ to be negative. $f_t(158, 21, 9)$

represents the rate of change of temperature at the same time and location when only time varies. Since typically air

temperature increases from the morning to the afternoon as the sun warms it, we would expect $f_t(158, 21, 9)$ to be

positive.

3. (a) If we start at $(1, 2)$ and move in the positive x-direction, the graph of f increases. Thus $f_x(1, 2)$ is positive.

(b) If we start at $(1, 2)$ and move in the positive y-direction, the graph of f decreases. Thus $f_y(1, 2)$ is negative.

5. $f(x,y) = 16 - 4x^2 - y^2 \Rightarrow f_x(x,y) = -8x$ and $f_y(x,y) = -2y \Rightarrow f_x(1,2) = -8$ and $f_y(1,2) = -4$. The graph

of f is the paraboloid $z = 16 - 4x^2 - y^2$ and the vertical plane $y = 2$ intersects it in the parabola $z = 12 - 4x^2$, $y = 2$

(the curve C_1 in the first figure). The slope of the

tangent line to this parabola at $(1,2,8)$ is

$f_x(1,2) = -8$. Similarly the plane $x = 1$ intersects

the paraboloid in the parabola $z = 12 - y^2$, $x = 1$ (the

curve C_2 in the second figure) and the slope of the

tangent line at $(1,2,8)$ is $f_y(1,2) = -4$.

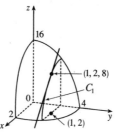
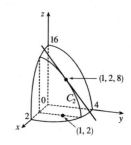

7. $f(x,y) = 3x - 2y^4 \Rightarrow f_x(x,y) = 3 - 0 = 3$, $f_y(x,y) = 0 - 8y^3 = -8y^3$

9. $z = xe^{3y} \Rightarrow \dfrac{\partial z}{\partial x} = e^{3y}$, $\dfrac{\partial z}{\partial y} = 3xe^{3y}$

11. $f(x,y) = \dfrac{x-y}{x+y} \Rightarrow f_x(x,y) = \dfrac{(1)(x+y) - (x-y)(1)}{(x+y)^2} = \dfrac{2y}{(x+y)^2}$,

$f_y(x,y) = \dfrac{(-1)(x+y) - (x-y)(1)}{(x+y)^2} = -\dfrac{2x}{(x+y)^2}$

13. $w = \sin\alpha\,\cos\beta \Rightarrow \dfrac{\partial w}{\partial \alpha} = \cos\alpha\,\cos\beta$, $\dfrac{\partial w}{\partial \beta} = -\sin\alpha\,\sin\beta$

15. $f(r,s) = r\ln(r^2 + s^2) \Rightarrow f_r(r,s) = r \cdot \dfrac{2r}{r^2 + s^2} + \ln(r^2 + s^2)\cdot 1 = \dfrac{2r^2}{r^2 + s^2} + \ln(r^2 + s^2)$,

$f_s(r,s) = r \cdot \dfrac{2s}{r^2 + s^2} + 0 = \dfrac{2rs}{r^2 + s^2}$

17. $u = te^{w/t} \Rightarrow \dfrac{\partial u}{\partial t} = t \cdot e^{w/t}(-wt^{-2}) + e^{w/t}\cdot 1 = e^{w/t} - \dfrac{w}{t}e^{w/t} = e^{w/t}\left(1 - \dfrac{w}{t}\right)$, $\dfrac{\partial u}{\partial w} = te^{w/t}\cdot\dfrac{1}{t} = e^{w/t}$

19. $f(x,y,z) = xy^2z^3 + 3yz \Rightarrow f_x(x,y,z) = y^2z^3$, $f_y(x,y,z) = 2xyz^3 + 3z$, $f_z(x,y,z) = 3xy^2z^2 + 3y$

21. $w = \ln(x + 2y + 3z) \Rightarrow \dfrac{\partial w}{\partial x} = \dfrac{1}{x+2y+3z}$, $\dfrac{\partial w}{\partial y} = \dfrac{2}{x+2y+3z}$, $\dfrac{\partial w}{\partial z} = \dfrac{3}{x+2y+3z}$

23. $u = xe^{-t}\sin\theta \Rightarrow \dfrac{\partial u}{\partial x} = e^{-t}\sin\theta$, $\dfrac{\partial u}{\partial t} = -xe^{-t}\sin\theta$, $\dfrac{\partial u}{\partial \theta} = xe^{-t}\cos\theta$

25. $f(x,y,z,t) = xyz^2\tan(yt) \Rightarrow f_x(x,y,z,t) = yz^2\tan(yt)$,

$f_y(x,y,z,t) = xyz^2 \cdot \sec^2(yt)\cdot t + xz^2\tan(yt) = xyz^2t\sec^2(yt) + xz^2\tan(yt)$,

$f_z(x,y,z,t) = 2xyz\tan(yt)$, $f_t(x,y,z,t) = xyz^2\sec^2(yt)\cdot y = xy^2z^2\sec^2(yt)$

27. $u = \sqrt{x_1^2 + x_2^2 + \cdots + x_n^2}$. For each $i = 1, \ldots, n$, $u_{x_i} = \frac{1}{2}\left(x_1^2 + x_2^2 + \cdots + x_n^2\right)^{-1/2}(2x_i) = \dfrac{x_i}{\sqrt{x_1^2 + x_2^2 + \cdots + x_n^2}}$.

29. $f(x,y) = \sqrt{x^2 + y^2} \Rightarrow f_x(x,y) = \frac{1}{2}(x^2 + y^2)^{-1/2}(2x) = \dfrac{x}{\sqrt{x^2 + y^2}}$, so $f_x(3,4) = \dfrac{3}{\sqrt{3^2 + 4^2}} = \dfrac{3}{5}$.

31. $f(x, y, z) = \dfrac{x}{y + z} = x(y + z)^{-1} \Rightarrow f_z(x, y, z) = x(-1)(y + z)^{-2} = -\dfrac{x}{(y + z)^2}$,

so $f_z(3, 2, 1) = -\dfrac{3}{(2 + 1)^2} = -\dfrac{1}{3}$.

33. $f(x, y) = xy^2 - x^3 y \Rightarrow$

$$f_x(x, y) = \lim_{h \to 0} \frac{f(x + h, y) - f(x, y)}{h} = \lim_{h \to 0} \frac{(x + h)y^2 - (x + h)^3 y - (xy^2 - x^3 y)}{h}$$

$$= \lim_{h \to 0} \frac{h(y^2 - 3x^2 y - 3xyh - yh^2)}{h} = \lim_{h \to 0}(y^2 - 3x^2 y - 3xyh - yh^2) = y^2 - 3x^2 y$$

$$f_y(x, y) = \lim_{h \to 0} \frac{f(x, y + h) - f(x, y)}{h} = \lim_{h \to 0} \frac{x(y + h)^2 - x^3(y + h) - (xy^2 - x^3 y)}{h} = \lim_{h \to 0} \frac{h(2xy + xh - x^3)}{h}$$

$$= \lim_{h \to 0}(2xy + xh - x^3) = 2xy - x^3$$

35. $f(x, y) = x^2 + y^2 + x^2 y \Rightarrow f_x = 2x + 2xy, \quad f_y = 2y + x^2$

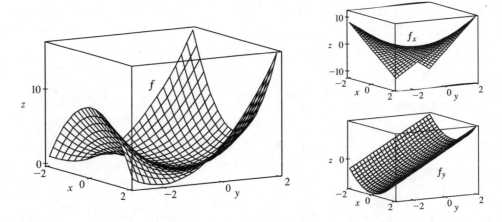

Note that the traces of f in planes parallel to the xz-plane are parabolas which open downward for $y < -1$ and upward for $y > -1$, and the traces of f_x in these planes are straight lines, which have negative slopes for $y < -1$ and positive slopes for $y > -1$. The traces of f in planes parallel to the yz-plane are parabolas which always open upward, and the traces of f_y in these planes are straight lines with positive slopes.

37. $x^2 + y^2 + z^2 = 3xyz \Rightarrow \dfrac{\partial}{\partial x}(x^2 + y^2 + z^2) = \dfrac{\partial}{\partial x}(3xyz) \Rightarrow 2x + 0 + 2z \dfrac{\partial z}{\partial x} = 3y\left(x \dfrac{\partial z}{\partial x} + z \cdot 1\right) \Leftrightarrow$

$2z \dfrac{\partial z}{\partial x} - 3xy \dfrac{\partial z}{\partial x} = 3yz - 2x \Leftrightarrow (2z - 3xy)\dfrac{\partial z}{\partial x} = 3yz - 2x$, so $\dfrac{\partial z}{\partial x} = \dfrac{3yz - 2x}{2z - 3xy}$.

$\dfrac{\partial}{\partial y}(x^2 + y^2 + z^2) = \dfrac{\partial}{\partial y}(3xyz) \Rightarrow 0 + 2y + 2z \dfrac{\partial z}{\partial y} = 3x\left(y \dfrac{\partial z}{\partial y} + z \cdot 1\right) \Leftrightarrow 2z \dfrac{\partial z}{\partial y} - 3xy \dfrac{\partial z}{\partial y} = 3xz - 2y \Leftrightarrow$

$(2z - 3xy)\dfrac{\partial z}{\partial y} = 3xz - 2y$, so $\dfrac{\partial z}{\partial y} = \dfrac{3xz - 2y}{2z - 3xy}$.

39. $x - z = \arctan(yz)$ $\Rightarrow$ $\dfrac{\partial}{\partial x}(x-z) = \dfrac{\partial}{\partial x}(\arctan(yz))$ $\Rightarrow$ $1 - \dfrac{\partial z}{\partial x} = \dfrac{1}{1+(yz)^2} \cdot y\,\dfrac{\partial z}{\partial x}$ $\Leftrightarrow$

$1 = \left(\dfrac{y}{1+y^2z^2} + 1\right)\dfrac{\partial z}{\partial x}$ $\Leftrightarrow$ $1 = \left(\dfrac{y+1+y^2z^2}{1+y^2z^2}\right)\dfrac{\partial z}{\partial x}$, so $\dfrac{\partial z}{\partial x} = \dfrac{1+y^2z^2}{1+y+y^2z^2}$.

$\dfrac{\partial}{\partial y}(x-z) = \dfrac{\partial}{\partial y}(\arctan(yz))$ $\Rightarrow$ $0 - \dfrac{\partial z}{\partial y} = \dfrac{1}{1+(yz)^2} \cdot \left(y\,\dfrac{\partial z}{\partial y} + z \cdot 1\right)$ $\Leftrightarrow$

$-\dfrac{z}{1+y^2z^2} = \left(\dfrac{y}{1+y^2z^2} + 1\right)\dfrac{\partial z}{\partial y}$ $\Leftrightarrow$ $-\dfrac{z}{1+y^2z^2} = \left(\dfrac{y+1+y^2z^2}{1+y^2z^2}\right)\dfrac{\partial z}{\partial y}$ $\Leftrightarrow$ $\dfrac{\partial z}{\partial y} = -\dfrac{z}{1+y+y^2z^2}$.

41. (a) $z = f(x) + g(y)$ $\Rightarrow$ $\dfrac{\partial z}{\partial x} = f'(x)$, $\dfrac{\partial z}{\partial y} = g'(y)$

(b) $z = f(x+y)$. Let $u = x + y$. Then $\dfrac{\partial z}{\partial x} = \dfrac{df}{du}\dfrac{\partial u}{\partial x} = \dfrac{df}{du}(1) = f'(u) = f'(x+y)$,

$\dfrac{\partial z}{\partial y} = \dfrac{df}{du}\dfrac{\partial u}{\partial y} = \dfrac{df}{du}(1) = f'(u) = f'(x+y)$.

43. $f(x,y) = x^4 - 3x^2y^3$ $\Rightarrow$ $f_x(x,y) = 4x^3 - 6xy^3$, $f_y(x,y) = -9x^2y^2$.

Then $f_{xx}(x,y) = 12x^2 - 6y^3$, $f_{xy}(x,y) = -18xy^2$, $f_{yx}(x,y) = -18xy^2$, $f_{yy}(x,y) = -18x^2y$.

45. $z = \dfrac{x}{x+y} = x(x+y)^{-1}$ $\Rightarrow$ $z_x = \dfrac{1(x+y) - 1(x)}{(x+y)^2} = \dfrac{y}{(x+y)^2}$, $z_y = x(-1)(x+y)^{-2} = -\dfrac{x}{(x+y)^2}$. Then

$z_{xx} = y(-2)(x+y)^{-3} = -\dfrac{2y}{(x+y)^3}$, $z_{xy} = \dfrac{1(x+y)^2 - y(2)(x+y)}{[(x+y)^2]^2} = \dfrac{x+y-2y}{(x+y)^3} = \dfrac{x-y}{(x+y)^3}$,

$z_{yx} = -\dfrac{1(x+y)^2 - x(2)(x+y)}{[(x+y)^2]^2} = -\dfrac{-x^2+xy+y^2}{(x+y)^2} = \dfrac{(x+y)(x-y)}{(x+y)^2} = \dfrac{x-y}{(x+y)^3}$,

$z_{yy} = -x(-2)(x+y)^{-3} = \dfrac{2x}{(x+y)^3}$.

47. $u = e^{-s}\sin t$ $\Rightarrow$ $u_s = -e^{-s}\sin t$, $u_t = e^{-s}\cos t$. Then $u_{ss} = e^{-s}\sin t$, $u_{st} = -e^{-s}\cos t$, $u_{ts} = -e^{-s}\cos t$,

$u_{tt} = -e^{-s}\sin t$.

49. $u = x\sin(x+2y)$ $\Rightarrow$ $u_x = x \cdot \cos(x+2y)(1) + \sin(x+2y) \cdot 1 = x\cos(x+2y) + \sin(x+2y)$,

$u_{xy} = x(-\sin(x+2y)(2)) + \cos(x+2y)(2) = 2\cos(x+2y) - 2x\sin(x+2y)$, $u_y = x\cos(x+2y)(2) = 2x\cos(x+2y)$,

$u_{yx} = 2x \cdot (-\sin(x+2y)(1)) + \cos(x+2y) \cdot 2 = 2\cos(x+2y) - 2x\sin(x+2y)$. Thus $u_{xy} = u_{yx}$.

51. $f(x,y) = 3xy^4 + x^3y^2$ $\Rightarrow$ $f_x = 3y^4 + 3x^2y^2$, $f_{xx} = 6xy^2$, $f_{xxy} = 12xy$ and

$f_y = 12xy^3 + 2x^3y$, $f_{yy} = 36xy^2 + 2x^3$, $f_{yyy} = 72xy$.

53. $f(x, y, z) = \cos(4x + 3y + 2z)$ ⇒

$f_x = -\sin(4x + 3y + 2z)(4) = -4\sin(4x + 3y + 2z)$,

$f_{xy} = -4\cos(4x + 3y + 2z)(3) = -12\cos(4x + 3y + 2z)$,

$f_{xyz} = -12(-\sin(4x + 3y + 2z))(2) = 24\sin(4x + 3y + 2z)$ and

$f_y = -\sin(4x + 3y + 2z)(3) = -3\sin(4x + 3y + 2z)$,

$f_{yz} = -3\cos(4x + 3y + 2z)(2) = -6\cos(4x + 3y + 2z)$,

$f_{yzz} = -6(-\sin(4x + 3y + 2z))(2) = 12\sin(4x + 3y + 2z)$.

55. $u = e^{r\theta}\sin\theta$ ⇒ $\dfrac{\partial u}{\partial \theta} = e^{r\theta}\cos\theta + \sin\theta \cdot e^{r\theta}(r) = e^{r\theta}(\cos\theta + r\sin\theta)$,

$\dfrac{\partial^2 u}{\partial r\, \partial\theta} = e^{r\theta}(\sin\theta) + (\cos\theta + r\sin\theta)e^{r\theta}(\theta) = e^{r\theta}(\sin\theta + \theta\cos\theta + r\theta\sin\theta)$,

$\dfrac{\partial^3 u}{\partial r^2\, \partial\theta} = e^{r\theta}(\theta\sin\theta) + (\sin\theta + \theta\cos\theta + r\theta\sin\theta)\cdot e^{r\theta}(\theta) = \theta e^{r\theta}(2\sin\theta + \theta\cos\theta + r\theta\sin\theta)$.

57. $u = e^{-\alpha^2 k^2 t}\sin kx$ ⇒ $u_x = ke^{-\alpha^2 k^2 t}\cos kx$, $u_{xx} = -k^2 e^{-\alpha^2 k^2 t}\sin kx$, and $u_t = -\alpha^2 k^2 e^{-\alpha^2 k^2 t}\sin kx$.

Thus $\alpha^2 u_{xx} = u_t$.

59. $u = \dfrac{1}{\sqrt{x^2 + y^2 + z^2}}$ ⇒ $u_x = \left(-\tfrac{1}{2}\right)(x^2 + y^2 + z^2)^{-3/2}(2x) = -x(x^2 + y^2 + z^2)^{-3/2}$ and

$u_{xx} = -(x^2 + y^2 + z^2)^{-3/2} - x\left(-\tfrac{3}{2}\right)(x^2 + y^2 + z^2)^{-5/2}(2x) = \dfrac{2x^2 - y^2 - z^2}{(x^2 + y^2 + z^2)^{5/2}}$.

By symmetry, $u_{yy} = \dfrac{2y^2 - x^2 - z^2}{(x^2 + y^2 + z^2)^{5/2}}$ and $u_{zz} = \dfrac{2z^2 - x^2 - y^2}{(x^2 + y^2 + z^2)^{5/2}}$.

Thus $u_{xx} + u_{yy} + u_{zz} = \dfrac{2x^2 - y^2 - z^2 + 2y^2 - x^2 - z^2 + 2z^2 - x^2 - y^2}{(x^2 + y^2 + z^2)^{5/2}} = 0$.

61. Let $v = x + at$, $w = x - at$. Then $u_t = \dfrac{\partial[f(v) + g(w)]}{\partial t} = \dfrac{df(v)}{dv}\dfrac{\partial v}{\partial t} + \dfrac{dg(w)}{dw}\dfrac{\partial w}{\partial t} = af'(v) - ag'(w)$ and

$u_{tt} = \dfrac{\partial[af'(v) - ag'(w)]}{\partial t} = a[af''(v) + ag''(w)] = a^2[f''(v) + g''(w)]$. Similarly, by using the Chain Rule we have

$u_x = f'(v) + g'(w)$ and $u_{xx} = f''(v) + g''(w)$. Thus $u_{tt} = a^2 u_{xx}$.

63. $z_x = e^y + ye^x$, $z_{xx} = ye^x$, $\partial^3 z/\partial x^3 = ye^x$. By symmetry $z_y = xe^y + e^x$, $z_{yy} = xe^y$, $\partial^3 z/\partial y^3 = xe^y$.

Then $\partial^3 z/\partial x\partial y^2 = e^y$ and $\partial^3 z/\partial x^2\partial y = e^x$. Thus $z = xe^y + ye^x$ satisfies the given partial differential equation.

65. By the Chain Rule, taking the partial derivative of both sides with respect to R_1 gives

$$\frac{\partial R^{-1}}{\partial R}\frac{\partial R}{\partial R_1} = \frac{\partial\left[(1/R_1)+(1/R_2)+(1/R_3)\right]}{\partial R_1} \text{ or } -R^{-2}\frac{\partial R}{\partial R_1} = -R_1^{-2}. \text{ Thus } \frac{\partial R}{\partial R_1} = \frac{R^2}{R_1^2}.$$

67. By Exercise 66, $PV = mRT$ $\Rightarrow$ $P = \dfrac{mRT}{V}$, so $\dfrac{\partial P}{\partial T} = \dfrac{mR}{V}$. Also, $PV = mRT$ $\Rightarrow$ $V = \dfrac{mRT}{P}$

and $\dfrac{\partial V}{\partial T} = \dfrac{mR}{P}$. Since $T = \dfrac{PV}{mR}$, we have $T\dfrac{\partial P}{\partial T}\dfrac{\partial V}{\partial T} = \dfrac{PV}{mR}\cdot\dfrac{mR}{V}\cdot\dfrac{mR}{P} = mR$.

69. $\dfrac{\partial K}{\partial m} = \frac{1}{2}v^2$, $\dfrac{\partial K}{\partial v} = mv$, $\dfrac{\partial^2 K}{\partial v^2} = m$. Thus $\dfrac{\partial K}{\partial m}\cdot\dfrac{\partial^2 K}{\partial v^2} = \frac{1}{2}v^2 m = K$.

71. $f_x(x,y) = x + 4y$ $\Rightarrow$ $f_{xy}(x,y) = 4$ and $f_y(x,y) = 3x - y$ $\Rightarrow$ $f_{yx}(x,y) = 3$. Since f_{xy} and f_{yx} are continuous everywhere but $f_{xy}(x,y) \neq f_{yx}(x,y)$, Clairaut's Theorem implies that such a function $f(x,y)$ does not exist.

73. By the geometry of partial derivatives, the slope of the tangent line is $f_x(1,2)$. By implicit differentiation of

$4x^2 + 2y^2 + z^2 = 16$, we get $8x + 2z\,(\partial z/\partial x) = 0$ $\Rightarrow$ $\partial z/\partial x = -4x/z$, so when $x = 1$ and $z = 2$ we have

$\partial z/\partial x = -2$. So the slope is $f_x(1,2) = -2$. Thus the tangent line is given by $z - 2 = -2(x-1)$, $y = 2$. Taking the

parameter to be $t = x - 1$, we can write parametric equations for this line: $x = 1 + t$, $y = 2$, $z = 2 - 2t$.

75. By Clairaut's Theorem, $f_{xyy} = (f_{xy})_y = (f_{yx})_y = f_{yxy} = (f_y)_{xy} = (f_y)_{yx} = f_{yyx}$.

77. Let $g(x) = f(x,0) = x(x^2)^{-3/2}e^0 = x\,|x|^{-3}$. But we are using the point $(1,0)$, so near $(1,0)$, $g(x) = x^{-2}$. Then

$g'(x) = -2x^{-3}$ and $g'(1) = -2$, so using (1) we have $f_x(1,0) = g'(1) = -2$.

79. (a)

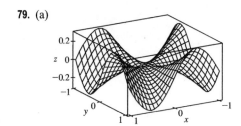

(b) For $(x,y) \neq (0,0)$, $f_x(x,y) = \dfrac{(3x^2 y - y^3)(x^2+y^2)-(x^3 y - xy^3)(2x)}{(x^2+y^2)^2} = \dfrac{x^4 y + 4x^2 y^3 - y^5}{(x^2+y^2)^2}$, and by symmetry

$f_y(x,y) = \dfrac{x^5 - 4x^3 y^2 - xy^4}{(x^2+y^2)^2}$.

(c) $f_x(0,0) = \lim\limits_{h\to 0}\dfrac{f(h,0)-f(0,0)}{h} = \lim\limits_{h\to 0}\dfrac{(0/h^2)-0}{h} = 0$ and $f_y(0,0) = \lim\limits_{h\to 0}\dfrac{f(0,h)-f(0,0)}{h} = 0$.

(d) By (3), $f_{xy}(0,0) = \dfrac{\partial f_x}{\partial y} = \lim\limits_{h\to 0}\dfrac{f_x(0,h)-f_x(0,0)}{h} = \lim\limits_{h\to 0}\dfrac{(-h^5-0)/h^4}{h} = -1$ while by (2),

$f_{yx}(0,0) = \dfrac{\partial f_y}{\partial x} = \lim\limits_{h\to 0}\dfrac{f_y(h,0)-f_y(0,0)}{h} = \lim\limits_{h\to 0}\dfrac{h^5/h^4}{h} = 1$.

(e) For $(x, y) \neq (0, 0)$, we use a CAS to compute

$$f_{xy}(x, y) = \frac{x^6 + 9x^4y^2 - 9x^2y^4 - y^6}{(x^2 + y^2)^3}.$$

Now as $(x, y) \to (0, 0)$ along the x-axis, $f_{xy}(x, y) \to 1$ while as

$(x, y) \to (0, 0)$ along the y-axis, $f_{xy}(x, y) \to -1$. Thus f_{xy} isn't

continuous at $(0, 0)$ and Clairaut's Theorem doesn't apply, so there is

no contradiction. The graphs of f_{xy} and f_{yx} are identical except at

the origin, where we observe the discontinuity.

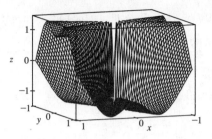

11.4 Tangent Planes and Linear Approximations

1. $z = f(x, y) = 4x^2 - y^2 + 2y \Rightarrow f_x(x, y) = 8x, f_y(x, y) = -2y + 2$, so $f_x(-1, 2) = -8, f_y(-1, 2) = -2$.

By Equation 2, an equation of the tangent plane is $z - 4 = f_x(-1, 2)[x - (-1)] + f_y(-1, 2)(y - 2) \Rightarrow$

$z - 4 = -8(x + 1) - 2(y - 2)$ or $z = -8x - 2y$.

3. $z = f(x, y) = \sqrt{4 - x^2 - 2y^2} \Rightarrow f_x(x, y) = \frac{1}{2}(4 - x^2 - 2y^2)^{-1/2}(-2x) = -\frac{x}{\sqrt{4 - x^2 - 2y^2}}$,

$f_y(x, y) = \frac{1}{2}(4 - x^2 - 2y^2)^{-1/2}(-4y) = -\frac{2y}{\sqrt{4 - x^2 - 2y^2}}$, so $f_x(1, -1) = -1$ and $f_y(1, -1) = 2$. Thus, an equation

of the tangent plane is $z - 1 = f_x(1, -1)(x - 1) + f_y(1, -1)[y - (-1)] \Rightarrow z - 1 = -1(x - 1) + 2(y + 1)$

or $x - 2y + z = 4$.

5. $z = f(x, y) = y\cos(x - y) \Rightarrow f_x = y(-\sin(x - y)(1)) = -y\sin(x - y)$,

$f_y = y(-\sin(x - y)(-1)) + \cos(x - y) = y\sin(x - y) + \cos(x - y)$, so $f_x(2, 2) = -2\sin(0) = 0$,

$f_y(2, 2) = 2\sin(0) + \cos(0) = 1$ and an equation of the tangent plane is $z - 2 = 0(x - 2) + 1(y - 2)$ or $z = y$.

7. $z = f(x, y) = x^2 + xy + 3y^2$, so $f_x(x, y) = 2x + y \Rightarrow f_x(1, 1) = 3, f_y(x, y) = x + 6y \Rightarrow f_y(1, 1) = 7$ and an

equation of the tangent plane is $z - 5 = 3(x - 1) + 7(y - 1)$ or $z = 3x + 7y - 5$. After zooming in, the surface and the

tangent plane become almost indistinguishable. (Here, the tangent plane is below the surface.) If we zoom in farther, the

surface and the tangent plane will appear to coincide.

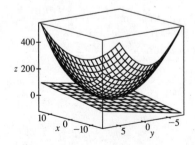
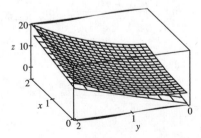

9. $f(x, y) = \dfrac{xy \sin (x - y)}{1 + x^2 + y^2}$. A CAS gives $f_x (x, y) = \dfrac{y \sin (x - y) + xy \cos (x - y)}{1 + x^2 + y^2} - \dfrac{2x^2 y \sin (x - y)}{(1 + x^2 + y^2)^2}$ and

$f_y (x, y) = \dfrac{x \sin (x - y) - xy \cos (x - y)}{1 + x^2 + y^2} - \dfrac{2xy^2 \sin (x - y)}{(1 + x^2 + y^2)^2}$. We use the CAS to evaluate these at $(1, 1)$, and then

substitute the results into Equation 2 to compute an equation of the tangent plane: $z = \frac{1}{3}x - \frac{1}{3}y$. The surface and tangent

plane are shown in the first graph below. After zooming in, the surface and the tangent plane become almost indistinguishable,

as shown in the second graph. (Here, the tangent plane is shown with fewer traces than the surface.) If we zoom in farther, the

surface and the tangent plane will appear to coincide.

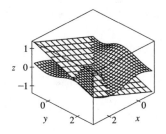
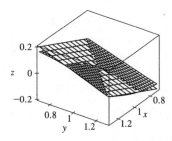

11. $f(x, y) = x \sqrt{y}$.   The partial derivatives are $f_x(x, y) = \sqrt{y}$ and $f_y(x, y) = \dfrac{x}{2 \sqrt{y}}$, so $f_x(1, 4) = 2$ and $f_y(1, 4) = \frac{1}{4}$. Both

f_x and f_y are continuous functions for $y > 0$, so by Theorem 8, f is differentiable at $(1, 4)$. By Equation 3, the linearization of

f at $(1, 4)$ is given by $L(x, y) = f(1, 4) + f_x(1, 4)(x - 1) + f_y(1, 4)(y - 4) = 2 + 2(x - 1) + \frac{1}{4}(y - 4) = 2x + \frac{1}{4}y - 1$.

13. $f(x, y) = \tan^{-1}(x + 2y)$.   The partial derivatives are $f_x(x, y) = \dfrac{1}{1 + (x + 2y)^2}$ and $f_y(x, y) = \dfrac{2}{1 + (x + 2y)^2}$, so

$f_x(1, 0) = \frac{1}{2}$ and $f_y(1, 0) = 1$. Both f_x and f_y are continuous functions, so f is differentiable at $(1, 0)$, and the linearization

of f at $(1, 0)$ is $L(x, y) = f(1, 0) + f_x(1, 0)(x - 1) + f_y(1, 0)(y - 0) = \frac{\pi}{4} + \frac{1}{2}(x - 1) + 1(y) = \frac{1}{2}x + y + \frac{\pi}{4} - \frac{1}{2}$.

15. $f(x, y) = \sqrt{20 - x^2 - 7y^2}$ $\Rightarrow$ $f_x(x, y) = -\dfrac{x}{\sqrt{20 - x^2 - 7y^2}}$ and $f_y(x, y) = -\dfrac{7y}{\sqrt{20 - x^2 - 7y^2}}$,

so $f_x(2, 1) = -\frac{2}{3}$ and $f_y(2, 1) = -\frac{7}{3}$. Then the linear approximation of f at $(2, 1)$ is given by

$f(x, y) \approx f(2, 1) + f_x(2, 1)(x - 2) + f_y(2, 1)(y - 1) = 3 - \frac{2}{3}(x - 2) - \frac{7}{3}(y - 1) = -\frac{2}{3}x - \frac{7}{3}y + \frac{20}{3}$.

Thus $f(1.95, 1.08) \approx -\frac{2}{3}(1.95) - \frac{7}{3}(1.08) + \frac{20}{3} = 2.84\overline{6}$.

17. $f(x, y, z) = \sqrt{x^2 + y^2 + z^2}$ $\Rightarrow$ $f_x(x, y, z) = \dfrac{x}{\sqrt{x^2 + y^2 + z^2}}$, $f_y(x, y, z) = \dfrac{y}{\sqrt{x^2 + y^2 + z^2}}$, and

$f_z(x, y, z) = \dfrac{z}{\sqrt{x^2 + y^2 + z^2}}$, so $f_x(3, 2, 6) = \frac{3}{7}$, $f_y(3, 2, 6) = \frac{2}{7}$, $f_z(3, 2, 6) = \frac{6}{7}$. Then the linear approximation of f at

$(3, 2, 6)$ is given by

$$f(x, y, z) \approx f(3, 2, 6) + f_x(3, 2, 6)(x - 3) + f_y(3, 2, 6)(y - 2) + f_z(3, 2, 6)(z - 6)$$
$$= 7 + \frac{3}{7}(x - 3) + \frac{2}{7}(y - 2) + \frac{6}{7}(z - 6) = \frac{3}{7}x + \frac{2}{7}y + \frac{6}{7}z$$

Thus $\sqrt{(3.02)^2 + (1.97)^2 + (5.99)^2} = f(3.02, 1.97, 5.99) \approx \frac{3}{7}(3.02) + \frac{2}{7}(1.97) + \frac{6}{7}(5.99) \approx 6.9914$.

19. $z = x^3 \ln(y^2) \quad \Rightarrow \quad dz = \dfrac{\partial z}{\partial x} dx + \dfrac{\partial z}{\partial y} dy = 3x^2 \ln(y^2)\, dx + x^3 \cdot \dfrac{1}{y^2}(2y)\, dy = 3x^2 \ln(y^2)\, dx + \dfrac{2x^3}{y} dy$

21. $R = \alpha\beta^2 \cos\gamma \quad \Rightarrow \quad dR = \dfrac{\partial R}{\partial \alpha} d\alpha + \dfrac{\partial R}{\partial \beta} d\beta + \dfrac{\partial R}{\partial \gamma} d\gamma = \beta^2 \cos\gamma\, d\alpha + 2\alpha\beta \cos\gamma\, d\beta - \alpha\beta^2 \sin\gamma\, d\gamma$

23. $dx = \Delta x = 0.05$, $dy = \Delta y = 0.1$, $z = 5x^2 + y^2$, $z_x = 10x$, $z_y = 2y$. Thus when $x = 1$ and $y = 2$,

$dz = z_x(1,2)\, dx + z_y(1,2)\, dy = (10)(0.05) + (4)(0.1) = 0.9$ while

$\Delta z = f(1.05, 2.1) - f(1,2) = 5(1.05)^2 + (2.1)^2 - 5 - 4 = 0.9225$.

25. $dA = \dfrac{\partial A}{\partial x} dx + \dfrac{\partial A}{\partial y} dy = y\, dx + x\, dy$ and $|\Delta x| \le 0.1$, $|\Delta y| \le 0.1$. We use $dx = 0.1$, $dy = 0.1$ with $x = 30$, $y = 24$; then

the maximum error in the area is about $dA = 24(0.1) + 30(0.1) = 5.4\text{ cm}^2$.

27. The volume of a can is $V = \pi r^2 h$ and $\Delta V \approx dV$ is an estimate of the amount of tin. Here $dV = 2\pi r h\, dr + \pi r^2\, dh$, so put

$dr = 0.04$, $dh = 0.08$ (0.04 on top, 0.04 on bottom) and then $\Delta V \approx dV = 2\pi(48)(0.04) + \pi(16)(0.08) \approx 16.08\text{ cm}^3$.

Thus the amount of tin is about 16 cm^3.

29. The errors in measurement are at most 2%, so $\left|\dfrac{\Delta w}{w}\right| \le 0.02$ and $\left|\dfrac{\Delta h}{h}\right| \le 0.02$. The relative error in the calculated surface

area is

$$\frac{\Delta S}{S} \approx \frac{dS}{S} = \frac{0.1091(0.425w^{0.425-1})h^{0.725}\, dw + 0.1091w^{0.425}(0.725h^{0.725-1})\, dh}{0.1091w^{0.425}h^{0.725}} = 0.425\frac{dw}{w} + 0.725\frac{dh}{h}$$

To estimate the maximum relative error, we use $\dfrac{dw}{w} = \left|\dfrac{\Delta w}{w}\right| = 0.02$ and $\dfrac{dh}{h} = \left|\dfrac{\Delta h}{h}\right| = 0.02 \quad \Rightarrow$

$\dfrac{dS}{S} = 0.425\,(0.02) + 0.725\,(0.02) = 0.023$. Thus the maximum percentage error is approximately 2.3%.

31. First we find $\dfrac{\partial R}{\partial R_1}$ implicitly by taking partial derivatives of both sides with respect to R_1:

$$\frac{\partial}{\partial R_1}\left(\frac{1}{R}\right) = \frac{\partial\,[(1/R_1) + (1/R_2) + (1/R_3)]}{\partial R_1} \quad \Rightarrow \quad -R^{-2}\frac{\partial R}{\partial R_1} = -R_1^{-2} \quad \Rightarrow \quad \frac{\partial R}{\partial R_1} = \frac{R^2}{R_1^2}.\ \text{Then by symmetry,}$$

$\dfrac{\partial R}{\partial R_2} = \dfrac{R^2}{R_2^2}$, $\dfrac{\partial R}{\partial R_3} = \dfrac{R^2}{R_3^2}$. When $R_1 = 25$, $R_2 = 40$ and $R_3 = 50$, $\dfrac{1}{R} = \dfrac{17}{200} \quad \Leftrightarrow \quad R = \frac{200}{17}\ \Omega$. Since the possible error

for each R_i is 0.5%, the maximum error of R is attained by setting $\Delta R_i = 0.005 R_i$. So

$$\Delta R \approx dR = \frac{\partial R}{\partial R_1}\Delta R_1 + \frac{\partial R}{\partial R_2}\Delta R_2 + \frac{\partial R}{\partial R_3}\Delta R_3 = (0.005)R^2\left(\frac{1}{R_1} + \frac{1}{R_2} + \frac{1}{R_3}\right) = (0.005)R = \tfrac{1}{17} \approx 0.059\ \Omega.$$

33. $\Delta z = f(a + \Delta x, b + \Delta y) - f(a,b) = (a + \Delta x)^2 + (b + \Delta y)^2 - (a^2 + b^2)$

$= a^2 + 2a\,\Delta x + (\Delta x)^2 + b^2 + 2b\,\Delta y + (\Delta y)^2 - a^2 - b^2 = 2a\,\Delta x + (\Delta x)^2 + 2b\,\Delta y + (\Delta y)^2$

But $f_x(a,b) = 2a$ and $f_y(a,b) = 2b$ and so $\Delta z = f_x(a,b)\,\Delta x + f_y(a,b)\,\Delta y + \Delta x\,\Delta x + \Delta y\,\Delta y$, which is Definition 7

with $\varepsilon_1 = \Delta x$ and $\varepsilon_2 = \Delta y$. Hence f is differentiable.

35. To show that f is continuous at (a, b) we need to show that $\displaystyle\lim_{(x,y)\to(a,b)} f(x, y) = f(a, b)$ or

equivalently $\displaystyle\lim_{(\Delta x, \Delta y)\to(0,0)} f(a + \Delta x, b + \Delta y) = f(a, b)$. Since f is differentiable at (a, b),

$f(a + \Delta x, b + \Delta y) - f(a, b) = \Delta z = f_x(a, b)\,\Delta x + f_y(a, b)\,\Delta y + \varepsilon_1\,\Delta x + \varepsilon_2\,\Delta y$, where ε_1 and $\varepsilon_2 \to 0$ as

$(\Delta x, \Delta y) \to (0, 0)$. Thus $f(a + \Delta x, b + \Delta y) = f(a, b) + f_x(a, b)\,\Delta x + f_y(a, b)\,\Delta y + \varepsilon_1\,\Delta x + \varepsilon_2\,\Delta y$. Taking the limit of

both sides as $(\Delta x, \Delta y) \to (0, 0)$ gives $\displaystyle\lim_{(\Delta x, \Delta y)\to(0,0)} f(a + \Delta x, b + \Delta y) = f(a, b)$. Thus f is continuous at (a, b).

11.5   The Chain Rule

1. $z = \sin x \cos y$, $x = \pi t$, $y = \sqrt{t}$ $\Rightarrow$

$$\frac{dz}{dt} = \frac{\partial z}{\partial x}\frac{dx}{dt} + \frac{\partial z}{\partial y}\frac{dy}{dt} = \cos x \cos y \cdot \pi + \sin x\,(-\sin y) \cdot \tfrac{1}{2}t^{-1/2} = \pi \cos x \cos y - \frac{1}{2\sqrt{t}}\sin x \sin y$$

3. $w = xe^{y/z}$, $x = t^2$, $y = 1 - t$, $z = 1 + 2t$ $\Rightarrow$

$$\frac{dw}{dt} = \frac{\partial w}{\partial x}\frac{dx}{dt} + \frac{\partial w}{\partial y}\frac{dy}{dt} + \frac{\partial w}{\partial z}\frac{dz}{dt} = e^{y/z} \cdot 2t + xe^{y/z}\left(\frac{1}{z}\right) \cdot (-1) + xe^{y/z}\left(-\frac{y}{z^2}\right) \cdot 2 = e^{y/z}\left(2t - \frac{x}{z} - \frac{2xy}{z^2}\right)$$

5. $z = x^2 + xy + y^2$, $x = s + t$, $y = st$ $\Rightarrow$

$$\frac{\partial z}{\partial s} = \frac{\partial z}{\partial x}\frac{\partial x}{\partial s} + \frac{\partial z}{\partial y}\frac{\partial y}{\partial s} = (2x + y)(1) + (x + 2y)(t) = 2x + y + xt + 2yt$$

$$\frac{\partial z}{\partial t} = \frac{\partial z}{\partial x}\frac{\partial x}{\partial t} + \frac{\partial z}{\partial y}\frac{\partial y}{\partial t} = (2x + y)(1) + (x + 2y)(s) = 2x + y + xs + 2ys$$

7. $z = e^r \cos\theta$, $r = st$, $\theta = \sqrt{s^2 + t^2}$ $\Rightarrow$

$$\frac{\partial z}{\partial s} = \frac{\partial z}{\partial r}\frac{\partial r}{\partial s} + \frac{\partial z}{\partial \theta}\frac{\partial \theta}{\partial s} = e^r \cos\theta \cdot t + e^r(-\sin\theta) \cdot \tfrac{1}{2}(s^2 + t^2)^{-1/2}(2s) = te^r \cos\theta - e^r \sin\theta \cdot \frac{s}{\sqrt{s^2 + t^2}}$$

$$= e^r\left(t\cos\theta - \frac{s}{\sqrt{s^2 + t^2}}\sin\theta\right)$$

$$\frac{\partial z}{\partial t} = \frac{\partial z}{\partial r}\frac{\partial r}{\partial t} + \frac{\partial z}{\partial \theta}\frac{\partial \theta}{\partial t} = e^r \cos\theta \cdot s + e^r(-\sin\theta) \cdot \tfrac{1}{2}(s^2 + t^2)^{-1/2}(2t) = se^r \cos\theta - e^r \sin\theta \cdot \frac{t}{\sqrt{s^2 + t^2}}$$

$$= e^r\left(s\cos\theta - \frac{t}{\sqrt{s^2 + t^2}}\sin\theta\right)$$

9. When $t = 3$, $x = g(3) = 2$ and $y = h(3) = 7$. By the Chain Rule (2),

$$\frac{dz}{dt} = \frac{\partial f}{\partial x}\frac{dx}{dt} + \frac{\partial f}{\partial y}\frac{dy}{dt} = f_x(2, 7)g'(3) + f_y(2, 7)\,h'(3) = (6)(5) + (-8)(-4) = 62.$$

11. $g(u,v) = f(x(u,v), y(u,v))$ where $x = e^u + \sin v$, $y = e^u + \cos v$ $\Rightarrow$

$\dfrac{\partial x}{\partial u} = e^u$, $\dfrac{\partial x}{\partial v} = \cos v$, $\dfrac{\partial y}{\partial u} = e^u$, $\dfrac{\partial y}{\partial v} = -\sin v$. By the Chain Rule (3), $\dfrac{\partial g}{\partial u} = \dfrac{\partial f}{\partial x}\dfrac{\partial x}{\partial u} + \dfrac{\partial f}{\partial y}\dfrac{\partial y}{\partial u}$. Then

$g_u(0,0) = f_x(x(0,0), y(0,0))\,x_u(0,0) + f_y(x(0,0), y(0,0))\,y_u(0,0) = f_x(1,2)(e^0) + f_y(1,2)(e^0) = 2(1) + 5(1) = 7.$

Similarly, $\dfrac{\partial g}{\partial v} = \dfrac{\partial f}{\partial x}\dfrac{\partial x}{\partial v} + \dfrac{\partial f}{\partial y}\dfrac{\partial y}{\partial v}$. Then

$g_v(0,0) = f_x(x(0,0), y(0,0))\,x_v(0,0) + f_y(x(0,0), y(0,0))\,y_v(0,0) = f_x(1,2)(\cos 0) + f_y(1,2)(-\sin 0)$

$\qquad = 2(1) + 5(0) = 2$

13.

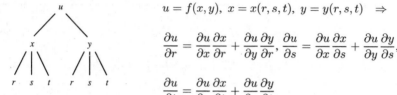

$u = f(x,y)$, $x = x(r,s,t)$, $y = y(r,s,t)$ $\Rightarrow$

$\dfrac{\partial u}{\partial r} = \dfrac{\partial u}{\partial x}\dfrac{\partial x}{\partial r} + \dfrac{\partial u}{\partial y}\dfrac{\partial y}{\partial r}$, $\dfrac{\partial u}{\partial s} = \dfrac{\partial u}{\partial x}\dfrac{\partial x}{\partial s} + \dfrac{\partial u}{\partial y}\dfrac{\partial y}{\partial s}$,

$\dfrac{\partial u}{\partial t} = \dfrac{\partial u}{\partial x}\dfrac{\partial x}{\partial t} + \dfrac{\partial u}{\partial y}\dfrac{\partial y}{\partial t}$

15.

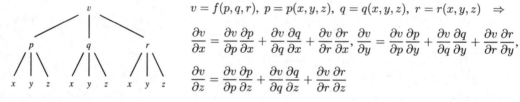

$v = f(p,q,r)$, $p = p(x,y,z)$, $q = q(x,y,z)$, $r = r(x,y,z)$ $\Rightarrow$

$\dfrac{\partial v}{\partial x} = \dfrac{\partial v}{\partial p}\dfrac{\partial p}{\partial x} + \dfrac{\partial v}{\partial q}\dfrac{\partial q}{\partial x} + \dfrac{\partial v}{\partial r}\dfrac{\partial r}{\partial x}$, $\dfrac{\partial v}{\partial y} = \dfrac{\partial v}{\partial p}\dfrac{\partial p}{\partial y} + \dfrac{\partial v}{\partial q}\dfrac{\partial q}{\partial y} + \dfrac{\partial v}{\partial r}\dfrac{\partial r}{\partial y}$,

$\dfrac{\partial v}{\partial z} = \dfrac{\partial v}{\partial p}\dfrac{\partial p}{\partial z} + \dfrac{\partial v}{\partial q}\dfrac{\partial q}{\partial z} + \dfrac{\partial v}{\partial r}\dfrac{\partial r}{\partial z}$

17. $z = x^2 + xy^3$, $x = uv^2 + w^3$, $y = u + ve^w$ $\Rightarrow$ $\dfrac{\partial z}{\partial u} = \dfrac{\partial z}{\partial x}\dfrac{\partial x}{\partial u} + \dfrac{\partial z}{\partial y}\dfrac{\partial y}{\partial u} = (2x + y^3)(v^2) + (3xy^2)(1),$

$\dfrac{\partial z}{\partial v} = \dfrac{\partial z}{\partial x}\dfrac{\partial x}{\partial v} + \dfrac{\partial z}{\partial y}\dfrac{\partial y}{\partial v} = (2x + y^3)(2uv) + (3xy^2)(e^w)$, $\dfrac{\partial z}{\partial w} = \dfrac{\partial z}{\partial x}\dfrac{\partial x}{\partial w} + \dfrac{\partial z}{\partial y}\dfrac{\partial y}{\partial w} = (2x + y^3)(3w^2) + (3xy^2)(ve^w).$

When $u = 2$, $v = 1$, and $w = 0$, we have $x = 2$, $y = 3$, so $\dfrac{\partial z}{\partial u} = (31)(1) + (54)(1) = 85,$

$\dfrac{\partial z}{\partial v} = (31)(4) + (54)(1) = 178$, $\dfrac{\partial z}{\partial w} = (31)(0) + (54)(1) = 54.$

19. $R = \ln(u^2 + v^2 + w^2)$, $u = x + 2y$, $v = 2x - y$, $w = 2xy$ $\Rightarrow$

$\dfrac{\partial R}{\partial x} = \dfrac{\partial R}{\partial u}\dfrac{\partial u}{\partial x} + \dfrac{\partial R}{\partial v}\dfrac{\partial v}{\partial x} + \dfrac{\partial R}{\partial w}\dfrac{\partial w}{\partial x} = \dfrac{2u}{u^2 + v^2 + w^2}(1) + \dfrac{2v}{u^2 + v^2 + w^2}(2) + \dfrac{2w}{u^2 + v^2 + w^2}(2y)$

$\quad = \dfrac{2u + 4v + 4wy}{u^2 + v^2 + w^2},$

$\dfrac{\partial R}{\partial y} = \dfrac{\partial R}{\partial u}\dfrac{\partial u}{\partial y} + \dfrac{\partial R}{\partial v}\dfrac{\partial v}{\partial y} + \dfrac{\partial R}{\partial w}\dfrac{\partial w}{\partial y} = \dfrac{2u}{u^2 + v^2 + w^2}(2) + \dfrac{2v}{u^2 + v^2 + w^2}(-1) + \dfrac{2w}{u^2 + v^2 + w^2}(2x)$

$\quad = \dfrac{4u - 2v + 4wx}{u^2 + v^2 + w^2}.$

When $x = y = 1$ we have $u = 3$, $v = 1$, and $w = 2$, so $\dfrac{\partial R}{\partial x} = \dfrac{9}{7}$ and $\dfrac{\partial R}{\partial y} = \dfrac{9}{7}.$

21. $u = x^2 + yz$, $x = pr\cos\theta$, $y = pr\sin\theta$, $z = p + r$ $\Rightarrow$

$$\frac{\partial u}{\partial p} = \frac{\partial u}{\partial x}\frac{\partial x}{\partial p} + \frac{\partial u}{\partial y}\frac{\partial y}{\partial p} + \frac{\partial u}{\partial z}\frac{\partial z}{\partial p} = (2x)(r\cos\theta) + (z)(r\sin\theta) + (y)(1) = 2xr\cos\theta + zr\sin\theta + y,$$

$$\frac{\partial u}{\partial r} = \frac{\partial u}{\partial x}\frac{\partial x}{\partial r} + \frac{\partial u}{\partial y}\frac{\partial y}{\partial r} + \frac{\partial u}{\partial z}\frac{\partial z}{\partial r} = (2x)(p\cos\theta) + (z)(p\sin\theta) + (y)(1) = 2xp\cos\theta + zp\sin\theta + y,$$

$$\frac{\partial u}{\partial \theta} = \frac{\partial u}{\partial x}\frac{\partial x}{\partial \theta} + \frac{\partial u}{\partial y}\frac{\partial y}{\partial \theta} + \frac{\partial u}{\partial z}\frac{\partial z}{\partial \theta} = (2x)(-pr\sin\theta) + (z)(pr\cos\theta) + (y)(0) = -2xpr\sin\theta + zpr\cos\theta.$$

When $p = 2$, $r = 3$, and $\theta = 0$ we have $x = 6$, $y = 0$, and $z = 5$, so $\dfrac{\partial u}{\partial p} = 36$, $\dfrac{\partial u}{\partial r} = 24$, and $\dfrac{\partial u}{\partial \theta} = 30$.

23. $\sqrt{xy} = 1 + x^2 y$, so let $F(x, y) = (xy)^{1/2} - 1 - x^2 y = 0$. Then by Equation 6

$$\frac{dy}{dx} = -\frac{F_x}{F_y} = -\frac{\frac{1}{2}(xy)^{-1/2}(y) - 2xy}{\frac{1}{2}(xy)^{-1/2}(x) - x^2} = -\frac{y - 4xy\sqrt{xy}}{x - 2x^2\sqrt{xy}} = \frac{4(xy)^{3/2} - y}{x - 2x^2\sqrt{xy}}.$$

25. $x^2 + y^2 + z^2 = 3xyz$, so let $F(x, y, z) = x^2 + y^2 + z^2 - 3xyz = 0$. Then by Equations 7

$$\frac{\partial z}{\partial x} = -\frac{F_x}{F_z} = -\frac{2x - 3yz}{2z - 3xy} = \frac{3yz - 2x}{2z - 3xy} \quad \text{and} \quad \frac{\partial z}{\partial y} = -\frac{F_y}{F_z} = -\frac{2y - 3xz}{2z - 3xy} = \frac{3xz - 2y}{2z - 3xy}.$$

27. $x - z = \arctan(yz)$, so let $F(x, y, z) = x - z - \arctan(yz) = 0$. Then

$$\frac{\partial z}{\partial x} = -\frac{F_x}{F_z} = -\frac{1}{-1 - \dfrac{1}{1 + (yz)^2}(y)} = \frac{1 + y^2 z^2}{1 + y + y^2 z^2} \quad \text{and}$$

$$\frac{\partial z}{\partial y} = -\frac{F_y}{F_z} = -\frac{-\dfrac{1}{1 + (yz)^2}(z)}{-1 - \dfrac{1}{1 + (yz)^2}(y)} = -\frac{\dfrac{z}{1 + y^2 z^2}}{\dfrac{1 + y^2 z^2 + y}{1 + y^2 z^2}} = -\frac{z}{1 + y + y^2 z^2}.$$

29. Since x and y are each functions of t, $T(x, y)$ is a function of t, so by the Chain Rule, $\dfrac{dT}{dt} = \dfrac{\partial T}{\partial x}\dfrac{dx}{dt} + \dfrac{\partial T}{\partial y}\dfrac{dy}{dt}$. After

3 seconds, $x = \sqrt{1 + t} = \sqrt{1 + 3} = 2$, $y = 2 + \frac{1}{3}t = 2 + \frac{1}{3}(3) = 3$, $\dfrac{dx}{dt} = \dfrac{1}{2\sqrt{1 + t}} = \dfrac{1}{2\sqrt{1 + 3}} = \dfrac{1}{4}$, and $\dfrac{dy}{dt} = \dfrac{1}{3}$.

Then $\dfrac{dT}{dt} = T_x(2, 3)\dfrac{dx}{dt} + T_y(2, 3)\dfrac{dy}{dt} = 4\left(\frac{1}{4}\right) + 3\left(\frac{1}{3}\right) = 2$. Thus the temperature is rising at a rate of $2°\text{C/s}$.

31. $C = 1449.2 + 4.6T - 0.055T^2 + 0.00029T^3 + 0.016D$, so $\dfrac{\partial C}{\partial T} = 4.6 - 0.11T + 0.00087T^2$ and $\dfrac{\partial C}{\partial D} = 0.016$.

According to the graph, the diver is experiencing a temperature of approximately $12.5°\text{C}$ at $t = 20$ minutes, so

$\dfrac{\partial C}{\partial T} = 4.6 - 0.11(12.5) + 0.00087(12.5)^2 \approx 3.36$. By sketching tangent lines at $t = 20$ to the graphs given, we estimate

$\dfrac{dD}{dt} \approx \dfrac{1}{2}$ and $\dfrac{dT}{dt} \approx -\dfrac{1}{10}$. Then, by the Chain Rule, $\dfrac{dC}{dt} = \dfrac{\partial C}{\partial T}\dfrac{dT}{dt} + \dfrac{\partial C}{\partial D}\dfrac{dD}{dt} \approx (3.36)\left(-\frac{1}{10}\right) + (0.016)\left(\frac{1}{2}\right) \approx -0.33$.

Thus the speed of sound experienced by the diver is decreasing at a rate of approximately 0.33 m/s per minute.

33. (a) $V = \ell w h$, so by the Chain Rule,

$$\frac{dV}{dt} = \frac{\partial V}{\partial \ell}\frac{d\ell}{dt} + \frac{\partial V}{\partial w}\frac{dw}{dt} + \frac{\partial V}{\partial h}\frac{dh}{dt} = wh\frac{d\ell}{dt} + \ell h\frac{dw}{dt} + \ell w\frac{dh}{dt} = 2\cdot 2\cdot 2 + 1\cdot 2\cdot 2 + 1\cdot 2\cdot(-3) = 6 \text{ m}^3/\text{s}.$$

(b) $S = 2(\ell w + \ell h + wh)$, so by the Chain Rule,

$$\frac{dS}{dt} = \frac{\partial S}{\partial \ell}\frac{d\ell}{dt} + \frac{\partial S}{\partial w}\frac{dw}{dt} + \frac{\partial S}{\partial h}\frac{dh}{dt} = 2(w+h)\frac{d\ell}{dt} + 2(\ell+h)\frac{dw}{dt} + 2(\ell+w)\frac{dh}{dt}$$

$$= 2(2+2)2 + 2(1+2)2 + 2(1+2)(-3) = 10 \text{ m}^2/\text{s}$$

(c) $L^2 = \ell^2 + w^2 + h^2 \;\Rightarrow\; 2L\frac{dL}{dt} = 2\ell\frac{d\ell}{dt} + 2w\frac{dw}{dt} + 2h\frac{dh}{dt} = 2(1)(2) + 2(2)(2) + 2(2)(-3) = 0 \;\Rightarrow$

$dL/dt = 0 \text{ m/s}.$

35. $\dfrac{dP}{dt} = 0.05$, $\dfrac{dT}{dt} = 0.15$, $V = 8.31\dfrac{T}{P}$ and $\dfrac{dV}{dt} = \dfrac{8.31}{P}\dfrac{dT}{dt} - 8.31\dfrac{T}{P^2}\dfrac{dP}{dt}$. Thus when $P = 20$ and $T = 320$,

$$\frac{dV}{dt} = 8.31\left[\frac{0.15}{20} - \frac{(0.05)(320)}{400}\right] \approx -0.27 \text{ L/s}.$$

37. (a) By the Chain Rule, $\dfrac{\partial z}{\partial r} = \dfrac{\partial z}{\partial x}\cos\theta + \dfrac{\partial z}{\partial y}\sin\theta$, $\dfrac{\partial z}{\partial\theta} = \dfrac{\partial z}{\partial x}(-r\sin\theta) + \dfrac{\partial z}{\partial y}r\cos\theta$.

(b) $\left(\dfrac{\partial z}{\partial r}\right)^2 = \left(\dfrac{\partial z}{\partial x}\right)^2\cos^2\theta + 2\dfrac{\partial z}{\partial x}\dfrac{\partial z}{\partial y}\cos\theta\sin\theta + \left(\dfrac{\partial z}{\partial y}\right)^2\sin^2\theta,$

$\left(\dfrac{\partial z}{\partial\theta}\right)^2 = \left(\dfrac{\partial z}{\partial x}\right)^2 r^2\sin^2\theta - 2\dfrac{\partial z}{\partial x}\dfrac{\partial z}{\partial y}r^2\cos\theta\sin\theta + \left(\dfrac{\partial z}{\partial y}\right)^2 r^2\cos^2\theta.$ Thus

$\left(\dfrac{\partial z}{\partial r}\right)^2 + \dfrac{1}{r^2}\left(\dfrac{\partial z}{\partial\theta}\right)^2 = \left[\left(\dfrac{\partial z}{\partial x}\right)^2 + \left(\dfrac{\partial z}{\partial y}\right)^2\right](\cos^2\theta + \sin^2\theta) = \left(\dfrac{\partial z}{\partial x}\right)^2 + \left(\dfrac{\partial z}{\partial y}\right)^2.$

39. Let $u = x - y$. Then $\dfrac{\partial z}{\partial x} = \dfrac{dz}{du}\dfrac{\partial u}{\partial x} = \dfrac{dz}{du}$ and $\dfrac{\partial z}{\partial y} = \dfrac{dz}{du}(-1)$. Thus $\dfrac{\partial z}{\partial x} + \dfrac{\partial z}{\partial y} = 0$.

41. Let $u = x + at$, $v = x - at$. Then $z = f(u) + g(v)$, so $\partial z/\partial u = f'(u)$ and $\partial z/\partial v = g'(v)$.

Thus $\dfrac{\partial z}{\partial t} = \dfrac{\partial z}{\partial u}\dfrac{\partial u}{\partial t} + \dfrac{\partial z}{\partial v}\dfrac{\partial v}{\partial t} = af'(u) - ag'(v)$ and

$\dfrac{\partial^2 z}{\partial t^2} = a\dfrac{\partial}{\partial t}[f'(u) - g'(v)] = a\left(\dfrac{df'(u)}{du}\dfrac{\partial u}{\partial t} - \dfrac{dg'(v)}{dv}\dfrac{\partial v}{\partial t}\right) = a^2 f''(u) + a^2 g''(v).$

Similarly $\dfrac{\partial z}{\partial x} = f'(u) + g'(v)$ and $\dfrac{\partial^2 z}{\partial x^2} = f''(u) + g''(v)$. Thus $\dfrac{\partial^2 z}{\partial t^2} = a^2\dfrac{\partial^2 z}{\partial x^2}$.

43. $\dfrac{\partial z}{\partial s} = \dfrac{\partial z}{\partial x}2s + \dfrac{\partial z}{\partial y}2r$. Then

$$\frac{\partial^2 z}{\partial r\,\partial s} = \frac{\partial}{\partial r}\left(\frac{\partial z}{\partial x}2s\right) + \frac{\partial}{\partial r}\left(\frac{\partial z}{\partial y}2r\right)$$

$$= \frac{\partial^2 z}{\partial x^2}\frac{\partial x}{\partial r}2s + \frac{\partial}{\partial y}\left(\frac{\partial z}{\partial x}\right)\frac{\partial y}{\partial r}2s + \frac{\partial z}{\partial x}\frac{\partial}{\partial r}2s + \frac{\partial^2 z}{\partial y^2}\frac{\partial y}{\partial r}2r + \frac{\partial}{\partial x}\left(\frac{\partial z}{\partial y}\right)\frac{\partial x}{\partial r}2r + \frac{\partial z}{\partial y}2$$

$$= 4rs\frac{\partial^2 z}{\partial x^2} + \frac{\partial^2 z}{\partial y\,\partial x}4s^2 + 0 + 4rs\frac{\partial^2 z}{\partial y^2} + \frac{\partial^2 z}{\partial x\,\partial y}4r^2 + 2\frac{\partial z}{\partial y}$$

By the continuity of the partials, $\dfrac{\partial^2 z}{\partial r\,\partial s} = 4rs\dfrac{\partial^2 z}{\partial x^2} + 4rs\dfrac{\partial^2 z}{\partial y^2} + (4r^2 + 4s^2)\dfrac{\partial^2 z}{\partial x\,\partial y} + 2\dfrac{\partial z}{\partial y}.$

45. $\dfrac{\partial z}{\partial r} = \dfrac{\partial z}{\partial x}\cos\theta + \dfrac{\partial z}{\partial y}\sin\theta$ and $\dfrac{\partial z}{\partial \theta} = -\dfrac{\partial z}{\partial x}\,r\sin\theta + \dfrac{\partial z}{\partial y}\,r\cos\theta$. Then

$$\frac{\partial^2 z}{\partial r^2} = \cos\theta\left(\frac{\partial^2 z}{\partial x^2}\cos\theta + \frac{\partial^2 z}{\partial y\,\partial x}\sin\theta\right) + \sin\theta\left(\frac{\partial^2 z}{\partial y^2}\sin\theta + \frac{\partial^2 z}{\partial x\,\partial y}\cos\theta\right)$$

$$= \cos^2\theta\,\frac{\partial^2 z}{\partial x^2} + 2\cos\theta\,\sin\theta\,\frac{\partial^2 z}{\partial x\,\partial y} + \sin^2\theta\,\frac{\partial^2 z}{\partial y^2}$$

and

$$\frac{\partial^2 z}{\partial \theta^2} = -r\cos\theta\,\frac{\partial z}{\partial x} + (-r\sin\theta)\left(\frac{\partial^2 z}{\partial x^2}(-r\sin\theta) + \frac{\partial^2 z}{\partial y\,\partial x}\,r\cos\theta\right)$$

$$\qquad -r\sin\theta\,\frac{\partial z}{\partial y} + r\cos\theta\left(\frac{\partial^2 z}{\partial y^2}\,r\cos\theta + \frac{\partial^2 z}{\partial x\,\partial y}(-r\sin\theta)\right)$$

$$= -r\cos\theta\,\frac{\partial z}{\partial x} - r\sin\theta\,\frac{\partial z}{\partial y} + r^2\sin^2\theta\,\frac{\partial^2 z}{\partial x^2} - 2r^2\cos\theta\,\sin\theta\,\frac{\partial^2 z}{\partial x\,\partial y} + r^2\cos^2\theta\,\frac{\partial^2 z}{\partial y^2}$$

Thus

$$\frac{\partial^2 z}{\partial r^2} + \frac{1}{r^2}\frac{\partial^2 z}{\partial \theta^2} + \frac{1}{r}\frac{\partial z}{\partial r} = (\cos^2\theta + \sin^2\theta)\frac{\partial^2 z}{\partial x^2} + (\sin^2\theta + \cos^2\theta)\frac{\partial^2 z}{\partial y^2}$$

$$\qquad -\frac{1}{r}\cos\theta\,\frac{\partial z}{\partial x} - \frac{1}{r}\sin\theta\,\frac{\partial z}{\partial y} + \frac{1}{r}\left(\cos\theta\,\frac{\partial z}{\partial x} + \sin\theta\,\frac{\partial z}{\partial y}\right)$$

$$= \frac{\partial^2 z}{\partial x^2} + \frac{\partial^2 z}{\partial y^2}\ \text{as desired.}$$

47. $F(x, y, z) = 0$ is assumed to define z as a function of x and y, that is, $z = f(x, y)$. So by (7), $\dfrac{\partial z}{\partial x} = -\dfrac{F_x}{F_z}$ since $F_z \neq 0$.

Similarly, it is assumed that $F(x, y, z) = 0$ defines x as a function of y and z, that is $x = h(x, z)$. Then $F(h(y, z), y, z) = 0$

and by the Chain Rule, $F_x\dfrac{\partial x}{\partial y} + F_y\dfrac{\partial y}{\partial y} + F_z\dfrac{\partial z}{\partial y} = 0$. But $\dfrac{\partial z}{\partial y} = 0$ and $\dfrac{\partial y}{\partial y} = 1$, so $F_x\dfrac{\partial x}{\partial y} + F_y = 0 \ \Rightarrow\ \dfrac{\partial x}{\partial y} = -\dfrac{F_y}{F_x}$.

A similar calculation shows that $\dfrac{\partial y}{\partial z} = -\dfrac{F_z}{F_y}$. Thus $\dfrac{\partial z}{\partial x}\dfrac{\partial x}{\partial y}\dfrac{\partial y}{\partial z} = \left(-\dfrac{F_x}{F_z}\right)\left(-\dfrac{F_y}{F_x}\right)\left(-\dfrac{F_z}{F_y}\right) = -1.$

11.6 Directional Derivatives and the Gradient Vector

1. $f(x, y) = \sqrt{5x - 4y} \ \Rightarrow\ f_x(x, y) = \tfrac{1}{2}(5x - 4y)^{-1/2}(5) = \dfrac{5}{2\sqrt{5x - 4y}}$ and

$f_y(x, y) = \tfrac{1}{2}(5x - 4y)^{-1/2}(-4) = -\dfrac{2}{\sqrt{5x - 4y}}$. If $\mathbf{u}$ is a unit vector in the direction of $\theta = -\tfrac{\pi}{6}$, then from Equation 6,

$D_{\mathbf{u}}\,f(4, 1) = f_x(4, 1)\cos\left(-\tfrac{\pi}{6}\right) + f_y(4, 1)\sin\left(-\tfrac{\pi}{6}\right) = \tfrac{5}{8}\cdot\tfrac{\sqrt{3}}{2} + \left(-\tfrac{1}{2}\right)\left(-\tfrac{1}{2}\right) = \tfrac{5\sqrt{3}}{16} + \tfrac{1}{4}.$

3. $f(x, y) = 5xy^2 - 4x^3 y$

(a) $\nabla f(x, y) = \langle f_x(x, y), f_y(x, y)\rangle = \langle 5y^2 - 12x^2 y, 10xy - 4x^3\rangle$

(b) $\nabla f(1, 2) = \langle 5(2)^2 - 12(1)^2(2), 10(1)(2) - 4(1)^3\rangle = \langle -4, 16\rangle$

(c) By Equation 9, $D_{\mathbf{u}}\,f(1, 2) = \nabla f(1, 2)\cdot\mathbf{u} = \langle -4, 16\rangle\cdot\left\langle \tfrac{5}{13}, \tfrac{12}{13}\right\rangle = (-4)\left(\tfrac{5}{13}\right) + (16)\left(\tfrac{12}{13}\right) = \tfrac{172}{13}.$

5. $f(x, y, z) = xe^{2yz}$

(a) $\nabla f(x, y, z) = \langle f_x(x, y, z), f_y(x, y, z), f_z(x, y, z) \rangle = \langle e^{2yz}, 2xze^{2yz}, 2xye^{2yz} \rangle$

(b) $\nabla f(3, 0, 2) = \langle 1, 12, 0 \rangle$

(c) By Equation 14, $D_{\mathbf{u}} f(3, 0, 2) = \nabla f(3, 0, 2) \cdot \mathbf{u} = \langle 1, 12, 0 \rangle \cdot \langle \frac{2}{3}, -\frac{2}{3}, \frac{1}{3} \rangle = \frac{2}{3} - \frac{24}{3} + 0 = -\frac{22}{3}$.

7. $f(x, y) = 1 + 2x \sqrt{y} \quad \Rightarrow \quad \nabla f(x, y) = \langle 2 \sqrt{y}, 2x \cdot \frac{1}{2}y^{-1/2} \rangle = \langle 2 \sqrt{y}, x/\sqrt{y} \rangle, \nabla f(3, 4) = \langle 4, \frac{3}{2} \rangle$, and a unit vector in

the direction of $\mathbf{v}$ is $\mathbf{u} = \dfrac{1}{\sqrt{4^2 + (-3)^2}} \langle 4, -3 \rangle = \langle \frac{4}{5}, -\frac{3}{5} \rangle$, so $D_{\mathbf{u}} f(3, 4) = \nabla f(3, 4) \cdot \mathbf{u} = \langle 4, \frac{3}{2} \rangle \cdot \langle \frac{4}{5}, -\frac{3}{5} \rangle = \frac{23}{10}$.

9. $g(s, t) = s^2 e^t \quad \Rightarrow \quad \nabla g(s, t) = 2se^t \mathbf{i} + s^2 e^t \mathbf{j}, \nabla g(2, 0) = 4\mathbf{i} + 4\mathbf{j}$, and a unit vector in the direction of $\mathbf{v}$ is

$\mathbf{u} = \frac{1}{\sqrt{2}}(\mathbf{i} + \mathbf{j})$, so $D_{\mathbf{u}} g(2, 0) = \nabla g(2, 0) \cdot \mathbf{u} = (4\mathbf{i} + 4\mathbf{j}) \cdot \frac{1}{\sqrt{2}}(\mathbf{i} + \mathbf{j}) = \frac{8}{\sqrt{2}} = 4 \sqrt{2}$.

11. $g(x, y, z) = (x + 2y + 3z)^{3/2} \quad \Rightarrow$

$\nabla g(x, y, z) = \langle \frac{3}{2}(x + 2y + 3z)^{1/2}(1), \frac{3}{2}(x + 2y + 3z)^{1/2}(2), \frac{3}{2}(x + 2y + 3z)^{1/2}(3) \rangle$

$\qquad = \langle \frac{3}{2}\sqrt{x + 2y + 3z}, 3 \sqrt{x + 2y + 3z}, \frac{9}{2} \sqrt{x + 2y + 3z} \rangle, \nabla g(1, 1, 2) = \langle \frac{9}{2}, 9, \frac{27}{2} \rangle,$

and a unit vector in the direction of $\mathbf{v} = 2\mathbf{j} - \mathbf{k}$ is $\mathbf{u} = \frac{2}{\sqrt{5}}\mathbf{j} - \frac{1}{\sqrt{5}}\mathbf{k}$, so

$D_{\mathbf{u}} g(1, 1, 2) = \langle \frac{9}{2}, 9, \frac{27}{2} \rangle \cdot \langle 0, \frac{2}{\sqrt{5}}, -\frac{1}{\sqrt{5}} \rangle = \frac{18}{\sqrt{5}} - \frac{27}{2\sqrt{5}} = \frac{9}{2\sqrt{5}}$.

13. $f(x, y) = \sqrt{xy} \quad \Rightarrow \quad \nabla f(x, y) = \langle \frac{1}{2}(xy)^{-1/2}(y), \frac{1}{2}(xy)^{-1/2}(x) \rangle = \langle \dfrac{y}{2 \sqrt{xy}}, \dfrac{x}{2 \sqrt{xy}} \rangle$, so $\nabla f(2, 8) = \langle 1, \frac{1}{4} \rangle$.

The unit vector in the direction of $\overrightarrow{PQ} = \langle 5 - 2, 4 - 8 \rangle = \langle 3, -4 \rangle$ is $\mathbf{u} = \langle \frac{3}{5}, -\frac{4}{5} \rangle$, so

$D_{\mathbf{u}} f(2, 8) = \nabla f(2, 8) \cdot \mathbf{u} = \langle 1, \frac{1}{4} \rangle \cdot \langle \frac{3}{5}, -\frac{4}{5} \rangle = \frac{2}{5}$.

15. $f(x, y) = y^2/x = y^2 x^{-1} \quad \Rightarrow \quad \nabla f(x, y) = \langle -y^2 x^{-2}, 2yx^{-1} \rangle = \langle -y^2/x^2, 2y/x \rangle$.

$\nabla f(2, 4) = \langle -4, 4 \rangle$, or equivalently $\langle -1, 1 \rangle$, is the direction of maximum rate of change, and the maximum rate

is $|\nabla f(2, 4)| = \sqrt{16 + 16} = 4 \sqrt{2}$.

17. $f(x, y, z) = \ln(xy^2 z^3) \quad \Rightarrow \quad \nabla f(x, y, z) = \langle \dfrac{y^2 z^3}{xy^2 z^3}, \dfrac{2xyz^3}{xy^2 z^3}, \dfrac{3xy^2 z^2}{xy^2 z^3} \rangle = \langle \dfrac{1}{x}, \dfrac{2}{y}, \dfrac{3}{z} \rangle$.

$\nabla f(1, -2, -3) = \langle 1, -1, -1 \rangle$ is the direction of maximum rate of change and the maximum rate is $|\nabla f(1, -2, -3)| = \sqrt{3}$.

19. (a) As in the proof of Theorem 15, $D_{\mathbf{u}} f = |\nabla f| \cos \theta$. Since the minimum value of $\cos \theta$ is -1 occurring when $\theta = \pi$, the

minimum value of $D_{\mathbf{u}} f$ is $-|\nabla f|$ occurring when $\theta = \pi$, that is when $\mathbf{u}$ is in the opposite direction of ∇f

(assuming $\nabla f \neq \mathbf{0}$).

(b) $f(x, y) = x^4 y - x^2 y^3 \quad \Rightarrow \quad \nabla f(x, y) = \langle 4x^3 y - 2xy^3, x^4 - 3x^2 y^2 \rangle$, so f decreases fastest at the point $(2, -3)$ in the

direction $-\nabla f(2, -3) = -\langle 12, -92 \rangle = \langle -12, 92 \rangle$.

21. The direction of fastest change is $\nabla f(x,y) = (2x-2)\,\mathbf{i} + (2y-4)\,\mathbf{j}$, so we need to find all points (x,y) where $\nabla f(x,y)$ is
parallel to $\mathbf{i} + \mathbf{j}$ $\Leftrightarrow$ $(2x-2)\,\mathbf{i} + (2y-4)\,\mathbf{j} = k\,(\mathbf{i}+\mathbf{j})$ $\Leftrightarrow$ $k = 2x-2$ and $k = 2y-4$. Then $2x-2 = 2y-4$ $\Rightarrow$
$y = x+1$, so the direction of fastest change is $\mathbf{i} + \mathbf{j}$ at all points on the line $y = x+1$.

23. $T = \dfrac{k}{\sqrt{x^2+y^2+z^2}}$ and $120 = T(1,2,2) = \dfrac{k}{3}$ so $k = 360$.

(a) $\mathbf{u} = \dfrac{\langle 1,-1,1\rangle}{\sqrt{3}}$,

$$D_{\mathbf{u}}T(1,2,2) = \nabla T(1,2,2)\cdot\mathbf{u} = \left[-360\left(x^2+y^2+z^2\right)^{-3/2}\langle x,y,z\rangle\right]_{(1,2,2)}\cdot\mathbf{u} = -\tfrac{40}{3}\langle 1,2,2\rangle\cdot\tfrac{1}{\sqrt{3}}\langle 1,-1,1\rangle = -\tfrac{40}{3\sqrt{3}}$$

(b) From (a), $\nabla T = -360\left(x^2+y^2+z^2\right)^{-3/2}\langle x,y,z\rangle$, and since $\langle x,y,z\rangle$ is the position vector of the point (x,y,z), the
vector $-\langle x,y,z\rangle$, and thus ∇T, always points toward the origin.

25. $\nabla V(x,y,z) = \langle 10x - 3y + yz,\, xz - 3x,\, xy\rangle$, $\nabla V(3,4,5) = \langle 38,6,12\rangle$

(a) $D_{\mathbf{u}}V(3,4,5) = \langle 38,6,12\rangle\cdot\tfrac{1}{\sqrt{3}}\langle 1,1,-1\rangle = \tfrac{32}{\sqrt{3}}$

(b) $\nabla V(3,4,5) = \langle 38,6,12\rangle$, or equivalently, $\langle 19,3,6\rangle$.

(c) $|\nabla V(3,4,5)| = \sqrt{38^2+6^2+12^2} = \sqrt{1624} = 2\sqrt{406}$

27. A unit vector in the direction of $\overrightarrow{AB}$ is $\mathbf{i}$ and a unit vector in the direction of $\overrightarrow{AC}$ is $\mathbf{j}$. Thus $D_{\overrightarrow{AB}}f(1,3) = f_x(1,3) = 3$ and
$D_{\overrightarrow{AC}}f(1,3) = f_y(1,3) = 26$. Therefore $\nabla f(1,3) = \langle f_x(1,3), f_y(1,3)\rangle = \langle 3,26\rangle$, and by definition,
$D_{\overrightarrow{AD}}f(1,3) = \nabla f\cdot\mathbf{u}$ where $\mathbf{u}$ is a unit vector in the direction of $\overrightarrow{AD}$, which is $\langle \tfrac{5}{13},\tfrac{12}{13}\rangle$. Therefore,
$D_{\overrightarrow{AD}}f(1,3) = \langle 3,26\rangle\cdot\langle \tfrac{5}{13},\tfrac{12}{13}\rangle = 3\cdot\tfrac{5}{13} + 26\cdot\tfrac{12}{13} = \tfrac{327}{13}$.

29. (a) $\nabla(au+bv) = \left\langle \dfrac{\partial(au+bv)}{\partial x}, \dfrac{\partial(au+bv)}{\partial y}\right\rangle = \left\langle a\dfrac{\partial u}{\partial x}+b\dfrac{\partial v}{\partial x}, a\dfrac{\partial u}{\partial y}+b\dfrac{\partial v}{\partial y}\right\rangle = a\left\langle\dfrac{\partial u}{\partial x},\dfrac{\partial u}{\partial y}\right\rangle + b\left\langle\dfrac{\partial v}{\partial x},\dfrac{\partial v}{\partial y}\right\rangle$
$= a\,\nabla u + b\,\nabla v$

(b) $\nabla(uv) = \left\langle v\dfrac{\partial u}{\partial x}+u\dfrac{\partial v}{\partial x}, v\dfrac{\partial u}{\partial y}+u\dfrac{\partial v}{\partial y}\right\rangle = v\left\langle\dfrac{\partial u}{\partial x},\dfrac{\partial u}{\partial y}\right\rangle + u\left\langle\dfrac{\partial v}{\partial x},\dfrac{\partial v}{\partial y}\right\rangle = v\,\nabla u + u\,\nabla v$

(c) $\nabla\left(\dfrac{u}{v}\right) = \left\langle \dfrac{v\dfrac{\partial u}{\partial x}-u\dfrac{\partial v}{\partial x}}{v^2}, \dfrac{v\dfrac{\partial u}{\partial y}-u\dfrac{\partial v}{\partial y}}{v^2}\right\rangle = \dfrac{v\left\langle\dfrac{\partial u}{\partial x},\dfrac{\partial u}{\partial y}\right\rangle - u\left\langle\dfrac{\partial v}{\partial x},\dfrac{\partial v}{\partial y}\right\rangle}{v^2} = \dfrac{v\,\nabla u - u\,\nabla v}{v^2}$

(d) $\nabla u^n = \left\langle \dfrac{\partial(u^n)}{\partial x}, \dfrac{\partial(u^n)}{\partial y}\right\rangle = \left\langle nu^{n-1}\dfrac{\partial u}{\partial x}, nu^{n-1}\dfrac{\partial u}{\partial y}\right\rangle = nu^{n-1}\,\nabla u$

31. Let $F(x, y, z) = x^2 - 2y^2 + z^2 + yz$. Then $x^2 - 2y^2 + z^2 + yz = 2$ is a level surface of F

and $\nabla F(x, y, z) = \langle 2x, -4y + z, 2z + y \rangle$.

(a) $\nabla F(2, 1, -1) = \langle 4, -5, -1 \rangle$ is a normal vector for the tangent plane at $(2, 1, -1)$, so an equation of the tangent plane

is $4(x - 2) - 5(y - 1) - 1(z + 1) = 0$ or $4x - 5y - z = 4$.

(b) The normal line has direction $\langle 4, -5, -1 \rangle$, so parametric equations are $x = 2 + 4t$, $y = 1 - 5t$, $z = -1 - t$, and

symmetric equations are $\dfrac{x - 2}{4} = \dfrac{y - 1}{-5} = \dfrac{z + 1}{-1}$.

33. $F(x, y, z) = -z + xe^y \cos z$ $\Rightarrow$ $\nabla F(x, y, z) = \langle e^y \cos z, xe^y \cos z, -1 - xe^y \sin z \rangle$ and $\nabla F(1, 0, 0) = \langle 1, 1, -1 \rangle$.

(a) $1(x - 1) + 1(y - 0) - 1(z - 0) = 0$ or $x + y - z = 1$

(b) $x - 1 = y = -z$

35. $F(x, y, z) = xy + yz + zx$, $\nabla F(x, y, z) = \langle y + z, x + z, y + x \rangle$, $\nabla F(1, 1, 1) = \langle 2, 2, 2 \rangle$, so an equation of the tangent

plane is $2x + 2y + 2z = 6$ or $x + y + z = 3$, and the normal line is given by $x - 1 = y - 1 = z - 1$ or $x = y = z$.

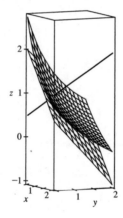

37. $\nabla f(x, y) = \langle 2x, 8y \rangle$, $\nabla f(2, 1) = \langle 4, 8 \rangle$.

The tangent line has equation $\nabla f(2, 1) \cdot \langle x - 2, y - 1 \rangle = 0$ $\Rightarrow$

$4(x - 2) + 8(y - 1) = 0$, which simplifies to $x + 2y = 4$.

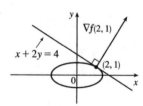

39. $\nabla F(x_0, y_0, z_0) = \left\langle \dfrac{2x_0}{a^2}, \dfrac{2y_0}{b^2}, \dfrac{2z_0}{c^2} \right\rangle$. Thus an equation of the tangent plane at (x_0, y_0, z_0) is

$\dfrac{2x_0}{a^2} x + \dfrac{2y_0}{b^2} y + \dfrac{2z_0}{c^2} z = 2\left(\dfrac{x_0^2}{a^2} + \dfrac{y_0^2}{b^2} + \dfrac{z_0^2}{c^2} \right) = 2(1) = 2$ since (x_0, y_0, z_0) is a point on the ellipsoid. Hence

$\dfrac{x_0}{a^2} x + \dfrac{y_0}{b^2} y + \dfrac{z_0}{c^2} z = 1$ is an equation of the tangent plane.

41. $\nabla f(x_0, y_0, z_0) = \langle 2x_0, -2y_0, 4z_0 \rangle$ and the given line has direction numbers $2, 4, 6$, so $\langle 2x_0, -2y_0, 4z_0 \rangle = k\langle 2, 4, 6 \rangle$ or

$x_0 = k$, $y_0 = -2k$ and $z_0 = \frac{3}{2}k$. But $x_0^2 - y_0^2 + 2z_0^2 = 1$ or $\left(1 - 4 + \frac{9}{2} \right)k^2 = 1$, so $k = \pm\sqrt{\frac{2}{3}} = \pm\frac{\sqrt{6}}{3}$ and there are two

such points: $\left(\pm\frac{\sqrt{6}}{3}, \mp\frac{2\sqrt{6}}{3}, \pm\frac{\sqrt{6}}{2} \right)$.

43. Let (x_0, y_0, z_0) be a point on the surface. Then an equation of the tangent plane at the point is

$$\frac{x}{2\sqrt{x_0}} + \frac{y}{2\sqrt{y_0}} + \frac{z}{2\sqrt{z_0}} = \frac{\sqrt{x_0} + \sqrt{y_0} + \sqrt{z_0}}{2}. \text{ But } \sqrt{x_0} + \sqrt{y_0} + \sqrt{z_0} = \sqrt{c}, \text{ so the equation is}$$

$\dfrac{x}{\sqrt{x_0}} + \dfrac{y}{\sqrt{y_0}} + \dfrac{z}{\sqrt{z_0}} = \sqrt{c}.$ The x-, y-, and z-intercepts are $\sqrt{cx_0}$, $\sqrt{cy_0}$ and $\sqrt{cz_0}$ respectively. (The x-intercept is found by

setting $y = z = 0$ and solving the resulting equation for x, and the y- and z-intercepts are found similarly.) So the sum of the

intercepts is $\sqrt{c}(\sqrt{x_0} + \sqrt{y_0} + \sqrt{z_0}) = c$, a constant.

45. If $f(x, y, z) = z - x^2 - y^2$ and $g(x, y, z) = 4x^2 + y^2 + z^2$, then the tangent line is perpendicular to both ∇f and ∇g

at $(-1, 1, 2)$. The vector $\mathbf{v} = \nabla f \times \nabla g$ will therefore be parallel to the tangent line.

We have $\nabla f(x, y, z) = \langle -2x, -2y, 1 \rangle \quad \Rightarrow \quad \nabla f(-1, 1, 2) = \langle 2, -2, 1 \rangle$, and $\nabla g(x, y, z) = \langle 8x, 2y, 2z \rangle \quad \Rightarrow$

$$\nabla g(-1, 1, 2) = \langle -8, 2, 4 \rangle. \text{ Hence } \mathbf{v} = \nabla f \times \nabla g = \begin{vmatrix} \mathbf{i} & \mathbf{j} & \mathbf{k} \\ 2 & -2 & 1 \\ -8 & 2 & 4 \end{vmatrix} = -10\,\mathbf{i} - 16\,\mathbf{j} - 12\,\mathbf{k}.$$

Parametric equations are: $x = -1 - 10t, \ y = 1 - 16t, \ z = 2 - 12t.$

47. (a) The direction of the normal line of F is given by ∇F, and that of G by ∇G. Assuming that

$\nabla F \neq 0 \neq \nabla G$, the two normal lines are perpendicular at P if $\nabla F \cdot \nabla G = 0$ at $P \quad \Leftrightarrow$

$\langle \partial F/\partial x, \partial F/\partial y, \partial F/\partial z \rangle \cdot \langle \partial G/\partial x, \partial G/\partial y, \partial G/\partial z \rangle = 0$ at $P \quad \Leftrightarrow \quad F_x G_x + F_y G_y + F_z G_z = 0$ at P.

(b) Here $F = x^2 + y^2 - z^2$ and $G = x^2 + y^2 + z^2 - r^2$, so

$\nabla F \cdot \nabla G = \langle 2x, 2y, -2z \rangle \cdot \langle 2x, 2y, 2z \rangle = 4x^2 + 4y^2 - 4z^2 = 4F = 0$, since the point (x, y, z) lies on the graph of

$F = 0$. To see that this is true without using calculus, note that $G = 0$ is the equation of a sphere centered at the origin and

$F = 0$ is the equation of a right circular cone with vertex at the origin (which is generated by lines through the origin). At

any point of intersection, the sphere's normal line (which passes through the origin) lies on the cone, and thus is

perpendicular to the cone's normal line. So the surfaces with equations $F = 0$ and $G = 0$ are everywhere orthogonal.

49. Let $\mathbf{u} = \langle a, b \rangle$ and $\mathbf{v} = \langle c, d \rangle$. Then we know that at the given point, $D_{\mathbf{u}} f = \nabla f \cdot \mathbf{u} = af_x + bf_y$ and

$D_{\mathbf{v}} f = \nabla f \cdot \mathbf{v} = cf_x + df_y$. But these are just two linear equations in the two unknowns f_x and f_y, and since $\mathbf{u}$ and $\mathbf{v}$ are

not parallel, we can solve the equations to find $\nabla f = \langle f_x, f_y \rangle$ at the given point. In fact,

$$\nabla f = \left\langle \frac{d\,D_{\mathbf{u}}\,f - b\,D_{\mathbf{v}}\,f}{ad - bc}, \frac{a\,D_{\mathbf{v}}\,f - c\,D_{\mathbf{u}}\,f}{ad - bc} \right\rangle.$$

11.7 Maximum and Minimum Values

1. (a) First we compute $D(1, 1) = f_{xx}(1, 1)\,f_{yy}(1, 1) - [f_{xy}(1, 1)]^2 = (4)(2) - (1)^2 = 7$. Since $D(1, 1) > 0$ and

$f_{xx}(1, 1) > 0$, f has a local minimum at $(1, 1)$ by the Second Derivatives Test.

(b) $D(1, 1) = f_{xx}(1, 1)\,f_{yy}(1, 1) - [f_{xy}(1, 1)]^2 = (4)(2) - (3)^2 = -1$. Since $D(1, 1) < 0$, f has a saddle point at $(1, 1)$ by

the Second Derivatives Test.

3. $f(x,y) = 9 - 2x + 4y - x^2 - 4y^2 \implies f_x = -2 - 2x, f_y = 4 - 8y,$

$f_{xx} = -2, \ f_{xy} = 0, f_{yy} = -8.$ Then $f_x = 0$ and $f_y = 0$ imply

$x = -1$ and $y = \frac{1}{2}$, and the only critical point is $\left(-1, \frac{1}{2}\right)$.

$D(x,y) = f_{xx}f_{yy} - (f_{xy})^2 = (-2)(-8) - 0^2 = 16$, and since

$D\left(-1, \frac{1}{2}\right) = 16 > 0$ and $f_{xx}\left(-1, \frac{1}{2}\right) = -2 < 0, f\left(-1, \frac{1}{2}\right) = 11$ is a

local maximum by the Second Derivatives Test.

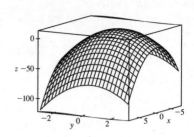

5. $f(x,y) = x^4 + y^4 - 4xy + 2 \implies f_x = 4x^3 - 4y, f_y = 4y^3 - 4x,$

$f_{xx} = 12x^2, \ f_{xy} = -4, f_{yy} = 12y^2.$ Then $f_x = 0$ implies $y = x^3$,

and substitution into $f_y = 0 \implies x = y^3$ gives $x^9 - x = 0 \implies$

$x(x^8 - 1) = 0 \implies x = 0$ or $x = \pm 1$. Thus the critical points are $(0,0)$,

$(1,1)$, and $(-1,-1)$. Now $D(0,0) = 0 \cdot 0 - (-4)^2 = -16 < 0$,

so $(0,0)$ is a saddle point. $D(1,1) = (12)(12) - (-4)^2 > 0$ and

$f_{xx}(1,1) = 12 > 0$, so $f(1,1) = 0$ is a local minimum. $D(-1,-1) = (12)(12) - (-4)^2 > 0$ and

$f_{xx} = (-1,-1) = 12 > 0$, so $f(-1,-1) = 0$ is also a local minimum.

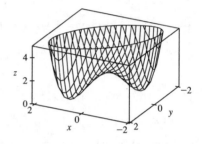

7. $f(x,y) = (1 + xy)(x + y) = x + y + x^2y + xy^2 \implies$

$f_x = 1 + 2xy + y^2, \ f_y = 1 + x^2 + 2xy, \ f_{xx} = 2y, \ f_{xy} = 2x + 2y,$

$f_{yy} = 2x.$ Then $f_x = 0$ implies $1 + 2xy + y^2 = 0$ and $f_y = 0$ implies

$1 + x^2 + 2xy = 0$. Subtracting the second equation from the first gives

$y^2 - x^2 = 0 \implies y = \pm x$, but if $y = x$ then $1 + 2xy + y^2 = 0 \implies$

$1 + 3x^2 = 0$ which has no real solution. If $y = -x$ then

$1 + 2xy + y^2 = 0 \implies 1 - x^2 = 0 \implies x = \pm 1$, so critical points

are $(1,-1)$ and $(-1,1)$. $D(1,-1) = (-2)(2) - 0 < 0$ and $D(-1,1) = (2)(-2) - 0 < 0$, so $(-1,1)$ and $(1,-1)$ are

saddle points.

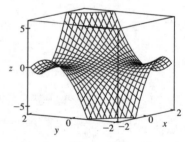

9. $f(x,y) = e^x \cos y \implies f_x = e^x \cos y, \ f_y = -e^x \sin y.$

Now $f_x = 0$ implies $\cos y = 0$ or $y = \frac{\pi}{2} + n\pi$ for n an integer.

But $\sin\left(\frac{\pi}{2} + n\pi\right) \neq 0$, so there are no critical points.

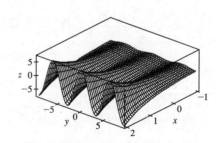

11. $f(x, y) = x \sin y \Rightarrow f_x = \sin y, \; f_y = x \cos y, \; f_{xx} = 0,$

$f_{yy} = -x \sin y, \; f_{xy} = \cos y.$ Then $f_x = 0$ if and only if $y = n\pi,$

n an integer, and substituting into $f_y = 0$ requires $x = 0$ for each of these

y-values. Thus the critical points are $(0, n\pi), \; n$ an integer. But

$D(0, n\pi) = -\cos^2(n\pi) < 0$ so each critical point is a saddle point.

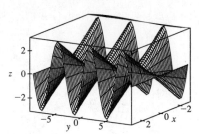

13. $f(x, y) = (x^2 + y^2)e^{y^2 - x^2} \Rightarrow$

$f_x = (x^2 + y^2)e^{y^2 - x^2}(-2x) + 2xe^{y^2 - x^2} = 2xe^{y^2 - x^2}(1 - x^2 - y^2),$

$f_y = (x^2 + y^2)e^{y^2 - x^2}(2y) + 2ye^{y^2 - x^2} = 2ye^{y^2 - x^2}(1 + x^2 + y^2),$

$f_{xx} = 2xe^{y^2 - x^2}(-2x) + (1 - x^2 - y^2)\left(2x\left(-2xe^{y^2 - x^2}\right) + 2e^{y^2 - x^2}\right) = 2e^{y^2 - x^2}((1 - x^2 - y^2)(1 - 2x^2) - 2x^2),$

$f_{xy} = 2xe^{y^2 - x^2}(-2y) + 2x(2y)e^{y^2 - x^2}(1 - x^2 - y^2) = -4xye^{y^2 - x^2}(x^2 + y^2),$

$f_{yy} = 2ye^{y^2 - x^2}(2y) + (1 + x^2 + y^2)\left(2y\left(2ye^{y^2 - x^2}\right) + 2e^{y^2 - x^2}\right) = 2e^{y^2 - x^2}((1 + x^2 + y^2)(1 + 2y^2) + 2y^2).$

$f_y = 0$ implies $y = 0$, and substituting into $f_x = 0$ gives

$2xe^{-x^2}(1 - x^2) = 0 \Rightarrow x = 0$ or $x = \pm 1$. Thus the critical points are

$(0, 0)$ and $(\pm 1, 0)$. Now $D(0, 0) = (2)(2) - 0 > 0$ and $f_{xx}(0, 0) = 2 > 0$,

so $f(0, 0) = 0$ is a local minimum. $D(\pm 1, 0) = (-4e^{-1})(4e^{-1}) - 0 < 0$

so $(\pm 1, 0)$ are saddle points.

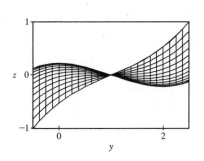

15. $f(x, y) = 3x^2y + y^3 - 3x^2 - 3y^2 + 2$

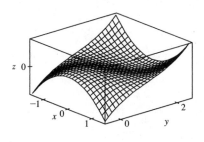

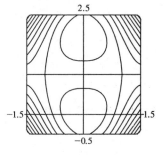

From the graphs, it appears that f has a local maximum $f(0, 0) \approx 2$ and a local minimum $f(0, 2) \approx -2$. There appear to be

saddle points near $(\pm 1, 1)$.

$f_x = 6xy - 6x, \; f_y = 3x^2 + 3y^2 - 6y$. Then $f_x = 0$ implies $x = 0$ or $y = 1$ and when $x = 0, \; f_y = 0$ implies $y = 0$

or $y = 2$; when $y = 1, \; f_y = 0$ implies $x^2 = 1$ or $x = \pm 1$. Thus the critical points are $(0, 0), \; (0, 2), \; (\pm 1, 1)$. Now

$f_{xx} = 6y - 6, \; f_{yy} = 6y - 6, \; f_{xy} = 6x$, so $D(0, 0) = D(0, 2) = 36 > 0$ while $D(\pm 1, 1) = -36 < 0$ and $f_{xx}(0, 0) = -6,$

$f_{xx}(0, 2) = 6$. Hence $(\pm 1, 1)$ are saddle points while $f(0, 0) = 2$ is a local maximum and $f(0, 2) = -2$ is a local minimum.

17. $f(x, y) = \sin x + \sin y + \sin(x + y)$, $0 \le x \le 2\pi$, $0 \le y \le 2\pi$

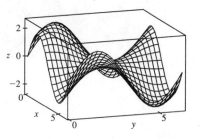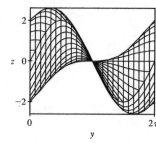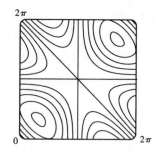

From the graphs it appears that f has a local maximum at about $(1, 1)$ with value approximately 2.6, a local minimum at about $(5, 5)$ with value approximately -2.6, and a saddle point at about $(3, 3)$.

$f_x = \cos x + \cos(x + y)$, $f_y = \cos y + \cos(x + y)$, $f_{xx} = -\sin x - \sin(x + y)$, $f_{yy} = -\sin y - \sin(x + y)$, $f_{xy} = -\sin(x + y)$. Setting $f_x = 0$ and $f_y = 0$ and subtracting gives $\cos x - \cos y = 0$ or $\cos x = \cos y$. Thus $x = y$ or $x = 2\pi - y$. If $x = y$, $f_x = 0$ becomes $\cos x + \cos 2x = 0$ or $2\cos^2 x + \cos x - 1 = 0$, a quadratic in $\cos x$. Thus $\cos x = -1$ or $\frac{1}{2}$ and $x = \pi$, $\frac{\pi}{3}$, or $\frac{5\pi}{3}$, yielding the critical points (π, π), $\left(\frac{\pi}{3}, \frac{\pi}{3}\right)$ and $\left(\frac{5\pi}{3}, \frac{5\pi}{3}\right)$. Similarly if $x = 2\pi - y$, $f_x = 0$ becomes $(\cos x) + 1 = 0$ and the resulting critical point is (π, π).

Now $D(x, y) = \sin x \sin y + \sin x \sin(x + y) + \sin y \sin(x + y)$. So $D(\pi, \pi) = 0$ and the Second Derivatives Test doesn't apply. $D\left(\frac{\pi}{3}, \frac{\pi}{3}\right) = \frac{9}{4} > 0$ and $f_{xx}\left(\frac{\pi}{3}, \frac{\pi}{3}\right) < 0$ so $f\left(\frac{\pi}{3}, \frac{\pi}{3}\right) = \frac{3\sqrt{3}}{2}$ is a local maximum while $D\left(\frac{5\pi}{3}, \frac{5\pi}{3}\right) = \frac{9}{4} > 0$ and $f_{xx}\left(\frac{5\pi}{3}, \frac{5\pi}{3}\right) > 0$, so $f\left(\frac{5\pi}{3}, \frac{5\pi}{3}\right) = -\frac{3\sqrt{3}}{2}$ is a local minimum.

19. $f(x, y) = x^4 - 5x^2 + y^2 + 3x + 2$ $\Rightarrow$ $f_x(x, y) = 4x^3 - 10x + 3$ and $f_y(x, y) = 2y$. $f_y = 0$ $\Rightarrow$ $y = 0$, and the graph of f_x shows that the roots of $f_x = 0$ are approximately $x = -1.714, 0.312$ and 1.402. (Alternatively, we could have used a calculator or a CAS to find these roots.) So to three decimal places, the critical points are $(-1.714, 0)$, $(1.402, 0)$, and $(0.312, 0)$. Now since $f_{xx} = 12x^2 - 10$, $f_{xy} = 0$, $f_{yy} = 2$, and $D = 24x^2 - 20$, we have $D(-1.714, 0) > 0$, $f_{xx}(-1.714, 0) > 0$, $D(1.402, 0) > 0$, $f_{xx}(1.402, 0) > 0$, and $D(0.312, 0) < 0$. Therefore $f(-1.714, 0) \approx -9.200$ and $f(1.402, 0) \approx 0.242$ are local minima, and $(0.312, 0)$ is a saddle point. The lowest point on the graph is approximately $(-1.714, 0, -9.200)$.

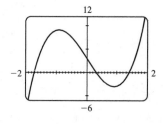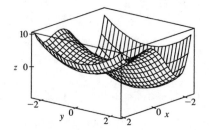

21. $f(x, y) = 2x + 4x^2 - y^2 + 2xy^2 - x^4 - y^4$ $\Rightarrow$ $f_x(x, y) = 2 + 8x + 2y^2 - 4x^3$, $f_y(x, y) = -2y + 4xy - 4y^3$.

Now $f_y = 0$ $\Leftrightarrow$ $2y(2y^2 - 2x + 1) = 0$ $\Leftrightarrow$ $y = 0$ or $y^2 = x - \frac{1}{2}$. The first of these implies that $f_x = -4x^3 + 8x + 2$, and the second implies that $f_x = 2 + 8x + 2\left(x - \frac{1}{2}\right) - 4x^3 = -4x^3 + 10x + 1$. From the graphs, we see that the first possibility for f_x has roots at approximately -1.267, -0.259, and 1.526, and the second has a root at approximately 1.629 (the negative roots do not give critical points, since $y^2 = x - \frac{1}{2}$ must be positive). So to three decimal places, f has critical points at $(-1.267, 0)$, $(-0.259, 0)$, $(1.526, 0)$, and $(1.629, \pm 1.063)$. Now since $f_{xx} = 8 - 12x^2$, $f_{xy} = 4y$, $f_{yy} = 4x - 12y^2$, and $D = (8 - 12x^2)(4x - 12y^2) - 16y^2$, we have $D(-1.267, 0) > 0$, $f_{xx}(-1.267, 0) > 0$, $D(-0.259, 0) < 0$, $D(1.526, 0) < 0$, $D(1.629, \pm 1.063) > 0$, and $f_{xx}(1.629, \pm 1.063) < 0$. Therefore, to three decimal places, $f(-1.267, 0) \approx 1.310$ and $f(1.629, \pm 1.063) \approx 8.105$ are local maxima, and $(-0.259, 0)$ and $(1.526, 0)$ are saddle points. The highest points on the graph are approximately $(1.629, \pm 1.063, 8.105)$.

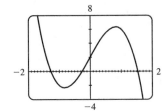
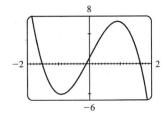
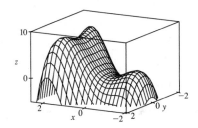

23. Since f is a polynomial it is continuous on D, so an absolute maximum and minimum exist. Here $f_x = 4$, $f_y = -5$ so there are no critical points inside D. Thus the absolute extrema must both occur on the boundary. Along L_1: $x = 0$ and $f(0, y) = 1 - 5y$ for $0 \le y \le 3$, a decreasing function in y, so the maximum value is $f(0, 0) = 1$ and the minimum value is $f(0, 3) = -14$. Along L_2: $y = 0$ and $f(x, 0) = 1 + 4x$ for $0 \le x \le 2$, an increasing function in x, so the minimum value is $f(0, 0) = 1$ and the maximum value is $f(2, 0) = 9$. Along L_3: $y = -\frac{3}{2}x + 3$ and $f\left(x, -\frac{3}{2}x + 3\right) = \frac{23}{2}x - 14$ for $0 \le x \le 2$, an increasing function in x, so the minimum value is $f(0, 3) = -14$ and the maximum value is $f(2, 0) = 9$. Thus the absolute maximum of f on D is $f(2, 0) = 9$ and the absolute minimum is $f(0, 3) = -14$.

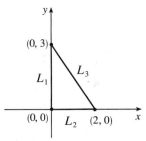

25. $f_x(x, y) = 2x + 2xy$, $f_y(x, y) = 2y + x^2$, and setting $f_x = f_y = 0$ gives $(0, 0)$ as the only critical point in D, with $f(0, 0) = 4$.

On L_1: $y = -1$, $f(x, -1) = 5$, a constant.

On L_2: $x = 1$, $f(1, y) = y^2 + y + 5$, a quadratic in y which attains its maximum at $(1, 1)$, $f(1, 1) = 7$ and its minimum at $\left(1, -\frac{1}{2}\right)$, $f\left(1, -\frac{1}{2}\right) = \frac{19}{4}$.

On L_3: $f(x, 1) = 2x^2 + 5$ which attains its maximum at $(-1, 1)$ and $(1, 1)$ with $f(\pm 1, 1) = 7$ and its minimum at $(0, 1)$, $f(0, 1) = 5$.

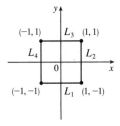

On L_4: $f(-1, y) = y^2 + y + 5$ with maximum at $(-1, 1)$, $f(-1, 1) = 7$ and minimum at $\left(-1, -\frac{1}{2}\right)$, $f\left(-1, -\frac{1}{2}\right) = \frac{19}{4}$. Thus the absolute maximum is attained at both $(\pm 1, 1)$ with $f(\pm 1, 1) = 7$ and the absolute minimum on D is attained at $(0, 0)$ with $f(0, 0) = 4$.

27. $f(x, y) = x^4 + y^4 - 4xy + 2$ is a polynomial and hence continuous on D, so it has an absolute maximum and minimum on D. In Exercise 5, we found the critical points of f; only $(1, 1)$ with $f(1, 1) = 0$ is inside D. On L_1: $y = 0$, $f(x, 0) = x^4 + 2$, $0 \le x \le 3$, a polynomial in x which attains its maximum at $x = 3$, $f(3, 0) = 83$, and its minimum at $x = 0$, $f(0, 0) = 2$.

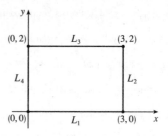

On L_2: $x = 3$, $f(3, y) = y^4 - 12y + 83$, $0 \le y \le 2$, a polynomial in y which attains its minimum at $y = \sqrt[3]{3}$, $f(3, \sqrt[3]{3}) = 83 - 9\sqrt[3]{3} \approx 70.0$, and its maximum at $y = 0$, $f(3, 0) = 83$.

On L_3: $y = 2$, $f(x, 2) = x^4 - 8x + 18$, $0 \le x \le 3$, a polynomial in x which attains its minimum at $x = \sqrt[3]{2}$, $f(\sqrt[3]{2}, 2) = 18 - 6\sqrt[3]{2} \approx 10.4$, and its maximum at $x = 3$, $f(3, 2) = 75$. On L_4: $x = 0$, $f(0, y) = y^4 + 2$, $0 \le y \le 2$, a polynomial in y which attains its maximum at $y = 2$, $f(0, 2) = 18$, and its minimum at $y = 0$, $f(0, 0) = 2$. Thus the absolute maximum of f on D is $f(3, 0) = 83$ and the absolute minimum is $f(1, 1) = 0$.

29. $f(x, y) = -(x^2 - 1)^2 - (x^2 y - x - 1)^2 \Rightarrow f_x(x, y) = -2(x^2 - 1)(2x) - 2(x^2 y - x - 1)(2xy - 1)$ and $f_y(x, y) = -2(x^2 y - x - 1)x^2$. Setting $f_y(x, y) = 0$ gives either $x = 0$ or $x^2 y - x - 1 = 0$.

There are no critical points for $x = 0$, since $f_x(0, y) = -2$, so we set $x^2 y - x - 1 = 0 \Leftrightarrow y = \dfrac{x+1}{x^2}$ $[x \ne 0]$,

so $f_x\left(x, \dfrac{x+1}{x^2}\right) = -2(x^2 - 1)(2x) - 2\left(x^2 \dfrac{x+1}{x^2} - x - 1\right)\left(2x \dfrac{x+1}{x^2} - 1\right) = -4x(x^2 - 1)$. Therefore

$f_x(x, y) = f_y(x, y) = 0$ at the points $(1, 2)$ and $(-1, 0)$. To classify these critical points, we calculate

$f_{xx}(x, y) = -12x^2 - 12x^2 y^2 + 12xy + 4y + 2$, $f_{yy}(x, y) = -2x^4$,

and $f_{xy}(x, y) = -8x^3 y + 6x^2 + 4x$. In order to use the Second Derivatives Test we calculate

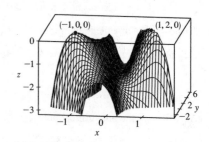

$D(-1, 0) = f_{xx}(-1, 0) f_{yy}(-1, 0) - [f_{xy}(-1, 0)]^2 = 16 > 0$,

$f_{xx}(-1, 0) = -10 < 0$, $D(1, 2) = 16 > 0$, and $f_{xx}(1, 2) = -26 < 0$, so both $(-1, 0)$ and $(1, 2)$ give local maxima.

31. Let d be the distance from $(2, 1, -1)$ to any point (x, y, z) on the plane $x + y - z = 1$, so $d = \sqrt{(x - 2)^2 + (y - 1)^2 + (z + 1)^2}$ where $z = x + y - 1$, and we minimize

$d^2 = f(x, y) = (x - 2)^2 + (y - 1)^2 + (x + y)^2$. Then $f_x(x, y) = 2(x - 2) + 2(x + y) = 4x + 2y - 4$,

$f_y(x, y) = 2(y - 1) + 2(x + y) = 2x + 4y - 2$. Solving $4x + 2y - 4 = 0$ and $2x + 4y - 2 = 0$ simultaneously gives $x = 1$,

$y = 0$. An absolute minimum exists (since there is a minimum distance from the point to the plane) and it must occur at a

critical point, so the shortest distance occurs for $x = 1$, $y = 0$ for which $d = \sqrt{(1 - 2)^2 + (0 - 1)^2 + (0 + 1)^2} = \sqrt{3}$.

33. Let d be the distance from the point $(4, 2, 0)$ to any point (x, y, z) on the cone, so $d = \sqrt{(x-4)^2 + (y-2)^2 + z^2}$ where

$z^2 = x^2 + y^2$, and we minimize $d^2 = (x-4)^2 + (y-2)^2 + x^2 + y^2 = f(x, y)$. Then

$f_x(x, y) = 2(x-4) + 2x = 4x - 8$, $f_y(x, y) = 2(y-2) + 2y = 4y - 4$, and the critical points occur when

$f_x = 0 \Rightarrow x = 2$, $f_y = 0 \Rightarrow y = 1$. Thus the only critical point is $(2, 1)$. An absolute minimum exists (since there

is a minimum distance from the cone to the point) which must occur at a critical point, so the points on the cone closest

to $(4, 2, 0)$ are $\left(2, 1, \pm\sqrt{5}\right)$.

35. $x + y + z = 100$, so maximize $f(x, y) = xy(100 - x - y)$. $f_x = 100y - 2xy - y^2$, $f_y = 100x - x^2 - 2xy$,

$f_{xx} = -2y$, $f_{yy} = -2x$, $f_{xy} = 100 - 2x - 2y$. Then $f_x = 0$ implies $y = 0$ or $y = 100 - 2x$. Substituting $y = 0$ into

$f_y = 0$ gives $x = 0$ or $x = 100$ and substituting $y = 100 - 2x$ into $f_y = 0$ gives $3x^2 - 100x = 0$ so $x = 0$ or $\frac{100}{3}$.

Thus the critical points are $(0, 0)$, $(100, 0)$, $(0, 100)$ and $\left(\frac{100}{3}, \frac{100}{3}\right)$.

$D(0, 0) = D(100, 0) = D(0, 100) = -10{,}000$ while $D\left(\frac{100}{3}, \frac{100}{3}\right) = \frac{10{,}000}{3}$ and $f_{xx}\left(\frac{100}{3}, \frac{100}{3}\right) = -\frac{200}{3} < 0$. Thus $(0, 0)$,

$(100, 0)$ and $(0, 100)$ are saddle points whereas $f\left(\frac{100}{3}, \frac{100}{3}\right)$ is a local maximum. Thus the numbers are $x = y = z = \frac{100}{3}$.

37. Maximize $f(x, y) = xy(36 - 9x^2 - 36y^2)^{1/2}/2$ with (x, y, z) in first octant. Then

$$f_x = \frac{y(36 - 9x^2 - 36y^2)^{1/2}}{2} + \frac{-9x^2y(36 - 9x^2 - 36y^2)^{-1/2}}{2} = \frac{(36y - 18x^2y - 36y^3)}{2(36 - 9x^2 - 36y^2)^{1/2}} \text{ and}$$

$$f_y = \frac{36x - 9x^3 - 72xy^2}{2(36 - 9x^2 - 36y^2)^{1/2}}. \text{ Setting } f_x = 0 \text{ gives } y = 0 \text{ or } y^2 = \frac{2 - x^2}{2} \text{ but } y > 0, \text{ so only the latter solution applies.}$$

Substituting this y into $f_y = 0$ gives $x^2 = \frac{4}{3}$ or $x = \frac{2}{\sqrt{3}}$, $y = \frac{1}{\sqrt{3}}$ and then $z^2 = (36 - 12 - 12)/4 = 3$. The fact

that this gives a maximum volume follows from the geometry. This maximum volume is

$$V = (2x)(2y)(2z) = 8\left(\frac{2}{\sqrt{3}}\right)\left(\frac{1}{\sqrt{3}}\right)\left(\sqrt{3}\right) = \frac{16}{\sqrt{3}}.$$

39. Maximize $f(x, y) = \frac{xy}{3}(6 - x - 2y)$, then the maximum volume is $V = xyz$.

$f_x = \frac{1}{3}(6y - 2xy - y^2) = \frac{1}{3}y(6 - 2x - 2y)$ and $f_y = \frac{1}{3}x(6 - x - 4y)$. Setting $f_x = 0$ and $f_y = 0$ gives the critical point

$(2, 1)$ which geometrically must yield a maximum. Thus the volume of the largest such box is $V = (2)(1)\left(\frac{2}{3}\right) = \frac{4}{3}$.

41. Let the dimensions be x, y, and z; then $4x + 4y + 4z = c$ and the volume is

$V = xyz = xy\left(\frac{1}{4}c - x - y\right) = \frac{1}{4}cxy - x^2y - xy^2$, $x > 0$, $y > 0$. Then $V_x = \frac{1}{4}cy - 2xy - y^2$ and $V_y = \frac{1}{4}cx - x^2 - 2xy$,

so $V_x = 0 = V_y$ when $2x + y = \frac{1}{4}c$ and $x + 2y = \frac{1}{4}c$. Solving, we get $x = \frac{1}{12}c$, $y = \frac{1}{12}c$ and $z = \frac{1}{4}c - x - y = \frac{1}{12}c$. From

the geometrical nature of the problem, this critical point must give an absolute maximum. Thus the box is a cube with edge

length $\frac{1}{12}c$.

43. Let the dimensions be x, y and z, then minimize $xy + 2(xz + yz)$ if $xyz = 32{,}000 \text{ cm}^3$. Then

$$f(x, y) = xy + [64{,}000(x + y)/xy] = xy + 64{,}000(x^{-1} + y^{-1}), \quad f_x = y - 64{,}000x^{-2}, \quad f_y = x - 64{,}000y^{-2}.$$

And $f_x = 0$ implies $y = 64{,}000/x^2$; substituting into $f_y = 0$ implies $x^3 = 64{,}000$ or $x = 40$ and then $y = 40$. Now

$D(x, y) = [(2)(64{,}000)]^2 x^{-3} y^{-3} - 1 > 0$ for $(40, 40)$ and $f_{xx}(40, 40) > 0$ so this is indeed a minimum. Thus the

dimensions of the box are $x = y = 40 \text{ cm}$, $z = 20 \text{ cm}$.

45. Let x, y, z be the dimensions of the rectangular box. Then the volume of the box is xyz and

$$L = \sqrt{x^2 + y^2 + z^2} \quad \Rightarrow \quad L^2 = x^2 + y^2 + z^2 \quad \Rightarrow \quad z = \sqrt{L^2 - x^2 - y^2}.$$

Substituting, we have volume $V(x, y) = xy \sqrt{L^2 - x^2 - y^2}$, $\ (x, y > 0)$.

$$V_x = xy \cdot \tfrac{1}{2}(L^2 - x^2 - y^2)^{-1/2}(-2x) + y\sqrt{L^2 - x^2 - y^2} = y\sqrt{L^2 - x^2 - y^2} - \frac{x^2 y}{\sqrt{L^2 - x^2 - y^2}},$$

$$V_y = x\sqrt{L^2 - x^2 - y^2} - \frac{xy^2}{\sqrt{L^2 - x^2 - y^2}}. \quad V_x = 0 \text{ implies } y(L^2 - x^2 - y^2) = x^2 y \ \Rightarrow \ y(L^2 - 2x^2 - y^2) = 0 \ \Rightarrow$$

$2x^2 + y^2 = L^2$ (since $y > 0$), and $V_y = 0$ implies $x(L^2 - x^2 - y^2) = xy^2 \ \Rightarrow \ x(L^2 - x^2 - 2y^2) = 0 \ \Rightarrow$

$x^2 + 2y^2 = L^2$ (since $x > 0$). Substituting $y^2 = L^2 - 2x^2$ into $x^2 + 2y^2 = L^2$ gives $x^2 + 2L^2 - 4x^2 = L^2 \ \Rightarrow$

$3x^2 = L^2 \ \Rightarrow \ x = L/\sqrt{3}$ (since $x > 0$) and then $y = \sqrt{L^2 - 2(L/\sqrt{3})^2} = L/\sqrt{3}$.

So the only critical point is $(L/\sqrt{3}, L/\sqrt{3})$ which, from the geometrical nature of the problem, must give an absolute

maximum. Thus the maximum volume is $V(L/\sqrt{3}, L/\sqrt{3}) = (L/\sqrt{3})^2 \sqrt{L^2 - (L/\sqrt{3})^2 - (L/\sqrt{3})^2} = L^3/(3\sqrt{3})$

cubic units.

47. Note that here the variables are m and b, and $f(m, b) = \sum_{i=1}^{n} [y_i - (mx_i + b)]^2$. Then $f_m = \sum_{i=1}^{n} -2x_i[y_i - (mx_i + b)] = 0$

implies $\sum_{i=1}^{n} (x_i y_i - mx_i^2 - bx_i) = 0$ or $\sum_{i=1}^{n} x_i y_i = m \sum_{i=1}^{n} x_i^2 + b \sum_{i=1}^{n} x_i$ and $f_b = \sum_{i=1}^{n} -2[y_i - (mx_i + b)] = 0$ implies

$\sum_{i=1}^{n} y_i = m \sum_{i=1}^{n} x_i + \sum_{i=1}^{n} b = m\left(\sum_{i=1}^{n} x_i \right) + nb$. Thus we have the two desired equations.

Now $f_{mm} = \sum_{i=1}^{n} 2x_i^2$, $f_{bb} = \sum_{i=1}^{n} 2 = 2n$ and $f_{mb} = \sum_{i=1}^{n} 2x_i$. And $f_{mm}(m, b) > 0$ always and

$$D(m, b) = 4n\left(\sum_{i=1}^{n} x_i^2 \right) - 4\left(\sum_{i=1}^{n} x_i \right)^2 = 4\left[n\left(\sum_{i=1}^{n} x_i^2 \right) - \left(\sum_{i=1}^{n} x_i \right)^2 \right] > 0 \text{ always so the solutions of these two}$$

equations do indeed minimize $\sum_{i=1}^{n} d_i^2$.

11.8 Lagrange Multipliers

1. $f(x, y) = x^2 + y^2$, $g(x, y) = xy = 1$, and $\nabla f = \lambda \nabla g$ $\Rightarrow$ $\langle 2x, 2y \rangle = \langle \lambda y, \lambda x \rangle$, so $2x = \lambda y$, $2y = \lambda x$, and $xy = 1$.

From the last equation, $x \neq 0$ and $y \neq 0$, so $2x = \lambda y$ $\Rightarrow$ $\lambda = 2x/y$. Substituting, we have $2y = (2x/y)\, x$ $\Rightarrow$

$y^2 = x^2$ $\Rightarrow$ $y = \pm x$. But $xy = 1$, so $x = y = \pm 1$ and the possible points for the extreme values of f are $(1, 1)$ and

$(-1, -1)$. Here there is no maximum value, since the constraint $xy = 1$ allows x or y to become arbitrarily large, and hence

$f(x, y) = x^2 + y^2$ can be made arbitrarily large. The minimum value is $f(1, 1) = f(-1, -1) = 2$.

3. $f(x, y) = x^2 y$, $g(x, y) = x^2 + 2y^2 = 6$ $\Rightarrow$ $\nabla f = \langle 2xy, x^2 \rangle$, $\lambda \nabla g = \langle 2\lambda x, 4\lambda y \rangle$. Then $2xy = 2\lambda x$ implies $x = 0$ or

$\lambda = y$. If $x = 0$, then $x^2 = 4\lambda y$ implies $\lambda = 0$ or $y = 0$. However, if $y = 0$ then $g(x, y) = 0$, a contradiction. So $\lambda = 0$ and

then $g(x, y) = 6$ $\Rightarrow$ $y = \pm\sqrt{3}$. If $\lambda = y$, then $x^2 = 4\lambda y$ implies $x^2 = 4y^2$, and so $g(x, y) = 6$ $\Rightarrow$ $4y^2 + 2y^2 = 6$

$\Rightarrow$ $y^2 = 1$ $\Rightarrow$ $y = \pm 1$. Thus f has possible extreme values at the points $(0, \pm\sqrt{3})$, $(\pm 2, 1)$, and $(\pm 2, -1)$. After

evaluating f at these points, we find the maximum value to be $f(\pm 2, 1) = 4$ and the minimum to be $f(\pm 2, -1) = -4$.

5. $f(x, y, z) = 2x + 6y + 10z$, $g(x, y, z) = x^2 + y^2 + z^2 = 35$ $\Rightarrow$ $\nabla f = \langle 2, 6, 10 \rangle$, $\lambda \nabla g = \langle 2\lambda x, 2\lambda y, 2\lambda z \rangle$. Then

$2\lambda x = 2$, $2\lambda y = 6$, $2\lambda z = 10$ imply $x = \dfrac{1}{\lambda}$, $y = \dfrac{3}{\lambda}$, and $z = \dfrac{5}{\lambda}$. But $35 = x^2 + y^2 + z^2 = \left(\dfrac{1}{\lambda}\right)^2 + \left(\dfrac{3}{\lambda}\right)^2 + \left(\dfrac{5}{\lambda}\right)^2$ $\Rightarrow$

$35 = \dfrac{35}{\lambda^2}$ $\Rightarrow$ $\lambda = \pm 1$, so f has possible extreme values at the points $(1, 3, 5)$, $(-1, -3, -5)$. The maximum value of f on

$x^2 + y^2 + z^2 = 35$ is $f(1, 3, 5) = 70$, and the minimum is $f(-1, -3, -5) = -70$.

7. $f(x, y, z) = xyz$, $g(x, y, z) = x^2 + 2y^2 + 3z^2 = 6$ $\Rightarrow$ $\nabla f = \langle yz, xz, xy \rangle$, $\lambda \nabla g = \langle 2\lambda x, 4\lambda y, 6\lambda z \rangle$. Then $\nabla f = \lambda \nabla g$

implies $\lambda = (yz)/(2x) = (xz)/(4y) = (xy)/(6z)$ or $x^2 = 2y^2$ and $z^2 = \frac{2}{3}y^2$. Thus $x^2 + 2y^2 + 3z^2 = 6$ implies $6y^2 = 6$

or $y = \pm 1$. Then the possible points are $\left(\sqrt{2}, \pm 1, \sqrt{\frac{2}{3}}\right)$, $\left(\sqrt{2}, \pm 1, -\sqrt{\frac{2}{3}}\right)$, $\left(-\sqrt{2}, \pm 1, \sqrt{\frac{2}{3}}\right)$, $\left(-\sqrt{2}, \pm 1, -\sqrt{\frac{2}{3}}\right)$. The

maximum value of f on the ellipsoid is $\frac{2}{\sqrt{3}}$, occurring when all coordinates are positive or exactly two are negative and the

minimum is $-\frac{2}{\sqrt{3}}$ occurring when 1 or 3 of the coordinates are negative.

9. $f(x, y, z) = x^2 + y^2 + z^2$, $g(x, y, z) = x^4 + y^4 + z^4 = 1$ $\Rightarrow$ $\nabla f = \langle 2x, 2y, 2z \rangle$, $\lambda \nabla g = \langle 4\lambda x^3, 4\lambda y^3, 4\lambda z^3 \rangle$.

Case 1: If $x \neq 0$, $y \neq 0$ and $z \neq 0$, then $\nabla f = \lambda \nabla g$ implies $\lambda = 1/(2x^2) = 1/(2y^2) = 1/(2z^2)$ or $x^2 = y^2 = z^2$ and

$3x^4 = 1$ or $x = \pm\frac{1}{\sqrt[4]{3}}$ giving the points $\left(\pm\frac{1}{\sqrt[4]{3}}, \frac{1}{\sqrt[4]{3}}, \frac{1}{\sqrt[4]{3}}\right)$, $\left(\pm\frac{1}{\sqrt[4]{3}}, -\frac{1}{\sqrt[4]{3}}, \frac{1}{\sqrt[4]{3}}\right)$, $\left(\pm\frac{1}{\sqrt[4]{3}}, \frac{1}{\sqrt[4]{3}}, -\frac{1}{\sqrt[4]{3}}\right)$, $\left(\pm\frac{1}{\sqrt[4]{3}}, -\frac{1}{\sqrt[4]{3}}, -\frac{1}{\sqrt[4]{3}}\right)$

all with an f-value of $\sqrt{3}$.

Case 2: If one of the variables equals zero and the other two are not zero, then the squares of the two nonzero coordinates are

equal with common value $\frac{1}{\sqrt{2}}$ and corresponding f value of $\sqrt{2}$.

Case 3: If exactly two of the variables are zero, then the third variable has value ± 1 with the corresponding f value of 1. Thus

on $x^4 + y^4 + z^4 = 1$, the maximum value of f is $\sqrt{3}$ and the minimum value is 1.

11. $f(x, y, z, t) = x + y + z + t$, $g(x, y, z, t) = x^2 + y^2 + z^2 + t^2 = 1$ $\Rightarrow$ $\langle 1, 1, 1, 1 \rangle = \langle 2\lambda x, 2\lambda y, 2\lambda z, 2\lambda t \rangle$, so

$\lambda = 1/(2x) = 1/(2y) = 1/(2z) = 1/(2t)$ and $x = y = z = t$. But $x^2 + y^2 + z^2 + t^2 = 1$, so the possible points are

$\left(\pm\frac{1}{2}, \pm\frac{1}{2}, \pm\frac{1}{2}, \pm\frac{1}{2} \right)$. Thus the maximum value of f is $f\left(\frac{1}{2}, \frac{1}{2}, \frac{1}{2}, \frac{1}{2} \right) = 2$ and the minimum value is

$f\left(-\frac{1}{2}, -\frac{1}{2}, -\frac{1}{2}, -\frac{1}{2} \right) = -2$.

13. $f(x, y, z) = x + 2y$, $g(x, y, z) = x + y + z = 1$, $h(x, y, z) = y^2 + z^2 = 4$ $\Rightarrow$ $\nabla f = \langle 1, 2, 0 \rangle$, $\lambda \nabla g = \langle \lambda, \lambda, \lambda \rangle$

and $\mu \nabla h = \langle 0, 2\mu y, 2\mu z \rangle$. Then $1 = \lambda$, $2 = \lambda + 2\mu y$ and $0 = \lambda + 2\mu z$ so $\mu y = \frac{1}{2} = -\mu z$ or $y = 1/(2\mu)$, $z = -1/(2\mu)$.

Thus $x + y + z = 1$ implies $x = 1$ and $y^2 + z^2 = 4$ implies $\mu = \pm\frac{1}{2\sqrt{2}}$. Then the possible points are $\left(1, \pm\sqrt{2}, \mp\sqrt{2} \right)$

and the maximum value is $f\left(1, \sqrt{2}, -\sqrt{2} \right) = 1 + 2\sqrt{2}$ and the minimum value is $f\left(1, -\sqrt{2}, \sqrt{2} \right) = 1 - 2\sqrt{2}$.

15. $f(x, y, z) = yz + xy$, $g(x, y, z) = xy = 1$, $h(x, y, z) = y^2 + z^2 = 1$ $\Rightarrow$ $\nabla f = \langle y, x + z, y \rangle$, $\lambda \nabla g = \langle \lambda y, \lambda x, 0 \rangle$,

$\mu \nabla h = \langle 0, 2\mu y, 2\mu z \rangle$. Then $y = \lambda y$ implies $\lambda = 1$ [$y \neq 0$ since $g(x, y, z) = 1$], $x + z = \lambda x + 2\mu y$ and $y = 2\mu z$. Thus

$\mu = z/(2y) = y/(2y)$ or $y^2 = z^2$, and so $y^2 + z^2 = 1$ implies $y = \pm\frac{1}{\sqrt{2}}$, $z = \pm\frac{1}{\sqrt{2}}$. Then $xy = 1$ implies $x = \pm\sqrt{2}$ and

the possible points are $\left(\pm\sqrt{2}, \pm\frac{1}{\sqrt{2}}, \frac{1}{\sqrt{2}} \right)$, $\left(\pm\sqrt{2}, \pm\frac{1}{\sqrt{2}}, -\frac{1}{\sqrt{2}} \right)$. Hence the maximum of f subject to the constraints is

$f\left(\pm\sqrt{2}, \pm\frac{1}{\sqrt{2}}, \pm\frac{1}{\sqrt{2}} \right) = \frac{3}{2}$ and the minimum is $f\left(\pm\sqrt{2}, \pm\frac{1}{\sqrt{2}}, \mp\frac{1}{\sqrt{2}} \right) = \frac{1}{2}$.

Note: Since $xy = 1$ is one of the constraints we could have solved the problem by solving $f(y, z) = yz + 1$ subject to

$y^2 + z^2 = 1$.

17. $f(x, y) = e^{-xy}$. For the interior of the region, we find the critical points: $f_x = -ye^{-xy}$, $f_y = -xe^{-xy}$, so the only

critical point is $(0, 0)$, and $f(0, 0) = 1$. For the boundary, we use Lagrange multipliers. $g(x, y) = x^2 + 4y^2 = 1$ $\Rightarrow$

$\lambda \nabla g = \langle 2\lambda x, 8\lambda y \rangle$, so setting $\nabla f = \lambda \nabla g$ we get $-ye^{-xy} = 2\lambda x$ and $-xe^{-xy} = 8\lambda y$. The first of these gives

$e^{-xy} = -2\lambda x/y$, and then the second gives $-x(-2\lambda x/y) = 8\lambda y$ $\Rightarrow$ $x^2 = 4y^2$. Solving this last equation with the

constraint $x^2 + 4y^2 = 1$ gives $x = \pm\frac{1}{\sqrt{2}}$ and $y = \pm\frac{1}{2\sqrt{2}}$. Now $f\left(\pm\frac{1}{\sqrt{2}}, \mp\frac{1}{2\sqrt{2}} \right) = e^{1/4} \approx 1.284$ and

$f\left(\pm\frac{1}{\sqrt{2}}, \pm\frac{1}{2\sqrt{2}} \right) = e^{-1/4} \approx 0.779$. The former are the maxima on the region and the latter are the minima.

19. At the extreme values of f, the level curves of f just touch the curve $g(x, y) = 8$ with a common tangent line. (See Figure 1

and the accompanying discussion.) We can observe several such occurrences on the contour map, but the level curve

$f(x, y) = c$ with the largest value of c which still intersects the curve $g(x, y) = 8$ is approximately $c = 59$, and the smallest

value of c corresponding to a level curve which intersects $g(x, y) = 8$ appears to be $c = 30$. Thus we estimate the maximum

value of f subject to the constraint $g(x, y) = 8$ to be about 59 and the minimum to be 30.

21. $P(L, K) = bL^\alpha K^{1-\alpha}$, $g(L, K) = mL + nK = p$ $\Rightarrow$ $\nabla P = \langle \alpha b L^{\alpha-1} K^{1-\alpha}, (1-\alpha)bL^\alpha K^{-\alpha}\rangle$, $\lambda \nabla g = \langle \lambda m, \lambda n\rangle$.

Then $\alpha b(K/L)^{1-\alpha} = \lambda m$ and $(1-\alpha)b(L/K)^\alpha = \lambda n$ and $mL + nK = p$, so $\alpha b(K/L)^{1-\alpha}/m = (1-\alpha)b(L/K)^\alpha/n$ or

$n\alpha/[m(1-\alpha)] = (L/K)^\alpha (L/K)^{1-\alpha}$ or $L = Kn\alpha/[m(1-\alpha)]$. Substituting into $mL + nK = p$ gives $K = (1-\alpha)p/n$

and $L = \alpha p/m$ for the maximum production.

23. Let the sides of the rectangle be x and y. Then $f(x, y) = xy$, $g(x, y) = 2x + 2y = p$ $\Rightarrow$ $\nabla f(x, y) = \langle y, x\rangle$,

$\lambda \nabla g = \langle 2\lambda, 2\lambda\rangle$. Then $\lambda = \frac{1}{2}y = \frac{1}{2}x$ implies $x = y$ and the rectangle with maximum area is a square with side length $\frac{1}{4}p$.

25. Let $f(x, y, z) = d^2 = (x-2)^2 + (y-1)^2 + (z+1)^2$, then we want to minimize f subject to the constraint

$g(x, y, z) = x + y - z = 1$. $\nabla f = \lambda \nabla g$ $\Rightarrow$ $\langle 2(x-2), 2(y-1), 2(z+1)\rangle = \lambda\langle 1, 1, -1\rangle$, so $x = (\lambda+4)/2$,

$y = (\lambda+2)/2$, $z = -(\lambda+2)/2$. Substituting into the constraint equation gives $\dfrac{\lambda+4}{2} + \dfrac{\lambda+2}{2} + \dfrac{\lambda+2}{2} = 1$ $\Rightarrow$

$3\lambda + 8 = 2$ $\Rightarrow$ $\lambda = -2$, so $x = 1$, $y = 0$, and $z = 0$. This must correspond to a minimum, so the shortest distance is

$d = \sqrt{(1-2)^2 + (0-1)^2 + (0+1)^2} = \sqrt{3}$.

27. Let $f(x, y, z) = d^2 = (x-4)^2 + (y-2)^2 + z^2$. Then we want to minimize f subject to the constraint

$g(x, y, z) = x^2 + y^2 - z^2 = 0$. $\nabla f = \lambda \nabla g$ $\Rightarrow$ $\langle 2(x-4), 2(y-2), 2z\rangle = \langle 2\lambda x, 2\lambda y, -2\lambda z\rangle$, so $x - 4 = \lambda x$,

$y - 2 = \lambda y$, and $z = -\lambda z$. From the last equation we have $z + \lambda z = 0$ $\Rightarrow$ $z(1+\lambda) = 0$, so either $z = 0$ or $\lambda = -1$.

But from the constraint equation we have $z = 0$ $\Rightarrow$ $x^2 + y^2 = 0$ $\Rightarrow$ $x = y = 0$ which is not possible from the first

two equations. So $\lambda = -1$ and $x - 4 = \lambda x$ $\Rightarrow$ $x = 2$, $y - 2 = \lambda y$ $\Rightarrow$ $y = 1$, and $x^2 + y^2 - z^2 = 0$ $\Rightarrow$

$4 + 1 - z^2 = 0$ $\Rightarrow$ $z = \pm\sqrt{5}$. This must correspond to a minimum, so the points on the cone closest to $(4, 2, 0)$

are $(2, 1, \pm\sqrt{5})$.

29. $f(x, y, z) = xyz$, $g(x, y, z) = x + y + z = 100$ $\Rightarrow$ $\nabla f = \langle yz, xz, xy\rangle = \lambda \nabla g = \langle \lambda, \lambda, \lambda\rangle$. Then $\lambda = yz = xz = xy$

implies $x = y = z = \frac{100}{3}$.

31. If the dimensions are $2x$, $2y$ and $2z$, then $f(x, y, z) = 8xyz$ and $g(x, y, z) = 9x^2 + 36y^2 + 4z^2 = 36$ $\Rightarrow$

$\nabla f = \langle 8yz, 8xz, 8xy\rangle = \lambda \nabla g = \langle 18\lambda x, 72\lambda y, 8\lambda z\rangle$. Thus $18\lambda x = 8yz$, $72\lambda y = 8xz$, $8\lambda z = 8xy$ so $x^2 = 4y^2$, $z^2 = 9y^2$

and $36y^2 + 36y^2 + 36y^2 = 36$ or $y = \frac{1}{\sqrt{3}}$ ($y > 0$). Thus the volume of the largest such box is $8\left(\frac{1}{\sqrt{3}}\right)\left(\frac{2}{\sqrt{3}}\right)\left(\frac{3}{\sqrt{3}}\right) = \frac{16}{\sqrt{3}}$.

33. $f(x, y, z) = xyz$, $g(x, y, z) = x + 2y + 3z = 6$ $\Rightarrow$ $\nabla f = \langle yz, xz, xy\rangle = \lambda \nabla g = \langle \lambda, 2\lambda, 3\lambda\rangle$.

Then $\lambda = yz = \frac{1}{2}xz = \frac{1}{3}xy$ implies $x = 2y$, $z = \frac{2}{3}y$. But $2y + 2y + 2y = 6$ so $y = 1$, $x = 2$, $z = \frac{2}{3}$ and the volume

is $V = \frac{4}{3}$.

35. $f(x,y,z) = xyz$, $g(x,y,z) = 4(x+y+z) = c$ $\Rightarrow$ $\nabla f = \langle yz, xz, xy \rangle$, $\lambda \nabla g = \langle 4\lambda, 4\lambda, 4\lambda \rangle$. Thus

$4\lambda = yz = xz = xy$ or $x = y = z = \frac{1}{12}c$ are the dimensions giving the maximum volume.

37. If the dimensions of the box are given by x, y, and z, then we need to find the maximum value of $f(x,y,z) = xyz$

$[x,y,z > 0]$ subject to the constraint $L = \sqrt{x^2 + y^2 + z^2}$ or $g(x,y,z) = x^2 + y^2 + z^2 = L^2$. $\nabla f = \lambda \nabla g$ $\Rightarrow$

$\langle yz, xz, xy \rangle = \lambda \langle 2x, 2y, 2z \rangle$, so $yz = 2\lambda x$ $\Rightarrow$ $\lambda = \dfrac{yz}{2x}$, $xz = 2\lambda y$ $\Rightarrow$ $\lambda = \dfrac{xz}{2y}$, and $xy = 2\lambda z$ $\Rightarrow$ $\lambda = \dfrac{xy}{2z}$.

Thus $\lambda = \dfrac{yz}{2x} = \dfrac{xz}{2y}$ $\Rightarrow$ $x^2 = y^2$ [since $z \ne 0$] $\Rightarrow$ $x = y$ and $\lambda = \dfrac{yz}{2x} = \dfrac{xy}{2z}$ $\Rightarrow$ $x = z$ [since $y \ne 0$].

Substituting into the constraint equation gives $x^2 + x^2 + x^2 = L^2$ $\Rightarrow$ $x^2 = L^2/3$ $\Rightarrow$ $x = L/\sqrt{3} = y = z$ and the

maximum volume is $\left(L/\sqrt{3} \right)^3 = L^3/(3\sqrt{3})$.

39. We need to find the extreme values of $f(x,y,z) = x^2 + y^2 + z^2$ subject to the two constraints $g(x,y,z) = x+y+2z = 2$

and $h(x,y,z) = x^2 + y^2 - z = 0$. $\nabla f = \langle 2x, 2y, 2z \rangle$, $\lambda \nabla g = \langle \lambda, \lambda, 2\lambda \rangle$ and $\mu \nabla h = \langle 2\mu x, 2\mu y, -\mu \rangle$. Thus we need

$2x = \lambda + 2\mu x$ **(1)**, $2y = \lambda + 2\mu y$ **(2)**, $2z = 2\lambda - \mu$ **(3)**, $x+y+2z = 2$ **(4)**, and $x^2 + y^2 - z = 0$ **(5)**.

From **(1)** and **(2)**, $2(x-y) = 2\mu(x-y)$, so if $x \ne y$, $\mu = 1$. Putting this in **(3)** gives $2z = 2\lambda - 1$ or $\lambda = z + \frac{1}{2}$, but putting

$\mu = 1$ into **(1)** says $\lambda = 0$. Hence $z + \frac{1}{2} = 0$ or $z = -\frac{1}{2}$. Then **(4)** and **(5)** become $x+y-3=0$ and $x^2 + y^2 + \frac{1}{2} = 0$. The

last equation cannot be true, so this case gives no solution. So we must have $x = y$. Then **(4)** and **(5)** become $2x + 2z = 2$ and

$2x^2 - z = 0$ which imply $z = 1 - x$ and $z = 2x^2$. Thus $2x^2 = 1 - x$ or $2x^2 + x - 1 = (2x-1)(x+1) = 0$ so $x = \frac{1}{2}$ or

$x = -1$. The two points to check are $\left(\frac{1}{2}, \frac{1}{2}, \frac{1}{2} \right)$ and $(-1, -1, 2)$: $f\left(\frac{1}{2}, \frac{1}{2}, \frac{1}{2} \right) = \frac{3}{4}$ and $f(-1, -1, 2) = 6$. Thus $\left(\frac{1}{2}, \frac{1}{2}, \frac{1}{2} \right)$ is

the point on the ellipse nearest the origin and $(-1, -1, 2)$ is the one farthest from the origin.

41. $f(x,y,z) = ye^{x-z}$, $g(x,y,z) = 9x^2 + 4y^2 + 36z^2 = 36$, $h(x,y,z) = xy + yz = 1$. $\nabla f = \lambda \nabla g + \mu \nabla h$ $\Rightarrow$

$\left\langle ye^{x-z}, e^{x-z}, -ye^{x-z} \right\rangle = \lambda \langle 18x, 8y, 72z \rangle + \mu \langle y, x+z, y \rangle$, so $ye^{x-z} = 18\lambda x + \mu y$, $e^{x-z} = 8\lambda y + \mu(x+z)$,

$-ye^{x-z} = 72\lambda z + \mu y$, $9x^2 + 4y^2 + 36z^2 = 36$, $xy + yz = 1$. Using a CAS to solve these 5 equations simultaneously for x,

y, z, λ, and μ (in Maple, use the `allvalues` command), we get 4 real-valued solutions:

$$x \approx 0.222444, \quad y \approx -2.157012, \quad z \approx -0.686049, \quad \lambda \approx -0.200401, \quad \mu \approx 2.108584$$
$$x \approx -1.951921, \quad y \approx -0.545867, \quad z \approx 0.119973, \quad \lambda \approx 0.003141, \quad \mu \approx -0.076238$$
$$x \approx 0.155142, \quad y \approx 0.904622, \quad z \approx 0.950293, \quad \lambda \approx -0.012447, \quad \mu \approx 0.489938$$
$$x \approx 1.138731, \quad y \approx 1.768057, \quad z \approx -0.573138, \quad \lambda \approx 0.317141, \quad \mu \approx 1.862675$$

Substituting these values into f gives $f(0.222444, -2.157012, -0.686049) \approx -5.3506$,

$f(-1.951921, -0.545867, 0.119973) \approx -0.0688$, $f(0.155142, 0.904622, 0.950293) \approx 0.4084$,

$f(1.138731, 1.768057, -0.573138) \approx 9.7938$. Thus the maximum is approximately 9.7938, and the mininum is

approximately -5.3506.

43. (a) We wish to maximize $f(x_1, x_2, \ldots, x_n) = \sqrt[n]{x_1 x_2 \cdots x_n}$ subject to

$g(x_1, x_2, \ldots, x_n) = x_1 + x_2 + \cdots + x_n = c$ and $x_i > 0$.

$$\nabla f = \left\langle \tfrac{1}{n}(x_1 x_2 \cdots x_n)^{\frac{1}{n}-1}(x_2 \cdots x_n), \tfrac{1}{n}(x_1 x_2 \cdots x_n)^{\frac{1}{n}-1}(x_1 x_3 \cdots x_n), \ldots, \tfrac{1}{n}(x_1 x_2 \cdots x_n)^{\frac{1}{n}-1}(x_1 \cdots x_{n-1}) \right\rangle$$

and $\lambda \nabla g = \langle \lambda, \lambda, \ldots, \lambda \rangle$, so we need to solve the system of equations

$$\frac{1}{n}(x_1 x_2 \cdots x_n)^{\frac{1}{n}-1}(x_2 \cdots x_n) = \lambda \quad \Rightarrow \quad x_1^{1/n} x_2^{1/n} \cdots x_n^{1/n} = n\lambda x_1$$

$$\frac{1}{n}(x_1 x_2 \cdots x_n)^{\frac{1}{n}-1}(x_1 x_3 \cdots x_n) = \lambda \quad \Rightarrow \quad x_1^{1/n} x_2^{1/n} \cdots x_n^{1/n} = n\lambda x_2$$

$$\vdots$$

$$\frac{1}{n}(x_1 x_2 \cdots x_n)^{\frac{1}{n}-1}(x_1 \cdots x_{n-1}) = \lambda \quad \Rightarrow \quad x_1^{1/n} x_2^{1/n} \cdots x_n^{1/n} = n\lambda x_n$$

This implies $n\lambda x_1 = n\lambda x_2 = \cdots = n\lambda x_n$. Note $\lambda \neq 0$, otherwise we can't have all $x_i > 0$. Thus $x_1 = x_2 = \cdots = x_n$.

But $x_1 + x_2 + \cdots + x_n = c \Rightarrow nx_1 = c \Rightarrow x_1 = \dfrac{c}{n} = x_2 = x_3 = \cdots = x_n$. Then the only point where f can

have an extreme value is $\left(\dfrac{c}{n}, \dfrac{c}{n}, \ldots, \dfrac{c}{n} \right)$. Since we can choose values for $(x_1, x_2, \ldots, x_n)$ that make f as close to

zero (but not equal) as we like, f has no minimum value. Thus the maximum value is

$$f\left(\frac{c}{n}, \frac{c}{n}, \ldots, \frac{c}{n} \right) = \sqrt[n]{\frac{c}{n} \cdot \frac{c}{n} \cdots \cdots \frac{c}{n}} = \frac{c}{n}.$$

(b) From part (a), $\dfrac{c}{n}$ is the maximum value of f. Thus $f(x_1, x_2, \ldots, x_n) = \sqrt[n]{x_1 x_2 \cdots x_n} \leq \dfrac{c}{n}$. But

$x_1 + x_2 + \cdots + x_n = c$, so $\sqrt[n]{x_1 x_2 \cdots x_n} \leq \dfrac{x_1 + x_2 + \cdots + x_n}{n}$. These two means are equal when f attains its

maximum value $\dfrac{c}{n}$, but this can occur only at the point $\left(\dfrac{c}{n}, \dfrac{c}{n}, \ldots, \dfrac{c}{n} \right)$ we found in part (a). So the means are equal only

when $x_1 = x_2 = x_3 = \cdots = x_n = \dfrac{c}{n}$.

11   Review

CONCEPT CHECK

1. (a) A function f of two variables is a rule that assigns to each ordered pair (x, y) of real numbers in its domain a unique real

number denoted by $f(x, y)$.

(b) One way to visualize a function of two variables is by graphing it, resulting in the surface $z = f(x, y)$. Another method for

visualizing a function of two variables is a contour map. The contour map consists of level curves of the function which are

horizontal traces of the graph of the function projected onto the xy-plane. Also, we can use an arrow diagram such as

Figure 1 in Section 11.1.

2. A function f of three variables is a rule that assigns to each ordered triple (x, y, z) in its domain a unique real number $f(x, y, z)$. We can visualize a function of three variables by examining its level surfaces $f(x, y, z) = k$, where k is a constant.

3. $\lim\limits_{(x,y)\to(a,b)} f(x, y) = L$ means the values of $f(x, y)$ approach the number L as the point (x, y) approaches the point (a, b) along any path that is within the domain of f. We can show that a limit at a point does not exist by finding two different paths approaching the point along which $f(x, y)$ has different limits.

4. (a) See Definition 11.2.4.

(b) If f is continuous on $\mathbb{R}^2$, its graph will appear as a surface without holes or breaks.

5. (a) See (2) and (3) in Section 11.3.

(b) See "Interpretations of Partial Derivatives" on page 610.

(c) To find f_x, regard y as a constant and differentiate $f(x, y)$ with respect to x. To find f_y, regard x as a constant and differentiate $f(x, y)$ with respect to y.

6. See the statement of Clairaut's Theorem on page 613.

7. (a) See (2) in Section 11.4.

(b) See (19) and the preceding discussion in Section 11.6.

8. See (3) and (4) and the accompanying discussion in Section 11.4. We can interpret the linearization of f at (a, b) geometrically as the linear function whose graph is the tangent plane to the graph of f at (a, b). Thus it is the linear function which best approximates f near (a, b).

9. (a) See Definition 11.4.7.

(b) Use Theorem 11.4.8.

10. See (10) and the associated discussion in Section 11.4.

11. See (2) and (3) in Section 11.5.

12. See (7) and the preceding discussion in Section 11.5.

13. (a) See Definition 11.6.2. We can interpret it as the rate of change of f at (x_0, y_0) in the direction of $\mathbf{u}$. Geometrically, if P is the point $(x_0, y_0, f(x_0, y_0))$ on the graph of f and C is the curve of intersection of the graph of f with the vertical plane that passes through P in the direction $\mathbf{u}$, the directional derivative of f at (x_0, y_0) in the direction of $\mathbf{u}$ is the slope of the tangent line to C at P. (See Figure 2 in Section 11.6.)

(b) See Theorem 11.6.3.

14. (a) See (8) and (13) in Section 11.6.

(b) $D_{\mathbf{u}} f(x, y) = \nabla f(x, y) \cdot \mathbf{u}$ or $D_{\mathbf{u}} f(x, y, z) = \nabla f(x, y, z) \cdot \mathbf{u}$

(c) The gradient vector of a function points in the direction of maximum rate of increase of the function. On a graph of the function, the gradient points in the direction of steepest ascent.

15. (a) f has a local maximum at (a, b) if $f(x, y) \leq f(a, b)$ when (x, y) is near (a, b).

(b) f has an absolute maximum at (a, b) if $f(x, y) \leq f(a, b)$ for all points (x, y) in the domain of f.

(c) f has a local minimum at (a, b) if $f(x, y) \geq f(a, b)$ when (x, y) is near (a, b).

(d) f has an absolute minimum at (a, b) if $f(x, y) \geq f(a, b)$ for all points (x, y) in the domain of f.

(e) f has a saddle point at (a, b) if $f(a, b)$ is a local maximum in one direction but a local minimum in another.

16. (a) By Theorem 11.7.2, if f has a local maximum at (a, b) and the first-order partial derivatives of f exist there, then

$$f_x(a, b) = 0 \text{ and } f_y(a, b) = 0.$$

(b) A critical point of f is a point (a, b) such that $f_x(a, b) = 0$ and $f_y(a, b) = 0$ or one of these partial derivatives does not exist.

17. See (3) in Section 11.7.

18. (a) See Figure 7 and the accompanying discussion in Section 11.7.

(b) See Theorem 11.7.4.

(c) See the procedure outlined in (5) in Section 11.7.

19. See the discussion beginning on page 653; see "Two Constraints" on page 656.

TRUE-FALSE QUIZ

1. True. $f_y(a, b) = \lim\limits_{h \to 0} \dfrac{f(a, b+h) - f(a, b)}{h}$ from Equation 11.3.3. Let $h = y - b$. As $h \to 0$, $y \to b$. Then by substituting,

we get $f_y(a, b) = \lim\limits_{y \to b} \dfrac{f(a, y) - f(a, b)}{y - b}$.

3. False. $f_{xy} = \dfrac{\partial^2 f}{\partial y \, \partial x}$.

5. False. See Example 11.2.3.

7. True. If f has a local minimum and f is differentiable at (a, b) then by Theorem 11.7.2, $f_x(a, b) = 0$ and $f_y(a, b) = 0$, so
$$\nabla f(a, b) = \langle f_x(a, b), f_y(a, b) \rangle = \langle 0, 0 \rangle = \mathbf{0}.$$

9. False. $\nabla f(x, y) = \langle 0, 1/y \rangle$.

11. True. $\nabla f = \langle \cos x, \cos y \rangle$, so $|\nabla f| = \sqrt{\cos^2 x + \cos^2 y}$. But $|\cos \theta| \leq 1$, so $|\nabla f| \leq \sqrt{2}$. Now
$$D_{\mathbf{u}} f(x, y) = \nabla f \cdot \mathbf{u} = |\nabla f| \, |\mathbf{u}| \cos \theta, \text{ but } \mathbf{u} \text{ is a unit vector, so } |D_{\mathbf{u}} f(x, y)| \leq \sqrt{2} \cdot 1 \cdot 1 = \sqrt{2}.$$

1. $\ln(x + y + 1)$ is defined only when $x + y + 1 > 0$ $\Rightarrow$ $y > -x - 1$,

so the domain of f is $\{(x, y) \mid y > -x - 1\}$, all those points above the

line $y = -x - 1$.

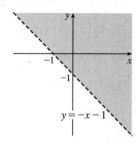

3. $z = f(x, y) = 1 - y^2$, a parabolic cylinder

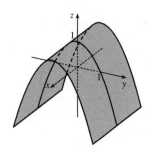

5. The level curves are $\sqrt{4x^2 + y^2} = k$ or $4x^2 + y^2 = k^2$,

$k \geq 0$, a family of ellipses.

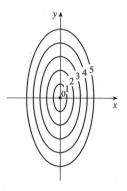

7.

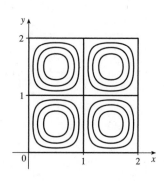

9. f is a rational function, so it is continuous on its

domain. Since f is defined at $(1, 1)$, we use direct

substitution to evaluate the limit:

$$\lim_{(x,y) \to (1,1)} \frac{2xy}{x^2 + 2y^2} = \frac{2(1)(1)}{1^2 + 2(1)^2} = \frac{2}{3}.$$

11. $f(x, y) = \sqrt{2x + y^2}$ $\Rightarrow$ $f_x = \frac{1}{2}(2x + y^2)^{-1/2}(2) = \dfrac{1}{\sqrt{2x + y^2}}$, $f_y = \frac{1}{2}(2x + y^2)^{-1/2}(2y) = \dfrac{y}{\sqrt{2x + y^2}}$

13. $g(u, v) = u \tan^{-1} v$ $\Rightarrow$ $g_u = \tan^{-1} v$, $g_v = \dfrac{u}{1 + v^2}$

15. $T(p, q, r) = p \ln(q + e^r)$ $\Rightarrow$ $T_p = \ln(q + e^r)$, $T_q = \dfrac{p}{q + e^r}$, $T_r = \dfrac{pe^r}{q + e^r}$

17. $f(x, y) = 4x^3 - xy^2$ $\Rightarrow$ $f_x = 12x^2 - y^2$, $f_y = -2xy$, $f_{xx} = 24x$, $f_{yy} = -2x$, $f_{xy} = f_{yx} = -2y$

19. $f(x,y,z) = x^k y^l z^m \Rightarrow f_x = kx^{k-1} y^l z^m,\ f_y = lx^k y^{l-1} z^m,\ f_z = mx^k y^l z^{m-1},\ f_{xx} = k(k-1)x^{k-2} y^l z^m,$

$f_{yy} = l(l-1)x^k y^{l-2} z^m,\ f_{zz} = m(m-1)x^k y^l z^{m-2},\ f_{xy} = f_{yx} = klx^{k-1} y^{l-1} z^m,\ f_{xz} = f_{zx} = kmx^{k-1} y^l z^{m-1},$

$f_{yz} = f_{zy} = lmx^k y^{l-1} z^{m-1}$

21. $z = xy + xe^{y/x} \Rightarrow \dfrac{\partial z}{\partial x} = y - \dfrac{y}{x}e^{y/x} + e^{y/x},\ \dfrac{\partial z}{\partial y} = x + e^{y/x}$ and

$x\dfrac{\partial z}{\partial x} + y\dfrac{\partial z}{\partial y} = x\left(y - \dfrac{y}{x}e^{y/x} + e^{y/x}\right) + y\left(x + e^{y/x}\right) = xy - ye^{y/x} + xe^{y/x} + xy + ye^{y/x} = xy + xy + xe^{y/x} = xy + z.$

23. (a) $z_x = 6x + 2 \Rightarrow z_x(1,-2) = 8$ and $z_y = -2y \Rightarrow z_y(1,-2) = 4$, so an equation of the tangent plane is

$z - 1 = 8(x-1) + 4(y+2)$ or $z = 8x + 4y + 1$.

(b) A normal vector to the tangent plane (and the surface) at $(1,-2,1)$ is $\langle 8, 4, -1 \rangle$. Then parametric equations for the normal

line there are $x = 1 + 8t,\ y = -2 + 4t,\ z = 1 - t$, and symmetric equations are $\dfrac{x-1}{8} = \dfrac{y+2}{4} = \dfrac{z-1}{-1}$.

25. (a) Let $F(x,y,z) = x^2 + 2y^2 - 3z^2$. Then $F_x = 2x,\ F_y = 4y,\ F_z = -6z$, so $F_x(2,-1,1) = 4,\ F_y(2,-1,1) = -4,$

$F_z(2,-1,1) = -6$. From Equation 11.6.19, an equation of the tangent plane is $4(x-2) - 4(y+1) - 6(z-1) = 0$

or, equivalently, $2x - 2y - 3z = 3$.

(b) From Equations 11.6.20, symmetric equations for the normal line are $\dfrac{x-2}{4} = \dfrac{y+1}{-4} = \dfrac{z-1}{-6}$.

27. (a) Let $F(x,y,z) = x + 2y + 3z - \sin(xyz)$. Then $F_x = 1 - yz\cos(xyz),\ F_y = 2 - xz\cos(xyz),\ F_z = 3 - xy\cos(xyz),$

so $F_x(2,-1,0) = 1,\ F_y(2,-1,0) = 2,\ F_z(2,-1,0) = 5$. From Equation 11.6.19, an equation of the tangent plane is

$1(x-2) + 2(y+1) + 5(z-0) = 0$ or $x + 2y + 5z = 0$.

(b) From Equations 11.6.20, symmetric equations for the normal line are $\dfrac{x-2}{1} = \dfrac{y+1}{2} = \dfrac{z}{5}$.

29. The hyperboloid is a level surface of the function $F(x,y,z) = x^2 + 4y^2 - z^2$, so a normal vector to the surface at (x_0, y_0, z_0)

is $\nabla F(x_0, y_0, z_0) = \langle 2x_0, 8y_0, -2z_0 \rangle$. A normal vector for the plane $2x + 2y + z = 5$ is $\langle 2, 2, 1 \rangle$. For the planes to be

parallel, we need the normal vectors to be parallel, so $\langle 2x_0, 8y_0, -2z_0 \rangle = k\langle 2, 2, 1 \rangle$, or $x_0 = k,\ y_0 = \frac{1}{4}k$, and $z_0 = -\frac{1}{2}k$.

But $x_0^2 + 4y_0^2 - z_0^2 = 4 \Rightarrow k^2 + \frac{1}{4}k^2 - \frac{1}{4}k^2 = 4 \Rightarrow k^2 = 4 \Rightarrow k = \pm 2$. So there are two such points:

$\left(2, \frac{1}{2}, -1\right)$ and $\left(-2, -\frac{1}{2}, 1\right)$.

31. $f(x,y,z) = x^3\sqrt{y^2 + z^2} \Rightarrow f_x(x,y,z) = 3x^2\sqrt{y^2 + z^2},\ f_y(x,y,z) = \dfrac{yx^3}{\sqrt{y^2 + z^2}},\ f_z(x,y,z) = \dfrac{zx^3}{\sqrt{y^2 + z^2}},$

so $f(2,3,4) = 8(5) = 40,\ f_x(2,3,4) = 3(4)\sqrt{25} = 60,\ f_y(2,3,4) = \frac{3(8)}{\sqrt{25}} = \frac{24}{5}$, and $f_z(2,3,4) = \frac{4(8)}{\sqrt{25}} = \frac{32}{5}$. Then the

linear approximation of f at $(2,3,4)$ is

$$f(x,y,z) \approx f(2,3,4) + f_x(2,3,4)(x-2) + f_y(2,3,4)(y-3) + f_z(2,3,4)(z-4)$$

$$= 40 + 60(x-2) + \tfrac{24}{5}(y-3) + \tfrac{32}{5}(z-4) = 60x + \tfrac{24}{5}y + \tfrac{32}{5}z - 120$$

Then $(1.98)^3\sqrt{(3.01)^2 + (3.97)^2} = f(1.98, 3.01, 3.97) \approx 60(1.98) + \tfrac{24}{5}(3.01) + \tfrac{32}{5}(3.97) - 120 = 38.656.$

33. $\dfrac{du}{dp} = \dfrac{\partial u}{\partial x}\dfrac{dx}{dp} + \dfrac{\partial u}{\partial y}\dfrac{dy}{dp} + \dfrac{\partial u}{\partial z}\dfrac{dz}{dp} = 2xy^3(1+6p) + 3x^2y^2(pe^p + e^p) + 4z^3(p\cos p + \sin p)$

35. By the Chain Rule, $\dfrac{\partial z}{\partial s} = \dfrac{\partial z}{\partial x}\dfrac{\partial x}{\partial s} + \dfrac{\partial z}{\partial y}\dfrac{\partial y}{\partial s}$. When $s = 1$ and $t = 2$, $x = g(1,2) = 3$ and $y = h(1,2) = 6$, so

$\dfrac{\partial z}{\partial s} = f_x(3,6)g_s(1,2) + f_y(3,6)h_s(1,2) = (7)(-1) + (8)(-5) = -47$. Similarly, $\dfrac{\partial z}{\partial t} = \dfrac{\partial z}{\partial x}\dfrac{\partial x}{\partial t} + \dfrac{\partial z}{\partial y}\dfrac{\partial y}{\partial t}$, so

$\dfrac{\partial z}{\partial t} = f_x(3,6)g_t(1,2) + f_y(3,6)h_t(1,2) = (7)(4) + (8)(10) = 108.$

37. $\dfrac{\partial z}{\partial x} = 2xf'(x^2 - y^2)$, $\quad \dfrac{\partial z}{\partial y} = 1 - 2yf'(x^2 - y^2)\quad \left[\text{where } f' = \dfrac{df}{d(x^2 - y^2)}\right]$. Then

$y\dfrac{\partial z}{\partial x} + x\dfrac{\partial z}{\partial y} = 2xyf'(x^2 - y^2) + x - 2xyf'(x^2 - y^2) = x.$

39. $\dfrac{\partial z}{\partial x} = \dfrac{\partial z}{\partial u}y + \dfrac{\partial z}{\partial v}\dfrac{-y}{x^2}$ and

$\dfrac{\partial^2 z}{\partial x^2} = y\dfrac{\partial}{\partial x}\left(\dfrac{\partial z}{\partial u}\right) + \dfrac{2y}{x^3}\dfrac{\partial z}{\partial v} + \dfrac{-y}{x^2}\dfrac{\partial}{\partial x}\left(\dfrac{\partial z}{\partial v}\right) = \dfrac{2y}{x^3}\dfrac{\partial z}{\partial v} + y\left(\dfrac{\partial^2 z}{\partial u^2}y + \dfrac{\partial^2 z}{\partial v\,\partial u}\dfrac{-y}{x^2}\right) + \dfrac{-y}{x^2}\left(\dfrac{\partial^2 z}{\partial v^2}\dfrac{-y}{x^2} + \dfrac{\partial^2 z}{\partial u\,\partial v}y\right)$

$= \dfrac{2y}{x^3}\dfrac{\partial z}{\partial v} + y^2\dfrac{\partial^2 z}{\partial u^2} - \dfrac{2y^2}{x^2}\dfrac{\partial^2 z}{\partial u\,\partial v} + \dfrac{y^2}{x^4}\dfrac{\partial^2 z}{\partial v^2}$

Also $\dfrac{\partial z}{\partial y} = x\dfrac{\partial z}{\partial u} + \dfrac{1}{x}\dfrac{\partial z}{\partial v}$ and

$\dfrac{\partial^2 z}{\partial y^2} = x\dfrac{\partial}{\partial y}\left(\dfrac{\partial z}{\partial u}\right) + \dfrac{1}{x}\dfrac{\partial}{\partial y}\left(\dfrac{\partial z}{\partial v}\right) = x\left(\dfrac{\partial^2 z}{\partial u^2}x + \dfrac{\partial^2 z}{\partial v\,\partial u}\dfrac{1}{x}\right) + \dfrac{1}{x}\left(\dfrac{\partial^2 z}{\partial v^2}\dfrac{1}{x} + \dfrac{\partial^2 z}{\partial u\,\partial v}x\right) = x^2\dfrac{\partial^2 z}{\partial u^2} + 2\dfrac{\partial^2 z}{\partial u\,\partial v} + \dfrac{1}{x^2}\dfrac{\partial^2 z}{\partial v^2}$

Thus

$x^2\dfrac{\partial^2 z}{\partial x^2} - y^2\dfrac{\partial^2 z}{\partial y^2} = \dfrac{2y}{x}\dfrac{\partial z}{\partial v} + x^2y^2\dfrac{\partial^2 z}{\partial u^2} - 2y^2\dfrac{\partial^2 z}{\partial u\,\partial v} + \dfrac{y^2}{x^2}\dfrac{\partial^2 z}{\partial v^2} - x^2y^2\dfrac{\partial^2 z}{\partial u^2} - 2y^2\dfrac{\partial^2 z}{\partial u\,\partial v} - \dfrac{y^2}{x^2}\dfrac{\partial^2 z}{\partial v^2}$

$= \dfrac{2y}{x}\dfrac{\partial z}{\partial v} - 4y^2\dfrac{\partial^2 z}{\partial u\,\partial v} = 2v\dfrac{\partial z}{\partial v} - 4uv\dfrac{\partial^2 z}{\partial u\,\partial v}$

since $y = xv = \dfrac{uv}{y}$ or $y^2 = uv$.

41. $\nabla f = \left\langle z^2\sqrt{y}\,e^{x\sqrt{y}}, \dfrac{xz^2 e^{x\sqrt{y}}}{2\sqrt{y}}, 2ze^{x\sqrt{y}}\right\rangle = ze^{x\sqrt{y}}\left\langle z\sqrt{y}, \dfrac{xz}{2\sqrt{y}}, 2\right\rangle$

43. $\nabla f = \langle 1/\sqrt{x}, -2y\rangle$, $\nabla f(1,5) = \langle 1, -10\rangle$, $\mathbf{u} = \frac{1}{5}\langle 3, -4\rangle$. Then $D_{\mathbf{u}}f(1,5) = \frac{43}{5}$.

45. $\nabla f = \langle 2xy, x^2 + 1/(2\sqrt{y})\rangle$, $|\nabla f(2,1)| = |\langle 4, \frac{9}{2}\rangle|$. Thus the maximum rate of change of f at $(2,1)$ is $\dfrac{\sqrt{145}}{2}$ in the

direction $\left\langle 4, \frac{9}{2}\right\rangle$.

47. $f(x,y) = x^2 - xy + y^2 + 9x - 6y + 10 \quad\Rightarrow\quad f_x = 2x - y + 9,$

$f_y = -x + 2y - 6$, $f_{xx} = 2 = f_{yy}$, $f_{xy} = -1$. Then $f_x = 0$ and $f_y = 0$ imply

$y = 1$, $x = -4$. Thus the only critical point is $(-4,1)$ and $f_{xx}(-4,1) > 0$,

$D(-4,1) = 3 > 0$, so $f(-4,1) = -11$ is a local minimum.

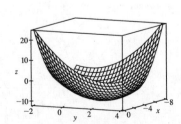

49. $f(x, y) = 3xy - x^2y - xy^2 \Rightarrow f_x = 3y - 2xy - y^2$, $f_y = 3x - x^2 - 2xy$,

$f_{xx} = -2y$, $f_{yy} = -2x$, $f_{xy} = 3 - 2x - 2y$. Then $f_x = 0$ implies

$y(3 - 2x - y) = 0$ so $y = 0$ or $y = 3 - 2x$. Substituting into $f_y = 0$ implies

$x(3 - x) = 0$ or $3x(-1 + x) = 0$. Hence the critical points are $(0, 0)$, $(3, 0)$,

$(0, 3)$ and $(1, 1)$. $D(0, 0) = D(3, 0) = D(0, 3) = -9 < 0$ so $(0, 0)$, $(3, 0)$, and

$(0, 3)$ are saddle points. $D(1, 1) = 3 > 0$ and $f_{xx}(1, 1) = -2 < 0$, so

$f(1, 1) = 1$ is a local maximum.

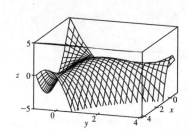

51. First solve inside D. Here $f_x = 4y^2 - 2xy^2 - y^3$, $f_y = 8xy - 2x^2y - 3xy^2$.

Then $f_x = 0$ implies $y = 0$ or $y = 4 - 2x$, but $y = 0$ isn't inside D. Substituting

$y = 4 - 2x$ into $f_y = 0$ implies $x = 0$, $x = 2$ or $x = 1$, but $x = 0$ isn't inside D,

and when $x = 2$, $y = 0$ but $(2, 0)$ isn't inside D. Thus the only critical point inside

D is $(1, 2)$ and $f(1, 2) = 4$. Secondly we consider the boundary of D.

On L_1: $f(x, 0) = 0$ and so $f = 0$ on L_1. On L_2: $x = -y + 6$ and

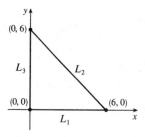

$f(-y + 6, y) = y^2(6 - y)(-2) = -2(6y^2 - y^3)$ which has critical points

at $y = 0$ and $y = 4$. Then $f(6, 0) = 0$ while $f(2, 4) = -64$. On L_3: $f(0, y) = 0$, so $f = 0$ on L_3. Thus on D the absolute

maximum of f is $f(1, 2) = 4$ while the absolute minimum is $f(2, 4) = -64$.

53. $f(x, y) = x^3 - 3x + y^4 - 2y^2$

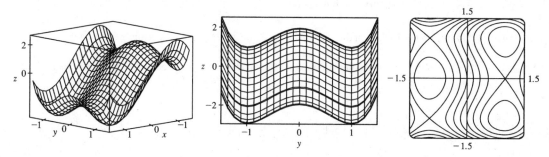

From the graphs, it appears that f has a local maximum $f(-1, 0) \approx 2$, local minima $f(1, \pm 1) \approx -3$, and saddle points at

$(-1, \pm 1)$ and $(1, 0)$.

To find the exact quantities, we calculate $f_x = 3x^2 - 3 = 0 \Leftrightarrow x = \pm 1$ and $f_y = 4y^3 - 4y = 0 \Leftrightarrow$

$y = 0, \pm 1$, giving the critical points estimated above. Also $f_{xx} = 6x$, $f_{xy} = 0$, $f_{yy} = 12y^2 - 4$, so using the Second

Derivatives Test, $D(-1, 0) = 24 > 0$ and $f_{xx}(-1, 0) = -6 < 0$ indicating a local maximum $f(-1, 0) = 2$;

$D(1, \pm 1) = 48 > 0$ and $f_{xx}(1, \pm 1) = 6 > 0$ indicating local minima $f(1, \pm 1) = -3$; and $D(-1, \pm 1) = -48$ and

$D(1, 0) = -24$, indicating saddle points.

55. $f(x,y) = x^2 y$, $g(x,y) = x^2 + y^2 = 1$ $\Rightarrow$ $\nabla f = \langle 2xy, x^2 \rangle = \lambda \nabla g = \langle 2\lambda x, 2\lambda y \rangle$. Then $2xy = 2\lambda x$ and $x^2 = 2\lambda y$

imply $\lambda = x^2/(2y)$ and $\lambda = y$ if $x \neq 0$ and $y \neq 0$. Hence $x^2 = 2y^2$. Then $x^2 + y^2 = 1$ implies $3y^2 = 1$ so $y = \pm\frac{1}{\sqrt{3}}$ and

$x = \pm\sqrt{\frac{2}{3}}$. [Note if $x = 0$ then $x^2 = 2\lambda y$ implies $y = 0$ and $f(0,0) = 0$.] Thus the possible points are $\left(\pm\sqrt{\frac{2}{3}}, \pm\frac{1}{\sqrt{3}}\right)$ and

the absolute maxima are $f\left(\pm\sqrt{\frac{2}{3}}, \frac{1}{\sqrt{3}}\right) = \frac{2}{3\sqrt{3}}$ while the absolute minima are $f\left(\pm\sqrt{\frac{2}{3}}, -\frac{1}{\sqrt{3}}\right) = -\frac{2}{3\sqrt{3}}$.

57. $f(x,y,z) = xyz$, $g(x,y,z) = x^2 + y^2 + z^2 = 3$. $\nabla f = \lambda \nabla g$ $\Rightarrow$ $\langle yz, xz, xy \rangle = \lambda \langle 2x, 2y, 2z \rangle$. If any of x, y, or z is

zero, then $x = y = z = 0$ which contradicts $x^2 + y^2 + z^2 = 3$. Then $\lambda = \dfrac{yz}{2x} = \dfrac{xz}{2y} = \dfrac{xy}{2z}$ $\Rightarrow$ $2y^2 z = 2x^2 z$ $\Rightarrow$

$y^2 = x^2$, and similarly $2yz^2 = 2x^2 y$ $\Rightarrow$ $z^2 = x^2$. Substituting into the constraint equation gives $x^2 + x^2 + x^2 = 3$ $\Rightarrow$

$x^2 = 1 = y^2 = z^2$. Thus the possible points are $(1, 1, \pm 1)$, $(1, -1, \pm 1)$, $(-1, 1, \pm 1)$, $(-1, -1, \pm 1)$. The absolute maximum

is $f(1, 1, 1) = f(1, -1, -1) = f(-1, 1, -1) = f(-1, -1, 1) = 1$ and the absolute

minimum is $f(1, 1, -1) = f(1, -1, 1) = f(-1, 1, 1) = f(-1, -1, -1) = -1$.

59. $f(x,y,z) = x^2 + y^2 + z^2$, $g(x,y,z) = xy^2 z^3 = 2$ $\Rightarrow$ $\nabla f = \langle 2x, 2y, 2z \rangle = \lambda \nabla g = \langle \lambda y^2 z^3, 2\lambda xyz^3, 3\lambda xy^2 z^2 \rangle$.

Since $xy^2 z^3 = 2$, $x \neq 0$, $y \neq 0$ and $z \neq 0$, so $2x = \lambda y^2 z^3$ **(1)**, $1 = \lambda xz^3 = xz^3 \lambda$ **(2)**, $2 = 3\lambda xy^2 z$ **(3)**. Then **(2)** and

(3) imply $\dfrac{1}{xz^3} = \dfrac{2}{3xy^2 z}$ or $y^2 = \frac{2}{3}z^2$ so $y = \pm z\sqrt{\frac{2}{3}}$. Similarly **(1)** and **(3)** imply $\dfrac{2x}{y^2 z^3} = \dfrac{2}{3xy^2 z}$ or $3x^2 = z^2$ so

$x = \pm\frac{1}{\sqrt{3}}z$. But $xy^2 z^3 = 2$ so x and z must have the same sign, that is, $x = \frac{1}{\sqrt{3}}z$. Thus $g(x,y,z) = 2$ implies

$\frac{1}{\sqrt{3}}z\left(\frac{2}{3}z^2\right)z^3 = 2$ or $z = \pm 3^{1/4}$ and the possible points are $(\pm 3^{-1/4}, 3^{-1/4}\sqrt{2}, \pm 3^{1/4})$, $(\pm 3^{-1/4}, -3^{-1/4}\sqrt{2}, \pm 3^{1/4})$.

However at each of these points f takes on the same value, $2\sqrt{3}$. But $(2, 1, 1)$ also satisfies $g(x,y,z) = 2$ and

$f(2, 1, 1) = 6 > 2\sqrt{3}$. Thus f has an absolute minimum value of $2\sqrt{3}$ and no absolute maximum subject to the constraint

$xy^2 z^3 = 2$.

Alternate solution: $g(x,y,z) = xy^2 z^3 = 2$ implies $y^2 = \dfrac{2}{xz^3}$, so minimize $f(x,z) = x^2 + \dfrac{2}{xz^3} + z^2$. Then

$f_x = 2x - \dfrac{2}{x^2 z^3}$, $f_z = -\dfrac{6}{xz^4} + 2z$, $f_{xx} = 2 + \dfrac{4}{x^3 z^3}$, $f_{zz} = \dfrac{24}{xz^5} + 2$ and $f_{xz} = \dfrac{6}{x^2 z^4}$. Now $f_x = 0$ implies

$2x^3 z^3 - 2 = 0$ or $z = 1/x$. Substituting into $f_y = 0$ implies $-6x^3 + 2x^{-1} = 0$ or $x = \frac{1}{\sqrt[4]{3}}$, so the two critical points are

$\left(\pm\frac{1}{\sqrt[4]{3}}, \pm\sqrt[4]{3}\right)$. Then $D\left(\pm\frac{1}{\sqrt[4]{3}}, \pm\sqrt[4]{3}\right) = (2 + 4)\left(2 + \frac{24}{3}\right) - \left(\frac{6}{\sqrt{3}}\right)^2 > 0$ and $f_{xx}\left(\pm\frac{1}{\sqrt[4]{3}}, \pm\sqrt[4]{3}\right) = 6 > 0$, so each point

is a minimum. Finally, $y^2 = \dfrac{2}{xz^3}$, so the four points closest to the origin are $\left(\pm\frac{1}{\sqrt[4]{3}}, \frac{\sqrt{2}}{\sqrt[4]{3}}, \pm\sqrt[4]{3}\right)$, $\left(\pm\frac{1}{\sqrt[4]{3}}, -\frac{\sqrt{2}}{\sqrt[4]{3}}, \pm\sqrt[4]{3}\right)$.

61.

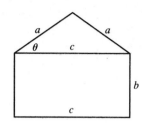

The area of the triangle is $\frac{1}{2}ca\sin\theta$ and the area of the rectangle is bc. Thus, the area of the whole object is $f(a,b,c) = \frac{1}{2}ca\sin\theta + bc$. The perimeter of the object is $g(a,b,c) = 2a + 2b + c = P$. To simplify $\sin\theta$ in terms of a, b, and c notice that $a^2\sin^2\theta + \left(\frac{1}{2}c\right)^2 = a^2 \;\Rightarrow\; \sin\theta = \dfrac{1}{2a}\sqrt{4a^2 - c^2}$. Thus

$$f(a,b,c) = \frac{c}{4}\sqrt{4a^2 - c^2} + bc.$$ (Instead of using θ, we could just have used the Pythagorean Theorem.) As a result, by Lagrange's method, we must find a, b, c, and λ by solving $\nabla f = \lambda \nabla g$ which gives the following equations: $ca(4a^2 - c^2)^{-1/2} = 2\lambda$ **(1)**, $c = 2\lambda$ **(2)**, $\frac{1}{4}(4a^2 - c^2)^{1/2} - \frac{1}{4}c^2(4a^2 - c^2)^{-1/2} + b = \lambda$ **(3)**, and $2a + 2b + c = P$ **(4)**. From **(2)**, $\lambda = \frac{1}{2}c$ and so **(1)** produces $ca(4a^2 - c^2)^{-1/2} = c \;\Rightarrow\; (4a^2 - c^2)^{1/2} = a \;\Rightarrow\; 4a^2 - c^2 = a^2 \;\Rightarrow\;$ **(5)** $c = \sqrt{3}\,a$. Similarly, since $\left(4a^2 - c^2\right)^{1/2} = a$ and $\lambda = \frac{1}{2}c$, **(3)** gives $\dfrac{a}{4} - \dfrac{c^2}{4a} + b = \dfrac{c}{2}$, so from **(5)**, $\dfrac{a}{4} - \dfrac{3a}{4} + b = \dfrac{\sqrt{3}\,a}{2} \;\Rightarrow\; -\dfrac{a}{2} - \dfrac{\sqrt{3}\,a}{2} = -b \;\Rightarrow\;$ **(6)** $b = \dfrac{a}{2}\left(1 + \sqrt{3}\right)$. Substituting **(5)** and **(6)** into **(4)** we get:

$$2a + a\left(1 + \sqrt{3}\right) + \sqrt{3}\,a = P \;\Rightarrow\; 3a + 2\sqrt{3}\,a = P \;\Rightarrow\; a = \frac{P}{3 + 2\sqrt{3}} = \frac{2\sqrt{3} - 3}{3}P$$ and thus

$$b = \frac{\left(2\sqrt{3} - 3\right)\left(1 + \sqrt{3}\right)}{6}P = \frac{3 - \sqrt{3}}{6}P \text{ and } c = \left(2 - \sqrt{3}\right)P.$$

12 □ MULTIPLE INTEGRALS

12.1 Double Integrals over Rectangles

1. (a) The subrectangles are shown in the figure.

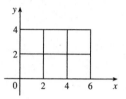

The surface is the graph of $f(x, y) = xy$ and $\Delta A = 4$, so we estimate

$$V \approx \sum_{i=1}^{3} \sum_{j=1}^{2} f(x_i, y_j) \, \Delta A$$

$$= f(2, 2) \, \Delta A + f(2, 4) \, \Delta A + f(4, 2) \, \Delta A + f(4, 4) \, \Delta A + f(6, 2) \, \Delta A + f(6, 4) \, \Delta A$$

$$= 4(4) + 8(4) + 8(4) + 16(4) + 12(4) + 24(4) = 288$$

(b) $V \approx \sum_{i=1}^{3} \sum_{j=1}^{2} f(\overline{x}_i, \overline{y}_j) \, \Delta A = f(1, 1) \, \Delta A + f(1, 3) \, \Delta A + f(3, 1) \, \Delta A + f(3, 3) \, \Delta A + f(5, 1) \, \Delta A + f(5, 3) \, \Delta A$

$$= 1(4) + 3(4) + 3(4) + 9(4) + 5(4) + 15(4) = 144$$

3. (a) The subrectangles are shown in the figure. Since $\Delta A = \pi^2/4$, we estimate

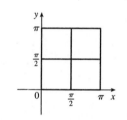

$$\iint_R \sin(x + y) \, dA \approx \sum_{i=1}^{2} \sum_{j=1}^{2} f\left(x_{ij}^*, y_{ij}^*\right) \Delta A$$

$$= f(0, 0) \, \Delta A + f\left(0, \tfrac{\pi}{2}\right) \Delta A + f\left(\tfrac{\pi}{2}, 0\right) \Delta A + f\left(\tfrac{\pi}{2}, \tfrac{\pi}{2}\right) \Delta A$$

$$= 0\left(\tfrac{\pi^2}{4}\right) + 1\left(\tfrac{\pi^2}{4}\right) + 1\left(\tfrac{\pi^2}{4}\right) + 0\left(\tfrac{\pi^2}{4}\right) = \tfrac{\pi^2}{2} \approx 4.935$$

(b) $\iint_R \sin(x + y) \, dA \approx \sum_{i=1}^{2} \sum_{j=1}^{2} f(\overline{x}_i, \overline{y}_j) \, \Delta A$

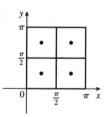

$$= f\left(\tfrac{\pi}{4}, \tfrac{\pi}{4}\right) \Delta A + f\left(\tfrac{\pi}{4}, \tfrac{3\pi}{4}\right) \Delta A + f\left(\tfrac{3\pi}{4}, \tfrac{\pi}{4}\right) \Delta A + f\left(\tfrac{3\pi}{4}, \tfrac{3\pi}{4}\right) \Delta A$$

$$= 1\left(\tfrac{\pi^2}{4}\right) + 0\left(\tfrac{\pi^2}{4}\right) + 0\left(\tfrac{\pi^2}{4}\right) + (-1)\left(\tfrac{\pi^2}{4}\right) = 0$$

(c) $\iint_R \sin(x + y) \, dA = \int_0^\pi \int_0^\pi \sin(x + y) \, dx \, dy = \int_0^\pi \left[-\cos(x + y)\right]_{x=0}^{x=\pi} dy = \int_0^\pi \left[\cos y - \cos(y + \pi)\right] dy$

$$= \sin y - \sin(y + \pi)\big]_0^\pi = 0$$

So the estimate from the Midpoint Rule in part (b) is same as the exact value.

5. With $m = n = 2$, we have $\Delta A = 4$. Using the contour map to estimate the value of f at the center of each subrectangle, we have

$$\iint_R f(x, y) \, dA \approx \sum_{i=1}^{2} \sum_{j=1}^{2} f(\overline{x}_i, \overline{y}_j) \, \Delta A = \Delta A[f(1, 1) + f(1, 3) + f(3, 1) + f(3, 3)] \approx 4(27 + 4 + 14 + 17) = 248.$$

7. $z = 3 > 0$, so we can interpret the integral as the volume of the solid S that lies below the plane $z = 3$ and above the rectangle $[-2, 2] \times [1, 6]$. S is a rectangular solid, thus $\iint_R 3 \, dA = 4 \cdot 5 \cdot 3 = 60$.

9. $z = f(x, y) = 4 - 2y \geq 0$ for $0 \leq y \leq 1$. Thus the integral represents the volume of that

part of the rectangular solid $[0, 1] \times [0, 1] \times [0, 4]$ which lies below the plane $z = 4 - 2y$.

So

$$\iint_R (4 - 2y)\, dA = (1)(1)(2) + \tfrac{1}{2}(1)(1)(2) = 3$$

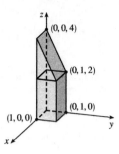

11. $\int_1^3 \int_0^1 (1 + 4xy)\, dx\, dy = \int_1^3 \left[x + 2x^2y\right]_{x=0}^{x=1} dy = \int_1^3 (1 + 2y)\, dy = \left[y + y^2\right]_1^3 = (3 + 9) - (1 + 1) = 10$

13. $\int_0^2 \int_0^{\pi/2} x \sin y\, dy\, dx = \int_0^2 x\, dx \int_0^{\pi/2} \sin y\, dy$ [as in Example 5] $= \left[\dfrac{x^2}{2}\right]_0^2 \left[-\cos y\right]_0^{\pi/2} = (2 - 0)(0 + 1) = 2$

15. $\int_0^2 \int_0^1 (2x + y)^8\, dx\, dy = \int_0^2 \left[\dfrac{1}{2}\dfrac{(2x + y)^9}{9}\right]_{x=0}^{x=1} dy$ [substitute $u = 2x + y \;\Rightarrow\; dx = \tfrac{1}{2}\, du$]

$$= \frac{1}{18}\int_0^2 \left[(2 + y)^9 - (0 + y)^9\right] dy = \frac{1}{18}\left[\frac{(2 + y)^{10}}{10} - \frac{y^{10}}{10}\right]_0^2$$

$$= \tfrac{1}{180}\left[(4^{10} - 2^{10}) - (2^{10} - 0^{10})\right] = \tfrac{1{,}046{,}528}{180} = \tfrac{261{,}632}{45}$$

17. $\displaystyle\int_1^4 \int_1^2 \left(\dfrac{x}{y} + \dfrac{y}{x}\right) dy\, dx = \int_1^4 \left[x \ln|y| + \dfrac{1}{x}\cdot\dfrac{1}{2}y^2\right]_{y=1}^{y=2} dx = \int_1^4 \left(x \ln 2 + \dfrac{3}{2x}\right) dx = \left[\tfrac{1}{2}x^2 \ln 2 + \tfrac{3}{2}\ln|x|\right]_1^4$

$$= 8 \ln 2 + \tfrac{3}{2}\ln 4 - \tfrac{1}{2}\ln 2 = \tfrac{15}{2}\ln 2 + 3 \ln 4^{1/2} = \tfrac{21}{2}\ln 2$$

19. $\int_0^{\ln 2} \int_0^{\ln 5} e^{2x - y}\, dx\, dy = \left(\int_0^{\ln 5} e^{2x}\, dx\right)\left(\int_0^{\ln 2} e^{-y}\, dy\right) = \left[\tfrac{1}{2}e^{2x}\right]_0^{\ln 5}\left[-e^{-y}\right]_0^{\ln 2} = \left(\tfrac{25}{2} - \tfrac{1}{2}\right)\left(-\tfrac{1}{2} + 1\right) = 6$

21. $\displaystyle\iint_R \dfrac{xy^2}{x^2 + 1}\, dA = \int_0^1 \int_{-3}^3 \dfrac{xy^2}{x^2 + 1}\, dy\, dx = \int_0^1 \dfrac{x}{x^2 + 1}\, dx \int_{-3}^3 y^2\, dy$

$$= \left[\tfrac{1}{2}\ln(x^2 + 1)\right]_0^1 \left[\tfrac{1}{3}y^3\right]_{-3}^3 = \tfrac{1}{2}(\ln 2 - \ln 1)\cdot\tfrac{1}{3}(27 + 27) = 9 \ln 2$$

23. $\int_0^{\pi/6} \int_0^{\pi/3} x \sin(x + y)\, dy\, dx$

$$= \int_0^{\pi/6} \left[-x \cos(x + y)\right]_{y=0}^{y=\pi/3} dx = \int_0^{\pi/6} \left[x \cos x - x \cos\left(x + \tfrac{\pi}{3}\right)\right] dx$$

$$= x\left[\sin x - \sin\left(x + \tfrac{\pi}{3}\right)\right]_0^{\pi/6} - \int_0^{\pi/6} \left[\sin x - \sin\left(x + \tfrac{\pi}{3}\right)\right] dx \qquad \text{[by integrating by parts separately for each term]}$$

$$= \tfrac{\pi}{6}\left[\tfrac{1}{2} - 1\right] - \left[-\cos x + \cos\left(x + \tfrac{\pi}{3}\right)\right]_0^{\pi/6} = -\tfrac{\pi}{12} - \left[-\tfrac{\sqrt{3}}{2} + 0 - \left(-1 + \tfrac{1}{2}\right)\right] = \tfrac{\sqrt{3} - 1}{2} - \tfrac{\pi}{12}$$

25. $\iint_R xye^{x^2 y}\, dA = \int_0^2 \int_0^1 xye^{x^2 y}\, dx\, dy = \int_0^2 \left[\tfrac{1}{2}e^{x^2 y}\right]_{x=0}^{x=1} dy = \tfrac{1}{2}\int_0^2 (e^y - 1)\, dy = \tfrac{1}{2}\left[e^y - y\right]_0^2$

$$= \tfrac{1}{2}\left[(e^2 - 2) - (1 - 0)\right] = \tfrac{1}{2}(e^2 - 3)$$

27. $V = \iint_R (12 - 3x - 2y)\, dA = \int_{-2}^3 \int_0^1 (12 - 3x - 2y)\, dx\, dy = \int_{-2}^3 \left[12x - \tfrac{3}{2}x^2 - 2xy\right]_{x=0}^{x=1} dy$

$$= \int_{-2}^3 \left(\tfrac{21}{2} - 2y\right) dy = \left[\tfrac{21}{2}y - y^2\right]_{-2}^3 = \tfrac{95}{2}$$

29. $V = \int_{-2}^{2} \int_{-1}^{1} \left(1 - \frac{1}{4}x^2 - \frac{1}{9}y^2\right) dx\, dy = 4 \int_{0}^{2} \int_{0}^{1} \left(1 - \frac{1}{4}x^2 - \frac{1}{9}y^2\right) dx\, dy$

$= 4 \int_{0}^{2} \left[x - \frac{1}{12}x^3 - \frac{1}{9}y^2 x\right]_{x=0}^{x=1} dy = 4 \int_{0}^{2} \left(\frac{11}{12} - \frac{1}{9}y^2\right) dy = 4\left[\frac{11}{12}y - \frac{1}{27}y^3\right]_{0}^{2} = 4 \cdot \frac{83}{54} = \frac{166}{27}$

31. Here we need the volume of the solid lying under the surface $z = x\sqrt{x^2 + y}$ and above the square $R = [0,1] \times [0,1]$ in the xy-plane.

$$V = \int_{0}^{1} \int_{0}^{1} x\sqrt{x^2 + y}\, dx\, dy = \int_{0}^{1} \frac{1}{3}\left[(x^2 + y)^{3/2}\right]_{x=0}^{x=1} dy = \frac{1}{3} \int_{0}^{1} \left[(1 + y)^{3/2} - y^{3/2}\right] dy$$

$$= \frac{1}{3} \cdot \frac{2}{5}\left[(1 + y)^{5/2} - y^{5/2}\right]_{0}^{1} = \frac{4}{15}\left(2\sqrt{2} - 1\right)$$

33. In the first octant, $z \geq 0 \quad \Rightarrow \quad y \leq 3$, so

$$V = \int_{0}^{3} \int_{0}^{2} (9 - y^2)\, dx\, dy = \int_{0}^{3} \left[9x - y^2 x\right]_{x=0}^{x=2} dy = \int_{0}^{3} (18 - 2y^2)\, dy = \left[18y - \frac{2}{3}y^3\right]_{0}^{3} = 36$$

35. In Maple, we can calculate the integral by defining the integrand as f and then using the command int(int(f,x=0..1),y=0..1);.

In Mathematica, we can use the command

Integrate[Integrate[f,{x,0,1}],{y,0,1}]. We find that

$\iint_R x^5 y^3 e^{xy}\, dA = 21e - 57 \approx 0.0839$. We can use plot3d (in Maple) or Plot3d (in Mathematica) to graph the function.

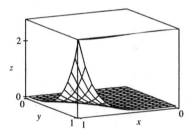

37. R is the rectangle $[-1,1] \times [0,5]$. Thus, $A(R) = 2 \cdot 5 = 10$ and

$f_{\text{ave}} = \dfrac{1}{A(R)} \iint_R f(x,y)\, dA = \frac{1}{10} \int_{0}^{5} \int_{-1}^{1} x^2 y\, dx\, dy = \frac{1}{10} \int_{0}^{5} \left[\frac{1}{3}x^3 y\right]_{x=-1}^{x=1} dy = \frac{1}{10} \int_{0}^{5} \frac{2}{3}y\, dy = \frac{1}{10}\left[\frac{1}{3}y^2\right]_{0}^{5} = \frac{5}{6}$.

39. If we divide R into mn subrectangles, $\iint_R k\, dA \approx \sum_{i=1}^{m} \sum_{j=1}^{n} f(x_{ij}^*, y_{ij}^*)\, \Delta A$ for any choice of sample points (x_{ij}^*, y_{ij}^*).

But $f(x_{ij}^*, y_{ij}^*) = k$ always and $\sum_{i=1}^{m} \sum_{j=1}^{n} \Delta A = \text{area of } R = (b-a)(d-c)$. Thus, no matter how we choose the sample

points, $\sum_{i=1}^{m} \sum_{j=1}^{n} f(x_{ij}^*, y_{ij}^*)\, \Delta A = k \sum_{i=1}^{m} \sum_{j=1}^{n} \Delta A = k(b-a)(d-c)$ and so

$\iint_R k\, dA = \lim\limits_{m,n \to \infty} \sum_{i=1}^{m} \sum_{j=1}^{n} f(x_{ij}^*, y_{ij}^*)\, \Delta A = \lim\limits_{m,n \to \infty} k \sum_{i=1}^{m} \sum_{j=1}^{n} \Delta A = \lim\limits_{m,n \to \infty} k(b-a)(d-c) = k(b-a)(d-c)$.

41. Let $f(x,y) = \dfrac{x - y}{(x + y)^3}$. Then a CAS gives $\int_{0}^{1} \int_{0}^{1} f(x,y)\, dy\, dx = \frac{1}{2}$ and $\int_{0}^{1} \int_{0}^{1} f(x,y)\, dx\, dy = -\frac{1}{2}$.

To explain the seeming violation of Fubini's Theorem, note that f has an infinite discontinuity at $(0,0)$ and thus does not satisfy the conditions of Fubini's Theorem. In fact, both iterated integrals involve improper integrals which diverge at their lower limits of integration.

12.2 Double Integrals over General Regions

1. $\int_0^1 \int_0^{x^2} (x + 2y)\, dy\, dx = \int_0^1 \left[xy + y^2\right]_{y=0}^{y=x^2} dx = \int_0^1 \left[x(x^2) + (x^2)^2 - 0 - 0\right] dx$

$\qquad\qquad = \int_0^1 (x^3 + x^4)\, dx = \left[\frac{1}{4}x^4 + \frac{1}{5}x^5\right]_0^1 = \frac{9}{20}$

3. $\int_0^1 \int_y^{e^y} \sqrt{x}\, dx\, dy = \int_0^1 \left[\frac{2}{3}x^{3/2}\right]_{x=y}^{x=e^y} dy = \frac{2}{3}\int_0^1 (e^{3y/2} - y^{3/2})\, dy = \frac{2}{3}\left[\frac{2}{3}e^{3y/2} - \frac{2}{5}y^{5/2}\right]_0^1$

$\qquad\qquad = \frac{2}{3}\left(\frac{2}{3}e^{3/2} - \frac{2}{5} - \frac{2}{3}e^0 + 0\right) = \frac{4}{9}e^{3/2} - \frac{32}{45}$

5. $\int_0^{\pi/2} \int_0^{\cos\theta} e^{\sin\theta}\, dr\, d\theta = \int_0^{\pi/2} \left[re^{\sin\theta}\right]_{r=0}^{r=\cos\theta} d\theta = \int_0^{\pi/2} (\cos\theta)\, e^{\sin\theta}\, d\theta = e^{\sin\theta}\Big]_0^{\pi/2} = e^{\sin(\pi/2)} - e^0 = e - 1$

7. $\iint_D x^3 y^2\, dA = \int_0^2 \int_{-x}^x x^3 y^2\, dy\, dx = \int_0^2 \left[\frac{1}{3}x^3 y^3\right]_{y=-x}^{y=x} dx = \frac{1}{3}\int_0^2 2x^6\, dx = \frac{2}{3}\left[\frac{1}{7}x^7\right]_0^2 = \frac{2}{21}\left[2^7 - 0\right] = \frac{256}{21}$

9. $\int_0^1 \int_0^{\sqrt{x}} \frac{2y}{x^2 + 1}\, dy\, dx = \int_0^1 \left[\frac{y^2}{x^2 + 1}\right]_{y=0}^{y=\sqrt{x}} dx = \int_0^1 \frac{x}{x^2 + 1}\, dx = \frac{1}{2}\ln\left|x^2 + 1\right|\Big]_0^1 = \frac{1}{2}(\ln 2 - \ln 1) = \frac{1}{2}\ln 2$

11. $\int_0^1 \int_0^{x^2} x\cos y\, dy\, dx = \int_0^1 \left[x\sin y\right]_{y=0}^{y=x^2} dx = \int_0^1 x\sin x^2\, dx = -\frac{1}{2}\cos x^2\Big]_0^1 = \frac{1}{2}(1 - \cos 1)$

13.

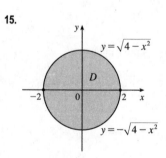

$\int_1^2 \int_{2-y}^{2y-1} y^3\, dx\, dy = \int_1^2 \left[xy^3\right]_{x=2-y}^{x=2y-1} dy = \int_1^2 \left[(2y-1) - (2-y)\right] y^3\, dy$

$\qquad\qquad = \int_1^2 (3y^4 - 3y^3)\, dy = \left[\frac{3}{5}y^5 - \frac{3}{4}y^4\right]_1^2$

$\qquad\qquad = \frac{96}{5} - 12 - \frac{3}{5} + \frac{3}{4} = \frac{147}{20}$

15.

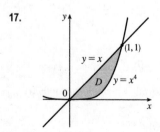

$\int_{-2}^2 \int_{-\sqrt{4-x^2}}^{\sqrt{4-x^2}} (2x - y)\, dy\, dx$

$\qquad = \int_{-2}^2 \left[2xy - \frac{1}{2}y^2\right]_{y=-\sqrt{4-x^2}}^{y=\sqrt{4-x^2}} dx$

$\qquad = \int_{-2}^2 \left[2x\sqrt{4-x^2} - \frac{1}{2}(4-x^2) + 2x\sqrt{4-x^2} + \frac{1}{2}(4-x^2)\right] dx$

$\qquad = \int_{-2}^2 4x\sqrt{4-x^2}\, dx = -\frac{4}{3}(4-x^2)^{3/2}\Big]_{-2}^2 = 0$

[Or, note that $4x\sqrt{4-x^2}$ is an odd function, so $\int_{-2}^2 4x\sqrt{4-x^2}\, dx = 0$.]

17.

$V = \int_0^1 \int_{x^4}^x (x + 2y)\, dy\, dx$

$\qquad = \int_0^1 \left[xy + y^2\right]_{y=x^4}^{y=x} dx = \int_0^1 (2x^2 - x^5 - x^8)\, dx$

$\qquad = \left[\frac{2}{3}x^3 - \frac{1}{6}x^6 - \frac{1}{9}x^9\right]_0^1 = \frac{2}{3} - \frac{1}{6} - \frac{1}{9} = \frac{7}{18}$

19.

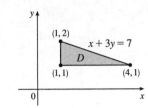

$$V = \int_1^2 \int_1^{7-3y} xy \, dx \, dy = \int_1^2 \left[\tfrac{1}{2} x^2 y \right]_{x=1}^{x=7-3y} dy$$

$$= \tfrac{1}{2} \int_1^2 (48y - 42y^2 + 9y^3) \, dy$$

$$= \tfrac{1}{2} \left[24y^2 - 14y^3 + \tfrac{9}{4} y^4 \right]_1^2 = \tfrac{31}{8}$$

21.

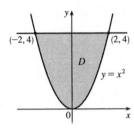

$$V = \int_0^1 \int_0^{1-x} (1 - x - y) \, dy \, dx = \int_0^1 \left[y - xy - \frac{y^2}{2} \right]_{y=0}^{y=1-x} dx$$

$$= \int_0^1 \left[(1-x)^2 - \tfrac{1}{2}(1-x)^2 \right] dx$$

$$= \int_0^1 \tfrac{1}{2} (1-x)^2 \, dx = \left[-\tfrac{1}{6}(1-x)^3 \right]_0^1 = \tfrac{1}{6}$$

23.

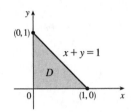

$$V = \int_{-2}^2 \int_{x^2}^4 x^2 \, dy \, dx$$

$$= \int_{-2}^2 x^2 \left[y \right]_{y=x^2}^{y=4} dx = \int_{-2}^2 (4x^2 - x^4) \, dx$$

$$= \left[\tfrac{4}{3} x^3 - \tfrac{1}{5} x^5 \right]_{-2}^2 = \tfrac{32}{3} - \tfrac{32}{5} + \tfrac{32}{3} - \tfrac{32}{5} = \tfrac{128}{15}$$

25.

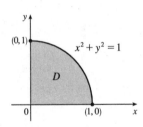

$$V = \int_0^1 \int_0^{\sqrt{1-x^2}} y \, dy \, dx = \int_0^1 \left[\frac{y^2}{2} \right]_{y=0}^{y=\sqrt{1-x^2}} dx$$

$$= \int_0^1 \frac{1-x^2}{2} \, dx = \tfrac{1}{2} \left[x - \tfrac{1}{3} x^3 \right]_0^1 = \tfrac{1}{3}$$

27. The two bounding curves $y = 1 - x^2$ and $y = x^2 - 1$ intersect at $(\pm 1, 0)$ with $1 - x^2 \geq x^2 - 1$ on $[-1, 1]$. Within this region, the plane $z = 2x + 2y + 10$ is above the plane $z = 2 - x - y$, so

$$V = \int_{-1}^1 \int_{x^2-1}^{1-x^2} (2x + 2y + 10) \, dy \, dx - \int_{-1}^1 \int_{x^2-1}^{1-x^2} (2 - x - y) \, dy \, dx$$

$$= \int_{-1}^1 \int_{x^2-1}^{1-x^2} (2x + 2y + 10 - (2 - x - y)) \, dy \, dx$$

$$= \int_{-1}^1 \int_{x^2-1}^{1-x^2} (3x + 3y + 8) \, dy \, dx = \int_{-1}^1 \left[3xy + \tfrac{3}{2} y^2 + 8y \right]_{y=x^2-1}^{y=1-x^2} dx$$

$$= \int_{-1}^1 \left[3x(1-x^2) + \tfrac{3}{2}(1-x^2)^2 + 8(1-x^2) - 3x(x^2-1) - \tfrac{3}{2}(x^2-1)^2 - 8(x^2-1) \right] dx$$

$$= \int_{-1}^1 (-6x^3 - 16x^2 + 6x + 16) \, dx = \left[-\tfrac{3}{2} x^4 - \tfrac{16}{3} x^3 + 3x^2 + 16x \right]_{-1}^1$$

$$= -\tfrac{3}{2} - \tfrac{16}{3} + 3 + 16 + \tfrac{3}{2} - \tfrac{16}{3} - 3 + 16 = \tfrac{64}{3}$$

29. The two surfaces intersect in the circle $x^2 + y^2 = 1$, $z = 0$ and the region of integration is the disk $D: x^2 + y^2 \leq 1$.

Using a CAS, the volume is $\displaystyle\iint_D (1 - x^2 - y^2)\, dA = \int_{-1}^{1} \int_{-\sqrt{1-x^2}}^{\sqrt{1-x^2}} (1 - x^2 - y^2)\, dy\, dx = \frac{\pi}{2}$.

31.

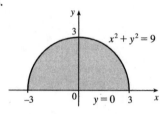

Because the region of integration is

$$D = \{(x, y) \mid 0 \leq y \leq \sqrt{x}, 0 \leq x \leq 4\} = \{(x, y) \mid y^2 \leq x \leq 4, 0 \leq y \leq 2\}$$

we have $\int_0^4 \int_0^{\sqrt{x}} f(x,y)\, dy\, dx = \iint_D f(x,y)\, dA = \int_0^2 \int_{y^2}^4 f(x,y)\, dx\, dy$.

33.

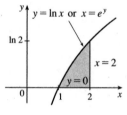

Because the region of integration is

$$D = \left\{(x, y) \mid -\sqrt{9 - y^2} \leq x \leq \sqrt{9 - y^2}, 0 \leq y \leq 3\right\}$$
$$= \{(x, y) \mid 0 \leq y \leq \sqrt{9 - x^2}, -3 \leq x \leq 3\}$$

we have

$$\int_0^3 \int_{-\sqrt{9-y^2}}^{\sqrt{9-y^2}} f(x,y)\, dx\, dy = \iint_D f(x,y)\, dA = \int_{-3}^{3} \int_0^{\sqrt{9-x^2}} f(x,y)\, dy\, dx.$$

35.

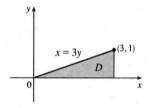

Because the region of integration is

$$D = \{(x, y) \mid 0 \leq y \leq \ln x, 1 \leq x \leq 2\} = \{(x, y) \mid e^y \leq x \leq 2, 0 \leq y \leq \ln 2\}$$

we have

$$\int_1^2 \int_0^{\ln x} f(x,y)\, dy\, dx = \iint_D f(x,y)\, dA = \int_0^{\ln 2} \int_{e^y}^2 f(x,y)\, dx\, dy$$

37.

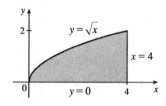

$$\int_0^1 \int_{3y}^3 e^{x^2}\, dx\, dy = \int_0^3 \int_0^{x/3} e^{x^2}\, dy\, dx = \int_0^3 \left[e^{x^2} y \right]_{y=0}^{y=x/3}\, dx$$

$$= \int_0^3 \left(\frac{x}{3}\right) e^{x^2}\, dx = \frac{1}{6}\, e^{x^2} \Big]_0^3 = \frac{e^9 - 1}{6}$$

39.

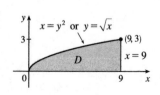

$$\int_0^3 \int_{y^2}^9 y \cos(x^2)\, dx\, dy = \int_0^9 \int_0^{\sqrt{x}} y \cos(x^2)\, dy\, dx$$

$$= \int_0^9 \cos(x^2) \left[\frac{y^2}{2}\right]_{y=0}^{y=\sqrt{x}}\, dx = \int_0^9 \frac{1}{2} x \cos(x^2)\, dx$$

$$= \frac{1}{4} \sin(x^2) \Big]_0^9 = \frac{1}{4} \sin 81$$

41.

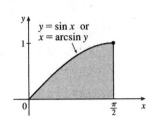

$y = \sin x$ or
$x = \arcsin y$

$$\int_0^1 \int_{\arcsin y}^{\pi/2} \cos x \sqrt{1 + \cos^2 x}\, dx\, dy$$

$$= \int_0^{\pi/2} \int_0^{\sin x} \cos x \sqrt{1 + \cos^2 x}\, dy\, dx$$

$$= \int_0^{\pi/2} \cos x \sqrt{1 + \cos^2 x}\, \Big[y \Big]_{y=0}^{y=\sin x}\, dx$$

$$= \int_0^{\pi/2} \cos x \sqrt{1 + \cos^2 x}\, \sin x\, dx \qquad \left[\begin{array}{l} \text{Let } u = \cos x,\ du = -\sin x\, dx, \\ \quad dx = du/(-\sin x) \end{array} \right]$$

$$= \int_1^0 -u\sqrt{1 + u^2}\, du = -\tfrac{1}{3}\big(1 + u^2\big)^{3/2}\Big]_1^0$$

$$= \tfrac{1}{3}\big(\sqrt{8} - 1\big) = \tfrac{1}{3}\big(2\sqrt{2} - 1\big)$$

43. $D = \{(x, y) \mid 0 \le x \le 1,\ -x + 1 \le y \le 1\} \cup \{(x, y) \mid -1 \le x \le 0,\ x + 1 \le y \le 1\}$

$\cup \{(x, y) \mid 0 \le x \le 1,\ -1 \le y \le x - 1\} \cup \{(x, y) \mid -1 \le x \le 0,\ -1 \le y \le -x - 1\}$, all type I.

$$\iint_D x^2\, dA = \int_0^1 \int_{1-x}^1 x^2\, dy\, dx + \int_{-1}^0 \int_{x+1}^1 x^2\, dy\, dx + \int_0^1 \int_{-1}^{x-1} x^2\, dy\, dx + \int_{-1}^0 \int_{-1}^{-x-1} x^2\, dy\, dx$$

$$= 4 \int_0^1 \int_{1-x}^1 x^2\, dy\, dx \qquad \text{[by symmetry of the regions and because } f(x, y) = x^2 \ge 0]$$

$$= 4 \int_0^1 x^3\, dx = 4\big[\tfrac{1}{4}x^4\big]_0^1 = 1$$

45. For $D = [0, 1] \times [0, 1]$, $0 \le \sqrt{x^3 + y^3} \le \sqrt{2}$ and $A(D) = 1$, so $0 \le \iint_D \sqrt{x^3 + y^3}\, dA \le \sqrt{2}$.

47. Since $m \le f(x, y) \le M$, $\iint_D m\, dA \le \iint_D f(x, y)\, dA \le \iint_D M\, dA$ by (8) $\Rightarrow$

$m \iint_D 1\, dA \le \iint_D f(x, y)\, dA \le M \iint_D 1\, dA$ by (7) $\Rightarrow$ $mA(D) \le \iint_D f(x, y)\, dA \le MA(D)$ by (10).

49. $\iint_D (x^2 \tan x + y^3 + 4)\, dA = \iint_D x^2 \tan x\, dA + \iint_D y^3\, dA + \iint_D 4\, dA$. But $x^2 \tan x$ is an odd function of x and D is

symmetric with respect to the y-axis, so $\iint_D x^2 \tan x\, dA = 0$. Similarly, y^3 is an odd function of y and D is symmetric with

respect to the x-axis, so $\iint_D y^3\, dA = 0$. Thus $\iint_D (x^2 \tan x + y^3 + 4)\, dA = 4 \iint_D dA = 4(\text{area of } D) = 4 \cdot \pi\big(\sqrt{2}\big)^2 = 8\pi$.

51. Since $\sqrt{1 - x^2 - y^2} \ge 0$, we can interpret $\iint_D \sqrt{1 - x^2 - y^2}\, dA$ as the volume of the solid that lies below the graph of

$z = \sqrt{1 - x^2 - y^2}$ and above the region D in the xy-plane. $z = \sqrt{1 - x^2 - y^2}$ is equivalent to $x^2 + y^2 + z^2 = 1$, $z \ge 0$

which meets the xy-plane in the circle $x^2 + y^2 = 1$, the boundary of D. Thus, the solid is an upper hemisphere of radius 1

which has volume $\tfrac{1}{2}\big[\tfrac{4}{3}\pi(1)^3\big] = \tfrac{2}{3}\pi$.

12.3 Double Integrals in Polar Coordinates

1. The region R is more easily described by polar coordinates: $R = \{(r, \theta) \mid 2 \le r \le 5,\ 0 \le \theta \le 2\pi\}$.

Thus $\iint_R f(x, y)\, dA = \int_0^{2\pi} \int_2^5 f(r \cos \theta, r \sin \theta)\, r\, dr\, d\theta$.

3. The region R is more easily described by rectangular coordinates: $R = \{(x, y) \mid -2 \le x \le 2,\ x \le y \le 2\}$.

Thus $\iint_R f(x, y)\, dA = \int_{-2}^2 \int_x^2 f(x, y)\, dy\, dx$.

5. The integral $\int_\pi^{2\pi} \int_4^7 r \, dr \, d\theta$ represents the area of the region

$R = \{(r, \theta) \mid 4 \le r \le 7, \pi \le \theta \le 2\pi\}$, the lower half of a ring.

$\int_\pi^{2\pi} \int_4^7 r \, dr \, d\theta = \left(\int_\pi^{2\pi} d\theta \right) \left(\int_4^7 r \, dr \right)$

$\qquad = \left[\theta \right]_\pi^{2\pi} \left[\frac{1}{2} r^2 \right]_4^7 = \pi \cdot \frac{1}{2} (49 - 16) = \frac{33\pi}{2}$

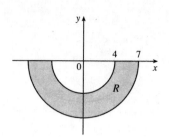

7. The disk D can be described in polar coordinates as $D = \{(r, \theta) \mid 0 \le r \le 3, 0 \le \theta \le 2\pi\}$. Then

$\iint_D xy \, dA = \int_0^{2\pi} \int_0^3 (r \cos \theta)(r \sin \theta) r \, dr \, d\theta = \left(\int_0^{2\pi} \sin \theta \cos \theta \, d\theta \right) \left(\int_0^3 r^3 \, dr \right) = \left[\frac{1}{2} \sin^2 \theta \right]_0^{2\pi} \left[\frac{1}{4} r^4 \right]_0^3 = 0.$

9. $\iint_R \cos(x^2 + y^2) \, dA = \int_0^\pi \int_0^3 \cos(r^2) r \, dr \, d\theta = \left(\int_0^\pi d\theta \right) \left(\int_0^3 r \cos(r^2) \, dr \right)$

$\qquad = \left[\theta \right]_0^\pi \left[\frac{1}{2} \sin(r^2) \right]_0^3 = \pi \cdot \frac{1}{2} (\sin 9 - \sin 0) = \frac{\pi}{2} \sin 9$

11. R is the region shown in the figure, and can be described

by $R = \{(r, \theta) \mid 0 \le \theta \le \pi/4, 1 \le r \le 2\}$. Thus

$\iint_R \arctan(y/x) \, dA = \int_0^{\pi/4} \int_1^2 \arctan(\tan \theta) r \, dr \, d\theta$ since $y/x = \tan \theta$.

Also, $\arctan(\tan \theta) = \theta$ for $0 \le \theta \le \pi/4$, so the integral becomes

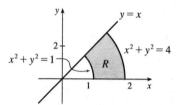

$\int_0^{\pi/4} \int_1^2 \theta r \, dr \, d\theta = \int_0^{\pi/4} \theta \, d\theta \int_1^2 r \, dr = \left[\frac{1}{2} \theta^2 \right]_0^{\pi/4} \left[\frac{1}{2} r^2 \right]_1^2 = \frac{\pi^2}{32} \cdot \frac{3}{2} = \frac{3}{64} \pi^2.$

13. $V = \iint_{x^2 + y^2 \le 4} \sqrt{x^2 + y^2} \, dA = \int_0^{2\pi} \int_0^2 \sqrt{r^2} \, r \, dr \, d\theta = \int_0^{2\pi} d\theta \int_0^2 r^2 \, dr = \left[\theta \right]_0^{2\pi} \left[\frac{1}{3} r^3 \right]_0^2 = 2\pi \left(\frac{8}{3} \right) = \frac{16}{3} \pi$

15. By symmetry,

$$V = 2 \iint_{x^2 + y^2 \le a^2} \sqrt{a^2 - x^2 - y^2} \, dA = 2 \int_0^{2\pi} \int_0^a \sqrt{a^2 - r^2} \, r \, dr \, d\theta = 2 \int_0^{2\pi} d\theta \int_0^a r \sqrt{a^2 - r^2} \, dr$$

$$= 2 \left[\theta \right]_0^{2\pi} \left[-\frac{1}{3} (a^2 - r^2)^{3/2} \right]_0^a = 2(2\pi) \left(0 + \frac{1}{3} a^3 \right) = \frac{4\pi}{3} a^3$$

17. The cone $z = \sqrt{x^2 + y^2}$ intersects the sphere $x^2 + y^2 + z^2 = 1$ when $x^2 + y^2 + \left(\sqrt{x^2 + y^2} \right)^2 = 1$ or $x^2 + y^2 = \frac{1}{2}$. So

$$V = \iint_{x^2 + y^2 \le 1/2} \left(\sqrt{1 - x^2 - y^2} - \sqrt{x^2 + y^2} \right) dA = \int_0^{2\pi} \int_0^{1/\sqrt{2}} \left(\sqrt{1 - r^2} - r \right) r \, dr \, d\theta$$

$$= \int_0^{2\pi} d\theta \int_0^{1/\sqrt{2}} \left(r \sqrt{1 - r^2} - r^2 \right) dr = \left[\theta \right]_0^{2\pi} \left[-\frac{1}{3} (1 - r^2)^{3/2} - \frac{1}{3} r^3 \right]_0^{1/\sqrt{2}} = 2\pi \left(-\frac{1}{3} \right) \left(\frac{1}{\sqrt{2}} - 1 \right) = \frac{\pi}{3} \left(2 - \sqrt{2} \right)$$

19. The given solid is the region inside the cylinder $x^2 + y^2 = 4$ between the surfaces $z = \sqrt{64 - 4x^2 - 4y^2}$

and $z = -\sqrt{64 - 4x^2 - 4y^2}$. So

$$V = \iint_{x^2 + y^2 \le 4} \left[\sqrt{64 - 4x^2 - 4y^2} - \left(-\sqrt{64 - 4x^2 - 4y^2} \right) \right] dA = \iint_{x^2 + y^2 \le 4} 2\sqrt{64 - 4x^2 - 4y^2} \, dA$$

$$= 4 \int_0^{2\pi} \int_0^2 \sqrt{16 - r^2} \, r \, dr \, d\theta = 4 \int_0^{2\pi} d\theta \int_0^2 r \sqrt{16 - r^2} \, dr = 4 \left[\theta \right]_0^{2\pi} \left[-\frac{1}{3} (16 - r^2)^{3/2} \right]_0^2$$

$$= 8\pi \left(-\frac{1}{3} \right) (12^{3/2} - 16^{2/3}) = \frac{8\pi}{3} \left(64 - 24\sqrt{3} \right)$$

21. One loop is given by the region

$D = \{(r, \theta) \,|\, -\pi/6 \le \theta \le \pi/6, 0 \le r \le \cos 3\theta \}$, so the area is

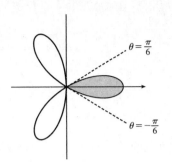

$\theta = \frac{\pi}{6}$

$\theta = -\frac{\pi}{6}$

$$\iint_D dA = \int_{-\pi/6}^{\pi/6} \int_0^{\cos 3\theta} r\, dr\, d\theta = \int_{-\pi/6}^{\pi/6} \left[\frac{1}{2} r^2 \right]_{r=0}^{r=\cos 3\theta} d\theta$$

$$= \int_{-\pi/6}^{\pi/6} \frac{1}{2} \cos^2 3\theta\, d\theta = 2 \int_0^{\pi/6} \frac{1}{2} \left(\frac{1 + \cos 6\theta}{2} \right) d\theta$$

$$= \frac{1}{2} \left[\theta + \frac{1}{6} \sin 6\theta \right]_0^{\pi/6} = \frac{\pi}{12}$$

23.

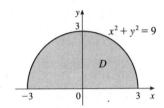

$x^2 + y^2 = 9$

D

$$\int_{-3}^3 \int_0^{\sqrt{9-x^2}} \sin(x^2 + y^2)\, dy\, dx = \int_0^\pi \int_0^3 \sin\left(r^2\right) r\, dr\, d\theta$$

$$= \int_0^\pi d\theta \int_0^3 r \sin\left(r^2\right)\, dr = [\theta]_0^\pi \left[-\frac{1}{2} \cos\left(r^2\right) \right]_0^3$$

$$= \pi \left(-\frac{1}{2}\right) (\cos 9 - 1) = \frac{\pi}{2} (1 - \cos 9)$$

25.

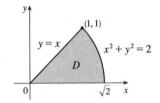

$(1, 1)$

$y = x$

$x^2 + y^2 = 2$

D

$\sqrt{2}$

$$\int_0^{\pi/4} \int_0^{\sqrt{2}} (r \cos\theta + r \sin\theta) r\, dr\, d\theta = \int_0^{\pi/4} (\cos\theta + \sin\theta)\, d\theta \int_0^{\sqrt{2}} r^2\, dr$$

$$= [\sin\theta - \cos\theta]_0^{\pi/4} \left[\frac{1}{3} r^3 \right]_0^{\sqrt{2}}$$

$$= \left[\frac{\sqrt{2}}{2} - \frac{\sqrt{2}}{2} - 0 + 1 \right] \cdot \frac{1}{3} \left(2\sqrt{2} - 0 \right) = \frac{2\sqrt{2}}{3}$$

27. The surface of the water in the pool is a circular disk D with radius 20 ft. If we place D on coordinate axes with the origin at

the center of D and define $f(x, y)$ to be the depth of the water at (x, y), then the volume of water in the pool is the volume of

the solid that lies above $D = \{(x, y) \,|\, x^2 + y^2 \le 400\}$ and below the graph of $f(x, y)$. We can associate north with the

positive y-direction, so we are given that the depth is constant in the x-direction and the depth increases linearly in the

y-direction from $f(0, -20) = 2$ to $f(0, 20) = 7$. The trace in the yz-plane is a line segment from $(0, -20, 2)$ to $(0, 20, 7)$.

The slope of this line is $\frac{7-2}{20-(-20)} = \frac{1}{8}$, so an equation of the line is $z - 7 = \frac{1}{8}(y - 20) \Rightarrow z = \frac{1}{8}y + \frac{9}{2}$. Since $f(x, y)$ is

independent of x, $f(x, y) = \frac{1}{8}y + \frac{9}{2}$. Thus the volume is given by $\iint_D f(x, y)\, dA$, which is most conveniently evaluated

using polar coordinates. Then $D = \{(r, \theta) \,|\, 0 \le r \le 20, 0 \le \theta \le 2\pi\}$ and substituting $x = r \cos\theta$, $y = r \sin\theta$ the integral

becomes

$$\int_0^{2\pi} \int_0^{20} \left(\frac{1}{8} r \sin\theta + \frac{9}{2} \right) r\, dr\, d\theta = \int_0^{2\pi} \left[\frac{1}{24} r^3 \sin\theta + \frac{9}{4} r^2 \right]_{r=0}^{r=20} d\theta = \int_0^{2\pi} \left(\frac{1000}{3} \sin\theta + 900 \right) d\theta$$

$$= \left[-\frac{1000}{3} \cos\theta + 900\theta \right]_0^{2\pi} = 1800\pi$$

Thus the pool contains $1800\pi \approx 5655$ ft^3 of water.

29. $\int_{1/\sqrt{2}}^{1} \int_{\sqrt{1-x^2}}^{x} xy \, dy \, dx + \int_{1}^{\sqrt{2}} \int_{0}^{x} xy \, dy \, dx + \int_{\sqrt{2}}^{2} \int_{0}^{\sqrt{4-x^2}} xy \, dy \, dx$

$$= \int_{0}^{\pi/4} \int_{1}^{2} r^3 \cos\theta \sin\theta \, dr \, d\theta = \int_{0}^{\pi/4} \left[\frac{r^4}{4} \cos\theta \sin\theta \right]_{r=1}^{r=2} d\theta$$

$$= \frac{15}{4} \int_{0}^{\pi/4} \sin\theta \cos\theta \, d\theta = \frac{15}{4} \left[\frac{\sin^2\theta}{2} \right]_{0}^{\pi/4} = \frac{15}{16}$$

31. (a) We integrate by parts with $u = x$ and $dv = xe^{-x^2} dx$. Then $du = dx$ and $v = -\frac{1}{2}e^{-x^2}$, so

$$\int_{0}^{\infty} x^2 e^{-x^2} dx = \lim_{t\to\infty} \int_{0}^{t} x^2 e^{-x^2} dx = \lim_{t\to\infty} \left(-\frac{1}{2}xe^{-x^2} \Big|_{0}^{t} + \int_{0}^{t} \frac{1}{2}e^{-x^2} dx \right)$$

$$= \lim_{t\to\infty} \left(-\frac{1}{2}te^{-t^2} \right) + \frac{1}{2} \int_{0}^{\infty} e^{-x^2} dx = 0 + \frac{1}{2} \int_{0}^{\infty} e^{-x^2} dx \qquad \text{[by l'Hospital's Rule]}$$

$$= \frac{1}{4} \int_{-\infty}^{\infty} e^{-x^2} dx \qquad \text{[since } e^{-x^2} \text{ is an even function]}$$

$$= \frac{1}{4}\sqrt{\pi} \qquad \text{[by Exercise 30(c)]}$$

(b) Let $u = \sqrt{x}$. Then $u^2 = x \implies dx = 2u \, du \implies$

$$\int_{0}^{\infty} \sqrt{x}e^{-x} dx = \lim_{t\to\infty} \int_{0}^{t} \sqrt{x}\,e^{-x} dx = \lim_{t\to\infty} \int_{0}^{\sqrt{t}} ue^{-u^2} 2u \, du = 2\int_{0}^{\infty} u^2 e^{-u^2} du = 2\left(\frac{1}{4}\sqrt{\pi}\right) \quad \text{[by part(a)]} \; = \frac{1}{2}\sqrt{\pi}.$$

12.4 Applications of Double Integrals

1. $Q = \iint_D \sigma(x,y) \, dA = \int_1^3 \int_0^2 (2xy + y^2) \, dy \, dx = \int_1^3 \left[xy^2 + \frac{1}{3}y^3 \right]_{y=0}^{y=2} dx$

$= \int_1^3 \left(4x + \frac{8}{3}\right) dx = \left[2x^2 + \frac{8}{3}x\right]_1^3 = 16 + \frac{16}{3} = \frac{64}{3}$ C

3. $m = \iint_D \rho(x,y) \, dA = \int_0^2 \int_{-1}^1 xy^2 \, dy \, dx = \int_0^2 x \, dx \int_{-1}^1 y^2 \, dy = \left[\frac{1}{2}x^2\right]_0^2 \left[\frac{1}{3}y^3\right]_{-1}^1 = 2 \cdot \frac{2}{3} = \frac{4}{3}$,

$\bar{x} = \frac{1}{m} \iint_D x\rho(x,y) \, dA = \frac{3}{4} \int_0^2 \int_{-1}^1 x^2 y^2 \, dy \, dx = \frac{3}{4} \int_0^2 x^2 \, dx \int_{-1}^1 y^2 \, dy = \frac{3}{4}\left[\frac{1}{3}x^3\right]_0^2 \left[\frac{1}{3}y^3\right]_{-1}^1 = \frac{3}{4} \cdot \frac{8}{3} \cdot \frac{2}{3} = \frac{4}{3}$,

$\bar{y} = \frac{1}{m} \iint_D y\rho(x,y) \, dA = \frac{3}{4} \int_0^2 \int_{-1}^1 xy^3 \, dy \, dx = \frac{3}{4} \int_0^2 x \, dx \int_{-1}^1 y^3 \, dy = \frac{3}{4}\left[\frac{1}{2}x^2\right]_0^2 \left[\frac{1}{4}y^4\right]_{-1}^1 = \frac{3}{4} \cdot 2 \cdot 0 = 0.$

Hence, $(\bar{x}, \bar{y}) = \left(\frac{4}{3}, 0\right)$.

5. $m = \int_0^2 \int_{x/2}^{3-x} (x + y) \, dy \, dx = \int_0^2 \left[xy + \frac{1}{2}y^2\right]_{y=x/2}^{y=3-x} dx = \int_0^2 \left[x\left(3 - \frac{3}{2}x\right) + \frac{1}{2}(3-x)^2 - \frac{1}{8}x^2\right] dx$

$= \int_0^2 \left(-\frac{9}{8}x^2 + \frac{9}{2}\right) dx = \left[-\frac{9}{8}\left(\frac{1}{3}x^3\right) + \frac{9}{2}x\right]_0^2 = 6,$

$M_y = \int_0^2 \int_{x/2}^{3-x} (x^2 + xy) \, dy \, dx = \int_0^2 \left[x^2 y + \frac{1}{2}xy^2\right]_{y=x/2}^{y=3-x} dx = \int_0^2 \left(\frac{9}{2}x - \frac{9}{8}x^3\right) dx = \frac{9}{2},$

$M_x = \int_0^2 \int_{x/2}^{3-y} (xy + y^2) \, dy \, dx = \int_0^2 \left[\frac{1}{2}xy^2 + \frac{1}{3}y^3\right]_{y=x/2}^{y=3-x} dx = \int_0^2 \left(9 - \frac{9}{2}x\right) dx = 9.$

Hence $m = 6$, $(\bar{x}, \bar{y}) = \left(\dfrac{M_y}{m}, \dfrac{M_x}{m}\right) = \left(\dfrac{3}{4}, \dfrac{3}{2}\right).$

7. $m = \int_0^1 \int_0^{e^x} y \, dy \, dx = \int_0^1 \left[\frac{1}{2}y^2\right]_{y=0}^{y=e^x} dx = \frac{1}{2}\int_0^1 e^{2x} \, dx = \frac{1}{4}e^{2x}\big]_0^1 = \frac{1}{4}(e^2 - 1),$

$M_y = \int_0^1 \int_0^{e^x} xy \, dy \, dx = \frac{1}{2}\int_0^1 xe^{2x} \, dx = \frac{1}{2}\left[\frac{1}{2}xe^{2x} - \frac{1}{4}e^{2x}\right]_0^1 = \frac{1}{8}(e^2 + 1),$

$M_x = \int_0^1 \int_0^{e^x} y^2 \, dy \, dx = \int_0^1 \left[\frac{1}{3}y^3\right]_{y=0}^{y=e^x} dx = \frac{1}{3}\int_0^1 e^{3x} \, dx = \frac{1}{3}\left[\frac{1}{3}e^{3x}\right]_0^1 = \frac{1}{9}(e^3 - 1).$

Hence $m = \frac{1}{4}(e^2 - 1)$, $(\overline{x}, \overline{y}) = \left(\dfrac{\frac{1}{8}(e^2 + 1)}{\frac{1}{4}(e^2 - 1)}, \dfrac{\frac{1}{9}(e^3 - 1)}{\frac{1}{4}(e^2 - 1)}\right) = \left(\dfrac{e^2 + 1}{2(e^2 - 1)}, \dfrac{4(e^3 - 1)}{9(e^2 - 1)}\right).$

9.

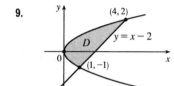

$m = \int_{-1}^2 \int_{y^2}^{y+2} 3 \, dx \, dy = \int_{-1}^2 (3y + 6 - 3y^2) \, dy = \frac{27}{2},$

$M_y = \int_{-1}^2 \int_{y^2}^{y+2} 3x \, dx \, dy = \int_{-1}^2 \frac{3}{2}\left[(y+2)^2 - y^4\right] dy$

$\quad = \left[\frac{1}{2}(y+2)^3 - \frac{3}{10}y^5\right]_{-1}^2 = \frac{108}{5}$

$M_x = \int_{-1}^2 \int_{y^2}^{y+2} 3y \, dx \, dy = \int_{-1}^2 (3y^2 + 6y - 3y^3) \, dy$

$\quad = \left[y^3 + 3y^2 - \frac{3}{4}y^4\right]_{-1}^2 = \frac{27}{4}$

Hence $m = \frac{27}{2}$, $(\overline{x}, \overline{y}) = \left(\frac{8}{5}, \frac{1}{2}\right).$

11. $\rho(x, y) = ky = kr\sin\theta$, $m = \int_0^{\pi/2}\int_0^1 kr^2 \sin\theta \, dr \, d\theta = \frac{1}{3}k \int_0^{\pi/2} \sin\theta \, d\theta = \frac{1}{3}k\left[-\cos\theta\right]_0^{\pi/2} = \frac{1}{3}k,$

$M_y = \int_0^{\pi/2}\int_0^1 kr^3 \sin\theta \cos\theta \, dr \, d\theta = \frac{1}{4}k \int_0^{\pi/2} \sin\theta \cos\theta \, d\theta = \frac{1}{8}k\left[-\cos 2\theta\right]_0^{\pi/2} = \frac{1}{8}k,$

$M_x = \int_0^{\pi/2}\int_0^1 kr^3 \sin^2\theta \, dr \, d\theta = \frac{1}{4}k \int_0^{\pi/2} \sin^2\theta \, d\theta = \frac{1}{8}k\left[\theta + \sin 2\theta\right]_0^{\pi/2} = \frac{\pi}{16}k.$

Hence $(\overline{x}, \overline{y}) = \left(\frac{3}{8}, \frac{3\pi}{16}\right).$

13. Placing the vertex opposite the hypotenuse at $(0, 0)$, $\rho(x, y) = k(x^2 + y^2)$. Then

$m = \int_0^a \int_0^{a-x} k(x^2 + y^2) \, dy \, dx = k\int_0^a \left[ax^2 - x^3 + \frac{1}{3}(a - x)^3\right] dx = k\left[\frac{1}{3}ax^3 - \frac{1}{4}x^4 - \frac{1}{12}(a - x)^4\right]_0^a = \frac{1}{6}ka^4.$

By symmetry,

$M_y = M_x = \int_0^a \int_0^{a-x} ky(x^2 + y^2) \, dy \, dx = k\int_0^a \left[\frac{1}{2}(a - x)^2 x^2 + \frac{1}{4}(a - x)^4\right] dx$

$\quad = k\left[\frac{1}{6}a^2 x^3 - \frac{1}{4}ax^4 + \frac{1}{10}x^5 - \frac{1}{20}(a - x)^5\right]_0^a = \frac{1}{15}ka^5$

Hence $(\overline{x}, \overline{y}) = \left(\frac{2}{5}a, \frac{2}{5}a\right).$

15. $I_x = \iint_D y^2 \rho(x, y) dA = \int_0^1 \int_0^{e^x} y^2 \cdot y \, dy \, dx = \int_0^1 \left[\frac{1}{4}y^4\right]_{y=0}^{y=e^x} dx = \frac{1}{4}\int_0^1 e^{4x} \, dx = \frac{1}{4}\left[\frac{1}{4}e^{4x}\right]_0^1 = \frac{1}{16}(e^4 - 1),$

$I_y = \iint_D x^2 \rho(x, y) \, dA = \int_0^1 \int_0^{e^x} x^2 y \, dy \, dx = \int_0^1 x^2 \left[\frac{1}{2}y^2\right]_{y=0}^{y=e^x} dx = \frac{1}{2}\int_0^1 x^2 e^{2x} \, dx$

$\quad = \frac{1}{2}\left[\left(\frac{1}{2}x^2 - \frac{1}{2}x + \frac{1}{4}\right)e^{2x}\right]_0^1$ [integrate by parts twice] $= \frac{1}{8}(e^2 - 1),$

and $I_0 = I_x + I_y = \frac{1}{16}(e^4 - 1) + \frac{1}{8}(e^2 - 1) = \frac{1}{16}(e^4 + 2e^2 - 3).$

17. $I_x = \int_{-1}^2 \int_{y^2}^{y+2} 3y^2 \, dx \, dy = \int_{-1}^2 (3y^3 + 6y^2 - 3y^4) \, dy = \left[\frac{3}{4}y^4 + 2y^3 - \frac{3}{5}y^5\right]_{-1}^2 = \frac{189}{20},$

$I_y = \int_{-1}^2 \int_{y^2}^{y+2} 3x^2 \, dx \, dy = \int_{-1}^2 \left[(y+2)^3 - y^6\right] dy = \left[\frac{1}{4}(y+2)^4 - \frac{1}{7}y^7\right]_{-1}^2 = \frac{1269}{28},$ and $I_0 = I_x + I_y = \frac{1917}{35}.$

19. Using a CAS, we find $m = \iint_D \rho(x,y)\,dA = \int_0^\pi \int_0^{\sin x} xy\,dy\,dx = \dfrac{\pi^2}{8}$. Then

$$\overline{x} = \frac{1}{m}\iint_D x\rho(x,y)\,dA = \frac{8}{\pi^2}\int_0^\pi \int_0^{\sin x} x^2 y\,dy\,dx = \frac{2\pi}{3} - \frac{1}{\pi} \text{ and}$$

$$\overline{y} = \frac{1}{m}\iint_D y\rho(x,y)\,dA = \frac{8}{\pi^2}\int_0^\pi \int_0^{\sin x} xy^2\,dy\,dx = \frac{16}{9\pi}, \text{ so } (\overline{x}, \overline{y}) = \left(\frac{2\pi}{3} - \frac{1}{\pi}, \frac{16}{9\pi}\right).$$

The moments of inertia are $I_x = \iint_D y^2\rho(x,y)\,dA = \int_0^\pi \int_0^{\sin x} xy^3\,dy\,dx = \dfrac{3\pi^2}{64}$,

$I_y = \iint_D x^2 \rho(x,y)\,dA = \int_0^\pi \int_0^{\sin x} x^3 y\,dy\,dx = \dfrac{\pi^2}{16}(\pi^2 - 3)$, and $I_0 = I_x + I_y = \dfrac{\pi^2}{64}(4\pi^2 - 9)$.

21. $I_x = \int_0^a \int_0^a \rho y^2\,dx\,dy = \rho \int_0^a dx \int_0^a y^2\,dy = \rho\big[x\big]_0^a \big[\tfrac{1}{3}y^3\big]_0^a = \rho a\big(\tfrac{1}{3}a^3\big) = \tfrac{1}{3}\rho a^4 = I_y$ by symmetry, and $m = \rho a^2$ since

the lamina is homogeneous. Hence $\overline{\overline{x}}^2 = \dfrac{I_y}{m} \;\Rightarrow\; \overline{\overline{x}} = \left[\big(\tfrac{1}{3}\rho a^4\big)/(\rho a^2)\right]^{1/2} = \tfrac{1}{\sqrt{3}}a$ and $\overline{\overline{y}}^2 = \dfrac{I_x}{m} \;\Rightarrow\; \overline{\overline{y}} = \tfrac{1}{\sqrt{3}}a.$

12.5 Triple Integrals

1. $\iiint_B xyz^2\,dV = \int_0^1 \int_{-1}^2 \int_0^3 xyz^2\,dz\,dx\,dy = \int_0^1 \int_{-1}^2 xy\big[\tfrac{1}{3}z^3\big]_{z=0}^{z=3}\,dx\,dy = \int_0^1 \int_{-1}^2 9xy\,dx\,dy$

$\qquad = \int_0^1 \big[\tfrac{9}{2}x^2 y\big]_{x=-1}^{x=2}\,dy = \int_0^1 \tfrac{27}{2}y\,dy = \tfrac{27}{4}y^2\big]_0^1 = \tfrac{27}{4}$

3. $\int_0^1 \int_0^z \int_0^{x+z} 6xz\,dy\,dx\,dz = \int_0^1 \int_0^z \big[6xyz\big]_{y=0}^{y=x+z}\,dx\,dz = \int_0^1 \int_0^z 6xz(x+z)\,dx\,dz$

$\qquad = \int_0^1 \big[2x^3 z + 3x^2 z^2\big]_{x=0}^{x=z}\,dz = \int_0^1 (2z^4 + 3z^4)\,dz = \int_0^1 5z^4\,dz = z^5\big]_0^1 = 1$

5. $\int_0^3 \int_0^1 \int_0^{\sqrt{1-z^2}} ze^y\,dx\,dz\,dy = \int_0^3 \int_0^1 \big[xze^y\big]_{x=0}^{x=\sqrt{1-z^2}}\,dz\,dy = \int_0^3 \int_0^1 ze^y\sqrt{1-z^2}\,dz\,dy$

$\qquad = \int_0^3 \big[-\tfrac{1}{3}(1-z^2)^{3/2}e^y\big]_{z=0}^{z=1}\,dy = \int_0^3 \tfrac{1}{3}e^y\,dy = \tfrac{1}{3}e^y\big]_0^3 = \tfrac{1}{3}(e^3 - 1)$

7. $\iiint_E 2x\,dV = \int_0^2 \int_0^{\sqrt{4-y^2}} \int_0^y 2x\,dz\,dx\,dy = \int_0^2 \int_0^{\sqrt{4-y^2}} \big[2xz\big]_{z=0}^{z=y}\,dx\,dy = \int_0^2 \int_0^{\sqrt{4-y^2}} 2xy\,dx\,dy$

$\qquad = \int_0^2 \big[x^2 y\big]_{x=0}^{x=\sqrt{4-y^2}}\,dy = \int_0^2 (4-y^2)y\,dy = \big[2y^2 - \tfrac{1}{4}y^4\big]_0^2 = 4$

9. Here $E = \{(x,y,z) \mid 0 \le x \le 1, 0 \le y \le \sqrt{x}, 0 \le z \le 1+x+y\}$, so

$$\iiint_E 6xy\,dV = \int_0^1 \int_0^{\sqrt{x}} \int_0^{1+x+y} 6xy\,dz\,dy\,dx = \int_0^1 \int_0^{\sqrt{x}} \big[6xyz\big]_{z=0}^{z=1+x+y}\,dy\,dx = \int_0^1 \int_0^{\sqrt{x}} 6xy(1+x+y)\,dy\,dx$$

$$= \int_0^1 \big[3xy^2 + 3x^2 y^2 + 2xy^3\big]_{y=0}^{y=\sqrt{x}}\,dx = \int_0^1 (3x^2 + 3x^3 + 2x^{5/2})\,dx = \big[x^3 + \tfrac{3}{4}x^4 + \tfrac{4}{7}x^{7/2}\big]_0^1 = \frac{65}{28}$$

11. Here E is the region that lies below the plane with x-, y-, and z-intercepts 1, 2, and 3 respectively, that is, below the plane

$2z + 6x + 3y = 6$ and above the region in the xy-plane bounded by the lines $x = 0$, $y = 0$ and $6x + 3y = 6$. So

$$\iiint_E xy\, dV = \int_0^1 \int_0^{2-2x} \int_0^{3-3x-3y/2} xy\, dz\, dy\, dx = \int_0^1 \int_0^{2-2x} \left(3xy - 3x^2y - \tfrac{3}{2}xy^2\right) dy\, dx$$

$$= \int_0^1 \left[\tfrac{3}{2}xy^2 - \tfrac{3}{2}x^2y^2 - \tfrac{1}{2}xy^3\right]_{y=0}^{y=2-2x} dx = \int_0^1 \left(2x - 6x^2 + 6x^3 - 2x^4\right) dx$$

$$= \left[x^2 - 2x^3 + \tfrac{3}{2}x^4 - \tfrac{2}{5}x^5\right]_0^1 = \tfrac{1}{10}$$

13.

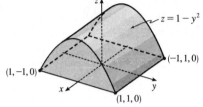

E is the region below the parabolic cylinder $z = 1 - y^2$ and above the

square $[-1, 1] \times [-1, 1]$ in the xy-plane.

$$\iiint_E x^2 e^y\, dV = \int_{-1}^1 \int_{-1}^1 \int_0^{1-y^2} x^2 e^y\, dz\, dy\, dx = \int_{-1}^1 \int_{-1}^1 x^2 e^y (1 - y^2)\, dy\, dx$$

$$= \int_{-1}^1 x^2\, dx \int_{-1}^1 (e^y - y^2 e^y)\, dy$$

$$= \left[\tfrac{1}{3}x^3\right]_{-1}^1 \left[e^y - (y^2 - 2y + 2)e^y\right]_{-1}^1 \quad \begin{bmatrix} \text{integrate by} \\ \text{parts twice} \end{bmatrix}$$

$$= \tfrac{1}{3}(2)[e - e - e^{-1} + 5e^{-1}] = \tfrac{8}{3e}$$

15.

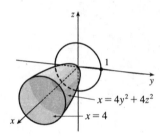

The projection E on the yz-plane is the disk $y^2 + z^2 \le 1$. Using polar

coordinates $y = r\cos\theta$ and $z = r\sin\theta$, we get

$$\iiint_E x\, dV = \iint_D \left[\int_{4y^2+4z^2}^4 x\, dx\right] dA = \tfrac{1}{2} \iint_D \left[4^2 - (4y^2 + 4z^2)^2\right] dA$$

$$= 8 \int_0^{2\pi} \int_0^1 (1 - r^4)\, r\, dr\, d\theta = 8 \int_0^{2\pi} d\theta \int_0^1 (r - r^5)\, dr$$

$$= 8(2\pi)\left[\tfrac{1}{2}r^2 - \tfrac{1}{6}r^6\right]_0^1 = \tfrac{16\pi}{3}$$

17. The plane $2x + y + z = 4$ intersects the xy-plane when

$2x + y + 0 = 4 \implies y = 4 - 2x$, so

$E = \{(x, y, z) \mid 0 \le x \le 2, 0 \le y \le 4 - 2x, 0 \le z \le 4 - 2x - y\}$ and

$V = \int_0^2 \int_0^{4-2x} \int_0^{4-2x-y} dz\, dy\, dx = \int_0^2 \int_0^{4-2x} (4 - 2x - y)\, dy\, dx$

$= \int_0^2 \left[4y - 2xy - \tfrac{1}{2}y^2\right]_{y=0}^{y=4-2x} dx$

$= \int_0^2 \left[4(4 - 2x) - 2x(4 - 2x) - \tfrac{1}{2}(4 - 2x)^2\right] dx$

$= \int_0^2 (2x^2 - 8x + 8)\, dx = \left[\tfrac{2}{3}x^3 - 4x^2 + 8x\right]_0^2 = \tfrac{16}{3}$

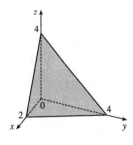

19. $V = \int_{-3}^{3} \int_{-\sqrt{9-x^2}}^{\sqrt{9-x^2}} \int_{1}^{5-y} dz\, dy\, dx = \int_{-3}^{3} \int_{-\sqrt{9-x^2}}^{\sqrt{9-x^2}} (5-y-1)\, dy\, dx = \int_{-3}^{3} \left[4y - \tfrac{1}{2}y^2 \right]_{y=-\sqrt{9-x^2}}^{y=\sqrt{9-x^2}} dx$

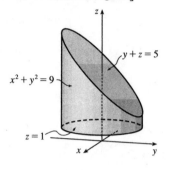

$= \int_{-3}^{3} 8\sqrt{9-x^2}\, dx = 8\left[\tfrac{x}{2}\sqrt{9-x^2} + \tfrac{9}{2}\sin^{-1}\left(\tfrac{x}{3}\right) \right]_{-3}^{3}$ $\qquad \left[\begin{array}{l} \text{using trigonometric substitution or} \\ \text{Formula 30 in the Table of Integrals} \end{array} \right]$

$= 8\left[\tfrac{9}{2}\sin^{-1}(1) - \tfrac{9}{2}\sin^{-1}(-1) \right] = 36\left(\tfrac{\pi}{2} - \left(-\tfrac{\pi}{2}\right) \right) = 36\pi$

Alternatively, use polar coordinates to evaluate the double integral:

$\int_{-3}^{3} \int_{-\sqrt{9-x^2}}^{\sqrt{9-x^2}} (4-y)\, dy\, dx = \int_{0}^{2\pi} \int_{0}^{3} (4 - r\sin\theta)\, r\, dr\, d\theta$

$= \int_{0}^{2\pi} \left[2r^2 - \tfrac{1}{3}r^3 \sin\theta \right]_{r=0}^{r=3} d\theta$

$= \int_{0}^{2\pi} (18 - 9\sin\theta)\, d\theta$

$= 18\theta + 9\cos\theta \Big|_{0}^{2\pi} = 36\pi$

21. (a) The wedge can be described as the region

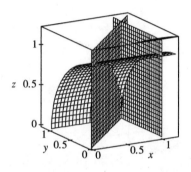

$D = \left\{ (x,y,z) \mid y^2 + z^2 \le 1,\, 0 \le x \le 1,\, 0 \le y \le x \right\}$
$= \left\{ (x,y,z) \mid 0 \le x \le 1,\, 0 \le y \le x,\, 0 \le z \le \sqrt{1-y^2} \right\}$

So the integral expressing the volume of the wedge is

$\iiint_D dV = \int_0^1 \int_0^x \int_0^{\sqrt{1-y^2}} dz\, dy\, dx.$

(b) A CAS gives $\int_0^1 \int_0^x \int_0^{\sqrt{1-y^2}} dz\, dy\, dx = \tfrac{\pi}{4} - \tfrac{1}{3}.$

(Or use Formulas 30 and 87 from the Table of Integrals.)

23. Here $f(x,y,z) = \dfrac{1}{\ln(1+x+y+z)}$ and $\Delta V = 2 \cdot 4 \cdot 2 = 16$, so the Midpoint Rule gives

$\iiint_B f(x,y,z)\, dV \approx \sum_{i=1}^{l} \sum_{j=1}^{m} \sum_{k=1}^{n} f(\bar{x}_i, \bar{y}_j, \bar{z}_k)\, \Delta V$

$= 16\left[f(1,2,1) + f(1,2,3) + f(1,6,1) + f(1,6,3) + f(3,2,1) + f(3,2,3) + f(3,6,1) + f(3,6,3) \right]$

$= 16\left[\tfrac{1}{\ln 5} + \tfrac{1}{\ln 7} + \tfrac{1}{\ln 9} + \tfrac{1}{\ln 11} + \tfrac{1}{\ln 7} + \tfrac{1}{\ln 9} + \tfrac{1}{\ln 11} + \tfrac{1}{\ln 13} \right] \approx 60.533$

25. $E = \left\{ (x,y,z) \mid 0 \le x \le 1,\, 0 \le z \le 1-x,\, 0 \le y \le 2-2z \right\}$,

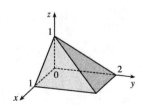

the solid bounded by the three coordinate planes and the planes

$z = 1 - x,\ y = 2 - 2z.$

27.

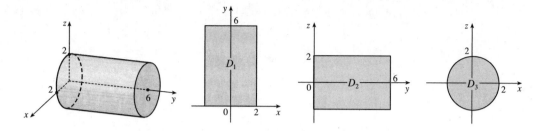

If D_1, D_2, D_3 are the projections of E on the xy-, yz-, and xz-planes, then

$$D_1 = \{(x,y) \mid -2 \leq x \leq 2, 0 \leq y \leq 6\}$$

$$D_2 = \{(y,z) \mid -2 \leq z \leq 2, 0 \leq y \leq 6\}$$

$$D_3 = \{(x,z) \mid x^2 + z^2 \leq 4\}$$

Therefore

$$E = \{(x,y,z) \mid -\sqrt{4-x^2} \leq z \leq \sqrt{4-x^2}, \ -2 \leq x \leq 2, 0 \leq y \leq 6\}$$

$$= \{(x,y,z) \mid -\sqrt{4-z^2} \leq x \leq \sqrt{4-z^2}, \ -2 \leq z \leq 2, 0 \leq y \leq 6\}$$

$$\iiint_E f(x,y,z)\, dV = \int_{-2}^{2} \int_{0}^{6} \int_{-\sqrt{4-x^2}}^{\sqrt{4-x^2}} f(x,y,z)\, dz\, dy\, dx = \int_{0}^{6} \int_{-2}^{2} \int_{-\sqrt{4-x^2}}^{\sqrt{4-x^2}} f(x,y,z)\, dz\, dx\, dy$$

$$= \int_{0}^{6} \int_{-2}^{2} \int_{-\sqrt{4-z^2}}^{\sqrt{4-z^2}} f(x,y,z)\, dx\, dz\, dy = \int_{-2}^{2} \int_{0}^{6} \int_{-\sqrt{4-z^2}}^{\sqrt{4-z^2}} f(x,y,z)\, dx\, dy\, dz$$

$$= \int_{-2}^{2} \int_{-\sqrt{4-x^2}}^{\sqrt{4-x^2}} \int_{0}^{6} f(x,y,z)\, dy\, dz\, dx = \int_{-2}^{2} \int_{-\sqrt{4-z^2}}^{\sqrt{4-z^2}} \int_{0}^{6} f(x,y,z)\, dy\, dx\, dz$$

29.

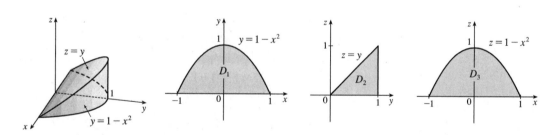

If D_1, D_2, and D_3 are the projections of E on the xy-, yz-, and xz-planes, then

$$D_1 = \{(x,y) \mid -1 \leq x \leq 1, 0 \leq y \leq 1 - x^2\} = \{(x,y) \mid 0 \leq y \leq 1, -\sqrt{1-y} \leq x \leq \sqrt{1-y}\},$$

$$D_2 = \{(y,z) \mid 0 \leq y \leq 1, 0 \leq z \leq y\} = \{(y,z) \mid 0 \leq z \leq 1, z \leq y \leq 1\}, \text{ and}$$

$$D_3 = \{(x,z) \mid -1 \leq x \leq 1, 0 \leq z \leq 1 - x^2\} = \{(x,z) \mid 0 \leq z \leq 1, -\sqrt{1-z} \leq x \leq \sqrt{1-z}\}$$

Therefore

$$E = \{(x,y,z) \mid -1 \le x \le 1, 0 \le y \le 1 - x^2, 0 \le z \le y\}$$
$$= \{(x,y,z) \mid 0 \le y \le 1, -\sqrt{1-y} \le x \le \sqrt{1-y}, 0 \le z \le y\}$$
$$= \{(x,y,z) \mid 0 \le y \le 1, 0 \le z \le y, -\sqrt{1-y} \le x \le \sqrt{1-y}\}$$
$$= \{(x,y,z) \mid 0 \le z \le 1, z \le y \le 1, -\sqrt{1-y} \le x \le \sqrt{1-y}\}$$
$$= \{(x,y,z) \mid -1 \le x \le 1, 0 \le z \le 1 - x^2, z \le y \le 1 - x^2\}$$
$$= \{(x,y,z) \mid 0 \le z \le 1, -\sqrt{1-z} \le x \le \sqrt{1-z}, z \le y \le 1 - x^2\}$$

Then

$$\iiint_E f(x,y,z)\,dV = \int_{-1}^{1}\int_{0}^{1-x^2}\int_{0}^{y} f(x,y,z)\,dz\,dy\,dx = \int_{0}^{1}\int_{-\sqrt{1-y}}^{\sqrt{1-y}}\int_{0}^{y} f(x,y,z)\,dz\,dx\,dy$$

$$= \int_{0}^{1}\int_{0}^{y}\int_{-\sqrt{1-y}}^{\sqrt{1-y}} f(x,y,z)\,dx\,dz\,dy = \int_{0}^{1}\int_{z}^{1}\int_{-\sqrt{1-y}}^{\sqrt{1-y}} f(x,y,z)\,dx\,dy\,dz$$

$$= \int_{-1}^{1}\int_{0}^{1-x^2}\int_{z}^{1-x^2} f(x,y,z)\,dy\,dz\,dx = \int_{0}^{1}\int_{-\sqrt{1-z}}^{\sqrt{1-z}}\int_{z}^{1-x^2} f(x,y,z)\,dy\,dx\,dz$$

31.

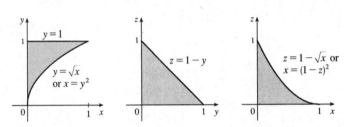

The diagrams show the projections of E on the xy-, yz-, and xz-planes. Therefore

$$\int_{0}^{1}\int_{\sqrt{x}}^{1}\int_{0}^{1-y} f(x,y,z)\,dz\,dy\,dx = \int_{0}^{1}\int_{0}^{y^2}\int_{0}^{1-y} f(x,y,z)\,dz\,dx\,dy = \int_{0}^{1}\int_{0}^{1-z}\int_{0}^{y^2} f(x,y,z)\,dx\,dy\,dz$$

$$= \int_{0}^{1}\int_{0}^{1-y}\int_{0}^{y^2} f(x,y,z)\,dx\,dz\,dy = \int_{0}^{1}\int_{0}^{1-\sqrt{x}}\int_{\sqrt{x}}^{1-z} f(x,y,z)\,dy\,dz\,dx$$

$$= \int_{0}^{1}\int_{0}^{(1-z)^2}\int_{\sqrt{x}}^{1-z} f(x,y,z)\,dy\,dx\,dz$$

33.

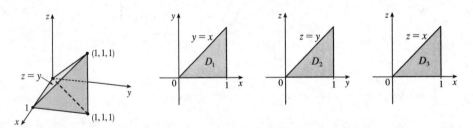

$\int_{0}^{1}\int_{y}^{1}\int_{0}^{y} f(x,y,z)\,dz\,dx\,dy = \iiint_E f(x,y,z)\,dV$ where $E = \{(x,y,z) \mid 0 \le z \le y, y \le x \le 1, 0 \le y \le 1\}$.

If D_1, D_2, and D_3 are the projections of E on the xy-, yz- and xz-planes then

$$D_1 = \{(x,y) \mid 0 \le y \le 1, y \le x \le 1\} = \{(x,y) \mid 0 \le x \le 1, 0 \le y \le x\},$$

$$D_2 = \{(y,z) \mid 0 \le y \le 1, 0 \le z \le y\} = \{(y,z) \mid 0 \le z \le 1, z \le y \le 1\}, \text{ and}$$

$$D_3 = \{(x, z) \mid 0 \le x \le 1, 0 \le z \le x\} = \{(x, z) \mid 0 \le z \le 1, z \le x \le 1\}.$$

Thus we also have

$$E = \{(x, y, z) \mid 0 \le x \le 1, 0 \le y \le x, 0 \le z \le y\} = \{(x, y, z) \mid 0 \le y \le 1, 0 \le z \le y, y \le x \le 1\}$$

$$= \{(x, y, z) \mid 0 \le z \le 1, z \le y \le 1, y \le x \le 1\} = \{(x, y, z) \mid 0 \le x \le 1, 0 \le z \le x, z \le y \le x\}$$

$$= \{(x, y, z) \mid 0 \le z \le 1, z \le x \le 1, z \le y \le x\}.$$

Then

$$\int_0^1 \int_y^1 \int_0^y f(x, y, z) \, dz \, dx \, dy = \int_0^1 \int_0^x \int_0^y f(x, y, z) \, dz \, dy \, dx = \int_0^1 \int_0^y \int_y^1 f(x, y, z) \, dx \, dz \, dy$$

$$= \int_0^1 \int_z^1 \int_y^1 f(x, y, z) \, dx \, dy \, dz = \int_0^1 \int_0^x \int_z^x f(x, y, z) \, dy \, dz \, dx$$

$$= \int_0^1 \int_z^1 \int_z^x f(x, y, z) \, dy \, dx \, dz.$$

35. $m = \iiint_E \rho(x, y, z) \, dV = \int_0^1 \int_0^{\sqrt{x}} \int_0^{1+x+y} 2 \, dz \, dy \, dx = \int_0^1 \int_0^{\sqrt{x}} 2(1 + x + y) \, dy \, dx$

$= \int_0^1 \left[2y + 2xy + y^2 \right]_{y=0}^{y=\sqrt{x}} dx = \int_0^1 \left(2\sqrt{x} + 2x^{3/2} + x \right) dx = \left[\frac{4}{3}x^{3/2} + \frac{4}{5}x^{5/2} + \frac{1}{2}x^2 \right]_0^1 = \frac{79}{30}$

$M_{yz} = \iiint_E x\rho(x, y, z) \, dV = \int_0^1 \int_0^{\sqrt{x}} \int_0^{1+x+y} 2x \, dz \, dy \, dx = \int_0^1 \int_0^{\sqrt{x}} 2x(1 + x + y) \, dy \, dx$

$= \int_0^1 \left[2xy + 2x^2y + xy^2 \right]_{y=0}^{y=\sqrt{x}} dx = \int_0^1 (2x^{3/2} + 2x^{5/2} + x^2) \, dx = \left[\frac{4}{5}x^{5/2} + \frac{4}{7}x^{7/2} + \frac{1}{3}x^3 \right]_0^1 = \frac{179}{105}$

$M_{xz} = \iiint_E y\rho(x, y, z) \, dV = \int_0^1 \int_0^{\sqrt{x}} \int_0^{1+x+y} 2y \, dz \, dy \, dx = \int_0^1 \int_0^{\sqrt{x}} 2y(1 + x + y) \, dy \, dx$

$= \int_0^1 \left[y^2 + xy^2 + \frac{2}{3}y^3 \right]_{y=0}^{y=\sqrt{x}} dx = \int_0^1 \left(x + x^2 + \frac{2}{3}x^{3/2} \right) dx = \left[\frac{1}{2}x^2 + \frac{1}{3}x^3 + \frac{4}{15}x^{5/2} \right]_0^1 = \frac{11}{10}$

$M_{xy} = \iiint_E z\rho(x, y, z) \, dV = \int_0^1 \int_0^{\sqrt{x}} \int_0^{1+x+y} 2z \, dz \, dy \, dx = \int_0^1 \int_0^{\sqrt{x}} \left[z^2 \right]_{z=0}^{z=1+x+y} dy \, dx = \int_0^1 \int_0^{\sqrt{x}} (1 + x + y)^2 \, dy \, dx$

$= \int_0^1 \int_0^{\sqrt{x}} (1 + 2x + 2y + 2xy + x^2 + y^2) \, dy \, dx = \int_0^1 \left[y + 2xy + y^2 + xy^2 + x^2y + \frac{1}{3}y^3 \right]_{y=0}^{y=\sqrt{x}} dx$

$= \int_0^1 \left(\sqrt{x} + \frac{7}{3}x^{3/2} + x + x^2 + x^{5/2} \right) dx = \left[\frac{2}{3}x^{3/2} + \frac{14}{15}x^{5/2} + \frac{1}{2}x^2 + \frac{1}{3}x^3 + \frac{2}{7}x^{7/2} \right]_0^1 = \frac{571}{210}$

Thus the mass is $\frac{79}{30}$ and the center of mass is $(\bar{x}, \bar{y}, \bar{z}) = \left(\dfrac{M_{yz}}{m}, \dfrac{M_{xz}}{m}, \dfrac{M_{xy}}{m} \right) = \left(\dfrac{358}{553}, \dfrac{33}{79}, \dfrac{571}{553} \right).$

37. $m = \int_0^a \int_0^a \int_0^a (x^2 + y^2 + z^2) \, dx \, dy \, dz = \int_0^a \int_0^a \left[\frac{1}{3}x^3 + xy^2 + xz^2 \right]_{x=0}^{x=a} dy \, dz = \int_0^a \int_0^a \left(\frac{1}{3}a^3 + ay^2 + az^2 \right) dy \, dz$

$= \int_0^a \left[\frac{1}{3}a^3 y + \frac{1}{3}ay^3 + ayz^2 \right]_{y=0}^{y=a} dz = \int_0^a \left(\frac{2}{3}a^4 + a^2z^2 \right) dz = \left[\frac{2}{3}a^4 z + \frac{1}{3}a^2 z^3 \right]_0^a = \frac{2}{3}a^5 + \frac{1}{3}a^5 = a^5$

$M_{yz} = \int_0^a \int_0^a \int_0^a \left[x^3 + x(y^2 + z^2) \right] dx \, dy \, dz = \int_0^a \int_0^a \left[\frac{1}{4}a^4 + \frac{1}{2}a^2(y^2 + z^2) \right] dy \, dz$

$= \int_0^a \left(\frac{1}{4}a^5 + \frac{1}{6}a^5 + \frac{1}{2}a^3 z^2 \right) dz = \frac{1}{4}a^6 + \frac{1}{3}a^6 = \frac{7}{12}a^6 = M_{xz} = M_{xy}$ by symmetry of E and $\rho(x, y, z)$

Hence $(\bar{x}, \bar{y}, \bar{z}) = \left(\frac{7}{12}a, \frac{7}{12}a, \frac{7}{12}a \right).$

39. (a) $m = \int_{-3}^{3} \int_{-\sqrt{9-x^2}}^{\sqrt{9-x^2}} \int_{1}^{5-y} \sqrt{x^2+y^2} \, dz \, dy \, dx$

(b) $(\bar{x}, \bar{y}, \bar{z}) = \left(\dfrac{M_{yz}}{m}, \dfrac{M_{xz}}{m}, \dfrac{M_{xy}}{m} \right)$ where

$M_{yz} = \int_{-3}^{3} \int_{-\sqrt{9-x^2}}^{\sqrt{9-x^2}} \int_{1}^{5-y} x \sqrt{x^2+y^2} \, dz \, dy \, dx$, $M_{xz} = \int_{-3}^{3} \int_{-\sqrt{9-x^2}}^{\sqrt{9-x^2}} \int_{1}^{5-y} y \sqrt{x^2+y^2} \, dz \, dy \, dx$, and

$M_{xy} = \int_{-3}^{3} \int_{-\sqrt{9-x^2}}^{\sqrt{9-x^2}} \int_{1}^{5-y} z \sqrt{x^2+y^2} \, dz \, dy \, dx$.

(c) $I_z = \int_{-3}^{3} \int_{-\sqrt{9-x^2}}^{\sqrt{9-x^2}} \int_{1}^{5-y} (x^2+y^2)\sqrt{x^2+y^2} \, dz \, dy \, dx = \int_{-3}^{3} \int_{-\sqrt{9-x^2}}^{\sqrt{9-x^2}} \int_{1}^{5-y} (x^2+y^2)^{3/2} \, dz \, dy \, dx$

41. (a) $m = \int_{0}^{1} \int_{0}^{\sqrt{1-x^2}} \int_{0}^{y} (1+x+y+z) \, dz \, dy \, dx = \dfrac{3\pi}{32} + \dfrac{11}{24}$

(b) $(\bar{x}, \bar{y}, \bar{z}) = \left(m^{-1} \int_{0}^{1} \int_{0}^{\sqrt{1-x^2}} \int_{0}^{y} x(1+x+y+z) \, dz \, dy \, dx, \right.$

$\qquad\qquad m^{-1} \int_{0}^{1} \int_{0}^{\sqrt{1-x^2}} \int_{0}^{y} y(1+x+y+z) \, dz \, dy \, dx,$

$\qquad\qquad\qquad \left. m^{-1} \int_{0}^{1} \int_{0}^{\sqrt{1-x^2}} \int_{0}^{y} z(1+x+y+z) \, dz \, dy \, dx \right)$

$\qquad = \left(\dfrac{28}{9\pi+44}, \dfrac{30\pi+128}{45\pi+220}, \dfrac{45\pi+208}{135\pi+660} \right)$

(c) $I_z = \int_{0}^{1} \int_{0}^{\sqrt{1-x^2}} \int_{0}^{y} (x^2+y^2)(1+x+y+z) \, dz \, dy \, dx = \dfrac{68+15\pi}{240}$

43. $I_x = \int_{0}^{L} \int_{0}^{L} \int_{0}^{L} k(y^2+z^2) \, dz \, dy \, dx = k \int_{0}^{L} \int_{0}^{L} \left(Ly^2 + \tfrac{1}{3}L^3 \right) dy \, dx = k \int_{0}^{L} \tfrac{2}{3}L^4 \, dx = \tfrac{2}{3}kL^5$.

By symmetry, $I_x = I_y = I_z = \tfrac{2}{3}kL^5$.

45. $V(E) = L^3$,

$$f_{\text{ave}} = \dfrac{1}{L^3} \int_{0}^{L} \int_{0}^{L} \int_{0}^{L} xyz \, dx \, dy \, dz = \dfrac{1}{L^3} \int_{0}^{L} x \, dx \int_{0}^{L} y \, dy \int_{0}^{L} z \, dz = \dfrac{1}{L^3} \left[\dfrac{x^2}{2} \right]_{0}^{L} \left[\dfrac{y^2}{2} \right]_{0}^{L} \left[\dfrac{z^2}{2} \right]_{0}^{L}$$

$$= \dfrac{1}{L^3} \dfrac{L^2}{2} \dfrac{L^2}{2} \dfrac{L^2}{2} = \dfrac{L^3}{8}$$

47. The triple integral will attain its maximum when the integrand $1 - x^2 - 2y^2 - 3z^2$ is positive in the region E and negative everywhere else. For if E contains some region F where the integrand is negative, the integral could be increased by excluding F from E, and if E fails to contain some part G of the region where the integrand is positive, the integral could be increased by including G in E. So we require that $x^2 + 2y^2 + 3z^2 \leq 1$. This describes the region bounded by the ellipsoid $x^2 + 2y^2 + 3z^2 = 1$.

12.6 Triple Integrals in Cylindrical Coordinates

1. (a)

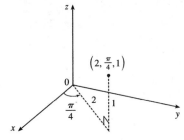

$x = 2\cos\dfrac{\pi}{4} = \sqrt{2}, y = 2\sin\dfrac{\pi}{4} = \sqrt{2}, z = 1,$

so the point is $\left(\sqrt{2}, \sqrt{2}, 1\right)$ in rectangular coordinates.

(b)

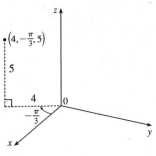

$x = 4\cos\left(-\tfrac{\pi}{3}\right) = 2, y = 4\sin\left(-\tfrac{\pi}{3}\right) = -2\sqrt{3},$

and $z = 5$, so the point is $\left(2, -2\sqrt{3}, 5\right)$ in rectangular coordinates.

3. (a) $r^2 = x^2 + y^2 = 1^2 + (-1)^2 = 2$ so $r = \sqrt{2}$; $\tan\theta = \dfrac{y}{x} = \dfrac{-1}{1} = -1$ and the point $(1, -1)$ is in the fourth quadrant of

the xy-plane, so $\theta = \tfrac{7\pi}{4} + 2n\pi$; $z = 4$. Thus, one set of cylindrical coordinates is $\left(\sqrt{2}, \tfrac{7\pi}{4}, 4\right)$.

(b) $r^2 = (-1)^2 + \left(-\sqrt{3}\right)^2 = 4$ so $r = 2$; $\tan\theta = \dfrac{-\sqrt{3}}{-1} = \sqrt{3}$ and the point $\left(-1, -\sqrt{3}\right)$ is in the third quadrant of the

xy-plane, so $\theta = \tfrac{4\pi}{3} + 2n\pi$; $z = 2$. Thus, one set of cylindrical coordinates is $\left(2, \tfrac{4\pi}{3}, 2\right)$.

5. Since $r = 3$, $x^2 + y^2 = 9$ and the surface is a circular cylinder with radius 3 and axis the z-axis.

7. $z = r^2 = x^2 + y^2$, so the surface is a circular paraboloid with vertex at the origin and axis the positive z-axis.

9. (a) $x^2 + y^2 = r^2$, so the equation becomes $z = r^2$.

(b) $r^2 = 2r\sin\theta$ or $r = 2\sin\theta$.

11.

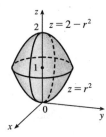

$z = r^2 = x^2 + y^2$ is a circular paraboloid with vertex $(0, 0, 0)$, opening upward.

$z = 2 - r^2 \;\Rightarrow\; z - 2 = -(x^2 + y^2)$ is a circular paraboloid with vertex $(0, 0, 2)$

opening downward. Thus $r^2 \le z \le 2 - r^2$ is the solid region enclosed by these two

surfaces.

13. We can position the cylindrical shell vertically so that its axis coincides with the z-axis and its base lies in the xy-plane. If we

use centimeters as the unit of measurement, then cylindrical coordinates conveniently describe the shell as $6 \le r \le 7$,

$0 \le \theta \le 2\pi, 0 \le z \le 20$.

15.

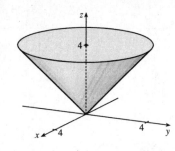

The region of integration is given in cylindrical coordinates by

$E = \{(r, \theta, z) \mid 0 \le \theta \le 2\pi, 0 \le r \le 4, r \le z \le 4\}$. This represents the solid

region bounded below by the cone $z = r$ and above by the horizontal plane $z = 4$.

$$\int_0^4 \int_0^{2\pi} \int_r^4 r \, dz \, d\theta \, dr = \int_0^4 \int_0^{2\pi} \left[rz\right]_{z=r}^{z=4} d\theta \, dr = \int_0^4 \int_0^{2\pi} r(4-r) \, d\theta \, dr$$

$$= \int_0^4 (4r - r^2) \, dr \int_0^{2\pi} d\theta = \left[2r^2 - \tfrac{1}{3}r^3\right]_0^4 \left[\theta\right]_0^{2\pi}$$

$$= \left(32 - \tfrac{64}{3}\right)(2\pi) = \tfrac{64\pi}{3}$$

17. In cylindrical coordinates, E is given by $\{(r, \theta, z) \mid 0 \le \theta \le 2\pi, 0 \le r \le 4, -5 \le z \le 4\}$. So

$$\iiint_E \sqrt{x^2 + y^2}\, dV = \int_0^{2\pi} \int_0^4 \int_{-5}^4 \sqrt{r^2}\, r \, dz \, dr \, d\theta = \int_0^{2\pi} d\theta \int_0^4 r^2 \, dr \int_{-5}^4 dz$$

$$= \left[\theta\right]_0^{2\pi} \left[\tfrac{1}{3}r^3\right]_0^4 \left[z\right]_{-5}^4 = (2\pi)\left(\tfrac{64}{3}\right)(9) = 384\pi$$

19. In cylindrical coordinates E is bounded by the paraboloid $z = 1 + r^2$, the cylinder $r^2 = 5$ or $r = \sqrt{5}$, and the xy-plane,

so E is given by $\{(r, \theta, z) \mid 0 \le \theta \le 2\pi, 0 \le r \le \sqrt{5}, 0 \le z \le 1 + r^2\}$. Thus

$$\iiint_E e^z \, dV = \int_0^{2\pi} \int_0^{\sqrt{5}} \int_0^{1+r^2} e^z \, r \, dz \, dr \, d\theta = \int_0^{2\pi} \int_0^{\sqrt{5}} r\left[e^z\right]_{z=0}^{z=1+r^2} dr \, d\theta = \int_0^{2\pi} \int_0^{\sqrt{5}} r(e^{1+r^2} - 1) \, dr \, d\theta$$

$$= \int_0^{2\pi} d\theta \int_0^{\sqrt{5}} \left(re^{1+r^2} - r\right) dr = 2\pi \left[\tfrac{1}{2}e^{1+r^2} - \tfrac{1}{2}r^2\right]_0^{\sqrt{5}} = \pi(e^6 - e - 5)$$

21. In cylindrical coordinates, E is bounded by the cylinder $r = 1$, the plane $z = 0$, and the cone $z = 2r$. So

$E = \{(r, \theta, z) \mid 0 \le \theta \le 2\pi, 0 \le r \le 1, 0 \le z \le 2r\}$ and

$$\iiint_E x^2 \, dV = \int_0^{2\pi} \int_0^1 \int_0^{2r} r^2 \cos^2 \theta \, r \, dz \, dr \, d\theta = \int_0^{2\pi} \int_0^1 \left[r^3 \cos^2 \theta \, z\right]_{z=0}^{z=2r} dr \, d\theta$$

$$= \int_0^{2\pi} \int_0^1 2r^4 \cos^2 \theta \, dr \, d\theta = \int_0^{2\pi} \left[\tfrac{2}{5}r^5 \cos^2 \theta\right]_{r=0}^{r=1} d\theta = \tfrac{2}{5} \int_0^{2\pi} \cos^2 \theta \, d\theta$$

$$= \frac{2}{5} \int_0^{2\pi} \frac{1 + \cos 2\theta}{2} \, d\theta = \frac{1}{5} \left[\theta + \frac{1}{2} \sin 2\theta\right]_0^{2\pi} = \frac{2\pi}{5}$$

23. (a) The paraboloids intersect when $x^2 + y^2 = 36 - 3x^2 - 3y^2 \;\Rightarrow\; x^2 + y^2 = 9$, so the region of integration

is $D = \{(x, y) \mid x^2 + y^2 \le 9\}$. Then, in cylindrical coordinates,

$E = \{(r, \theta, z) \mid r^2 \le z \le 36 - 3r^2, 0 \le r \le 3, 0 \le \theta \le 2\pi\}$ and

$$V = \int_0^{2\pi} \int_0^3 \int_{r^2}^{36 - 3r^2} r \, dz \, dr \, d\theta = \int_0^{2\pi} \int_0^3 \left(36r - 4r^3\right) dr \, d\theta = \int_0^{2\pi} \left[18r^2 - r^4\right]_{r=0}^{r=3} d\theta = \int_0^{2\pi} 81 \, d\theta = 162\pi.$$

(b) For constant density K, $m = KV = 162\pi K$ from part (a). Since the region is homogeneous and symmetric,

$M_{yz} = M_{xz} = 0$ and

$$M_{xy} = \int_0^{2\pi} \int_0^3 \int_{r^2}^{36-3r^2} (zK) r \, dz \, dr \, d\theta = K \int_0^{2\pi} \int_0^3 r\left[\tfrac{1}{2}z^2\right]_{z=r^2}^{z=36-3r^2} dr \, d\theta$$

$$= \tfrac{K}{2} \int_0^{2\pi} \int_0^3 r((36 - 3r^2)^2 - r^4) \, dr \, d\theta = \tfrac{K}{2} \int_0^{2\pi} d\theta \int_0^3 (8r^5 - 216r^3 + 1296r) \, dr$$

$$= \tfrac{K}{2} (2\pi) \left[\tfrac{8}{6}r^6 - \tfrac{216}{4}r^4 + \tfrac{1296}{2}r^2\right]_0^3 = \pi K(2430) = 2430\pi K$$

Thus $(\overline{x}, \overline{y}, \overline{z}) = \left(\dfrac{M_{yz}}{m}, \dfrac{M_{xz}}{m}, \dfrac{M_{xy}}{m}\right) = \left(0, 0, \dfrac{2430\pi K}{162\pi K}\right) = (0, 0, 15)$.

25. The paraboloid $z = 4x^2 + 4y^2$ intersects the plane $z = a$ when $a = 4x^2 + 4y^2$ or $x^2 + y^2 = \frac{1}{4}a$. So, in cylindrical coordinates, $E = \{(r, \theta, z) \mid 0 \le r \le \frac{1}{2}\sqrt{a}, 0 \le \theta \le 2\pi, 4r^2 \le z \le a\}$. Thus

$$m = \int_0^{2\pi} \int_0^{\sqrt{a}/2} \int_{4r^2}^a Kr\,dz\,dr\,d\theta = K \int_0^{2\pi} \int_0^{\sqrt{a}/2} (ar - 4r^3)\,dr\,d\theta$$

$$= K \int_0^{2\pi} \left[\frac{1}{2}ar^2 - r^4\right]_{r=0}^{r=\sqrt{a}/2} d\theta = K \int_0^{2\pi} \frac{1}{16}a^2\,d\theta = \frac{1}{8}a^2\pi K$$

Since the region is homogeneous and symmetric, $M_{yz} = M_{xz} = 0$ and

$$M_{xy} = \int_0^{2\pi} \int_0^{\sqrt{a}/2} \int_{4r^2}^a Krz\,dz\,dr\,d\theta = K \int_0^{2\pi} \int_0^{\sqrt{a}/2} \left(\frac{1}{2}a^2 r - 8r^5\right)dr\,d\theta$$

$$= K \int_0^{2\pi} \left[\frac{1}{4}a^2 r^2 - \frac{4}{3}r^6\right]_{r=0}^{r=\sqrt{a}/2} d\theta = K \int_0^{2\pi} \frac{1}{24}a^3\,d\theta = \frac{1}{12}a^3\pi K$$

Hence $(\overline{x}, \overline{y}, \overline{z}) = (0, 0, \frac{2}{3}a)$.

27. The region of integration is the region above the cone $z = \sqrt{x^2 + y^2}$, or $z = r$, and below the plane $z = 2$. Also, we have $-2 \le y \le 2$ with $-\sqrt{4 - y^2} \le x \le \sqrt{4 - y^2}$ which describes a circle of radius 2 in the xy-plane centered at $(0, 0)$. Thus,

$$\int_{-2}^{2} \int_{-\sqrt{4-y^2}}^{\sqrt{4-y^2}} \int_{\sqrt{x^2+y^2}}^{2} xz\,dz\,dx\,dy = \int_0^{2\pi} \int_0^2 \int_r^2 (r\cos\theta)\,zr\,dz\,dr\,d\theta = \int_0^{2\pi} \int_0^2 \int_r^2 r^2(\cos\theta)\,z\,dz\,dr\,d\theta$$

$$= \int_0^{2\pi} \int_0^2 r^2(\cos\theta)\left[\frac{1}{2}z^2\right]_{z=r}^{z=2} dr\,d\theta = \frac{1}{2}\int_0^{2\pi} \int_0^2 r^2(\cos\theta)\left(4 - r^2\right)dr\,d\theta$$

$$= \frac{1}{2}\int_0^{2\pi} \cos\theta\,d\theta \int_0^2 \left(4r^2 - r^4\right)dr = \frac{1}{2}[\sin\theta]_0^{2\pi} \left[\frac{4}{3}r^3 - \frac{1}{5}r^5\right]_0^2 = 0$$

29. (a) The mountain comprises a solid conical region C. The work done in lifting a small volume of material ΔV with density $g(P)$ to a height $h(P)$ above sea level is $h(P)g(P)\,\Delta V$. Summing over the whole mountain we get $W = \iiint_C h(P)g(P)\,dV$.

(b) Here C is a solid right circular cone with radius $R = 62{,}000$ ft, height $H = 12{,}400$ ft, and density $g(P) = 200$ lb/ft^3 at all points P in C. We use cylindrical coordinates:

$$W = \int_0^{2\pi} \int_0^H \int_0^{R(1-z/H)} z \cdot 200r\,dr\,dz\,d\theta = 2\pi \int_0^H 200z\left[\frac{1}{2}r^2\right]_{r=0}^{r=R(1-z/H)} dz$$

$$= 400\pi \int_0^H z\,\frac{R^2}{2}\left(1 - \frac{z}{H}\right)^2 dz = 200\pi R^2 \int_0^H \left(z - \frac{2z^2}{H} + \frac{z^3}{H^2}\right)dz$$

$$= 200\pi R^2 \left[\frac{z^2}{2} - \frac{2z^3}{3H} + \frac{z^4}{4H^2}\right]_0^H = 200\pi R^2 \left(\frac{H^2}{2} - \frac{2H^2}{3} + \frac{H^2}{4}\right)$$

$$= \frac{50}{3}\pi R^2 H^2 = \frac{50}{3}\pi(62{,}000)^2(12{,}400)^2 \approx 3.1 \times 10^{19} \text{ ft-lb}$$

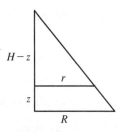

$$\frac{r}{R} = \frac{H-z}{H} = 1 - \frac{z}{H}$$

12.7   Triple Integrals in Spherical Coordinates

1. (a)

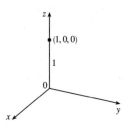

$x = \rho \sin\phi \cos\theta = (1)\sin 0 \cos 0 = 0,$

$y = \rho \sin\phi \sin\theta = (1)\sin 0 \sin 0 = 0,$ and

$z = \rho \cos\phi = (1)\cos 0 = 1$ so the point is

$(0, 0, 1)$ in rectangular coordinates.

(b)

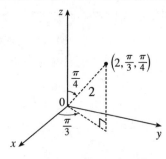

$x = 2\sin\frac{\pi}{4}\cos\frac{\pi}{3} = \frac{\sqrt{2}}{2}, y = 2\sin\frac{\pi}{4}\sin\frac{\pi}{3} = \frac{\sqrt{6}}{2},$

$z = 2\cos\frac{\pi}{4} = \sqrt{2}$ so the point is $\left(\frac{\sqrt{2}}{2}, \frac{\sqrt{6}}{2}, \sqrt{2}\right)$ in

rectangular coordinates.

3. (a) $\rho = \sqrt{x^2 + y^2 + z^2} = \sqrt{1 + 3 + 12} = 4, \cos\phi = \dfrac{z}{\rho} = \dfrac{2\sqrt{3}}{4} = \dfrac{\sqrt{3}}{2} \;\Rightarrow\; \phi = \dfrac{\pi}{6},$ and

$\cos\theta = \dfrac{x}{\rho \sin\phi} = \dfrac{1}{4\sin(\pi/6)} = \dfrac{1}{2} \;\Rightarrow\; \theta = \dfrac{\pi}{3}$ [since $y > 0$]. Thus spherical coordinates are $\left(4, \dfrac{\pi}{3}, \dfrac{\pi}{6}\right).$

(b) $\rho = \sqrt{0 + 1 + 1} = \sqrt{2}, \cos\phi = \dfrac{-1}{\sqrt{2}} \;\Rightarrow\; \phi = \dfrac{3\pi}{4},$ and $\cos\theta = \dfrac{0}{\sqrt{2}\sin(3\pi/4)} = 0 \;\Rightarrow\; \theta = \dfrac{3\pi}{2}$

[since $y < 0$]. Thus spherical coordinates are $\left(\sqrt{2}, \dfrac{3\pi}{2}, \dfrac{3\pi}{4}\right).$

5. Since $\phi = \frac{\pi}{3}$, the surface is the top half of the right circular cone with vertex at the origin and axis the positive z-axis.

7. Since $\rho \sin\phi = 2$ and $x = \rho \sin\phi \cos\theta$, $x = 2\cos\theta$. Also $y = \rho \sin\phi \sin\theta$ so $y = 2\sin\theta$. Then

$x^2 + y^2 = 4\cos^2\theta + 4\sin^2\theta = 4$, a circular cylinder of radius 2 about the z-axis.

9. (a) $x = \rho \sin\phi \cos\theta, y = \rho \sin\phi \sin\theta,$ and $z = \rho \cos\phi,$ so the equation becomes

$\rho \cos\phi = (\rho \sin\phi \cos\theta)^2 + (\rho \sin\phi \sin\theta)^2$ or $\rho \cos\phi = \rho^2 \sin^2\phi$ or $\rho \sin^2\phi = \cos\phi.$

(b) $\rho^2 \sin^2\phi(\cos^2\theta + \sin^2\theta) = 2\rho \sin\phi \sin\theta$ or $\rho \sin^2\phi = 2\sin\phi \sin\theta$ or $\rho \sin\phi = 2\sin\theta.$

11. $\rho = 2$ represents a sphere of radius 2, centered at the origin, so $\rho \le 2$ is this

sphere and its interior. $0 \le \phi \le \frac{\pi}{2}$ restricts the solid to that portion of the

region that lies on or above the xy-plane, and $0 \le \theta \le \frac{\pi}{2}$ further restricts

the solid to the first octant. Thus the solid is the portion in the first octant of

the solid ball centered at the origin with radius 2.

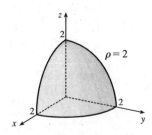

13. $-\frac{\pi}{2} \le \theta \le \frac{\pi}{2}$ restricts the solid to the 4 octants in which x is positive.

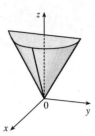

$\rho = \sec\phi \;\Rightarrow\; \rho\cos\phi = z = 1$, which is the equation of a horizontal

plane. $0 \le \phi \le \frac{\pi}{6}$ describes a cone, opening upward. So the solid lies

above the cone $\phi = \frac{\pi}{6}$ and below the plane $z = 1$.

15. $z \ge \sqrt{x^2 + y^2}$ because the solid lies above the cone. Squaring both sides of this inequality gives $z^2 \ge x^2 + y^2 \;\Rightarrow\;$

$2z^2 \ge x^2 + y^2 + z^2 = \rho^2 \;\Rightarrow\; z^2 = \rho^2\cos^2\phi \ge \frac{1}{2}\rho^2 \;\Rightarrow\; \cos^2\phi \ge \frac{1}{2}$. The cone opens upward so that the inequality is

$\cos\phi \ge \frac{1}{\sqrt{2}}$, or equivalently $0 \le \phi \le \frac{\pi}{4}$. In spherical coordinates the sphere $z = x^2 + y^2 + z^2$ is $\rho\cos\phi = \rho^2 \;\Rightarrow\;$

$\rho = \cos\phi$. $0 \le \rho \le \cos\phi$ because the solid lies below the sphere. The solid can therefore be described as the region in

spherical coordinates satisfying $0 \le \rho \le \cos\phi, 0 \le \phi \le \frac{\pi}{4}$.

17.

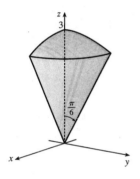

The region of integration is given in spherical coordinates by

$E = \{(\rho, \theta, \phi) \mid 0 \le \rho \le 3, 0 \le \theta \le \pi/2, 0 \le \phi \le \pi/6\}$. This represents the solid

region in the first octant bounded above by the sphere $\rho = 3$ and below by the cone

$\phi = \pi/6$.

$$\int_0^{\pi/6}\int_0^{\pi/2}\int_0^3 \rho^2\sin\phi\,d\rho\,d\theta\,d\phi = \int_0^{\pi/6}\sin\phi\,d\phi\int_0^{\pi/2}d\theta\int_0^3\rho^2\,d\rho$$

$$= \left[-\cos\phi\right]_0^{\pi/6}\left[\theta\right]_0^{\pi/2}\left[\tfrac{1}{3}\rho^3\right]_0^3$$

$$= \left(1 - \frac{\sqrt{3}}{2}\right)\left(\frac{\pi}{2}\right)(9) = \frac{9\pi}{4}\left(2 - \sqrt{3}\right)$$

19. The solid E is most conveniently described if we use cylindrical coordinates:

$E = \{(r, \theta, z) \mid 0 \le \theta \le \frac{\pi}{2}, 0 \le r \le 3, 0 \le z \le 2\}$. Then

$\iiint_E f(x, y, z)\,dV = \int_0^{\pi/2}\int_0^3\int_0^2 f(r\cos\theta, r\sin\theta, z)\,r\,dz\,dr\,d\theta$.

21. In spherical coordinates, B is represented by $\{(\rho, \theta, \phi) \mid 0 \le \rho \le 1, 0 \le \theta \le 2\pi, 0 \le \phi \le \pi\}$. Thus

$\iiint_B (x^2 + y^2 + z^2)\,dV = \int_0^\pi\int_0^{2\pi}\int_0^1 (\rho^2)\rho^2\sin\phi\,d\rho\,d\theta\,d\phi = \int_0^\pi\sin\phi\,d\phi\int_0^{2\pi}d\theta\int_0^1\rho^4\,d\rho$

$$= \left[-\cos\phi\right]_0^\pi \left[\theta\right]_0^{2\pi}\left[\tfrac{1}{5}\rho^5\right]_0^1 = (2)(2\pi)\left(\tfrac{1}{5}\right) = \tfrac{4\pi}{5}$$

23. In spherical coordinates, E is represented by $\{(\rho, \theta, \phi) \mid 1 \le \rho \le 2, 0 \le \theta \le \frac{\pi}{2}, 0 \le \phi \le \frac{\pi}{2}\}$. Thus

$\iiint_E z\,dV = \int_0^{\pi/2}\int_0^{\pi/2}\int_1^2 (\rho\cos\phi)\,\rho^2\sin\phi\,d\rho\,d\theta\,d\phi = \int_0^{\pi/2}\cos\phi\sin\phi\,d\phi\int_0^{\pi/2}d\theta\int_1^2\rho^3\,d\rho$

$$= \left[\tfrac{1}{2}\sin^2\phi\right]_0^{\pi/2}\left[\theta\right]_0^{\pi/2}\left[\tfrac{1}{4}\rho^4\right]_1^2 = \left(\tfrac{1}{2}\right)\left(\tfrac{\pi}{2}\right)\left(\tfrac{15}{4}\right) = \tfrac{15\pi}{16}$$

25. $\iiint_E x^2 \, dV = \int_0^\pi \int_0^\pi \int_3^4 (\rho \sin\phi \cos\theta)^2 \, \rho^2 \sin\phi \, d\rho \, d\phi \, d\theta = \int_0^\pi \cos^2\theta \, d\theta \int_0^\pi \sin^3\phi \, d\phi \int_3^4 \rho^4 \, d\rho$

$= \left[\frac{1}{2}\theta + \frac{1}{4}\sin 2\theta\right]_0^\pi \left[-\frac{1}{3}(2+\sin^2\phi)\cos\phi\right]_0^\pi \left[\frac{1}{5}\rho^5\right]_3^4 = \left(\frac{\pi}{2}\right)\left(\frac{2}{3}+\frac{2}{3}\right)\frac{1}{5}(4^5 - 3^5) = \frac{1562}{15}\pi$

27. (a) Since $\rho = 4\cos\phi$ implies $\rho^2 = 4\rho\cos\phi$, the equation is that of a sphere of radius 2 with center at $(0,0,2)$. Thus

$V = \int_0^{2\pi} \int_0^{\pi/3} \int_0^{4\cos\phi} \rho^2 \sin\phi \, d\rho \, d\phi \, d\theta = \int_0^{2\pi} \int_0^{\pi/3} \left[\frac{1}{3}\rho^3\right]_{\rho=0}^{\rho=4\cos\phi} \sin\phi \, d\phi \, d\theta = \int_0^{2\pi} \int_0^{\pi/3} \left(\frac{64}{3}\cos^3\phi\right) \sin\phi \, d\phi \, d\theta$

$= \int_0^{2\pi} \left[-\frac{16}{3}\cos^4\phi\right]_{\phi=0}^{\phi=\pi/3} d\theta = \int_0^{2\pi} -\frac{16}{3}\left(\frac{1}{16} - 1\right) d\theta = 5\theta\Big]_0^{2\pi} = 10\pi$

(b) By the symmetry of the problem $M_{yz} = M_{xz} = 0$. Then

$M_{xy} = \int_0^{2\pi} \int_0^{\pi/3} \int_0^{4\cos\phi} \rho^3 \cos\phi \sin\phi \, d\rho \, d\phi \, d\theta = \int_0^{2\pi} \int_0^{\pi/3} \cos\phi \sin\phi \left(64\cos^4\phi\right) d\phi \, d\theta$

$= \int_0^{2\pi} 64\left[-\frac{1}{6}\cos^6\phi\right]_{\phi=0}^{\phi=\pi/3} d\theta = \int_0^{2\pi} \frac{21}{2} \, d\theta = 21\pi$

Hence $(\overline{x}, \overline{y}, \overline{z}) = (0, 0, 2.1)$.

29. (a) The density function is $\rho(x, y, z) = K$, a constant, and by the symmetry of the problem $M_{xz} = M_{yz} = 0$. Then

$M_{xy} = \int_0^{2\pi} \int_0^{\pi/2} \int_0^a K\rho^3 \sin\phi \cos\phi \, d\rho \, d\phi \, d\theta = \frac{1}{2}\pi K a^4 \int_0^{\pi/2} \sin\phi \cos\phi \, d\phi = \frac{1}{8}\pi K a^4$. But the mass is

$K(\text{volume of the hemisphere}) = \frac{2}{3}\pi K a^3$, so the centroid is $\left(0, 0, \frac{3}{8}a\right)$.

(b) Place the center of the base at $(0, 0, 0)$; the density function is $\rho(x, y, z) = K$. By symmetry, the moments of inertia about any two such diameters will be equal, so we just need to find I_x:

$I_x = \int_0^{2\pi} \int_0^{\pi/2} \int_0^a \left(K\rho^2 \sin\phi\right) \rho^2 \left(\sin^2\phi \sin^2\theta + \cos^2\phi\right) d\rho \, d\phi \, d\theta$

$= K \int_0^{2\pi} \int_0^{\pi/2} \left(\sin^3\phi \sin^2\theta + \sin\phi \cos^2\phi\right)\left(\frac{1}{5}a^5\right) d\phi \, d\theta$

$= \frac{1}{5}K a^5 \int_0^{2\pi} \left[\sin^2\theta \left(-\cos\phi + \frac{1}{3}\cos^3\phi\right) + \left(-\frac{1}{3}\cos^3\phi\right)\right]_{\phi=0}^{\phi=\pi/2} d\theta = \frac{1}{5}K a^5 \int_0^{2\pi} \left[\frac{2}{3}\sin^2\theta + \frac{1}{3}\right] d\theta$

$= \frac{1}{5}K a^5 \left[\frac{2}{3}\left(\frac{1}{2}\theta - \frac{1}{4}\sin 2\theta\right) + \frac{1}{3}\theta\right]_0^{2\pi} = \frac{1}{5}K a^5 \left[\frac{2}{3}(\pi - 0) + \frac{1}{3}(2\pi - 0)\right] = \frac{4}{15}K a^5 \pi$

31. In spherical coordinates $z = \sqrt{x^2 + y^2}$ becomes $\cos\phi = \sin\phi$ or $\phi = \frac{\pi}{4}$. Then

$V = \int_0^{2\pi} \int_0^{\pi/4} \int_0^1 \rho^2 \sin\phi \, d\rho \, d\phi \, d\theta = \int_0^{2\pi} d\theta \int_0^{\pi/4} \sin\phi \, d\phi \int_0^1 \rho^2 \, d\rho = \frac{1}{3}\pi\left(2 - \sqrt{2}\right)$,

$M_{xy} = \int_0^{2\pi} \int_0^{\pi/4} \int_0^1 \rho^3 \sin\phi \cos\phi \, d\rho \, d\phi \, d\theta = 2\pi\left[-\frac{1}{4}\cos 2\phi\right]_0^{\pi/4}\left(\frac{1}{4}\right) = \frac{\pi}{8}$ and by symmetry $M_{yz} = M_{xz} = 0$.

Hence $(\overline{x}, \overline{y}, \overline{z}) = \left(0, 0, \dfrac{3}{8\left(2 - \sqrt{2}\right)}\right)$.

33. In cylindrical coordinates the paraboloid is given by $z = r^2$ and the plane by $z = 2r\sin\theta$ and they intersect in the circle $r = 2\sin\theta$. Then $\iiint_E z \, dV = \int_0^\pi \int_0^{2\sin\theta} \int_{r^2}^{2r\sin\theta} rz \, dz \, dr \, d\theta = \frac{5\pi}{6}$ [using a CAS].

35. The region E of integration is the region above the cone $z = \sqrt{x^2 + y^2}$ and below the sphere $x^2 + y^2 + z^2 = 2$ in the first

octant. Because E is in the first octant we have $0 \le \theta \le \frac{\pi}{2}$. The cone has equation $\phi = \frac{\pi}{4}$ (as in Example 4), so $0 \le \phi \le \frac{\pi}{4}$,

and $0 \le \rho \le \sqrt{2}$. So the integral becomes

$\int_0^{\pi/4} \int_0^{\pi/2} \int_0^{\sqrt{2}} (\rho \sin \phi \cos \theta)(\rho \sin \phi \sin \theta) \rho^2 \sin \phi \, d\rho \, d\theta \, d\phi$

$$= \int_0^{\pi/4} \sin^3 \phi \, d\phi \int_0^{\pi/2} \sin \theta \cos \theta \, d\theta \int_0^{\sqrt{2}} \rho^4 \, d\rho = \left(\int_0^{\pi/4} (1 - \cos^2 \phi) \sin \phi \, d\phi \right) \left[\frac{1}{2} \sin^2 \theta \right]_0^{\pi/2} \left[\frac{1}{5} \rho^5 \right]_0^{\sqrt{2}}$$

$$= \left[\frac{1}{3} \cos^3 \phi - \cos \phi \right]_0^{\pi/4} \cdot \frac{1}{2} \cdot \frac{1}{5} \left(\sqrt{2} \right)^5 = \left[\frac{\sqrt{2}}{12} - \frac{\sqrt{2}}{2} - \left(\frac{1}{3} - 1 \right) \right] \cdot \frac{2\sqrt{2}}{5} = \frac{4\sqrt{2} - 5}{15}$$

37. In cylindrical coordinates, the equation of the cylinder is $r = 3, 0 \le z \le 10$.

The hemisphere is the upper part of the sphere radius 3, center $(0, 0, 10)$, equation

$r^2 + (z - 10)^2 = 3^2, z \ge 10$. In Maple, we can use either the `coords=cylindrical`

option in a regular `plot` command, or the `plots[cylinderplot]` command. In

Mathematica, we can use `ParametricPlot3d`.

39. If E is the solid enclosed by the surface $\rho = 1 + \frac{1}{5} \sin 6\theta \sin 5\phi$, it can be described in spherical coordinates as

$E = \left\{ (\rho, \theta, \phi) \mid 0 \le \rho \le 1 + \frac{1}{5} \sin 6\theta \sin 5\phi, 0 \le \theta \le 2\pi, 0 \le \phi \le \pi \right\}$. Its volume is given by

$V(E) = \iiint_E dV = \int_0^\pi \int_0^{2\pi} \int_0^{1 + (\sin 6\theta \sin 5\phi)/5} \rho^2 \sin \phi \, d\rho \, d\theta \, d\phi = \frac{136\pi}{99}$ [using a CAS].

41. (a) From the diagram, $z = r \cot \phi_0$ to $z = \sqrt{a^2 - r^2}, r = 0$

to $r = a \sin \phi_0$ (or use $a^2 - r^2 = r^2 \cot^2 \phi_0$). Thus

$V = \int_0^{2\pi} \int_0^{a \sin \phi_0} \int_{r \cot \phi_0}^{\sqrt{a^2 - r^2}} r \, dz \, dr \, d\theta$

$= 2\pi \int_0^{a \sin \phi_0} \left(r \sqrt{a^2 - r^2} - r^2 \cot \phi_0 \right) dr$

$= \frac{2\pi}{3} \left[-(a^2 - r^2)^{3/2} - r^3 \cot \phi_0 \right]_0^{a \sin \phi_0}$

$= \frac{2\pi}{3} \left[-\left(a^2 - a^2 \sin^2 \phi_0 \right)^{3/2} - a^3 \sin^3 \phi_0 \cot \phi_0 + a^3 \right]$

$= \frac{2}{3} \pi a^3 \left[1 - \left(\cos^3 \phi_0 + \sin^2 \phi_0 \cos \phi_0 \right) \right] = \frac{2}{3} \pi a^3 (1 - \cos \phi_0)$

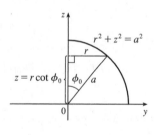

(b) The wedge in question is the shaded area rotated from $\theta = \theta_1$ to $\theta = \theta_2$.

Letting

V_{ij} = volume of the region bounded by the sphere of radius ρ_i

and the cone with angle ϕ_j ($\theta = \theta_1$ to θ_2)

and letting V be the volume of the wedge, we have

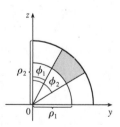

$$V = (V_{22} - V_{21}) - (V_{12} - V_{11})$$

$$= \tfrac{1}{3}(\theta_2 - \theta_1)\big[\rho_2^3(1 - \cos\phi_2) - \rho_2^3(1 - \cos\phi_1) - \rho_1^3(1 - \cos\phi_2) + \rho_1^3(1 - \cos\phi_1)\big]$$

$$= \tfrac{1}{3}(\theta_2 - \theta_1)\big[(\rho_2^3 - \rho_1^3)(1 - \cos\phi_2) - (\rho_2^3 - \rho_1^3)(1 - \cos\phi_1)\big] = \tfrac{1}{3}(\theta_2 - \theta_1)\big[(\rho_2^3 - \rho_1^3)(\cos\phi_1 - \cos\phi_2)\big]$$

Or: Show that $V = \displaystyle\int_{\theta_1}^{\theta_2} \int_{\rho_1 \sin\phi_1}^{\rho_2 \sin\phi_2} \int_{r\cot\phi_2}^{r\cot\phi_1} r\,dz\,dr\,d\theta.$

(c) By the Mean Value Theorem with $f(\rho) = \rho^3$ there exists some $\tilde\rho$ with $\rho_1 \le \tilde\rho \le \rho_2$ such that

$$f(\rho_2) - f(\rho_1) = f'(\tilde\rho)(\rho_2 - \rho_1) \text{ or } \rho_1^3 - \rho_2^3 = 3\tilde\rho^2\,\Delta\rho. \text{ Similarly there exists } \phi \text{ with } \phi_1 \le \tilde\phi \le \phi_2$$

such that $\cos\phi_2 - \cos\phi_1 = \left(-\sin\tilde\phi\right)\Delta\phi.$ Substituting into the result from (b) gives

$$\Delta V = (\tilde\rho^2\,\Delta\rho)(\theta_2 - \theta_1)(\sin\tilde\phi)\,\Delta\phi = \tilde\rho^2 \sin\tilde\phi\,\Delta\rho\,\Delta\phi\,\Delta\theta.$$

12.8 Change of Variables in Multiple Integrals

1. $x = u + 4v,\ y = 3u - 2v.$ The Jacobian is $\dfrac{\partial(x,y)}{\partial(u,v)} = \begin{vmatrix} \partial x/\partial u & \partial x/\partial v \\ \partial y/\partial u & \partial y/\partial v \end{vmatrix} = \begin{vmatrix} 1 & 4 \\ 3 & -2 \end{vmatrix} = 1(-2) - 4(3) = -14.$

3. $\dfrac{\partial(x,y)}{\partial(u,v)} = \begin{vmatrix} \dfrac{\partial x}{\partial u} & \dfrac{\partial x}{\partial v} \\[2mm] \dfrac{\partial y}{\partial u} & \dfrac{\partial y}{\partial v} \end{vmatrix} = \begin{vmatrix} \dfrac{v}{(u+v)^2} & -\dfrac{u}{(u+v)^2} \\[3mm] -\dfrac{v}{(u-v)^2} & \dfrac{u}{(u-v)^2} \end{vmatrix} = \dfrac{uv}{(u+v)^2(u-v)^2} - \dfrac{uv}{(u+v)^2(u-v)^2} = 0$

5. $\dfrac{\partial(x,y,z)}{\partial(u,v,w)} = \begin{vmatrix} \partial x/\partial u & \partial x/\partial v & \partial x/\partial w \\ \partial y/\partial u & \partial y/\partial v & \partial y/\partial w \\ \partial z/\partial u & \partial z/\partial v & \partial z/\partial w \end{vmatrix} = \begin{vmatrix} v & u & 0 \\ 0 & w & v \\ w & 0 & u \end{vmatrix} = v\begin{vmatrix} w & v \\ 0 & u \end{vmatrix} - u\begin{vmatrix} 0 & v \\ w & u \end{vmatrix} + 0\begin{vmatrix} 0 & w \\ w & 0 \end{vmatrix}$

$$= v(uw - 0) - u(0 - vw) = 2uvw$$

7. The transformation maps the boundary of S to the boundary of the image R, so we first look at side S_1 in the uv-plane. S_1 is

described by $v = 0\ \ (0 \le u \le 3),$ so $x = 2u + 3v = 2u$ and $y = u - v = u.$ Eliminating $u,$ we have $x = 2y,\ 0 \le x \le 6.$ S_2

is the line segment $u = 3,\ 0 \le v \le 2,$ so $x = 6 + 3v$ and $y = 3 - v.$ Then $v = 3 - y \ \Rightarrow \ x = 6 + 3(3 - y) = 15 - 3y,$

$6 \le x \le 12.$ S_3 is the line segment $v = 2,\ 0 \le u \le 3,$ so $x = 2u + 6$ and $y = u - 2,$ giving $u = y + 2 \ \Rightarrow \ x = 2y + 10,$

$6 \le x \le 12.$ Finally, S_4 is the segment $u = 0,\ 0 \le v \le 2,$ so $x = 3v$ and $y = -v \ \Rightarrow \ x = -3y,\ 0 \le x \le 6.$ The image of

set S is the region R shown in the xy-plane, a parallelogram bounded by these four segments.

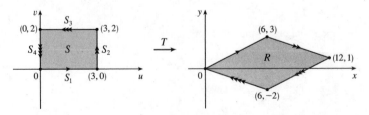

9. S_1 is the line segment $u = v$, $0 \leq u \leq 1$, so $y = v = u$ and $x = u^2 = y^2$. Since $0 \leq u \leq 1$, the image is the portion of the parabola $x = y^2$, $0 \leq y \leq 1$. S_2 is the segment $v = 1$, $0 \leq u \leq 1$, thus $y = v = 1$ and $x = u^2$, so $0 \leq x \leq 1$. The image is the line segment $y = 1$, $0 \leq x \leq 1$. S_3 is the segment $u = 0$, $0 \leq v \leq 1$, so $x = u^2 = 0$ and $y = v$ $\Rightarrow$ $0 \leq y \leq 1$. The image is the segment $x = 0$, $0 \leq y \leq 1$. Thus, the image of S is the region R in the first quadrant bounded by the parabola $x = y^2$, the y-axis, and the line $y = 1$.

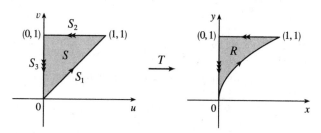

11. $\dfrac{\partial(x,y)}{\partial(u,v)} = \begin{vmatrix} 2 & 1 \\ 1 & 2 \end{vmatrix} = 3$ and $x - 3y = (2u + v) - 3(u + 2v) = -u - 5v$. To find the region S in the uv-plane that

corresponds to R we first find the corresponding boundary under the given transformation. The line through $(0,0)$ and $(2,1)$ is $y = \frac{1}{2}x$ which is the image of $u + 2v = \frac{1}{2}(2u + v)$ $\Rightarrow$ $v = 0$; the line through $(2,1)$ and $(1,2)$ is $x + y = 3$ which is the image of $(2u + v) + (u + 2v) = 3$ $\Rightarrow$ $u + v = 1$; the line through $(0,0)$ and $(1,2)$ is $y = 2x$ which is the image of $u + 2v = 2(2u + v)$ $\Rightarrow$ $u = 0$. Thus S is the triangle $0 \leq v \leq 1 - u$, $0 \leq u \leq 1$ in the uv-plane and

$$\iint_R (x - 3y)\, dA = \int_0^1 \int_0^{1-u} (-u - 5v)\, |3|\, dv\, du = -3 \int_0^1 \left[uv + \frac{5}{2}v^2 \right]_{v=0}^{v=1-u}\, du$$

$$= -3 \int_0^1 \left(u - u^2 + \frac{5}{2}(1 - u)^2 \right) du = -3 \left[\frac{1}{2}u^2 - \frac{1}{3}u^3 - \frac{5}{6}(1 - u)^3 \right]_0^1 = -3 \left(\frac{1}{2} - \frac{1}{3} + \frac{5}{6} \right) = -3$$

13. $\dfrac{\partial(x,y)}{\partial(u,v)} = \begin{vmatrix} 2 & 0 \\ 0 & 3 \end{vmatrix} = 6$, $x^2 = 4u^2$ and the planar ellipse $9x^2 + 4y^2 \leq 36$ is the image of the disk $u^2 + v^2 \leq 1$. Thus

$$\iint_R x^2\, dA = \iint_{u^2 + v^2 \leq 1} (4u^2)(6)\, du\, dv = \int_0^{2\pi} \int_0^1 (24r^2 \cos^2 \theta)\, r\, dr\, d\theta = 24 \int_0^{2\pi} \cos^2 \theta\, d\theta \int_0^1 r^3\, dr$$

$$= 24 \left[\frac{1}{2}x + \frac{1}{4} \sin 2x \right]_0^{2\pi} \left[\frac{1}{4}r^4 \right]_0^1 = 24(\pi)\left(\frac{1}{4} \right) = 6\pi$$

15. $\dfrac{\partial(x,y)}{\partial(u,v)} = \begin{vmatrix} 1/v & -u/v^2 \\ 0 & 1 \end{vmatrix} = \dfrac{1}{v}$, $xy = u$, $y = x$ is the image of the parabola $v^2 = u$, $y = 3x$ is the image of the parabola

$v^2 = 3u$, and the hyperbolas $xy = 1$, $xy = 3$ are the images of the lines $u = 1$ and $u = 3$ respectively. Thus

$$\iint_R xy\, dA = \int_1^3 \int_{\sqrt{u}}^{\sqrt{3u}} u\left(\frac{1}{v} \right) dv\, du = \int_1^3 u\left(\ln \sqrt{3u} - \ln \sqrt{u} \right) du = \int_1^3 u \ln \sqrt{3}\, du = 4 \ln \sqrt{3} = 2 \ln 3.$$

17. (a) $\dfrac{\partial(x,y,z)}{\partial(u,v,w)} = \begin{vmatrix} a & 0 & 0 \\ 0 & b & 0 \\ 0 & 0 & c \end{vmatrix} = abc$ and since $u = \dfrac{x}{a}$, $v = \dfrac{y}{b}$, $w = \dfrac{z}{c}$ the solid enclosed by the ellipsoid is the image of the

ball $u^2 + v^2 + w^2 \leq 1$. So

$$\iiint_E dV = \iiint_{u^2+v^2+w^2 \leq 1} abc\, du\, dv\, dw = (abc)(\text{volume of the ball}) = \tfrac{4}{3}\pi abc$$

(b) If we approximate the surface of the Earth by the ellipsoid $\dfrac{x^2}{6378^2} + \dfrac{y^2}{6378^2} + \dfrac{z^2}{6356^2} = 1$, then we can estimate

the volume of the Earth by finding the volume of the solid E enclosed by the ellipsoid. From part (a), this is

$$\iiint_E dV = \tfrac{4}{3}\pi(6378)(6378)(6356) \approx 1.083 \times 10^{12} \text{ km}^3.$$

19. Letting $u = x - 2y$ and $v = 3x - y$, we have $x = \tfrac{1}{5}(2v - u)$ and $y = \tfrac{1}{5}(v - 3u)$. Then $\dfrac{\partial(x,y)}{\partial(u,v)} = \begin{vmatrix} -1/5 & 2/5 \\ -3/5 & 1/5 \end{vmatrix} = \dfrac{1}{5}$

and R is the image of the rectangle enclosed by the lines $u = 0$, $u = 4$, $v = 1$, and $v = 8$. Thus

$$\iint_R \frac{x - 2y}{3x - y}\, dA = \int_0^4 \int_1^8 \frac{u}{v}\left|\frac{1}{5}\right| dv\, du = \frac{1}{5}\int_0^4 u\, du \int_1^8 \frac{1}{v}\, dv = \tfrac{1}{5}\left[\tfrac{1}{2}u^2\right]_0^4 \left[\ln|v|\right]_1^8 = \tfrac{8}{5}\ln 8.$$

21. Letting $u = y - x$, $v = y + x$, we have $y = \tfrac{1}{2}(u + v)$, $x = \tfrac{1}{2}(v - u)$. Then $\dfrac{\partial(x,y)}{\partial(u,v)} = \begin{vmatrix} -1/2 & 1/2 \\ 1/2 & 1/2 \end{vmatrix} = -\dfrac{1}{2}$ and R is the

image of the trapezoidal region with vertices $(-1,1)$, $(-2,2)$, $(2,2)$, and $(1,1)$. Thus

$$\iint_R \cos\tfrac{y-x}{y+x}\, dA = \int_1^2 \int_{-v}^{v} \cos\tfrac{u}{v}\left|-\tfrac{1}{2}\right| du\, dv = \tfrac{1}{2}\int_1^2 \left[v\sin\tfrac{u}{v}\right]_{u=-v}^{u=v} dv$$

$$= \tfrac{1}{2}\int_1^2 2v\sin(1)\, dv = \tfrac{3}{2}\sin 1$$

23. Let $u = x + y$ and $v = -x + y$. Then $u + v = 2y \;\Rightarrow\; y = \tfrac{1}{2}(u + v)$ and

$u - v = 2x \;\Rightarrow\; x = \tfrac{1}{2}(u - v)$. $\dfrac{\partial(x,y)}{\partial(u,v)} = \begin{vmatrix} 1/2 & -1/2 \\ 1/2 & 1/2 \end{vmatrix} = \dfrac{1}{2}.$

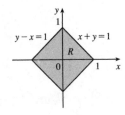

Now $|u| = |x + y| \leq |x| + |y| \leq 1 \;\Rightarrow\; -1 \leq u \leq 1$, and

$|v| = |-x + y| \leq |x| + |y| \leq 1 \;\Rightarrow\; -1 \leq v \leq 1$. R is the image of

the square region with vertices $(1,1)$, $(1,-1)$, $(-1,-1)$, and $(-1,1)$.

So $\iint_R e^{x+y}\, dA = \tfrac{1}{2}\int_{-1}^1 \int_{-1}^1 e^u\, du\, dv = \tfrac{1}{2}\left[e^u\right]_{-1}^1 \left[v\right]_{-1}^1 = e - e^{-1}.$

12   Review

CONCEPT CHECK

1. (a) A double Riemann sum of f is $\sum_{i=1}^{m} \sum_{j=1}^{n} f\left(x_{ij}^{*}, y_{ij}^{*}\right) \Delta A_{ij}$, where ΔA_{ij} is the area of the ij-th subrectangle and $\left(x_{ij}^{*}, y_{ij}^{*}\right)$

   is a sample point in each subrectangle. If $f(x, y) \geq 0$, this sum represents an approximation to the volume of the solid that lies above the rectangle R and below the graph of f.

   (b) $\iint_{R} f(x, y)\, dA = \lim_{\max \Delta x_i, \Delta y_j \to 0} \sum_{i=1}^{m} \sum_{j=1}^{n} f\left(x_{ij}^{*}, y_{ij}^{*}\right) \Delta A_{ij}$

   (c) If $f(x, y) \geq 0$, $\iint_{R} f(x, y)\, dA$ represents the volume of the solid that lies above the rectangle R and below the surface $z = f(x, y)$. If f takes on both positive and negative values, $\iint_{R} f(x, y)\, dA$ is the difference of the volume above R but below the surface $z = f(x, y)$ and the volume below R but above the surface $z = f(x, y)$.

   (d) We usually evaluate $\iint_{R} f(x, y)\, dA$ as an iterated integral according to Fubini's Theorem (see Theorem 12.1.10).

   (e) The Midpoint Rule for Double Integrals says that we approximate the double integral $\iint_{R} f(x, y)\, dA$ by the double

   Riemann sum $\sum_{i=1}^{m} \sum_{j=1}^{n} f\left(\overline{x}_i, \overline{y}_j\right) \Delta A$ where the sample points $\left(\overline{x}_i, \overline{y}_j\right)$ are the centers of the subrectangles.

2. (a) See (1) and (2) and the accompanying discussion in Section 12.2.

   (b) See (3) and the accompanying discussion in Section 12.2.

   (c) See (5) and the preceding discussion in Section 12.2.

   (d) See (6)–(11) in Section 12.2.

3. We may want to change from rectangular to polar coordinates in a double integral if the region R of integration is more easily described in polar coordinates. To accomplish this, we use $\iint_{R} f(x, y)\, dA = \int_{\alpha}^{\beta} \int_{a}^{b} f(r \cos \theta, r \sin \theta)\, r\, dr\, d\theta$ where R is given by $0 \leq a \leq r \leq b$, $\alpha \leq \theta \leq \beta$.

4. (a) $m = \iint_{D} \rho(x, y)\, dA$

   (b) $M_x = \iint_{D} y\rho(x, y)\, dA$, $M_y = \iint_{D} x\rho(x, y)\, dA$

   (c) The center of mass is $(\overline{x}, \overline{y})$ where $\overline{x} = \dfrac{M_y}{m}$ and $\overline{y} = \dfrac{M_x}{m}$.

   (d) $I_x = \iint_{D} y^2 \rho(x, y)\, dA$, $I_y = \iint_{D} x^2 \rho(x, y)\, dA$, $I_0 = \iint_{D} (x^2 + y^2)\rho(x, y)\, dA$

5. (a) $\iiint_{B} f(x, y, z)\, dV = \lim_{\max \Delta x_i, \Delta y_j, \Delta z_k \to 0} \sum_{i=1}^{l} \sum_{j=1}^{m} \sum_{k=1}^{n} f\left(x_{ijk}^{*}, y_{ijk}^{*}, z_{ijk}^{*}\right) \Delta V_{ijk}$

(b) We usually evaluate $\iiint_B f(x, y, z) \, dV$ as an iterated integral according to Fubini's Theorem for Triple Integrals (see Theorem 12.5.4).

(c) See the paragraph following Example 12.5.1.

(d) See (5) and (6) and the accompanying discussion in Section 12.5.

(e) See (10) and the accompanying discussion in Section 12.5.

(f) See (11) and the preceding discussion in Section 12.5.

6. (a) $m = \iiint_E \rho(x, y, z) \, dV$

(b) $M_{yz} = \iiint_E x\rho(x, y, z) \, dV$, $M_{xz} = \iiint_E y\rho(x, y, z) \, dV$, $M_{xy} = \iiint_E z\rho(x, y, z) \, dV$.

(c) The center of mass is $(\overline{x}, \overline{y}, \overline{z})$ where $\overline{x} = \dfrac{M_{yz}}{m}$, $\overline{y} = \dfrac{M_{xz}}{m}$, and $\overline{z} = \dfrac{M_{xy}}{m}$.

(d) $I_x = \iiint_E (y^2 + z^2)\rho(x, y, z) \, dV$, $I_y = \iiint_E (x^2 + z^2)\rho(x, y, z) \, dV$, $I_z = \iiint_E (x^2 + y^2)\rho(x, y, z) \, dV$.

7. (a) $x = r \cos \theta$, $y = r \sin \theta$, $z = z$. See the discussion after Example 1 in Section 12.6.

(b) $x = \rho \sin \phi \cos \theta$, $y = \rho \sin \phi \sin \theta$, $z = \rho \cos \phi$. See Figure 1 and the accompanying discussion in Section 12.7.

8. (a) See Formula 12.6.4 and the accompanying discussion.

(b) See Formula 12.7.3 and the accompanying discussion.

(c) We may want to change from rectangular to cylindrical or spherical coordinates in a triple integral if the region E of integration is more easily described in cylindrical or spherical coordinates or if the triple integral is easier to evaluate using cylindrical or spherical coordinates.

9. (a) $\dfrac{\partial (x, y)}{\partial (u, v)} = \begin{vmatrix} \partial x/\partial u & \partial x/\partial v \\ \partial y/\partial u & \partial y/\partial v \end{vmatrix} = \dfrac{\partial x}{\partial u} \dfrac{\partial y}{\partial v} - \dfrac{\partial x}{\partial v} \dfrac{\partial y}{\partial u}$

(b) See (9) and the accompanying discussion in Section 12.8.

(c) See (13) and the accompanying discussion in Section 12.8.

TRUE-FALSE QUIZ

1. This is true by Fubini's Theorem.

3. True by Equation 12.1.11.

5. True: $\iint_D \sqrt{4 - x^2 - y^2} \, dA =$ the volume under the surface $x^2 + y^2 + z^2 = 4$ and above the xy-plane

$$= \tfrac{1}{2} \left(\text{the volume of the sphere } x^2 + y^2 + z^2 = 4\right) = \tfrac{1}{2} \cdot \tfrac{4}{3}\pi(2)^3 = \tfrac{16}{3}\pi$$

7. The volume enclosed by the cone $z = \sqrt{x^2 + y^2}$ and the plane $z = 2$ is, in cylindrical coordinates,

$$V = \int_0^{2\pi} \int_0^2 \int_r^2 r \, dz \, dr \, d\theta \neq \int_0^{2\pi} \int_0^2 \int_r^2 dz \, dr \, d\theta, \text{ so the assertion is false.}$$

EXERCISES

1. As shown in the contour map, we divide R into 9 equally sized subsquares, each with area $\Delta A = 1$. Then we approximate

$\iint_R f(x,y)\, dA$ by a Riemann sum with $m = n = 3$ and the sample points the upper right corners of each square, so

$$\iint_R f(x,y)\, dA \approx \sum_{i=1}^{3} \sum_{j=1}^{3} f(x_i, y_j)\, \Delta A$$

$$= \Delta A\, [f(1,1) + f(1,2) + f(1,3) + f(2,1) + f(2,2) + f(2,3) + f(3,1) + f(3,2) + f(3,3)]$$

Using the contour lines to estimate the function values, we have

$$\iint_R f(x,y)\, dA \approx 1[2.7 + 4.7 + 8.0 + 4.7 + 6.7 + 10.0 + 6.7 + 8.6 + 11.9] \approx 64.0$$

3. $\int_1^2 \int_0^2 (y + 2xe^y)\, dx\, dy = \int_1^2 \left[xy + x^2 e^y\right]_{x=0}^{x=2} dy = \int_1^2 (2y + 4e^y)\, dy = \left[y^2 + 4e^y\right]_1^2$
$$= 4 + 4e^2 - 1 - 4e = 4e^2 - 4e + 3$$

5. $\int_0^1 \int_0^x \cos(x^2)\, dy\, dx = \int_0^1 \left[\cos(x^2)y\right]_{y=0}^{y=x} dx = \int_0^1 x\cos(x^2)\, dx = \frac{1}{2}\sin(x^2)\big]_0^1 = \frac{1}{2}\sin 1$

7. $\int_0^\pi \int_0^1 \int_0^{\sqrt{1-y^2}} y\sin x\, dz\, dy\, dx = \int_0^\pi \int_0^1 \left[(y\sin x)z\right]_{z=0}^{z=\sqrt{1-y^2}} dy\, dx = \int_0^\pi \int_0^1 y\sqrt{1-y^2}\sin x\, dy\, dx$
$$= \int_0^\pi \left[-\frac{1}{3}(1-y^2)^{3/2}\sin x\right]_{y=0}^{y=1} dx = \int_0^\pi \frac{1}{3}\sin x\, dx = -\frac{1}{3}\cos x\big]_0^\pi = \frac{2}{3}$$

9. The region R is more easily described by polar coordinates: $R = \{(r,\theta) \mid 2 \le r \le 4, 0 \le \theta \le \pi\}$. Thus

$\iint_R f(x,y)\, dA = \int_0^\pi \int_2^4 f(r\cos\theta, r\sin\theta)\, r\, dr\, d\theta$.

11.

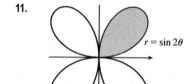

$r = \sin 2\theta$

The region whose area is given by $\int_0^{\pi/2} \int_0^{\sin 2\theta} r\, dr\, d\theta$ is

$\left\{(r,\theta) \mid 0 \le \theta \le \frac{\pi}{2}, 0 \le r \le \sin 2\theta\right\}$, which is the region contained in the

loop in the first quadrant of the four-leaved rose $r = \sin 2\theta$.

13.

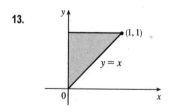

$(1,1)$

$y = x$

$\int_0^1 \int_x^1 \cos(y^2)\, dy\, dx = \int_0^1 \int_0^y \cos(y^2)\, dx\, dy$

$$= \int_0^1 \cos(y^2)[x]_{x=0}^{x=y}\, dy = \int_0^1 y\cos(y^2)\, dy$$

$$= \left[\frac{1}{2}\sin(y^2)\right]_0^1 = \frac{1}{2}\sin 1$$

15. $\iint_R ye^{xy}\, dA = \int_0^3 \int_0^2 ye^{xy}\, dx\, dy = \int_0^3 \left[e^{xy}\right]_{x=0}^{x=2} dy = \int_0^3 (e^{2y} - 1)\, dy = \left[\frac{1}{2}e^{2y} - y\right]_0^3$
$$= \frac{1}{2}e^6 - 3 - \frac{1}{2} = \frac{1}{2}e^6 - \frac{7}{2}$$

17.

$$\iint_D \frac{y}{1+x^2}\, dA = \int_0^1 \int_0^{\sqrt{x}} \frac{y}{1+x^2}\, dy\, dx = \int_0^1 \frac{1}{1+x^2}\left[\tfrac{1}{2}y^2\right]_{y=0}^{y=\sqrt{x}}\, dx$$

$$= \tfrac{1}{2}\int_0^1 \frac{x}{1+x^2}\, dx = \left[\tfrac{1}{4}\ln(1+x^2)\right]_0^1 = \tfrac{1}{4}\ln 2$$

19.

$$\iint_D y\, dA = \int_0^2 \int_{y^2}^{8-y^2} y\, dx\, dy$$

$$= \int_0^2 y[x]_{x=y^2}^{x=8-y^2}\, dy = \int_0^2 y(8-y^2-y^2)\, dy$$

$$= \int_0^2 (8y-2y^3)\, dy = \left[4y^2-\tfrac{1}{2}y^4\right]_0^2 = 8$$

21.

$$\iint_D (x^2+y^2)^{3/2}\, dA = \int_0^{\pi/3}\int_0^3 (r^2)^{3/2}r\, dr\, d\theta$$

$$= \int_0^{\pi/3} d\theta \int_0^3 r^4\, dr = \left[\theta\right]_0^{\pi/3}\left[\tfrac{1}{5}r^5\right]_0^3$$

$$= \frac{\pi}{3}\frac{3^5}{5} = \frac{81\pi}{5}$$

23. $\iiint_E xy\, dV = \int_0^3 \int_0^x \int_0^{x+y} xy\, dz\, dy\, dx = \int_0^3\int_0^x xy\left[z\right]_{z=0}^{z=x+y}\, dy\, dx = \int_0^3\int_0^x xy(x+y)\, dy\, dx$

$\quad = \int_0^3\int_0^x (x^2y+xy^2)\, dy\, dx = \int_0^3 \left[\tfrac{1}{2}x^2y^2+\tfrac{1}{3}xy^3\right]_{y=0}^{y=x}\, dx = \int_0^3 \left(\tfrac{1}{2}x^4+\tfrac{1}{3}x^4\right)\, dx$

$\quad = \tfrac{5}{6}\int_0^3 x^4\, dx = \left[\tfrac{1}{6}x^5\right]_0^3 = \tfrac{81}{2} = 40.5$

25. $\iiint_E y^2z^2\, dV = \int_{-1}^1 \int_{-\sqrt{1-y^2}}^{\sqrt{1-y^2}} \int_0^{1-y^2-z^2} y^2z^2\, dx\, dz\, dy = \int_{-1}^1 \int_{-\sqrt{1-y^2}}^{\sqrt{1-y^2}} y^2z^2(1-y^2-z^2)\, dz\, dy$

$\quad = \int_0^{2\pi}\int_0^1 (r^2\cos^2\theta)(r^2\sin^2\theta)(1-r^2)\, r\, dr\, d\theta = \int_0^{2\pi}\int_0^1 \tfrac{1}{4}\sin^2 2\theta(r^5-r^7)\, dr\, d\theta$

$\quad = \int_0^{2\pi}\tfrac{1}{8}(1-\cos 4\theta)\left[\tfrac{1}{6}r^6-\tfrac{1}{8}r^8\right]_{r=0}^{r=1}\, d\theta = \tfrac{1}{192}\left[\theta-\tfrac{1}{4}\sin 4\theta\right]_0^{2\pi} = \tfrac{2\pi}{192} = \tfrac{\pi}{96}$

27. $\iiint_E yz\, dV = \int_{-2}^2 \int_0^{\sqrt{4-x^2}} \int_0^y yz\, dz\, dy\, dx = \int_{-2}^2\int_0^{\sqrt{4-x^2}} \tfrac{1}{2}y^3\, dy\, dx = \int_0^\pi \int_0^2 \tfrac{1}{2}r^3(\sin^3\theta)\, r\, dr\, d\theta$

$\quad = \tfrac{16}{5}\int_0^\pi \sin^3\theta\, d\theta = \tfrac{16}{5}\left[-\cos\theta+\tfrac{1}{3}\cos^3\theta\right]_0^\pi = \tfrac{64}{15}$

29. $V = \int_0^2\int_1^4 (x^2+4y^2)\, dy\, dx = \int_0^2 \left[x^2y+\tfrac{4}{3}y^3\right]_{y=1}^{y=4}\, dx = \int_0^2 (3x^2+84)\, dx = 176$

31.

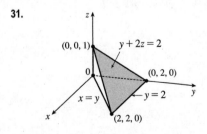

$$V = \int_0^2\int_0^y \int_0^{(2-y)/2} dz\, dx\, dy$$

$$= \int_0^2\int_0^y \left(1-\tfrac{1}{2}y\right)\, dx\, dy$$

$$= \int_0^2 \left(y-\tfrac{1}{2}y^2\right)\, dy = \tfrac{2}{3}$$

33. Using the wedge above the plane $z = 0$ and below the plane $z = mx$ and noting that we have the same volume for $m < 0$ as

for $m > 0$ (so use $m > 0$), we have

$$V = 2 \int_0^{a/3} \int_0^{\sqrt{a^2 - 9y^2}} mx \, dx \, dy = 2 \int_0^{a/3} \tfrac{1}{2} m(a^2 - 9y^2) \, dy = m \left[a^2 y - 3y^3 \right]_0^{a/3} = m \left(\tfrac{1}{3} a^3 - \tfrac{1}{9} a^3 \right) = \tfrac{2}{9} ma^3.$$

35. (a) $m = \int_0^1 \int_0^{1-y^2} y \, dx \, dy = \int_0^1 (y - y^3) \, dy = \tfrac{1}{2} - \tfrac{1}{4} = \tfrac{1}{4}$

(b) $M_y = \int_0^1 \int_0^{1-y^2} xy \, dx \, dy = \int_0^1 \tfrac{1}{2} y (1 - y^2)^2 \, dy = -\tfrac{1}{12} (1 - y^2)^3 \Big]_0^1 = \tfrac{1}{12}$,

$\quad M_x = \int_0^1 \int_0^{1-y^2} y^2 \, dx \, dy = \int_0^1 (y^2 - y^4) \, dy = \tfrac{2}{15}$. Hence $(\overline{x}, \overline{y}) = \left(\tfrac{1}{3}, \tfrac{8}{15} \right)$.

(c) $I_x = \int_0^1 \int_0^{1-y^2} y^3 \, dx \, dy = \int_0^1 (y^3 - y^5) \, dy = \tfrac{1}{12}$,

$\quad I_y = \int_0^1 \int_0^{1-y^2} yx^2 \, dx \, dy = \int_0^1 \tfrac{1}{3} y (1 - y^2)^3 \, dy = -\tfrac{1}{24} (1 - y^2)^4 \Big]_0^1 = \tfrac{1}{24}$,

$\quad I_0 = I_x + I_y = \tfrac{1}{8}, \overline{\overline{y}}^2 = \tfrac{1/12}{1/4} = \tfrac{1}{3} \Rightarrow \overline{\overline{y}} = \tfrac{1}{\sqrt{3}}, \text{ and } \overline{\overline{x}}^2 = \tfrac{1/24}{1/4} = \tfrac{1}{6} \Rightarrow \overline{\overline{x}} = \tfrac{1}{\sqrt{6}}.$

37. (a) The equation of the cone with the suggested orientation is $(h - z) = \tfrac{h}{a} \sqrt{x^2 + y^2}, 0 \le z \le h$. Then $V = \tfrac{1}{3} \pi a^2 h$ is the

volume of one frustum of a cone; by symmetry $M_{yz} = M_{xz} = 0$; and

$$M_{xy} = \iint\limits_{x^2+y^2 \le a^2} \int_0^{h - (h/a)\sqrt{x^2+y^2}} z \, dz \, dA = \int_0^{2\pi} \int_0^a \int_0^{(h/a)(a-r)} rz \, dz \, dr \, d\theta = \pi \int_0^a r \frac{h^2}{a^2} (a - r)^2 \, dr$$

$$= \frac{\pi h^2}{a^2} \int_0^a (a^2 r - 2ar^2 + r^3) \, dr = \frac{\pi h^2}{a^2} \left(\frac{a^4}{2} - \frac{2a^4}{3} + \frac{a^4}{4} \right) = \frac{\pi h^2 a^2}{12}$$

Hence the centroid is $(\overline{x}, \overline{y}, \overline{z}) = \left(0, 0, \tfrac{1}{4} h \right)$.

(b) $I_z = \int_0^{2\pi} \int_0^a \int_0^{(h/a)(a-r)} r^3 \, dz \, dr \, d\theta = 2\pi \int_0^a \frac{h}{a} (ar^3 - r^4) \, dr = \frac{2\pi h}{a} \left(\frac{a^5}{4} - \frac{a^5}{5} \right) = \frac{\pi a^4 h}{10}$

39. $x = r \cos \theta = 2\sqrt{3} \cos \tfrac{\pi}{3} = 2\sqrt{3} \cdot \tfrac{1}{2} = \sqrt{3}, y = r \sin \theta = 2\sqrt{3} \sin \tfrac{\pi}{3} = 2\sqrt{3} \cdot \tfrac{\sqrt{3}}{2} = 3, z = 2$, so in rectangular

coordinates the point is $(\sqrt{3}, 3, 2)$. $\rho = \sqrt{r^2 + z^2} = \sqrt{12 + 4} = 4, \theta = \tfrac{\pi}{3}$, and $\cos \phi = \tfrac{z}{\rho} = \tfrac{1}{2}$, so $\phi = \tfrac{\pi}{3}$ and spherical

coordinates are $\left(4, \tfrac{\pi}{3}, \tfrac{\pi}{3} \right)$.

41. $x = \rho \sin \phi \cos \theta = 8 \sin \tfrac{\pi}{6} \cos \tfrac{\pi}{4} = 8 \cdot \tfrac{1}{2} \cdot \tfrac{\sqrt{2}}{2} = 2\sqrt{2}, y = \rho \sin \phi \sin \theta = 8 \sin \tfrac{\pi}{6} \sin \tfrac{\pi}{4} = 2\sqrt{2}$, and

$z = \rho \cos \phi = 8 \cos \tfrac{\pi}{6} = 8 \cdot \tfrac{\sqrt{3}}{2} = 4\sqrt{3}$. Thus rectangular coordinates for the point are $(2\sqrt{2}, 2\sqrt{2}, 4\sqrt{3})$.

$r^2 = x^2 + y^2 = 8 + 8 = 16 \Rightarrow r = 4, \theta = \tfrac{\pi}{4}$, and $z = 4\sqrt{3}$, so cylindrical coordinates are $\left(4, \tfrac{\pi}{4}, 4\sqrt{3} \right)$.

43. $x^2 + y^2 + z^2 = 4$. In cylindrical coordinates, this becomes $r^2 + z^2 = 4$. In spherical coordinates, it becomes $\rho^2 = 4$ or

$\rho = 2$.

45.

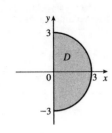

$$\int_0^3 \int_{-\sqrt{9-x^2}}^{\sqrt{9-x^2}} (x^3 + xy^2)\, dy\, dx = \int_0^3 \int_{-\sqrt{9-x^2}}^{\sqrt{9-x^2}} x(x^2 + y^2)\, dy\, dx$$

$$= \int_{-\pi/2}^{\pi/2} \int_0^3 (r\cos\theta)(r^2)\, r\, dr\, d\theta$$

$$= \int_{-\pi/2}^{\pi/2} \cos\theta\, d\theta \int_0^3 r^4\, dr$$

$$= \left[\sin\theta\right]_{-\pi/2}^{\pi/2} \left[\tfrac{1}{5}r^5\right]_0^3 = 2 \cdot \tfrac{1}{5}(243) = \tfrac{486}{5} = 97.2$$

47.

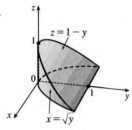
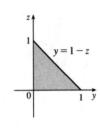

$$\int_{-1}^1 \int_{x^2}^1 \int_0^{1-y} f(x,y,z)\, dz\, dy\, dx = \int_0^1 \int_0^{1-z} \int_{-\sqrt{y}}^{\sqrt{y}} f(x,y,z)\, dx\, dy\, dz$$

49. Since $u = x - y$ and $v = x + y$, $x = \tfrac{1}{2}(u + v)$ and $y = \tfrac{1}{2}(v - u)$.

Thus $\dfrac{\partial(x,y)}{\partial(u,v)} = \begin{vmatrix} 1/2 & 1/2 \\ -1/2 & 1/2 \end{vmatrix} = \dfrac{1}{2}$ and $\displaystyle\iint_R \dfrac{x-y}{x+y}\, dA = \int_2^4 \int_{-2}^0 \dfrac{u}{v}\left(\dfrac{1}{2}\right) du\, dv = -\int_2^4 \dfrac{dv}{v} = -\ln 2.$

51. Let $u = y - x$ and $v = y + x$ so $x = y - u = (v - x) - u \;\Rightarrow\; x = \tfrac{1}{2}(v - u)$ and $y = v - \tfrac{1}{2}(v - u) = \tfrac{1}{2}(v + u)$.

$\left|\dfrac{\partial(x,y)}{\partial(u,v)}\right| = \left|\dfrac{\partial x}{\partial u}\dfrac{\partial y}{\partial v} - \dfrac{\partial x}{\partial v}\dfrac{\partial y}{\partial u}\right| = \left|-\tfrac{1}{2}\left(\tfrac{1}{2}\right) - \tfrac{1}{2}\left(\tfrac{1}{2}\right)\right| = \left|-\tfrac{1}{2}\right| = \tfrac{1}{2}$. R is the image under this transformation of the square

with vertices $(u, v) = (0, 0)$, $(-2, 0)$, $(0, 2)$, and $(-2, 2)$. So

$$\iint_R xy\, dA = \int_0^2 \int_{-2}^0 \dfrac{v^2 - u^2}{4}\left(\dfrac{1}{2}\right) du\, dv = \tfrac{1}{8}\int_0^2 \left[v^2 u - \tfrac{1}{3}u^3\right]_{u=-2}^{u=0} dv = \tfrac{1}{8}\int_0^2 \left(2v^2 - \tfrac{8}{3}\right) dv = \tfrac{1}{8}\left[\tfrac{2}{3}v^3 - \tfrac{8}{3}v\right]_0^2 = 0$$

This result could have been anticipated by symmetry, since the integrand is an odd function of y and R is symmetric about

the x-axis.

13 □ VECTOR CALCULUS

13.1 Vector Fields

1. $\mathbf{F}(x, y) = \frac{1}{2}(\mathbf{i} + \mathbf{j})$

All vectors in this field are identical, with length $\frac{1}{\sqrt{2}}$ and

direction parallel to the line $y = x$.

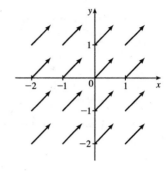

3. $\mathbf{F}(x, y) = y\,\mathbf{i} + \frac{1}{2}\mathbf{j}$

The length of the vector $y\,\mathbf{i} + \frac{1}{2}\mathbf{j}$ is $\sqrt{y^2 + \frac{1}{4}}$. Vectors

are tangent to parabolas opening about the x-axis.

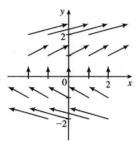

5. $\mathbf{F}(x, y) = \dfrac{y\,\mathbf{i} + x\,\mathbf{j}}{\sqrt{x^2 + y^2}}$

The length of the vector $\dfrac{y\,\mathbf{i} + x\,\mathbf{j}}{\sqrt{x^2 + y^2}}$ is 1.

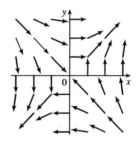

7. $\mathbf{F}(x, y, z) = \mathbf{k}$

All vectors in this field are parallel to the z-axis and have

length 1.

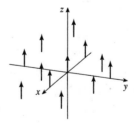

9. $\mathbf{F}(x, y, z) = x\,\mathbf{k}$

At each point (x, y, z), $\mathbf{F}(x, y, z)$ is a vector of length $|x|$.

For $x > 0$, all point in the direction of the positive z-axis,

while for $x < 0$, all are in the direction of the negative

z-axis. In each plane $x = k$, all the vectors are identical.

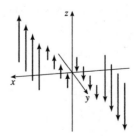

11. $\mathbf{F}(x, y) = \langle y, x \rangle$ corresponds to graph II. In the first quadrant all the vectors have positive x- and y-components, in the second quadrant all vectors have positive x-components and negative y-components, in the third quadrant all vectors have negative x- and y-components, and in the fourth quadrant all vectors have negative x-components and positive y-components. In addition, the vectors get shorter as we approach the origin.

13. $\mathbf{F}(x, y) = \langle x - 2, x + 1 \rangle$ corresponds to graph I since the vectors are independent of y (vectors along vertical lines are identical) and, as we move to the right, both the x- and the y-components get larger.

15. $\mathbf{F}(x, y, z) = \mathbf{i} + 2\mathbf{j} + 3\mathbf{k}$ corresponds to graph IV, since all vectors have identical length and direction.

17. $\mathbf{F}(x, y, z) = x\mathbf{i} + y\mathbf{j} + 3\mathbf{k}$ corresponds to graph III; the projection of each vector onto the xy-plane is $x\mathbf{i} + y\mathbf{j}$, which points away from the origin, and the vectors point generally upward because their z-components are all 3.

19.

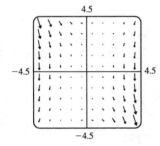

The vector field seems to have very short vectors near the line $y = 2x$.

For $\mathbf{F}(x, y) = \langle 0, 0 \rangle$ we must have $y^2 - 2xy = 0$ and $3xy - 6x^2 = 0$.

The first equation holds if $y = 0$ or $y = 2x$, and the second holds if $x = 0$ or $y = 2x$. So both equations hold [and thus $\mathbf{F}(x, y) = \mathbf{0}$] along the line $y = 2x$.

21. $\nabla f(x, y) = f_x(x, y)\mathbf{i} + f_y(x, y)\mathbf{j} = \dfrac{1}{x + 2y}\mathbf{i} + \dfrac{2}{x + 2y}\mathbf{j}$

23. $\nabla f(x, y, z) = f_x(x, y, z)\mathbf{i} + f_y(x, y, z)\mathbf{j} + f_z(x, y, z)\mathbf{k} = \dfrac{x}{\sqrt{x^2 + y^2 + z^2}}\mathbf{i} + \dfrac{y}{\sqrt{x^2 + y^2 + z^2}}\mathbf{j} + \dfrac{z}{\sqrt{x^2 + y^2 + z^2}}\mathbf{k}$

25. $f(x, y) = xy - 2x \quad \Rightarrow \quad \nabla f(x, y) = (y - 2)\mathbf{i} + x\mathbf{j}$.

The length of $\nabla f(x, y)$ is $\sqrt{(y - 2)^2 + x^2}$ and $\nabla f(x, y)$

terminates on the line $y = x + 2$ at the point

$(x + y - 2, x + y)$.

27. We graph ∇f along with a contour map of f.

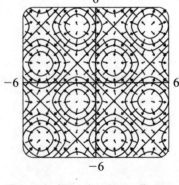

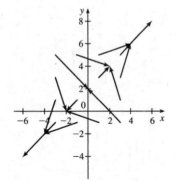

The graph shows that the gradient vectors are perpendicular to the level curves. Also, the gradient vectors point in the direction in which f is increasing and are longer where the level curves are closer together.

29. At $t = 3$ the particle is at $(2, 1)$ so its velocity is $\mathbf{V}(2, 1) = \langle 4, 3 \rangle$. After 0.01 units of time, the particle's change in location should be approximately $0.01\,\mathbf{V}(2, 1) = 0.01\,\langle 4, 3 \rangle = \langle 0.04, 0.03 \rangle$, so the particle should be approximately at the point $(2.04, 1.03)$.

31. (a) We sketch the vector field $\mathbf{F}(x, y) = x\,\mathbf{i} - y\,\mathbf{j}$ along with several approximate flow lines. The flow lines appear to be hyperbolas with shape similar to the graph of $y = \pm 1/x$, so we might guess that the flow lines have equations $y = C/x$.

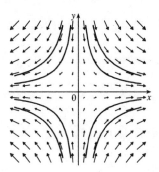

(b) If $x = x(t)$ and $y = y(t)$ are parametric equations of a flow line, then the velocity vector of the flow line at the point (x, y) is $x'(t)\,\mathbf{i} + y'(t)\,\mathbf{j}$. Since the velocity vectors coincide with the vectors in the vector field, we have

$x'(t)\,\mathbf{i} + y'(t)\,\mathbf{j} = x\,\mathbf{i} - y\,\mathbf{j} \quad\Rightarrow\quad dx/dt = x,\ dy/dt = -y$. To solve these differential equations, we know

$dx/dt = x \quad\Rightarrow\quad dx/x = dt \quad\Rightarrow\quad \ln|x| = t + C \quad\Rightarrow\quad x = \pm e^{t+C} = Ae^t$ for some constant A, and

$dy/dt = -y \quad\Rightarrow\quad dy/y = -dt \quad\Rightarrow\quad \ln|y| = -t + K \quad\Rightarrow\quad y = \pm e^{-t+K} = Be^{-t}$ for some constant B. Therefore

$xy = Ae^t Be^{-t} = AB = $ constant. If the flow line passes through $(1, 1)$ then $(1)(1) = $ constant $= 1 \quad\Rightarrow\quad xy = 1 \quad\Rightarrow\quad y = 1/x,\ x > 0$.

13.2 Line Integrals

1. $x = t^2$ and $y = t$, $0 \le t \le 2$, so by Formula 3

$$\int_C y\,ds = \int_0^2 t\sqrt{\left(\tfrac{dx}{dt}\right)^2 + \left(\tfrac{dy}{dt}\right)^2}\,dt = \int_0^2 t\sqrt{(2t)^2 + (1)^2}\,dt$$
$$= \int_0^2 t\sqrt{4t^2 + 1}\,dt = \tfrac{1}{12}\left(4t^2 + 1\right)^{3/2}\Big]_0^2 = \tfrac{1}{12}\left(17\sqrt{17} - 1\right)$$

3. Parametric equations for C are $x = 4\cos t$, $y = 4\sin t$, $-\tfrac{\pi}{2} \le t \le \tfrac{\pi}{2}$. Then

$$\int_C xy^4\,ds = \int_{-\pi/2}^{\pi/2} (4\cos t)(4\sin t)^4\sqrt{(-4\sin t)^2 + (4\cos t)^2}\,dt = \int_{-\pi/2}^{\pi/2} 4^5\cos t \sin^4 t \sqrt{16\left(\sin^2 t + \cos^2 t\right)}\,dt$$
$$= 4^5\int_{-\pi/2}^{\pi/2} \left(\sin^4 t \cos t\right)(4)\,dt = (4)^6\left[\tfrac{1}{5}\sin^5 t\right]_{-\pi/2}^{\pi/2} = \tfrac{2 \cdot 4^6}{5} = 1638.4$$

5.

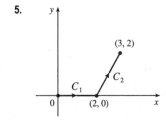

$C = C_1 + C_2$

On C_1: $x = x$, $y = 0 \quad\Rightarrow\quad dy = 0\,dx$, $0 \le x \le 2$.

On C_2: $x = x$, $y = 2x - 4 \quad\Rightarrow\quad dy = 2\,dx$, $2 \le x \le 3$.

Then

$$\int_C xy\,dx + (x - y)\,dy = \int_{C_1} xy\,dx + (x - y)\,dy + \int_{C_2} xy\,dx + (x - y)\,dy$$
$$= \int_0^2 (0 + 0)\,dx + \int_2^3 \left[(2x^2 - 4x) + (-x + 4)(2)\right]dx$$
$$= \int_2^3 (2x^2 - 6x + 8)\,dx = \tfrac{17}{3}$$

7. $x = 4\sin t$, $y = 4\cos t$, $z = 3t$, $0 \le t \le \frac{\pi}{2}$. Then by Formula 9,

$$\int_C xy^3\, ds = \int_0^{\pi/2} (4\sin t)(4\cos t)^3 \sqrt{\left(\frac{dx}{dt}\right)^2 + \left(\frac{dy}{dt}\right)^2 + \left(\frac{dz}{dt}\right)^2}\, dt$$

$$= \int_0^{\pi/2} 4^4 \cos^3 t \sin t \sqrt{(4\cos t)^2 + (-4\sin t)^2 + (3)^2}\, dt = \int_0^{\pi/2} 256\cos^3 t \sin t \sqrt{16(\cos^2 t + \sin^2 t) + 9}\, dt$$

$$= 1280 \int_0^{\pi/2} \cos^3 t \sin t\, dt = -320\cos^4 t \Big]_0^{\pi/2} = 320$$

9. Parametric equations for C are $x = t$, $y = 2t$, $z = 3t$, $0 \le t \le 1$. Then

$$\int_C xe^{yz}\, ds = \int_0^1 te^{(2t)(3t)} \sqrt{1^2 + 2^2 + 3^2}\, dt = \sqrt{14}\int_0^1 te^{6t^2}\, dt = \sqrt{14}\left[\frac{1}{12}e^{6t^2}\right]_0^1 = \frac{\sqrt{14}}{12}(e^6 - 1).$$

11. $\int_C x^2 y \sqrt{z}\, dz = \int_0^1 (t^3)^2 (t)\sqrt{t^2} \cdot 2t\, dt = \int_0^1 2t^9\, dt = \frac{1}{5}t^{10}\Big]_0^1 = \frac{1}{5}$

13.

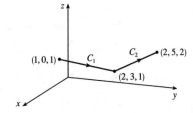

On C_1: $x = 1 + t \ \Rightarrow\ dx = dt$, $y = 3t \ \Rightarrow\ dy = 3\, dt$, $z = 1$
$\Rightarrow\ dz = 0\, dt$, $0 \le t \le 1$.

On C_2: $x = 2 \ \Rightarrow\ dx = 0\, dt$, $y = 3 + 2t \ \Rightarrow$
$dy = 2\, dt$, $z = 1 + t \ \Rightarrow\ dz = dt$, $0 \le t \le 1$.

Then

$$\int_C (x + yz)\, dx + 2x\, dy + xyz\, dz$$

$$= \int_{C_1} (x + yz)\, dx + 2x\, dy + xyz\, dz + \int_{C_2}(x + yz)\, dx + 2x\, dy + xyz\, dz$$

$$= \int_0^1 (1 + t + (3t)(1))\, dt + 2(1 + t) \cdot 3\, dt + (1 + t)(3t)(1) \cdot 0\, dt$$

$$+ \int_0^1 (2 + (3 + 2t)(1 + t)) \cdot 0\, dt + 2(2) \cdot 2\, dt + (2)(3 + 2t)(1 + t)\, dt$$

$$= \int_0^1 (10t + 7)\, dt + \int_0^1 (4t^2 + 10t + 14)\, dt = \left[5t^2 + 7t\right]_0^1 + \left[\frac{4}{3}t^3 + 5t^2 + 14t\right]_0^1 = 12 + \frac{61}{3} = \frac{97}{3}$$

15. (a) Along the line $x = -3$, the vectors of $\mathbf{F}$ have positive y-components, so since the path goes upward, the integrand $\mathbf{F} \cdot \mathbf{T}$ is always positive. Therefore $\int_{C_1} \mathbf{F} \cdot d\mathbf{r} = \int_{C_1} \mathbf{F} \cdot \mathbf{T}\, ds$ is positive.

(b) All of the (nonzero) field vectors along the circle with radius 3 are pointed in the clockwise direction, that is, opposite the direction to the path. So $\mathbf{F} \cdot \mathbf{T}$ is negative, and therefore $\int_{C_2} \mathbf{F} \cdot d\mathbf{r} = \int_{C_2} \mathbf{F} \cdot \mathbf{T}\, ds$ is negative.

17. $\mathbf{r}(t) = t^2\,\mathbf{i} - t^3\,\mathbf{j}$, so $\mathbf{F}(\mathbf{r}(t)) = (t^2)^2(-t^3)^3\,\mathbf{i} - (-t^3)\sqrt{t^2}\,\mathbf{j} = -t^{13}\,\mathbf{i} + t^4\,\mathbf{j}$ and $\mathbf{r}'(t) = 2t\,\mathbf{i} - 3t^2\,\mathbf{j}$.

Thus $\int_C \mathbf{F} \cdot d\mathbf{r} = \int_0^1 \mathbf{F}(\mathbf{r}(t)) \cdot \mathbf{r}'(t)\, dt = \int_0^1 (-2t^{14} - 3t^6)\, dt = \left[-\frac{2}{15}t^{15} - \frac{3}{7}t^7\right]_0^1 = -\frac{59}{105}$.

19. $\int_C \mathbf{F} \cdot d\mathbf{r} = \int_0^1 \langle \sin t^3, \cos(-t^2), t^4\rangle \cdot \langle 3t^2, -2t, 1\rangle\, dt$

$$= \int_0^1 (3t^2 \sin t^3 - 2t\cos t^2 + t^4)\, dt = \left[-\cos t^3 - \sin t^2 + \frac{1}{5}t^5\right]_0^1 = \frac{6}{5} - \cos 1 - \sin 1$$

21. We graph $\mathbf{F}(x, y) = (x - y)\mathbf{i} + xy\mathbf{j}$ and the curve C. We see that most of the vectors starting on C point in roughly the same direction as C, so for these portions of C the tangential component $\mathbf{F} \cdot \mathbf{T}$ is positive. Although some vectors in the third quadrant which start on C point in roughly the opposite direction, and hence give negative tangential components, it seems reasonable that the effect of these portions of C is outweighed by the positive tangential components. Thus, we would expect $\int_C \mathbf{F} \cdot d\mathbf{r} = \int_C \mathbf{F} \cdot \mathbf{T} \, ds$ to be positive.

To verify, we evaluate $\int_C \mathbf{F} \cdot d\mathbf{r}$. The curve C can be represented by $\mathbf{r}(t) = 2\cos t \, \mathbf{i} + 2\sin t \, \mathbf{j}$, $0 \le t \le \frac{3\pi}{2}$, so $\mathbf{F}(\mathbf{r}(t)) = (2\cos t - 2\sin t)\mathbf{i} + 4\cos t \sin t \, \mathbf{j}$ and $\mathbf{r}'(t) = -2\sin t \, \mathbf{i} + 2\cos t \, \mathbf{j}$. Then

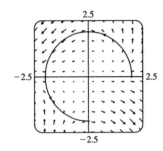

$$\int_C \mathbf{F} \cdot d\mathbf{r} = \int_0^{3\pi/2} \mathbf{F}(\mathbf{r}(t)) \cdot \mathbf{r}'(t) \, dt$$
$$= \int_0^{3\pi/2} [-2\sin t(2\cos t - 2\sin t) + 2\cos t(4\cos t \sin t)] \, dt$$
$$= 4 \int_0^{3\pi/2} (\sin^2 t - \sin t \cos t + 2\sin t \cos^2 t) \, dt$$
$$= 3\pi + \tfrac{2}{3} \quad \text{[using a CAS]}$$

23. (a) $\int_C \mathbf{F} \cdot d\mathbf{r} = \int_0^1 \left\langle e^{t^2 - 1}, t^5 \right\rangle \cdot \left\langle 2t, 3t^2 \right\rangle dt = \int_0^1 \left(2t e^{t^2 - 1} + 3t^7\right) dt = \left[e^{t^2 - 1} + \tfrac{3}{8}t^8\right]_0^1 = \tfrac{11}{8} - 1/e$

(b) $\mathbf{r}(0) = \mathbf{0}$, $\mathbf{F}(\mathbf{r}(0)) = \left\langle e^{-1}, 0 \right\rangle$;

$\mathbf{r}\!\left(\tfrac{1}{\sqrt{2}}\right) = \left\langle \tfrac{1}{2}, \tfrac{1}{2\sqrt{2}} \right\rangle$, $\mathbf{F}\!\left(\mathbf{r}\!\left(\tfrac{1}{\sqrt{2}}\right)\right) = \left\langle e^{-1/2}, \tfrac{1}{4\sqrt{2}} \right\rangle$;

$\mathbf{r}(1) = \langle 1, 1 \rangle$, $\mathbf{F}(\mathbf{r}(1)) = \langle 1, 1 \rangle$.

In order to generate the graph with Maple, we use the PLOT command (not to be confused with the plot command) to define each of the vectors. For example,

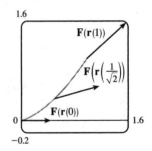

```
v1:=PLOT(CURVES([[0,0],[evalf(1/exp(1)),0]]));
```

generates the vector from the vector field at the point $(0, 0)$ (but without an arrowhead) and gives it the name v1. To show everything on the same screen, we use the display command. In Mathematica, we use ListPlot (with the PlotJoined - > True option) to generate the vectors, and then Show to show everything on the same screen.

25. The part of the astroid that lies in the quadrant is parametrized by $x = \cos^3 t$, $y = \sin^3 t$, $0 \le t \le \frac{\pi}{2}$.

Now $\dfrac{dx}{dt} = 3\cos^2 t\,(-\sin t)$ and $\dfrac{dy}{dt} = 3\sin^2 t \cos t$, so

$$\sqrt{\left(\dfrac{dx}{dt}\right)^2 + \left(\dfrac{dy}{dt}\right)^2} = \sqrt{9\cos^4 t \sin^2 t + 9\sin^4 t \cos^2 t} = 3\cos t \sin t \sqrt{\cos^2 t + \sin^2 t} = 3\cos t \sin t.$$

Therefore $\int_C x^3 y^5 \, ds = \int_0^{\pi/2} \cos^9 t \sin^{15} t \,(3\cos t \sin t)\, dt = \dfrac{945}{16{,}777{,}216}\pi.$

27. We use the parametrization $x = 2\cos t$, $y = 2\sin t$, $-\frac{\pi}{2} \le t \le \frac{\pi}{2}$. Then

$$ds = \sqrt{\left(\frac{dx}{dt}\right)^2 + \left(\frac{dy}{dt}\right)^2}\, dt = \sqrt{(-2\sin t)^2 + (2\cos t)^2}\, dt = 2\, dt, \text{ so } m = \int_C k\, ds = 2k\int_{-\pi/2}^{\pi/2} dt = 2k(\pi),$$

$\overline{x} = \frac{1}{2\pi k}\int_C xk\, ds = \frac{1}{2\pi}\int_{-\pi/2}^{\pi/2}(2\cos t)2\, dt = \frac{1}{2\pi}\left[4\sin t\right]_{-\pi/2}^{\pi/2} = \frac{4}{\pi}, \overline{y} = \frac{1}{2\pi k}\int_C yk\, ds = \frac{1}{2\pi}\int_{-\pi/2}^{\pi/2}(2\sin t)2\, dt = 0.$ Hence

$(\overline{x}, \overline{y}) = \left(\frac{4}{\pi}, 0\right).$

29. (a) $\overline{x} = \frac{1}{m}\int_C x\rho(x,y,z)\, ds$, $\overline{y} = \frac{1}{m}\int_C y\rho(x,y,z)\, ds$, $\overline{z} = \frac{1}{m}\int_C z\rho(x,y,z)\, ds$ where $m = \int_C \rho(x,y,z)\, ds$.

(b) $m = \int_C k\, ds = k\int_0^{2\pi}\sqrt{4\sin^2 t + 4\cos^2 t + 9}\, dt = k\sqrt{13}\int_0^{2\pi} dt = 2\pi k\sqrt{13}$,

$$\overline{x} = \frac{1}{2\pi k\sqrt{13}}\int_0^{2\pi} 2k\sqrt{13}\,\sin t\, dt = 0, \overline{y} = \frac{1}{2\pi k\sqrt{13}}\int_0^{2\pi} 2k\sqrt{13}\,\cos t\, dt = 0,$$

$$\overline{z} = \frac{1}{2\pi k\sqrt{13}}\int_0^{2\pi}\left(k\sqrt{13}\right)(3t)\, dt = \frac{3}{2\pi}\left(2\pi^2\right) = 3\pi. \text{ Hence } (\overline{x}, \overline{y}, \overline{z}) = (0, 0, 3\pi).$$

31. From Example 3, $\rho(x,y) = k(1-y)$, $x = \cos t$, $y = \sin t$, and $ds = dt$, $0 \le t \le \pi$ $\Rightarrow$

$$I_x = \int_C y^2\rho(x,y)\, ds = \int_0^\pi \sin^2 t\, [k(1-\sin t)]\, dt = k\int_0^\pi(\sin^2 t - \sin^3 t)\, dt$$

$$= \tfrac{1}{2}k\int_0^\pi(1-\cos 2t)\, dt - k\int_0^\pi\left(1-\cos^2 t\right)\sin t\, dt \qquad \left[\begin{array}{l}\text{Let } u = \cos t,\, du = -\sin t\, dt \\ \text{in the second integral}\end{array}\right]$$

$$= k\left[\tfrac{\pi}{2} + \int_1^{-1}(1-u^2)\, du\right] = k\left(\tfrac{\pi}{2} - \tfrac{4}{3}\right)$$

$$I_y = \int_C x^2\rho(x,y)\, ds = k\int_0^\pi \cos^2 t\,(1-\sin t)\, dt = \tfrac{k}{2}\int_0^\pi(1+\cos 2t)\, dt - k\int_0^\pi\cos^2 t\sin t\, dt$$

$$= k\left(\tfrac{\pi}{2} - \tfrac{2}{3}\right), \text{ using the same substitution as above.}$$

33. $W = \int_C \mathbf{F}\cdot d\mathbf{r} = \int_0^{2\pi}\langle t - \sin t, 3 - \cos t\rangle\cdot\langle 1 - \cos t, \sin t\rangle\, dt$

$= \int_0^{2\pi}(t - t\cos t - \sin t + \sin t\cos t + 3\sin t - \sin t\cos t)\, dt$

$= \int_0^{2\pi}(t - t\cos t + 2\sin t)\, dt = \left[\tfrac{1}{2}t^2 - (t\sin t + \cos t) - 2\cos t\right]_0^{2\pi} \qquad \left[\begin{array}{l}\text{by integrating by parts}\\ \text{in the second term}\end{array}\right]$

$= 2\pi^2$

35. $\mathbf{r}(t) = \langle 1 + 2t, 4t, 2t\rangle$, $0 \le t \le 1$,

$$W = \int_C \mathbf{F}\cdot d\mathbf{r} = \int_0^1\langle 6t, 1 + 4t, 1 + 6t\rangle\cdot\langle 2, 4, 2\rangle\, dt = \int_0^1(12t + 4(1 + 4t) + 2(1 + 6t))\, dt$$

$= \int_0^1(40t + 6)\, dt = \left[20t^2 + 6t\right]_0^1 = 26$

37. Let $\mathbf{F} = 185\,\mathbf{k}$. To parametrize the staircase, let

$x = 20\cos t$, $y = 20\sin t$, $z = \frac{90}{6\pi}t = \frac{15}{\pi}t$, $0 \le t \le 6\pi$ $\Rightarrow$

$$W = \int_C \mathbf{F}\cdot d\mathbf{r} = \int_0^{6\pi}\langle 0, 0, 185\rangle\cdot\langle -20\sin t, 20\cos t, \tfrac{15}{\pi}\rangle\, dt = (185)\tfrac{15}{\pi}\int_0^{6\pi} dt = (185)(90) \approx 1.67\times 10^4 \text{ ft-lb}$$

39. (a) $\mathbf{r}(t) = \langle \cos t, \sin t \rangle$, $0 \le t \le 2\pi$, and let $\mathbf{F} = \langle a, b \rangle$. Then

$$W = \int_C \mathbf{F} \cdot d\mathbf{r} = \int_0^{2\pi} \langle a, b \rangle \cdot \langle -\sin t, \cos t \rangle \, dt = \int_0^{2\pi} (-a \sin t + b \cos t) \, dt = \left[a \cos t + b \sin t \right]_0^{2\pi}$$
$$= a + 0 - a + 0 = 0$$

(b) Yes. $\mathbf{F}(x, y) = k\,\mathbf{x} = \langle kx, ky \rangle$ and

$$W = \int_C \mathbf{F} \cdot d\mathbf{r} = \int_0^{2\pi} \langle k \cos t, k \sin t \rangle \cdot \langle -\sin t, \cos t \rangle \, dt = \int_0^{2\pi} (-k \sin t \cos t + k \sin t \cos t) \, dt = \int_0^{2\pi} 0 \, dt = 0$$

13.3 The Fundamental Theorem for Line Integrals

1. C appears to be a smooth curve, and since ∇f is continuous, we know f is differentiable. Then Theorem 2 says that the value
of $\int_C \nabla f \cdot d\mathbf{r}$ is simply the difference of the values of f at the terminal and initial points of C. From the graph, this is
$50 - 10 = 40$.

3. $\partial(6x + 5y)/\partial y = 5 = \partial(5x + 4y)/\partial x$ and the domain of $\mathbf{F}$ is $\mathbb{R}^2$ which is open and simply-connected, so by Theorem 6
$\mathbf{F}$ is conservative. Thus, there exists a function f such that $\nabla f = \mathbf{F}$, that is, $f_x(x, y) = 6x + 5y$ and $f_y(x, y) = 5x + 4y$.
But $f_x(x, y) = 6x + 5y$ implies $f(x, y) = 3x^2 + 5xy + g(y)$ and differentiating both sides of this equation with respect to
y gives $f_y(x, y) = 5x + g'(y)$. Thus $5x + 4y = 5x + g'(y)$ so $g'(y) = 4y$ and $g(y) = 2y^2 + K$ where K is a constant.
Hence $f(x, y) = 3x^2 + 5xy + 2y^2 + K$ is a potential function for $\mathbf{F}$.

5. $\partial(xe^y)/\partial y = xe^y$, $\partial(ye^x)/\partial x = ye^x$. Since these are not equal, $\mathbf{F}$ is not conservative.

7. $\partial(2x \cos y - y \cos x)/\partial y = -2x \sin y - \cos x = \partial(-x^2 \sin y - \sin x)/\partial x$ and the domain of $\mathbf{F}$ is $\mathbb{R}^2$. Hence $\mathbf{F}$ is
conservative so there exists a function f such that $\nabla f = \mathbf{F}$. Then $f_x(x, y) = 2x \cos y - y \cos x$ implies
$f(x, y) = x^2 \cos y - y \sin x + g(y)$ and $f_y(x, y) = -x^2 \sin y - \sin x + g'(y)$. But $f_y(x, y) = -x^2 \sin y - \sin x$ so
$g'(y) = 0 \ \Rightarrow \ g(y) = K$. Then $f(x, y) = x^2 \cos y - y \sin x + K$ is a potential function for $\mathbf{F}$.

9. $\partial(ye^x + \sin y)/\partial y = e^x + \cos y = \partial(e^x + x \cos y)/\partial x$ and the domain of $\mathbf{F}$ is $\mathbb{R}^2$. Hence $\mathbf{F}$ is conservative so there
exists a function f such that $\nabla f = \mathbf{F}$. Then $f_x(x, y) = ye^x + \sin y$ implies $f(x, y) = ye^x + x \sin y + g(y)$ and
$f_y(x, y) = e^x + x \cos y + g'(y)$. But $f_y(x, y) = e^x + x \cos y$ so $g(y) = K$ and $f(x, y) = ye^x + x \sin y + K$ is a potential
function for $\mathbf{F}$.

11. (a) $f_x(x, y) = x^3 y^4$ implies $f(x, y) = \frac{1}{4}x^4 y^4 + g(y)$ and $f_y(x, y) = x^4 y^3 + g'(y)$. But $f_y(x, y) = x^4 y^3$ so
$g'(y) = 0 \ \Rightarrow \ g(y) = K$, a constant. We can take $K = 0$, so $f(x, y) = \frac{1}{4}x^4 y^4$.

(b) The initial point of C is $\mathbf{r}(0) = (0, 1)$ and the terminal point is $\mathbf{r}(1) = (1, 2)$, so
$$\int_C \mathbf{F} \cdot d\mathbf{r} = f(1, 2) - f(0, 1) = 4 - 0 = 4.$$

13. (a) $f_x(x, y, z) = yz$ implies $f(x, y, z) = xyz + g(y, z)$ and so $f_y(x, y, z) = xz + g_y(y, z)$. But $f_y(x, y, z) = xz$ so

$g_y(y, z) = 0 \;\Rightarrow\; g(y, z) = h(z)$. Thus $f(x, y, z) = xyz + h(z)$ and $f_z(x, y, z) = xy + h'(z)$. But

$f_z(x, y, z) = xy + 2z$, so $h'(z) = 2z \;\Rightarrow\; h(z) = z^2 + K$. Hence $f(x, y, z) = xyz + z^2$ (taking $K = 0$).

(b) $\int_C \mathbf{F} \cdot d\mathbf{r} = f(4, 6, 3) - f(1, 0, -2) = 81 - 4 = 77$.

15. (a) $f_x(x, y, z) = y^2 \cos z$ implies $f(x, y, z) = xy^2 \cos z + g(y, z)$ and so $f_y(x, y, z) = 2xy \cos z + g_y(y, z)$. But

$f_y(x, y, z) = 2xy \cos z$ so $g_y(y, z) = 0 \;\Rightarrow\; g(y, z) = h(z)$. Thus $f(x, y, z) = xy^2 \cos z + h(z)$ and

$f_z(x, y, z) = -xy^2 \sin z + h'(z)$. But $f_z(x, y, z) = -xy^2 \sin z$, so $h'(z) = 0 \;\Rightarrow\; h(z) = K$. Hence

$f(x, y, z) = xy^2 \cos z$ (taking $K = 0$).

(b) $\mathbf{r}(0) = \langle 0, 0, 0 \rangle, \mathbf{r}(\pi) = \langle \pi^2, 0, \pi \rangle$ so $\int_C \mathbf{F} \cdot d\mathbf{r} = f(\pi^2, 0, \pi) - f(0, 0, 0) = 0 - 0 = 0$.

17. Here $\mathbf{F}(x, y) = \tan y \, \mathbf{i} + x \sec^2 y \, \mathbf{j}$. Then $f(x, y) = x \tan y$ is a potential function for $\mathbf{F}$, that is, $\nabla f = \mathbf{F}$ so

$\mathbf{F}$ is conservative and thus its line integral is independent of path. Hence

$\int_C \tan y \, dx + x \sec^2 y \, dy = \int_C \mathbf{F} \cdot d\mathbf{r} = f\left(2, \frac{\pi}{4}\right) - f(1, 0) = 2 \tan \frac{\pi}{4} - \tan 0 = 2$.

19. $\mathbf{F}(x, y) = 2y^{3/2} \mathbf{i} + 3x \sqrt{y} \mathbf{j}$, $W = \int_C \mathbf{F} \cdot d\mathbf{r}$. Since $\partial(2y^{3/2})/\partial y = 3\sqrt{y} = \partial(3x\sqrt{y})/\partial x$, there exists a function f such

that $\nabla f = \mathbf{F}$. In fact, $f_x(x, y) = 2y^{3/2} \;\Rightarrow\; f(x, y) = 2xy^{3/2} + g(y) \;\Rightarrow\; f_y(x, y) = 3xy^{1/2} + g'(y)$. But

$f_y(x, y) = 3x\sqrt{y}$ so $g'(y) = 0$ or $g(y) = K$. We can take $K = 0 \;\Rightarrow\; f(x, y) = 2xy^{3/2}$. Thus

$W = \int_C \mathbf{F} \cdot d\mathbf{r} = f(2, 4) - f(1, 1) = 2(2)(8) - 2(1) = 30$.

21. We know that if the vector field (call it $\mathbf{F}$) is conservative, then around any closed path C, $\int_C \mathbf{F} \cdot d\mathbf{r} = 0$. But take C to be a

circle centered at the origin, oriented counterclockwise. All of the field vectors that start on C are roughly in the direction of

motion along C, so the integral around C will be positive. Therefore the field is not conservative.

23.

From the graph, it appears that $\mathbf{F}$ is conservative, since around all closed paths, the

number and size of the field vectors pointing in directions similar to that of the path

seem to be roughly the same as the number and size of the vectors pointing in the

opposite direction. To check, we calculate

$$\frac{\partial}{\partial y} (\sin y) = \cos y = \frac{\partial}{\partial x} (1 + x \cos y). \text{ Thus } \mathbf{F} \text{ is conservative, by Theorem 6.}$$

25. Since $\mathbf{F}$ is conservative, there exists a function f such that $\mathbf{F} = \nabla f$, that is, $P = f_x, Q = f_y$, and $R = f_z$. Since P, Q and R

have continuous first order partial derivatives, Clairaut's Theorem says that $\partial P/\partial y = f_{xy} = f_{yx} = \partial Q/\partial x$,

$\partial P/\partial z = f_{xz} = f_{zx} = \partial R/\partial x$, and $\partial Q/\partial z = f_{yz} = f_{zy} = \partial R/\partial y$.

27. $D = \{(x, y) \mid x > 0, y > 0\}$ = the first quadrant (excluding the axes).

(a) D is open because around every point in D we can put a disk that lies in D.

(b) D is connected because the straight line segment joining any two points in D lies in D.

(c) D is simply-connected because it's connected and has no holes.

29. $D = \{(x, y) \mid 1 < x^2 + y^2 < 4\}$ = the annular region between the circles with center $(0, 0)$ and radii 1 and 2.

(a) D is open.

(b) D is connected.

(c) D is not simply-connected. For example, $x^2 + y^2 = (1.5)^2$ is simple and closed and lies within D but encloses points that are not in D. (Or we can say, D has a hole, so is not simply-connected.)

31. (a) $P = -\dfrac{y}{x^2 + y^2}$, $\dfrac{\partial P}{\partial y} = \dfrac{y^2 - x^2}{(x^2 + y^2)^2}$ and $Q = \dfrac{x}{x^2 + y^2}$, $\dfrac{\partial Q}{\partial x} = \dfrac{y^2 - x^2}{(x^2 + y^2)^2}$. Thus $\dfrac{\partial P}{\partial y} = \dfrac{\partial Q}{\partial x}$.

(b) C_1: $x = \cos t$, $y = \sin t$, $0 \le t \le \pi$, C_2: $x = \cos t$, $y = \sin t$, $t = 2\pi$ to $t = \pi$. Then

$$\int_{C_1} \mathbf{F} \cdot d\mathbf{r} = \int_0^\pi \frac{(-\sin t)(-\sin t) + (\cos t)(\cos t)}{\cos^2 t + \sin^2 t} \, dt = \int_0^\pi dt = \pi \text{ and } \int_{C_2} \mathbf{F} \cdot d\mathbf{r} = \int_{2\pi}^\pi dt = -\pi$$

Since these aren't equal, the line integral of $\mathbf{F}$ isn't independent of path. (Or notice that $\int_{C_3} \mathbf{F} \cdot d\mathbf{r} = \int_0^{2\pi} dt = 2\pi$ where C_3 is the circle $x^2 + y^2 = 1$, and apply the contrapositive of Theorem 3.) This doesn't contradict Theorem 6, since the domain of $\mathbf{F}$, which is $\mathbb{R}^2$ except the origin, isn't simply-connected.

13.4   Green's Theorem

1. (a)

C_1: $x = t$ $\Rightarrow$ $dx = dt$, $y = 0$ $\Rightarrow$ $dy = 0 \, dt$, $0 \le t \le 2$.

C_2: $x = 2$ $\Rightarrow$ $dx = 0 \, dt$, $y = t$ $\Rightarrow$ $dy = dt$, $0 \le t \le 3$.

C_3: $x = 2 - t$ $\Rightarrow$ $dx = -dt$, $y = 3$ $\Rightarrow$ $dy = 0 \, dt$, $0 \le t \le 2$.

C_4: $x = 0$ $\Rightarrow$ $dx = 0 \, dt$, $y = 3 - t$ $\Rightarrow$ $dy = -dt$, $0 \le t \le 3$.

Thus $\oint_C xy^2 \, dx + x^3 \, dy = \oint_{C_1 + C_2 + C_3 + C_4} xy^2 \, dx + x^3 \, dy$

$$= \int_0^2 0 \, dt + \int_0^3 8 \, dt + \int_0^2 -9(2 - t) \, dt + \int_0^3 0 \, dt$$

$$= 0 + 24 - 18 + 0 = 6$$

(b) $\oint_C xy^2 \, dx + x^3 \, dy = \iint_D \left[\frac{\partial}{\partial x}(x^3) - \frac{\partial}{\partial y}(xy^2) \right] dA = \int_0^2 \int_0^3 (3x^2 - 2xy) \, dy \, dx = \int_0^2 (9x^2 - 9x) \, dx = 24 - 18 = 6$

3. (a)

$C_1: x = t \;\Rightarrow\; dx = dt, y = 0 \;\Rightarrow\; dy = 0\,dt, 0 \le t \le 1.$

$C_2: x = 1 \;\Rightarrow\; dx = 0\,dt, y = t \;\Rightarrow\; dy = dt, 0 \le t \le 2.$

$C_3: x = 1 - t \;\Rightarrow\; dx = -dt, y = 2 - 2t \;\Rightarrow\; dy = -2\,dt, 0 \le t \le 1.$

Thus
$$\oint_C xy\,dx + x^2 y^3\,dy = \oint_{C_1 + C_2 + C_3} xy\,dx + x^2 y^3\,dy$$
$$= \int_0^1 0\,dt + \int_0^2 t^3\,dt + \int_0^1 \left[-(1-t)(2-2t) - 2(1-t)^2(2-2t)^3\right] dt$$
$$= 0 + \left[\tfrac{1}{4}t^4\right]_0^2 + \left[\tfrac{2}{3}(1-t)^3 + \tfrac{8}{3}(1-t)^6\right]_0^1 = 4 - \tfrac{10}{3} = \tfrac{2}{3}$$

(b) $\oint_C xy\,dx + x^2 y^3\,dy = \iint_D \left[\frac{\partial}{\partial x}(x^2 y^3) - \frac{\partial}{\partial y}(xy)\right] dA = \int_0^1 \int_0^{2x} (2xy^3 - x)\,dy\,dx$

$\qquad = \int_0^1 \left[\tfrac{1}{2}xy^4 - xy\right]_{y=0}^{y=2x} dx = \int_0^1 (8x^5 - 2x^2)\,dx = \tfrac{4}{3} - \tfrac{2}{3} = \tfrac{2}{3}$

5. We can parametrize C as $x = \cos\theta$, $y = \sin\theta$, $0 \le \theta \le 2\pi$. Then the line integral is

$\oint_C P\,dx + Q\,dy = \int_0^{2\pi} \cos^4\theta \sin^5\theta\,(-\sin\theta)\,d\theta + \int_0^{2\pi} (-\cos^7\theta \sin^6\theta)\cos\theta\,d\theta = -\frac{29\pi}{1024}$, according to a CAS.

The double integral is

$$\iint_D \left(\frac{\partial Q}{\partial x} - \frac{\partial P}{\partial y}\right) dA = \int_{-1}^1 \int_{-\sqrt{1-x^2}}^{\sqrt{1-x^2}} (-7x^6 y^6 - 5x^4 y^4)\,dy\,dx = -\frac{29\pi}{1024},$$

verifying Green's Theorem in this case.

7. The region D enclosed by C is $[0, 1] \times [0, 1]$, so

$$\int_C e^y\,dx + 2xe^y\,dy = \iint_D \left[\frac{\partial}{\partial x}(2xe^y) - \frac{\partial}{\partial y}(e^y)\right] dA = \int_0^1 \int_0^1 (2e^y - e^y)\,dy\,dx$$
$$= \int_0^1 dx \int_0^1 e^y\,dy = (1)(e^1 - e^0) = e - 1$$

9. $\int_C \left(y + e^{\sqrt{x}}\right) dx + \left(2x + \cos y^2\right) dy = \iint_D \left[\frac{\partial}{\partial x}\left(2x + \cos y^2\right) - \frac{\partial}{\partial y}\left(y + e^{\sqrt{x}}\right)\right] dA$

$\qquad = \int_0^1 \int_{y^2}^{\sqrt{y}} (2 - 1)\,dx\,dy = \int_0^1 (y^{1/2} - y^2)\,dy = \tfrac{1}{3}$

11. $\int_C y^3\,dx - x^3\,dy = \iint_D \left[\frac{\partial}{\partial x}(-x^3) - \frac{\partial}{\partial y}(y^3)\right] dA = \iint_D (-3x^2 - 3y^2)\,dA = \int_0^{2\pi} \int_0^2 (-3r^2)\,r\,dr\,d\theta$

$\qquad = -3 \int_0^{2\pi} d\theta \int_0^2 r^3\,dr = -3(2\pi)(4) = -24\pi$

13. $\mathbf{F}(x, y) = \langle \sqrt{x} + y^3, x^2 + \sqrt{y} \rangle$ and the region D enclosed by C is given by $\{(x, y) \mid 0 \le x \le \pi, 0 \le y \le \sin x\}$.

C is traversed clockwise, so $-C$ gives the positive orientation.

$$\int_C \mathbf{F} \cdot d\mathbf{r} = -\int_{-C} \left(\sqrt{x} + y^3 \right) dx + \left(x^2 + \sqrt{y} \right) dy = -\iint_D \left[\frac{\partial}{\partial x} \left(x^2 + \sqrt{y} \right) - \frac{\partial}{\partial y} \left(\sqrt{x} + y^3 \right) \right] dA$$

$$= -\int_0^\pi \int_0^{\sin x} (2x - 3y^2)\, dy\, dx = -\int_0^\pi \left[2xy - y^3 \right]_{y=0}^{y=\sin x} dx$$

$$= -\int_0^\pi (2x \sin x - \sin^3 x)\, dx = -\int_0^\pi (2x \sin x - (1 - \cos^2 x) \sin x)\, dx$$

$$= -\left[2 \sin x - 2x \cos x + \cos x - \tfrac{1}{3} \cos^3 x \right]_0^\pi \qquad \text{[integrate by parts in the first term]}$$

$$= -\left(2\pi - 2 + \tfrac{2}{3} \right) = \tfrac{4}{3} - 2\pi$$

15. $\mathbf{F}(x, y) = \langle e^x + x^2 y, e^y - xy^2 \rangle$ and the region D enclosed by C is the disk $x^2 + y^2 \le 25$.

C is traversed clockwise, so $-C$ gives the positive orientation.

$$\int_C \mathbf{F} \cdot d\mathbf{r} = -\int_{-C} (e^x + x^2 y)\, dx + (e^y - xy^2)\, dy = -\iint_D \left[\frac{\partial}{\partial x} (e^y - xy^2) - \frac{\partial}{\partial y} (e^x + x^2 y) \right] dA$$

$$= -\iint_D (-y^2 - x^2)\, dA = \iint_D (x^2 + y^2)\, dA = \int_0^{2\pi} \int_0^5 (r^2)\, r\, dr\, d\theta = \int_0^{2\pi} d\theta \int_0^5 r^3\, dr = 2\pi \left[\tfrac{1}{4} r^4 \right]_0^5 = \tfrac{625}{2} \pi$$

17. By Green's Theorem, $W = \int_C \mathbf{F} \cdot d\mathbf{r} = \int_C x(x + y)\, dx + xy^2\, dy = \iint_D (y^2 - x)\, dy\, dx$ where C is the path described in the question and D is the triangle bounded by C. So

$$W = \int_0^1 \int_0^{1-x} (y^2 - x)\, dy\, dx = \int_0^1 \left[\tfrac{1}{3} y^3 - xy \right]_{y=0}^{y=1-x} dx = \int_0^1 \left(\tfrac{1}{3} (1 - x)^3 - x(1 - x) \right) dx$$

$$= \left[-\tfrac{1}{12} (1 - x)^4 - \tfrac{1}{2} x^2 + \tfrac{1}{3} x^3 \right]_0^1 = \left(-\tfrac{1}{2} + \tfrac{1}{3} \right) - \left(-\tfrac{1}{12} \right) = -\tfrac{1}{12}$$

19. Let C_1 be the arch of the cycloid from $(0, 0)$ to $(2\pi, 0)$, which corresponds to $0 \le t \le 2\pi$, and let C_2 be the segment from $(2\pi, 0)$ to $(0, 0)$, so C_2 is given by $x = 2\pi - t$, $y = 0$, $0 \le t \le 2\pi$. Then $C = C_1 \cup C_2$ is traversed clockwise, so $-C$ is oriented positively. Thus $-C$ encloses the area under one arch of the cycloid and from (5) we have

$$A = -\oint_{-C} y\, dx = \int_{C_1} y\, dx + \int_{C_2} y\, dx = \int_0^{2\pi} (1 - \cos t)(1 - \cos t)\, dt + \int_0^{2\pi} 0\, (-dt)$$

$$= \int_0^{2\pi} (1 - 2\cos t + \cos^2 t)\, dt + 0 = \left[t - 2 \sin t + \tfrac{1}{2} t + \tfrac{1}{4} \sin 2t \right]_0^{2\pi} = 3\pi$$

21. (a) Using Equation 13.2.8, we write parametric equations of the line segment as $x = (1 - t)x_1 + tx_2$, $y = (1 - t)y_1 + ty_2$, $0 \le t \le 1$. Then $dx = (x_2 - x_1)\, dt$ and $dy = (y_2 - y_1)\, dt$, so

$$\int_C x\, dy - y\, dx = \int_0^1 [(1 - t)x_1 + tx_2](y_2 - y_1)\, dt + [(1 - t)y_1 + ty_2](x_2 - x_1)\, dt$$

$$= \int_0^1 (x_1(y_2 - y_1) - y_1(x_2 - x_1) + t[(y_2 - y_1)(x_2 - x_1) - (x_2 - x_1)(y_2 - y_1)])\, dt$$

$$= \int_0^1 (x_1 y_2 - x_2 y_1)\, dt = x_1 y_2 - x_2 y_1$$

(b) We apply Green's Theorem to the path $C = C_1 \cup C_2 \cup \cdots \cup C_n$, where C_i is the line segment that joins (x_i, y_i) to

(x_{i+1}, y_{i+1}) for $i = 1, 2, \ldots, n-1$, and C_n is the line segment that joins (x_n, y_n) to (x_1, y_1). From (5),

$\frac{1}{2}\int_C x\,dy - y\,dx = \iint_D dA$, where D is the polygon bounded by C. Therefore

$$\text{area of polygon} = A(D) = \iint_D dA = \frac{1}{2}\int_C x\,dy - y\,dx$$
$$= \frac{1}{2}\left(\int_{C_1} x\,dy - y\,dx + \int_{C_2} x\,dy - y\,dx + \cdots + \int_{C_{n-1}} x\,dy - y\,dx + \int_{C_n} x\,dy - y\,dx\right)$$

To evaluate these integrals we use the formula from (a) to get

$$A(D) = \frac{1}{2}[(x_1 y_2 - x_2 y_1) + (x_2 y_3 - x_3 y_2) + \cdots + (x_{n-1} y_n - x_n y_{n-1}) + (x_n y_1 - x_1 y_n)].$$

(c) $A = \frac{1}{2}[(0 \cdot 1 - 2 \cdot 0) + (2 \cdot 3 - 1 \cdot 1) + (1 \cdot 2 - 0 \cdot 3) + (0 \cdot 1 - (-1) \cdot 2) + (-1 \cdot 0 - 0 \cdot 1)]$

$= \frac{1}{2}(0 + 5 + 2 + 2) = \frac{9}{2}$

23. Here $A = \frac{1}{2}(1)(1) = \frac{1}{2}$ and $C = C_1 + C_2 + C_3$, where C_1: $x = x$, $y = 0$, $0 \le x \le 1$;

C_2: $x = x$, $y = 1 - x$, $x = 1$ to $x = 0$; and C_3: $x = 0$, $y = 1$ to $y = 0$. Then

$\bar{x} = \frac{1}{2A}\int_C x^2\,dy = \int_{C_1} x^2\,dy + \int_{C_2} x^2\,dy + \int_{C_3} x^2\,dy = 0 + \int_1^0 (x^2)(-dx) + 0 = \frac{1}{3}$. Similarly,

$\bar{y} = -\frac{1}{2A}\int_C y^2\,dx = \int_{C_1} y^2\,dx + \int_{C_2} y^2\,dx + \int_{C_3} y^2\,dx = 0 + \int_1^0 (1-x)^2(-dx) + 0 = \frac{1}{3}$. Therefore $(\bar{x}, \bar{y}) = \left(\frac{1}{3}, \frac{1}{3}\right)$.

25. By Green's Theorem, $-\frac{1}{3}\rho\oint_C y^3\,dx = -\frac{1}{3}\rho\iint_D(-3y^2)\,dA = \iint_D y^2\rho\,dA = I_x$ and

$\frac{1}{3}\rho\oint_C x^3\,dy = \frac{1}{3}\rho\iint_D(3x^2)\,dA = \iint_D x^2\rho\,dA = I_y$.

27. Since C is a simple closed path which doesn't pass through or enclose the origin, there exists an open region that doesn't

contain the origin but does contain D. Thus $P = -y/(x^2 + y^2)$ and $Q = x/(x^2 + y^2)$ have continuous partial derivatives on

this open region containing D and we can apply Green's Theorem. But by Exercise 13.3.31(a), $\partial P/\partial y = \partial Q/\partial x$, so

$\oint_C \mathbf{F} \cdot d\mathbf{r} = \iint_D 0\,dA = 0$.

29. Using the first part of (5), we have that $\iint_R dx\,dy = A(R) = \int_{\partial R} x\,dy$. But $x = g(u, v)$, and $dy = \frac{\partial h}{\partial u}\,du + \frac{\partial h}{\partial v}\,dv$, and we

orient ∂S by taking the positive direction to be that which corresponds, under the mapping, to the positive direction along ∂R,

so

$$\int_{\partial R} x\,dy = \int_{\partial S} g(u, v)\left(\frac{\partial h}{\partial u}\,du + \frac{\partial h}{\partial v}\,dv\right) = \int_{\partial S} g(u, v)\frac{\partial h}{\partial u}\,du + g(u, v)\frac{\partial h}{\partial v}\,dv$$
$$= \pm\iint_S \left[\frac{\partial}{\partial u}\left(g(u, v)\frac{\partial h}{\partial v}\right) - \frac{\partial}{\partial v}\left(g(u, v)\frac{\partial h}{\partial u}\right)\right]dA \quad \text{[using Green's Theorem in the } uv\text{-plane]}$$
$$= \pm\iint_S \left(\frac{\partial g}{\partial u}\frac{\partial h}{\partial v} + g(u, v)\frac{\partial^2 h}{\partial u\,\partial v} - \frac{\partial g}{\partial v}\frac{\partial h}{\partial u} - g(u, v)\frac{\partial^2 h}{\partial v\,\partial u}\right)dA \quad \text{[using the Chain Rule]}$$
$$= \pm\iint_S \left(\frac{\partial x}{\partial u}\frac{\partial y}{\partial v} - \frac{\partial x}{\partial v}\frac{\partial y}{\partial u}\right)dA \quad \text{[by the equality of mixed partials]} = \pm\iint_S \frac{\partial(x, y)}{\partial(u, v)}\,du\,dv$$

The sign is chosen to be positive if the orientation that we gave to ∂S corresponds to the usual positive orientation, and it is

negative otherwise. In either case, since $A(R)$ is positive, the sign chosen must be the same as the sign of $\dfrac{\partial(x, y)}{\partial(u, v)}$. Therefore

$$A(R) = \iint_R dx\,dy = \iint_S \left|\frac{\partial(x, y)}{\partial(u, v)}\right|du\,dv.$$

13.5   Curl and Divergence

1. (a) curl $\mathbf{F} = \nabla \times \mathbf{F} = \begin{vmatrix} \mathbf{i} & \mathbf{j} & \mathbf{k} \\ \partial/\partial x & \partial/\partial y & \partial/\partial z \\ xyz & 0 & -x^2 y \end{vmatrix} = (-x^2 - 0)\,\mathbf{i} - (-2xy - xy)\,\mathbf{j} + (0 - xz)\,\mathbf{k}$

$\qquad = -x^2\,\mathbf{i} + 3xy\,\mathbf{j} - xz\,\mathbf{k}$

(b) div $\mathbf{F} = \nabla \cdot \mathbf{F} = \dfrac{\partial}{\partial x}(xyz) + \dfrac{\partial}{\partial y}(0) + \dfrac{\partial}{\partial z}(-x^2 y) = yz + 0 + 0 = yz$

3. (a) curl $\mathbf{F} = \nabla \times \mathbf{F} = \begin{vmatrix} \mathbf{i} & \mathbf{j} & \mathbf{k} \\ \partial/\partial x & \partial/\partial y & \partial/\partial z \\ 1 & x + yz & xy - \sqrt{z} \end{vmatrix} = (x - y)\,\mathbf{i} - (y - 0)\,\mathbf{j} + (1 - 0)\,\mathbf{k}$

$\qquad = (x - y)\,\mathbf{i} - y\,\mathbf{j} + \mathbf{k}$

(b) div $\mathbf{F} = \nabla \cdot \mathbf{F} = \dfrac{\partial}{\partial x}(1) + \dfrac{\partial}{\partial y}(x + yz) + \dfrac{\partial}{\partial z}(xy - \sqrt{z}) = z - \dfrac{1}{2\sqrt{z}}$

5. (a) curl $\mathbf{F} = \nabla \times \mathbf{F} = \begin{vmatrix} \mathbf{i} & \mathbf{j} & \mathbf{k} \\ \partial/\partial x & \partial/\partial y & \partial/\partial z \\ e^x \sin y & e^x \cos y & z \end{vmatrix} = (0 - 0)\,\mathbf{i} - (0 - 0)\,\mathbf{j} + (e^x \cos y - e^x \cos y)\,\mathbf{k} = \mathbf{0}$

(b) div $\mathbf{F} = \nabla \cdot \mathbf{F} = \dfrac{\partial}{\partial x}(e^x \sin y) + \dfrac{\partial}{\partial y}(e^x \cos y) + \dfrac{\partial}{\partial z}(z) = e^x \sin y - e^x \sin y + 1 = 1$

7. (a) curl $\mathbf{F} = \nabla \times \mathbf{F} = \begin{vmatrix} \mathbf{i} & \mathbf{j} & \mathbf{k} \\ \partial/\partial x & \partial/\partial y & \partial/\partial z \\ \ln x & \ln(xy) & \ln(xyz) \end{vmatrix} = \left(\dfrac{xz}{xyz} - 0\right)\mathbf{i} - \left(\dfrac{yz}{xyz} - 0\right)\mathbf{j} + \left(\dfrac{y}{xy} - 0\right)\mathbf{k}$

$\qquad = \left\langle \dfrac{1}{y}, -\dfrac{1}{x}, \dfrac{1}{x} \right\rangle$

(b) div $\mathbf{F} = \nabla \cdot \mathbf{F} = \dfrac{\partial}{\partial x}(\ln x) + \dfrac{\partial}{\partial y}(\ln(xy)) + \dfrac{\partial}{\partial z}(\ln(xyz)) = \dfrac{1}{x} + \dfrac{x}{xy} + \dfrac{xy}{xyz} = \dfrac{1}{x} + \dfrac{1}{y} + \dfrac{1}{z}$

9. If the vector field is $\mathbf{F} = P\,\mathbf{i} + Q\,\mathbf{j} + R\,\mathbf{k}$, then we know $R = 0$. In addition, the y-component of each vector of $\mathbf{F}$ is 0, so

$Q = 0$, hence $\dfrac{\partial Q}{\partial x} = \dfrac{\partial Q}{\partial y} = \dfrac{\partial Q}{\partial z} = \dfrac{\partial R}{\partial x} = \dfrac{\partial R}{\partial y} = \dfrac{\partial R}{\partial z} = 0$. P increases as y increases, so $\dfrac{\partial P}{\partial y} > 0$, but P doesn't change in

the x- or z-directions, so $\dfrac{\partial P}{\partial x} = \dfrac{\partial P}{\partial z} = 0$.

(a) div $\mathbf{F} = \dfrac{\partial P}{\partial x} + \dfrac{\partial Q}{\partial y} + \dfrac{\partial R}{\partial z} = 0 + 0 + 0 = 0$

(b) curl $\mathbf{F} = \left(\dfrac{\partial R}{\partial y} - \dfrac{\partial Q}{\partial z}\right)\mathbf{i} + \left(\dfrac{\partial P}{\partial z} - \dfrac{\partial R}{\partial x}\right)\mathbf{j} + \left(\dfrac{\partial Q}{\partial x} - \dfrac{\partial P}{\partial y}\right)\mathbf{k} = (0 - 0)\,\mathbf{i} + (0 - 0)\,\mathbf{j} + \left(0 - \dfrac{\partial P}{\partial y}\right)\mathbf{k} = -\dfrac{\partial P}{\partial y}\mathbf{k}$

Since $\dfrac{\partial P}{\partial y} > 0$, $-\dfrac{\partial P}{\partial y}\mathbf{k}$ is a vector pointing in the negative z-direction.

11. $\operatorname{curl}\mathbf{F} = \nabla \times \mathbf{F} = \begin{vmatrix} \mathbf{i} & \mathbf{j} & \mathbf{k} \\ \partial/\partial x & \partial/\partial y & \partial/\partial z \\ yz & xz & xy \end{vmatrix} = (x - x)\mathbf{i} - (y - y)\mathbf{j} + (z - z)\mathbf{k} = \mathbf{0}$

and $\mathbf{F}$ is defined on all of $\mathbb{R}^3$ with component functions which have continuous partial derivatives, so by Theorem 4, $\mathbf{F}$ is

conservative. Thus, there exists a function f such that $\mathbf{F} = \nabla f$. Then $f_x(x, y, z) = yz$ implies $f(x, y, z) = xyz + g(y, z)$

and $f_y(x, y, z) = xz + g_y(y, z)$. But $f_y(x, y, z) = xz$, so $g(y, z) = h(z)$ and $f(x, y, z) = xyz + h(z)$. Thus

$f_z(x, y, z) = xy + h'(z)$ but $f_z(x, y, z) = xy$ so $h(z) = K$, a constant. Hence a potential function for $\mathbf{F}$ is

$f(x, y, z) = xyz + K$.

13. $\operatorname{curl}\mathbf{F} = \nabla \times \mathbf{F} = \begin{vmatrix} \mathbf{i} & \mathbf{j} & \mathbf{k} \\ \partial/\partial x & \partial/\partial y & \partial/\partial z \\ 2xy & x^2 + 2yz & y^2 \end{vmatrix} = (2y - 2y)\mathbf{i} - (0 - 0)\mathbf{j} + (2x - 2x)\mathbf{k} = \mathbf{0}$, $\mathbf{F}$ is defined on all of $\mathbb{R}^3$,

and the partial derivatives of the component functions are continuous, so $\mathbf{F}$ is conservative. Thus there exists a function f such

that $\nabla f = \mathbf{F}$. Then $f_x(x, y, z) = 2xy$ implies $f(x, y, z) = x^2 y + g(y, z)$ and $f_y(x, y, z) = x^2 + g_y(y, z)$. But

$f_y(x, y, z) = x^2 + 2yz$, so $g(y, z) = y^2 z + h(z)$ and $f(x, y, z) = x^2 y + y^2 z + h(z)$. Thus $f_z(x, y, z) = y^2 + h'(z)$ but

$f_z(x, y, z) = y^2$ so $h(z) = K$ and $f(x, y, z) = x^2 y + y^2 z + K$.

15. $\operatorname{curl}\mathbf{F} = \nabla \times \mathbf{F} = \begin{vmatrix} \mathbf{i} & \mathbf{j} & \mathbf{k} \\ \partial/\partial x & \partial/\partial y & \partial/\partial z \\ ye^{-x} & e^{-x} & 2z \end{vmatrix} = (0 - 0)\mathbf{i} - (0 - 0)\mathbf{j} + (-e^{-x} - e^{-x})\mathbf{k} = -2e^{-x}\mathbf{k} \neq \mathbf{0}$,

so $\mathbf{F}$ is not conservative.

17. No. Assume there is such a $\mathbf{G}$. Then $\operatorname{div}(\operatorname{curl}\mathbf{G}) = y^2 + z^2 + x^2 \neq 0$, which contradicts Theorem 11.

19. $\operatorname{curl}\mathbf{F} = \begin{vmatrix} \mathbf{i} & \mathbf{j} & \mathbf{k} \\ \partial/\partial x & \partial/\partial y & \partial/\partial z \\ f(x) & g(y) & h(z) \end{vmatrix} = (0 - 0)\mathbf{i} + (0 - 0)\mathbf{j} + (0 - 0)\mathbf{k} = \mathbf{0}$.

Hence $\mathbf{F} = f(x)\mathbf{i} + g(y)\mathbf{j} + h(z)\mathbf{k}$ is irrotational.

For Exercises 21–27, let $\mathbf{F}(x, y, z) = P_1\mathbf{i} + Q_1\mathbf{j} + R_1\mathbf{k}$ and $\mathbf{G}(x, y, z) = P_2\mathbf{i} + Q_2\mathbf{j} + R_2\mathbf{k}$.

21. $\operatorname{div}(\mathbf{F} + \mathbf{G}) = \dfrac{\partial(P_1 + P_2)}{\partial x} + \dfrac{\partial(Q_1 + Q_2)}{\partial y} + \dfrac{\partial(R_1 + R_2)}{\partial z}$

$= \left(\dfrac{\partial P_1}{\partial x} + \dfrac{\partial Q_1}{\partial y} + \dfrac{\partial R_1}{\partial z}\right) + \left(\dfrac{\partial P_2}{\partial x} + \dfrac{\partial Q_2}{\partial y} + \dfrac{\partial R_3}{\partial z}\right) = \operatorname{div}\mathbf{F} + \operatorname{div}\mathbf{G}$

23. $\operatorname{div}(f\mathbf{F}) = \dfrac{\partial(fP_1)}{\partial x} + \dfrac{\partial(fQ_1)}{\partial y} + \dfrac{\partial(fR_1)}{\partial z} = \left(f\dfrac{\partial P_1}{\partial x} + P_1\dfrac{\partial f}{\partial x}\right) + \left(f\dfrac{\partial Q_1}{\partial y} + Q_1\dfrac{\partial f}{\partial y}\right) + \left(f\dfrac{\partial R_1}{\partial z} + R_1\dfrac{\partial f}{\partial z}\right)$

$= f\left(\dfrac{\partial P_1}{\partial x} + \dfrac{\partial Q_1}{\partial y} + \dfrac{\partial R_1}{\partial z}\right) + \langle P_1, Q_1, R_1 \rangle \cdot \left\langle \dfrac{\partial f}{\partial x}, \dfrac{\partial f}{\partial y}, \dfrac{\partial f}{\partial z} \right\rangle = f\operatorname{div}\mathbf{F} + \mathbf{F} \cdot \nabla f$

25. $\operatorname{div}(\mathbf{F} \times \mathbf{G}) = \nabla \cdot (\mathbf{F} \times \mathbf{G}) = \begin{vmatrix} \partial/\partial x & \partial/\partial y & \partial/\partial z \\ P_1 & Q_1 & R_1 \\ P_2 & Q_2 & R_2 \end{vmatrix} = \frac{\partial}{\partial x}\begin{vmatrix} Q_1 & R_1 \\ Q_2 & R_2 \end{vmatrix} - \frac{\partial}{\partial y}\begin{vmatrix} P_1 & R_1 \\ P_2 & R_2 \end{vmatrix} + \frac{\partial}{\partial z}\begin{vmatrix} P_1 & Q_1 \\ P_2 & Q_2 \end{vmatrix}$

$$= \left[Q_1 \frac{\partial R_2}{\partial x} + R_2 \frac{\partial Q_1}{\partial x} - Q_2 \frac{\partial R_1}{\partial x} - R_1 \frac{\partial Q_2}{\partial x}\right] - \left[P_1 \frac{\partial R_2}{\partial y} + R_2 \frac{\partial P_1}{\partial y} - P_2 \frac{\partial R_1}{\partial y} - R_1 \frac{\partial P_2}{\partial y}\right]$$

$$+ \left[P_1 \frac{\partial Q_2}{\partial z} + Q_2 \frac{\partial P_1}{\partial z} - P_2 \frac{\partial Q_1}{\partial z} - Q_1 \frac{\partial P_2}{\partial z}\right]$$

$$= \left[P_2\left(\frac{\partial R_1}{\partial y} - \frac{\partial Q_1}{\partial z}\right) + Q_2\left(\frac{\partial P_1}{\partial z} - \frac{\partial R_1}{\partial x}\right) + R_2\left(\frac{\partial Q_1}{\partial x} - \frac{\partial P_1}{\partial y}\right)\right]$$

$$- \left[P_1\left(\frac{\partial R_2}{\partial y} - \frac{\partial Q_2}{\partial z}\right) + Q_1\left(\frac{\partial P_2}{\partial z} - \frac{\partial R_2}{\partial x}\right) + R_1\left(\frac{\partial Q_2}{\partial x} - \frac{\partial P_2}{\partial y}\right)\right]$$

$$= \mathbf{G} \cdot \operatorname{curl}\mathbf{F} - \mathbf{F} \cdot \operatorname{curl}\mathbf{G}$$

27. $\operatorname{curl}(\operatorname{curl}\mathbf{F}) = \nabla \times (\nabla \times \mathbf{F}) = \begin{vmatrix} \mathbf{i} & \mathbf{j} & \mathbf{k} \\ \partial/\partial x & \partial/\partial y & \partial/\partial z \\ \partial R_1/\partial y - \partial Q_1/\partial z & \partial P_1/\partial z - \partial R_1/\partial x & \partial Q_1/\partial x - \partial P_1/\partial y \end{vmatrix}$

$$= \left(\frac{\partial^2 Q_1}{\partial y \partial x} - \frac{\partial^2 P_1}{\partial y^2} - \frac{\partial^2 P_1}{\partial z^2} + \frac{\partial^2 R_1}{\partial z \partial x}\right)\mathbf{i} + \left(\frac{\partial^2 R_1}{\partial z \partial y} - \frac{\partial^2 Q_1}{\partial z^2} - \frac{\partial^2 Q_1}{\partial x^2} + \frac{\partial^2 P_1}{\partial x \partial y}\right)\mathbf{j}$$

$$+ \left(\frac{\partial^2 P_1}{\partial x \partial z} - \frac{\partial^2 R_1}{\partial x^2} - \frac{\partial^2 R_1}{\partial y^2} + \frac{\partial^2 Q_1}{\partial y \partial z}\right)\mathbf{k}$$

Now let's consider $\operatorname{grad}(\operatorname{div}\mathbf{F}) - \nabla^2\mathbf{F}$ and compare with the above.
(Note that $\nabla^2\mathbf{F}$ is defined on page 762.)

$$\operatorname{grad}(\operatorname{div}\mathbf{F}) - \nabla^2\mathbf{F} = \left[\left(\frac{\partial^2 P_1}{\partial x^2} + \frac{\partial^2 Q_1}{\partial x \partial y} + \frac{\partial^2 R_1}{\partial x \partial z}\right)\mathbf{i} + \left(\frac{\partial^2 P_1}{\partial y \partial x} + \frac{\partial^2 Q_1}{\partial y^2} + \frac{\partial^2 R_1}{\partial y \partial z}\right)\mathbf{j} + \left(\frac{\partial^2 P_1}{\partial z \partial x} + \frac{\partial^2 Q_1}{\partial z \partial y} + \frac{\partial^2 R_1}{\partial z^2}\right)\mathbf{k}\right]$$

$$- \left[\left(\frac{\partial^2 P_1}{\partial x^2} + \frac{\partial^2 P_1}{\partial y^2} + \frac{\partial^2 P_1}{\partial z^2}\right)\mathbf{i} + \left(\frac{\partial^2 Q_1}{\partial x^2} + \frac{\partial^2 Q_1}{\partial y^2} + \frac{\partial^2 Q_1}{\partial z^2}\right)\mathbf{j}\right.$$

$$\left. + \left(\frac{\partial^2 R_1}{\partial x^2} + \frac{\partial^2 R_1}{\partial y^2} + \frac{\partial^2 R_1}{\partial z^2}\right)\mathbf{k}\right]$$

$$= \left(\frac{\partial^2 Q_1}{\partial x \partial y} + \frac{\partial^2 R_1}{\partial x \partial z} - \frac{\partial^2 P_1}{\partial y^2} - \frac{\partial^2 P_1}{\partial z^2}\right)\mathbf{i} + \left(\frac{\partial^2 P_1}{\partial y \partial x} + \frac{\partial^2 R_1}{\partial y \partial z} - \frac{\partial^2 Q_1}{\partial x^2} - \frac{\partial^2 Q_1}{\partial z^2}\right)\mathbf{j}$$

$$+ \left(\frac{\partial^2 P_1}{\partial z \partial x} + \frac{\partial^2 Q_1}{\partial z \partial y} - \frac{\partial^2 R_1}{\partial x^2} - \frac{\partial^2 R_2}{\partial y^2}\right)\mathbf{k}$$

Then applying Clairaut's Theorem to reverse the order of differentiation in the second partial derivatives as needed and comparing, we have $\operatorname{curl}\operatorname{curl}\mathbf{F} = \operatorname{grad}\operatorname{div}\mathbf{F} - \nabla^2\mathbf{F}$ as desired.

29. (a) $\nabla r = \nabla \sqrt{x^2 + y^2 + z^2} = \dfrac{x}{\sqrt{x^2 + y^2 + z^2}}\,\mathbf{i} + \dfrac{y}{\sqrt{x^2 + y^2 + z^2}}\,\mathbf{j} + \dfrac{z}{\sqrt{x^2 + y^2 + z^2}}\,\mathbf{k} = \dfrac{x\,\mathbf{i} + y\,\mathbf{j} + z\,\mathbf{k}}{\sqrt{x^2 + y^2 + z^2}} = \dfrac{\mathbf{r}}{r}$

(b) $\nabla \times \mathbf{r} = \begin{vmatrix} \mathbf{i} & \mathbf{j} & \mathbf{k} \\[4pt] \dfrac{\partial}{\partial x} & \dfrac{\partial}{\partial y} & \dfrac{\partial}{\partial z} \\[8pt] x & y & z \end{vmatrix} = \left[\dfrac{\partial}{\partial y}(z) - \dfrac{\partial}{\partial z}(y)\right]\mathbf{i} + \left[\dfrac{\partial}{\partial z}(x) - \dfrac{\partial}{\partial x}(z)\right]\mathbf{j} + \left[\dfrac{\partial}{\partial x}(y) - \dfrac{\partial}{\partial y}(x)\right]\mathbf{k} = \mathbf{0}$

(c) $\nabla\left(\dfrac{1}{r}\right) = \nabla\left(\dfrac{1}{\sqrt{x^2 + y^2 + z^2}}\right)$

$\qquad = \dfrac{-\dfrac{1}{2\sqrt{x^2 + y^2 + z^2}}(2x)}{x^2 + y^2 + z^2}\,\mathbf{i} - \dfrac{\dfrac{1}{2\sqrt{x^2 + y^2 + z^2}}(2y)}{x^2 + y^2 + z^2}\,\mathbf{j} - \dfrac{\dfrac{1}{2\sqrt{x^2 + y^2 + z^2}}(2z)}{x^2 + y^2 + z^2}\,\mathbf{k}$

$\qquad = -\dfrac{x\,\mathbf{i} + y\,\mathbf{j} + z\,\mathbf{k}}{(x^2 + y^2 + z^2)^{3/2}} = -\dfrac{\mathbf{r}}{r^3}$

(d) $\nabla \ln r = \nabla \ln\left(x^2 + y^2 + z^2\right)^{1/2} = \tfrac{1}{2}\nabla \ln\left(x^2 + y^2 + z^2\right)$

$\qquad = \dfrac{x}{x^2 + y^2 + z^2}\,\mathbf{i} + \dfrac{y}{x^2 + y^2 + z^2}\,\mathbf{j} + \dfrac{z}{x^2 + y^2 + z^2}\,\mathbf{k} = \dfrac{x\,\mathbf{i} + y\,\mathbf{j} + z\,\mathbf{k}}{x^2 + y^2 + z^2} = \dfrac{\mathbf{r}}{r^2}$

31. By (13), $\oint_C f(\nabla g) \cdot \mathbf{n}\,ds = \iint_D \operatorname{div}(f\nabla g)\,dA = \iint_D [f\operatorname{div}(\nabla g) + \nabla g \cdot \nabla f]\,dA$ by Exercise 23. But $\operatorname{div}(\nabla g) = \nabla^2 g$. Hence $\iint_D f\nabla^2 g\,dA = \oint_C f(\nabla g) \cdot \mathbf{n}\,ds - \iint_D \nabla g \cdot \nabla f\,dA$.

33. Let $f(x, y) = 1$. Then $\nabla f = \mathbf{0}$ and Green's first identity (see Exercise 31) says

$\iint_D \nabla^2 g\,dA = \oint_C (\nabla g) \cdot \mathbf{n}\,ds - \iint_D \mathbf{0} \cdot \nabla g\,dA \quad \Rightarrow \quad \iint_D \nabla^2 g\,dA = \oint_C \nabla g \cdot \mathbf{n}\,ds$. But g is harmonic on D, so $\nabla^2 g = 0 \quad \Rightarrow \quad \oint_C \nabla g \cdot \mathbf{n}\,ds = 0$ and $\oint_C D_{\mathbf{n}}g\,ds = \oint_C (\nabla g \cdot \mathbf{n})\,ds = 0$.

35. (a) We know that $\omega = v/d$, and from the diagram $\sin\theta = d/r \quad \Rightarrow \quad v = d\omega = (\sin\theta)r\omega = |\mathbf{w} \times \mathbf{r}|$. But $\mathbf{v}$ is perpendicular to both $\mathbf{w}$ and $\mathbf{r}$, so that $\mathbf{v} = \mathbf{w} \times \mathbf{r}$.

(b) From (a), $\mathbf{v} = \mathbf{w} \times \mathbf{r} = \begin{vmatrix} \mathbf{i} & \mathbf{j} & \mathbf{k} \\ 0 & 0 & \omega \\ x & y & z \end{vmatrix} = (0 \cdot z - \omega y)\mathbf{i} + (\omega x - 0 \cdot z)\mathbf{j} + (0 \cdot y - x \cdot 0)\mathbf{k} = -\omega y\,\mathbf{i} + \omega x\,\mathbf{j}$

(c) $\operatorname{curl}\mathbf{v} = \nabla \times \mathbf{v} = \begin{vmatrix} \mathbf{i} & \mathbf{j} & \mathbf{k} \\[4pt] \partial/\partial x & \partial/\partial y & \partial/\partial z \\[4pt] -\omega y & \omega x & 0 \end{vmatrix}$

$\qquad = \left[\dfrac{\partial}{\partial y}(0) - \dfrac{\partial}{\partial z}(\omega x)\right]\mathbf{i} + \left[\dfrac{\partial}{\partial z}(-\omega y) - \dfrac{\partial}{\partial x}(0)\right]\mathbf{j} + \left[\dfrac{\partial}{\partial x}(\omega x) - \dfrac{\partial}{\partial y}(-\omega y)\right]\mathbf{k}$

$\qquad = [\omega - (-\omega)]\mathbf{k} = 2\omega\,\mathbf{k} = 2\mathbf{w}$

13.6 Parametric Surfaces and Their Areas

1. $\mathbf{r}(u, v) = (u + v)\,\mathbf{i} + (3 - v)\,\mathbf{j} + (1 + 4u + 5v)\,\mathbf{k} = \langle 0, 3, 1 \rangle + u \langle 1, 0, 4 \rangle + v \langle 1, -1, 5 \rangle$. From Example 3, we recognize

this as a vector equation of a plane through the point $(0, 3, 1)$ and containing vectors $\mathbf{a} = \langle 1, 0, 4 \rangle$ and $\mathbf{b} = \langle 1, -1, 5 \rangle$. If we

wish to find a more conventional equation for the plane, a normal vector to the plane is $\mathbf{a} \times \mathbf{b} = \begin{vmatrix} \mathbf{i} & \mathbf{j} & \mathbf{k} \\ 1 & 0 & 4 \\ 1 & -1 & 5 \end{vmatrix} = 4\,\mathbf{i} - \mathbf{j} - \mathbf{k}$

and an equation of the plane is $4(x - 0) - (y - 3) - (z - 1) = 0$ or $4x - y - z = -4$.

3. $\mathbf{r}(s, t) = \langle s, t, t^2 - s^2 \rangle$, so the corresponding parametric equations for the surface are $x = s$, $y = t$, $z = t^2 - s^2$. For any

point (x, y, z) on the surface, we have $z = y^2 - x^2$. With no restrictions on the parameters, the surface is $z = y^2 - x^2$, which

we recognize as a hyperbolic paraboloid.

5. $\mathbf{r}(u, v) = \langle u^2 + 1, v^3 + 1, u + v \rangle$, $-1 \le u \le 1$, $-1 \le v \le 1$.

The surface has parametric equations $x = u^2 + 1$, $y = v^3 + 1$, $z = u + v$,

$-1 \le u \le 1$, $-1 \le v \le 1$. If we keep u constant at u_0, $x = u_0^2 + 1$, a

constant, so the corresponding grid curves must be the curves parallel to the

yz-plane. If v is constant, we have $y = v_0^3 + 1$, a constant, so these grid

curves are the curves parallel to the xz-plane.

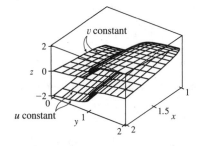

7. $\mathbf{r}(u, v) = \langle \cos^3 u \cos^3 v, \sin^3 u \cos^3 v, \sin^3 v \rangle$.

The surface has parametric equations $x = \cos^3 u \cos^3 v$, $y = \sin^3 u \cos^3 v$,

$z = \sin^3 v$, $0 \le u \le \pi$, $0 \le v \le 2\pi$. Note that if $v = v_0$ is constant then

$z = \sin^3 v_0$ is constant, so the corresponding grid curves must be the curves

parallel to the xy-plane. The vertically oriented grid curves, then, correspond

to $u = u_0$ being held constant, giving $x = \cos^3 u_0 \cos^3 v$,

$y = \sin^3 u_0 \cos^3 v$, $z = \sin^3 v$. These curves lie in vertical planes that

contain the z-axis.

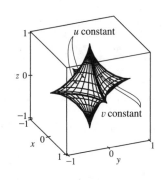

9. $x = \cos u \sin 2v$, $y = \sin u \sin 2v$, $z = \sin v$.

The complete graph of the surface is given by the parametric domain

$0 \le u \le \pi$, $0 \le v \le 2\pi$. Note that if $v = v_0$ is constant, the parametric

equations become $x = \cos u \sin 2v_0$, $y = \sin u \sin 2v_0$, $z = \sin v_0$ which

represent a circle of radius $\sin 2v_0$ in the plane $z = \sin v_0$. So the circular

grid curves we see lying horizontally are the grid curves which have

v constant. The vertical grid curves, then, correspond to $u = u_0$ being held

constant, giving $x = \cos u_0 \sin 2v$ and $y = \sin u_0 \sin 2v$ with $z = \sin v$

which has a "figure-eight" shape.

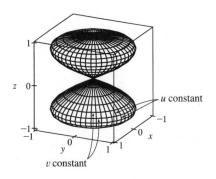

11. $\mathbf{r}(u, v) = u \cos v\, \mathbf{i} + u \sin v\, \mathbf{j} + v\, \mathbf{k}$. The parametric equations for the surface are $x = u \cos v$, $y = u \sin v$, $z = v$. We look at the grid curves first; if we fix v, then x and y parametrize a straight line in the plane $z = v$ which intersects the z-axis. If u is held constant, the projection onto the xy-plane is circular; with $z = v$, each grid curve is a helix. The surface is a spiraling ramp, graph I.

13. $x = (u - \sin u) \cos v$, $y = (1 - \cos u) \sin v$, $z = u$. If u is held constant, x and y give an equation of an ellipse in the plane $z = u$, thus the grid curves are horizontally oriented ellipses. Note that when $u = 0$, the "ellipse" is the single point $(0, 0, 0)$, and when $u = \pi$, we have $y = 0$ while x ranges from $-\pi$ to π, a line segment parallel to the x-axis in the plane $z = \pi$. This is the upper "seam" we see in graph II. When v is held constant, $z = u$ is free to vary, so the corresponding grid curves are the curves we see running up and down along the surface.

15. From Example 3, parametric equations for the plane through the point $(1, 2, -3)$ that contains the vectors $\mathbf{a} = \langle 1, 1, -1 \rangle$ and $\mathbf{b} = \langle 1, -1, 1 \rangle$ are $x = 1 + u(1) + v(1) = 1 + u + v$, $y = 2 + u(1) + v(-1) = 2 + u - v$, $z = -3 + u(-1) + v(1) = -3 - u + v$.

17. Solving the equation for y gives $y^2 = 1 - x^2 + z^2$ $\Rightarrow$ $y = \sqrt{1 - x^2 + z^2}$. (We choose the positive root since we want the part of the hyperboloid that corresponds to $y \geq 0$.) If we let x and z be the parameters, parametric equations are $x = x$, $z = z$, $y = \sqrt{1 - x^2 + z^2}$.

19. Since the cone intersects the sphere in the circle $x^2 + y^2 = 2$, $z = \sqrt{2}$ and we want the portion of the sphere above this, we can parametrize the surface as $x = x$, $y = y$, $z = \sqrt{4 - x^2 - y^2}$ where $x^2 + y^2 \leq 2$.

Alternate solution: Using spherical coordinates, $x = 2 \sin \phi \cos \theta$, $y = 2 \sin \phi \sin \theta$, $z = 2 \cos \phi$ where $0 \leq \phi \leq \frac{\pi}{4}$ and $0 \leq \theta \leq 2\pi$.

21. Parametric equations are $x = x$, $y = 4 \cos \theta$, $z = 4 \sin \theta$, $0 \leq x \leq 5$, $0 \leq \theta \leq 2\pi$.

23. The surface appears to be a portion of a circular cylinder of radius 3 with axis the x-axis. An equation of the cylinder is $y^2 + z^2 = 9$, and we can impose the restrictions $0 \leq x \leq 5$, $y \leq 0$ to obtain the portion shown.

To graph the surface on a CAS, we can use parametric equations $x = u$, $y = 3 \cos v$, $z = 3 \sin v$ with the parameter domain $0 \leq u \leq 5$, $\frac{\pi}{2} \leq v \leq \frac{3\pi}{2}$. Alternatively, we can regard x and z as parameters. Then parametric equations are $x = x$, $z = z$, $y = -\sqrt{9 - z^2}$, where $0 \leq x \leq 5$ and $-3 \leq z \leq 3$.

25. Using Equations 3, we have the parametrization $x = x$, $y = e^{-x} \cos \theta$, $z = e^{-x} \sin \theta$, $0 \leq x \leq 3$, $0 \leq \theta \leq 2\pi$.

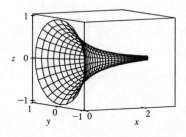

27. (a) Replacing $\cos u$ by $\sin u$ and $\sin u$ by $\cos u$ gives parametric equations

$x = (2 + \sin v) \sin u$, $y = (2 + \sin v) \cos u$, $z = u + \cos v$. From the graph, it

appears that the direction of the spiral is reversed. We can verify this observation

by noting that the projection of the spiral grid curves onto the xy-plane, given by

$x = (2 + \sin v) \sin u$, $y = (2 + \sin v) \cos u$, $z = 0$, draws a circle in the

clockwise direction for each value of v. The original equations, on the other hand,

give circular projections drawn in the counterclockwise direction. The equation for

z is identical in both surfaces, so as z increases, these grid curves spiral up in

opposite directions for the two surfaces.

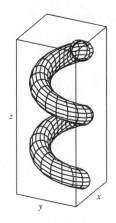

(b) Replacing $\cos u$ by $\cos 2u$ and $\sin u$ by $\sin 2u$ gives parametric equations

$x = (2 + \sin v) \cos 2u$, $y = (2 + \sin v) \sin 2u$, $z = u + \cos v$. From the graph, it

appears that the number of coils in the surface doubles within the same parametric

domain. We can verify this observation by noting that the projection of the spiral

grid curves onto the xy-plane, given by $x = (2 + \sin v) \cos 2u$,

$y = (2 + \sin v) \sin 2u$, $z = 0$ (where v is constant), complete circular revolutions

for $0 \le u \le \pi$ while the original surface requires $0 \le u \le 2\pi$ for a complete

revolution. Thus, the new surface winds around twice as fast as the original

surface, and since the equation for z is identical in both surfaces, we observe twice

as many circular coils in the same z-interval.

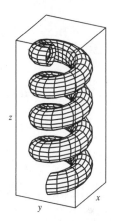

29. $\mathbf{r}(u, v) = (u + v)\,\mathbf{i} + 3u^2\,\mathbf{j} + (u - v)\,\mathbf{k}$.

$\mathbf{r}_u = \mathbf{i} + 6u\,\mathbf{j} + \mathbf{k}$ and $\mathbf{r}_v = \mathbf{i} - \mathbf{k}$, so $\mathbf{r}_u \times \mathbf{r}_v = -6u\,\mathbf{i} + 2\,\mathbf{j} - 6u\,\mathbf{k}$.

Since the point $(2, 3, 0)$ corresponds to $u = 1$, $v = 1$, a normal vector

to the surface at $(2, 3, 0)$ is $-6\,\mathbf{i} + 2\,\mathbf{j} - 6\,\mathbf{k}$, and an equation of the

tangent plane is $-6x + 2y - 6z = -6$ or $3x - y + 3z = 3$.

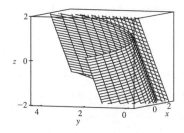

31. $\mathbf{r}(u, v) = u\,\mathbf{i} + \ln(uv)\,\mathbf{j} + v\,\mathbf{k}$ $\Rightarrow$ $\mathbf{r}_u(u, v) = \mathbf{i} + \frac{1}{u}\,\mathbf{j}$,

$\mathbf{r}_v(u, v) = \frac{1}{v}\,\mathbf{j} + \mathbf{k}$. $\mathbf{r}(1, 1) = \mathbf{i} + \mathbf{k}$, so the point corresponding to

$u = 1$, $v = 1$ is $(1, 0, 1)$. A normal vector for the tangent plane is

$\mathbf{r}_u(1, 1) \times \mathbf{r}_v(1, 1) = (\mathbf{i} + \mathbf{j}) \times (\mathbf{j} + \mathbf{k}) = \mathbf{i} - \mathbf{j} + \mathbf{k}$, so an equation

of the tangent plane is $(x - 1) - (y - 0) + (z - 1) = 0$

or $x - y + z = 2$.

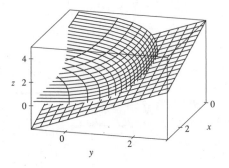

33. $z = f(x, y) = 6 - 3x - 2y$ which intersects the xy-plane in the line $3x + 2y = 6$, so D is the triangular region given

by $\{(x, y) \mid 0 \le x \le 2, 0 \le y \le 3 - \frac{3}{2}x\}$. Thus

$$A(S) = \iint_D \sqrt{1 + (-3)^2 + (-2)^2}\, dA = \sqrt{14} \iint_D dA = \sqrt{14}\, A(D) = \sqrt{14}\left(\tfrac{1}{2} \cdot 2 \cdot 3\right) = 3\sqrt{14}.$$

35. $z = f(x, y) = \frac{2}{3}(x^{3/2} + y^{3/2})$ and $D = \{(x, y) \mid 0 \le x \le 1, 0 \le y \le 1\}$. Then $f_x = x^{1/2}$, $f_y = y^{1/2}$ and

$$A(S) = \iint_D \sqrt{1 + \left(\sqrt{x}\right)^2 + \left(\sqrt{y}\right)^2}\, dA = \int_0^1 \int_0^1 \sqrt{1 + x + y}\, dy\, dx$$

$$= \int_0^1 \left[\tfrac{2}{3}(x + y + 1)^{3/2}\right]_{y=0}^{y=1} dx = \tfrac{2}{3} \int_0^1 \left[(x + 2)^{3/2} - (x + 1)^{3/2}\right] dx$$

$$= \tfrac{2}{3}\left[\tfrac{2}{5}(x + 2)^{5/2} - \tfrac{2}{5}(x + 1)^{5/2}\right]_0^1 = \tfrac{4}{15}(3^{5/2} - 2^{5/2} - 2^{5/2} + 1) = \tfrac{4}{15}(3^{5/2} - 2^{7/2} + 1)$$

37. $z = f(x, y) = xy$ with $0 \le x^2 + y^2 \le 1$, so $f_x = y$, $f_y = x$ $\Rightarrow$

$$A(S) = \iint_D \sqrt{1 + y^2 + x^2}\, dA = \int_0^{2\pi} \int_0^1 \sqrt{r^2 + 1}\, r\, dr\, d\theta = \int_0^{2\pi} \left[\tfrac{1}{3}(r^2 + 1)^{3/2}\right]_{r=0}^{r=1} d\theta$$

$$= \int_0^{2\pi} \tfrac{1}{3}\left(2\sqrt{2} - 1\right) d\theta = \tfrac{2\pi}{3}\left(2\sqrt{2} - 1\right)$$

39. $z = f(x, y) = y^2 - x^2$ with $1 \le x^2 + y^2 \le 4$. Then

$$A(S) = \iint_D \sqrt{1 + 4x^2 + 4y^2}\, dA = \int_0^{2\pi} \int_1^2 \sqrt{1 + 4r^2}\, r\, dr\, d\theta = \int_0^{2\pi} d\theta \int_1^2 r\sqrt{1 + 4r^2}\, dr$$

$$= \left[\theta\right]_0^{2\pi} \left[\tfrac{1}{12}(1 + 4r^2)^{3/2}\right]_1^2 = \tfrac{\pi}{6}\left(17\sqrt{17} - 5\sqrt{5}\right)$$

41. A parametric representation of the surface is $x = x$, $y = 4x + z^2$, $z = z$ with $0 \le x \le 1, 0 \le z \le 1$.

Hence $\mathbf{r}_x \times \mathbf{r}_z = (\mathbf{i} + 4\mathbf{j}) \times (2z\,\mathbf{j} + \mathbf{k}) = 4\mathbf{i} - \mathbf{j} + 2z\,\mathbf{k}$.

Note: In general, if $y = f(x, z)$ then $\mathbf{r}_x \times \mathbf{r}_z = \dfrac{\partial f}{\partial x}\mathbf{i} - \mathbf{j} + \dfrac{\partial f}{\partial z}\mathbf{k}$ and $A(S) = \iint_D \sqrt{1 + \left(\dfrac{\partial f}{\partial x}\right)^2 + \left(\dfrac{\partial f}{\partial z}\right)^2}\, dA$. Then

$$A(S) = \int_0^1 \int_0^1 \sqrt{17 + 4z^2}\, dx\, dz = \int_0^1 \sqrt{17 + 4z^2}\, dz$$

$$= \tfrac{1}{2}\left(z\sqrt{17 + 4z^2} + \tfrac{17}{2}\ln|2z + \sqrt{4z^2 + 17}|\right)\Big]_0^1 = \tfrac{\sqrt{21}}{2} + \tfrac{17}{4}\left[\ln\left(2 + \sqrt{21}\right) - \ln\sqrt{17}\right]$$

43. $\mathbf{r}_u = \langle v, 1, 1\rangle$, $\mathbf{r}_v = \langle u, 1, -1\rangle$ and $\mathbf{r}_u \times \mathbf{r}_v = \langle -2, u + v, v - u\rangle$. Then

$$A(S) = \iint_{u^2 + v^2 \le 1} \sqrt{4 + 2u^2 + 2v^2}\, dA = \int_0^{2\pi} \int_0^1 r\sqrt{4 + 2r^2}\, dr\, d\theta = \int_0^{2\pi} d\theta \int_0^1 r\sqrt{4 + 2r^2}\, dr$$

$$= 2\pi\left[\tfrac{1}{6}(4 + 2r^2)^{3/2}\right]_0^1 = \tfrac{\pi}{3}\left(6\sqrt{6} - 8\right) = \pi\left(2\sqrt{6} - \tfrac{8}{3}\right)$$

45. $z = f(x, y) = e^{-x^2 - y^2}$ with $x^2 + y^2 \le 4$.

$$A(S) = \iint_D \sqrt{1 + \left(-2xe^{-x^2-y^2}\right)^2 + \left(-2ye^{-x^2-y^2}\right)^2}\, dA = \iint_D \sqrt{1 + 4(x^2 + y^2)e^{-2(x^2+y^2)}}\, dA$$

$$= \int_0^{2\pi} \int_0^2 \sqrt{1 + 4r^2 e^{-2r^2}}\, r\, dr\, d\theta = \int_0^{2\pi} d\theta \int_0^2 r\sqrt{1 + 4r^2 e^{-2r^2}}\, dr = 2\pi \int_0^2 r\sqrt{1 + 4r^2 e^{-2r^2}}\, dr \approx 13.9783$$

47. (a) The midpoints of the four squares are $\left(\frac{1}{4},\frac{1}{4}\right)$, $\left(\frac{1}{4},\frac{3}{4}\right)$, $\left(\frac{3}{4},\frac{1}{4}\right)$, and $\left(\frac{3}{4},\frac{3}{4}\right)$. Here $f(x,y)=x^2+y^2$, so the Midpoint Rule gives

$$A(S)=\iint_D \sqrt{[f_x(x,y)]^2+[f_y(x,y)]^2+1}\,dA=\iint_D \sqrt{(2x)^2+(2y)^2+1}\,dA$$

$$\approx \tfrac{1}{4}\left(\sqrt{\left[2\left(\tfrac{1}{4}\right)\right]^2+\left[2\left(\tfrac{1}{4}\right)\right]^2+1}+\sqrt{\left[2\left(\tfrac{1}{4}\right)\right]^2+\left[2\left(\tfrac{3}{4}\right)\right]^2+1}\right.$$

$$\left.+\sqrt{\left[2\left(\tfrac{3}{4}\right)\right]^2+\left[2\left(\tfrac{1}{4}\right)\right]^2+1}+\sqrt{\left[2\left(\tfrac{3}{4}\right)\right]^2+\left[2\left(\tfrac{3}{4}\right)\right]^2+1}\right)$$

$$=\tfrac{1}{4}\left(\sqrt{\tfrac{3}{2}}+2\sqrt{\tfrac{7}{2}}+\sqrt{\tfrac{11}{2}}\right)\approx 1.8279$$

(b) A CAS estimates the integral to be $A(S)=\iint_D \sqrt{1+(2x)^2+(2y)^2}\,dA=\int_0^1\int_0^1 \sqrt{1+4x^2+4y^2}\,dy\,dx\approx 1.8616$.

This agrees with the Midpoint estimate only in the first decimal place.

49. $z=1+2x+3y+4y^2$, so

$$A(S)=\iint_D \sqrt{1+\left(\frac{\partial z}{\partial x}\right)^2+\left(\frac{\partial z}{\partial y}\right)^2}\,dA=\int_1^4\int_0^1 \sqrt{1+4+(3+8y)^2}\,dy\,dx=\int_1^4\int_0^1 \sqrt{14+48y+64y^2}\,dy\,dx.$$

Using a CAS, we have

$$\int_1^4\int_0^1 \sqrt{14+48y+64y^2}\,dy\,dx=\tfrac{45}{8}\sqrt{14}+\tfrac{15}{16}\ln\left(11\sqrt{5}+3\sqrt{14}\sqrt{5}\right)-\tfrac{15}{16}\ln\left(3\sqrt{5}+\sqrt{14}\sqrt{5}\right)$$

or $\tfrac{45}{8}\sqrt{14}+\tfrac{15}{16}\ln\frac{11\sqrt{5}+3\sqrt{70}}{3\sqrt{5}+\sqrt{70}}$.

51. (a) $x=a\sin u\cos v$, $y=b\sin u\sin v$, $z=c\cos u$ $\Rightarrow$

(b)

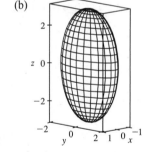

$$\frac{x^2}{a^2}+\frac{y^2}{b^2}+\frac{z^2}{c^2}=(\sin u\cos v)^2+(\sin u\sin v)^2+(\cos u)^2$$

$$=\sin^2 u+\cos^2 u=1$$

and since the ranges of u and v are sufficient to generate the entire graph,

the parametric equations represent an ellipsoid.

(c) From the parametric equations (with $a=1$, $b=2$, and $c=3$),

we calculate $\mathbf{r}_u=\cos u\cos v\,\mathbf{i}+2\cos u\sin v\,\mathbf{j}-3\sin u\,\mathbf{k}$ and

$\mathbf{r}_v=-\sin u\sin v\,\mathbf{i}+2\sin u\cos v\,\mathbf{j}$. So $\mathbf{r}_u\times\mathbf{r}_v=6\sin^2 u\cos v\,\mathbf{i}+3\sin^2 u\sin v\,\mathbf{j}+2\sin u\cos u\,\mathbf{k}$, and the surface

area is given by

$$A(S)=\int_0^{2\pi}\int_0^\pi |\mathbf{r}_u\times\mathbf{r}_v|\,du\,dv$$

$$=\int_0^{2\pi}\int_0^\pi \sqrt{36\sin^4 u\cos^2 v+9\sin^4 u\sin^2 v+4\cos^2 u\sin^2 u}\,du\,dv$$

53. To find the region D: $z = x^2 + y^2$ implies $z + z^2 = 4z$ or $z^2 - 3z = 0$. Thus $z = 0$ or $z = 3$ are the planes where the

surfaces intersect. But $x^2 + y^2 + z^2 = 4z$ implies $x^2 + y^2 + (z-2)^2 = 4$, so $z = 3$ intersects the upper hemisphere.

Thus $(z-2)^2 = 4 - x^2 - y^2$ or $z = 2 + \sqrt{4 - x^2 - y^2}$. Therefore D is the region inside the circle $x^2 + y^2 + (3-2)^2 = 4$,

that is, $D = \{(x, y) \mid x^2 + y^2 \le 3\}$.

$$A(S) = \iint_D \sqrt{1 + [(-x)(4 - x^2 - y^2)^{-1/2}]^2 + [(-y)(4 - x^2 - y^2)^{-1/2}]^2} \, dA$$

$$= \int_0^{2\pi} \int_0^{\sqrt{3}} \sqrt{1 + \frac{r^2}{4 - r^2}} \, r \, dr \, d\theta = \int_0^{2\pi} \int_0^{\sqrt{3}} \frac{2r \, dr}{\sqrt{4 - r^2}} \, d\theta = \int_0^{2\pi} \left[-2(4 - r^2)^{1/2} \right]_{r=0}^{r=\sqrt{3}} d\theta$$

$$= \int_0^{2\pi} (-2 + 4) \, d\theta = 2\theta \Big]_0^{2\pi} = 4\pi$$

55. If we revolve the curve $y = f(x)$, $a \le x \le b$ about the x-axis, where $f(x) \ge 0$, then from Equations 3 we know we can

parametrize the surface using $x = x$, $y = f(x) \cos\theta$, and $z = f(x) \sin\theta$, where $a \le x \le b$ and $0 \le \theta \le 2\pi$. Thus we can

say the surface is represented by $\mathbf{r}(x, \theta) = x\,\mathbf{i} + f(x) \cos\theta\,\mathbf{j} + f(x) \sin\theta\,\mathbf{k}$, with $a \le x \le b$ and $0 \le \theta \le 2\pi$. Then by (6),

the surface area is given by $A(S) = \iint_D |\mathbf{r}_x \times \mathbf{r}_\theta| \, dA$ where D is the rectangular parameter region $[a, b] \times [0, 2\pi]$. Here,

$\mathbf{r}_x(x, \theta) = \mathbf{i} + f'(x) \cos\theta\,\mathbf{j} + f'(x) \sin\theta\,\mathbf{k}$ and $\mathbf{r}_\theta(x) = -f(x) \sin\theta\,\mathbf{j} + f(x) \cos\theta\,\mathbf{k}$. So

$$\mathbf{r}_x \times \mathbf{r}_\theta = \begin{vmatrix} \mathbf{i} & \mathbf{j} & \mathbf{k} \\ 1 & f'(x) \cos\theta & f'(x) \sin\theta \\ 0 & -f(x) \sin\theta & f(x) \cos\theta \end{vmatrix} = \left[f(x) f'(x) \cos^2\theta + f(x) f'(x) \sin^2\theta \right] \mathbf{i} - f(x) \cos\theta\,\mathbf{j} - f(x) \sin\theta\,\mathbf{k}$$

$$= f(x) f'(x) \mathbf{i} - f(x) \cos\theta\,\mathbf{j} - f(x) \sin\theta\,\mathbf{k} \text{ and}$$

$$|\mathbf{r}_x \times \mathbf{r}_\theta| = \sqrt{[f(x) f'(x)]^2 + [f(x)]^2 \cos^2\theta + [f(x)]^2 \sin^2\theta}$$

$$= \sqrt{[f(x)]^2 \left([f'(x)]^2 + 1\right)} = f(x) \sqrt{1 + [f'(x)]^2} \text{ [since } f(x) \ge 0]. \text{ Thus}$$

$$A(S) = \iint_D |\mathbf{r}_x \times \mathbf{r}_\theta| \, dA = \int_a^b \int_0^{2\pi} f(x) \sqrt{1 + [f'(x)]^2} \, d\theta \, dx$$

$$= \int_a^b f(x) \sqrt{1 + [f'(x)]^2} \, [\theta]_0^{2\pi} \, dx = 2\pi \int_a^b f(x) \sqrt{1 + [f'(x)]^2} \, dx$$

57. $y = \sqrt{x} \;\Rightarrow\; 1 + \left(\dfrac{dy}{dx}\right)^2 = 1 + \left(\dfrac{1}{2\sqrt{x}}\right)^2 = 1 + \dfrac{1}{4x}$. So

$$S = \int_4^9 2\pi y \sqrt{1 + \left(\frac{dy}{dx}\right)^2} \, dx = \int_4^9 2\pi \sqrt{x} \sqrt{1 + \frac{1}{4x}} \, dx = 2\pi \int_4^9 \left(x + \tfrac{1}{4}\right) dx$$

$$= 2\pi \left[\tfrac{2}{3} \left(x + \tfrac{1}{4}\right)^{3/2} \right]_4^9 = \tfrac{4\pi}{3} \left[\tfrac{1}{8}(4x + 1)^{3/2} \right]_4^9 = \tfrac{\pi}{6} \left(37\sqrt{37} - 17\sqrt{17} \right)$$

13.7   Surface Integrals

1. Each face of the cube has surface area $2^2 = 4$, and the points P_{ij}^* are the points where the cube intersects the coordinate axes.

Here, $f(x, y, z) = \sqrt{x^2 + 2y^2 + 3z^2}$, so by Definition 1,

$$\iint_S f(x, y, z)\, dS \approx [f(1, 0, 0)](4) + [f(-1, 0, 0)](4) + [f(0, 1, 0)](4) + [f(0, -1, 0)](4)$$
$$+ [f(0, 0, 1)](4) + [f(0, 0, -1)](4)$$
$$= 4\left(1 + 1 + 2\sqrt{2} + 2\sqrt{3}\right) = 8\left(1 + \sqrt{2} + \sqrt{3}\right) \approx 33.170$$

3. We can use the xz- and yz-planes to divide H into four patches of equal size, each with surface area equal to $\frac{1}{8}$ the surface

area of a sphere with radius $\sqrt{50}$, so $\Delta S = \frac{1}{8}(4)\pi\left(\sqrt{50}\right)^2 = 25\pi$. Then $(\pm 3, \pm 4, 5)$ are sample points in the four patches,

and using a Riemann sum as in Definition 1, we have

$$\iint_H f(x, y, z)\, dS \approx f(3, 4, 5)\, \Delta S + f(3, -4, 5)\, \Delta S + f(-3, 4, 5)\, \Delta S + f(-3, -4, 5)\, \Delta S$$
$$= (7 + 8 + 9 + 12)(25\pi) = 900\pi \approx 2827$$

5. $\mathbf{r}(u, v) = u^2\,\mathbf{i} + u\sin v\,\mathbf{j} + u\cos v\,\mathbf{k}$, $0 \le u \le 1$, $0 \le v \le \pi/2$ and

$\mathbf{r}_u \times \mathbf{r}_v = (2u\,\mathbf{i} + \sin v\,\mathbf{j} + \cos v\,\mathbf{k}) \times (u\cos v\,\mathbf{j} - u\sin v\,\mathbf{k}) = -u\,\mathbf{i} + 2u^2\sin v\,\mathbf{j} + 2u^2\cos v\,\mathbf{k}$ and

$|\mathbf{r}_u \times \mathbf{r}_v| = \sqrt{u^2 + 4u^4\sin^2 v + 4u^4\cos^2 v} = \sqrt{u^2 + 4u^4(\sin^2 v + \cos^2 v)} = u\sqrt{1 + 4u^2}$ (since $u \ge 0$).

Then $\iint_S yz\, dS = \int_0^{\pi/2}\int_0^1 (u\sin v)(u\cos v) \cdot u\sqrt{1 + 4u^2}\, du\, dv = \int_0^1 u^3\sqrt{1 + 4u^2}\, du \int_0^{\pi/2}\sin v\cos v\, dv$

$$[\text{let } t = 1 + 4u^2 \quad \Rightarrow \quad u^2 = \tfrac{1}{4}(t - 1) \text{ and } \tfrac{1}{8}\, dt = u\, du]$$

$$= \int_1^5 \tfrac{1}{8} \cdot \tfrac{1}{4}(t - 1)\sqrt{t}\, dt \int_0^{\pi/2}\sin v\cos v\, dv = \tfrac{1}{32}\int_1^5\left(t^{3/2} - \sqrt{t}\right) dt \int_0^{\pi/2}\sin v\cos v\, dv$$

$$= \tfrac{1}{32}\left[\tfrac{2}{5}t^{5/2} - \tfrac{2}{3}t^{3/2}\right]_1^5 \left[\tfrac{1}{2}\sin^2 v\right]_0^{\pi/2} = \tfrac{1}{32}\left(\tfrac{2}{5}(5)^{5/2} - \tfrac{2}{3}(5)^{3/2} - \tfrac{2}{5} + \tfrac{2}{3}\right) \cdot \tfrac{1}{2}(1 - 0) = \tfrac{5}{48}\sqrt{5} + \tfrac{1}{240}$$

7. $z = 1 + 2x + 3y$ so $\dfrac{\partial z}{\partial x} = 2$ and $\dfrac{\partial z}{\partial y} = 3$. Then by Formula 2,

$$\iint_S x^2 yz\, dS = \iint_D x^2 yz\sqrt{\left(\dfrac{\partial z}{\partial x}\right)^2 + \left(\dfrac{\partial z}{\partial y}\right)^2 + 1}\, dA = \int_0^3\int_0^2 x^2 y(1 + 2x + 3y)\sqrt{4 + 9 + 1}\, dy\, dx$$

$$= \sqrt{14}\int_0^3\int_0^2 (x^2 y + 2x^3 y + 3x^2 y^2)\, dy\, dx = \sqrt{14}\int_0^3\left[\tfrac{1}{2}x^2 y^2 + x^3 y^2 + x^2 y^3\right]_{y=0}^{y=2}\, dx$$

$$= \sqrt{14}\int_0^3 (10x^2 + 4x^3)\, dx = \sqrt{14}\left[\tfrac{10}{3}x^3 + x^4\right]_0^3 = 171\sqrt{14}$$

9. S is the part of the plane $z = 1 - x - y$ over the region $D = \{(x, y) \mid 0 \le x \le 1, 0 \le y \le 1 - x\}$. Thus

$$\iint_S yz\, dS = \iint_D y(1 - x - y)\sqrt{(-1)^2 + (-1)^2 + 1}\, dA = \sqrt{3}\int_0^1\int_0^{1-x}\left(y - xy - y^2\right) dy\, dx$$

$$= \sqrt{3}\int_0^1\left[\tfrac{1}{2}y^2 - \tfrac{1}{2}xy^2 - \tfrac{1}{3}y^3\right]_{y=0}^{y=1-x}\, dx = \sqrt{3}\int_0^1 \tfrac{1}{6}(1 - x)^3\, dx = -\tfrac{\sqrt{3}}{24}(1 - x)^4\Big|_0^1 = \tfrac{\sqrt{3}}{24}$$

11. S is the portion of the cone $z^2 = x^2 + y^2$ for $1 \le z \le 3$, or equivalently, S is the part of the surface $z = \sqrt{x^2 + y^2}$ over the

region $D = \{(x, y) \mid 1 \le x^2 + y^2 \le 9\}$. Thus

$$\iint_S x^2 z^2 \, dS = \iint_D x^2(x^2 + y^2) \sqrt{\left(\frac{x}{\sqrt{x^2 + y^2}}\right)^2 + \left(\frac{y}{\sqrt{x^2 + y^2}}\right)^2 + 1} \, dA$$

$$= \iint_D x^2(x^2 + y^2) \sqrt{\frac{x^2 + y^2}{x^2 + y^2} + 1} \, dA = \iint_D \sqrt{2} \, x^2(x^2 + y^2) \, dA = \sqrt{2} \int_0^{2\pi} \int_1^3 (r \cos\theta)^2 (r^2) \, r \, dr \, d\theta$$

$$= \sqrt{2} \int_0^{2\pi} \cos^2\theta \, d\theta \int_1^3 r^5 \, dr = \sqrt{2} \left[\tfrac{1}{2}\theta + \tfrac{1}{4}\sin 2\theta\right]_0^{2\pi} \left[\tfrac{1}{6}r^6\right]_1^3 = \sqrt{2}\,(\pi) \cdot \tfrac{1}{6}(3^6 - 1) = \frac{364\sqrt{2}}{3}\pi$$

13. Using x and z as parameters, we have $\mathbf{r}(x, z) = x\,\mathbf{i} + (x^2 + z^2)\,\mathbf{j} + z\,\mathbf{k}$, $x^2 + z^2 \le 4$. Then

$$\mathbf{r}_x \times \mathbf{r}_z = (\mathbf{i} + 2x\,\mathbf{j}) \times (2z\,\mathbf{j} + \mathbf{k}) = 2x\,\mathbf{i} - \mathbf{j} + 2z\,\mathbf{k} \text{ and } |\mathbf{r}_x \times \mathbf{r}_z| = \sqrt{4x^2 + 1 + 4z^2} = \sqrt{1 + 4(x^2 + z^2)}. \text{ Thus}$$

$$\iint_S y \, dS = \iint_{x^2+z^2 \le 4} (x^2 + z^2)\sqrt{1 + 4(x^2 + z^2)} \, dA = \int_0^{2\pi} \int_0^2 r^2 \sqrt{1 + 4r^2}\, r \, dr \, d\theta$$

$$= \int_0^{2\pi} d\theta \int_0^2 r^2 \sqrt{1 + 4r^2}\, r \, dr = 2\pi \int_0^2 r^2 \sqrt{1 + 4r^2}\, r \, dr$$

$$[\text{let } u = 1 + 4r^2 \quad \Rightarrow \quad r^2 = \tfrac{1}{4}(u - 1) \text{ and } \tfrac{1}{8}\,du = r \, dr]$$

$$= 2\pi \int_1^{17} \tfrac{1}{4}(u - 1)\sqrt{u} \cdot \tfrac{1}{8}\,du = \tfrac{1}{16}\pi \int_1^{17} (u^{3/2} - u^{1/2})\, du$$

$$= \tfrac{1}{16}\pi \left[\tfrac{2}{5}u^{5/2} - \tfrac{2}{3}u^{3/2}\right]_1^{17} = \tfrac{1}{16}\pi\left[\tfrac{2}{5}(17)^{5/2} - \tfrac{2}{3}(17)^{3/2} - \tfrac{2}{5} + \tfrac{2}{3}\right] = \frac{\pi}{60}\left(391\sqrt{17} + 1\right)$$

15. Using spherical coordinates and Example 13.6.4 we have $\mathbf{r}(\phi, \theta) = 2\sin\phi\cos\theta\,\mathbf{i} + 2\sin\phi\sin\theta\,\mathbf{j} + 2\cos\phi\,\mathbf{k}$ and

$|\mathbf{r}_\phi \times \mathbf{r}_\theta| = 4\sin\phi$. Then $\iint_S (x^2 z + y^2 z)\, dS = \int_0^{2\pi} \int_0^{\pi/2} (4\sin^2\phi)(2\cos\phi)(4\sin\phi)\, d\phi\, d\theta = 16\pi\sin^4\phi\big]_0^{\pi/2} = 16\pi$.

17. Using cylindrical coordinates, we have $\mathbf{r}(\theta, z) = 3\cos\theta\,\mathbf{i} + 3\sin\theta\,\mathbf{j} + z\,\mathbf{k}$, $0 \le \theta \le 2\pi$, $0 \le z \le 2$,

and $|\mathbf{r}_\theta \times \mathbf{r}_z| = 3$.

$$\iint_S (x^2 y + z^2)\, dS = \int_0^{2\pi} \int_0^2 (27\cos^2\theta\sin\theta + z^2)\, 3\, dz\, d\theta = \int_0^{2\pi} (162\cos^2\theta\sin\theta + 8)\, d\theta = 16\pi$$

19. $\mathbf{F}(x, y, z) = xy\,\mathbf{i} + yz\,\mathbf{j} + zx\,\mathbf{k}$, $z = g(x, y) = 4 - x^2 - y^2$, and D is the square $[0, 1] \times [0, 1]$, so by Equation 10

$$\iint_S \mathbf{F} \cdot d\mathbf{S} = \iint_D [-xy(-2x) - yz(-2y) + zx]\, dA = \int_0^1 \int_0^1 [2x^2 y + 2y^2(4 - x^2 - y^2) + x(4 - x^2 - y^2)]\, dy\, dx$$

$$= \int_0^1 \left(\tfrac{1}{3}x^2 + \tfrac{11}{3}x - x^3 + \tfrac{34}{15}\right) dx = \frac{713}{180}$$

21. $\mathbf{F}(x, y, z) = xze^y\,\mathbf{i} - xze^y\,\mathbf{j} + z\,\mathbf{k}$, $z = g(x, y) = 1 - x - y$, and $D = \{(x, y) \mid 0 \le x \le 1, 0 \le y \le 1 - x\}$. Since S has

downward orientation, we have

$$\iint_S \mathbf{F} \cdot d\mathbf{S} = -\iint_D [-xze^y(-1) - (-xze^y)(-1) + z]\, dA = -\int_0^1 \int_0^{1-x} (1 - x - y)\, dy\, dx$$

$$= -\int_0^1 \left(\tfrac{1}{2}x^2 - x + \tfrac{1}{2}\right) dx = -\tfrac{1}{6}$$

23. $\mathbf{F}(x, y, z) = x\,\mathbf{i} - z\,\mathbf{j} + y\,\mathbf{k}$, $z = g(x, y) = \sqrt{4 - x^2 - y^2}$ and D is the quarter disk

$\{(x, y) \mid 0 \le x \le 2, 0 \le y \le \sqrt{4 - x^2}\}$. S has downward orientation, so by Formula 10,

$$\iint_S \mathbf{F} \cdot d\mathbf{S} = -\iint_D \left[-x \cdot \tfrac{1}{2}(4 - x^2 - y^2)^{-1/2}(-2x) - (-z) \cdot \tfrac{1}{2}(4 - x^2 - y^2)^{-1/2}(-2y) + y\right] dA$$

$$= -\iint_D \left(\frac{x^2}{\sqrt{4 - x^2 - y^2}} - \sqrt{4 - x^2 - y^2} \cdot \frac{y}{\sqrt{4 - x^2 - y^2}} + y\right) dA$$

$$= -\iint_D x^2(4 - (x^2 + y^2))^{-1/2}\, dA = -\int_0^{\pi/2} \int_0^2 (r\cos\theta)^2(4 - r^2)^{-1/2}\, r\, dr\, d\theta$$

$$= -\int_0^{\pi/2} \cos^2\theta\, d\theta \int_0^2 r^3(4 - r^2)^{-1/2}\, dr \quad [\text{let } u = 4 - r^2 \Rightarrow r^2 = 4 - u \text{ and } -\tfrac{1}{2}\, du = r\, dr]$$

$$= -\int_0^{\pi/2} \left(\tfrac{1}{2} + \tfrac{1}{2}\cos 2\theta\right) d\theta \int_4^0 -\tfrac{1}{2}(4 - u)(u)^{-1/2}\, du$$

$$= -\left[\tfrac{1}{2}\theta + \tfrac{1}{4}\sin 2\theta\right]_0^{\pi/2} \left(-\tfrac{1}{2}\right)\left[8\sqrt{u} - \tfrac{2}{3}u^{3/2}\right]_4^0 = -\tfrac{\pi}{4}\left(-\tfrac{1}{2}\right)\left(-16 + \tfrac{16}{3}\right) = -\tfrac{4}{3}\pi$$

25. Let S_1 be the paraboloid $y = x^2 + z^2$, $0 \le y \le 1$ and S_2 the disk $x^2 + z^2 \le 1$, $y = 1$. Since S is a closed surface, we use the outward orientation. On S_1: $\mathbf{F}(\mathbf{r}(x, z)) = (x^2 + z^2)\,\mathbf{j} - z\,\mathbf{k}$ and $\mathbf{r}_x \times \mathbf{r}_z = 2x\,\mathbf{i} - \mathbf{j} + 2z\,\mathbf{k}$ (since the $\mathbf{j}$-component must be negative on S_1). Then

$$\iint_{S_1} \mathbf{F} \cdot d\mathbf{S} = \iint_{x^2+z^2 \le 1} \left[-(x^2 + z^2) - 2z^2\right] dA = -\int_0^{2\pi} \int_0^1 (r^2 + 2r^2\cos^2\theta)\, r\, dr\, d\theta$$

$$= -\int_0^{2\pi} \tfrac{1}{4}(1 + 2\cos^2\theta)\, d\theta = -\left(\tfrac{\pi}{2} + \tfrac{\pi}{2}\right) = -\pi$$

On S_2: $\mathbf{F}(\mathbf{r}(x, z)) = \mathbf{j} - z\,\mathbf{k}$ and $\mathbf{r}_z \times \mathbf{r}_x = \mathbf{j}$. Then $\iint_{S_2} \mathbf{F} \cdot d\mathbf{S} = \iint_{x^2+z^2 \le 1} (1)\, dA = \pi$. Hence $\iint_S \mathbf{F} \cdot d\mathbf{S} = -\pi + \pi = 0$.

27. Here S consists of four surfaces: S_1, the top surface (a portion of the circular cylinder $y^2 + z^2 = 1$); S_2, the bottom surface (a portion of the xy-plane); S_3, the front half-disk in the plane $x = 2$, and S_4, the back half-disk in the plane $x = 0$.

On S_1: The surface is $z = \sqrt{1 - y^2}$ for $0 \le x \le 2$, $-1 \le y \le 1$ with upward orientation, so

$$\iint_{S_1} \mathbf{F} \cdot d\mathbf{S} = \int_0^2 \int_{-1}^1 \left[-x^2(0) - y^2\left(-\frac{y}{\sqrt{1 - y^2}}\right) + z^2\right] dy\, dx = \int_0^2 \int_{-1}^1 \left(\frac{y^3}{\sqrt{1 - y^2}} + 1 - y^2\right) dy\, dx$$

$$= \int_0^2 \left[-\sqrt{1 - y^2} + \tfrac{1}{3}(1 - y^2)^{3/2} + y - \tfrac{1}{3}y^3\right]_{y=-1}^{y=1} dx = \int_0^2 \tfrac{4}{3}\, dx = \tfrac{8}{3}$$

On S_2: The surface is $z = 0$ with downward orientation, so

$$\iint_{S_2} \mathbf{F} \cdot d\mathbf{S} = \int_0^2 \int_{-1}^1 (-z^2)\, dy\, dx = \int_0^2 \int_{-1}^1 (0)\, dy\, dx = 0$$

On S_3: The surface is $x = 2$ for $-1 \le y \le 1$, $0 \le z \le \sqrt{1 - y^2}$, oriented in the positive x-direction. Regarding y and z as parameters, we have $\mathbf{r}_y \times \mathbf{r}_z = \mathbf{i}$ and

$$\iint_{S_3} \mathbf{F} \cdot d\mathbf{S} = \int_{-1}^{1} \int_{0}^{\sqrt{1-y^2}} x^2 \, dz \, dy = \int_{-1}^{1} \int_{0}^{\sqrt{1-y^2}} 4 \, dz \, dy = 4A\,(S_3) = 2\pi$$

On S_4: The surface is $x = 0$ for $-1 \le y \le 1$, $0 \le z \le \sqrt{1-y^2}$, oriented in the negative x-direction. Regarding y and z as

parameters, we use $-(\mathbf{r}_y \times \mathbf{r}_z) = -\mathbf{i}$ and

$$\iint_{S_4} \mathbf{F} \cdot d\mathbf{S} = \int_{-1}^{1} \int_{0}^{\sqrt{1-y^2}} x^2 \, dz \, dy = \int_{-1}^{1} \int_{0}^{\sqrt{1-y^2}} (0) \, dz \, dy = 0$$

Thus $\iint_S \mathbf{F} \cdot d\mathbf{S} = \frac{8}{3} + 0 + 2\pi + 0 = 2\pi + \frac{8}{3}$.

29. We use Formula 4 with $z = 3 - 2x^2 - y^2$ $\Rightarrow$ $\partial z/\partial x = -4x$, $\partial z/\partial y = -2y$. The boundaries of the region

$3 - 2x^2 - y^2 \ge 0$ are $-\sqrt{\frac{3}{2}} \le x \le \sqrt{\frac{3}{2}}$ and $-\sqrt{3-2x^2} \le y \le \sqrt{3-2x^2}$, so we use a CAS (with precision reduced to

seven or fewer digits; otherwise the calculation takes a very long time) to calculate

$$\iint_S x^2 y^2 z^2 \, dS = \int_{-\sqrt{3/2}}^{\sqrt{3/2}} \int_{-\sqrt{3-2x^2}}^{\sqrt{3-2x^2}} x^2 y^2 (3 - 2x^2 - y^2)^2 \sqrt{16x^2 + 4y^2 + 1} \, dy \, dx \approx 3.4895$$

31. If S is given by $y = h(x, z)$, then S is also the level surface $f(x, y, z) = y - h(x, z) = 0$.

$\mathbf{n} = \dfrac{\nabla f(x,y,z)}{|\nabla f(x,y,z)|} = \dfrac{-h_x \mathbf{i} + \mathbf{j} - h_z \mathbf{k}}{\sqrt{h_x^2 + 1 + h_z^2}}$, and $-\mathbf{n}$ is the unit normal that points to the left. Now we proceed as in the

derivation of (10), using Formula 4 to evaluate

$$\iint_S \mathbf{F} \cdot d\mathbf{S} = \iint_S \mathbf{F} \cdot \mathbf{n} \, dS = \iint_D (P\mathbf{i} + Q\mathbf{j} + R\mathbf{k}) \, \frac{\dfrac{\partial h}{\partial x}\mathbf{i} - \mathbf{j} + \dfrac{\partial h}{\partial z}\mathbf{k}}{\sqrt{\left(\dfrac{\partial h}{\partial x}\right)^2 + 1 + \left(\dfrac{\partial h}{\partial z}\right)^2}} \sqrt{\left(\dfrac{\partial h}{\partial x}\right)^2 + 1 + \left(\dfrac{\partial h}{\partial z}\right)^2} \, dA$$

where D is the projection of $\mathbf{S}$ onto the xz-plane. Therefore $\displaystyle\iint_S \mathbf{F} \cdot d\mathbf{S} = \iint_D \left(P\frac{\partial h}{\partial x} - Q + R\frac{\partial h}{\partial z}\right) dA$.

33. $m = \iint_S K \, dS = K \cdot 4\pi\left(\frac{1}{2}a^2\right) = 2\pi a^2 K$; by symmetry $M_{xz} = M_{yz} = 0$, and

$M_{xy} = \iint_S zK \, dS = K\int_0^{2\pi}\int_0^{\pi/2}(a\cos\phi)(a^2\sin\phi)\,d\phi\,d\theta = 2\pi Ka^3\left[-\frac{1}{4}\cos 2\phi\right]_0^{\pi/2} = \pi Ka^3$.

Hence $(\overline{x}, \overline{y}, \overline{z}) = \left(0, 0, \frac{1}{2}a\right)$.

35. (a) $I_z = \iint_S (x^2 + y^2)\rho(x, y, z)\, dS$

(b) $I_z = \iint_S (x^2 + y^2)\left(10 - \sqrt{x^2 + y^2}\right) dS = \iint\limits_{1 \le x^2 + y^2 \le 16} (x^2 + y^2)\left(10 - \sqrt{x^2 + y^2}\right)\sqrt{2}\, dA$

$\quad = \int_0^{2\pi}\int_1^4 \sqrt{2}\,(10r^3 - r^4)\, dr\, d\theta = 2\sqrt{2}\,\pi\left(\frac{4329}{10}\right) = \frac{4329}{5}\sqrt{2}\,\pi$

37. The rate of flow through the cylinder is the flux $\iint_S \rho \mathbf{v} \cdot \mathbf{n}\, dS = \iint_S \rho \mathbf{v} \cdot d\mathbf{S}$. We use the parametric representation

$\mathbf{r}(u, v) = 2\cos u\, \mathbf{i} + 2\sin u\, \mathbf{j} + v\, \mathbf{k}$ for S, where $0 \le u \le 2\pi$, $0 \le v \le 1$, so $\mathbf{r}_u = -2\sin u\, \mathbf{i} + 2\cos u\, \mathbf{j}$, $\mathbf{r}_v = \mathbf{k}$, and the

outward orientation is given by $\mathbf{r}_u \times \mathbf{r}_v = 2\cos u\, \mathbf{i} + 2\sin u\, \mathbf{j}$. Then

$$\iint_S \rho \mathbf{v} \cdot d\mathbf{S} = \rho \int_0^{2\pi} \int_0^1 \left(v\, \mathbf{i} + 4\sin^2 u\, \mathbf{j} + 4\cos^2 u\, \mathbf{k} \right) \cdot \left(2\cos u\, \mathbf{i} + 2\sin u\, \mathbf{j} \right) dv\, du$$

$$= \rho \int_0^{2\pi} \int_0^1 \left(2v\cos u + 8\sin^3 u \right) dv\, du = \rho \int_0^{2\pi} \left(\cos u + 8\sin^3 u \right) du$$

$$= \rho \left[\sin u + 8\left(-\tfrac{1}{3}\right)\left(2 + \sin^2 u\right)\cos u \right]_0^{2\pi} = 0 \text{ kg/s}$$

39. S consists of the hemisphere S_1 given by $z = \sqrt{a^2 - x^2 - y^2}$ and the disk S_2 given by $0 \le x^2 + y^2 \le a^2$, $z = 0$. On S_1:

$\mathbf{E} = a\sin\phi\cos\theta\, \mathbf{i} + a\sin\phi\sin\theta\, \mathbf{j} + 2a\cos\phi\, \mathbf{k}$, $\mathbf{T}_\phi \times \mathbf{T}_\theta = a^2\sin^2\phi\cos\theta\, \mathbf{i} + a^2\sin^2\phi\sin\theta\, \mathbf{j} + a^2\sin\phi\cos\phi\, \mathbf{k}$. Thus

$$\iint_{S_1} \mathbf{E} \cdot d\mathbf{S} = \int_0^{2\pi} \int_0^{\pi/2} (a^3\sin^3\phi + 2a^3\sin\phi\cos^2\phi)\, d\phi\, d\theta$$

$$= \int_0^{2\pi} \int_0^{\pi/2} (a^3\sin\phi + a^3\sin\phi\cos^2\phi)\, d\phi\, d\theta = (2\pi)a^3\left(1 + \tfrac{1}{3}\right) = \tfrac{8}{3}\pi a^3$$

On S_2: $\mathbf{E} = x\, \mathbf{i} + y\, \mathbf{j}$, and $\mathbf{r}_y \times \mathbf{r}_x = -\mathbf{k}$ so $\iint_{S_2} \mathbf{E} \cdot d\mathbf{S} = 0$. Hence the total charge is $q = \varepsilon_0 \iint_S \mathbf{E} \cdot d\mathbf{S} = \tfrac{8}{3}\pi a^3 \varepsilon_0$.

41. $K\nabla u = 6.5(4y\, \mathbf{j} + 4z\, \mathbf{k})$. S is given by $\mathbf{r}(x, \theta) = x\, \mathbf{i} + \sqrt{6}\cos\theta\, \mathbf{j} + \sqrt{6}\sin\theta\, \mathbf{k}$ and since we want the inward heat flow, we

use $\mathbf{r}_x \times \mathbf{r}_\theta = -\sqrt{6}\cos\theta\, \mathbf{j} - \sqrt{6}\sin\theta\, \mathbf{k}$. Then the rate of heat flow inward is given by

$$\iint_S (-K\,\nabla u) \cdot d\mathbf{S} = \int_0^{2\pi} \int_0^4 -(6.5)(-24)\, dx\, d\theta = (2\pi)(156)(4) = 1248\pi.$$

43. Let S be a sphere of radius a centered at the origin. Then $|\mathbf{r}| = a$ and $\mathbf{F}(\mathbf{r}) = c\mathbf{r}/|\mathbf{r}|^3 = (c/a^3)(x\, \mathbf{i} + y\, \mathbf{j} + z\, \mathbf{k})$. A

parametric representation for S is $\mathbf{r}(\phi, \theta) = a\sin\phi\cos\theta\, \mathbf{i} + a\sin\phi\sin\theta\, \mathbf{j} + a\cos\phi\, \mathbf{k}$, $0 \le \phi \le \pi$, $0 \le \theta \le 2\pi$. Then

$\mathbf{r}_\phi = a\cos\phi\cos\theta\, \mathbf{i} + a\cos\phi\sin\theta\, \mathbf{j} - a\sin\phi\, \mathbf{k}$, $\mathbf{r}_\theta = -a\sin\phi\sin\theta\, \mathbf{i} + a\sin\phi\cos\theta\, \mathbf{j}$, and the outward orientation is given

by $\mathbf{r}_\phi \times \mathbf{r}_\theta = a^2\sin^2\phi\cos\theta\, \mathbf{i} + a^2\sin^2\phi\sin\theta\, \mathbf{j} + a^2\sin\phi\cos\phi\, \mathbf{k}$. The flux of $\mathbf{F}$ across S is

$$\iint_S \mathbf{F} \cdot d\mathbf{S} = \int_0^{\pi} \int_0^{2\pi} \frac{c}{a^3} \left(a\sin\phi\cos\theta\, \mathbf{i} + a\sin\phi\sin\theta\, \mathbf{j} + a\cos\phi\, \mathbf{k} \right)$$

$$\cdot \left(a^2\sin^2\phi\cos\theta\, \mathbf{i} + a^2\sin^2\phi\sin\theta\, \mathbf{j} + a^2\sin\phi\cos\phi\, \mathbf{k} \right) d\theta\, d\phi$$

$$= \frac{c}{a^3} \int_0^{\pi} \int_0^{2\pi} a^3 \left(\sin^3\phi + \sin\phi\cos^2\phi \right) d\theta\, d\phi = c \int_0^{\pi} \int_0^{2\pi} \sin\phi\, d\theta\, d\phi = 4\pi c$$

Thus the flux does not depend on the radius a.

13.8 Stokes' Theorem

1. The boundary curve C is the circle $x^2 + y^2 = 4$, $z = 0$ oriented in the counterclockwise direction. The vector

   equation is $\mathbf{r}(t) = 2\cos t\,\mathbf{i} + 2\sin t\,\mathbf{j}$, $0 \le t \le 2\pi$, so $\mathbf{r}'(t) = -2\sin t\,\mathbf{i} + 2\cos t\,\mathbf{j}$ and

   $\mathbf{F}(\mathbf{r}(t)) = (2\cos t)^2 e^{(2\sin t)(0)}\,\mathbf{i} + (2\sin t)^2 e^{(2\cos t)(0)}\,\mathbf{j} + (0)^2 e^{(2\cos t)(2\sin t)}\,\mathbf{k} = 4\cos^2 t\,\mathbf{i} + 4\sin^2 t\,\mathbf{j}$. Then, by Stokes'

   Theorem,

   $$\iint_S \operatorname{curl} \mathbf{F} \cdot d\mathbf{S} = \int_C \mathbf{F} \cdot d\mathbf{r} = \int_0^{2\pi} \mathbf{F}(\mathbf{r}(t)) \cdot \mathbf{r}'(t)\, dt = \int_0^{2\pi} (-8\cos^2 t \sin t + 8\sin^2 t \cos t)\, dt$$
   $$= 8\left[\tfrac{1}{3}\cos^3 t + \tfrac{1}{3}\sin^3 t\right]_0^{2\pi} = 0$$

3. C is the square in the plane $z = -1$. By (3), $\iint_{S_1} \operatorname{curl} \mathbf{F} \cdot d\mathbf{S} = \oint_C \mathbf{F} \cdot d\mathbf{r} = \iint_{S_2} \operatorname{curl} \mathbf{F} \cdot d\mathbf{S}$ where S_1 is the original cube

   without the bottom and S_2 is the bottom face of the cube. $\operatorname{curl} \mathbf{F} = x^2 z\,\mathbf{i} + (xy - 2xyz)\,\mathbf{j} + (y - xz)\,\mathbf{k}$. For S_2, we choose

   $\mathbf{n} = \mathbf{k}$ so that C has the same orientation for both surfaces. Then $\operatorname{curl} \mathbf{F} \cdot \mathbf{n} = y - xz = x + y$ on S_2, where $z = -1$. Thus

   $\iint_{S_2} \operatorname{curl} \mathbf{F} \cdot d\mathbf{S} = \int_{-1}^{1} \int_{-1}^{1} (x + y)\, dx\, dy = 0$ so $\iint_{S_1} \operatorname{curl} \mathbf{F} \cdot d\mathbf{S} = 0$.

5. $\operatorname{curl} \mathbf{F} = -2z\,\mathbf{i} - 2x\,\mathbf{j} - 2y\,\mathbf{k}$ and we take the surface S to be the planar region enclosed by C, so S is the portion of the plane

   $x + y + z = 1$ over $D = \{(x, y) \mid 0 \le x \le 1, 0 \le y \le 1 - x\}$. Since C is oriented counterclockwise, we orient S upward.

   Using Equation 13.7.10, we have $z = g(x, y) = 1 - x - y$, $P = -2z$, $Q = -2x$, $R = -2y$, and

   $$\int_C \mathbf{F} \cdot d\mathbf{r} = \iint_S \operatorname{curl} \mathbf{F} \cdot d\mathbf{S} = \iint_D [-(-2z)(-1) - (-2x)(-1) + (-2y)]\, dA$$
   $$= \int_0^1 \int_0^{1-x} (-2)\, dy\, dx = -2 \int_0^1 (1 - x)\, dx = -1$$

7. $\operatorname{curl} \mathbf{F} = (xe^{xy} - 2x)\,\mathbf{i} - (ye^{xy} - y)\,\mathbf{j} + (2z - z)\,\mathbf{k}$ and we take S to be the disk $x^2 + y^2 \le 16$, $z = 5$. Since C is oriented

   counterclockwise (from above), we orient S upward. Then $\mathbf{n} = \mathbf{k}$ and $\operatorname{curl} \mathbf{F} \cdot \mathbf{n} = 2z - z$ on S, where $z = 5$. Thus

   $\oint \mathbf{F} \cdot d\mathbf{r} = \iint_S \operatorname{curl} \mathbf{F} \cdot \mathbf{n}\, dS = \iint_S (2z - z)\, dS = \iint_S (10 - 5)\, dS = 5(\text{area of } S) = 5(\pi \cdot 4^2) = 80\pi$.

9. (a) The curve of intersection is an ellipse in the plane $x + y + z = 1$ with unit normal $\mathbf{n} = \frac{1}{\sqrt{3}}(\mathbf{i} + \mathbf{j} + \mathbf{k})$,

   $\operatorname{curl} \mathbf{F} = x^2\,\mathbf{j} + y^2\,\mathbf{k}$, and $\operatorname{curl} \mathbf{F} \cdot \mathbf{n} = \frac{1}{\sqrt{3}}(x^2 + y^2)$. Then

   $$\oint_C \mathbf{F} \cdot d\mathbf{r} = \iint_S \frac{1}{\sqrt{3}}(x^2 + y^2)\, dS = \iint_{x^2 + y^2 \le 9} (x^2 + y^2)\, dx\, dy = \int_0^{2\pi} \int_0^3 r^3\, dr\, d\theta = 2\pi\left(\tfrac{81}{4}\right) = \tfrac{81\pi}{2}$$

(b)

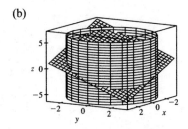

(c) One possible parametrization is $x = 3\cos t$, $y = 3\sin t$,

   $z = 1 - 3\cos t - 3\sin t$, $0 \le t \le 2\pi$.

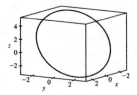

11. The boundary curve C is the circle $x^2 + y^2 = 1$, $z = 1$ oriented in the counterclockwise direction as viewed from above.

We can parametrize C by $\mathbf{r}(t) = \cos t \,\mathbf{i} + \sin t \,\mathbf{j} + \mathbf{k}$, $0 \le t \le 2\pi$, and then $\mathbf{r}'(t) = -\sin t \,\mathbf{i} + \cos t \,\mathbf{j}$. Thus

$\mathbf{F}(\mathbf{r}(t)) = \sin^2 t \,\mathbf{i} + \cos t \,\mathbf{j} + \mathbf{k}$, $\mathbf{F}(\mathbf{r}(t)) \cdot \mathbf{r}'(t) = \cos^2 t - \sin^3 t$, and

$$\oint_C \mathbf{F} \cdot d\mathbf{r} = \int_0^{2\pi} (\cos^2 t - \sin^3 t)\, dt = \int_0^{2\pi} \tfrac{1}{2}(1 + \cos 2t)\, dt - \int_0^{2\pi} (1 - \cos^2 t)\sin t \, dt$$
$$= \tfrac{1}{2}\left[t + \tfrac{1}{2}\sin 2t\right]_0^{2\pi} - \left[-\cos t + \tfrac{1}{3}\cos^3 t\right]_0^{2\pi} = \pi$$

Now curl $\mathbf{F} = (1 - 2y)\,\mathbf{k}$, and the projection D of S on the xy-plane is the disk

$x^2 + y^2 \le 1$, so by Equation 13.7.10 with $z = g(x, y) = x^2 + y^2$ we have

$$\iint_S \text{curl } \mathbf{F} \cdot d\mathbf{S} = \iint_D (1 - 2y)\, dA = \int_0^{2\pi} \int_0^1 (1 - 2r\sin\theta)\, r \, dr \, d\theta = \int_0^{2\pi} \left(\tfrac{1}{2} - \tfrac{2}{3}\sin\theta\right) d\theta = \pi.$$

13. It is easier to use Stokes' Theorem than to compute the work directly. Let S be the planar region enclosed by the path of the

particle, so S is the portion of the plane $z = \tfrac{1}{2}y$ for $0 \le x \le 1$, $0 \le y \le 2$, with upward orientation.

curl $\mathbf{F} = 8y\,\mathbf{i} + 2z\,\mathbf{j} + 2y\,\mathbf{k}$ and

$$\oint_C \mathbf{F} \cdot d\mathbf{r} = \iint_S \text{curl } \mathbf{F} \cdot d\mathbf{S} = \iint_D \left[-8y\,(0) - 2z\left(\tfrac{1}{2}\right) + 2y\right] dA = \int_0^1 \int_0^2 \left(2y - \tfrac{1}{2}y\right) dy \, dx$$
$$= \int_0^1 \int_0^2 \tfrac{3}{2}y \, dy \, dx = \int_0^1 \left[\tfrac{3}{4}y^2\right]_{y=0}^{y=2} dx = \int_0^1 3\, dx = 3$$

15. Assume S is centered at the origin with radius a and let H_1 and H_2 be the upper and lower hemispheres, respectively, of S.

Then $\iint_S \text{curl } \mathbf{F} \cdot d\mathbf{S} = \iint_{H_1} \text{curl } \mathbf{F} \cdot d\mathbf{S} + \iint_{H_2} \text{curl } \mathbf{F} \cdot d\mathbf{S} = \oint_{C_1} \mathbf{F} \cdot d\mathbf{r} + \oint_{C_2} \mathbf{F} \cdot d\mathbf{r}$ by Stokes' Theorem. But C_1 is the

circle $x^2 + y^2 = a^2$ oriented in the counterclockwise direction while C_2 is the same circle oriented in the clockwise direction.

Hence $\oint_{C_2} \mathbf{F} \cdot d\mathbf{r} = -\oint_{C_1} \mathbf{F} \cdot d\mathbf{r}$ so $\iint_S \text{curl } \mathbf{F} \cdot d\mathbf{S} = 0$ as desired.

13.9   The Divergence Theorem

1. div $\mathbf{F} = 3 + x + 2x = 3 + 3x$, so

$\iiint_E \text{div } \mathbf{F} \, dV = \int_0^1 \int_0^1 \int_0^1 (3x + 3)\, dx \, dy \, dz = \tfrac{9}{2}$ (notice the triple integral is

three times the volume of the cube plus three times $\bar{x}$).

To compute $\iint_S \mathbf{F} \cdot d\mathbf{S}$, on S_1: $\mathbf{n} = \mathbf{i}$, $\mathbf{F} = 3\,\mathbf{i} + y\,\mathbf{j} + 2z\,\mathbf{k}$, and

$\iint_{S_1} \mathbf{F} \cdot d\mathbf{S} = \iint_{S_1} 3\, dS = 3$;

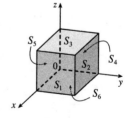

S_2: $\mathbf{F} = 3x\,\mathbf{i} + x\,\mathbf{j} + 2xz\,\mathbf{k}$, $\mathbf{n} = \mathbf{j}$ and $\iint_{S_2} \mathbf{F} \cdot d\mathbf{S} = \iint_{S_2} x \, dS = \tfrac{1}{2}$;

S_3: $\mathbf{F} = 3x\,\mathbf{i} + xy\,\mathbf{j} + 2x\,\mathbf{k}$, $\mathbf{n} = \mathbf{k}$ and $\iint_{S_3} \mathbf{F} \cdot d\mathbf{S} = \iint_{S_3} 2x \, dS = 1$;

S_4: $\mathbf{F} = \mathbf{0}$, $\iint_{S_4} \mathbf{F} \cdot d\mathbf{S} = 0$; S_5: $\mathbf{F} = 3x\,\mathbf{i} + 2x\,\mathbf{k}$, $\mathbf{n} = -\mathbf{j}$ and $\iint_{S_5} \mathbf{F} \cdot d\mathbf{S} = \iint_{S_5} 0\, dS = 0$;

S_6: $\mathbf{F} = 3x\,\mathbf{i} + xy\,\mathbf{j}$, $\mathbf{n} = -\mathbf{k}$ and $\iint_{S_6} \mathbf{F} \cdot d\mathbf{S} = \iint_{S_6} 0\, dS = 0$. Thus $\iint_S \mathbf{F} \cdot d\mathbf{S} = \tfrac{9}{2}$.

3. div $\mathbf{F} = x + y + z$, so

$$\iiint_E \text{div } \mathbf{F}\, dV = \int_0^{2\pi} \int_0^1 \int_0^1 (r\cos\theta + r\sin\theta + z)\, r\, dz\, dr\, d\theta = \int_0^{2\pi} \int_0^1 \left(r^2\cos\theta + r^2\sin\theta + \tfrac{1}{2}r\right) dr\, d\theta$$
$$= \int_0^{2\pi} \left(\tfrac{1}{3}\cos\theta + \tfrac{1}{3}\sin\theta + \tfrac{1}{4}\right) d\theta = \tfrac{1}{4}(2\pi) = \tfrac{\pi}{2}$$

Let S_1 be the top of the cylinder, S_2 the bottom, and S_3 the vertical edge. On S_1, $z = 1$, $\mathbf{n} = \mathbf{k}$, and $\mathbf{F} = xy\,\mathbf{i} + y\,\mathbf{j} + x\,\mathbf{k}$, so

$$\iint_{S_1} \mathbf{F} \cdot d\mathbf{S} = \iint_{S_1} \mathbf{F} \cdot \mathbf{n}\, dS = \iint_{S_1} x\, dS = \int_0^{2\pi} \int_0^1 (r\cos\theta)\, r\, dr\, d\theta = \left[\sin\theta\right]_0^{2\pi} \left[\tfrac{1}{3}r^3\right]_0^1 = 0. \text{ On } S_2, z = 0, \mathbf{n} = -\mathbf{k}, \text{ and}$$

$\mathbf{F} = xy\,\mathbf{i}$ so $\iint_{S_2} \mathbf{F} \cdot d\mathbf{S} = \iint_{S_2} 0\, dS = 0$. S_3 is given by $\mathbf{r}(\theta, z) = \cos\theta\,\mathbf{i} + \sin\theta\,\mathbf{j} + z\,\mathbf{k}$, $0 \le \theta \le 2\pi$, $0 \le z \le 1$. Then

$\mathbf{r}_\theta \times \mathbf{r}_z = \cos\theta\,\mathbf{i} + \sin\theta\,\mathbf{j}$ and

$$\iint_{S_3} \mathbf{F} \cdot d\mathbf{S} = \iint_D \mathbf{F} \cdot (\mathbf{r}_\theta \times \mathbf{r}_z)\, dA = \int_0^{2\pi} \int_0^1 (\cos^2\theta\sin\theta + z\sin^2\theta)\, dz\, d\theta$$
$$= \int_0^{2\pi} \left(\cos^2\theta\sin\theta + \tfrac{1}{2}\sin^2\theta\right) d\theta = \left[-\tfrac{1}{3}\cos^3\theta + \tfrac{1}{4}\left(\theta - \tfrac{1}{2}\sin 2\theta\right)\right]_0^{2\pi} = \tfrac{\pi}{2}$$

Thus $\iint_S \mathbf{F} \cdot d\mathbf{S} = 0 + 0 + \tfrac{\pi}{2} = \tfrac{\pi}{2}$.

5. div $\mathbf{F} = \frac{\partial}{\partial x}(e^x\sin y) + \frac{\partial}{\partial y}(e^x\cos y) + \frac{\partial}{\partial z}(yz^2) = e^x\sin y - e^x\sin y + 2yz = 2yz$, so by the Divergence Theorem,

$$\iint_S \mathbf{F} \cdot d\mathbf{S} = \iiint_E \text{div } \mathbf{F}\, dV = \int_0^1 \int_0^1 \int_0^2 2yz\, dz\, dy\, dx = 2\int_0^1 dx \int_0^1 y\, dy \int_0^1 z\, dz = 2\left[x\right]_0^1 \left[\tfrac{1}{2}y^2\right]_0^1 \left[\tfrac{1}{2}z^2\right]_0^2 = 2.$$

7. div $\mathbf{F} = 3y^2 + 0 + 3z^2$, so using cylindrical coordinates with $y = r\cos\theta$, $z = r\sin\theta$, $x = x$ we have

$$\iint_S \mathbf{F} \cdot d\mathbf{S} = \iiint_E (3y^2 + 3z^2)\, dV = \int_0^{2\pi} \int_0^1 \int_{-1}^2 (3r^2\cos^2\theta + 3r^2\sin^2\theta)\, r\, dx\, dr\, d\theta$$
$$= 3\int_0^{2\pi} d\theta \int_0^1 r^3\, dr \int_{-1}^2 dx = 3(2\pi)\left(\tfrac{1}{4}\right)(3) = \tfrac{9\pi}{2}$$

9. div $\mathbf{F} = y\sin z + 0 - y\sin z = 0$, so by the Divergence Theorem, $\iint_S \mathbf{F} \cdot d\mathbf{S} = \iiint_E 0\, dV = 0$.

11. div $\mathbf{F} = y^2 + 0 + x^2 = x^2 + y^2$ so

$$\iint_S \mathbf{F} \cdot d\mathbf{S} = \iiint_E (x^2 + y^2)\, dV = \int_0^{2\pi} \int_0^2 \int_{r^2}^4 r^2 \cdot r\, dz\, dr\, d\theta = \int_0^{2\pi} \int_0^2 r^3(4 - r^2)\, dr\, d\theta$$
$$= \int_0^{2\pi} d\theta \int_0^2 (4r^3 - r^5)\, dr = 2\pi\left[r^4 - \tfrac{1}{6}r^6\right]_0^2 = \tfrac{32}{3}\pi$$

13. div $\mathbf{F} = 12x^2z + 12y^2z + 12z^3$ so

$$\iint_S \mathbf{F} \cdot d\mathbf{S} = \iiint_E 12z(x^2 + y^2 + z^2)\, dV = \int_0^{2\pi} \int_0^\pi \int_0^R 12(\rho\cos\phi)(\rho^2)\rho^2\sin\phi\, d\rho\, d\phi\, d\theta$$
$$= 12\int_0^{2\pi} d\theta \int_0^\pi \sin\phi\cos\phi\, d\phi \int_0^R \rho^5\, d\rho = 12(2\pi)\left[\tfrac{1}{2}\sin^2\phi\right]_0^\pi \left[\tfrac{1}{6}\rho^6\right]_0^R = 0$$

15. $\iint_S \mathbf{F} \cdot d\mathbf{S} = \iiint_E \sqrt{3 - x^2}\, dV = \int_{-1}^1 \int_{-1}^1 \int_0^{2 - x^4 - y^4} \sqrt{3 - x^2}\, dz\, dy\, dx = \tfrac{341}{60}\sqrt{2} + \tfrac{81}{20}\sin^{-1}\left(\tfrac{\sqrt{3}}{3}\right)$

17. For S_1 we have $\mathbf{n} = -\mathbf{k}$, so $\mathbf{F} \cdot \mathbf{n} = \mathbf{F} \cdot (-\mathbf{k}) = -x^2z - y^2 = -y^2$ (since $z = 0$ on S_1). So if D is the unit disk, we get

$\iint_{S_1} \mathbf{F} \cdot d\mathbf{S} = \iint_{S_1} \mathbf{F} \cdot \mathbf{n}\, dS = \iint_D (-y^2)\, dA = -\int_0^{2\pi} \int_0^1 r^2(\sin^2\theta)\, r\, dr\, d\theta = -\tfrac{1}{4}\pi$. Now since S_2 is closed, we can use

the Divergence Theorem. Since div $\mathbf{F} = \frac{\partial}{\partial x}(z^2x) + \frac{\partial}{\partial y}\left(\tfrac{1}{3}y^3 + \tan z\right) + \frac{\partial}{\partial z}(x^2z + y^2) = z^2 + y^2 + x^2$, we use spherical

coordinates to get $\iint_{S_2} \mathbf{F} \cdot d\mathbf{S} = \iiint_E \text{div } \mathbf{F}\, dV = \int_0^{2\pi} \int_0^{\pi/2} \int_0^1 \rho^2 \cdot \rho^2\sin\phi\, d\rho\, d\phi\, d\theta = \tfrac{2}{5}\pi$. Finally

$\iint_S \mathbf{F} \cdot d\mathbf{S} = \iint_{S_2} \mathbf{F} \cdot d\mathbf{S} - \iint_{S_1} \mathbf{F} \cdot d\mathbf{S} = \tfrac{2}{5}\pi - \left(-\tfrac{1}{4}\pi\right) = \tfrac{13}{20}\pi$.

19. The vectors that end near P_1 are longer than the vectors that start near P_1, so the net flow is inward near P_1 and div $\mathbf{F}(P_1)$ is negative. The vectors that end near P_2 are shorter than the vectors that start near P_2, so the net flow is outward near P_2 and div $\mathbf{F}(P_2)$ is positive.

21.

From the graph it appears that for points above the x-axis, vectors starting near a particular point are longer than vectors ending there, so divergence is positive. The opposite is true at points below the x-axis, where divergence is negative.

$\mathbf{F}(x, y) = \langle xy, x + y^2 \rangle \quad \Rightarrow$

div $\mathbf{F} = \frac{\partial}{\partial x}(xy) + \frac{\partial}{\partial y}(x + y^2) = y + 2y = 3y$. Thus div $\mathbf{F} > 0$ for $y > 0$, and div $\mathbf{F} < 0$ for $y < 0$.

23. Since $\dfrac{\mathbf{x}}{|\mathbf{x}|^3} = \dfrac{x\,\mathbf{i} + y\,\mathbf{j} + z\,\mathbf{k}}{(x^2 + y^2 + z^2)^{3/2}}$ and $\dfrac{\partial}{\partial x}\left(\dfrac{x}{(x^2 + y^2 + z^2)^{3/2}}\right) = \dfrac{(x^2 + y^2 + z^2) - 3x^2}{(x^2 + y^2 + z^2)^{5/2}}$ with similar expressions for

$\dfrac{\partial}{\partial y}\left(\dfrac{y}{(x^2 + y^2 + z^2)^{3/2}}\right)$ and $\dfrac{\partial}{\partial z}\left(\dfrac{z}{(x^2 + y^2 + z^2)^{3/2}}\right)$, we have

div $\left(\dfrac{\mathbf{x}}{|\mathbf{x}|^3}\right) = \dfrac{3(x^2 + y^2 + z^2) - 3(x^2 + y^2 + z^2)}{(x^2 + y^2 + z^2)^{5/2}} = 0$, except at $(0, 0, 0)$ where it is undefined.

25. $\iint_S \mathbf{a} \cdot \mathbf{n}\,dS = \iiint_E \operatorname{div} \mathbf{a}\,dV = 0$ since div $\mathbf{a} = 0$.

27. $\iint_S \operatorname{curl} \mathbf{F} \cdot d\mathbf{S} = \iiint_E \operatorname{div}(\operatorname{curl} \mathbf{F})\,dV = 0$ by Theorem 13.5.11.

29. $\iint_S (f\nabla g) \cdot \mathbf{n}\,dS = \iiint_E \operatorname{div}(f\nabla g)\,dV = \iiint_E (f\nabla^2 g + \nabla g \cdot \nabla f)\,dV$ by Exercise 13.5.23.

13   Review

CONCEPT CHECK

1. See Definitions 1 and 2 in Section 13.1. A vector field can represent, for example, the wind velocity at any location in space, the speed and direction of the ocean current at any location, or the force vectors of Earth's gravitational field at a location in space.

2. (a) A conservative vector field $\mathbf{F}$ is a vector field which is the gradient of some scalar function f.

(b) The function f in part (a) is called a potential function for $\mathbf{F}$, that is, $\mathbf{F} = \nabla f$.

3. (a) See Definition 13.2.2.

(b) We normally evaluate the line integral using Formula 13.2.3.

(c) The mass is $m = \int_C \rho(x, y)\,ds$, and the center of mass is $(\overline{x}, \overline{y})$ where $\overline{x} = \frac{1}{m}\int_C x\rho(x, y)\,ds$, $\overline{y} = \frac{1}{m}\int_C y\rho(x, y)\,ds$.

(d) See (5) and (6) in Section 13.2 for plane curves; we have similar definitions when C is a space curve

[see the equation preceding (10) in Section 13.2].

(e) For plane curves, see Equations 13.2.7. We have similar results for space curves

[see the equation preceding (10) in Section 13.2].

4. (a) See Definition 13.2.13.

(b) If $\mathbf{F}$ is a force field, $\int_C \mathbf{F} \cdot d\mathbf{r}$ represents the work done by $\mathbf{F}$ in moving a particle along the curve C.

(c) $\int_C \mathbf{F} \cdot d\mathbf{r} = \int_C P\,dx + Q\,dy + R\,dz$

5. See Theorem 13.3.2.

6. (a) $\int_C \mathbf{F} \cdot d\mathbf{r}$ is independent of path if the line integral has the same value for any two curves that have the same initial and terminal points.

(b) See Theorem 13.3.4.

7. See the statement of Green's Theorem on page 751.

8. See Equations 13.4.5.

9. (a) $\operatorname{curl} \mathbf{F} = \left(\dfrac{\partial R}{\partial y} - \dfrac{\partial Q}{\partial z} \right) \mathbf{i} + \left(\dfrac{\partial P}{\partial z} - \dfrac{\partial R}{\partial x} \right) \mathbf{j} + \left(\dfrac{\partial Q}{\partial x} - \dfrac{\partial P}{\partial y} \right) \mathbf{k} = \nabla \times \mathbf{F}$

(b) $\operatorname{div} \mathbf{F} = \dfrac{\partial P}{\partial x} + \dfrac{\partial Q}{\partial y} + \dfrac{\partial R}{\partial z} = \nabla \cdot \mathbf{F}$

(c) For curl $\mathbf{F}$, see the discussion accompanying Figure 1 on page 760 as well as Figure 6 and the accompanying discussion on page 789. For div $\mathbf{F}$, see the discussion following Example 5 on page 761 as well as the discussion preceding (8) on page 795.

10. See Theorem 13.3.6; see Theorem 13.5.4.

11. (a) See (1) and (2) and the accompanying discussion in Section 13.6; See Figure 4 and the accompanying discussion on page 766.

(b) See Definition 13.6.6 .

(c) See Equation 13.6.9 .

12. (a) See (1) in Section 13.7.

(b) We normally evaluate the surface integral using Formula 13.7.2.

(c) See Formula 13.7.4.

(d) The mass is $m = \iint_S \rho(x, y, z)\,dS$ and the center of mass is $(\bar{x}, \bar{y}, \bar{z})$ where $\bar{x} = \frac{1}{m} \iint_S x\rho(x, y, z)\,dS,$

$\bar{y} = \frac{1}{m} \iint_S y\rho(x, y, z)\,dS, \bar{z} = \frac{1}{m} \iint_S z\rho(x, y, z)\,dS.$

13. (a) See Figures 6 and 7 and the accompanying discussion in Section 13.7. A Möbius strip is a nonorientable surface; see Figures 4 and 5 and the accompanying discussion on page 779.

(b) See Definition 13.7.8.

(c) See Formula 13.7.9.

(d) See Formula 13.7.10.

14. See the statement of Stokes' Theorem on page 786.

15. See the statement of the Divergence Theorem on page 791.

16. In each theorem, we have an integral of a "derivative" over a region on the left side, while the right side involves the values of the original function only on the boundary of the region.

TRUE-FALSE QUIZ

1. False; div $\mathbf{F}$ is a scalar field.

3. True, by Theorem 13.5.3 and the fact that div $\mathbf{0} = 0$.

5. False. See Exercise 13.3.31. (But the assertion is true if D is simply-connected; see Theorem 13.3.6.)

7. True. Apply the Divergence Theorem and use the fact that div $\mathbf{F} = 0$.

EXERCISES

1. (a) Vectors starting on C point in roughly the direction opposite to C, so the tangential component $\mathbf{F} \cdot \mathbf{T}$ is negative. Thus

$\int_C \mathbf{F} \cdot d\mathbf{r} = \int_C \mathbf{F} \cdot \mathbf{T}\, ds$ is negative.

(b) The vectors that end near P are shorter than the vectors that start near P, so the net flow is outward near P and div $\mathbf{F}\,(P)$ is positive.

3. $\int_C yz \cos x\, ds = \int_0^\pi (3\cos t)(3\sin t)\cos t \sqrt{(1)^2 + (-3\sin t)^2 + (3\cos t)^2}\, dt = \int_0^\pi (9\cos^2 t \sin t)\sqrt{10}\, dt$

$= 9\sqrt{10}\left(-\frac{1}{3}\cos^3 t\right)\Big]_0^\pi = -3\sqrt{10}\,(-2) = 6\sqrt{10}$

5. $\int_C y^3\, dx + x^2\, dy = \int_{-1}^1 \left[y^3(-2y) + (1-y^2)^2\right] dy = \int_{-1}^1 (-y^4 - 2y^2 + 1)\, dy$

$= \left[-\frac{1}{5}y^5 - \frac{2}{3}y^3 + y\right]_{-1}^1 = -\frac{1}{5} - \frac{2}{3} + 1 - \frac{1}{5} - \frac{2}{3} + 1 = \frac{4}{15}$

7. $C: x = 1 + 2t \quad \Rightarrow \quad dx = 2\, dt,\ y = 4t \quad \Rightarrow \quad dy = 4\, dt,\ z = -1 + 3t \quad \Rightarrow \quad dz = 3\, dt, 0 \le t \le 1.$

$\int_C xy\, dx + y^2\, dy + yz\, dz = \int_0^1 \left[(1+2t)(4t)(2) + (4t)^2(4) + (4t)(-1+3t)(3)\right] dt$

$= \int_0^1 (116t^2 - 4t)\, dt = \left[\frac{116}{3}t^3 - 2t^2\right]_0^1 = \frac{116}{3} - 2 = \frac{110}{3}$

9. $\mathbf{F}(\mathbf{r}(t)) = e^{-t}\mathbf{i} + t^2(-t)\mathbf{j} + (t^2 + t^3)\mathbf{k},\ \mathbf{r}'(t) = 2t\mathbf{i} + 3t^2\mathbf{j} - \mathbf{k}$ and

$\int_C \mathbf{F} \cdot d\mathbf{r} = \int_0^1 (2te^{-t} - 3t^5 - (t^2 + t^3))\, dt = \left[-2te^{-t} - 2e^{-t} - \frac{1}{2}t^6 - \frac{1}{3}t^3 - \frac{1}{4}t^4\right]_0^1 = \frac{11}{12} - \frac{4}{e}.$

11. $\frac{\partial}{\partial y}\left[(1+xy)e^{xy}\right] = 2xe^{xy} + x^2ye^{xy} = \frac{\partial}{\partial x}\left[e^y + x^2e^{xy}\right]$ and the domain of $\mathbf{F}$ is $\mathbb{R}^2$, so $\mathbf{F}$ is conservative. Thus there exists a function f such that $\mathbf{F} = \nabla f$. Then $f_y(x,y) = e^y + x^2e^{xy}$ implies $f(x,y) = e^y + xe^{xy} + g(x)$ and then $f_x(x,y) = xye^{xy} + e^{xy} + g'(x) = (1+xy)e^{xy} + g'(x)$. But $f_x(x,y) = (1+xy)e^{xy}$, so $g'(x) = 0 \quad \Rightarrow \quad g(x) = K$. Thus $f(x,y) = e^y + xe^{xy} + K$ is a potential function for $\mathbf{F}$.

13. Since $\frac{\partial}{\partial y}(4x^3y^2 - 2xy^3) = 8x^3y - 6xy^2 = \frac{\partial}{\partial x}(2x^4y - 3x^2y^2 + 4y^3)$ and the domain of $\mathbf{F}$ is $\mathbb{R}^2$, $\mathbf{F}$ is conservative. Furthermore $f(x,y) = x^4y^2 - x^2y^3 + y^4$ is a potential function for $\mathbf{F}$. $t = 0$ corresponds to the point $(0,1)$ and $t = 1$ corresponds to $(1,1)$, so $\int_C \mathbf{F} \cdot d\mathbf{r} = f(1,1) - f(0,1) = 1 - 1 = 0.$

15. $C_1: \mathbf{r}(t) = t\,\mathbf{i} + t^2\,\mathbf{j}, \, -1 \leq t \leq 1;$

$C_2: \mathbf{r}(t) = -t\,\mathbf{i} + \mathbf{j}, \, -1 \leq t \leq 1.$

Then

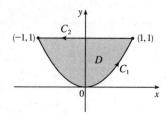

$\int_C xy^2\,dx - x^2 y\,dy = \int_{-1}^{1}(t^5 - 2t^5)\,dt + \int_{-1}^{1} t\,dt = \left[-\frac{1}{6}t^6\right]_{-1}^{1} + \left[\frac{1}{2}t^2\right]_{-1}^{1} = 0.$

Using Green's Theorem, we have

$$\int_C xy^2\,dx - x^2 y\,dy = \iint_D \left[\frac{\partial}{\partial x}(-x^2 y) - \frac{\partial}{\partial y}(xy^2)\right]dA = \iint_D (-2xy - 2xy)\,dA = \int_{-1}^{1}\int_{x^2}^{1} -4xy\,dy\,dx$$

$$= \int_{-1}^{1}\left[-2xy^2\right]_{y=x^2}^{y=1}\,dx = \int_{-1}^{1}(2x^5 - 2x)\,dx = \left[\frac{1}{3}x^6 - x^2\right]_{-1}^{1} = 0$$

17. $\int_C x^2 y\,dx - xy^2\,dy = \iint_{x^2+y^2 \leq 4} \left[\frac{\partial}{\partial x}(-xy^2) - \frac{\partial}{\partial y}(x^2 y)\right]dA = \iint_{x^2+y^2 \leq 4}(-y^2 - x^2)\,dA = -\int_0^{2\pi}\int_0^2 r^3\,dr\,d\theta = -8\pi$

19. If we assume there is such a vector field $\mathbf{G}$, then $\operatorname{div}(\operatorname{curl}\mathbf{G}) = 2 + 3z - 2xz$. But $\operatorname{div}(\operatorname{curl}\mathbf{F}) = 0$ for all vector fields $\mathbf{F}$.

Thus such a $\mathbf{G}$ cannot exist.

21. For any piecewise-smooth simple closed plane curve C bounding a region D, we can apply Green's Theorem to

$\mathbf{F}(x,y) = f(x)\,\mathbf{i} + g(y)\,\mathbf{j}$ to get $\int_C f(x)\,dx + g(y)\,dy = \iint_D \left[\frac{\partial}{\partial x}g(y) - \frac{\partial}{\partial y}f(x)\right]dA = \iint_D 0\,dA = 0.$

23. $\nabla^2 f = 0$ means that $\dfrac{\partial^2 f}{\partial x^2} + \dfrac{\partial^2 f}{\partial y^2} = 0$. Now if $\mathbf{F} = f_y\,\mathbf{i} - f_x\,\mathbf{j}$ and C is any closed path in D, then applying Green's

Theorem, we get

$$\int_C \mathbf{F}\cdot d\mathbf{r} = \int_C f_y\,dx - f_x\,dy = \iint_D \left[\frac{\partial}{\partial x}(-f_x) - \frac{\partial}{\partial y}(f_y)\right]dA = -\iint_D (f_{xx} + f_{yy})\,dA = -\iint_D 0\,dA = 0.$$

Therefore the line integral is independent of path, by Theorem 13.3.3.

25. $z = f(x,y) = x^2 + 2y$ with $0 \leq x \leq 1, \, 0 \leq y \leq 2x$. Thus

$$A(S) = \iint_D \sqrt{1 + 4x^2 + 4}\,dA = \int_0^1 \int_0^{2x} \sqrt{5 + 4x^2}\,dy\,dx = \int_0^1 2x\sqrt{5+4x^2}\,dx = \frac{1}{6}(5+4x^2)^{3/2}\Big]_0^1 = \frac{1}{6}\left(27 - 5\sqrt{5}\right).$$

27. $z = f(x,y) = x^2 + y^2$ with $0 \leq x^2 + y^2 \leq 4$ so $\mathbf{r}_x \times \mathbf{r}_y = -2x\,\mathbf{i} - 2y\,\mathbf{j} + \mathbf{k}$ (using upward orientation). Then

$$\iint_S z\,dS = \iint_{x^2+y^2 \leq 4} (x^2+y^2)\sqrt{4x^2 + 4y^2 + 1}\,dA = \int_0^{2\pi}\int_0^2 r^3\sqrt{1+4r^2}\,dr\,d\theta = \frac{1}{60}\pi\left(391\sqrt{17} + 1\right)$$

(Substitute $u = 1 + 4r^2$ and use tables.)

29. Since the sphere bounds a simple solid region, the Divergence Theorem applies and

$\iint_S \mathbf{F}\cdot d\mathbf{S} = \iiint_E (z - 2)\,dV = \iiint_E z\,dV - 2\iiint_E dV = m\bar{z} - 2\left(\frac{4}{3}\pi 2^3\right) = -\frac{64}{3}\pi.$

Alternate solution: $\mathbf{F}(\mathbf{r}(\phi,\theta)) = 4\sin\phi\cos\theta\cos\phi\,\mathbf{i} - 4\sin\phi\sin\theta\,\mathbf{j} + 6\sin\phi\cos\theta\,\mathbf{k},$

$\mathbf{r}_\phi \times \mathbf{r}_\theta = 4\sin^2\phi\cos\theta\,\mathbf{i} + 4\sin^2\phi\sin\theta\,\mathbf{j} + 4\sin\phi\cos\phi\,\mathbf{k},$ and

$\mathbf{F}\cdot(\mathbf{r}_\phi \times \mathbf{r}_\theta) = 16\sin^3\phi\cos^2\theta\cos\phi - 16\sin^3\phi\sin^2\theta + 24\sin^2\phi\cos\phi\cos\theta.$ Then

$$\iint_S \mathbf{F}\cdot d\mathbf{S} = \int_0^{2\pi}\int_0^{\pi}(16\sin^3\phi\cos\phi\cos^2\theta - 16\sin^3\phi\sin^2\theta + 24\sin^2\phi\cos\phi\cos\theta)\,d\phi\,d\theta$$

$$= \int_0^{2\pi}\frac{4}{3}(-16\sin^2\theta)\,d\theta = -\frac{64}{3}\pi$$

31. Since $\operatorname{curl}\mathbf{F} = \mathbf{0}$, $\iint_S (\operatorname{curl}\mathbf{F}) \cdot d\mathbf{S} = 0$. We parametrize C: $\mathbf{r}(t) = \cos t\,\mathbf{i} + \sin t\,\mathbf{j}$, $0 \leq t \leq 2\pi$ and

$$\oint_C \mathbf{F} \cdot d\mathbf{r} = \int_0^{2\pi}(-\cos^2 t \sin t + \sin^2 t \cos t)\,dt = \tfrac{1}{3}\cos^3 t + \tfrac{1}{3}\sin^3 t\Big]_0^{2\pi} = 0.$$

33. The surface is given by $x + y + z = 1$ or $z = 1 - x - y$, $0 \leq x \leq 1$, $0 \leq y \leq 1 - x$ and $\mathbf{r}_x \times \mathbf{r}_y = \mathbf{i} + \mathbf{j} + \mathbf{k}$. Then

$$\oint_C \mathbf{F} \cdot d\mathbf{r} = \iint_S \operatorname{curl}\mathbf{F} \cdot d\mathbf{S} = \iint_D (-y\,\mathbf{i} - z\,\mathbf{j} - x\,\mathbf{k}) \cdot (\mathbf{i} + \mathbf{j} + \mathbf{k})\,dA = \iint_D (-1)\,dA = -(\text{area of } D) = -\tfrac{1}{2}$$

35. $\iiint_E \operatorname{div}\mathbf{F}\,dV = \iiint\limits_{x^2 + y^2 + z^2 \leq 1} 3\,dV = 3(\text{volume of sphere}) = 4\pi$. Then

$$\mathbf{F}(\mathbf{r}(\phi, \theta)) \cdot (\mathbf{r}_\phi \times \mathbf{r}_\theta) = \sin^3 \phi \cos^2 \theta + \sin^3 \phi \sin^2 \theta + \sin \phi \cos^2 \phi = \sin \phi \text{ and}$$

$$\iint_S \mathbf{F} \cdot d\mathbf{S} = \int_0^{2\pi}\int_0^\pi \sin \phi\,d\phi\,d\theta = (2\pi)(2) = 4\pi.$$

37. By the Divergence Theorem, $\iint_S \mathbf{F} \cdot \mathbf{n}\,dS = \iiint_E \operatorname{div}\mathbf{F}\,dV = 3(\text{volume of } E) = 3(8 - 1) = 21$.

☐ APPENDIXES

A TRIGONOMETRY

1. $210° = 210\left(\frac{\pi}{180}\right) = \frac{7\pi}{6}$ rad

3. $9° = 9\left(\frac{\pi}{180}\right) = \frac{\pi}{20}$ rad

5. $900° = 900\left(\frac{\pi}{180}\right) = 5\pi$ rad

7. 4π rad $= 4\pi\left(\frac{180}{\pi}\right) = 720°$

9. $\frac{5\pi}{12}$ rad $= \frac{5\pi}{12}\left(\frac{180}{\pi}\right) = 75°$

11. $-\frac{3\pi}{8}$ rad $= -\frac{3\pi}{8}\left(\frac{180}{\pi}\right) = -67.5°$

13. Using Formula 3, $a = r\theta = 36 \cdot \frac{\pi}{12} = 3\pi$ cm.

15. Using Formula 3, $\theta = a/r = \frac{1}{1.5} = \frac{2}{3}$ rad $= \frac{2}{3}\left(\frac{180}{\pi}\right) = \left(\frac{120}{\pi}\right)° \approx 38.2°$.

17.

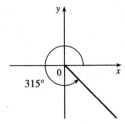

19.

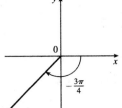

21.

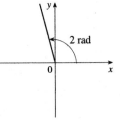

23.

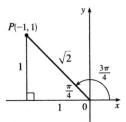

From the diagram we see that a point on the terminal side is $P(-1, 1)$. Therefore, taking $x = -1$, $y = 1$, $r = \sqrt{2}$ in the definitions of the trigonometric ratios, we have $\sin\frac{3\pi}{4} = \frac{1}{\sqrt{2}}$, $\cos\frac{3\pi}{4} = -\frac{1}{\sqrt{2}}$, $\tan\frac{3\pi}{4} = -1$, $\csc\frac{3\pi}{4} = \sqrt{2}$, $\sec\frac{3\pi}{4} = -\sqrt{2}$, and $\cot\frac{3\pi}{4} = -1$.

25.

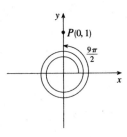

From the diagram we see that a point on the terminal line is $P(0, 1)$. Therefore taking $x = 0$, $y = 1$, $r = 1$ in the definitions of the trigonometric ratios, we have $\sin\frac{9\pi}{2} = 1$, $\cos\frac{9\pi}{2} = 0$, $\tan\frac{9\pi}{2} = y/x$ is undefined since $x = 0$, $\csc\frac{9\pi}{2} = 1$, $\sec\frac{9\pi}{2} = r/x$ is undefined since $x = 0$, and $\cot\frac{9\pi}{2} = 0$.

27.

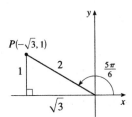

Using Figure 8 we see that a point on the terminal line is $P\left(-\sqrt{3}, 1\right)$. Therefore taking $x = -\sqrt{3}$, $y = 1$, $r = 2$ in the definitions of the trigonometric ratios, we have $\sin\frac{5\pi}{6} = \frac{1}{2}$, $\cos\frac{5\pi}{6} = -\frac{\sqrt{3}}{2}$, $\tan\frac{5\pi}{6} = -\frac{1}{\sqrt{3}}$, $\csc\frac{5\pi}{6} = 2$, $\sec\frac{5\pi}{6} = -\frac{2}{\sqrt{3}}$, and $\cot\frac{5\pi}{6} = -\sqrt{3}$.

29. $\sin\theta = y/r = \frac{3}{5}$ $\Rightarrow$ $y = 3, r = 5$, and $x = \sqrt{r^2 - y^2} = 4$ (since $0 < \theta < \frac{\pi}{2}$). Therefore taking $x = 4, y = 3, r = 5$ in

the definitions of the trigonometric ratios, we have $\cos\theta = \frac{4}{5}$, $\tan\theta = \frac{3}{4}$, $\csc\theta = \frac{5}{3}$, $\sec\theta = \frac{5}{4}$, and $\cot\theta = \frac{4}{3}$.

31. $\frac{\pi}{2} < \phi < \pi$ $\Rightarrow$ ϕ is in the second quadrant, where x is negative and y is positive. Therefore $\sec\phi = r/x = -1.5 = -\frac{3}{2}$

$\Rightarrow$ $r = 3, x = -2$, and $y = \sqrt{r^2 - x^2} = \sqrt{5}$. Taking $x = -2, y = \sqrt{5}$, and $r = 3$ in the definitions of the trigonometric

ratios, we have $\sin\phi = \frac{\sqrt{5}}{3}$, $\cos\phi = -\frac{2}{3}$, $\tan\phi = -\frac{\sqrt{5}}{2}$, $\csc\phi = \frac{3}{\sqrt{5}}$, and $\cot\theta = -\frac{2}{\sqrt{5}}$.

33. $\pi < \beta < 2\pi$ means that β is in the third or fourth quadrant where y is negative. Also since $\cot\beta = x/y = 3$ which is

positive, x must also be negative. Therefore $\cot\beta = x/y = \frac{3}{1}$ $\Rightarrow$ $x = -3, y = -1$, and $r = \sqrt{x^2 + y^2} = \sqrt{10}$. Taking

$x = -3, y = -1$ and $r = \sqrt{10}$ in the definitions of the trigonometric ratios, we have $\sin\beta = -\frac{1}{\sqrt{10}}$, $\cos\beta = -\frac{3}{\sqrt{10}}$,

$\tan\beta = \frac{1}{3}$, $\csc\beta = -\sqrt{10}$, and $\sec\beta = -\frac{\sqrt{10}}{3}$.

35. $\sin 35° = \dfrac{x}{10}$ $\Rightarrow$ $x = 10\sin 35° \approx 5.73576$ cm   **37.** $\tan\frac{2\pi}{5} = \dfrac{x}{8}$ $\Rightarrow$ $x = 8\tan\frac{2\pi}{5} \approx 24.62147$ cm

39.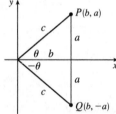

(a) From the diagram we see that $\sin\theta = \dfrac{y}{r} = \dfrac{a}{c}$, and $\sin(-\theta) = \dfrac{-a}{c} = -\dfrac{a}{c} = -\sin\theta$.

(b) Again from the diagram we see that $\cos\theta = \dfrac{x}{r} = \dfrac{b}{c} = \cos(-\theta)$.

41. (a) Using (12a) and (13a), we have

$$\tfrac{1}{2}\left[\sin(x+y) + \sin(x-y)\right] = \tfrac{1}{2}\left[\sin x \cos y + \cos x \sin y + \sin x \cos y - \cos x \sin y\right]$$
$$= \tfrac{1}{2}(2\sin x \cos y) = \sin x \cos y$$

(b) This time, using (12b) and (13b), we have

$$\tfrac{1}{2}\left[\cos(x+y) + \cos(x-y)\right] = \tfrac{1}{2}\left[\cos x \cos y - \sin x \sin y + \cos x \cos y + \sin x \sin y\right]$$
$$= \tfrac{1}{2}(2\cos x \cos y) = \cos x \cos y$$

(c) Again using (12b) and (13b), we have

$$\tfrac{1}{2}\left[\cos(x-y) - \cos(x+y)\right] = \tfrac{1}{2}\left[\cos x \cos y + \sin x \sin y - \cos x \cos y + \sin x \sin y\right]$$
$$= \tfrac{1}{2}(2\sin x \sin y) = \sin x \sin y$$

43. Using (12a), we have $\sin\left(\frac{\pi}{2} + x\right) = \sin\frac{\pi}{2}\cos x + \cos\frac{\pi}{2}\sin x = 1\cdot\cos x + 0\cdot\sin x = \cos x$.

45. Using (6), we have $\sin\theta\cot\theta = \sin\theta\cdot\dfrac{\cos\theta}{\sin\theta} = \cos\theta$.

47. $\sec y - \cos y = \dfrac{1}{\cos y} - \cos y$ [by (6)] $= \dfrac{1 - \cos^2 y}{\cos y} = \dfrac{\sin^2 y}{\cos y}$ [by (7)] $= \dfrac{\sin y}{\cos y}\sin y = \tan y \sin y$ [by (6)]

49. $\cot^2\theta + \sec^2\theta = \dfrac{\cos^2\theta}{\sin^2\theta} + \dfrac{1}{\cos^2\theta}$ [by (6)] $= \dfrac{\cos^2\theta\cos^2\theta + \sin^2\theta}{\sin^2\theta\cos^2\theta}$

$$= \frac{\left(1-\sin^2\theta\right)\left(1-\sin^2\theta\right)+\sin^2\theta}{\sin^2\theta\cos^2\theta} \quad \text{[by (7)]} = \frac{1-\sin^2\theta+\sin^4\theta}{\sin^2\theta\cos^2\theta}$$

$$= \frac{\cos^2\theta+\sin^4\theta}{\sin^2\theta\cos^2\theta} \quad \text{[by (7)]} = \frac{1}{\sin^2\theta} + \frac{\sin^2\theta}{\cos^2\theta} = \csc^2\theta + \tan^2\theta \quad \text{[by (6)]}$$

51. Using (14a), we have $\tan 2\theta = \tan(\theta + \theta) = \dfrac{\tan\theta + \tan\theta}{1-\tan\theta\tan\theta} = \dfrac{2\tan\theta}{1-\tan^2\theta}.$

53. Using (15a) and (16a),

$$\sin x \sin 2x + \cos x \cos 2x = \sin x\,(2\sin x\cos x) + \cos x\,(2\cos^2 x - 1) = 2\sin^2 x\cos x + 2\cos^3 x - \cos x$$

$$= 2\left(1-\cos^2 x\right)\cos x + 2\cos^3 x - \cos x \quad \text{[by (7)]}$$

$$= 2\cos x - 2\cos^3 x + 2\cos^3 x - \cos x = \cos x$$

Or: $\sin x \sin 2x + \cos x \cos 2x = \cos\,(2x - x)$ [by 13(b)] $= \cos x$

55. $\dfrac{\sin\phi}{1-\cos\phi} = \dfrac{\sin\phi}{1-\cos\phi}\cdot\dfrac{1+\cos\phi}{1+\cos\phi} = \dfrac{\sin\phi\,(1+\cos\phi)}{1-\cos^2\phi} = \dfrac{\sin\phi\,(1+\cos\phi)}{\sin^2\phi}$ [by (7)]

$$= \frac{1+\cos\phi}{\sin\phi} = \frac{1}{\sin\phi} + \frac{\cos\phi}{\sin\phi} = \csc\phi + \cot\phi \quad \text{[by (6)]}$$

57. Using (12a),

$$\sin 3\theta + \sin\theta = \sin\,(2\theta + \theta) + \sin\theta = \sin 2\theta\cos\theta + \cos 2\theta\sin\theta + \sin\theta$$

$$= \sin 2\theta\cos\theta + \left(2\cos^2\theta - 1\right)\sin\theta + \sin\theta \quad \text{[by (16a)]}$$

$$= \sin 2\theta\cos\theta + 2\cos^2\theta\sin\theta - \sin\theta + \sin\theta = \sin 2\theta\cos\theta + \sin 2\theta\cos\theta \quad \text{[by (15a)]}$$

$$= 2\sin 2\theta\cos\theta$$

59. Since $\sin x = \frac{1}{3}$ we can label the opposite side as having length 1, the hypotenuse as having length 3, and use the Pythagorean Theorem to get that the adjacent side has length $\sqrt{8}$. Then, from the diagram, $\cos x = \frac{\sqrt{8}}{3}$. Similarly we have that $\sin y = \frac{3}{5}$. Now use (12a):

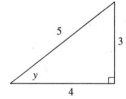

$$\sin(x+y) = \sin x\cos y + \cos x\sin y = \tfrac{1}{3}\cdot\tfrac{4}{5} + \tfrac{\sqrt{8}}{3}\cdot\tfrac{3}{5} = \tfrac{4}{15} + \tfrac{3\sqrt{8}}{15} = \tfrac{4+6\sqrt{2}}{15}.$$

61. Using (13b) and the values for $\cos x$ and $\sin y$ obtained in Exercise 59, we have

$$\cos\,(x-y) = \cos x\cos y + \sin x\sin y = \tfrac{\sqrt{8}}{3}\cdot\tfrac{4}{5} + \tfrac{1}{3}\cdot\tfrac{3}{5} = \tfrac{8\sqrt{2}+3}{15}$$

63. Using (15a) and the values for $\sin y$ and $\cos y$ obtained in Exercise 59, we have

$$\sin 2y = 2\sin y\cos y = 2\cdot\tfrac{3}{5}\cdot\tfrac{4}{5} = \tfrac{24}{25}$$

65. $2\cos x - 1 = 0 \iff \cos x = \frac{1}{2} \implies x = \frac{\pi}{3}, \frac{5\pi}{3}$ for $x \in [0, 2\pi]$.

67. $2\sin^2 x = 1 \iff \sin^2 x = \frac{1}{2} \iff \sin x = \pm\frac{1}{\sqrt{2}} \implies x = \frac{\pi}{4}, \frac{3\pi}{4}, \frac{5\pi}{4}, \frac{7\pi}{4}.$

69. Using (15a), we have $\sin 2x = \cos x \iff 2\sin x\cos x - \cos x = 0 \iff \cos x(2\sin x - 1) = 0 \iff \cos x = 0$ or $2\sin x - 1 = 0 \implies x = \frac{\pi}{2}, \frac{3\pi}{2}$ or $\sin x = \frac{1}{2} \implies x = \frac{\pi}{6}$ or $\frac{5\pi}{6}$. Therefore, the solutions are $x = \frac{\pi}{6}, \frac{\pi}{2}, \frac{5\pi}{6}, \frac{3\pi}{2}.$

71. $\sin x = \tan x \iff \sin x - \tan x = 0 \iff \sin x - \dfrac{\sin x}{\cos x} = 0 \iff \sin x\left(1 - \dfrac{1}{\cos x}\right) = 0 \iff \sin x = 0$ or

$1 - \dfrac{1}{\cos x} = 0 \implies x = 0, \pi, 2\pi$ or $1 = \dfrac{1}{\cos x} \implies \cos x = 1 \implies x = 0, 2\pi$. Therefore the solutions are $x = 0, \pi, 2\pi$.

73. We know that $\sin x = \frac{1}{2}$ when $x = \frac{\pi}{6}$ or $\frac{5\pi}{6}$, and from Figure 13(a), we see that $\sin x \le \frac{1}{2}$ $\Rightarrow$ $0 \le x \le \frac{\pi}{6}$ or

$\frac{5\pi}{6} \le x \le 2\pi$ for $x \in [0, 2\pi]$.

75. $\tan x = -1$ when $x = \frac{3\pi}{4}$, $\frac{7\pi}{4}$, and $\tan x = 1$ when $x = \frac{\pi}{4}$ or $\frac{5\pi}{4}$. From Figure 14 we see that $-1 < \tan x < 1$ $\Rightarrow$

$0 \le x < \frac{\pi}{4}$, $\frac{3\pi}{4} < x < \frac{5\pi}{4}$, and $\frac{7\pi}{4} < x \le 2\pi$.

77. $y = \cos\left(x - \frac{\pi}{3}\right)$. We start with the graph of $y = \cos x$ and shift it $\frac{\pi}{3}$ units to the right.

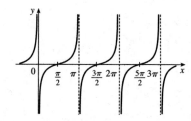

79. $y = \frac{1}{3}\tan\left(x - \frac{\pi}{2}\right)$. We start with the graph of $y = \tan x$, shift it $\frac{\pi}{2}$ units to the right and compress it to $\frac{1}{3}$ of its original

vertical size.

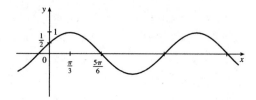

81. $y = |\sin x|$. We start with the graph of $y = \sin x$ and reflect the parts below the x-axis about the x-axis.

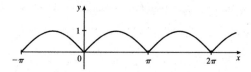

83. From the figure in the text, we see that $x = b\cos\theta$, $y = b\sin\theta$, and from the distance formula we have that the

distance c from (x, y) to $(a, 0)$ is $c = \sqrt{(x-a)^2 + (y-0)^2}$ $\Rightarrow$

$$c^2 = (b\cos\theta - a)^2 + (b\sin\theta)^2 = b^2\cos^2\theta - 2ab\cos\theta + a^2 + b^2\sin^2\theta$$
$$= a^2 + b^2(\cos^2\theta + \sin^2\theta) - 2ab\cos\theta = a^2 + b^2 - 2ab\cos\theta \quad [\text{by (7)}]$$

85. Using the Law of Cosines, we have $c^2 = 1^2 + 1^2 - 2(1)(1)\cos(\alpha - \beta) = 2[1 - \cos(\alpha - \beta)]$. Now, using the distance

formula, $c^2 = |AB|^2 = (\cos\alpha - \cos\beta)^2 + (\sin\alpha - \sin\beta)^2$. Equating these two expressions for c^2, we get

$2[1 - \cos(\alpha - \beta)] = \cos^2\alpha + \sin^2\alpha + \cos^2\beta + \sin^2\beta - 2\cos\alpha\cos\beta - 2\sin\alpha\sin\beta$ $\Rightarrow$

$1 - \cos(\alpha - \beta) = 1 - \cos\alpha\cos\beta - \sin\alpha\sin\beta$ $\Rightarrow$ $\cos(\alpha - \beta) = \cos\alpha\cos\beta + \sin\alpha\sin\beta$.

87. In Exercise 86 we used the subtraction formula for cosine to prove the addition formula for cosine. Using that formula with

$x = \frac{\pi}{2} - \alpha$, $y = \beta$, we get $\cos\left[\left(\frac{\pi}{2} - \alpha\right) + \beta\right] = \cos\left(\frac{\pi}{2} - \alpha\right)\cos\beta - \sin\left(\frac{\pi}{2} - \alpha\right)\sin\beta$ $\Rightarrow$

$\cos\left[\frac{\pi}{2} - (\alpha - \beta)\right] = \cos\left(\frac{\pi}{2} - \alpha\right)\cos\beta - \sin\left(\frac{\pi}{2} - \alpha\right)\sin\beta$. Now we use the identities given in the problem,

$\cos\left(\frac{\pi}{2} - \theta\right) = \sin\theta$ and $\sin\left(\frac{\pi}{2} - \theta\right) = \cos\theta$, to get $\sin(\alpha - \beta) = \sin\alpha\cos\beta - \cos\alpha\sin\beta$.

89. Using $A = \frac{1}{2}ab\sin\theta$, the area of the triangle is $\frac{1}{2}(10)(3)\sin 107° \approx 14.34457$ cm^2.

C   SIGMA NOTATION

1. $\displaystyle\sum_{i=1}^{5} \sqrt{i} = \sqrt{1} + \sqrt{2} + \sqrt{3} + \sqrt{4} + \sqrt{5}$

3. $\displaystyle\sum_{i=4}^{6} 3^i = 3^4 + 3^5 + 3^6$

5. $\displaystyle\sum_{k=0}^{4} \frac{2k-1}{2k+1} = -1 + \frac{1}{3} + \frac{3}{5} + \frac{5}{7} + \frac{7}{9}$

7. $\displaystyle\sum_{i=1}^{n} i^{10} = 1^{10} + 2^{10} + 3^{10} + \cdots + n^{10}$

9. $\displaystyle\sum_{j=0}^{n-1} (-1)^j = 1 - 1 + 1 - 1 + \cdots + (-1)^{n-1}$

11. $1 + 2 + 3 + 4 + \cdots + 10 = \displaystyle\sum_{i=1}^{10} i$

13. $\dfrac{1}{2} + \dfrac{2}{3} + \dfrac{3}{4} + \dfrac{4}{5} + \cdots + \dfrac{19}{20} = \displaystyle\sum_{i=1}^{19} \dfrac{i}{i+1}$

15. $2 + 4 + 6 + 8 + \cdots + 2n = \displaystyle\sum_{i=1}^{n} 2i$

17. $1 + 2 + 4 + 8 + 16 + 32 = \displaystyle\sum_{i=0}^{5} 2^i$

19. $x + x^2 + x^3 + \cdots + x^n = \displaystyle\sum_{i=1}^{n} x^i$

21. $\displaystyle\sum_{i=4}^{8} (3i - 2) = [3(4) - 2] + [3(5) - 2] + [3(6) - 2] + [3(7) - 2] + [3(8) - 2] = 10 + 13 + 16 + 19 + 22 = 80$

23. $\displaystyle\sum_{j=1}^{6} 3^{j+1} = 3^2 + 3^3 + 3^4 + 3^5 + 3^6 + 3^7 = 9 + 27 + 81 + 243 + 729 + 2187 = 3276$

(For a more general method, see Exercise 47.)

25. $\displaystyle\sum_{n=1}^{20} (-1)^n = -1 + 1 - 1 + 1 - 1 + 1 - 1 + 1 - 1 + 1 - 1 + 1 - 1 + 1 - 1 + 1 - 1 + 1 - 1 + 1 = 0$

27. $\displaystyle\sum_{i=0}^{4} \left(2^i + i^2\right) = (1 + 0) + (2 + 1) + (4 + 4) + (8 + 9) + (16 + 16) = 61$

29. $\displaystyle\sum_{i=1}^{n} 2i = 2 \sum_{i=1}^{n} i = 2 \cdot \frac{n(n+1)}{2}$ [by Theorem 3(c)] $= n(n+1)$

31. $\displaystyle\sum_{i=1}^{n} \left(i^2 + 3i + 4\right) = \sum_{i=1}^{n} i^2 + 3\sum_{i=1}^{n} i + \sum_{i=1}^{n} 4 = \frac{n(n+1)(2n+1)}{6} + \frac{3n(n+1)}{2} + 4n$

$\qquad = \frac{1}{6}\left[\left(2n^3 + 3n^2 + n\right) + \left(9n^2 + 9n\right) + 24n\right] = \frac{1}{6}\left(2n^3 + 12n^2 + 34n\right)$

$\qquad = \frac{1}{3}n\left(n^2 + 6n + 17\right)$

33. $\displaystyle\sum_{i=1}^{n}(i+1)(i+2) = \sum_{i=1}^{n}\left(i^2+3i+2\right) = \sum_{i=1}^{n}i^2 + 3\sum_{i=1}^{n}i + \sum_{i=1}^{n}2$

$$= \frac{n(n+1)(2n+1)}{6} + \frac{3n(n+1)}{2} + 2n = \frac{n(n+1)}{6}\left[(2n+1)+9\right] + 2n$$

$$= \frac{n(n+1)}{3}(n+5) + 2n = \frac{n}{3}\left[(n+1)(n+5)+6\right] = \frac{n}{3}\left(n^2+6n+11\right)$$

35. $\displaystyle\sum_{i=1}^{n}\left(i^3-i-2\right) = \sum_{i=1}^{n}i^3 - \sum_{i=1}^{n}i - \sum_{i=1}^{n}2 = \left[\frac{n(n+1)}{2}\right]^2 - \frac{n(n+1)}{2} - 2n$

$$= \tfrac{1}{4}n(n+1)\left[n(n+1)-2\right] - 2n = \tfrac{1}{4}n(n+1)(n+2)(n-1) - 2n$$

$$= \tfrac{1}{4}n\left[(n+1)(n-1)(n+2)-8\right] = \tfrac{1}{4}n\left[(n^2-1)(n+2)-8\right] = \tfrac{1}{4}n\left(n^3+2n^2-n-10\right)$$

37. By Theorem 2(a) and Example 3, $\displaystyle\sum_{i=1}^{n}c = c\sum_{i=1}^{n}1 = cn.$

39. $\displaystyle\sum_{i=1}^{n}\left[(i+1)^4-i^4\right] = \left(2^4-1^4\right) + \left(3^4-2^4\right) + \left(4^4-3^4\right) + \cdots + \left[(n+1)^4-n^4\right]$

$$= (n+1)^4 - 1^4 = n^4 + 4n^3 + 6n^2 + 4n$$

On the other hand,

$$\sum_{i=1}^{n}\left[(i+1)^4-i^4\right] = \sum_{i=1}^{n}\left(4i^3+6i^2+4i+1\right) = 4\sum_{i=1}^{n}i^3 + 6\sum_{i=1}^{n}i^2 + 4\sum_{i=1}^{n}i + \sum_{i=1}^{n}1$$

$$= 4S + n(n+1)(2n+1) + 2n(n+1) + n \quad \left[\text{where } S = \sum_{i=1}^{n}i^3\right]$$

$$= 4S + 2n^3 + 3n^2 + n + 2n^2 + 2n + n = 4S + 2n^3 + 5n^2 + 4n$$

Thus, $n^4 + 4n^3 + 6n^2 + 4n = 4S + 2n^3 + 5n^2 + 4n$, from which it follows that

$$4S = n^4 + 2n^3 + n^2 = n^2\left(n^2+2n+1\right) = n^2(n+1)^2 \text{ and } S = \left[\frac{n(n+1)}{2}\right]^2.$$

41. (a) $\displaystyle\sum_{i=1}^{n}\left[i^4-(i-1)^4\right] = \left(1^4-0^4\right) + \left(2^4-1^4\right) + \left(3^4-2^4\right) + \cdots + \left[n^4-(n-1)^4\right] = n^4 - 0 = n^4$

(b) $\displaystyle\sum_{i=1}^{100}\left(5^i-5^{i-1}\right) = \left(5^1-5^0\right) + \left(5^2-5^1\right) + \left(5^3-5^2\right) + \cdots + \left(5^{100}-5^{99}\right) = 5^{100} - 5^0 = 5^{100} - 1$

(c) $\displaystyle\sum_{i=3}^{99}\left(\frac{1}{i}-\frac{1}{i+1}\right) = \left(\frac{1}{3}-\frac{1}{4}\right) + \left(\frac{1}{4}-\frac{1}{5}\right) + \left(\frac{1}{5}-\frac{1}{6}\right) + \cdots + \left(\frac{1}{99}-\frac{1}{100}\right) = \frac{1}{3} - \frac{1}{100} = \frac{97}{300}$

(d) $\displaystyle\sum_{i=1}^{n}(a_i-a_{i-1}) = (a_1-a_0) + (a_2-a_1) + (a_3-a_2) + \cdots + (a_n-a_{n-1}) = a_n - a_0$

43. $\displaystyle\lim_{n\to\infty}\sum_{i=1}^{n}\frac{1}{n}\left(\frac{i}{n}\right)^2 = \lim_{n\to\infty}\frac{1}{n^3}\sum_{i=1}^{n}i^2 = \lim_{n\to\infty}\frac{1}{n^3}\frac{n(n+1)(2n+1)}{6} = \lim_{n\to\infty}\frac{1}{6}\left(1+\frac{1}{n}\right)\left(2+\frac{1}{n}\right)$

$$= \tfrac{1}{6}(1)(2) = \tfrac{1}{3}$$

45. $\lim\limits_{n\to\infty}\sum\limits_{i=1}^{n}\dfrac{2}{n}\left[\left(\dfrac{2i}{n}\right)^3+5\left(\dfrac{2i}{n}\right)\right]=\lim\limits_{n\to\infty}\sum\limits_{i=1}^{n}\left[\dfrac{16}{n^4}i^3+\dfrac{20}{n^2}i\right]=\lim\limits_{n\to\infty}\left[\dfrac{16}{n^4}\sum\limits_{i=1}^{n}i^3+\dfrac{20}{n^2}\sum\limits_{i=1}^{n}i\right]$

$\qquad=\lim\limits_{n\to\infty}\left[\dfrac{16}{n^4}\dfrac{n^2(n+1)^2}{4}+\dfrac{20}{n^2}\dfrac{n(n+1)}{2}\right]=\lim\limits_{n\to\infty}\left[\dfrac{4(n+1)^2}{n^2}+\dfrac{10n(n+1)}{n^2}\right]$

$\qquad=\lim\limits_{n\to\infty}\left[4\left(1+\dfrac{1}{n}\right)^2+10\left(1+\dfrac{1}{n}\right)\right]=4\cdot1+10\cdot1=14$

47. Let $S=\sum\limits_{i=1}^{n}ar^{i-1}=a+ar+ar^2+\cdots+ar^{n-1}$. Multiplying both sides by r gives us

$rS=ar+ar^2+\cdots+ar^{n-1}+ar^n$. Subtracting the first equation from the second, we find

$(r-1)S=ar^n-a=a(r^n-1)$, so $S=\dfrac{a(r^n-1)}{r-1}$ (since $r\neq1$).

49. $\sum\limits_{i=1}^{n}\left(2i+2^i\right)=2\sum\limits_{i=1}^{n}i+\sum\limits_{i=1}^{n}2\cdot2^{i-1}=2\dfrac{n(n+1)}{2}+\dfrac{2(2^n-1)}{2-1}=2^{n+1}+n^2+n-2.$

For the first sum we have used Theorem 3(c), and for the second, Exercise 47 with $a=r=2$.

D   THE LOGARITHM DEFINED AS AN INTEGRAL

1. (a)

We interpret $\ln 1.5$ as the area under the curve $y=1/x$ from $x=1$ to $x=1.5$. The area of the rectangle $BCDE$ is $\frac{1}{2}\cdot\frac{2}{3}=\frac{1}{3}$. The area of the trapezoid $ABCD$ is $\frac{1}{2}\cdot\frac{1}{2}\left(1+\frac{2}{3}\right)=\frac{5}{12}$. Thus, by comparing areas, we observe that $\frac{1}{3}<\ln 1.5<\frac{5}{12}$.

(b) With $f(t)=1/t$, $n=10$, and $\Delta x=0.05$, we have

$$\ln 1.5=\int_1^{1.5}(1/t)\,dt\approx(0.05)[f(1.025)+f(1.075)+\cdots+f(1.475)]$$
$$=(0.05)\left[\tfrac{1}{1.025}+\tfrac{1}{1.075}+\cdots+\tfrac{1}{1.475}\right]\approx0.4054$$

3.

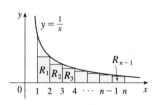

The area of R_i is $\dfrac{1}{i+1}$ and so $\dfrac{1}{2}+\dfrac{1}{3}+\cdots+\dfrac{1}{n}<\displaystyle\int_1^n\dfrac{1}{t}\,dt=\ln n.$

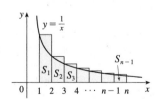

The area of S_i is $\dfrac{1}{i}$ and so $1+\dfrac{1}{2}+\cdots+\dfrac{1}{n-1}>\displaystyle\int_1^n\dfrac{1}{t}\,dt=\ln n.$

Thus, $\dfrac{1}{2}+\dfrac{1}{3}+\cdots+\dfrac{1}{n}<\ln n<1+\dfrac{1}{2}+\cdots+\dfrac{1}{n-1}.$

5. If $f(x) = \ln(x^r)$, then $f'(x) = (1/x^r)(rx^{r-1}) = r/x$. But if $g(x) = r \ln x$, then $g'(x) = r/x$. So f and g must differ by a constant: $\ln(x^r) = r \ln x + C$. Put $x = 1$: $\ln(1^r) = r \ln 1 + C \;\Rightarrow\; C = 0$, so $\ln(x^r) = r \ln x$.

7. Using the third law of logarithms and Equation 10, we have $\ln e^{rx} = rx = r \ln e^x = \ln(e^x)^r$. Since $\ln$ is a one-to-one function, it follows that $e^{rx} = (e^x)^r$.

9. Using Definition 13, the first law of logarithms, and the first law of exponents for e^x, we have
$$(ab)^x = e^{x \ln(ab)} = e^{x(\ln a + \ln b)} = e^{x \ln a + x \ln b} = e^{x \ln a} e^{x \ln b} = a^x b^x.$$